D1611005

BIOLOGY OF THE SAUROPOD DINOSAURS

LIFE OF THE PAST
James O. Farlow, editor

BIOLOGY OF THE SAUROPOD DINOSAURS

Understanding the Life of Giants

EDITED BY

NICOLE KLEIN

KRISTIAN REMES

CAROLE T. GEE

P. MARTIN SANDER

Indiana University Press
Bloomington and Indianapolis

This book is a publication of

Indiana University Press
601 North Morton Street
Bloomington, IN 47404-3797 USA

iupress.indiana.edu

Telephone orders	800-842-6796
Fax orders	812-855-7931
Orders by e-mail	iuporder@indiana.edu

⊗ The paper used in this publication meets the minimum requirements of the American National Standard for Information Sciences—Permanence of Paper for Printed Library Materials, ANSI Z39.48-1992.

Manufactured in the United States of America

Library of Congress Cataloging-in-Publication Data

Biology of the sauropod dinosaurs : understanding the life of giants / edited by Nicole Klein ... [et al.].
p.cm. — (Life of the past)
Includes bibliographical references and index.
ISBN 978-0-253-35508-9 (cloth : alk. paper) 1. Saurischia.
I. Klein, Nicole, [date]
QE862.S3B56 2011
567.913—dc22
2010046735
1 2 3 4 5 16 15 14 13 12 11

CONTENTS

CONTRIBUTORS

Andras Borbély
Abteilung für Werkstoffdiagnostik und Technologie der Stähle, Max-Planck-Institut für Eisenforschung, Max-Planck-Str. 1, 40237 Düsseldorf, Germany; *a.borbely@mpie.de*

Thomas Breuer
Institut für Zoologie, Rheinische Friedrich-Wilhelms-Universität Bonn, Poppelsdorfer Schloss, 53115 Bonn, Germany; *thomas-breuer@gmx.net*

Andreas Christian
Institut für Biologie und Sachunterricht und ihre Didaktik, Universität Flensburg, Auf dem Campus 1, 24943 Flensburg, Germany; *christian@uni-flensburg.de*

Marcus Clauss
Klinik für Zoo-, Heim- und Wildtiere, Vetsuisse-Fakultät, Universität Zürich, Winterthurerstr. 260, 8057 Zürich, Switzerland; *mclauss@vetclinics.unizh.ch*

Maïtena Dumont
Abteilung für Werkstoffdiagnostik und Technologie der Stähle, Max-Planck-Institut für Eisenforschung, Max-Planck-Str. 1, 40237 Düsseldorf, Germany; *m.dumont@mpie.de*

Gordon Dzemski
Institut für Biologie und Sachunterricht und ihre Didaktik, Universität Flensburg, Auf dem Campus 1, 24943 Flensburg, Germany; *dzemski@uni-flensburg.de*

Regina Fechner
Institut für Zoologie und Neurobiologie, Ruhr-Universität Bochum, 44780 Bochum, Germany; *r.fechner@lrz.uni-muenchen.de*

Bergita Ganse
Schwerpunkt Unfallchirurgie, Klinik für Orthopädie und Unfallchirurgie, Universität zu Köln, Kerpener Str. 62, 50924 Köln, Germany; *bergita.ganse@uk-koeln.de*

Carole T. Gee
Steinmann-Institut für Geologie, Mineralogie und Paläontologie, Rheinische Friedrich-Wilhelms-Universität Bonn, Nussallee 8, 53115 Bonn, Germany; *cgee@uni-bonn.de*

Rainer Goessling
Arbeitsgruppe Biomechanik, Fakultät für Maschinenbau, Ruhr-Universität Bochum, 44780 Bochum, Germany; *rainer.goessling@ruhr-uni-bochum.de*

Eva Maria Griebeler
Abteilung Ökologie, Zoologisches Institut, Universität Mainz, Postfach 3980, 55099 Mainz, Germany; *em.griebeler@uni-mainz.de*

Hanns-Christian Gunga
Institut für Physiologie, Charité—Universitätsmedizin Berlin, Campus Benjamin Franklin, Arnimallee 22, 14195 Berlin, Germany; *hanns-christian.gunga@charite.de*

Oliver Hampe
Museum für Naturkunde—Leibniz-Institut für Evolutions- und Biodiversitätsforschung an der Humboldt-Universität zu Berlin, Invalidenstr. 43, 10115 Berlin, Germany; *oliver.hampe@mfn-berlin.de*

Wolf-Dieter Heinrich
Museum für Naturkunde—Leibniz-Institut für Evolutions- und Biodiversitätsforschung an der Humboldt-Universität zu Berlin, Invalidenstr. 43, 10115 Berlin, Germany; *wolf-dieter.heinrich@mfn-berlin.de*

Bianca Hohn
Institut für Zoologie und Neurobiologie, Ruhr-Universität Bochum, 44780 Bochum, Germany; *bianca.hohn@web.de*

Jürgen Hummel
Institut für Tierwissenschaften, Rheinische Friedrich-Wilhelms-Universität Bonn, Endenicher Allee 15, 53115 Bonn, Germany; *jhum@itz.uni-bonn.de*

Anke Kaysser-Pyzalla
Helmholtz-Zentrum Berlin für Materialien und Energie, Lise-Meitner Campus, Glienicker Straße 100, 14109 Berlin, Germany; *anke.pyzalla@helmholtz-berlin.de*

Nicole Klein
Steinmann-Institut für Geologie, Mineralogie und Paläontologie, Rheinische Friedrich-Wilhelms-University of Bonn, Nussallee 8, 53115 Bonn, Germany; *nklein@uni-bonn.de*

Aleksander Kostka
Abteilung für Werkstoffdiagnostik und Technologie der Stähle, Max-Planck-Institut für Eisenforschung, Max-Planck-Str. 1, 40237 Düsseldorf, Germany; *a.kostka@mpie.de*

Heinrich Mallison
Museum für Naturkunde—Leibniz-Institut für Evolutions- und Biodiversitätsforschung an der Humboldt-Universität zu Berlin, Invalidenstr. 43, 10115 Berlin, Germany; *heinrich.mallison@googlemail.com*

Julia Mannhardt
Arbeitsgruppe Biomechanik, Fakultät für Maschinenbau, Ruhr-Universität Bochum, 44780 Bochum, Germany; *julia.mannhardt@gmx.de*

Pascal de Micheli
Arbeitsgruppe Biomechanik, Fakultät für Maschinenbau, Ruhr-Universität Bochum, 44780 Bochum, Germany; *pascal.demicheli@free.fr*

Nadine Pajor
Institut für Zoologie, Rheinische Friedrich-Wilhelms-Universität Bonn, Poppelsdorfer Schloss, 53115 Bonn, Germany; *pajor@uni-bonn.de*

Steven F. Perry
Institut für Zoologie, Rheinische Friedrich-Wilhelms-Universität Bonn, Poppelsdorfer Schloss, 53115 Bonn, Germany; *perry@uni-bonn.de*

Holger Preuschoft
Anatomisches Institut, Medizinische Fakultät, Ruhr-Universität Bochum, Universitätsstraße 150, 44780 Bochum, Germany; *holger.preuschoft@rub.de*

Oliver W. M. Rauhut
Bayerische Staatssammlung für Paläontologie und Geologie, Richard-Wagner-Str. 10, 80333 München, Germany; *o.rauhut@lrz.uni-muenchen.de*

Katrin Reis
Bayerische Staatssammlung für Paläontologie und Geologie, Richard-Wagner-Str. 10, 80333 München, Germany, *k.moser@lrz.uni-muenchen.de*

Kristian Remes
Steinmann-Institut für Geologie, Mineralogie und Paläontologie, Rheinische Friedrich-Wilhelms-University of Bonn, Nussallee 8, 53115 Bonn, Germany; current address: DFG, 53175 Bonn, Germany; *kristian.remes@dfg.de*

P. Martin Sander
Steinmann-Institut für Geologie, Mineralogie und Paläontologie, Rheinische Friedrich-Wilhelms-University of Bonn, Nussallee 8, 53115 Bonn, Germany; *martin.sander@uni-bonn.de*

Alexander Stahn
Institut für Physiologie, Charité—Universitätsmedizin Berlin, Campus Benjamin Franklin, Arnimallee 22, 14195 Berlin, Germany; *alexander.stahn@charite.de*

Koen Stein
Steinmann-Institut für Geologie, Mineralogie und Paläontologie, Rheinische Friedrich-Wilhelms-University of Bonn, Nussallee 8, 53115 Bonn, Germany; *koen.stein@uni-bonn.de*

Stefan Stoinski
Computer Vision and Remote Sensing, Technische Universität Berlin, Franklinstr. 28/29, 10587 Berlin, Germany; *stoinski@fpk.tu-berlin.de*

Tim Suthau
Director of Imaging, MÖLLER-WEDEL GmbH, Rosengarten 10, 22880 Wedel, Germany; *suthau@arcor.de*

Thomas Tütken
Steinmann-Institut für Geologie, Mineralogie und Paläontologie, Rheinische Friedrich-Wilhelms-University of Bonn, Poppelsdorfer Schloß, 53115 Bonn, Germany; *tuetken@uni-bonn.de*

David M. Unwin
Department of Museum Studies, University of Leicester, 105 Princess Road East, Leicester LE1 2LG, United Kingdom; *dmu1@le.ac.uk*

Jan Werner
Abteilung Ökologie, Zoologisches Institut, Universität Mainz, P.O. Box 3980, 55099 Mainz, Germany; *wernerja@uni-mainz.de*

Oliver Wings
Museum für Naturkunde—Leibniz-Institut für Evolutions- und Biodiversitätsforschung an der Humboldt-Universität zu Berlin, Invalidenstr. 43, 10115 Berlin, Germany; *oliver.wings@mfn-berlin.de*

Ulrich Witzel
Arbeitsgruppe Biomechanik, Fakultät für Maschinenbau, Ruhr-Universität Bochum, 44780 Bochum, Germany; *Ulrich.Witzel@ruhr-uni-bochum.de*

PREFACE

"The never-since-surpassed size of the largest dinosaurs remains unexplained." This resigned conclusion voiced a decade ago (Burness et al. 2001, 14523) has inspired us, a highly diverse group of researchers in Germany and Switzerland, to join forces in an attempt to understand why and how the largest of the large, the long-necked sauropod dinosaurs attained their gargantuan proportions. Dinosaur gigantism is a scientific problem that has puzzled evolutionary biologists since the earliest discoveries of sauropod dinosaur bones almost 160 years ago, which were aptly named *Cetiosaurus*, the "whale lizard." In terms of body mass, sauropod dinosaurs are second in size only to the large baleen whales that evolved some 180 million years later in the Tertiary. However, whales and sauropods cannot really be compared to one another because the rules of the game in regard to body size are so different on land and in the water. What does bind whales and the "whale lizards" together, though, is their evolutionary trend toward ever larger body sizes. This raises the question of how sauropods achieved their gigantic sizes and, more importantly, what ultimately stopped them from getting even bigger.

These fascinating issues were the driving force behind the formation of a research consortium, Research Unit 533 "Biology of the Sauropod Dinosaurs: The Evolution of Gigantism," funded by the German Research Foundation (Deutsche Forschungsgemeinschaft, or DFG). We, the members of the Research Unit, feel that only a better understanding of the biology of the sauropods and their role in Mesozoic ecosystems can bring us closer to an understanding of their gigantism. For the most part, geological reasons for sauropod gigantism can be discounted because none of the environmental parameters of the Mesozoic, for example, atmospheric oxygen content (Sander et al. 2010), are reflected in changes in sauropod body size. This then leaves mainly biological reasons behind the success story of these huge animals that ruled the Earth for 145 million years.

Our Research Unit consists of experts from all walks of scientific life. Indeed, there are not many dinosaur research projects in which paleontologists are outnumbered by non-paleontologists, but working on the issue of gigantism required just that. The 38 authors who have helped to put together the latest knowledge on sauropod dinosaur biology in this volume are specialists in animal nutrition, biomechanics, bone histology, computer modeling, dinosaur anatomy, evolutionary ecology, geochemistry, materials science, paleobotany, physiology, veterinary medicine, and zoology. Listed here in alphabetical order, each and every one of these fields has contributed to our basic research on sauropod gigantism, bringing new ideas and fresh approaches to the problem.

Our shared journey down the road of scientific discovery has been exciting and productive. When we first started out, however, we had to learn to speak in a common language. Intense, three-day workshops every six months have taught us how to effectively communicate with one another—not a trivial task if an isotope geochemist is to exchange research results with a functional morphologist, or an animal nutritionist is to discuss profound interpretations with a materials scientist. But from the very beginning, the enthusiasm of all members of our research group and their willingness to trade ideas and reach out to one another have been so great that even in the early stages, the cross-fertilization between disparate fields succeeded. Observations or data that puzzled one specialist were readily explained by an expert in another field. Now, at the start of our third and final funding period, our collaborative efforts are bringing forth the fruit of knowledge that we have been searching for.

Our inquiries regarding sauropods began in a manner fairly similar to those of most research scientists. Initially, each question or line of investigation about sauropod biology was tackled by formulating a hypothesis on how it could have made sauropod gigantism possible; this hypothesis was then tested with experimentation and research. Take, as a case in point, the mechanical properties of sauropod bone tissue. In this example, the hypothesis tested by our materials scientist (see Chapter 5) was that sauropod bone has superior mechanical properties compared to large-mammal bone which would have resulted in stronger skeletons in sauropods with relatively less bone material. The hypothesis was proven wrong because sauropod and cow bone tissues have the same strength. Although this result was a surprise to some, it brought us another step closer to a better understanding of sauropod gigantism; the uniqueness of sauropods does not lie in the mechanical properties of their bones, but somewhere else in their biological make-up. And thus our search had to move on to other avenues of investigation.

In our seven-year journey together, we have discovered that it is not one single feature that sets sauropods apart. The small head, wide grin, cheekless face, ridiculously long neck, barrel-shaped chest, spacious abdomen, and fancy tail all make up the iconic sauropod of a child's picture book. Evolutionarily, however, they represent a suite of primitive characters and key innovations in sauropods. This suite of features coupled with others that are not obvious in a sauropod's appearance is proving to be the key to why gigantism could have been developed to such extremes in this particular group of plant-eating animals (see Chapter 1). It is this research—our studies and how they contribute to the understanding of sauropod gigantism—that we have compiled in this book. To offer the reader a more integrated view of our work, we approach the biology of the sauropods from four major perspectives: nutrition, physiology, construction, and growth. If you want to find out more about the biology and gigantism of the sauropod dinosaurs, we suggest consulting the list of scientific papers by our Research Unit at www.sauropod-dinosaurs.uni-bonn.de.

Putting together this book has clearly been a Herculean teamwork effort, and on behalf of my three co-editors, I profusely thank all of our authors for their contributions and the DFG for the funding, assistance, and support that has made this project an amazing success. For technical assistance, we are indebted to Kay Heitplatz, Maren Jansen, Anja Königs, Dorothea Krantz, Jean Sebastian Marpmann, and Katja Waskow (listed here in alphabetical order and all at the University of Bonn). Our special thanks go to following reviewers, whose efforts and constructive comments were invaluable: Jörg Albertz (Technische Universität Berlin), R. McNeill Alexander (University of Leeds), Ronan Allain (Muséum National d'Histoire Naturelle, Paris), Karl Bates (University of Manchester), David Berman (Carnegie Museum of Natural History, Pittsburgh), Matt Bonnan (Western Illinois University, Macomb), Vivian de Buffrénil (Université Pierre et Marie Curie, Paris), Chris Carbone (Zoological Society, London), William G. Chaloner (Royal Holloway University of London), Leon Claessens (Harvard University, Cambridge), James O. Farlow (Indiana University–Purdue University, Fort Wayne), Henry Fricke (Colorado College, Colorado Springs), Hartmut Haubold (Martin-Luther-Universität Halle-Wittenberg), Donald Henderson (Royal Tyrrell Museum, Drumheller), John Hutchinson (University of London, Hatfield), Frankie Jackson (Montana State University, Bozeman), Ivan Lonardelli (University of Trento), Mehran Moazen (University of Hull), Daniela Schwarz-Wings (Museum für Naturkunde, Berlin), Roger Seymour (University of Adelaide), Mike Taylor (University of Portsmouth), Clive Trueman (University of Southampton), Paul Upchurch (University College, London), Peter van Soest (Cornell University, Ithaca), and Ray Wilhite (Louisiana State University, Baton Rouge).

Finally, we express our gratitude to Robert Sloan, editorial director at Indiana University Press, and to James O. Farlow, editor of the press's *Life of the Past* series, for their support of this volume.

P. Martin Sander
Speaker of DFG Research Unit 533
Bonn, Germany

References

Burness, G. P., Diamond, J. & Flannery, T. 2001. Dinosaurs, dragons, and dwarfs: the evolution of maximal body size.—*Proceedings of the National Academy of Sciences of the United States of America* 98: 14518–14523.

Sander, P. M., Christian, A., Clauss, M., Fechner, R., Gee, C. T., Griebeler, E. M., Gunga, H.-C., Hummel, J., Mallison, H., Perry, S., Preuschoft, H., Rauhut, O., Remes, K., Tütken, T., Wings, O. & Witzel, U. 2010. Biology of the sauropod dinosaurs: the evolution of gigantism.—*Biological Reviews of the Cambridge Philosophical Society.* doi: 10.1111/j.1469=185X .2010.00137.x.

INSTITUTIONAL ABBREVIATIONS

AMNH	American Museum of Natural History, New York, USA
BPI	Bernard Price Institute for Palaeontological Research, Johannesburg, South Africa
BSP	Bayerische Staatssammlung für Paläontologie und Geologie, Munich, Germany
BYU	BYU Museum of Paleontology, Brigham Young University, Provo, Utah, USA (formerly BYU Earth Science Museum)
CM	Carnegie Museum of Natural History, Pittsburgh, Pennsylvania, USA
DFMMh/FV	Dinosaurier-Freilichtmuseum Münchehagen/Verein zur Förderung der Niedersächsischen Paläontologie e.V., Münchehagen, Germany
FGGUB	Faculty of Geology and Geophysics, University of Bucharest, Bucharest, Romania
GPIT	*See* IFG
IFG	Institut für Geowissenschaften, Eberhard-Karls-Universität Tübingen, Tübingen, Germany (formerly Geologisch-Paläontologisches Institut Tübingen, abbreviated GPIT)
IPB	Steinmann-Institut für Geologie, Mineralogie und Paläontologie, Rheinische Friedrich-Wilhelms-Universität Bonn, Germany
IVPP	Institute for Vertebrate Paleontology and Paleoanthropology, Beijing, China
LMC	Musée de Cruzy, Association Culturelle Archéologique et Paléontologique, Cruzy, Hérault, France
MACN	Museo Argentino de Ciencias Naturales "Bernardino Rivadavia," Buenos Aires, Argentina
MAFI	Geological Survey of Hungary (MAFI), Budapest, Hungary
MB	*See* MFN
MCP	Museu de Ciências e Tecnologia, Porto Alegre, Brazil
MDE	Musée des Dinosaures, Espéraza, Aude, France
MFN	Museum für Naturkunde—Leibniz-Institut für Evolutions- und Biodiversitätsforschung an der Humboldt-Universität zu Berlin, Germany
NAA	Naturama Naturmuseum Aargau, Aarau, Aargau, Switzerland
NM	National Museum, Bloemfontein, South Africa
OMNH	Oklahoma Museum of Natural History, Norman, Oklahoma, USA
P.DMR	Palaeontological collection, Department of Mineral Ressorces, Khon Kaen, Kalasin, Thailand
PMU	Museum of Evolution, University of Uppsala, Uppsala, Sweden
PVL	Fundación Miguel Lillo, Tucumán, Argentina
PVSJ	Universidad Nacional de San Juan, Argentina
SMA	Sauriermuseum Aathal, Aathal, Switzerland
SMF	Sauriermuseum Frick, Frick, Switzerland
SMNS	Staatliches Museum für Naturkunde Stuttgart, Germany
UCMP	University of California, Museum of Paleontology, Berkeley, USA
YPM	Yale Peabody Museum, New Haven, USA
ZDM	Zigong Dinosaur Museum, Zigong, China

INTRODUCTION

1

Sauropod Biology and the Evolution of Gigantism: What Do We Know?

MARCUS CLAUSS

Life scientists are concerned with the description of the life forms that exist and how they work—an inventory of *what is.* Additionally, life scientists want to understand why life forms are what they are—from both a historical and functional perspective. Evolutionary theory offers a link between both perspectives via the sequence of organisms that have evolved and are constantly adapting to their environment by natural selection. But, still unsatisfied, life scientists want to discover why selection acts in a certain way. We want to understand *what is* within the framework of *what is possible,* by distilling universal rules from our inventories to understand the limitations of *what could be.* Only if we understand what is possible will we be ready to accept historical reasons for the absence of a life form. "It just didn't happen" will only sound plausible and satisfying if we know whether it could have.

With this approach, any expansion of the inventory of *what is* will automatically lead to a reevaluation of those theories that explain *what is possible.* Every discovery of a new species or a new ecosystem will make such a reevaluation necessary; the more the new discovery deviates from what has been recorded so far, the more necessary the reevaluation. In this respect, dinosaurs are invaluable to us. They expand the inventory of life forms that have developed at some stage during the existence of our planet and evidently must have been subjected to a similar set of constraints that we assume for extant life forms. Yet because they are different enough, they are a challenge to our concepts—an outgroup against which our biological understanding must be tested. Therefore, as Dodson (1990) put it, advancing our understanding of dinosaurs also means understanding the world we live in.

Sauropods are the ultimate outgroup among terrestrial vertebrates simply because of their size. Their vast dimensions and sheer existence in the history of life oblige us to evaluate any potential limits of body size in terrestrial vertebrates. Whereas many other fossil life forms can fit comparatively easily within existing frameworks, sauropods appear so far out of the range that they are a definite challenge.

However, before the riddle of sauropod size can be solved, the seemingly more profane task of reconstructing these organisms from the fossil record must be carried out (Sander et al. 2010a). This alone can be demanding, as can be nicely traced in the history of sauropod research—for example, the conceptual shift from an aquatic to a terrestrial lifestyle, the shift from a viviparous to an oviparous reproduction, the shift from a sprawling to a columnar stance (McIntosh 1997), or the shift from a digestive system with a gizzard full of gastroliths to a digestive tract with no particle size reduction at all (Wings & Sander 2007). Knowledge about sauropod morphology, systematics, diversity, and evolution has been summarized by Upchurch et al. (2004) and Curry Rogers & Wilson (2005). The latter reference also deals with some aspects of sauropod biology.

With particular reference to the chapters of this book and to the work of our research group, the current knowledge about the history, form, and function of sauropods can be briefly summarized as follows (see also Sander et al. 2010a).

Sauropods evolved from basal sauropodomorphs, although exact phylogenetic relationships are not resolved (Chapter 8). They are characterized by a quadrupedal stance with columnar legs, a long neck and tail, and a comparatively small head (Chapters 8, 11, 15). Regardless of the large taxonomic diversity of sauropods, this basic body plan hardly varies (Chapter 8). Sauropods can be broadly grouped into forms with longer front legs, a presumably upright neck, and a rather cranial center of gravity, and forms with longer hind legs, a presumably more horizontal neck, and a rather caudal center of gravity (Chapters 8, 14). The reconstruction of their muscular and skeletal anatomy reflects biomechanical particularities of their body shape and size at the macroscopic as well as the microscopic level (Chapters 8–11, 15). Niche diversification in sauropod taxa can be inferred from differences in dental and cranial anatomy, neck length, and posture (Chapters 2, 10, 14, 15) as well as from isotope studies (Chapter 4). Sauropods were herbivores that did not chew their food and most likely did not possess other means of food particle size reduction, such as a gizzard with gastroliths (Chapter 2). They probably relied on symbiotic microflora in a massive hindgut to ferment plant material (Chapter 2), using the available plant resources of their time, as extant herbivores do today (Chapters 3, 4). They probably had heterogeneous "bird-like" lungs with air sacs and pneumatization of various bony structures, in particular the neck vertebrae (Chapter 5). It is generally thought that sauropods had a metabolic rate higher than that of extant ectotherms, although the difference in rates is still under debate; an ontogenetic decrease of metabolic rate has been suggested (Sander & Clauss 2008). Even more controversially debated is their cardiovascular system, for which consensus has not been reached, apart from assuming that they had four-chambered hearts (Chapter 7). Sauropods were oviparous and laid hard-shelled eggs, probably in numerous small clutches (Sander et al. 2008), which also facilitated fast population regrowth (Chapter 16). The young grew rapidly and reached sexual maturity in their second decade of life (Chapter 17). Parental care was probably absent and juvenile mortality high (Chapter 16), with different predators of the time feeding on the various ontogenetic stages of sauropods (Hummel & Clauss 2008). Although evidence has been hard to come by, it is likely that sauropods lived in groups or herds, some of which appear to have been age segregated (Coombs 1990; Myers & Fiorillo 2009).

All of the above does not sound particularly exceptional. However, sauropods did all this while achieving adult body masses between 15 and 100 metric tons (Chapter 6; Appendix). No other groups of terrestrial vertebrates have ever reached such a size. Because the advantages of a large body size (Chapter 12) apply to terrestrial vertebrates in general, the obvious question haunting life scientists (including paleontologists) is: what factors allowed the sauropods—and the sauropods alone—to become so large?

The easy way out is to simply answer that it is the combination of all the factors mentioned above that allowed these terrestrial vertebrates to become giants. In other words, to become as large as a sauropod, you have to be a sauropod. In a historical sense, this is probably true. In a functional approach, however, characteristics that are independent of body mass, characteristics that just follow body mass, and characteristics that truly facilitate gigantism should be differentiated from one another.

For example, many biomechanical adaptations of sauropods were a precondition for, and a consequence of, their large body size (Chapters 8–11, 15), but these adaptations appear easy to achieve by other vertebrates in the sense of convergent evolution and thus do not appear to be the crucial factors triggering gigantism. Actually, it is the universal applicability of the laws of static and dynamic mechanics that facilitates our understanding of these convergent adaptations. The origin of these adaptations, according to mechanical principles, makes them particularly suitable for investigations by computer modeling based on these principles (Chapters 10, 11, 13, 14). These studies are crucial for our understanding of how a giant works, but they cannot explain the origin—and the uniqueness—of sauropod gigantism. Similarly, the botanical (Chapters 3, 4) and nutritional (Chapter 2) composition of potential sauropod food and the presumably enormous digestive tract of sauropods (Chapter 2) can be described, but again, these factors do not set sauropods apart from other vertebrates. Unless we are thinking of absolute limits to skeletal static due to gravity (Hokkanen 1986; Alexander 1989), it seems that both the vertebrate musculoskeletal and the digestive system can accommodate any given body size, whether large or not.

Whether the same can be assumed for the cardiovascular system is a topic of intensive scientific debate (Seymour 2009a; Sander et al. 2009; Chapter 7). The peculiar neck of sauropods, which has been suggested to have been held in many sauropods in an upright, distinctly inclined or curved posture based on skeletal reconstructions and in analogy with extant amniotes (Taylor et al. 2009; Chapter 15), poses a dramatic conceptual problem in terms of the mechanics and energetics of the cardiovascular system (Seymour 2009a, 2009b). To me—an animal nutritionist and digestive physiologist with no background in cardiovascular physiology or musculoskeletal reconstructions—both sets of arguments appear convincing; the resolution of this scientific issue is a major challenge for future studies on sauropod paleobiology. However, it is noteworthy that the posture of a particular body part, not the giant body size in general, is the bone of contention here.

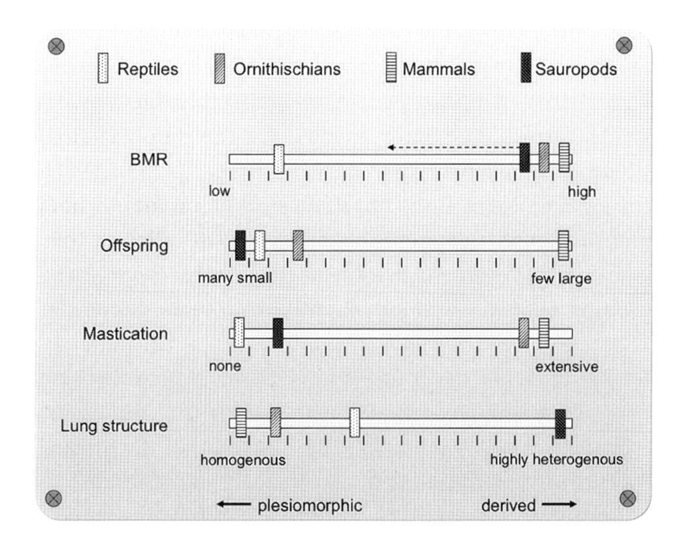

FIGURE 1.1. Respective sets of selected morphophysiological characteristics of sauropods, ornithischians, terrestrial mammals, and nondinosaurian reptiles, visualized as a slider panel. BMR, basal metabolic rate; the dashed arrow indicates a hypothetical ontogenetic reduction in BMR. The gigantism of sauropods is explained by their peculiar combination of plesiomorphic and derived characteristics. *Adapted after Sander & Clauss (2008).*

Whether the topic of thermoregulation in sauropods (Chapter 7) is only interesting for the biology of these particular animals, or whether thermoregulation—and hence metabolism—is crucial for the evolution of gigantism is also subject to ongoing scientific debate. Different authors have claimed that gigantic body size poses a constraint on heat dissipation and hence the level of metabolism at which a giant can operate. However, at the First International Workshop on Sauropod Biology and Gigantism held by our research group in Bonn in 2008, Roger Seymour explained that data on the body temperature (Clarke & Rothery 2008), metabolic rate (Paladino et al. 1981), and geographic distribution of elephants do not point out particular problems with overheating in these animals and suggested that in previous models, the immediate transport of heat to the body surface via the vascular system had not been appropriately considered. Evidently, more elaborate models are needed to understand the potential implications of heat production and heat loss in giant organisms. At the gigantic size of sauropods, thermal inertia will have undoubtedly guaranteed a comparatively constant core body temperature. Analogy with extant "mass homoiotherms," such as giant tortoises, however, might raise doubts that their high level of activity (as inferred from sauropod trackways, for example) can be accounted for by mass homoiothermy alone. Growth rates assumed for sauropods (Chapter 17) are hard to imagine without high metabolic rates, and it has been suggested that gigantism as observed in sauropods is not possible for ectothermic animals (Head et al. 2009). As a convenient compromise between the different aspects of sauropod metabolism, we could consider the ideas of Farlow (1990) and Sander & Clauss (2008), who suggest an ontogenetic drop in metabolic rate (indicated by the dashed arrow in Fig. 1.1) that facilitated the rapid growth of juveniles but eased heat stress and nutritional requirements in adults. This hypothesis awaits further corroboration.

When comparing sauropods to giant terrestrial mammals, anatomical and physiological features set sauropods apart—in particular their mode of reproduction, long neck, respiratory system, and lack of mastication. In contrast, growth rates and possibly metabolism were similar enough to some degree in these groups (Chapters 2, 5, 15–17). Therefore, the hypothesis that it was a combination of these factors that made sauropod gigantism possible comes to mind (Sander & Clauss 2008; Fig. 1.1). However, each of these factors will have to be scrutinized for plausibility and, if possible, tested.

Testing physiological features in extinct animals is obviously problematic. More precise concepts of niche partitioning are difficult to evaluate because the fossil record does not provide sufficient resolution to associate specific dinosaurs with specific plants (Butler et al. 2009, 2010; Chapter 4). Bone and dental tissue can yield information on growth through histological analyses (Chapter 17), as well as additional information on diet, thermoregulation, and migration through isotope analyses (Tütken et al. 2004; Amiot et al. 2006; Fricke et al. 2009; Chapter 4). Isotopic studies in particular have the advantage that they present alternative approaches to questions that have previously been answered with other methods; in this respect, they represent true tests. So far, such tests appear to be in accordance with our hypotheses.

Unfortunately, generating hypotheses that are based on skeletal features that can be tested by other skeletal features alone is rarely possible. The association of features that facilitate the rearing of a sauropod on its hind legs and the mobility of its neck (Chapter 14) represents such a rare example. For other hypotheses—such as the possible role of a long neck, the presence of a bird-like respiratory system, and the absence of mastication—more theoretical approaches, often involving allometric extrapolations, have to be used. Because sauropods invariably lie outside the range of the data from which the allometric regressions have been derived, such an approach must always remain speculative. Only the qualitative difference between vivipary and ovipary is so evident that its relevance for population survival can be immediately understood (Chapter 16).

Whether a long neck represents an energetic advantage, as suggested in Chapter 12, that might have enhanced the evolution of giant body size, or whether it simply represents a feature that most nonchewing herbivores could evolve independently of body size has been hotly debated within our research group. As long as model calculations on the energetic costs and benefits of long necks over the entire body size range covered by juvenile to adult sauropods are lacking, this issue will remain unresolved (Seymour 2009a, 2009b; Sander et al. 2009). Similarly, the potential advantage of bird-like lungs remains speculative as long as physiological models that take a comparative approach in quantifying particular lung functions—for example, that of heat exchange—for "mammal-like" and "bird-like" systems are lacking. However, even if the direct link between bird-like lungs and gigantism is not yet compelling, its absence in both terrestrial mammals and the Ornithischia (Wedel 2006; Fig. 1.1), which both did not attain the giant sizes of sauropods, is a strong indication for the relevance of such a system in the evolution of gigantism. Unfortunately, we still lack a model demonstrating that mammal-like lungs constrain body size.

However, for another sauropod characteristic, such a constraint can be comfortably assumed, and it is with the narrow-mindedness of a researcher trapped in his own research field that I state here that its connection to gigantism can be considered relatively obvious: the absence of mastication (Sander & Clauss 2008; Sander et al. 2010a, 2010b). As with the respiratory system, sauropods differ from both mammalian and ornithischian herbivores in this respect (Fig. 1.1). Among terrestrial mammalian herbivores, which all display formidable adaptations for masticatory particle-size reduction of their food, the percentage of time spent feeding increases in an allometric fashion with body mass that would require feeding for more than 100% of the day (Owen-Smith 1988; Chapter 2) in animals weighing more than approximately 18 metric tons. Because this threshold coincides with mass estimates for the largest terrestrial mammal (*Indricotherium;* Fortelius & Kappelman 1993), the largest ornithischian (*Shantungosaurus;* Horner et al. 2004), and roughly with the lower body-mass range of the adults of many sauropod taxa, the interpretation appears attractive that herbivores, once they had evolved the very efficient adaptation of mastication, were prevented from evolving giant body size because this would have necessitated a secondary loss of mastication. Thus, it seems that a primitive feature of sauropods—the absence of mastication—allowed them to enter the niche of giants. From a certain body size onward, food particle size will be determined by plant morphology alone and hence will remain rather constant, while gut capacity will further increase with increasing body size. Therefore, sauropods might represent a rare example of herbivores that actually benefit from an increase in body size in terms of a larger gut and a longer retention of food in that gut without incurring the disadvantage of decreasing chewing efficiency (Chapter 2).

Ultimately, though, body mass will be constrained by the resources available. Biomass availability depends on climatic factors and habitat quality, and at giant size, it is restricted mainly by land mass. Evidence suggests that sauropods—somehow—followed this pattern (Burness et al. 2001). The question of whether the number and diversity of smaller herbivores had an impact on the resources available for these giants is difficult to answer using the fossil record, and it is not resolved for recent ecosystems either. However, in parallel to the argument that an increased diversity of carnivores could indicate a higher amount of biomass available for secondary consumers in dinosaur ecosystems (Hummel & Clauss 2008), it might be possible to test whether the diversity of regular-sized and giant-sized herbivores is reciprocal across ecosystems, indicating that the presence of giant herbivores can reinforce their own dominance via interference competition (Persson 1985). In the end, it will be through the understanding of their ecosystems, as has been repeatedly advocated, for example, by Farlow (2007), that the full dimension of gigantism will be understood. Until then, we will continue our integrated studies on the reconstruction of sauropod physiology, life history, and population biology in our attempt at understanding the life of giants (Sander et al. 2010a).

Acknowledgments

I thank Martin Sander (University of Bonn) for inviting me into, and for efficiently leading, this fascinating research unit; my colleagues of this research unit; the Deutsche Forschungsgemeinschaft (DFG) for supporting our work; Jürgen Hummel (University of Bonn) for his companionship in the "Sauropod Nutrition Squad"; Martin Sander (University of Bonn) for drawing the figure; and Nicole Klein, Kristian Remes, Carole Gee, and Martin Sander (all at the University of Bonn) for editing this book and inviting me to write this introduction. This is contribution number 60 of the DFG Research Unit 533 "Biology of the Sauropod Dinosaurs: The Evolution of Gigantism."

References

Alexander, R. M. 1989. *Dynamics of Dinosaurs and Other Extinct Giants.* Columbia University Press, New York.

Amiot, R., Lécuyer, C., Buffetaut, E., Escarguel, G., Fluteau, F. & Martineau, F. 2006. Oxygen isotopes from biogenic apatites suggest widespread endothermy in Cretaceous dinosaurs.—*Earth and Planetary Science Letters* 246: 41–54.

Burness, G. P., Diamond, J. & Flannery, T. 2001. Dinosaurs, dragons, and dwarfs: the evolution of maximal body size.—*Proceedings of the National Academy of Sciences of the United States of America* 98: 14518–14523.

Butler, R. J., Barrett, P. M., Kenrick, P. & Penn, M. G. 2009. Diversity patterns amongst herbivorous dinosaurs and plants during the Cretaceous: implications for hypotheses of dinosaur/angiosperm co-evolution.—*Journal of Evolutionary Biology* 22: 446–459.

Butler, R. J., Barrett, P. M., Penn, M. G. & Kenrick, P. 2010. Testing evolutionary hypotheses over geological time scales: interactions between Cretaceous dinosaurs and plants.—*Biological Journal of the Linnean Society.* 100: 1–15.

Clarke, A. & Rothery, P. 2008. Scaling of body temperature in mammals and birds.—*Functional Ecology* 22: 58–67.

Coombs, W. P. 1990. Behavior patterns of dinosaurs. *In* Weishampel, D. B., Dodson, P. & Osmólska, H. (eds.). *The Dinosauria.* University of California Press, Berkeley: pp. 32–42.

Curry Rogers, K. & Wilson, J. A. (eds.). 2005. *The Sauropods: Evolution and Paleobiology.* University of California Press, Berkeley.

Dodson, P. 1990. Dinosaur paleobiology. *In* Weishampel, D. B., Dodson, P. & Osmólska, H. (eds.). *The Dinosauria.* University of California Press, Berkeley: p. 31.

Farlow, J. O. 1990. Dinosaur energetics and thermal biology. *In* Weishampel, D. B., Dodson, P. & Osmólska, H. (eds.). *The Dinosauria.* University of California Press, Berkeley: pp. 43–55.

Farlow, J. O. 2007. A speculative look at the paleoecology of large dinosaurs of the Morrison Formation, or Life with *Camarasaurus* and *Allosaurus*. *In* Kvale, E. P., Brett-Surman, M. K. & Farlow, J. O. (eds.). *Dinosaur Paleoecology and Geology: The Life and Times of Wyoming's Jurassic Dinosaurs and Marine Reptiles.* Geoscience Adventure Workshop, Shell: pp. 98–151.

Fortelius, M. & Kappelman, J. 1993. The largest land mammal ever imagined.—*Zoological Journal of the Linnean Society of London* 107: 85–101.

Fricke, H. C., Rogers, R. R. & Gates, T. A. 2009. Hadrosaurid migration: inferences based on stable isotope comparisons among Late Cretaceous dinosaur localities.—*Paleobiology* 35: 270–288.

Head, J. J., Bloch, J. I., Hastings, A. K., Bourque, J. R., Cadena, E. A., Herrera, F. A., Polly, P. D. & Jaramillo, C. A. 2009. Giant boid snake from the Palaeocene neotropics reveals hotter past equatorial temperatures.—*Nature* 457: 715–717.

Hokkanen, J. E. I. 1986. The size of the largest land animal.—*Journal of Theoretical Biology* 118: 491–499.

Horner, J. R., Weishampel, D. B. & Forster, C. A. 2004. Hadrosauridae. *In* Weishampel, D. B., Dodson, P. & Osmólska, H. (eds.). *The Dinosauria. 2nd edition.* University of California Press, Berkeley: pp. 438–463.

Hummel, J. & Clauss, M. 2008. Megaherbivores as pacemakers of carnivore diversity and biomass: distributing or sinking trophic energy?—*Evolutionary Ecology Research* 10: 925–930.

McIntosh, J. S. 1997. Sauropoda. *In* Currie, P. J. & Padian, K. (eds.). *Encyclopedia of Dinosaurs.* Academic Press, San Diego: pp. 654–658.

Myers, T. S. & Fiorillo, A. R. 2009. Evidence for gregarious behavior and age segregation in sauropod dinosaurs.—*Palaeogeography, Palaeoclimatology, Palaeoecology* 274: 96–104.

Owen-Smith, N. 1988. *Megaherbivores: The Influence of Very Large Body Size on Ecology.* Cambridge University Press, Cambridge.

Paladino, F. V., Spotila, J. R. & Pendergast, D. 1981. Respiratory variables of Indian and African elephants.—*American Zoologist* 21: 1043.

Persson, L. 1985. Asymmetrical competition: are larger animals competitively superior?—*American Naturalist* 126: 261–266.

Sander, P. M. & Clauss, M. 2008. Sauropod gigantism.—*Science* 322: 200–201.

Sander, P. M., Peitz, C., Jackson, F. D. & Chiappe, L. M. 2008. Upper Cretaceous titanosaur nesting sites and their implications for sauropod dinosaur reproductive biology.—*Palaeontographica Abt.* A 284: 69–107.

Sander, P. M., Christian, A. & Gee, C. T. 2009. Sauropods kept their heads down. Response.—*Science* 323: 1671–1672.

Sander, P. M., Christian, A., Clauss, M., Fechner, R., Gee, C. T., Griebeler, E. M., Gunga, H.-C., Hummel, J., Mallison, H., Perry, S., Preuschoft, H., Rauhut, O., Remes, K., Tütken, T., Wings, O. & Witzel, U. 2010a. Biology of the sauropod dinosaurs: the evolution of gigantism. *Biological Reviews of the Cambridge Philosophical Society.* doi:10.1111/j.1469-185X.2010.00137.x.

Sander, P. M., Gee, C. T., Hummel, J. & Clauss, M. 2010b. Mesozoic plants and dinosaur herbivory. *In* Gee, C. T. (ed.). *Plants in Mesozoic Time: Morphological Innovations, Phylogeny, Ecosystems.* Indiana University Press, Bloomington: 331–359.

Seymour, R. S. 2009a. Sauropods kept their heads down.—*Science* 323: 1671.

Seymour, R. S. 2009b. Raising the sauropod neck: it costs more to get less.—*Biology Letters* 5: 317–319.

Taylor, M. P., Wedel, M. J. & Naish, D. 2009. Head and neck posture in sauropod dinosaurs inferred from extant animals.—*Acta Palaeontologica Polonica* 54: 213–220.

Tütken, T., Pfretzschner, H. U., Vennemann, T. W., Sun, G. & Wang, Y. D. 2004. Paleobiology and skeletochronology of Jurassic dinosaurs: implications from the histology and oxygen isotope compositions of bones.—*Palaeogeography, Palaeoclimatology, Palaeoecology* 206: 217–238.

Upchurch, P., Barrett, P. M. & Dodson, P. 2004. Sauropoda. *In* Weishampel, D. B., Dodson, P. & Osmólska, H. (eds.). *The Dinosauria. 2nd edition.* University of California Press, Berkeley: pp. 259–322.

Wedel, M. J. 2006. Origin of postcranial skeletal pneumaticity in dinosaurs.—*Integrative Zoology* 2: 80–85.

Wings, O. & Sander, P. M. 2007. No gastric mill in sauropod dinosaurs: new evidence from analysis of gastrolith mass and function in ostriches.—*Proceedings of the Royal Society B: Biological Sciences* 274: 635–640.

PART ONE

NUTRITION

2

Sauropod Feeding and Digestive Physiology

JÜRGEN HUMMEL AND MARCUS CLAUSS

SAUROPOD DINOSAURS DOMINATED the large herbivore niche in many Mesozoic ecosystems. On the basis of evidence from extant herbivores, significant symbiotic gut microbe activity can safely be inferred for these animals. A hindgut fermentation chamber as in horses or elephants appears more likely than a foregut system. Sauropods are unusual in several herbivore-relevant features such as their large foraging range (due to a long neck), apparent lack of food comminution (which is highly untypical for large extant herbivores), and their extremely high body weights (which is likely linked to several key features of herbivore foraging and digestion). On the basis of regressions on extant herbivores, their gut capacity can be safely assumed to have been highly comprehensive in relation to energy requirements. This can, but need not necessarily, imply extremely long food retention times. Besides these animal features, the spectrum of food plants available for sauropods in sufficient quantity (sphenophytes, pteridophytes, and gymnosperms) was completely different from that of extant herbivores (mostly angiosperms), which has some potential implications for the respective harvesters of these plants. Gymnosperms have a tendency to facilitate rather large cropping sizes (measured in kilograms of dry matter per bite) and therefore large intakes. In vitro digestibility of several living representatives of potential sauropod food plants was estimated to be better than expected, and at least comparable to the level of extant browse. Although sauropods are different from extant large herbivores in several aspects, they must be considered one of the greatest success stories in the long history of large animal herbivory.

Introduction

The longnecks, as sauropods are sometimes called by young dinosaur enthusiasts, are still often perceived as gigantic but strange creatures with a funny body shape, rather than as evolutionary successful animals. However, because they are the largest herbivores ever, as well as the terrestrial vertebrates that dominated the megaherbivore niche of most land masses from the end of the Triassic until the end of the Cretaceous for an incredible 135 million years, they should instead be regarded as the most successful vertebrate herbivores ever known. When referring to them as "the sauropods," one must not forget that this group is made up of a large group of diverse herbivores that should probably be no more regarded as uniform in their digestive physiology than, for example, "the primates," which also utilize a great variety of digestive strategies.

The differences repeatedly demonstrated in the skull anatomy and dentition between different sauropod clades (Calvo 1994b; Christiansen 2000; Upchurch & Barrett 2000), for example, exceed in their complexity those observed between artiodactyls and perissodactyls. It seems likely that some taxa were specialized at least to some degree on certain groups of plants—just like many megaherbivores today that can feed quite selectively on certain plant types such as grass (hippo, white rhino, large bovids), browse (giraffe, black rhino, Sumatran rhino), or use a combination of grass and browse (Indian rhino, with some tendency to include more grass in its diet, or elephants, of which African elephants have the tendency to include less grass in their diet compared to the Asian elephant) (Clauss et al. 2008a). When speculating on the digestive physiology of sauropods, one has to be aware that we rely almost exclusively on extrapolations from extant organisms; in other words, we have to extrapolate far beyond the body mass range from which our knowledge on digestive processes is derived, and we have to use as modern analogs another clade (mammals) that currently occupies the megaherbivore niche. Almost all information on the digestive physiology is in the soft tissue of the stomach and intestines, which do not occur in the fossil record of dinosaurs. Although a variety of coprolites and fossilized gut contents have been described from herbivorous dinosaurs (Stokes 1964; Chin & Gill 1996; Hollocher et al. 2001; Ghosh et al. 2003; Prasad et al. 2005), hardly any of these can be safely considered of sauropod origin (Sander et al. 2010); therefore, this source of information is at the moment not available for sauropod research. Instead, we have to speculate on the digestion of sauropods by means of educated guesses that are in part based on extrapolations from extant herbivores.

Just about everything has been said about dinosaur feeding (Coombs 1975; Bakker 1978; Krassilov 1981; Weaver 1983; Coe et al. 1987; Farlow 1987; Dunham et al. 1989; Weishampel & Norman 1989; Dodson 1990; Taggart & Cross 1997; Tiffney 1997; Upchurch & Barrett 2000; Magnol 2003). Any attempt on our part to outline the physiological characteristics of dinosaurs will therefore by necessity reiterate statements that can be found somewhere in the scientific literature. Especially because we are dealing with speculation rather than hard data, it is often difficult to properly honor all those who have already published a similar thought. For this reason, our review is not meant to be a conclusive history of all citations and ideas but instead a selective presentation. However, we want to specifically mention here the insightful works by Farlow (1987) and Paul (1998) that touch on sauropod feeding, nutrition, and digestive physiology.

Sauropods are different. Apart from our fascination in reconstructing these giants, they show alternative evolutionary strategies in vertebrate digestive physiology that we would have not thought of, and by doing so, they elucidate constraints under which extant herbivores operate that we would not have noticed as constraints but rather would have taken as a matter of course.

Feeding and Food Processing in Sauropods

On the basis of analyses on their dentition, all sauropods appear to have been exclusively herbivorous as adults (Upchurch & Barrett 2000; Weishampel & Jianu 2000; Barrett & Upchurch 2005; Stevens & Parrish 2005; Sander et al. 2009, 2010a), which does not exclude the occasional use of arthropods or other small animals by hatchlings (Barrett 2000).

NECKS

Among the most remarkable features of sauropods are their large body size and very long necks. Neck types and body forms vary in sauropods, with different types such as *Brachiosaurus* (long neck with long front legs), *Diplodocus* (long neck but rather short front legs), or *Dicraeosaurus* (rather short neck). The biomechanical function of the neck (Stevens & Parrish 1999, 2005; Christian 2002; Christian & Dzemski 2007; Christian & Dzemski, this volume) and evolutionary causes behind the evolution of a long neck (such as that hypothesized for giraffe by Simmons & Scheepers 1996; e.g., sexual selection, Senter 2007) have received considerable attention and discussion (for a review see Sander et al. 2010b). Despite its apparent obviousness, it has only recently been shown explicitly for the giraffe that its long neck is most likely an outcome of feeding competition within the browsing guild (Cameron & Du Toit 2007). Du Toit (1990) also described a clear stratification of feeding height between giraffe and other African browsing ruminants. It appears most likely that this interpretation applies to sauropods as well—a result that sorts sauropod taxa according to their feeding height (Upchurch & Barrett 2000). Given their enormous neck, the feeding range of most sauropods has to be regarded as extremely large, a characteristic also found in the elephant, the largest living herbivore, as a result of its trunk (Colbert 1993).

SKULL AND TEETH

The pencil-shaped teeth restricted to the front of the snout in diplodocoids and titanosaurs versus the more massive dentitions of spoon-shaped teeth with wear facets in basal sauropods and basal macronarians can be regarded as extremes of sauropod dentition and skull types (Fig. 2.1). Although the former type suggests a raking type of plant cropping (the animal raking off the leaves of a twig, leaving behind the stripped, less digestible woody shoots), the latter type of teeth allows some biting off and potentially a limited degree of mastication in the sense of puncturing or even crushing the material—or at least damaging the leaf cuticle, the major barrier for microbial access—during ingestion. Differences in the microwear of teeth have been demonstrated (Fiorillo 1998), and corre-

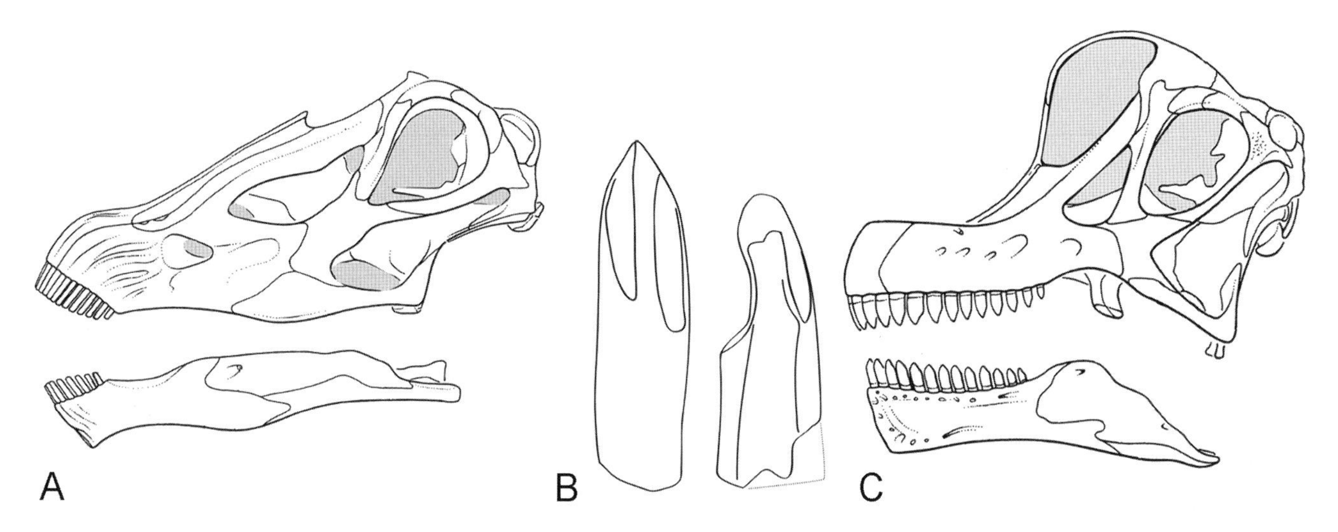

FIGURE 2.1. Sketch of two different sauropod skulls and their characteristic dentition. (A) Skull of *Diplodocus* sp. Note the typical small pencil-like teeth restricted to the anterior part of the skull. This kind of dentition allows only cropping and raking off plant material and no further mastication. (B) The spoon-shaped teeth of *Brachiosaurus* sp. are larger and show wear facets with a typical abrasion pattern. (C) Skull of *Brachiosaurus* sp. Note the far posterior reaching dentition. *Modified after Wilson & Sereno (1998).*

sponding differences in selected food type have been proposed (Bakker 1986; Galton 1986). Recent investigations and finds reveal an unexpected diversity of dentitions (Barrett & Upchurch 2005), and a more detailed separation of skull types can be applied that produced varying degrees of oral processing of food (Calvo 1994b; Christiansen 2000; Upchurch & Barrett 2000; Barrett & Upchurch 2005). In our opinion, the term "oral processing" should be avoided because it is ambiguous; it is not clear to what extent biting off/cropping of forage, or masticating/comminution of the cropped forage material is meant. More descriptive terms, such as "biting off, " "stripping off, " "chewing," "mastication," and "particle size reduction," and even the more self-evident components of oral processing such as swallowing and lubrication, would facilitate a better understanding. Whatever component of oral processing is referred to in the literature, compared to mammals, sauropods are exceptional herbivores insofar as their teeth lack any adaptation for masticating and grinding food. In concert with different feeding heights, differences in dentition and the extrapolated way of cropping food are commonly thought to have contributed significantly to niche separation of sympatric sauropod taxa, although a concrete interpretation of what the different feeding niches might have consisted of remains vague.

GASTRIC MILL

The alternative efficient option of food processing realized in extant vertebrates is the gastric mill of birds. Although gizzards with some grinding function have been described in a variety of invertebrates (Morton 1979; Dall & Moriarty 1983), among vertebrates, only some herbivorous fish such as mullets, have been reported, apart from the birds, to have a functional food particle size reduction device within their guts (Guillaume et al. 1999). Many authors have favored the existence of such a device in sauropods (Janensch 1929; Bakker 1986; Galton 1986; Farlow 1987; Weishampel & Norman 1989; Wing et al. 1992; Christiansen 1996; Taggart & Cross 1997; Bonaparte & Mateus 1999; Upchurch & Barrett 2000; Sanders et al. 2001). However, others deny the existence of an avian-like gastric mill, arguing that pebble aggregations interpreted as gastric mills are a sedimentological phenomenon (Calvo 1994a; Lucas 2000; Wings 2003, 2005) and that the amounts recovered are far too small to be regarded as functional in animals with the body size of sauropods (Wings & Sander 2007). On the basis of a critical review and evaluation of the fossil record and comparative studies on the gastric mill in ostriches, Wings & Sander (2007) arrived at the conclusion that there is to date no evidence for an avian-style gastric mill in sauropods.

The relevance of this assumption cannot be overestimated. The food that terrestrial herbivores ingest is basically reduced in size in one way: mechanical breakdown. Having passed the site of particle reduction—either the oral cavity with its dental apparatus, or the gastric mill with its gastroliths—there is generally little further breakdown of ingesta particles in terrestrial herbivores (Pearce 1967; Poppi et al. 1980; Murphy & Nicoletti 1984; Mcleod & Minson 1988; Freudenberger 1992; Spalinger & Robbins 1992; Moore 1999). This means that if both a masticatory apparatus and a gastric mill are supposed to be absent in sauropods, it is unlikely that there was any other significant means of ingesta particle size reduction. In particular, long ingesta retention times in the gut, and therefore a long exposure to microbial fermentation, might well have compensated for the lack of particle breakdown (see below), but this does not represent other means by which ingesta particle size was actually reduced. Thus, sauropods appear to be the ultimate herbivore nonchewers.

It should be added here that it is thought that by penetrating lignified cell walls during initial colonization, rumen fungi increase the degradation rate of coarse forage by helping other microbes obtain access to cell walls (Van Soest 1994; see Bjorndal 1997 for a discussion of a similar role of nematodes in the digestive tract of herbivorous reptiles). Such action would partially lessen the decreasing effect of large particle size on degradation rate.

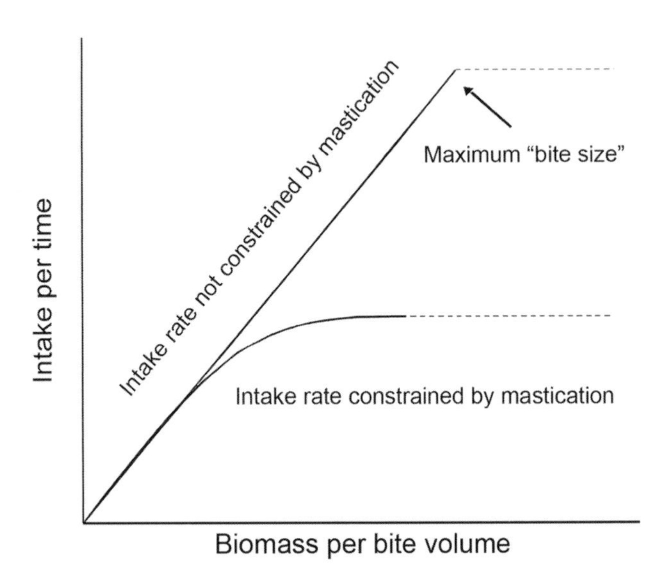

FIGURE 2.2. Potential differences in the functional response of masticating (lower curve) and nonmasticating herbivores (upper curve) (sensu Holling 1959).

CROPPING EFFICIENCY

Animal Factors

Much speculation has been caused by the supposedly small head of sauropods (Russell et al. 1980; Weaver 1983; Coe et al. 1987; Farlow 1987; Dodson 1990; Colbert 1993), which is assumed to be too small to allow a sufficient intake for an endotherm-like metabolism. Although this hypothesis is still sometimes referred to, two separate data collections (Paul 1998; Christiansen 1999) argue against this scenario. Although the skull of sauropods may appear small, this is mainly due to its shortness compared to those of extant herbivores (as a result of its complete lack of chewing teeth); the width of its mouth opening is within the range expected if extrapolated from mammalian herbivores.

Paul (1998) states that the skull width of indricotheres and sauropods of the same body mass did not differ or can even be narrower in the mammalian herbivore, and Christiansen (1999) based his conclusion on measurements on sauropodomorphs (11 species) and mammals (88 species, including 27 species of ungulates). It should be added that applying the respective regressions set up by Christiansen (1999) for artiodactyls/perissodactyls and sauropodomorphs to 10,000 kg animals results in a considerably higher value for the mammalian compared to the dinosaurian herbivore (29.5 vs. 17.7 cm muzzle width). However, any difference in skull dimensions should not be overemphasized in its predictive value for metabolic rate because factors such as plant morphology may determine upper limits for an extensive constant increase in muzzle width with body weight.

Feeding time could still be considered as a limiting factor for sauropod-sized herbivores, given the allometric increase of time devoted to feeding in the activity budget of herbivores (Owen-Smith 1988; see below) and the extremely long feeding times of, for example, elephants (up to 80% of their 24 hour budget). In models of the intake rates of herbivores, there is usually a trade-off between bite size (the amount of food taken in one bite, a function of biomass availability and biomass structure) and bite rate (the number of bites taken per unit time), because larger bite size usually implies longer mastication and hence more time elapsed before the next bite can be taken. Intake rate therefore increases with bite size, but in an asymptotic function (Spalinger et al. 1988; Shipley & Spalinger 1992; Spalinger & Hobbs 1992; Gross et al. 1993a, 1993b; Ginnett & Demment 1995; Bergman et al. 2000; Illius et al. 2002); this function is usually referred to as a type II functional response curve of a species (sensu Holling 1959) (Fig. 2.2). The derived intake models are all based on the assumption that food intake rate is ultimately limited by the rate of oral processing—that is, mainly mastication (Yearsley et al. 2001).

In animals that do not masticate their food, such as extant birds, but also sauropods or stegosaurs, such a trade-off should not exist, and any gain in bite size should instead more directly translate into a gain in foraging rate. In such animals, oral processing consists only of cropping and swallowing, both of which are processes that can be considered to be much less dependent on bite size in terms of the time they require. Thus, the increase in intake rate with increasing bite size might be either linear (type I functional response) or should contain a longer linear component before the limiting effect of cropping/swallowing sets in. Maximum intake would then not be dependent on the process of mastication but on the maximum amount such animals could crop and/or swallow (Fig. 2.2).

Type I functional response curves have actually been found in herbivorous birds (Rowcliffe et al. 1999), but type II curves with an assumed limitation due to the increasing swallowing effort with increasing bite size have also been described (Durant et al. 2003). Actually, it has even been suggested that the oral handling of cropped food in birds may be a time-consuming process similar to chewing in mammals (Van Gils et al. 2007). Studies on the functional response of herbivorous reptiles are lacking.

Plant Factors

An additional important factor that is rarely considered in the discussion of cropping efficiency or intake rate is the morphology of the plants that are actually cropped. In theory, animals should favor such food plants that allow grasping a high amount of biomass per bite (and hence make cropping more efficient) over plants that are structurally formed such that only a small amount can be harvested per bite. Plant distribution expressed as a density measure in terms of biomass per area correlates to intake in mammalian herbivores (Stobbs 1973; Trudell & White 1981; Wickstrom et al. 1984; Spalinger et al. 1988; Wallis de Vries & Daleboudt 1994; Heartley et al. 1997; Shipley et al. 1998; Illius et al. 2002). Although these studies dealt mainly with the effect of moving from one feeding place to the other, Shipley et al. (1998) demonstrated that for moose, conifers offered up to 20 times the available mass in twigs and leaves as compared to dicotyledonous trees. Although the study was performed in winter, when the deciduous dicots would be leafless anyhow, the authors stated that this effect was particularly due to the fundamental difference in plant architecture between the groups. In feeding trials with captive deer (*Odocoileus virginianus*), it was demonstrated that plant morphology, measured as the leaf mass of the terminal 20 cm of a twig, correlated positively with foraging efficiency in terms of intake rate (Koerth & Stuth 1991). Therefore, the assumption that a higher twig biomass will result in higher intakes is supported empirically.

Such considerations can influence our perceptions on sauropod feeding. On the basis of the subjective experience that comparably sized twigs are considerably heavier in a conifer than in a deciduous broad-leaved tree and therefore provide more biomass per bite to an herbivore, we wanted to test whether this impression reflected a real parameter. Twenty individual shoots from different individuals of 14 plant species including different conifers, *Ginkgo,* and *Metasequoia* were investigated for their biomass per sauropod bite (Table 2.1). The depth to which a shoot could be cropped was estimated at 30 cm with a *Diplodocus* skull. Therefore, the first 30 cm of a shoot was clipped, and the resulting piece was divided into two 15 cm parts, representing the first and the second to be encountered by a potential herbivore. The leaves were stripped from the twig, dried to a constant weight, and weighed separately.

For all parameters (leaf weight and twig weight at 15 and 30 cm, respectively, and total weight at 30 cm), there were significant differences between plant groups (Table 2.1). Differences between conifers, angiosperms, and the *Ginkgo/Metasequoia* group were always significant, except for twig weight at 30 cm between angiosperms and *Ginkgo/Metasequoia.* The general pattern was that of higher weights in conifers, intermediate weights in *Ginkgo/Metasequoia,* and lower weights in the deciduous angiosperm trees (Table 2.1).

The clearest differences were apparent in the category of total biomass of the first 30 cm of a twig. However, similar patterns were found in twig and leaf portions, which support these differences, and can be considered as equally relevant for cropping and raking/stripping feeding types. Given this effect of plant morphology on cropping efficiency, differences in cropping efficiency should be one parameter used in models exploring dinosaur foraging. For example, should larger herbivorous dinosaurs be considered time-limited insofar as they had to ingest larger amounts of food, then a tendency for such species to select habitats rich in forage plants of a beneficial morphology would be expected. There are herbivore groups with less efficient forage particle size reduction such as stegosaurs and sauropods, and herbivore groups with a more efficient forage particle size reduction such as ornithopods. Should differences in the degree of forage particle size reduction (comminution/mastication) force herbivore groups that were less effective in this respect to ingest more plant material to compensate for the lower digestibility of larger particles, then again, such species could be expected to cluster in plant communities where plant morphology enhances cropping efficiency.

This explicitly does not rule out the use of conifer vegetation by ornithopods, as, for example, indicated by coprolites (Chin & Gill 1996; Taggart & Cross 1997).

A Potential Factor in Sauropod–Ornithopod Competition?

Several authors have speculated that sauropods occur particularly often in association with conifers in the fossil record, whereas ornithopod dinosaurs, which were notably smaller, occurred mostly in association with angiosperms during the Cretaceous (Coe et al. 1987; Wing & Tiffney 1987; Weishampel & Norman 1989; Dodson 1990; Wing et al. 1992; Leckey 2004; Rees et al. 2004). However, both a convincing demonstration

Table 2.1. Dry Matter (DM) Weights of Leaves and Twigs of Different Plant Species at Different Cropping Lengths

	First 15 cm (g DM)		First 30 cm (g DM)		
Species	Leaf mass	Twig mass	Leaf mass	Twig mass	Total mass
Deciduous angiosperms					
Salix caprea	2.7 ± 0.9	1.0 ± 0.3	6.4 ± 1.6	3.7 ± 1.1	10.1 ± 2.6
Salix alba	1.3 ± 0.2	0.3 ± 0.1	3.1 ± 0.4	1.1 ± 0.1	4.2 ± 0.5
Quercus rubra	2.2 ± 1.1	0.7 ± 0.4	5.4 ± 2.2	2.9 ± 1.6	8.4 ± 3.7
Crataegus laevigata	1.1 ± 0.6	0.9 ± 0.5	2.6 ± 1.1	3.0 ± 1.5	5.6 ± 2.6
Fagus sylvaticus	1.0 ± 0.4	0.3 ± 0.1	2.6 ± 0.7	1.2 ± 0.3	3.8 ± 0.8
Acer pseudoplatanus	2.0 ± 0.6	0.3 ± 0.1	5.1 ± 1.5	1.5 ± 0.4	6.7 ± 1.9
Mean	**1.7** ± 0.7[a]	**0.6** ± 0.3[a]	**4.2** ± 1.6[a]	**2.2** ± 1.1[a]	**6.4** ± 2.4[a]
Conifers					
Cedrus atlantica	4.3 ± 1.7	1.1 ± 0.6	10.8 ± 3.8	4.0 ± 1.4	14.8 ± 4.9
Picea abies	2.2 ± 0.8	1.3 ± 0.6	11.0 ± 3.5	6.1 ± 3.2	17.1 ± 6.4
Abies alba	3.9 ± 1.2	1.8 ± 1.0	14.8 ± 5.5	6.5 ± 3.4	21.4 ± 8.7
Pinus sylvestris	16.9 ± 5.1	2.9 ± 1.2	33.7 ± 10.1	7.9 ± 2.8	41.7 ± 12.5
Mean	**6.8** ± 6.8[b]	**1.8** ± 0.8[b]	**17.6** ± 10.9[b]	**6.1** ± 1.6[b]	**23.7** ± 12.3[b]
Deciduous conifer					
Larix decidua	**0.8**[c] ± 0.3	**0.5**[a] ± 0.2	**2.1**[c] ± 0.8	**2.0**[a] ± 1.1	**4.1**[c] ± 1.8
Extant relatives of potential sauropod food plants					
Ginkgo biloba	4.1 ± 1.5	1.8 ± 0.9	8.8 ± 3.0	4.6 ± 1.2	13.4 ± 3.6
Metasequoia glyptostroboides	2.1 ± 1.1	0.3 ± 0.2	6.9 ± 3.0	1.5 ± 0.8	8.3 ± 3.8
Mean	**3.1**[d]	**1.0**[a]	**7.8**[d]	**3.0**[a]	**10.8**[d]
Fern					
Dryopteris filixmas	0.5[c] ± 0.1	0.1[c] ± 0.0	1.9[c] ± 0.4	0.5[c] ± 0.2	2.4[e] ± 0.6

Data are presented as mean ± standard deviation. Parameters were tested by nested ANOVA by SPSS 12.0 (SPSS, Chicago, IL), with plant species nested into the respective plant groups. Because plant groups and not species were the target of this investigation, statistically significant differences were not pursued among species, only among plant groups, by post hoc tests. Different superscripts indicate significant differences between plant groups within cropping length. Boldface type indicates the mean.

of the parallel rise of angiosperms and ornithopods and a convincing theory for differential adaptation to conifer and angiosperm foliage are presently lacking (Butler et al. 2009, 2010). In particular, the suggestion that the diversification of the more sophisticated ornithopods was driven by the radiation of angiosperms (Weishampel & Norman 1989) lacks a causative connection. The theory of Bakker (1986) focuses on feeding heights but offers little in connection with the newer information on chewing mechanisms.

Our findings suggest that in coniferous habitats, ornithopods would have faced a higher degree of resource competition with sauropods. It is hypothesized that angiosperms, with their decreased biomass density, may have created a competition-reduced refuge for those herbivorous dinosaurs less dependent on high intakes. Thus, aware of the lack of evidence for a connection between angiosperm evolution and major events in the evolution of dinosaur herbivory (Sereno 1997; Taggart & Cross 1997; Weishampel & Jianu 2000; Barrett & Willis 2001; Butler et al. 2009, 2010) and of the historical overemphasis on plant–animal interactions (Midgley & Bond 1991), we suggest that some dinosaur groups could have better thrived feeding on the angiosperms.

Digestive Strategies of Herbivores

SELECTIVITY AND DIET QUALITY

Vertebrate herbivores make use of the cell walls of the plants they ingest by the help of symbiotic gut microbes. These bacteria (with some fungi and protozoa) ferment the structural carbohydrates of plants (such as cellulose, hemicellulose, and pectin), producing the short-chained fatty acids in this process. These are absorbed by the vertebrate host and used as an energy source and precursors for long-chained fatty acids (i.e., adipose tissue) (Stevens & Hume 1998). Extant large herbivores and particularly megaherbivores all have to focus their feeding on cell wall-rich vegetative plant parts such as leaves, stems, and young twigs and bark. Although they will obviously ingest fruits or seeds if available (but mostly accidentally), the spatial and seasonal availability of such high-quality feeds is much too low to meet the quantitative daily requirements of large herbivores (Demment & Van Soest 1985). A diet very high in cell walls can therefore be safely postulated for sauropods. Any scenario considering an extraordinarily high quality diet as a trigger of sauropod gigantism (ironically referred to as the power-bar theory) is extremely unlikely. Although it is obviously true that a good food supply helps an animal to take advantage of its maximal growth potential, this only influences the development of large body size on an individual ontogenetic level, and this relation must not be confused with phylogenetic questions. Some authors even postulate that the evolution of large body size is triggered by an extraordinarily low food quality (Midgley et al. 2002). However, although this is an interesting hypothesis, there is little hard evidence to date to which degree low food quality is more than simply a consequence of poor selectivity associated with large body size.

FERMENTATION VERSUS CELL CONTENT DIGESTION

Among extant herbivores faced with the same digestive challenges as sauropods, different strategies are used to extract nutrients from cell wall-rich plant parts, a food resource with large proportions of material that cannot be digested autoenzymatically (with the enzymes of the animal), but only with the help of gut microbial populations (alloenzymatically) (Langer 1987). In theory, even strict herbivores could adopt a strategy to extract only easily digestible nutrients from plants that can be digested by the enzymes produced by the vertebrate itself, excreting the fiber more or less undigested. However, this strategy has many limitations. Most importantly, it requires a drastically increased food intake as compared to animals that use gut bacteria to ferment the fiber. Few specialized herbivores employ an extreme strategy of high intake/low fiber digestibility. Among the most successful are geese, which (at least seasonally) only extract the easily available nutrients from their food, excreting most fiber (Prop & Vulink 1992). One may speculate that the development of a capacious fermentation chamber is not a good option in (at least seasonal) long-distance flyers such as geese (Klasing 1998).

Among mammals, the giant panda should be mentioned here as a herbivore without a significant symbiotic gut flora (Dierenfeld et al. 1982), but it is obviously not a good example of a very successful herbivore. The concept that sauropods followed a comparable strategy (high intake, very low fiber digestion) means that one has to assume an extremely high food intake—higher than that observed in any other large herbivore of more than 200 kg of body mass today, which is not the most likely option for a megaherbivore, given the special time budget constraints of megaherbivores as outlined above. Among the extant homeotherm megaherbivores, many of which forage for the better part of the day, there is no organism that does *not* relying on fiber fermentation.

Evidence for a Functional Gut Flora in Sauropods

Any skepticism about the existence of a functional gut flora in sauropods may be countered by the following arguments:

1. For members of major extant large herbivore lineages, the development of a fermentation chamber within their guts is the rule rather than the exception. This development occurred in taxa as diverse as artiodactyls, perissodactyls, proboscidians, lagomorphs, rodents, various marsupials, and others among mammals (Stevens & Hume 1995); in birds such as ostrich, rhea, and galliformes (Klasing 1998); in tortoises, iguanids, and agamids (Bjorndal 1997); and even in tadpoles (Pryor & Bjorndal 2005) and several fish lineages (Clements 1997). Among the extant large herbivores, there are still differences in the degree to which fiber is digested. The elephant is the classic example of a herbivore that is dependent on its gut fauna but that nevertheless pursues a strategy of high intake and low digestibility as compared to other herbivores (Clauss et al. 2003), but still with a much higher digestive efficiency than geese or panda.
2. The groups of microbes responsible for the degradation of fiber are among the evolutionarily oldest, which existed some 1,000 million years before dinosaurs appeared (Hume & Warner 1980; Van Soest 1994). Terrestrial vertebrates with symbiotic, fiber-degrading gut microbiota are thought to have appeared in the Carboniferous/Permian. In these animals, the ingestion of detritus or herbivorous insects facilitated colonization of the gut with fiber-degrading microbes (Hotton et al. 1997; Sues & Reisz 1998; Reisz & Sues 2000). Thus, the prerequisites for establishing a functional symbiotic gut flora existed long before sauropods became the ruling large herbivores.
3. Evidence from ruminants shows that isolation of calves may prevent the colonization of their guts by protozoa (which are not considered an essential part of a functional gut flora of the host) (Van Soest 1994), but colonization of the gut with fiber-degrading bacteria occurs even without direct contact with other animals. Even under extremely severe isolation (including sterilized feeds), colonization of the rumen/gut by bacteria cannot be completely prevented (Males 1973). Significant atypical populations may develop under such extremely unnatural circumstances that still

exhibit relevant fiber degrading capacities (dry matter digestion was decreased by 2–10%, and cellulose digestion by 15–40%, according to Males 1973, cited in Dehority & Orpin 1997).

Acquiring Gut Microbes

Consequently, this means that acquiring a functional gut flora is much less of a problem on an evolutionary level than often perceived. For example, the acquisition of symbiotic gut microbes is considered to have occurred independently in several lineages in the late Paleozoic (Sues & Reisz 1998). In line with Hotton et al. (1997), the inoculation with suitable gut microbes need not be considered as limiting, as long as the anatomical prerequisite in the form of a voluminous chamber in the gut is available.

This does not mean that in the ontogenetic development of an individual, active inoculation by gut microbes from conspecifics is not beneficial. If a young animal can acquire a microbial flora from its mother or from conspecifics by mouth-to-mouth contact or by the ingestion of feces, this will represent a digestive advantage, because this flora is probably already adapted to the respective food sources. But particular behavioral adaptations for the acquisition of a gut fauna should be considered more as an improvement of the system rather than a prerequisite (Troyer 1982 is often cited here as supportive evidence for obligatory sociality in herbivorous dinosaurs).

In conclusion, given the broad distribution of symbiotic gut microbes among extant specialized herbivores, it is safe to consider herbivorous dinosaurs as also harboring a symbiotic fiber digesting gut flora (Farlow 1987; Van Soest 1994).

FERMENTATIVE HEAT

The existence of an extensive, active microbial population in a large fermentation chamber has been hypothesized to contribute significantly to temperature regulation. Indeed, sauropods have been compared to giant compost heaps (Farlow 1987). Whether fermentative heat represents a significant contribution to the thermoregulation of herbivores has not yet been analyzed in detail. However, the limited evidence that exists contradicts this idea: Clarke & Rothery (2008) analyzed body temperature across a large variety of mammalian species and concluded that no general pattern of either increasing or decreasing body temperature with increasing or decreasing body mass among herbivores was evident, leading to the conclusion that the contribution of fermentation heat to overall temperature regulation does not follow a consistent pattern. However, this does not mean that such compensation could not occur in other groups such as herbivorous dinosaurs.

FOREGUT VERSUS HINDGUT FERMENTATION

Among vertebrate herbivores, basically two principal sites for fermentation chambers are known (Stevens & Hume 1995). The most basic site to host a microbial population is the hindgut because it is here that some degree of fermentation occurs, in as lightly specialized herbivores such as humans. Taxa employing this strategy cannot make use of the huge amount of microbial mass developing in their gut (these microbes, which are a significant source of protein, cannot be digested and are only excreted), but only of the products of their fermentation (absorbing the short-chained fatty acids). Hindgut fermentation occurs in such diverse groups as tortoises, iguanas, agamids, sea turtles, herbivorous skinks, perissodactyls (horses, rhinos, tapirs), elephants, sirenians, koalas, and wombats, which is concentrated in all of these animals to a considerable extent in the colon. Other sites in the hindgut used as a fermentation chamber are the paired blindsacs in birds such as ostrich and grouse and the cecum of rodents such as capybaras, nutria, guinea pigs, and many others, and lagomorphs such as rabbits. In these mammalian taxa, the strategy of hindgut fermentation is often coupled to coprophagy, allowing the animals to make use of the microbial protein built up in the gut.

In another group of herbivores, the microbial fermentation chamber is located in the foregut. These animals are called foregut fermenters. Apart from producing short-chained fatty acids, the microbes also serve as an important source of protein in these animals; as they are washed out of the foregut, they enter the stomach and small intestine where they can be digested. This setup can be considered a slightly more complicated solution, occurring almost exclusively in specialized mammalian herbivores such as all ruminants, camelids, hippos, peccaries, colobus monkeys, sloths, kangaroos, and to a certain extent in hamsters and voles (Langer 1988). Although it has been assumed to be a strategy restricted to mammals, at least one bird, the hoatzin, has been shown to carry out intensive fermentation in its crop and is therefore counted as a foregut fermenter (Grajal et al. 1989). In reptiles, however, evidence for foregut fermentation is lacking.

But which strategy did sauropods adopt? Because elephants and rhinoceroses use a strategy of hindgut fermentation, most authors have taken this as the most likely option. Given their relatedness with birds, large blindsacs like the paired cecum of ostriches could be envisioned. However, the solution of a foregut cannot be discarded completely on this rough basis of analogy because even among the mammalian megaherbivores, there is the common hippo, with its extensive forestomach fermentation (Clauss et al. 2004). The example of primates shows that even within one closely related taxonomic unit, both systems (hindgut fermentation, as in, e.g., howler monkeys, and foregut fermentation, as in, e.g., colobines) may evolve (Chivers & Hladik 1980). A foregut system seems to be the more complicated system to evolve, and it should be noted that the majority of herbivorous life forms among mammals, fossil or extant, is considered to be or have been hindgut fermenters (Langer 1991). Only when coupled with the physiological mechanism of rumination (regurgitating sorted forestomach contents and rechewing them) did the foregut fermentation system lead to a high degree of species diversity (the camelids and ruminants) (Langer 1994; Schwarm et al. 2009). Without an efficient mastication system, this option is far less likely in sauropods.

ARGUMENTS AGAINST FOREGUT FERMENTATION IN SAUROPODS

It seems that the question of foregut versus hindgut fermentation in dinosaurs has a certain potential fascination (Farlow 1987; Marshall & Stevens 2000). Therefore, we want to present an additional set of arguments that support, in our view, the conclusion that foregut fermentation is a particularly unlikely option for sauropod dinosaurs.

Foreguts Only Function at Low Intake Levels

The foregut fermentation system represents an important constraint that is linked to the differential speed at which plant fiber on the one hand and soluble carbohydrates and other nutrients such as protein or fat on the other hand can be digested. Enzymatic digestion of soluble carbohydrates, protein, and fat is a speedy process, as is the bacterial fermentation of these substances (Hummel et al. 2006a). In energetic terms, however, the bacterial fermentation of these substances represents a loss as compared to autoenzymatic digestion (Stevens & Hume 1998). In contrast to these comparatively quick processes, bacterial fermentation of plant fiber—the major energy source of strict herbivores—requires more time; therefore, a long ingesta retention time is the characteristic of most herbivorous species (Stevens & Hume 1998; Hummel et al. 2006b).

For any given gut system, the ingesta retention time is a function of food intake and the indigestible fraction: the more food ingested, the faster the ingesta is propelled through the gut (Clauss et al. 2007a, 2007b). A hindgut fermenting system is flexible in this respect and allows for a low (e.g., rhinoceros) or high (e.g., elephant) food intake (Clauss et al. 2008b). Autoenzymatic digestion of soluble carbohydrates, proteins, and fat in the small intestine will occur efficiently at any intake level, and only plant fiber digestion in the large intestine will be affected by intake level—higher at lower intake levels (longer retention) or lower at higher intake levels (shorter retention). In the latter case, the lower fiber digestibility can be compensated for by the generally higher food intake. A foregut fermenting system, by contrast, is limited to a comparatively low food intake. Any nutrient ingested will be fermented by the forestomach bacteria first. In the case of soluble carbohydrates, proteins, or fat, this results in a reduced energetic efficiency as compared to autoenzymatic digestion.

Because these easily digestible components are fermented quickly, comparative energetic losses will always occur, regardless of a high or low food intake. However, given a high food intake and hence a shorter retention in the forestomach, plant fiber will be fermented less efficiently. A foregut fermenter with a high food intake would have the worst of both worlds: easily digestible substrates are lost to the less efficient foregut fermentation, and plant fiber is also used less efficiently as a result of the short retention time. Therefore, a comparatively low food intake is the only logical option for foregut fermenters. Although hindgut fermentation allows for the flexibility of either strategy (high intake and less efficient fiber digestion, or low intake and efficient fiber digestion), foregut fermentation is restricted to one of these options: low intake and efficient use of fiber (Clauss et al. 2008b, 2010). This theoretical assumption is supported by the available empirical data for mammalian herbivores. Only those foregut fermenters that have additionally evolved rumination can achieve comparatively high food intakes (Clauss et al. 2007a; Schwarm et al. 2009) because their forestomach can clear the fine (digested) particles selectively while the larger particles are still being digested.

Fats Are Saturated in Foreguts Before They Are Absorbed

Another important effect of forestomach fermentation is the saturation of the ingested fat (Clauss et al. 2009a). Herbivores consume diets high in polyunsaturated fatty acids. If these are absorbed in the small intestine, they are incorporated into body tissue. As a result of the high unsaturation of the absorbed fats in hindgut-fermenting herbivores, their fat is soft. In domestic pigs (hindgut fermenters), there are limitations for the amount of polyunsaturated fats that their diet should contain, because otherwise, their fat becomes too soft and oily for the taste of human consumers. In foregut fermenters, ingested fats are modified by the forestomach bacteria before they are absorbed in the small intestine. This modification is a process of partial saturation, so that foregut fermenters absorb mostly saturated fats. Therefore, their adipose tissue is harder ("lard"). This is the reason why ruminant milk can be turned into butter that is firm at room temperature ("horse butter" would be a fluid oil at room temperature). Because polyunsaturated fatty acids are essential to vertebrates, one often wonders how foregut fermenters can meet their respective nutritional requirements, and the question is still considered unsolved (Karasov & Martínez del Rio 2007). Probably just enough of these polyunsaturated fatty acids can escape the forestomach intact.

These effects would become prohibitive at the retention times necessary in animals that do not chew their food. The bacterial fermentation of plant fiber depends not only on the time available for this fermentation (the ingesta retention time), but also on the size of the fiber particles. As ingesta particles become smaller, they have a higher surface-to-volume ratio, allowing for more bacterial or enzymatic attack per unit volume (reviewed in Clauss & Hummel 2005). In other words, the digestive process is speeded up by ingesta particle size reduction. This fact itself could be an indication why foregut fermentation is rare among reptiles and birds but more common among the chewing mammals. Chewing reduces the retention time necessary for thorough fiber fermentation, so that in chewers, a foregut fermentation strategy can be supported. Although the ingesta throughput is low (when compared to other chewers), the steady outflow of foregut ingesta containing bacterial protein and escaping unsaturated fat is sufficient to maintain body functions. In a hypothetical nonchewing foregut fermenter, given the much longer ingesta retention times required for thorough fiber fermentation in the forestomach, the extremely low ingesta throughput and hence low supply of protein and unsaturated fat may be prohibitive. Actual tests of these hypotheses are lacking. It should be noted, however, that among the primates, foregut fermenters appear to have evolved the more efficient dentition (Fritz 2007), which perhaps alleviates the described effect.

Foreguts Are Problematic for Ontogenetic Diet Shifts

Another important aspect of the fermentation system is its flexibility in terms of ontogenetic diet shifts, as one would expect in organisms that span an enormous body size range during their growth. Ontogenetic diet shifts occur in all mammals (from an animal-derived milk diet to any of the typical mammalian diets) but have also been reported for bird and reptile species (e.g., Bouchard & Bjorndal 2006); they occur in any direction, from carnivory to herbivory, as in lizards and turtles, or from herbivory to carnivory, as in amphibians. Even in groups such as iguanas, which consume a herbivorous diet during all life stages, a diet shift in terms of the fiber content of the ingested diet takes place between juveniles and adults (Troyer 1984; Wikelski et al. 1993). Ontogenetic diet shift has been suggested in sauropods on the basis of the absence of pits in juvenile *Camarasaurus* teeth as opposed to those of adults (Fiorillo 1991). One of the important advantages of a hindgut fermentation system is that it allows an ontogenetic diet shift (as from a carnivorous to an herbivorous diet) and can even be rendered, over time, more efficient in terms of fiber digestion (e.g., if the diet becomes more fibrous over time) by reducing the relative food intake. In contrast, a foregut system only serves for the digestion of plant matter, and other food is not digested efficiently; apart from some marine carnivores, the baleen whales, which feed on animal matter with a certain proportion of chitin (which is chemically similar to cellulose), no extant foregut fermenter is a regular carnivore. In mammalian foregut fermenters, the massive ontogenetic diet shift typical in herbivorous mammals (from an animal-based food —milk—to plant matter) is facilitated by a special anatomical structure, the gastric groove that channels the milk past the foregut and thus prevents malfermentation at this site (Langer 1988, 1993). Malfermentation in this context means that if animal protein such as milk or meat is digested by microbial processes, it becomes a rotting or decay process that can lead to intoxication of the host animal. However, this bypass mechanism depends on the fact that the channeled food is liquid. No bypass of solid food has as yet been described in foregut fermenters. Therefore, ontogenetic diet shifts such as described in nonmammals, which all comprise a shift between two different solid diets, are easy in a hindgut fermentation system, but may be more difficult in a foregut fermentation system.

Given these considerations, it appears more likely that animals that did not masticate or grind their food and for which a huge ontogenetic shift in food composition and digestive strategy must be inferred—such as the sauropods—would have relied on a hindgut fermentation system. It is only in the chewing ornithopods that the evolution of a foregut system would appear more likely.

Allometry of Digestion and Food Selection

INTRODUCTION TO THE METHOD AND USE OF ALLOMETRIC EXTRAPOLATIONS

Although most traits of mammalian anatomy and physiology correlate with body size (mostly expressed as body mass), this relationship is hardly ever linear. A large animal species that has twice the body mass (BM) of a smaller species does not have twice the energy requirements, but only $2 \times BM^{0.75}$ times these requirements; hence $BM^{0.75}$ is called the metabolic body size. Similarly, allometric equations (with a typical exponent) exist for a huge variety of anatomical and physiological and even life-history parameters (Calder 1996).

A common method used to infer characteristics of fossil life forms such as dinosaurs is allometric extrapolation. A mathematical equation is generated on the basis of data from extant animals; one of the factors in this equation is a measure that can also be conveniently derived from fossil material. Subsequently, the data for the fossil animals are entered into the equation to calculate the respective character of interest. This is often done in sequence. For example, first a correlation between bone length (or bone circumference) and body mass is used to estimate body mass of a fossil organism. Next, body

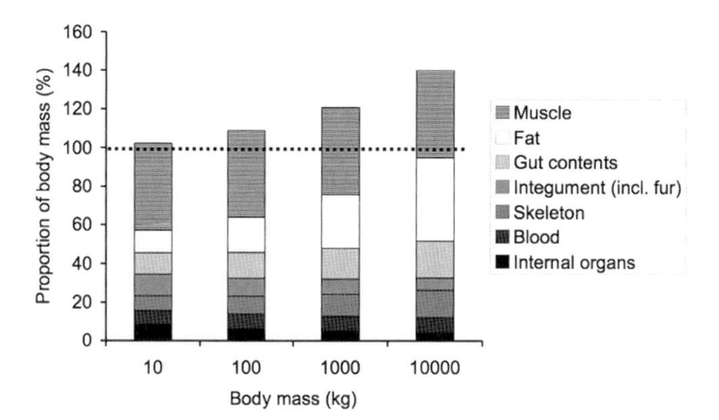

FIGURE 2.3. Calculation of the proportion of mammalian body tissues of the total assumed body mass using allometric equations from Calder (1996) and Parra (1978). Note that for convenience, most data are obtained from small mammals; at small body mass, the extrapolation therefore remains accurate, but at larger body masses, significant deviations from reality are possible.

mass is used to estimate a huge variety of other parameters, such as organ mass, home range, and reproductive turnover time. Because body mass is in practice a predictor of most other physiological variables (although the accuracy of these predictions remains debatable), it is usually at the center of these considerations (Calder 1996).

It is important to understand the limits of this technique. A classic example of the limits of extrapolation is body size ranges that are not covered by the original data set from which the predictive equations are made (Fig. 2.3). We may use allometric equations on organ and tissue mass from Calder (1996) and gut capacity from Parra (1978) to calculate the hypothetical body composition of a very large herbivorous mammal. If expressed as a percentage of the assumed body mass, it becomes obvious that if summed together, the individually calculated parts add up to more than the body mass entered into the individual equations, which is a physical impossibility. This example should remind everyone to remain careful about extrapolations derived from allometric equations.

ALLOMETRY

Choice of Data Sets

One important choice in the application of an allometric equation is what data set the equation used is derived from. For example, when extrapolating the organ size of an organism the size of a sauropod, the question whether the extrapolation is based on organ allometry of mammals or reptiles will have a considerable effect on the results (Franz et al. 2009). Because it has been speculated that the metabolism of sauropods might have changed during ontogeny with a trend towards a low, "ectothermic" metabolism (mass homoiothermy) in adulthood (Sander & Clauss 2008), the use of a reptile-based equation for the extrapolation of the gastrointestinal tissue mass of an adult 38 metric ton sauropod could be considered reasonable and would result in a gut tissue mass of 1,670 kg (or 4.4% of the assumed body mass) less than if the equation for mammals was used (Franz et al. 2009).

Relevance of the Allometric Exponent

The relevance of the allometric exponent has been extensively debated, for example, for the metabolic rate: whether metabolism scales to $BM^{0.67}$ or $BM^{0.75}$ has been considered critical in terms of the biological explanation underlying this pattern (e.g., White & Seymour 2003, 2005; Savage et al. 2004). When extrapolating to very large body masses, slight (potentially spurious) differences in the numerical value of the allometric exponent will overrule a (potentially biologically meaningful) ranking between groups on the basis of the factor a in the allometric equation $a \times BM^b$. For example, if two data sets on a hypothetical parameter Y that is biologically linked to metabolism yielded the equations

$$Y_{\text{mammal}} = 4 \times BM^{0.73}$$

$$Y_{\text{reptile}} = 0.4 \times BM^{0.85},$$

then this result would support the general concept that the mammalian endothermic metabolism represents an increase in reptile ectotherm metabolism by a factor of 10. The difference in the exponent is due to the particular data sets but is, most likely, not significant; the 95% confidence intervals for the exponent probably overlap. However, if these two equations are used to investigate potential differences in Y between an "endotherm" and an "ectotherm" giant, the astonishing result is that at a body mass of 50 metric tons, the metabolic parameter of an endotherm or ectotherm would be identical. Unless a sound biological reason for the difference in scaling the exponent is evident, it appears prudent not to link too much interpretation to a finding most likely generated by spurious differences in the allometric exponent of the equations used (Franz et al. 2009).

HOW MASTICATION MIGHT CONSTRAIN GIGANTISM

Applying allometric equations derived from extant (mostly mammalian) herbivores to the body masses of sauropod dinosaurs is fun and can often serve to illustrate that the animal groups in question—those from which the equation was derived, and those to which it is applied—are subject to different sets of constraints. For example, applying the equation to the correlation between the proportion of time spent foraging (out of the total 24 hour time budget) by mammalian herbivores from Owen-Smith (1988) results in the following:

Foraging budget (in % of day) = $19.0 \times BM^{0.17}$.

It appears that from a body mass of 18 metric tons onward, animals would become limited by the fact that they cannot put more than 24 hours of feeding in a day. Interestingly, this is about the size estimated for the largest terrestrial mammal ever found to date, the *Indricotherium* (Fortelius & Kappelman 1993). This consideration can now serve to highlight that somehow, sauropods must have been different from mammals. The question is, in which respect? A quick conclusion such as the sauropods could not have been endothermic cannot be corroborated by such an allometric consideration. The allometric consideration can only serve as additional evidence. Other solutions—for instance, that a body size threshold of 18 metric tons appears to apply not only to the largest terrestrial mammals, but also to the largest herbivorous dinosaurs with a dental apparatus evolved for thorough mastication. The Ornithopoda and Ceratopsia are well below this threshold (Paul 1997) and must also be considered, which opens a new hypothesis that *mastication* is the size-limiting factor, not endothermy. Indeed, the absence of mastication has been suggested as a major factor facilitating gigantism in sauropods (Sander & Clauss 2008).

ALLOMETRY AND SAUROPOD DIGESTIVE PHYSIOLOGY

With respect to feeding and digestive physiology, different allometric predictions have been proposed for sauropods. Perhaps the most prominent of these investigations are the ones already mentioned that were carried out by Paul (1998) and Christiansen (1999), who independently came to the conclusion that the snout width of sauropods was not smaller than one would expect for animals of their body size. This was an important refutation of the general idea that sauropods have small heads for their body size and therefore should have been comparatively limited to a low-intake (and hence ectothermic) metabolic strategy.

An important result in the reconstruction of the sauropod organism is that current estimates of the volume of the coelomic cavity in a sauropod and the estimated volume of its organs, whether based on mammalian or on reptilian equations, do not match. It appears that the reconstructed sauropod coelomic cavity offers much more space than we would consider necessary to harbor the reconstructed organs (Gunga et al. 2008; Franz et al. 2009). This means that current body size reconstructions do not indicate organismal size constraints. On the contrary, they allow conceptual leeway in the reconstruction of sauropods, for example, a disproportionately larger gut in these animals. On the bases of estimates of food intake and food digestibility in sauropods (Hummel et al. 2008) and on estimates of the gut capacity of sauropods, the mean ingesta retention time can also be estimated (Franz et al. 2009).

An intermediate metabolism and a food of medium quality were assumed for a 38 metric ton sauropod. The apparent energy digestibility was assumed at 44% with an intake of 96–140 kg dry matter per day, resulting in a mean ingesta retention time between six and eight days in this case. However, presuming a "regular" gut capacity (as observed in mammals and reptiles) and that the above-mentioned space leeway in the coelomic cavity of sauropods would even allow for a doubling of this gut capacity, resulting mean ingesta retention times could range between 11 and 16 days. Thus, estimated retention times fall within the range of the 11 days measured in Galapagos tortoises (*Geochelone nigra*) (Hatt et al. 2002), which are living reptiles that do not chew their food. We suggest that one should not overemphasize such numerical estimates, attractive as they may appear. Yet these values appear to indicate that even in the absence of mastication, the potentially feasible ingesta retention times in sauropods would be sufficient to allow for a reasonable digestion of plant matter. Yet far more interesting in our view are some general considerations on herbivore digestive physiology facilitated by the study of sauropods.

BODY MASS

Relationship between Body Mass and Gut Capacity, Food Intake, and Ingesta Retention

Perhaps the most prominent set of allometric considerations in general herbivore digestive physiology is the Jarman–Bell principle, which states that larger animals tolerate food of lower quality (Bell 1971; Geist 1974; Jarman 1974; elaborated by Parra 1978; Demment & Van Soest 1983, 1985; Illius & Gordon 1992). This concept is based on a discrepancy between the allometric scaling of gut capacity and gut fill rate (food intake rate). Gut capacity (measured as the wet weight of digestive tract contents) of mammalian herbivores generally scales linearly (or isometrically) with body mass, that is, at body mass$^{1.0}$ (read: body mass to the power of one). Although the largest terrestrial herbivore, the elephant, appeared to be a weak outlier in the original data set, the investigation of a specimen that was allowed to feed just before it was humanely killed for medical reasons indicated that elephants are no exception to the general pattern (Clauss et al. 2005b). Finally, a more expanded data set corroborated the pattern (Clauss et al. 2007a). This means that in mammalian herbivores, if body mass increases by a factor of 2, so does gut capacity. However, gut fill rate or food intake (usually measured as dry matter intake) does not scale linearly with body mass, but—just like metabolism and energy requirements—with metabolic body weight or $BM^{0.75}$. Again, this relationship has been corrobo-

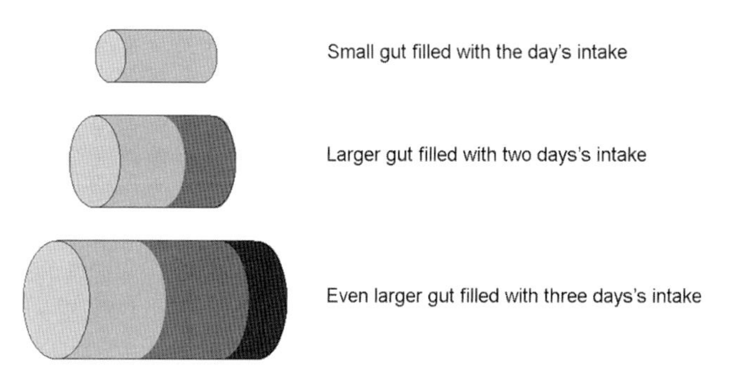

FIGURE 2.4. Schematic representation of the theoretical difference in the scaling of gut capacity (symbolized by the tube) and food intake (symbolized by the gray filling). At a given body size (top), it is assumed that the daily food intake fills out the complete gut. This translates into a food retention time of 24 hours, because on the next day, this food is pushed out by the next meal. As body size increases (middle and bottom), gut capacity increases (linearly with body mass, i.e., body mass^1), but gut fill (= food intake) does not increase at the same rate (only at $BM^{0.75}$). Therefore, not all of the food from the previous day is pushed out (food from previous days indicated by darker shades of gray), which translates into longer residence of food in the digestive tract. Note that differences in food digestibility due to differences in food selection in accordance with body mass are not part of this scheme; if larger animals ingest less digestible food, then their indigestible gut fill (the pushing portion of the contents; the digestible part will not portion but be absorbed) will be disproportionately higher than in smaller animals.

rated in numerous different studies (reviewed in Clauss et al. 2007a). If body mass increases by a factor of 2, food intake only increases by a factor of $2^{0.75} = 1.68$. Therefore, as animals become larger, their gut capacity should increase more than their gut fill rate; this should translate into a longer retention time of food in the gut, because more gut is available per unit ingested food (Fig. 2.4).

This has been considered a major digestive advantage of increasing body size and an evolutionary incentive for body size increase in herbivores (Demment & Van Soest 1985). Actually, given the two relationships,

gut capacity approximately $BM^{1.00}$

food intake rate approximately $BM^{0.75}$,

it can be concluded theoretically that the time food stays in the gut (the ingesta passage or ingesta retention time) scales to $BM^{(1.00-0.75)}$, or $BM^{0.25}$ (reviewed in Clauss et al. 2007a). This concept has been used to explain or claim the following: first, larger herbivores can use food of lower quality (because longer retention time allows for more thorough digestion); and second, on similar diets, larger herbivores achieve higher digestibilities (because the same diet is exposed to a longer digestion time).

Empirical data, however, do not necessarily corroborate these predictions. Evidence for an increase of ingesta retention time with body mass is poor (Clauss et al. 2007a, 2008b). Similarly, evidence for an increase in digestive efficiency with increasing body mass is also poor (Pérez-Barberìa et al. 2004; Clauss & Hummel 2005). Actually, the data rather appear to support the concept that among large herbivores, both digestive efficiency and ingesta retention are relatively independent of body size. This is most likely due to two different mechanisms: physiological consequences of large body size on one hand, and consequences of large body size on the quality of food that can be efficiently cropped on the other.

Relationship between Body Mass and Ingesta Particle Size

Digestive "disadvantages" of large body size that have received little attention have been reviewed in Clauss & Hummel (2005). Maybe the most evident of these disadvantages is that with increasing body mass, ingesta particle size increases (Fritz et al. 2009). Mice chew their food into smaller particles than elephants. Because larger particles require a longer time to be digested to the same extent as smaller particles, any potential advantage of a longer retention time bestowed by a larger body size could be annihilated in comparison among "particle size reducers" by the concomitant higher proportion of large particles. In individual comparisons between species, such as between the horse, rhinoceros, and elephant (Clauss et al. 2005a), or between buffalo and hippopotamus (Schwarm et al. 2009), variation in chewing efficiency has been invoked to explain differences in digestive efficiency that could not be explained by variation in ingesta retention. The concept that an increased ingesta retention can compensate for a lack of ingestive particle size reduction has been proposed for the comparison of reptilian and mammalian herbivores (Karasov et al. 1986), and potentially long ingesta retention times have been evoked as a compensatory mechanism in gigantic herbivorous dinosaurs that lacked mechanisms of particle size reduction (Farlow 1987). Actually, among some grazing mammals, ingesta retention time and ingesta particle size have a compensatory effect on digestive efficiency (Clauss et al. 2009b). Thus, the theoretical assumptions on the digestive advantage of larger body size will probably not apply directly to a guild of herbivores that evolved adaptations for ingesta particle size reduction. Even without grinding teeth or a gizzard, animals can influence ingesta particle size, for example by adjusting bite size (that is, taking more, smaller bites) (Bjorndal & Bolten 1992; Fritz et al. 2010). However, above a certain body size threshold, adjustment of ingesta particle size by small bite sizes will no longer be a feasible option due to time constraints.

Therefore, the general predictions of the advantages of large body size based on gut capacity and ingesta retention alone might indeed apply to groups of large, herbivorous, nonchewing dinosaurs such as ankylosaurs, stegosaurs, and sauropods. Especially in sauropods, the only way to influence ingesta particle size was probably the choice of forage, whether the cropped food had large or small leaves (note that leaf surface is often difficult to penetrate for gut microbes, except for perhaps fungi; Van Soest 1994). Sauropods might have actually achieved what comparative digestive physiologists dream of: such a dramatic increase in gut capacity due to their enormous body size that this capacity alone, irrespective of complicating considerations of particle size, served to achieve reasonable digestive efficiencies.

Within the extant fauna, the level of metabolism is roughly correlated to adaptations for a speedy digestive process (Reilly et al. 2001; Lucas 2004; Franz et al. 2010). The most remarkable of adaptations in this respect are those aiming at ingesta particle size reduction in mammals (grinding teeth) and birds (gizzard). In competitive situations, animals less adapted to speedy digestion occur in specific niches of sedentary lifestyles (e.g., herbivorous reptiles vs. herbivorous mammals; hippos as compared to ruminants; sedentary arboreal folivores such as sloths or koalas). It is tempting to assume a similar difference in lifestyle between the chewing ornithopods and ceratopsians on one hand and ankylosaurs and stegosaurs on the other hand, as has been done in the paleontological literature (Farlow 1987). Sauropods, we think, are peculiar in this respect because they evolved body sizes that liberated them from any potential constraints of ingesta particle size and from chewing competitors.

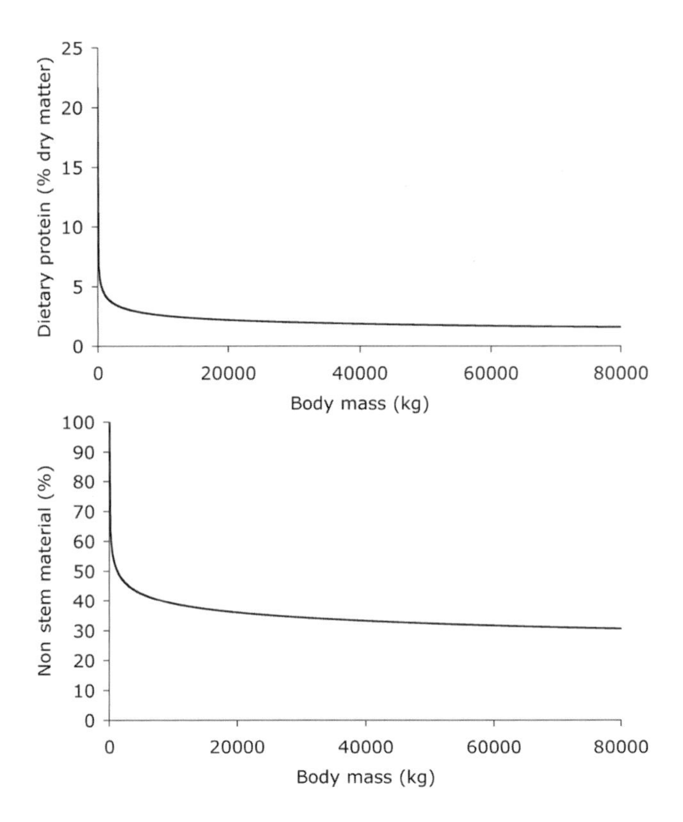

FIGURE 2.5. Extrapolation of characteristics of diet quality to very large body sizes in herbivores, based on regression equations from Owen-Smith (1988) for dietary crude protein content (regression derived from wild ruminant species) and for the proportion of nonstem material in the diet (regression derived from ruminants, hippos, and hindgut fermenters). Note that at very large body sizes, the decline of diet quality with increasing body size appears negligible.

Relationship between Body Mass and Food Quality

The other mechanism that will lessen the predicted positive effect of large body size on digestive performance is a decrease in food quality with increasing body size. It is generally believed that larger animals ingest a diet of lesser quality (Bell 1971; Jarman 1974; Demment & Van Soest 1983, 1985); for herbivores, this translates into a diet of higher fiber and lower protein content, and probably also a higher indigestible proportion in the fiber fraction. Empirical data collections that actually test this assumption are rare, with the exception of data on protein levels in stomach contents, fermentation rate, and proportion of nonstem material in large African herbivores (Owen-Smith 1988). A conceptual question is whether large body size should be considered as an adaptation to low-quality forage, as for example was done by Midgley et al. (2002), or whether low forage quality is simply a (necessary) consequence of large body size that evolved for other reasons (Renecker & Hudson 1992). Larger animals have higher absolute energy requirements and therefore need to ingest larger absolute amounts of food. They do not have the time to be very selective in their food intake but must take whatever is available in large batches.

Additionally, as a result of their larger ingestive organs (e.g., larger snout width), larger animals cannot feed as selectively as smaller ones. Thus, the quality of the food they ingest depends on what is available in large, easily accessible batches. As with most precious things on this planet, high-quality food is mostly rare and dispersed, whereas low-quality food is more easily obtained and available in larger batches. Note, however, that there are evident exceptions in marine ecosystems: the largest mammals, the baleen whales, consume high-quality diets in the form of krill and fish; this high quality is available in an aggregate that makes foraging by these large predators efficient. On land, large body size mostly implies a low-quality diet for logistical reasons, but it does not oblige animals to consume such diets if higher-quality food is available in reasonable amounts and batches. Note that domestic large herbivores do survive well on feeds of much higher dietary quality than the forages they originally evolved to feed on.

However, if larger herbivores consume forage of lesser quality, this will then also reduce the potential advantage outlined in Fig. 2.4. The portion of the ingested food that will actually push the ingesta through the gut and hence be the major determinant of ingesta retention is the indigestible portion. If food quality declines with increasing body size, then the proportion of indigestible material in the gut will increase, making the difference between gut capacity and gut fill rate less distinct, and hence any digestive advantage conferred by large body size less effective. However, although investigations on this topic are missing, the question is whether the decrease in dietary quality is an effect within a limited range of body sizes only, or whether it is a consistent correlation that can be expected to continue far into the body size range of sauropods.

Actually, in the data collections of Owen-Smith (1988), any effect that demonstrates decreasing diet quality with increasing body mass is significant because of the inclusion of animals from the 10–500 kg range. If only animals from 500 kg onward were regarded, a decline would be difficult to prove, and if the regression equations given by this author are used to extrapolate the protein content or the proportion of nonstem material in the diet of very large herbivores, it becomes obvious that at very large body sizes, further reductions in dietary quality become minimal (Fig. 2.5).

Therefore, one could speculate that at the body size of megaherbivores, including the sauropods, the effects of decreasing diet quality with increasing body mass should not play an important role and hence should not annihilate the advantage of increasing body mass outlined in Fig. 2.4. Another problem that deserves further investigation is at which body size diet quality will not deteriorate significantly, and how this is influenced by botanical characteristics. Again, sauropods might have freed themselves by their sheer size from any factors complicating the easy digestive concept of Fig. 2.4.

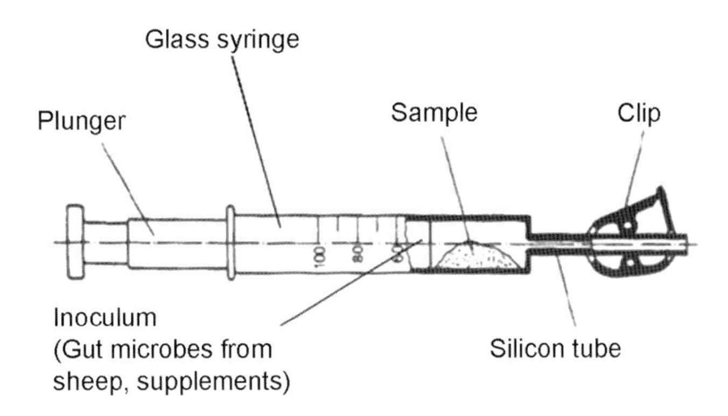

FIGURE 2.6. Glass syringe used in Hohenheim gas test. Gas produced during fermentation is taken as a measure of the degradation of the incubated plant material.

Potential Food Plants

Besides their gigantic body size, the exclusive use of nonangiosperm flora as food plants is among the particularities of sauropod feeding, at least before the late Early Cretaceous, because few vertebrates today make use of these resources—plants such as ferns, horsetails, *Ginkgo,* and conifers (Gee, this volume). Generally they are regarded as food plants of extremely low quality (Coe et al. 1987; Wing & Tiffney 1987; Van Soest 1994; Taggart & Cross 1997; Midgley et al. 2002; Farlow 2007); however, little quantification of this assumption is available. Most paleobotanists have hypothesized that soft-tissue plants like ferns or *Ginkgo* were selected over spinier and less palatable conifers by herbivorous dinosaurs (Coe et al. 1987; Dodson 1990; Taggart & Cross 1997; Tiffney 1997). On the basis of the peculiarities of the respective skull and teeth, Krassilov (1981) considered ferns and horsetails as diplodocid and conifers as camasaurid food, while Stevens & Parrish (2005) concluded from the neck postures of sauropod taxa that only brachiosaurids and camarasaurids were capable of feeding on high-growing conifers, with all other forms relying on low-growing vegetation such as ferns and horsetails. On the basis of the investigations of Weaver (1983), who measured gross energy contents in samples of extent ferns and conifers, Fiorillo (1998) dismissed ferns and horsetails as favorable food plants, while Engelmann et al. (2004) favored ferns and horsetails as sauropod fodder, irrespective of their presumably low energy content.

Given a lack of quantitative information, we began a project aimed at quantifying the digestibility of extant relatives of potential dinosaur food plants (Hummel et al. 2008). Sufficient quantities of different plant taxa were available from the Botanical Gardens of the University of Bonn and the Botanical Gardens of Cologne. To get an estimate of the digestibility of the plants, an in vitro fermentation test (Hohenheim gas test; Menke et al. 1979; Menke & Steingass 1988) was used that simulates in the laboratory microbial digestive processes that occur in herbivorous animals. This test is regularly used to estimate the energy content of fodder for herbivores such as ruminants. Small samples of the plants are digested by microbes from the gut of herbivores in gas-tight glass syringes under conditions favorable for their growth, and the degradation of the samples is quantified over time by recording the gas produced during fermentation (Fig. 2.6). It is general practice to use inoculum adapted to a standardized diet in this test rather than inoculum adapted to a peculiar feed source. A general criticism to the use of such a standard test is that the inoculum is taken from individuals of an animal species not adapted to the tested forages—both in terms of individual and evolutionary adaptations. However, our basic conclusion will be that the potential feeding plants performed better than expected (see below), and any inoculum adapted to the particular feeding plants would just have given even better results.

Some of the fern and gymnosperm foliage yielded levels of energy only moderately lower than food of extant herbivores such as forbs or grasses. Many of the extant relatives of potential sauropod food plants performed on a level comparable to moderate browse (Fig. 2.7), with some exceptions such as the Podocarpaceae and the cycads, which both gave rather low yields. Among the ferns, samples from *Angiopteris* and *Osmunda* yielded comparatively high amounts of energy, while *Dicksonia* was very low in energy (Table 2.2).

The results for the conifer genus *Araucaria,* known for its widespread occurrence in the Mesozoic, were especially interesting in having a fermentative behavior of being digested rather slow initially, but then to a high extent, if given enough time. This fermentative behavior would require a long retention time in the respective herbivore—potentially not a problem for a sauropod, as discussed above. In combination with their widespread occurrence, this gives *Araucaria* some potential as food plant for high-browsing dinosaurs. Another plant group of particular interest are the horsetails, which performed particularly well in the in vitro test, resulting in rather high energy yields despite their high silica content.

In summary, our study arrived at a slightly different ranking of potential food plants than that of Weaver (1983), who judged cycads as best and horsetails as worst (Table 2.2). However, in Weaver's study, the gross energy content of plants was investigated, which is of limited value in estimating the energy actually available from plant material to a herbivore (Gfe 2003): although a piece of solid wood is nearly indigestible to a herbivore, its combustion energy (=gross energy) is on a level comparable to young leaves.

Although the in vitro test looks at the overall digestibility of the samples, which is considered as the best overall measure of diet quality by Owen-Smith (1988), it is obvious that other characteristics of the food plants must not be overlooked. Nutrients such as protein are also of considerable interest. The level of this nutrient was low in all *Araucaria* samples, making the use of this group by young, growing animals at least less likely. However, it has also been shown that requirements of animals may evolve in concert with their respective food resources to some extent (Grubb 1992; Midgley 2005).

Our results obviously raise the question why these plants do not form a staple food resource for extant herbivores. Several arguments can be made in this respect. In some instances, the extant range of plants represents a relict distribution, in which the species could only survive because of a low predation level by herbivores. Other plant groups investigated here are in fact regularly used as feeding plants by some herbivores; for example, deer obviously feed on various types of conifers, and horsetails are an important food source for geese (see also Gee, this volume). However, the extreme abrasiveness of horsetails (the result of their high silica content) makes them a challenging food for animals that masticate their food extensively, but far less so for birds like geese and nonchewing dinosaurs like sauropods. Obviously defense chemicals (toxic or/and digestion inhibiting) are present in considerable amounts in

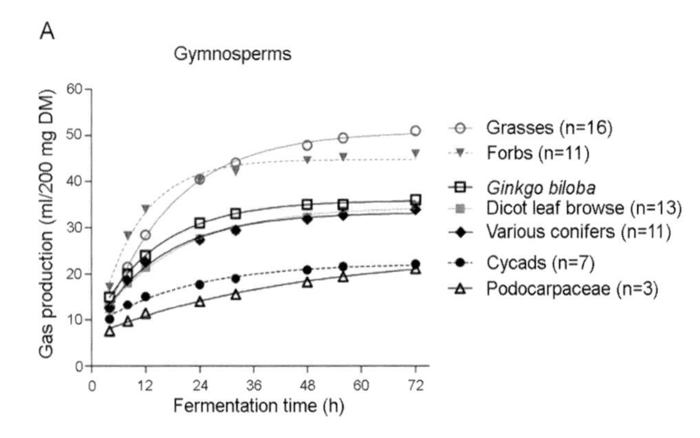

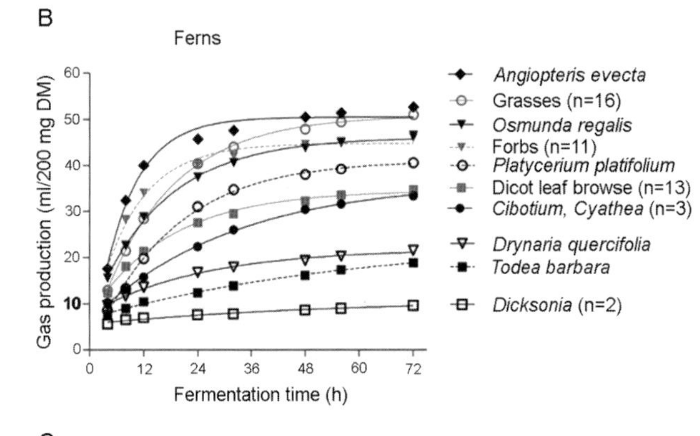

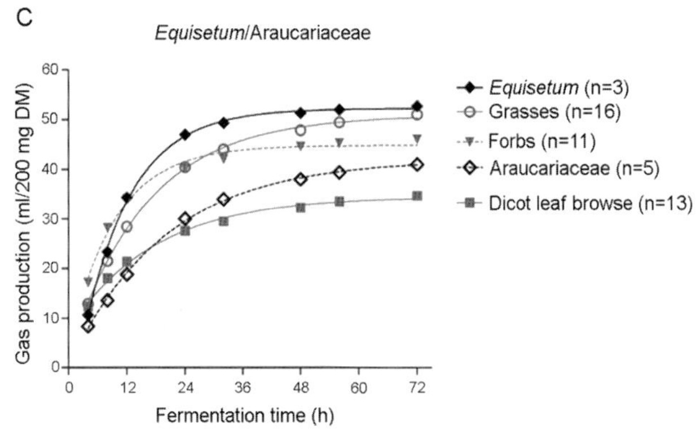

FIGURE 2.7. Fermentative behavior of potential dinosaur food plants compared to that of angiosperms. Gas production in the Hohenheim gas test (Menke et al. 1979) is plotted versus fermentation time. (A) Various gymnosperms compared to angiosperms. Note that *Ginkgo* and some conifers (Cephalotaxaceae, Taxodiaceae, Pinaceae, and Taxaceae) performed at the level of angiosperm browse, whereas podocarp conifers and cycads fared poorly. (B) Ferns compared to angiosperms. Note the great variability among ferns, including the very poor performance of the tree fern *Dicksonia*. (C) Araucariaceae and horsetails (*Equisetum* spp.) compared to angiosperms. Note that horsetails even surpass grasses and that araucarias outperform browse after 72 hours. DM, dry matter; means and standard error of the mean (SEM) are indicated. *Figure from Hummel et al. (2008).*

Table 2.2. Nutrient and Energy (Measured as Metabolizable Energy, ME) Content of Potential Dinosaur Food Plants

Sample type (no. of spp.)	ME (Gp72h) (MJ/kg DM)	ME (Weaver 1983) (MJ/kg DM)	Crude protein (% DM)	NDF (% DM)
Grasses (16)	11.3 (9.3–13.6)		15.3	62.8
Forbs (11)	10.4 (9.1–11.9)		19.8	37.8
Dicot browse (13)	7.5 (5.5–10.0)		20.7	43.2
Ginkgo (1)	8.6	6.7	15.6	27.5
Araucariaceae (5)	9.4 (8.0–11.6)	7.0	4.4	65.2
Podocarpaceae (3)	5.9 (5.0–6.1)		6.6	62.3
Various conifers (13)	8.3 (6.3–10.8)	7.0 (6.4–7.5)	10.0	51.3
Cycads (7)	6.1 (4.4–7.7)	7.6 (7.1–8.6)	11.4	65.3
Various ferns (9)	7.7 (4.7–11.7)	6.6 (5.4–7.4)	11.5	62.8
Tree ferns (5)	6.4 (3.6–9.3)	6.9 (6.6–7.2)	11.3	63.6
Equisetum (3)	11.6 (10.8–12.9)	5.3	11.7	48.4

Gp, gas production; DM, dry matter; NDF, neutral detergent fiber. ME for the data of Weaver (1983) was calculated by multiplying gross energy with a factor of 0.5 (digestible energy, according to Weaver 1983), and consequently with a factor of 0.76 to get ME (according to Robbins 1993). *Table from Hummel et al. (2008).*

many, if not most, of the plants investigated here (see Swain 1976, 1978, for overview). However, the competition with angiosperms and the high predatory pressure by extant herbivores may have led to the development of a higher level of protection in living members of these plant groups as compared to their Mesozoic relatives. The same may be true for the often considerable structural protection, as seen in some species of *Araucaria*. In addition, the presence of toxins should not be regarded as precluding their use by herbivores per se, but instead limiting the amount of incorporation of these plants into the diets of animals specialized to feed on them. Black rhinos are the best example of a hindgut-fermenting ungulate feeding on plants well known for their toxicity, such as the Euphorbiaceae (Adcock & Emslie 1997).

Regardless of their use by extant herbivores, sauropods must have fed at least on some of these plants, simply because angiosperm foliage was not available for most of their existence. The results of our study indicate that the food supply of sauropods was less problematic than usually thought (see also Gee, this volume) and may help explain the diversity of herbivorous dinosaurs in Mesozoic ecosystems.

Conclusions

In summary, sauropods are fascinating because they are different from any other vertebrate model organism we know. Just like other herbivores, they evolved into a variety of forms that most likely represent a variety of feeding niches. Nevertheless, the fermentative digestion of plant matter in a hindgut fermentation system was most likely common to all of them. The plants that were potentially available to them would have yielded reasonable amounts of energy, comparable to at least those of extant browse leaves. When trying to understand the ecological relevance of their potential digestive strategy, we must resort to concepts developed for the understanding of comparable body sizes for niche differentiation in mammalian herbivores. When evaluating these concepts, we find them insufficient because they do not incorporate the effects of body size on chewing efficiency, and because they do not consider the effect of increasing body size on the quality of the ingested diet (while trying to explain how larger animals can ingest lower-quality food). Sauropods did not chew their food or grind it in a gizzard, and they reached adult body sizes that were most likely far beyond the point at where size has an influence on diet quality. In this respect, sauropods appear conceptually simple—and ironically, they may represent exactly the logical outcome of selection pressures assumed to work on mammalian herbivores. In mammals, these selection pressures could not produce such gigantic life forms as a result of the mechanical and time-constraining effects of mastication, ontogenetically inflexible metabolic rates, and the population-limiting effects of vivipary (Sander & Clauss 2008; Sander et al. 2010a). Further work on the digestive physiology of sauropods should strive to reconcile their dramatic ontogenetic metabolic requirements with their potential food sources.

Acknowledgments

We are grateful to Peter Van Soest (Cornell University, Ithaca, New York) and Jim Farlow (Purdue University, Fort Wayne, Indiana) for their comments; and Stephan Anhalt (Botanical Gardens, Cologne) and Wolfgang Lobin (Botanical Garden, University of Bonn) for the generous supply of plant samples. We thank the Deutsche Forschungsgemeinschaft (DFG) for facilitating our participation in the DFG Research Unit 533 "Biology of the Sauropod Dinosaurs: The Evolution of Gigantism"; and Martin Sander (University of Bonn) for his organization, motivation, and enthusiasm. This paper is contribu-

tion number 61 of the DFG Research Unit 533 "Biology of the Sauropod Dinosaurs: The Evolution of Gigantism."

References

Adcock, K. & Emslie, R. H. 1997. Biologie, Verhalten und Ökologie des Spitzmaul-Nashorns. *In* Emslie, R. & van Strien, N. (eds.). *Die Nashörner.* Filander Verlag, Fürth: pp. 115–137.

Bakker, R. 1986. *The Dinosaur Heresies.* Longman Scientific & Technical, Harlow, UK.

Bakker, R. T. 1978. Dinosaur feeding behaviour and the origin of flowering plants.—*Nature* 274: 661–663.

Barrett, P. M. 2000. Prosauropod dinosaurs and iguanas: speculations on diets of extinct reptiles. *In* Sues, H.-D. (ed.).—*Evolution of Herbivory in Terrestrial Vertebrates. Perspectives from the Fossil Record.* Cambridge University Press, Cambridge: pp. 42–78.

Barrett, P. M. & Upchurch, P. 2005. Sauropod diversity through time: possible macroevolutionary and palaeoecological implications. *In* Curry Rogers, K. A. & Wilson, J. (eds.). *The Sauropods: Evolution and Paleobiology.* University of California Press, Berkeley: pp. 125–156.

Barrett, P. M. & Willis, K. J. 2001. Did dinosaurs invent flowers? Dinosaur–angiosperm coevolution revisited.—*Biological Reviews of the Cambridge Philosophical Society* 76: 411–447.

Bell, R. H. V. 1971. A grazing ecosystem in the Serengeti.—*Scientific American* 225: 86–93.

Bergman, C. M., Fryxell, J. M. & Gates, C. C. 2000. The effect of tissue complexity and sward height on the functional response of wood bison.—*Functional Ecology* 14: 61–69.

Bjorndal, K. A. 1997. Fermentation in reptiles and amphibians. *In* Mackie, R. I. & White, B. A. (eds.). *Gastrointestinal Microbiology, Vol. 1: Gastrointestinal Ecosystems and Fermentations.* ITP, New York: pp. 199–230.

Bjorndal, K. A. & Bolten, A. B. 1992. Body size and digestive efficiency in a herbivorous freshwater turtle: advantages of small bite size.—*Physiological Zoology* 65: 1028–1039.

Bonaparte, J. F. & Mateus, O. 1999. A new diplodocid, *Dinheirosaurus lourinhanensis* gen. et sp. nov., from the Late Jurassic beds of Portugal.—*Revista del Museo Argentino de Ciencias Naturales* 5: 13–29.

Bouchard, S. S. & Bjorndal, K. A. 2006. Ontogenetic diet shifts and digestive constraints in the omnivorous freshwater turtle *Trachemys scripta.*—*Physiological and Biochemical Zoology* 79: 150–158.

Butler, R. J., Barrett, P. M., Kenrick, P. & Penn, M. G. 2009. Diversity patterns amongst herbivorous dinosaurs and plants during the Cretaceous: implications for hypotheses of dinosaur/angiosperm co-evolution.—*Journal of Evolutionary Biology* 22: 446–459.

Butler, R. J., Barrett, P. M., Penn, M. G. & Kenrick, P. 2010. Testing evolutionary hypotheses over geological time scales; interactions between Cretaceous dinosaurs and plants.—*Biological Journal of the Linnean Society* 100: 1–15.

Calder, W. A. 1996. *Size, Function and Life History.* Harvard University Press, Cambridge.

Calvo, J. O. 1994a. Gastroliths in sauropod dinosaurs.—*Gaia* 10: 205–208.

Calvo, J. O. 1994b. Jaw mechanics in sauropod dinosaurs.—*Gaia* 10: 183–193.

Cameron, E. Z. & Du Toit, J. T. 2007. Winning by a neck: tall giraffes avoid competing with shorter browsers.—*American Naturalist* 169: 130–135.

Chin, K. & Gill, B. D. 1996. Dinosaurs, dung beetles, and conifers: participants in a Cretaceous food web.—*Palaios* 11: 280–285.

Chivers, D. J. & Hladik, C. M. 1980. Morphology of the gastrointestinal tract in primates: comparisons with other mammals in relation to diet.—*Journal of Morphology* 166: 337–386.

Christian, A. 2002. Neck posture and overall body design in sauropods.—*Fossil Record* 5: 269–279.

Christian, A. & Dzemski, G. 2007. Reconstruction of the cervical skeleton posture of *Brachiosaurus brancai* Janensch, 1914 by an analysis of the intervertebral stress along the neck and a comparison with the results of different approaches.—*Fossil Record* 10: 38–49.

Christian, A. & Dzemski, G. This volume. Neck posture in sauropods. *In* Klein, N., Remes, K., Gee, C. T. & Sander, P. M. (eds.). *Biology of the Sauropod Dinosaurs: Understanding the Life of Giants.* Indiana University Press, Bloomington: pp. 251–260.

Christiansen, P. 1996. The evidence for implications of gastroliths in sauropods.—*Gaia* 12: 1–7.

Christiansen, P. 1999. On the head size of sauropodomorph dinosaurs: implications for ecology and physiology.—*Historical Biology* 13: 269–297.

Christiansen, P. 2000. Feeding mechanisms of the sauropod dinosaurs *Brachiosaurus, Camarasaurus, Diplodocus,* and *Dicraeosaurus.*—*Historical Biology* 14: 137–152.

Clarke, A. & Rothery, P. 2008. Scaling of body temperature in mammals and birds.—*Functional Ecology* 22: 58–67.

Clauss, M. & Hummel, J. 2005. The digestive performance of mammalian herbivores: why big may not be *that* much better.—*Mammal Review* 35: 174–187.

Clauss, M., Löhlein, W., Kienzle, E. & Wiesner, H. 2003. Studies on feed digestibilities in captive Asian elephants (*Elephas maximus*).—*Journal of Animal Physiology and Animal Nutrition* 87: 160–173.

Clauss, M., Schwarm, A., Ortmann, S., Alber, D., Flach, E. J., Kühne, R., Hummel, J., Streich, W. J. & Hofer, H. 2004. Intake, ingesta retention, particle size distribution and digestibility in the Hippopotamidae.—*Comparative Biochemistry and Physiology A* 139: 449–459.

Clauss, M., Polster, C., Kienzle, E., Wiesner, H., Baumgartner, K., von Houwald, F., Ortmann, S., Streich, W. J. & Dierenfeld, E. S. 2005a. Studies on digestive physiology and feed digestibilities in captive Indian rhinoceros (*Rhinoceros unicornis*).—*Journal of Animal Physiology and Animal Nutrition* 89: 229–237.

Clauss, M., Robert, N., Walzer, C., Vitaud, C. & Hummel, J. 2005b. Testing predictions on body mass and gut contents: dissection of an African elephant (*Loxodonta africana*).—*European Journal of Wildlife Research* 51: 291–294.

Clauss, M., Schwarm, A., Ortmann, S., Streich, W. J. & Hummel, J. 2007a. A case of non-scaling in mammalian physiology? Body size, digestive capacity, food intake, and ingesta passage in mammalian herbivores.—*Comparative Biochemistry and Physiology A* 148: 249–265.

Clauss, M., Streich, W. J., Schwarm, A., Ortmann, S. & Hummel, J. 2007b. The relationship of food intake and ingesta passage predicts feeding ecology in two different megaherbivore groups.—*Oikos* 116: 209–216.

Clauss, M., Kaiser, T. & Hummel, J. 2008a. The morphophysiological adaptations of browsing and grazing mammals. *In* Gordon, I. J. & Prins, H. H. T. (eds.). *The Ecology of Browsing and Grazing*. Springer, Heidelberg: pp. 47–88.

Clauss, M., Streich, W. J., Nunn, C. L., Ortmann, S., Hohmann, G., Schwarm, A. & Hummel, J. 2008b. The influence of natural diet composition, food intake level, and body size on ingesta passage in primates.—*Comparative Biochemistry and Physiology A* 150: 274–281.

Clauss, M., Grum, C. & Hatt, J. M. 2009a. Polyunsaturated fatty acid content in adipose tissue in foregut and hindgut fermenting mammalian herbivores: a literature survey.—*Mammalian Biology* 74: 153–158.

Clauss, M., Nunn, C., Fritz, J. & Hummel, J. 2009b. Evidence for a tradeoff between retention time and chewing efficiency in large mammalian herbivores.—*Comparative Biochemistry and Physiology A* 154: 376–382.

Clauss, M., Hume, I. D. & Hummel, J. 2010. Evolutionary adaptations of ruminants and their potential relevance for modern production systems.—*Animal* 4: 979–992.

Clements, K. D. 1997. Fermentation and gastrointestinal microorganisms in fishes. *In* Mackie, R. I. & White, B. A. (eds.). *Gastrointestinal Microbiology*. ITP, New York: pp. 156–198.

Coe, M. J., Dilcher, D. L., Farlow, J. O., Jarzen, D. M. & Russell, D. A. 1987. Dinosaurs and land plants. *In* Friis, E. M., Chaloner, W. G. & Crane, R. (eds.). *The Origins of Angiosperms and Their Biological Consequences*. Cambridge University Press, New York: pp. 225–258.

Colbert, E. H. 1993. Feeding strategies and metabolism in elephants and sauropod dinosaurs.—*American Journal of Science* 293A: 1–19.

Coombs, W. 1975. Sauropod habits and habitats.—*Palaeogeography, Palaeoclimatology, Palaeoecology* 17: 1–33.

Dall, W. & Moriarty, D. J. W. 1983. Functional aspects of nutrition and digestion. *In* Mantel, L. H. & Bliss, D. E. (eds.). *The Biology of Crustacea Internal Anatomy and Physiological Regulation*. Academic Press, New York: pp. 215–261.

Dehority, B. A. & Orpin, C. G. 1997. Development of, and natural fluctuations in, rumen microbial populations. *In* Hobson, P. N. & Stewart, C. S. (eds.). *The Rumen Microbial Ecosystem*. Blackie Academic and Professional, London: pp. 196–245.

Demment, M. W. & Van Soest, P. J. 1983. *Body Size, Digestive Capacity, and Feeding Strategies of Herbivores*. Winrock International Livestock Research & Training Center, Morrilton.

Demment, M. W. & Van Soest, P. J. 1985. A nutritional explanation for body-size patterns of ruminant and nonruminant herbivores.—*American Naturalist* 125: 641–672.

Dierenfeld, E. S., Hintz, H. F., Robertson, J. B., Van Soest, P. J. & Oftedal, O. T. 1982. Utilization of bamboo by the giant panda.—*Journal of Nutrition* 112: 636–641.

Dodson, P. 1990. Sauropod paleoecology. *In* Weishampel, D. B., Dodson, P. & Osmólska, A. (eds.). *The Dinosauria*. University of California Press, Berkeley: pp. 402–407.

Dunham, A. E., Overall, K. L., Porter, W. P. & Forster, C. A. 1989. Implications of ecological energetics and biophysiological and development constraints for life-history variation in dinosaurs.—*Geological Society of America, Special Paper* 238: 1–19.

Durant, D., Fritz, H., Blais, S. & Duncan, P. 2003. The functional response in three species of herbivorous Anatidae: effects of sward height, body mass and bill size.—*Journal of Animal Ecology* 72: 220–231.

Du Toit, J. T. 1990. Feeding-height stratification among African browsing ruminants.—*African Journal of Ecology* 28: 55–61.

Engelmann, G. F., Chure, D. J. & Fiorillo, A. R. 2004. The implications of dry climate for the paleoecology of the fauna of the Upper Jurassic Morrison Formation.—*Sedimentary Geology* 167: 297–308.

Farlow, J. O. 1987. Speculations about the diet and digestive physiology of herbivorous dinosaurs.—*Paleobiology* 13: 60–72.

Farlow, J. O. 2007. A speculative look at the paleoecology of large dinosaurs at the Morrison Formation, or, life with *Camarasaurus* and *Allosaurus*. *In* Kvale, E. P., Brett-Surman, M. K. & Farlow, J. O. (eds.). *Dinosaur Paleoecology and Geology: The Life and Times of Wyoming's Jurassic Dinosaurs and Marine Reptiles*. Geoscience Adventure Workshop, Shell: pp. 98–151.

Fiorillo, A. R. 1991. Dental microwear on the teeth of *Camarasaurus* and *Diplodocus:* implications for sauropod paleoecology.—*Contributions of the Paleontological Museum, University of Oslo* 364: 23–24.

Fiorillo, A. R. 1998. Dental microwear of sauropod dinosaurs *Camarasaurus* and *Diplodocus:* evidence for resource partitioning in the Late Jurassic of North America.—*Historical Biology* 13: 1–16.

Fortelius, M. & Kappelman, J. 1993. The largest land mammal ever imagined.—*Zoological Journal of the Linnean Society of London* 107: 85–101.

Franz, R., Hummel, J., Kienzle, E., Kölle, P., Gunga, H. C. & Clauss, M. 2009. Allometry of visceral organs in living amniotes and its implications for sauropod dinosaurs.—*Proceedings of the Royal Society B: Biological Sciences* 276: 1731–1736.

Franz, R., Hummel, J., Müller, J. W. H., Bauert, M., Hatt, J. M. & Clauss, M. 2010. Herbivorous reptiles and body mass effects on food intake, digesta retention, digestibility and gut capacity, and a comparison with mammals.—*Comparative Biochemistry and Physiology*. doi: 10.1016/j.cbpa.2010.09.007.

Freudenberger, D. O. 1992. Gut capacity, functional allocation of gut volume and size distributions of digesta particles in two macropodid marsupials (*Macropus robustus robustus* and *M. r. erubescens*) and the feral goat (*Capra hircus*).—*Australian Journal of Zoology* 40: 551–561.

Fritz, J., 2007. *Allometrie der Kotpartikelgröße von pflanzenfressenden Säugern, Reptilien und Vögeln*. Ph.D. Dissertation. Ludwig-Maximilian-Universität München, Munich.

Fritz, J., Hummel, J., Kienzle, E., Arnold, C., Nunn, C. & Clauss, M. 2009. Comparative chewing efficiency in mammalian herbivores.—*Oikos* 118: 1623–1632.

Fritz, J., Hummel, J., Kienzie, E., Streich, W. J. & Clauss, M. 2010. To chew or not to chew: faecal particle size in herbivorous reptiles and mammals.—*Journal of Experimental Zoology A* 313: 579–586.

Galton, P. M. 1986. Herbivorous adaptations of Late Triassic and Early Jurassic dinosaurs. *In* Padian, K. (ed.). *The Beginning of the Age of Dinosaurs*. Cambridge University Press, Cambridge: pp. 203–221.

Gee, C. T. This volume. Dietary options for the sauropod dinosaurs from an integrated botanical and paleobotanical perspective. *In* Klein, N., Remes, K., Gee, C. T. & Sander, P. M. (eds.). *Biology of the Sauropod Dinosaurs: Understanding the Life of Giants*. Indiana University Press, Bloomington: pp. 34–56.

Geist, V. 1974. On the relationship of social evolution and ecology in ungulates.—*American Zoologist* 14: 205–220.

Gesellschaft für Ernährungswissenschaft (GfE). 2003. *Recommendations for the Supply of Energy and Nutrients to Goats*. DLG-Verlag, Frankfurt.

Ghosh, P., Bhattacharya, S. K., Sahni, A., Kar, R. K., Mohabey, D. M. & Ambwani, K. 2003. Dinosaur coprolites from the Late Cretaceous (Maastrichtian) Lameta Formation of India: isotopic and other markers suggesting a C_3 plant diet.—*Cretaceous Research* 24: 743–750.

Ginnett, T. F. & Demment, M. W. 1995. The functional response of herbivores: analysis and test of a simple mechanistic model. —*Functional Ecology* 9: 376–384.

Grajal, A., Strahl, S. D., Parra, R., Dominguez, M. G. & Neher, A. 1989. Foregut fermentation in the hoatzin, a neotropical leaf-eating bird.—*Science* 245: 1236–1238.

Gross, J. E., Hobbs, N. T. & Wunder, B. A. 1993a. Independent variables for predicting intake rate in mammalian herbivores: biomass density, plant density, or bite size?—*Oikos* 68: 75–81.

Gross, J. E., Shipley, L. A., Hobbs, N. T., Spalinger, D. E. & Wunder, B. A. 1993b. Functional response of herbivores in food-concentrated patches: tests of a mechanistic model.—*Ecology* 74: 778–791.

Grubb, P. J. 1992. A positive distrust in simplicity: lessons from plant defences and from competition among plants and among animals.—*Journal of Ecology* 80: 585–610.

Guillaume, J., Kaushik, S., Bergot, P. & Métailler, R. 1999. *Nutrition and Feeding of Fish and Crustaceans*. Springer, London.

Gunga, H., Suthau, T., Bellmann, A., Stoinski, S., Friedrich, A., Trippel, T., Kirsch, K. & Hellwich, O. 2008. A new body mass estimation of *Brachiosaurus brancai* Janensch, 1914 mounted and exhibited at the Museum of Natural History (Berlin, Germany).—*Fossil Record* 11: 28–33.

Hatt, J. M., Gisler, R., Mayes, R., Lechner-Doll, M., Clauss, M., Liesegang, A. & Wanner, M. 2002. The use of dosed and herbage n-alkanes as markers for the determination of intake, digestibility, mean retention time and diet selection in Galapagos tortoises.—*Herpetological Journal* 12: 45–54.

Heartley, S. E., Iason, G. R., Duncan, A. J. & Hitchcock, D. 1997. Feeding behaviour of red deer (*Cervus elaphus*) offered sitka spruce saplings (*Picea sitchensis*) grown under different light and nutrient regimes.—*Functional Ecology* 11: 348–357.

Holling, C. S. 1959. The components of predation as revealed by a study of small mammal predation of the European Pine sawfly.—*Canadian Entomology* 91: 293–332.

Hollocher, T., Chin, K., Hollocher, K. & Kruge, M. A. 2001. Bacterial residues in coprolite of herbivorous dinosaurs: role of bacteria in mineralization of feces.—*Palaios* 16: 547–565.

Hotton, N., Olson, E. C. & Beerbower, R. 1997. Amniote origins and the discovery of herbivory. *In* Sumida, S. S. & Martin, K. L. M. (eds.). *Amniote Origins. Completing the Transition to Land*. Academic Press, San Diego: pp. 207–264.

Hume, I. D. & Warner, A. C. I. 1980. Evolution of microbial digestion in mammals. *In* Ruckebusch, Y. & Thievend, P. (eds.). *Digestive Physiology and Metabolism in Ruminants*. MTP Press, Lancaster: pp. 665–684.

Hummel, J., Nogge, G., Clauss, M., Norgaard, C., Johanson, K., Nijboer, J. & Pfeffer, E. 2006a. Energetic nutrition of the okapi in captivity: fermentation characteristics of feedstuffs.—*Zoo Biology* 25: 251–266.

Hummel, J., Südekum, K. H., Streich, W. J. & Clauss, M. 2006b. Forage fermentation patterns and their implications for herbivore ingesta retention times.—*Functional Ecology* 20: 989–1002.

Hummel, J., Gee, C. T., Südekum, K. H., Sander, P. M., Nogge, G. & Clauss, M. 2008. In vitro digestibility of fern and gymnosperm foliage: implications for sauropod feeding ecology and diet selection.—*Proceedings of the Royal Society B: Biological Sciences* 275: 1015–1021.

Illius, A. W. & Gordon, I. J. 1992. Modelling the nutritional ecology of ungulate herbivores: evolution of body size and competitive interactions.—*Oecologia* 89: 428–434.

Illius, A. W., Duncan, P., Richard, C. & Mesochina, P. 2002. Mechanisms of functional response and resource exploitation in browsing roe deer.—*Journal of Animal Ecology* 71: 723–734.

Janensch, W. 1929. Magensteine bei Sauropoden der Tendaguru-Schichten.—*Palaentographica Supplement* 7 (2) 1: 137–144.

Jarman, P. J. 1974. The social organization of antelope in relation to their ecology.—*Behaviour* 48: 215–266.

Karasov, W. H. & Martínez del Rio, C. 2007. *Physiological Ecology: How Animals Process Energy, Nutrients, and Toxins*. Princeton University Press, Princeton.

Karasov, W. H., Petrossian, E., Rosenberg, L. & Diamond, J. M. 1986. How do food passage rate and assimilation differ between herbivorous lizards and non-ruminant mammals?—*Journal of Comparative Physiology B* 156: 599–609.

Klasing, K. 1998. *Comparative Avian Nutrition.* CAB International, Wallingford.

Koerth, B. H. & Stuth, J. W. 1991. Instantaneous intake rates of 9 browse species by white-tailed deer.—*Journal of Range Management* 44: 614–618.

Krassilov, V. A. 1981. Changes of Mesozoic vegetation and the extinction of dinosaurs.—*Palaeography, Palaeoclimatology, Palaeoecology* 34: 207–234.

Langer, P. 1987. Der Verdauungstrakt bei pflanzenfressenden Säugetieren.—*Biologie in unserer Zeit* 17: 9–14.

Langer, P. 1988. *The Mammalian Herbivore Stomach.* Gustav Fischer Verlag, Stuttgart, New York.

Langer, P. 1991. Evolution of the digestive tract in mammals.—*Verhandlungen der Deutschen Zoologischen Gesellschaft* 84: 169–193.

Langer, P. 1993. The gastric groove, a specific adaptation to nutritional ontogeny in eutherian mammals.—*Growth Development & Aging* 57: 139–146.

Langer, P. 1994. Food and digestion of Cenozoic mammals in Europe. *In* Chivers, D. J. & Langer, P. (eds.). *The Digestive System of Mammals: Food, Form, and Function.* Cambridge University Press, Cambridge: pp. 9–24.

Leckey, E. H., 2004. *Co-evolution in Herbivorous Dinosaurs and Land Plants: the Use of Fossil Locality Data in an Examination of Spatial Co-occurrence.* M.Sc. Thesis. University of California, Santa Barbara.

Lucas, S. G. 2000. The gastromyths of "*Seismosaurus,*" a Late Jurassic dinosaur from New Mexico.—*New Mexico Museum of Natural History and Science Bulletin* 17: 61–68.

Lucas, P. W. 2004. *Dental Functional Morphology: How Teeth Work.* Cambridge University Press, Cambridge.

Magnol, J. P. 2003. Appareil digestif, appareil locomoteur et stratégies de prise alimentaire chez les dinosaures.—*Revue de Médecine Vétérinaire* 154: 543–563.

Males, J. R. 1973. *Ration Digestibility, Rumen Bacteria and Several Rumen Parameters in Sheep Born and Reared in Isolation.* Ph.D. Dissertation. Ohio State University, Columbus.

Marshall, C. L. & Stevens, C. E. 2000. Yes Virginia, there were foregut fermenting dinosaurs.—*Proceedings of the Comparative Nutrition Society* 3: 138–142.

McLeod, M. N. & Minson, D. J. 1988. Large particle breakdown by cattle eating ryegrass and alfalfa.—*Journal of Animal Science* 66: 992–999.

Menke, K. H. & Steingaß, H. 1988. Estimation of the energetic feed value obtained from chemical analysis and in vitro gas production using rumen fluid.—*Animal Research and Development* 28: 7–55.

Menke, K. H., Raab, L., Salewski, A., Steingaß, H., Fritz, D. & Schneider, W. 1979. The estimation of the digestibility and metabolizable energy content of ruminant feedingstuffs from the gas production when they are incubated with rumen liquor in vitro.—*Journal of Agricultural Science* 93: 217–222.

Midgley, J. J. 2005. Why don't leaf-eating animals prevent the formation of vegetation? Relative vs. absolute dietary requirements.—*New Phytologist* 168: 271–273.

Midgley, J. J. & Bond, W. J. 1991. Ecological aspects of the rise of angiosperms: a challenge to the reproductive superiority hypotheses.—*Biological Journal of the Linnean Society* 44: 81–92.

Midgley, J. J., Midgley, G. & Bond, W. J. 2002. Why were dinosaurs so large? A food quality hypothesis.—*Evolutionary Ecology Research* 4: 1093–1095.

Moore, S. J. 1999. Food breakdown in an avian herbivore: who needs teeth?—*Australian Journal of Zoology* 47: 625–632.

Morton, J. 1979. *Guts: The Form and Function of the Digestive System.* Edward Arnold, Southampton.

Murphy, M. R. & Nicoletti, J. M. 1984. Potential reduction of forage and rumen digesta particle size by microbial action.—*Journal of Dairy Science* 67: 1221–1226.

Owen-Smith, N. 1988. *Megaherbivores: The Influence of Very Large Body Size on Ecology.* Cambridge University Press, Cambridge.

Parra, R. 1978. Comparison of foregut and hindgut fermentation in herbivores. *In* Montgomery, G. G. (ed.). *The Ecology of Arboreal Folivores.* Smithsonian Institution Press, Washington, D.C.: pp. 205–229.

Paul, G. S. 1997. Dinosaur models: the good, the bad, and using them to estimate the mass of dinosaurs. *In* Wolberg, D. L., Stumps, E. & Rosenberg, G. D. (eds.). *Dinofest.* International Academy of Natural Sciences, Philadelphia: pp. 129–154.

Paul, G. S. 1998. Terramegathermy and Cope's rule in the land of titans.—*Modern Geology* 23: 179–217.

Pearce, G. R. 1967. Changes in particle size in the reticulorumen of sheep.—*Australian Journal of Agricultural Research* 18: 119–125.

Pérez-Barberìa, F. J., Elston, D. A., Gordon, I. J. & Illius, A. W. 2004. The evolution of phylogenetic differences in the efficiency of digestion in ruminants.—*Proceedings of the Royal Society B: Biological Sciences* 271: 1081–1090.

Poppi, D. P., Norton, B. W., Minson, D. J. & Hendricksen, R. E. 1980. The validity of the critical size theory for particles leaving the rumen.—*Journal of Agricultural Science* 94: 275–280.

Prasad, V., Strömberg, C. A. E., Alimohammadian, H. & Sahni, A. 2005. Dinosaur coprolites and the early evolution of grasses and grazers.—*Science* 310: 1177–1180.

Prop, J. & Vulink, T. 1992. Digestion by barnacle geese in the annual cycle—the interplay between retention time and food quality.—*Functional Ecology* 6: 180–189.

Pryor, G. & Bjorndal, K. A. 2005. Symbiotic fermentation, digesta passage, and gastrointestinal morphology in bullfrog tadpoles (*Rana catesbeiana*).—*Physiological and Biochemical Zoology* 78: 201–215.

Rees, P. M., Noto, C. R., Parrish, J. M. & Parrish, J. T. 2004. Late Jurassic climates, vegetation, and dinosaur distribution.—*Journal of Geology* 112: 643–653.

Reilly, S. M., McBrayer, L. D. & White, T. D. 2001. Prey processing in amniotes: biomechanical and behavioral patterns of food reduction.—*Comparative Biochemistry and Physiology A* 128: 397–415.

Reisz, R. R. & Sues, H.-D. 2000. Herbivory in late Paleozoic and Triassic terrestrial vertebrates. *In* Sues, H.-D. (ed.). *Evolution of Herbivory in Terrestrial Vertebrates: Perspectives from the Fossil Record.* Cambridge University Press, Cambridge: pp. 9–41.

Renecker, L. A. & Hudson, R. J. 1992. Thermoregulatory and behavioral response of moose: is large body size an adaptation or constraint?—*Alces Supplement* 1: 52–64.

Robbins, C. T. 1993. *Wildlife Feeding and Nutrition*. Academic Press, San Diego.

Rowcliffe, J. M., Sutherland, W. J. & Watkinson, A. R. 1999. The functional and aggregative responses of a herbivore: underlying mechanisms and the spatial implications for plant depletion.—*Journal of Animal Ecology* 68: 853–868.

Russell, D. A., Beland, P. & McIntosh, J. S. 1980. Paleoecology of the dinosaurs of Tendaguru (Tanzania).—*Memoirs de la Societe of Geologique de France* 59: 169–176.

Sander, P. M. & Clauss, M. 2008. Sauropod gigantism.—*Science* 322: 200–201.

Sander, P. M., Christian, A. & Gee, C. T. 2009. Sauropods kept their heads down. Response.—*Science* 323: 1671–1672.

Sander, P. M., Christian, A., Clauss, M., Fechner, R., Gee, C. T., Griebeler, E. M., Gunga, H.-C., Hummel, J., Mallison, H., Perry, S., Preuschoft, H., Rauhut, O., Remes, K., Tütken, T., Wings, O. & Witzel, U. 2010a. Biology of the sauropod dinosaurs: the evolution of gigantism.—*Biological Reviews of the Cambridge Philosophical Society*. doi: 10.1111/j.1469=185X.2010.00137.x.

Sander, P. M., Gee, C. T., Hummel, J. & Clauss, M. 2010b. Mesozoic plants and dinosaur herbivory. *In* Gee, C. T. (ed). *Plants in Mesozoic Time: Morphological Innovations, Phylogeny, Ecosystems*. Indiana University Press, Bloomington: pp. 331–359.

Sanders, F., Manley, K. & Carpenter, K. 2001. Gastroliths from the Lower Cretaceous sauropod *Cedarosaurus weiskopfae*. *In* Tanke, D. H. & Carpenter, K. (eds.). *Mesozoic Vertebrate Life*. Indiana University Press, Bloomington: pp. 166–180.

Savage, V. M., Gillooly, J. F., Woodruff, W. H., West, G. B., Alen, A. P., Enquist, B. J. & Brown, J. H. 2004. The predominance of quarter-power scaling in biology.—*Functional Ecology* 18: 257–282.

Schwarm, A., Ortmann, S., Wolf, C., Streich, W. J. & Clauss, M. 2009. More efficient mastication allows increasing intake without compromising digestibility or necessitating a larger gut: comparative feeding trials in banteng (*Bos javanicus*) and pygmy hippopotamus (*Hexaprotodon liberiensis*).—*Comparative Biochemistry and Physiology A* 152: 504–512.

Senter, P. 2007. Necks for sex: sexual selection as an explanation for sauropod dinosaur neck elongation.—*Journal of Zoology* 271: 45–53.

Sereno, P. C. 1997. The origin and evolution of dinosaurs.—*Annual Review of Earth and Planetary Science* 25: 435–489.

Shipley, L. A. & Spalinger, D. E. 1992. Mechanics of browsing in dense food patches: effects of plant and animal morphology on intake rate.—*Canadian Journal of Zoology* 70: 1743–1752.

Shipley, L. A., Blomquist, S. & Danell, K. 1998. Diet choices made by free-ranging moose in northern Sweden in relation to plant distribution, chemistry, and morphology.—*Canadian Journal of Zoology* 76: 1722–1733.

Simmons, R. & Scheepers, L. 1996. Winning by a neck: sexual selection in the evolution of giraffe.—*American Naturalist* 148: 771–786.

Spalinger, D. E. & Hobbs, N. T. 1992. Mechanisms of foraging in mammalian herbivores: new models of functional response.—*American Naturalist* 140: 325–348.

Spalinger, D. E. & Robbins, C. T. 1992. The dynamics of particle flow in the rumen of mule deer and elk.—*Physiological Zoology* 65: 379–402.

Spalinger, D. E., Hanley, T. A. & Robbins, C. T. 1988. Analysis of the functional response in foraging in the Sitka black-tailed deer.—*Ecology* 69: 1166–1175.

Stevens, C. E. & Hume, I. D. 1995. *Comparative Physiology of the Vertebrate Digestive System*. Cambridge University Press, New York.

Stevens, C. E. & Hume, I. D. 1998. Contributions of microbes in vertebrate gastrointestinal tract to production and conservation of nutrients.—*Physiological Reviews* 78: 393–427.

Stevens, K. A. & Parrish, J. M. 1999. Neck posture and feeding habits of two Jurassic sauropod dinosaurs.—*Science* 284: 798–800.

Stevens, K. A. & Parrish, J. M. 2005. Neck posture, dentition, and feeding strategies in Jurassic sauropod dinosaurs. *In* Tidwell, V. & Carpenter, K. (eds.). *Thunder-lizards: The Sauropodomorph Dinosaurs*. Indiana University Press, Bloomington: pp. 212–232.

Stobbs, T. H. 1973. The effect of plant structure on the intake of tropical pastures. II. Differences in sward structure, nutritive value, and bite size of animals grazing *Setaria anceps* and *Chloris gayana* at various stages of growth.—*Australian Journal of Agricultural Research* 24: 821–829.

Stokes, W. L. 1964. Fossilized stomach contents of a sauropod dinosaur.—*Science* 143: 576–577.

Sues, H.-D. & Reisz, R. R. 1998. Origins and early evolution of herbivory in tetrapods.—*Trends in Ecology and Evolution* 13: 141–145.

Swain, T. 1976. Angiosperm-reptile co-evolution. *In* Bellairs, A. d. A. & Barry Cox, C. (eds.). *Morphology and Biology of Reptiles*. Academic Press, London: pp. 107–122.

Swain, T. 1978. Plant-animal coevolution: a synoptic view of the Paleozoic and Mesozoic. *In* Harborne, J. B. (ed.). *Biochemical Aspects of Plant and Animal Coevolution*. Academic Press, London: pp. 3–19.

Taggart, R. E. & Cross, A. T. 1997. The relationship between land plant diversity and productivity and patterns of dinosaur herbivory. *In* Wolberg, D. L., Stump, E. & Rosenberg, G. (eds.). *Dinofest*. International Academy of Natural Sciences, Philadelphia: pp. 403–416.

Tiffney, B. H. 1997. Land plants as food and habitat in the age of dinosaurs. *In* Farlow, J. O. & Brett-Surman, M. K. (eds.). *The Complete Dinosaur*. Indiana University Press, Bloomington: pp. 352–370.

Troyer, K. 1982. Transfer of fermentative microbes between generations in a herbivorous lizard.—*Science* 216: 540–542.

Troyer, K. 1984. Diet selection and digestion in *Iguana iguana:* the importance of age and nutrient requirements.—*Oecologia* 61: 201–207.

Trudell, J. & White, R. G. 1981. The effect of forage structure and availability on food intake, biting rate, bite size and daily eating time of reindeer.—*Journal of Applied Ecology* 18: 63–81.

Upchurch, P. & Barrett, P. M. 2000. The evolution of sauropod feeding mechanisms. *In* Sues, H.-D. (ed.). *Evolution of Herbivory in Terrestrial Vertebrates. Perspectives from the Fossil Record.* Cambridge University Press, Cambridge: pp. 79–122.

Van Gils, J. A., Gyimesi, A. & Van Lith, B. 2007. Avian herbivory: an experiment, a field test, and an allometric comparison with mammals.—*Ecology* 88: 2926–2935.

Van Soest, P. J. 1994. *Nutritional Ecology of the Ruminant.* Cornell University Press, Ithaca.

Wallis de Vries, M. F. & Daleboudt, C. 1994. Foraging strategy of cattle in patchy grassland.—*Oecologia* 100: 98–106.

Weaver, J. C. 1983. The improbable endotherm: the energetics of the sauropod dinosaur *Brachiosaurus*.—*Paleobiology* 9: 173–182.

Weishampel, D. B. & Jianu, C. M. 2000. Plant-eaters and ghost lineages: dinosaurian herbivory revisited. *In* Sues, H.-D. (ed.). *Evolution of Herbivory in Terrestrial Vertebrates. Perspectives from the Fossil Record.* Cambridge University Press, Cambridge: pp. 123–143.

Weishampel, D. B. & Norman, D. B. 1989. Vertebrate herbivory in the Mesozoic: jaws, plants, and evolutionary metrics. *In* Farlow, J. O. (ed.). *Paleobiology of the Dinosaurs.* Geological Society of America 238, Boulder, Colorado: pp. 87–100.

White, C. R. & Seymour, R. S. 2003. Mammalian basal metabolic rate is proportional to body mass$^{2/3}$.—*Proceedings of the National Academy of Sciences of the United States of America* 100: 4046–4049.

White, C. R. & Seymour, R. S. 2005. Allometric scaling of mammalian metabolism.—*Journal of Experimental Biology* 208: 1611–1619.

Wickstrom, M. L., Robbins, C. T., Hanley, T. A., Spalinger, D. E. & Parish, S. M. 1984. Food intake and foraging energetics of elk and mule deer.—*Journal of Wildlife Management* 48: 1285–1301.

Wilson, J. A. & Sereno, P. C. 1998. Early evolution and higher-level phylogeny of sauropod dinosaurs.—*Society of Vertebrate Paleontology Memoir* 5: 1–68.

Wikelski, M., Gall, B. & Trillmich, F. 1993. Ontogenetic changes in food intake and digestion rate of the herbivorous marine iguana (*Amblyrhynchus cristatus*).—*Oecologia* 94: 373–379.

Wing, S. L. & Tiffney, B. H. 1987. The reciprocal interaction of angiosperm evolution and tetrapod herbivory. *Review of Paleobotany and Palynology* 50: 179–210.

Wing, S. L., Sues, H.-D., Tiffney, B. H., Stucky, R. K., Weishampel, D. B., Spicer, R. A., Jablonski, D., Badgley, C. E., Wilson, M. V. H. & Kovach, W. L. 1992. Mesozoic and early Cenozoic terrestrial ecosystems. *In* Behrensmeyer, A. K., Damuth, J. D., DiMichele, W. A., Potts, R., Sues, H.-D. & Wing, S. L. (eds.). *Terrestrial Ecosystems through Time: Evolutionary Paleoecology of Plants and Animals.* University of Chicago Press, Chicago: pp. 327–416.

Wings, O. 2003. The function of gastroliths in dinosaurs: new considerations following studies on extant birds.—*Journal of Vertebrate Paleontology* 23: 111A.

Wings, O. 2005. Taphonomy, gastroliths, and the lithophagic behavior of sauropodomorph dinosaurs.—*Journal of Vertebrate Paleontology* 25: 131A.

Wings, O. & Sander, P. M. 2007. No gastric mill in sauropod dinosaurs: new evidence from analysis of gastrolith mass and function in ostriches.—*Proceedings of the Royal Society B: Biological Sciences* 274: 635–640.

Yearsley, J., Tolkamp, B. J. & Illius, A. W. 2001. Theoretical developments in the study and prediction of food intake.—*Proceedings of the Nutrition Society* 60: 145–156.

3

Dietary Options for the Sauropod Dinosaurs from an Integrated Botanical and Paleobotanical Perspective

CAROLE T. GEE

DURING THE MAJORITY OF THE MESOZOIC, from the Triassic to the mid Cretaceous, the food plants of the sauropod dinosaurs were virtually limited to ferns, fern allies, and gymnosperms because the diversification of the angiosperms, which include the broad-leaved trees and grasses of today, only began in the Late Cretaceous. In this chapter, the preferences of the sauropods for one or more of these Mesozoic plant groups are evaluated by means of a survey approach that integrates botanical and paleobotanical data. These data include the growth habits of the nearest living relatives of these plant groups, their habitat, the amount of biomass produced, and the ability to regrow shoots, branches, and leaves after injury through herbivory. The relative quantities of energy and essential nutrients yielded to herbivores with hindgut fermentation, the consumption of the various plant groups by modern herbivores, and the coeval occurrence of sauropods and individual plant groups in the fossil record are other major factors taken into consideration here. As a result of this extensive survey, it appears that *Araucaria, Equisetum,* the Cheirolepidiaceae (an extinct conifer family), and *Ginkgo* would have been most accessible, sustaining, and/or preferred sources of food for the sauropods. Moderately accessible, sustaining, and/or commonly encountered plants would have been other conifers such as the Podocarpaceae, Cupressaceae, and Pinaceae. Less commonly browsed by the sauropods, especially by large, fully grown individuals, would have been forest-dwelling ferns such as *Angiopteris* and *Osmunda*. The least frequently eaten plants were probably the cycads and bennettitaleans.

Introduction

Botanically, it seems that the thick-cuticle conifers, toxic cycads, and low-biomass ferns would have offered little in terms of palatable, sustaining fodder to the early and mid Mesozoic sauropods, yet we know that giant sauropods did exist and must have thrived on these plant groups. Indeed, ferns and gymnosperms dominated the flora during the better part of the Mesozoic, specifically, some 140 million years altogether. The great diversification of the flowering plants, and hence the onset of the modern flora, took place in the mid Cretaceous, after about four fifths of the Mesozoic had already passed by. Angiosperms, the producers of broad leaves, flowers, and fruits in the present-day world, thus fed the herbivorous dinosaurs for only a relatively short span of time.

The plant groups that greened the Earth from the Early Triassic (251 million years ago) to the mid Cretaceous (100 million years ago) included ferns and fern allies, cycads and bennettitaleans, seed ferns, ginkgophytes, and conifers. Like the dinosaurs, some of these taxa—major groups such as the bennettitaleans, seed ferns, and the cheirolepidiaceans—went extinct at the end of the Mesozoic, but others have survived to this day.

Indeed, a few plants that have remained unchanged at the genus level since the Mesozoic, for example, the horsetail *Equisetum*, the tropical ferns *Angiopteris* and *Marattia*, the maidenhair tree *Ginkgo*, and the southern conifer *Araucaria*. The living species of these genera can be looked on as the embodiment of close relatives that nourished the dinosaurs.

Although it is clear that herbivorous dinosaurs must have fed on and attained gigantic sizes using nonangiospermous plants, it is less obvious which of these plants—if any—they might have preferred as a food source. Were some plant groups more nutritious than others? Did the dinosaurs preferentially live (and feed) in certain kinds of vegetation? Furthermore, after a herd of dinosaurs or perhaps just a particularly hungry individual had decimated an area of its flora, were some plants better able to recover and regenerate their foliage than others, thus surviving another season to grow, reproduce, and possibly withstand another onslaught of herbivory?

This chapter tackles the question of dietary options for the herbivorous dinosaurs, focusing on sauropods, by conducting a survey of the preangiospermous Mesozoic flora, especially in light of any nearest living relatives, to see how suitably they would have served as dinosaur fodder in regard to their growth habit, the habitat in which they grew, the amount of biomass they were able to produce, their ability to regrow shoots, branches, and leaves, and the quantity of energy and important nutrients they would have offered herbivores. Although these nonangiospermous plant groups are not usually considered forage plants for modern plant eaters, reports of consumption involving living animals are noted here as well.

In addition to analyzing living plants, the fossil record can be gleaned for clues to the food preferences of the sauropod dinosaurs. For example, the coeval occurrence of both sauropods and specific plant groups at the same fossil localities increases the possibility that these dinosaurs fed on these plants, assuming that these death assemblages represent the living biota to a reasonable extent.

Finally, on the basis of data from living plants, extant herbivores, and the fossil record, each Mesozoic plant group is rated comparatively for its likelihood as a food option for the sauropod dinosaurs.

PREVIOUS APPROACHES

A number of approaches have been developed to try to find out what sorts of plants herbivorous dinosaurs preferred feeding on. These include using trace fossils, dinosaur morphology, and plant factors such as morphology and physiology.

Trace Fossils

The most direct approach would seem to be the study of plant remains in dinosaur coprolites or digestive-tract remains. Yet this approach is not as straightforward as generally assumed for several reasons, which were recently discussed in great detail (Sander et al. 2010). In regard to coprolites, it is difficult to positively and unequivocally trace fossilized fecal material to a particular genus or even type of plant-eating dinosaur. Even in the case of the putative sauropod coprolites found in the same horizon as titanosaurs in the Upper Cretaceous Lameta Formation of India, there is some doubt that such gigantic animals would have produced such small coprolites (Sander et al. 2010). Furthermore, it is questionable whether the "soft" plant tissues, especially of fern fronds, reported from these nodule-shaped structures (Mohabey 2005) would have survived the three to four days of chemical and mechanical processing in a large herbivore's digestive system (cf. Hummel & Clauss, this volume).

In the case of stomach remains and other reports of digestive-tract remains, even if fossilized plant material is found inside a dinosaur's body cavity, it is still unclear whether this material truly represents the digestive remains or a postmortem accumulation. In one well studied case of a mummified hadrosaur, the highly diverse pollen flora in the plant material, coupled with the uniform particle size of the plant megafossils, plus the occurrence of an unusual mixture of charcoal, dinocysts, and other algae, led researchers to suggest that these plant remains were washed into the body after death (Currie et al. 1995).

Indeed, in a critical look at coprolite and digestive-tract remains in herbivorous dinosaurs, Sander et al. (2010a) reported that most cases are problematic in their authenticity, with the exception of one well documented specimen of an apparently frugivorous bird, *Jeholornis prima* from the Lower Cretaceous of China (Zhou & Zhang 2002), with seeds in its gut, and of another case consisting of clusters of pellet-like coprolites with bennettitalean leaf cuticle from an English dinosaur (Hill 1976). It should be noted that in both of these instances, the plant remains were monotypic in regard to plant parts (i.e., seeds or leaves) and generally monospecific in regard to plant taxon (i.e., *Carpolithes* or bennettitaleans). This would be in line with a feeding behavior in which the animal feeds on only one type of plant part of a single species, depending on what is ripe or available at the time. Thus, such cases would depict the last meal eaten by the dinosaur immediately before death and burial, and would be less likely to represent the spectrum of the animal's overall diet.

Dinosaur Morphology

Other approaches to deciphering herbivorous dinosaur feeding habits involve the analysis of teeth, jaw mechanics, skull shape, neck length, and other food-processing structures such as gastric mills. Because this topic has also been subject to a recent critical discussion (e.g., Sander et al. 2010a; Hummel & Clauss, this volume), it will be not repeated here. However, it should be noted that it is thought that the success of the sauropod dinosaurs in regard to gigantism was based in part on the fact that they did not chew, which enabled them to ingest large mouthfuls of food rapidly and continuously (Sander & Clauss 2008; Sander et al. 2010b; Hummel & Clauss, this volume). Other probable factors included egg laying, a high growth rate, avian-style respiration, an ontogenetically flexible metabolic rate, and a long neck (Sander & Clauss 2008; Sander et al. 2010a, 2010b; Griebeler & Werner, this volume; Sander et al., this volume).

Plant Factors

The giant body size of sauropods, along with their need to consume huge amounts of plant material for growth and maintenance, necessarily means that they were bulk feeders. As revealed by the cropping efficiency experiments on conifer and angiosperm twigs carried out by Hummel & Clauss (this volume, and references therein), sauropods would have been most efficient when foraging on conifers, and not on angiosperms (i.e., broad-leaved trees) or ferns because they would have received more biomass with each bite of conifer foliage.

Qualitatively, there are also differences in energy and nutrient yield between different plant groups. Although it had been assumed for many years that the preangiosperm Mesozoic plants could have provided only low-quality fodder for herbivorous dinosaurs (Coe et al. 1987; Taggart & Cross 1997; Tiffney 1997; Midgley et al. 2002), recent empirical laboratory experiments estimating the amount of energy yielded by the nearest living relatives of the Mesozoic flora (Hummel et al. 2008; Sander et al. 2010a; Hummel & Clauss, this volume) have shown that a number of ferns, conifers, and other gymnosperms actually offer herbivores relatively large amounts of energy, comparable to or even surpassing grasses or broad-leaved trees (see Hummel & Clauss, this volume). These high-energy plants include horsetails such as *Equisetum,* ferns such as *Osmunda* and *Angiopteris,* conifers such as *Araucaria, Torreya, Taxus,* and *Cephalotaxus,* and the maidenhair tree *Ginkgo.* Conversely, there are other plants, namely the cycads and podocarps, that prove to be poor sources of energy. These new perspectives reveal another factor that can be used in the evaluation of the various Mesozoic plant groups for their potential as sauropod fodder and are incorporated here into the botanical and paleobotanical data on each plant group.

Parameters Considered

CHRONOSTRATIGRAPHY AND PALEOGEOGRAPHY

To control for floral and faunal differences dependent on geological time, this chapter focuses on the Late Jurassic sites or regions that have produced well documented sauropod faunas and Mesozoic floras. Specifically, this encompasses the Morrison Formation of the Western Interior of North America, the Patagonian region of southern South America, and the Tendaguru Beds in Tanzania, East Africa (Fig. 3.1).

FERMENTATION VALUES

Because the fermentation curves of Hummel et al. (2008) are reproduced in another chapter (Hummel & Clauss, this volume; see also Sander et al. 2010b), they will not be duplicated here. However, I propose here a five-point scale categorizing the amounts of gas production in order to describe qualitatively the relative amount of energy released from each food plant type during the 72 hour long fermentation trials of Hummel et al. (2008) and to facilitate comparison between plant groups. Hence, excellent is over 45 ml/200 mg dry matter; very good is 35–45 ml/200 mg dry matter; good is 25–35 ml/200 mg dry matter; poor is 15–25 ml/200 mg dry matter; and very poor is under 15 ml/200 mg dry matter.

Analysis of Plant Taxa

ARAUCARIACEAE

Habit and Habitat

The family Araucariaceae is represented today by three genera—*Araucaria* (Plate 3.1), *Agathis,* and *Wollemia*—all of which form tall evergreen trees. On average, the 20 or so species of *Araucaria* reach heights of about 30 m, but *A. heterophylla,* which is commonly grown in northern temperate regions as a garden or house plant (i.e., the Norfolk Island pine), can top out at 70 m (Krüssmann 1972). Mature trees of *Araucaria* (Plate 3.1F) or *Agathis* often occur singly as canopy dominants in the forest, although *Araucaria* can also make up relatively dense, monospecific groves of trees (Schütt et al. 2004). All three genera occur in tropical and subtropical forests. While *Araucaria* is native to both South America and Australasia, *Agathis* is widespread in Australasia (Krüssmann 1972), and *Wollemia* is restricted to a single rain forest gorge near Sydney, Australia (Jones et al. 1995).

Biomass Production

Like present-day members of the family, fossil Araucariaceae formed tall trees, and their spreading branches probably constituted quite a bit of biomass. In the past, as in the present, these trees were likely slow growers; *Araucaria* plantations in Northern Queensland, for example, take twice as long to reach maturity as commercial timber like those with nonnative pines (Hanrahan, pers. comm., 2007). When compared to other tree species in its native habitat, *Araucaria araucana* is also a slower grower. In a mixed *Nothofagus* forest in South America, for example, the average increment of growth of *Araucaria araucana* trees each year was only 5–8.2 cm in height

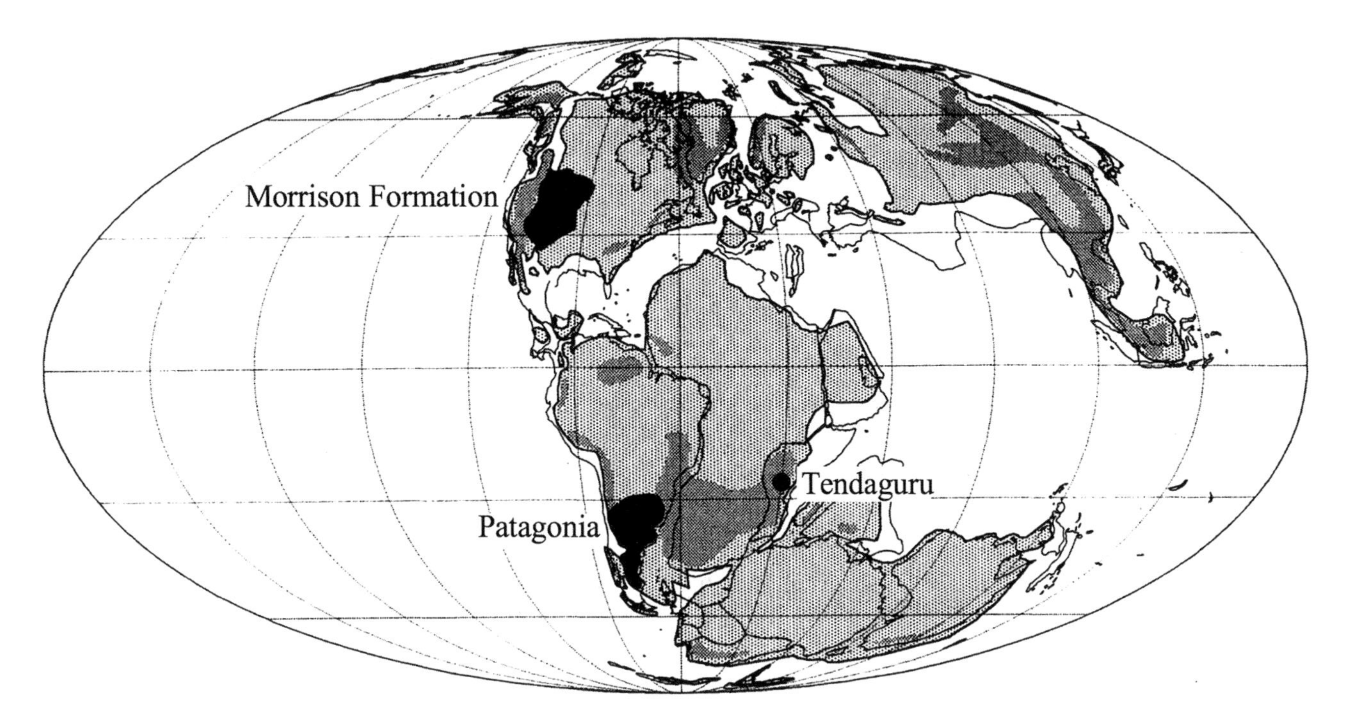

FIGURE 3.1. Paleogeographic map of the Late Jurassic (Kimmeridgian). Shown in black are the approximate locations of the three major sauropod faunas discussed in this chapter. (1) Morrison Formation, Western Interior of North America. (2) Patagonia, southern South America. (3) Tendaguru, Tanzania, Africa. Light gray indicates land; medium gray, highlands; and white areas within the black lines, flooded shelves. *Map modified after Smith et al. (2004).*

and 2.3–2.7 mm in diameter (Donoso et al. 2004). Among the southern conifers—that is, the Podocarpaceae, Araucariaceae, and half of the Cupressaceae—*Araucaria* and *Agathis* are moderate in their growth rate (Enright & Ogden 1995).

Potential for Recovery

All extant members of the Araucariaceae have the ability to regenerate branches or treetops that have been broken off by way of epicormic and coppice buds (e.g., Burrows 1990; Burrows et al. 2003), which are, respectively, dormant buds on the trunk or at the base of the tree that are triggered into producing new growth after damage (Plate 3.1G, H). This trait provides a way to continuously regenerate new organs resulting from damage that may occur through drought, small-scale fires, or blowdowns caused by tropical cyclones (e.g., Rigg et al. 1998; Burrows et al. 2003; Gee, pers. obs.). In the Mesozoic, this would have also been advantageous to the individual trees after intense feeding on leafy branches and twigs by tall, voracious sauropods, or damage by natural causes such as volcanic activity or fire. Indeed, fossil structures representing the dormant woody buds of conifers such as *Araucaria mirabilis* (Spegazzini) Windhausen or *Pararaucaria patagonica* Wieland have been collected from the Middle Jurassic Cerro Cuadrado Petrified Forest for many decades (Stockey 2002). It is thought that these buds, called aerial lignotubers, may have developed in the axils of leaves of conifers in this area in response to fire caused by volcanism, which was a repeated occurrence in the Cerro Cuadrado sequence (Stockey 2002).

Digestibility

The in vitro digestibility of *Araucaria* foliage, based on several trials with five different species, is very good (Hummel et al. 2008). Leaves and pollen cones (Plate 3.1B, D, E) of *Araucaria* are moderately good in their digestibility, while foliage of *Agathis* is weakly digestible (Hummel, pers. comm.). In contrast to other plants, *Araucaria* foliage provides little in the form of protein (Hummel et al. 2008; Hummel & Clauss, this volume) and thus would be less appropriate for young, growing animals requiring a high protein intake. It should be noted that herbivore gut fermentation behavior of *Araucaria* foliage rises slowly in the first 30 hours of digestion, but finishes at a relatively higher rate over the course of the next four days. Such an energy release would be most advantageous to a large animal with a long hindgut retention time, such as a giant-sized adult sauropod dinosaur.

Consumption by Modern Herbivores

Although few vertebrates are known to feed on *Araucaria*, there are reports of girdling (also known as ring barking) by cockatoos in Northern Queensland, Australia (Hanrahan, pers. comm. 2007). *Araucaria* seeds of the species that produce large seeds, such as *A. bidwillii, A. araucana,* or *A. angustifolia,* are commonly eaten by rats, by large birds such as cockatoos, parrots, crows, or jays, and by humans.

Co-occurrence with Major Late Jurassic Sauropod Faunas

Fossils pertaining to *Araucaria* occur throughout most of the Mesozoic, starting from the Late Triassic and extending beyond the Late Cretaceous; they were globally widespread as well. Species diversity within just the genus *Araucaria* was high during this era, judging from the more than 30 different species of seed cones that have been described (Gee & Tidwell 2010). In the Morrison Formation, araucarian plant compressions (e.g., Plate 3.1C) and wood are abundant and in fact are by far the most common plant fossils in a sauropod bonebed, the Howe-Stephens Quarry on the Howe Ranch in north-central Wyoming (Gee & Tidwell 2010). Palynologically, araucariaceous pollen is a major element throughout the Morrison Formation, second only to the Cheirolepidiaceae in terms of frequency (Hotton & Baghai-Riding 2010). In Patagonia, araucarian cones and trees are locally very common, for example at the Cerro Cuadrado Petrified Forest in Patagonia (e.g., Calder 1953; Stockey 1975, 1978; Zamuner & Falaschi 2005; Falaschi 2009). In Tendaguru, araucarian cuticle occurs as a minor element of the flora, but the family also shows up throughout the Dinosaur Beds in the form of wood, as well as several species of abundant pollen (Schrank 1999; Aberhan et al. 2002) that suggest the widespread dominance of araucarians, especially in the Middle Saurian Bed (Schrank 2010).

Remarks and Rating

In the Mesozoic, the Araucariaceae, especially *Araucaria,* would have been a good source of food for large herbivores such as adult sauropods. Not only were araucarian trees common all over the world during most of the Mesozoic, but as a consequence of their arborescent growth habit and probable occurrence in forests, they would have provided large amounts of biomass for consumption. By extrapolating from fermentation experiments with extant *Araucaria* species, it can be inferred that the foliage and cones of the Mesozoic araucarians would have been high in energy and most suitable for large animals with a long hindgut retention time. On the basis of these parameters and because of the intimate association of *Araucaria* with sauropods in bonebeds or in sediments coeval with the major sauropod faunas, it is quite likely that the Araucariaceae were a frequent and attractive source of food for the sauropods.

EQUISETUM

Habit and Habitat

The genus *Equisetum* (Plate 3.2A), commonly known as the scouring rush or horsetail, is represented by about 15 species today. All species are perennial, although those in temperate areas often die back to the ground in the winter. Most temperate species are low-growing, attaining a height of up to about 1 m, although one species in Chile, *E. giganteum,* reaches 5 m (Husby 2003). Characteristic of all species in the genus is their occurrence in large monospecific stands, which they achieve by spreading underground with horizontal stems called rhizomes. Because *Equisetum* can produce extensive root systems to tap sources of deep groundwater, it grows in seemingly parched, disturbed habitats, such as gravels along railroad tracks or dry, sandy riverbeds, although it is common to find extensive thickets of *Equisetum* in wet, marshy areas around ponds or lakes, or along rivers.

Morphologically and anatomically, *Equisetum* has changed very little since the Middle Triassic, down to the four strap-shaped bands called elaters around its spores (Schwendemann et al. 2010), which strongly suggests that it likely had the same sort of growth habit and may have occupied the same sort of moist habitats in the geological past as it does in the present day. It is, for example, not uncommon to find numerous axes of Jurassic *Equisetum,* such as *E. laterale* (Plate 3.2B), amassed on one slab, which also suggests that this species grew in pure stands in the Mesozoic as it still does today (Harris 1961; Gee 1989; Cantrill & Hunter 2005; Gee & Sander, pers. obs.).

Biomass Production and Potential for Recovery

Because the stem of *Equisetum* is basically a hollow cylinder, each individual shoot is composed of relatively little biomass. However, taking into account the low, thicket-like growth of *Equisetum,* a large colony around a lake or several populations on a floodplain could amount to quite a significant supply of biomass for low-grazing herbivores.

Equisetum has a fast growth rate. Temperate species that die back in the winter can produce a 1 m high shoot in a single growing season. Gardeners are very familiar with the ability of this plant to spread laterally in a relatively short period of time, as well as with the difficulty of eradicating this weed because of its extensive rhizome system. Similarly, damage to *Equisetum* shoots—for example, due to herbivory—would not affect the plant's underground rhizomes and roots, which are protected underground and would merely resprout new upright stems.

Digestibility and Consumption by Modern Herbivores

The energy yield of *Equisetum,* based on three species, is excellent, exceeding that of all other plant groups, including grasses (Hummel et al. 2008; Hummel and Clauss, this volume). Although silica is thought to hinder the digestibility of cell walls (Van Soest 1994), the surface of *Equisetum* shoots, which are rich in silica (e.g., Holzhüter et al. 2003), did not seem to have much of a deleterious effect in the trials of Hummel et al. (2008) and Hummel & Clauss (this volume). Furthermore, the digestibility of *Equisetum* rises very quickly within the first 24 hours, which suggests that smaller herbivores with a shorter gut retention time would have especially benefited from its consumption.

Despite possible negative effects of silica on digestion and the purported abrasive effect of silica on mammalian teeth (Sander, pers. comm.), several large herbivorous mammals include *Equisetum* as a key food plant in their diets. These include caribou, moose, musk ox, dall sheep, and buffalo (Palmer 1944). Musk oxen, for example, commonly feed on horsetails in the summertime, when they spend most of their time in moist habitats (WMAC(NS) 2007). Similarly, many types of waterfowl, such as Canada geese, lesser snow geese, pink-footed geese, barnacle geese, and trumpeter swans, depend on various species of *Equisetum,* especially during egg incubation and after hatching (Thomas & Prevett 1982; Grant et al. 1994). Even young birds, such as Icelandic pink-footed goslings, feed extensively on *E. arvense* (Gardarsson & Sigurdsson 1972). Trumpeter swan cygnets spend more time feeding on *Equisetum* than on submerged aquatic plants, as compared with their parents (Grant et al. 1994). This is no wonder given the high protein (22% dry weight) and high phosphorous content of the rhizome tips and upright shoots of *Equisetum*—that is, of *E. fluvatile*—in late spring and summer (Thomas & Prevett 1982), when the needs for energy, protein, and minerals are at their highest levels in geese (Thomas & Prevett 1982). In tropical regions, domestic cattle have been observed to graze on *Equisetum giganteum* with relish (Hauke 1969).

Co-occurrence with Major Late Jurassic Sauropod Faunas

Like *Araucaria, Equisetum* had a cosmopolitan distribution in the Mesozoic from the Late Triassic onward, although there are reports of the genus from the Carboniferous (Taylor et al. 2009). In the Late Jurassic of the Western Interior of North America, several species of *Equisetum* have been reported from Utah, Colorado, Wyoming, Montana, British Columbia, and Alberta, including occurrences in sauropod bonebeds, such as in the Mygatt-Moore Quarry in Colorado (Tidwell et al. 1998) and the Howe-Stephens Quarry in Wyoming (Plate 3C; Gee, unpublished data), or in beds that have yielded a typical assemblage of Morrison sauropods, namely, in the Como Bluff "Member" in Wyoming (Tidwell et al. 2006). At the last site, *Equisetum* shoots are abundant and are represented by two different species. In all cases, these shoots are slender and presumably pertain to short stems, attaining less than 1 m in height. Several species of small-stature *Equisetum* have also been described from Early Jurassic to mid Cretaceous floras from both Patagonia (see recent summary by Villar de Seoane 2005; Falaschi et al. 2009) and the Antarctic Peninsula (e.g., Halle 1913; Gee 1989; Rees & Cleal 2004; Cantrill & Hunter 2005). Evidence of *Equisetum* has not yet been recovered from the Tendaguru flora (cf. Aberhan et al. 2002; Schrank 2010).

Remarks and Rating

In the Mesozoic, *Equisetum* (or *Equisetites*, as it is also known), would have been a good source of food for all low-browsing herbivores, especially young sauropods. Given their global distribution, these plants were probably common in freshwater wetland areas, where they likely covered large areas along shorelines and in this way provided much biomass for consumption. The plant's characteristic vigorous growth of aerial shoots, protection of its main stem underground, and deep root system would have helped to quickly regenerate stands of *Equisetum* after intense feeding by herbivores. Fermentation experiments with extant *Equisetum* species by Hummel et al. (2008) show that the aerial shoots as well as the tips of the rhizome offer herbivores a high-energy source of high protein, phosphorous, and other nutrient content that would especially benefit young, fast-growing animals such as young sauropods. The rapid digestibility of *Equisetum* would have also been most advantageous to small or young dinosaurs with a short gut retention time. Moreover, the short stature of most horsetails in the past and their proximity to sources of water would have enhanced their attractiveness as a commonly sought-after food plant for young and smaller sauropods.

CHEIROLEPIDIACEAE

Habit and Habitat

The family Cheirolepidiaceae is an extinct group of conifers that was an important constituent of global floras during the Mesozoic from the Late Triassic onward (Stewart & Rothwell 1993). In comparison to other conifer families, whether past or present, the Cheirolepidiaceae show the greatest diversity in regard to habit, habitat, and morphology. Some members of the family were tall trees, for example, with a trunk diameter up to about 1 m in one species (Francis 1983) or a height of at least 23.4 m in another (Axsmith & Jacobs 2005). On the basis of fossil wood studies on an in situ stand of cheirolepidiacean trees from the Late Jurassic of England, many of the trees attained at least 200 years of age; the largest tree was probably over 700 years old, indicating that there was a lengthy history of continuity in this long-lived forest (Francis 1983). Although a nonarborescent habit is unusual among conifers, other members of the Cheirolepidiaceae were herbaceous or scrubby, growing, for example, in low, dense stands in a salt marsh setting (Jung 1974; Daghlian & Person 1977). Cheirolepidiaceans grew in a variety of plant communities, ranging from monospecific or low-diversity floras on hypersaline substrates or in brackish coastal swamps, to species-rich assemblages in mesic, riparian settings (e.g., Daghlian & Person 1977; Francis 1983; Gomez et al. 2002; Axsmith & Jacobs

2005); they thrived in warm habitats under semiarid or even arid conditions, as well as in strongly seasonal climates (Francis 1983), especially at low paleolatitudes (<40°) during the Cretaceous (Taylor et al. 2009 and references therein).

There are two major kinds of foliage in the Cheirolepidiaceae: leaves that are spirally arranged with either scale-like or spreading leaves (*Brachyphyllum* or *Pagiophyllum* type) and leaves that clasp around the shoot with a jointed appearance (*Frenelopsis* and *Pseudofrenelopsis* type) (Watson 1988). Despite this diversity in leaf morphology, traits that unify the family (Watson 1988) are the distinctive pollen *Classopollis* and, to a lesser extent, thick cuticles with sunken stomata and papillae that extend over the stomata (Plate 3.2E). It is thought that some species bore fleshy, succulent leaves (Watson 1988) and that some species may have been deciduous (Behrensmeyer et al. 1992).

Biomass Production and Potential for Recovery

The arboresent cheirolepidiaceans may have dominated the woody vegetation in some areas, forming monospecific groves or forests. In this case, their spreading branches and foliage would have offered taller sauropods a large amount of biomass. Similarly, the dense colonies of low-growing, halophytic cheirolepidiaceans in the salt marshes of the Early Cretaceous of Texas (Daghlian & Person 1977), which have a modern ecological analog in the form of *Salicornia* (pickleweed), would have also provided a plentiful source of food for small and large sauropods alike. It is unknown how quickly cheirolepidiaceans grew or could regenerate after damage.

Digestibility and Consumption by Modern Herbivores

Because this group of plants does not have any close living relatives, it is impossible to test foliage for digestibility, nor is it possible to relate any accounts of consumption by modern herbivores.

Co-occurrence with Major Late Jurassic Sauropod Faunas

In the palynoflora of the Morrison Formation, *Classopollis* pollen is extremely abundant in the southern states (New Mexico and Arizona) and becomes increasingly less common in the northern region (Hotton & Baghai-Riding 2010). *Classopollis* pollen (also called *Corollina*) occurs throughout the Salt Wash Member and older sediments at Dinosaur National Monument in Utah (Litwin et al. 1998). Shoots with *Brachyphyllum*-type leaves bearing cuticle with papillae overhanging the stomata occur at two sauropod bonebeds, the Mygatt-Moore Quarry in Colorado (Tidwell et al. 1998) and the Howe-Stephens Quarry in Wyoming (Gee, unpublished data); these shoots likely pertain to the Cheirolepidiaceae. At Tendaguru, both cuticle and pollen floras are dominated by the Cheirolepidiaceae throughout most of the section, at times forming a monotypical assemblage (Schrank 1999, 2010; Aberhan et al. 2002). The Cheirolepidiaceae have not yet been found in the Jurassic of Patagonia, although the family does occur in the Early Cretaceous of Argentina and Brazil as frenelopsid and nonfrenelopsid foliage (Archangelsky 1963, 1966, 1968; Kunzmann et al. 2006).

Although the Cretaceous is technically beyond the scope of this survey of co-occurrences, it is interesting to note that Early Cretaceous cheirolepidiaceans occur in coastal sediments in Texas in which sauropod trackways and bonebeds have been discovered. One species (*Frenelopsis varians*) was collected from the Glen Rose Formation at a site northwest of Austin, Texas, and grew in low colonies in salt marshes near a hypersaline lagoon or bay depositional system (Daghlian & Person 1977). A second species of *Frenelopsis, F. ramosissima,* was found in a sauropod bonebed in the Twin Mountains Formation on the Jones Ranch southwest of Fort Worth and, in contrast, formed a monospecific stand of massive trees in a semiarid coastal forest (Axsmith & Jacobs 2005).

Remarks and Rating

Although the Cheirolepidiaceae no longer have any close relatives on which we can run fermentation experiments or measure biomass productivity, their habit as arborescent or scrubby plants, dominance in xeric or saline habitats, and general co-occurrence with sauropod during the mid and late Mesozoic suggest that the members of this family have constituted a major portion of a large sauropod's diet. Furthermore, as first pointed out by Tiffney (1997), their leaves, which have a succulent appearance, may have been quite palatable to the herbivorous dinosaurs.

GINKGOPHYTES

Habit and Habitat

Ginkgo biloba (Plate 3.2F, G), the maidenhair tree, is the sole surviving member of this group of plants, which once flourished in the Northern Hemisphere during the Mesozoic and Paleogene (Stewart & Rothwell 1993). Ginkgos are long-lived trees that can survive up to 3,000 years (Del Tredici 1991) and usually attain heights between 20 and 30 m, but can reach 60 m in height (Del Tredici 1991; Schütt et al. 2004). Some shrubby forms may have existed in the Mesozoic (Green 2005, 2007). A distinctive trait of the maidenhair tree is its fan-shaped leaves, which turn a brilliant golden color in the fall (Plate 3.2G). Today, *Ginkgo* is deciduous, and it is commonly assumed that ginkgophytes were deciduous in ancient times too (e.g., Spicer & Parrish 1986). The natural distribution of *Ginkgo biloba* is limited to a small, refugial area in southeastern China, which has a mesic, warm-temperate climate (Del Tredici 1992b). However, *G. biloba* is widely cultivated in areas with cold temperate, warm temperate, and Mediterranean climates (Del Tredici 2007), and is thus well known for its broad environmental tolerance in regard to moisture, temperature, and topography.

Biomass Production

Because of its economic importance in medicine, the harvesting of *Ginkgo biloba* leaves has received some attention, especially in Asia. It was found on a 15 year old plantation in central Korea that the above-ground biomass of *Ginkgo,* which includes stem wood, stem bark, branches, and foliage, equaled 23,780 kg/ha (Son & Kim 1998). This falls within the range of biomass values for 10–20 year old stands of conifers, which vary from 15,000–70,000 kg/ha (Kimmins et al. 1985). The foliage-only biomass on the Korean *Ginkgo* plantation made up 10% of the above-ground tree biomass, which is considered a relatively large proportion (Son & Kim 1998).

In the leaf biomass experiments of Hummel & Clauss (this volume) in which the distal 30 cm of foliage of various tree species were stripped, dried, and weighed, *Ginkgo biloba* produced 8.8 ± 3.0 g in dry matter, compared to an average of 4.2 ± 1.6 for six different broad-leaved trees, an increase of more than twice as much biomass for *Ginkgo* over the angiosperms.

Potential for Recovery

As a deciduous gymnosperm, *Ginkgo biloba* sheds its leaves every year, which means that it will renew its foliage annually in any case. It also has two kinds of lignotubers—sometimes called basal chichi and aerial chichi—that will propagate new trunks or branches from the parent plant vegetatively in the event of traumatic damage or changes in substrate stability (Del Tredici 1992a, 1992b). In fact, a *Ginkgo biloba* growing in the center of the atomic blast over Hiroshima, which had its trunk completely destroyed in 1945, is survived by a new tree that sprouted from its base by way of its basal lignotubers (Del Tredici 1991). These dormant woody buds may be the key to ginkgo's longevity through the centuries as well as through geological time, as reproduction by seeds appears to be mostly unsuccessful, especially in closed-canopy forests (Del Tredici 1992a, 1992b, 2007). This is due in great part to seed predation (see below) and to the low-light conditions in a closed forest.

It has been observed that the small natural populations of *Ginkgo* on Tian Mu Shan near Hangzhou, China, occur today on disturbance-generated microsites with soil erosion such as on stream banks, rocky slopes, and edges of exposed cliffs. A preference for disturbed habitats would have been advantageous for the germination and establishment of new ginkgo trees in the wake of any trampling, soil-churning sauropods feeding on older ginkgo stands, although it would have taken some decades before the ginkgo seeds and saplings grew into good-sized trees.

Ginkgo biloba is a slow grower. In plantations in its native habitat on Tian Mu Shan, China, and in Virginia, USA, the average growth rate was 21 cm/yr and 34 cm/yr for trees 25 and 35 years old, respectively (Del Tredici 2004). A more vigorous average growth rate of 48 cm/yr was measured on the 15 year old plantation in Korea mentioned above (Son & Kim 1998).

Digestibility and Consumption by Living Animals and Humans

In laboratory fermentation experiments (Hummel et al. 2008), the energy yield of *Ginkgo biloba* was found to be good. Among the many plant groups tested by Hummel et al. (2008), *Ginkgo biloba* leaves yielded by far the most crude protein, surpassing the percentage in dry matter of *Equisetum* (the next best source of crude protein) by 1.3 times and that of araucariaceous leaves (the worst source of crude protein among the Mesozoic plant types tested) by 3.5 times (Hummel, pers. comm.).

The nutritious seeds of the extant ginkgo tree are consumed by a number of different animals. Tree squirrels in North America and China, such as the red-bellied squirrel on Tian Mu Shan, feed on the ginkgo "nuts," and humans, especially in Asia, have eaten the boiled seeds for centuries (Del Tredici 1991). In addition, three omnivorous members of the Carnivora, the leopard cat and the masked palm civet in China and the raccoon dog in Japan (Del Tredici 1992b, 2008), feed on ginkgo "berries"—the inner seed surrounded by a fleshy, foul-smelling seed coat. Indeed, the droppings of the raccoon dog have been found to contain intact seeds, which then germinated the next spring (Rothwell & Holt 1997). The feeding of these nocturnal scavengers on ginkgo nuts has led Del Tredici (1992b) to speculate that the foul-smelling seed coat of *Ginkgo* attracts animal dispersers by posing as a carrion mimic.

In regard to its foliage, *Ginkgo biloba* is reputed to be quite resistant to damage from insects, fungi, bacteria, and viruses (Del Tredici 2004). Although there are some modern insects that feed on *Ginkgo* leaves, their number is extremely small compared to that attacking other gymnosperms (Honda 1997).

Consumption by Fossil Animals

In the Cretaceous–Paleogene boundary fossil flora in North Dakota, the leaves of *Ginkgo adiantoides* show a few types of insect damage (Labandeira et al. 2002).

Co-occurrence with Major Late Jurassic Sauropod Faunas

Several species of *Ginkgo* leaves (Brown 1975; Ash & Tidwell 1998), as well as pollen (Hotton & Baghai-Riding 2010), occur throughout the Morrison Formation, from Canada to Colorado and Utah. *Ginkgo* seeds have been found in Utah and Montana (Tidwell 1990b), while *Ginkgo* foliage co-occurs with sauropod remains at the Mygatt-Moore Quarry in western

Colorado (Tidwell et al. 1998). At Tendaguru, ginkgo cuticle does occur, but it is less commonly preserved than conifer cuticle (Kahlert et al. 1999; Aberhan et al. 2002).

Remarks and Rating

The ginkgophytes would have been a good source of energy and protein for medium-sized to large herbivores in the Northern Hemisphere throughout the Mesozoic. As a consequence of their arborescent growth habit and production of abundant leaves, they would have provided much biomass for consumption during times of the year when the trees bore leaves. Moreover, mature trees with ginkgo nuts would have provided taller sauropods with additional nutrition when fruiting. The trees may have occupied more open or disturbed habitats, which might have enabled larger animals more room for maneuvering, and any damage to branches, the trunk, or the base of the tree during feeding might have not permanently harmed the tree, but instead activated its dormant growth buds. Extrapolating from the good energy content of *Ginkgo biloba,* as well as from the frequent association of *Ginkgo* fossils and sauropod remains in the Late Jurassic Morrison Formation, it is likely that *Ginkgo* leaves and its fructifications may have been a good, attractive source of nutrition for larger herbivores in the Northern Hemisphere.

PODOCARPACEAE

Habit and Habitat

The Podocarpaceae are a large family of mostly evergreen conifers, consisting of about 18 genera and 170–200 species (Hill 1995). *Podocarpus* is the largest genus in the family and forms either trees from 20–30 m or occasionally 40 m in height, or shorter, single- or multistemmed shrubs from 4–12 m in height (Krüssmann 1972). Most members of the family are native to the warm temperate and subtropical zones of the Southern Hemisphere, although a few species do occur in Japan, China, Malaysia, and the Philippines (Krüssmann 1972). The taller trees can dominate the canopy layer in mid to upper montane forests and can live longer than 1,000 years. Podocarps are forest forming, as in New Zealand, where they make up dense podocarp-dominated forests, as well as mixed podocarp–hardwood forests (Ogden & Stewart 1995).

This family has a long history of plant megafossils and pollen extending back to the Early Triassic (Taylor et al. 2009). Nearly all fossils occur in the Southern Hemisphere (Hill 1995), although there are a few reports of megafossils (e.g., from China; Zhou 1983) and pollen (e.g., from the Morrison Formation, USA; Hotton & Baghai-Riding 2010) from the Mesozoic of the Northern Hemisphere as well. Although extant podocarps have free, spreading leaves, it should be noted that podocarpaceous foliage described from the Mesozoic and Cenozoic of Australasia resembles *Brachyphyllum* and *Pagiophyllum* (Gee, pers. obs. based on specimens figured by Hill 1995), which are two form genera with short leaves closely appressed to the shoot axis that are common in the Mesozoic floras all over the world.

Biomass Production and Potential for Recovery

In general, most members of the Podocarpaceae are very slow growers, even when compared to other relatively slow-growing conifers (Enright & Ogden 1995) or when growing under benign conditions. For example, the net primary production of a 35 year old plantation of *Podocarpus imbricatus* on the tropical island of Hainan, China, averaged 10.3 metric tons per hectare and year, reaching a maximum of 14 metric tons per hectare and year (Chen et al. 2004). The former value is much lower than the usual net primary production of tropical forests, and is instead roughly equivalent to that of warm temperate forests (cf. Lieth 1975).

Some species of *Podocarpus* are known to possess epicormic shoots, similar to those in the Araucariaceae.

Digestibility and Consumption by Modern Animals

The digestibility of three different genera (*Podocarpus, Dacrydium, Phyllocladus*) in this family proved to be poor (Hummel, pers. comm.), and the family Podocarpaceae was one of the worst plant groups in the experimental trials of Hummel et al. (2008).

A fleshy tissue around podocarpaceous seeds called the epimatium (Plate 3.2H) adds to their attractiveness as a food option; such seeds are part of the normal diet of animals such as brushtail opossums (*Podocarpus hallii* and *Dacrydium cupressinum* seeds) or ship rats (*Prumnopitys ferruginea* seeds) in mixed podocarp–hardwood forests in New Zealand (Sweetapple & Nugent 2007). The opossums rely on *P. hallii* as their main food on both North Island and South Island of New Zealand (Nugent et al. 1997 and references therein; Rogers 1997; Bellingham et al. 1999). In fact, opossums bulk feed on the *P. hallii* leaves all night when feeding in the canopy layer (Rogers 1997). Strangely, in the same mixed podocarp–hardwood forests, *P. hallii* is avoided by browsing red deer. Indeed, red deer in a temperate forest heavily dominated by *Podocarpus nagii* also avoid eating podocarp leaves of this species, even in feeding trials (Ohmae et al. 1996). This is thought to be due to the antiherbivory effect of nagilactones in the leaves (Ohmae et al. 1996).

Co-occurrence with Major Late Jurassic Sauropod Faunas

As mentioned earlier, fossil shoots with leaves that have been unequivocally identified as podocarpaceous elsewhere resemble the form genera of *Brachyphyllum* and *Pagiophyllum* that pertain to the foliage of the Araucariaceae and Cheirolepidiaceae. Several species of *Brachyphyllum* and *Pagiophyllum* are known from the Morrison Formation (Tidwell 1990b; Ash & Tidwell 1998; Tidwell et al. 1998, 2006), but none of them have been assigned to the Podocarpaceae. In contrast, 16 form taxa pertaining to podocarpaceous pollen have been described from many parts of the Morrison Formation (Hotton & Baghai-Riding 2010). The Podocarpaceae are also a common element in the Tendaguru flora, appearing in all or most of the units as wood, cuticle, or pollen (Aberhahn et al. 2002; Schrank 2010). In Patagonia, the Podocarpaceae first appear in the Triassic (Troncoso et al. 2000) and continue to show up as pollen, wood, leaves, and pollen and seed cones throughout the Mesozoic (e.g., Del Fueyo 1996 on pollen; Gnaedinger 2007 on wood; Taylor et al. 2009 and references therein on compression fossils).

Remarks and Rating

Assuming the Podocarpaceae had the same woody habits and forest habitats in the Mesozoic as they do now, and considering their co-occurrence with the sauropod faunas, especially in the Southern Hemisphere, they would have been a common food plant for herbivorous dinosaurs. However, in view of their poor fermentation values, which translate into comparatively low amounts of energy in each bite, slow growth rates, and documented unpalatability to some large present-day herbivores (red deer), the Podocarpaceae may have been a less sought-after source of food compared to other gymnosperms such as *Araucaria,* the Cheirolepidiaceae, and ginkgophytes.

OTHER CONIFERS

Living conifers make up a large group of plants, called the Pinophyta or Coniferae, consisting of 7 families, 69 genera, and roughly 600 species (Earle 2009a). Two extant families (Araucariaceae and Podocarpaceae) have been discussed here separately as a result of their dominance in Mesozoic ecosystems. The remaining families are the Cupressaceae, Pinaceae, Taxaceae, Cephalotaxaceae, and Sciadopityaceae. It should also be noted that the Taxodiaceae are now generally regarded as part of the Cupressaceae, with the exception of *Sciadopitys,* which is now commonly put into a separate family of its own. Because the Taxaceae and Cephalotaxaceae have poor fossil records and the Sciadopityaceae first appears in the Late Cretaceous (Taylor et al. 2009), these families will not be treated here.

CUPRESSACEAE

Habit and Habitat

The Cupressaceae comprise a large family that includes about 27 genera and 127 species (Mabberley 1993) of shrubs or, more commonly, trees, which can attain heights up to 112 m (*Sequoia sempervirens,* the coast redwood; Lanner 2002). The trees in this family are also notable for being the largest (*Sequoiadendron giganteum,* the giant sequoia), the stoutest (*Taxodium mucronatum,* the Montezuma cypress or ahuehuete), and the second longest lived (*Fitzroya cupressoides,* the alerce) in the world. The Cupressaceae are the most widely distributed family of conifers and can be found on all continents except for Antarctica (Earle 2009b). Accordingly, they are found in a variety of habitats ranging from coastal settings, floodplains, freshwater swamps (Plate 3.3A), riverbanks, and mountains, and they thrive under various climatic regimes, which include tropical, subtropical, warm temperate, and semiarid conditions (Burns & Honkala 1990).

The oldest fossil generally accepted as pertaining to the Cupressaceae occurs in the Middle Triassic (Yao et al. 1997). A more solid fossil record of the family shows up in the Jurassic, and the worldwide distribution of the Cupressaceae becomes apparent in the Cretaceous (Taylor et al. 2009).

Biomass Production and Potential for Recovery

As large trees with spreading branches, like the other conifer families described in this chapter, the members of the Cupressaceae offer quite a bit of biomass. Some members of the family, such as *Sequoia sempervirens,* have such a high growth rate that, of all the world's vegetation types, a mature coast redwood forest produces the greatest biomass per unit area, even exceeding that of tropical forests (Lanner 2002). In its first year, a coast redwood sapling can grow up to 1.8 m in height (Lanner 2002). *Sequoiadendron giganteum, Metasequoia glyptostroboides,* and *Glyptostrobus pensilis* are other examples of fast-growing members of the family (Schütt et al. 2004).

Like araucarians and ginkgoes, some cupressaceous genera have the capacity to regenerate from lignotubers, known in many conifers as burls, after injury or death to the parent tree. In the case of the coast redwood, for example, these dormant growth buds are located along its roots, allowing it to form lines of clones up to 30 m long after damage by fire (Lanner 2002).

Digestibility and Consumption by Modern Animals

The in vitro fermentation of the Cupressaceae in the laboratory experiments of Hummel et al. (2008) was good. On the basis of 11 samples from a variety of genera, the average amount of digestibility nearly matches that of *Ginkgo.*

The wood of some cupressaceous trees, for instance, *Juniperus virginia, Sequoia sempervirens,* and *Callitris glaucophylla,* can contain a compound that smells like camphor and deters insects, especially termites. However, the foliage of the trees in this family seems to be palatable to large herbivores. *Juniperus communis, J. occidentalis, J. californica,* and *Austrocedrus chilensis* are, for example, commonly heavily browsed by livestock such as goats (e.g., Zanoni & Adams 1973; Torrano & Valderrábano 2005) and deer, especially red deer (Relva & Veblen 1998).

Co-occurrence with Major Late Jurassic Sauropod Faunas

Foliage and wood of the Cupressaceae occurs in the Morrison Formation (e.g., Tidwell 1990b; Ash & Tidwell 1998; Tidwell et al. 1998). Wood assigned to the morphogenus *Glyptostroboxylon,* which may pertain to the Cupressaceae/Taxodiaceae, is also found at Tendaguru (Kahlert et al. 1999; Süss & Schultka 2001; Aberhan et al. 2002). In Argentina, cupressaceous wood has been reported from the Jurassic (Gnaedinger 2004), and leafy twigs with seed cones occur in the mid Cretaceous of Argentina (Halle 1913; Archangelsky 1963; Villar de Seoane 1998; Llorens & Del Fueyo 2003; Del Fueyo et al. 2008). The first convincing evidence of the Cupressaceae from the Jurassic—a new genus and species (*Austrohamia minuta*) of leafy twigs and branches bearing seed and pollen cones—was recently described from Patagonia (Escapa et al. 2008).

Remarks and Rating

With an arborescent habit, spreading branches, the ability to thrive in a variety of habitats, and good digestibility, the trees of the Cupressaceae in the widest sense (that is, including the basal members of the former Taxodiaceae) would have been a good source of nutrition for the sauropod dinosaurs. Taking into consideration their general co-occurrence with the major sauropod faunas in both hemispheres, as well as the palatability of cupressaceous leaves to extant herbivores, the foliage of this conifer family may have comprised a good portion of a herbivorous dinosaur's diet.

PINACEAE

Habit and Habitat

Like most conifers, the Pinaceae are predominantly evergreen trees and are rarely deciduous or shrubs (Plate 3.3B); they bear needle-like foliage and woody seed cones (Plate 3.3C). There are 9 genera and 194 species in this family, nearly all of which are concentrated in the Northern Hemisphere (Krüssmann 1972; Mabberley 1993). The Pinaceae form forests, with most trees growing to a maximum height of 30 to 40 m, although a few can reach 87 m (Krüssmann 1972). This family dominates the boreal forest—the world's largest biome—as well as most temperate and boreal mountain forests and semiarid woodlands (Earle & Frankis 2009). Like the Cupressaceae, the members of this family can also be found in coastal settings (Plate 3.3B), in freshwater swamps, and on floodplains (Burns & Honkala 1990). Pinaceous trees have great longevity; the longest lived organisms on earth, the bristlecone pines (*Pinus longaeva* and *P. aristata*), pertain to the Pinaceae (Lanner 2002). The genera *Abies* (fir), *Cedrus* (cedar), *Larix* (larch), *Picea* (spruce), *Pinus* (pine), *Pseudotsuga* (Douglas fir), and *Tsuga* (hemlock spruce) are especially well known because they are important sources of timber and pulp, turpentine, resins, cultivated ornamentals, and edible seeds (Mabberley 1993). Indeed, the family Pinaceae are economically and ecologically the most important gymnosperm family on earth (Earle & Frankis 2009).

From the diversity of seed cones in the Cretaceous, it is thought that the Pinaceae was well established early in the Mesozoic (Taylor et al. 2009). One of the oldest members of this family is represented by the seed cone *Compsostrobus* from the Late Triassic of North Carolina (Delevoryas & Hope 1973, 1987).

Biomass Production and Potential for Recovery

Many trees in the Pinaceae exhibit a classical Christmas tree shape, with long, downward-drooping branches. Combined with their ability to form forests, this would offer much biomass for browsing herbivores. Growth rates are variable in the family. Early growth in *Tsuga canadensis,* for example, is extremely slow, and trees with a d.b.h. (diameter at breast height, a standard forester's measurement) of less than 2.5 cm (1 inch) may be as old as 100 years (Godman & Lancaster 1990). On the other hand, *Pinus halepensis* is a fast grower and can attain a height of 30 cm in its first year (Schütt et al. 2004).

Regeneration in the Pinaceae is also variable. Although some members of the family (e.g., *Pinus virginiana*) do not show any resprouting (Carter & Snow 1990), many more resprout readily in response to injury, particularly fire, sending up new shoots from epicormic buds in the needle fascicles and leaf axils, along the trunk, or from the roots. This latter group includes species of *Picea, Pinus, Pseudotsuga,* and *Tsuga* (Earle & Frankis 2009).

Digestibility and Consumption by Modern Herbivores

The digestibility of the Pinaceae in the in vitro fermentation experiments of Hummel et al. (2008) is good.

Contrary to common belief, large herbivores such as forest ungulates commonly feed on these conifers; they eat leaves, strip bark, and tend to decimate saplings between 10 to 40 cm high (e.g., Bergström & Bergqvist 1997; Kupferschmid & Bugmann 2005). For example, red deer, roe deer, and chamois feed on Norway spruce (*Picea abies*) in European forests (Kupferschmid & Bugmann 2005), while sika deer browse on young Japanese larches (*Larix kaempferi*) in Japan (e.g., Akashi 2006). A wide range of animals have been documented as feeding on eastern hemlock (*Tsuga canadensis*), causing serious damage, loss of vigor, slowing of growth rate, or death to the tree;

these animals include white-tailed deer, snowshoe hares, New England cottontails, mice, voles, squirrels and other rodents, porcupines, and sapsuckers (Godman & Lancaster 1990). White-tailed deer and rabbits are also known to browse on young sprouts and seedlings of pitch pine (*Pinus rigida;* Little & Garrett 1990). Meadow voles girdle young trees of Virginia pine (*Pinus virginiana*), preferring it over other species in the area (Carter & Snow 1990), and were responsible for devastating seedling plantations of Norway spruce (*Picea abies*) and Norway pine (*Pinus resinosa*) in Canada during a vole population density peak in 1987–1988 (Bucyanayandi et al. 1990).

A number of species of *Pinus* produce large, edible seeds, namely the pinyon pines and stone pine, which are eaten by squirrels, a variety of birds, and humans. For instance, the nutcracker and pinyon jays cache huge numbers of pinyon pine seeds each year, which are eaten later by the birds in the winter or left to germinate the coming spring (Lanner 2002). In whitebark pine, uneaten seeds cached by Clark's nutcrackers are the only reliable means for the species to regenerate itself (Lanner 2002). Even species of *Pinus* with smaller seeds, such as *P. rigida,* are an important food for squirrels, quail, and small birds such as the pine warbler, pine grosbeak, and black-capped chickadee (Little & Garrett 1990).

Insect herbivory on seed cones is evident in the recent as well as in the fossil record. Tunneled borings filled with frass of boring beetles in a mid Cretaceous pinaceous seed cone resemble the infestation of boring beetles of the genus *Conophthorus* on seed cones of living *Pinus* spp., which eat through the nutritive tissues of the vascular cambium, phloem, and cortex of the cone (Falder et al. 1998).

Co-occurrence with Major Late Jurassic Sauropod Faunas

Pinaceous foliage (commonly called *Pityocladus* or *Pityophyllum*) and wood are known from the Late Jurassic Morrison Formation (Tidwell 1990b; Ash & Tidwell 1998). Bisaccate pollen grains typical of this family are found throughout the Morrison Formation as well (Hotton & Baghai-Riding 2010). Fossil remains of the Pinaceae do not occur at Tendaguru or in Patagonia, nor would it be expected to find them there because this family had its main distribution in the Northern Hemisphere during the Mesozoic, as it does today (Taylor et al. 2009).

Remarks and Rating

For the same reasons as in the Cupressaceae—a tree habit with long, spreading branches, the ability to thrive in a number of different habitats, and good digestibility—the Pinaceae would have been good food plants in the Mesozoic, albeit only in the Northern Hemisphere.

FERNS

Today, the ferns comprise a large division of plants, known as the Filicophyta, containing some 20,000 species. Although ferns can occur as epiphytes on trees or as floating macrophytes in freshwater, the ferns of interest here are ground-dwelling forms with a long fossil history, such as the families Marattiaceae and Osmundaceae.

MARATTIACEAE

Habit and Habitat

The Marattiaceae are a family of tropical ferns, with either leaves arising near ground level from an underground rhizome or elevated to the top of a tree-like trunk (Kramer et al. 1995). The 6 genera and roughly 260 species of this family grow in rain forests under year-round uniform conditions of high temperature and high humidity (Christenhusz 2009), often in wet soils and shady spots (Jones 1987). The center of diversity of the family today is in the Asian tropics (cf. Christenhusz et al. 2008).

The order Marattiales has an extensive fossil history that extends back to the Early Carboniferous, about 300 million years ago (Taylor et al. 2009), while fossil leaves identical to those of the extant genera *Marattia, Danaea,* and *Angiopteris* have been recorded as far back as the Late Triassic, Early Jurassic, and Middle Jurassic, respectively (e.g., Harris 1931; Hill 1987; Stewart & Rothwell 1993; Yang et al. 2008).

Biomass Production and Potential for Recovery

The largest member of the extant Marattiaceae, the king fern *Angiopteris evecta,* produces large, robust fronds that can reach lengths of 8 m. However, admittedly little is known about its growth rate and longevity in its native habitat, although it is under protection as an endangered species in parts of Australia (NSW National Parks and Wildlife Service 2001). Once established, the fronds of *Angiopteris* are massive and robust, although growth is presumably slow. Even under ideal horticultural conditions, the propagation of *Marattia* by spores, for example, proceeds at an extremely leisurely pace, taking up to four years to produce a plant about 5 cm high (Large & Braggins 2004).

Digestibility and Consumption by Modern Herbivores

The digestibility of *Angiopteris evecta* leaves is excellent, the best of all the ferns tested by Hummel & Clauss (this volume), and after 72 hours, it reaches the energy yield of grasses (Hummel et al. 2008). Like *Equisetum,* the rate of fermentation of *Angiopteris* is greatest within the first 24 hours of digestion.

There are reports of insect damage on living leaves of *Marat-*

tia and *Angiopteris* (Beck & Labandeira 1998), but it is not known whether vertebrates also feed on these ferns today.

Co-occurrence with Major Late Jurassic Sauropod Faunas

As mentioned above, fossils of *Marattia, Danaea,* and *Angiopteris* are known from the Mesozoic, but they have not been found in association with the major sauropod faunas from the Morrison Formation, Tendaguru, or Patagonia.

Remarks and Rating

Although *Angiopteris* yields a remarkably high amount of energy, especially for a fern, it is doubtful whether this genus, along with *Marattia* and *Danaea,* formed a major part of a giant sauropod's diet. Not only would the general habitat of marattiaceous ferns in dense, closed-canopy rain forests have been less accessible to the larger sauropods, but the slow growth and propagation of living marattiaceans also suggests that there would have been a poor response of their Mesozoic relatives to intense herbivore feeding pressure. Furthermore, the lack of intimate association with major sauropod faunas in the fossil record also suggests a lack of opportunities for plant–herbivore interactions.

OSMUNDACEAE

Habit and Habitat

The Osmundaceae are represented today by three genera that form fronds near ground level (*Osmunda,* Plate 3.3D) or sometimes at the top of tree-like stems (*Todea,* Plate 3.3E, and *Leptopteris*). These ferns prefer moist, poorly drained conditions in open or closed-canopy habitats such as stream banks, damp woods, moist forests, and acidic swamps in temperate and subtropical areas (Jones 1987; Kramer et al. 1995).

The family has an extensive fossil record, extending back to the Permian (Stewart & Rothwell 1993). In Jurassic sediments, osmundaceous rhizomes are abundant and occur all over the world, but are most common in the Southern Hemisphere (Tian et al. 2008). The existence of *Osmunda* and *Todea* in the Middle Triassic and Late Cretaceous, respectively, document the longevity of these genera (Jud et al. 2008). Indeed, one living species (*O. cinnamomea,* the cinnamon fern in the eastern North America) has remained unchanged since the Late Cretaceous (Serbet & Rothwell 1999).

Biomass Production and Potential for Recovery

The fronds of osmundaceous ferns range from relatively small (less than 1 m long; *Leptopteris*) to moderately large (2–4 m long; *Osmunda* and *Todea*). *O. cinnamomea* commonly grows in monospecific colonies and, like *Equisetum,* can form dense thickets.

Osmunda cinnamomea readily resprouts from its underground rhizomes after its aerial portions have been destroyed by fire, exhibiting vigorous rhizome growth after fire damage, and in fact does best in areas that regularly experience burning (Walsh 1994). Once established, individual plants of *Osmunda* are reported to grow relatively fast, but it has also been noted that the rate of vegetative spreading in *Osmunda, Todea,* and *Leptopteris* is slow (Walsh 1994; Large & Braggins 2004).

Digestibility and Consumption by Modern Herbivores and Humans

Osmunda and *Todea* have vastly different fermentation curves. *Osmunda* is a good/excellent producer of energy, similar to grasses and forbs, while *Todea* is a poor producer of energy and has the second to worst fermentation curves (Hummel, pers. comm.). Hence, a herbivore would be wiser in regard to energy intake to graze on *Osmunda* than on *Todea.*

Livestock and white-tailed deer like to feed on *Osmunda cinnamomea,* especially on tender fronds that are no older than a month (Walsh 1994). In fact, cattle prefer to browse cinnamon ferns second only to cane (*Arundinaria gigantea*). Young fronds of *O. cinnamomea* can also be steamed or boiled and eaten by humans (Elias & Dykeman 1990).

Co-occurrence with Major Late Jurassic Sauropod Faunas

Two genera of the Osmundaceae, the tree ferns *Osmundacaulis* and *Ashicaulis* (Tidwell & Rushfort 1970; Tidwell 1990a, 1994), occur in the Morrison Formation and are abundant near Ferro and Moab in Utah (Tidwell 1990b), although the plant remains are not directly associated with any sauropod remains. Nevertheless, the spores of the family (*Baculatisporites, Osmundacites,* and *Todisporites*) occur laterally and vertically throughout the Morrison Formation (Litwin et al. 1998; Hotton & Baghai-Riding 2010), indicating that they were part of the regional vegetation. At Tendaguru, the spores of *Todisporites* and *Osmundacites* show up in the palynoflora (Schrank 1999, 2010).

Although several species of osmundaceous rhizomes have been reported from southernmost Patagonia, none of them occur in the Late Jurassic (Tian et al. 2008). However, the form genus *Cladophlebis,* which is thought to pertain to the Osmundaceae, occurs throughout the Triassic and Jurassic of Argentina and the Antarctic Peninsula (e.g., Herbst 1971; Gee 1989). Fossil *Cladophlebis* leaves (Plate 3.3F) are twice pinnate and robust, closely resembling the fronds of extant *Osmunda* (Plate 3.3D) and *Todea* (Plate 3.3E). Fertile pinnules known as *Todites* bearing small, round, densely packed sporangia are similar to the sporangia-bearing pinnules of living *Todea.*

Remarks and Rating

The Osmundaceae, especially *Osmunda,* may have formed a recurrent but minor part of the sauropod diet in mesic habitats, given its high energy content, dense thicket-forming habit, high palatability to some grazers and browsers, and coeval occurrence in the Morrison and Tendaguru sediments. If a parallel can be drawn between the rapid regeneration of fronds and the new colonization of disturbed areas after fire in recent environments and sauropod herbivory in Mesozoic times, Jurassic *Osmunda* rhizomes may have responded with vigorous regrowth of their fronds after being cropped.

CYCADS AND BENNETTITALEANS

The Cycadales (the true cycads; Plate 3.3G) and Bennettitales (also called the Cycadeoidales; Plate 3.3H) are two different groups of gymnosperms with similar growth forms and leaf morphology. These plants, which are commonly treated together as the cycadophytes, have pinnately compound leaves that are so similar that at times they can only be distinguished from one another by details of epidermal features such as stomata. The organization of their cones was very different, however, and this supports the continued separation of the enigmatic Bennettitales from the Cycadales (cf. Crepet & Stevenson 2010). Although cycads first showed up in the Paleozoic and continue to the present day, bennettitaleans occurred exclusively in the Mesozoic.

Habit and Habitat

Cycads have long, evergreen, pinnately compound fronds. Their stems commonly form stout or tall, upright trunks and are covered with a mantle of hard, woody leaf bases. There are 12 genera and over 300 species in the family (Chaw et al. 2005), and individual plants can live several hundred years. Cycads occur today in the tropical, subtropical, and warm temperate regions of both the Northern and Southern Hemispheres, ranging northward from the southern islands of the Japanese archipelago and southward to southern parts of Australia, while the center of diversity of the cycads is in Central America (Jones 1993). They can be found in mesic habitats, such as rain forests, as well as in semiarid to xeric environments, such as grasslands and sparse woods, and on rocky escarpments and in gorges (Jones 1993). Cycads have a long fossil history that stretches back to the Carboniferous and continues until today. They reached their heyday during the Mesozoic in regard to geographic distribution and number of taxa (Taylor et al. 2009).

The fronds of the bennettitaleans, on the other hand, look similar to those of the cycads, but are commonly much shorter in length. Their stems can appear stout and trunk-like, similar to those of the cycads, or massively globose in shape. Like the cycads, they were a characteristic feature of the landscape in the Mesozoic. In Patagonia, for example, bennettitaleans were shrubby plants that formed the understory vegetation alongside corystosperms (a type of seed fern; see below) in a variety of forest types (evergreen, deciduous, sclerophyllous) or grew in open areas as shrubs in the Triassic (Artabe et al. 2001; Cúneo et al. 2010). In the Cretaceous of Patagonia, bennettitaleans, as well as cycads, show morphological features adapted to warm and seasonally dry climates, due at least in part to the constant volcanic activity in the region (Archangelsky et al. 1995; Archangelsky 2003; Cúneo et al. 2010). Thus, like cycads, bennettitaleans could grow in a variety of environments ranging from mesic to xeric habitats.

Biomass Production and Potential for Recovery

Among gardeners and horticulturalists, cycads are notorious for being slow growers. Cycads do not develop new leaves continuously but produce a burst or flush of leaves at irregular intervals in a tuft at the top of the trunk. Thus, young plants often have few leaves and offer little in terms of biomass. The amount of biomass contained in the leaves of an older plant may be significantly higher, as cycad leaves tend to be persistent, remaining on the plant for a long time.

As mentioned above, the leaves of the bennettitaleans were shorter than those of modern cycads and would have thus offered even less in terms of biomass. However, nothing is known about the rate of growth in the bennettitaleans.

One cycad genus, *Cycas,* can reproduce itself vegetatively by producing basal suckers, also known as bulbils or offsets, but these also grow extremely slowly (Gee, pers. obs.). Other genera, such as *Encephalartos* and *Lepidozamia,* have epicormic buds in their trunks that will generate new growth if the plant is damaged (Jones 1993).

As with most other extinct plant groups, nothing is known about the potential for recovery and regrowth in the Bennettiales.

Digestibility and Consumption by Modern Herbivores and Humans

The fermentation experiments of Hummel et al. (2008) show that cycad foliage is a poor producer of energy, vying with the Podocarpaceae for the third-to-worst source of nutrition for herbivores. Fermentation experiments could not be run on the bennettitaleans because they do not have any living relatives.

A major problem for herbivores feeding on cycads is the presence of several strong toxins—macrozamin, cycasin, and neocycasin, to mention a few—in various parts of the plants (Hegnauer 1962). In Australia, the coordination of livestock ingesting large quantities of the foliage or seeds of *Bowenia, Cycas, Lepidozamia,* or *Macrozamia* becomes debilitated, and the animals lose control of their hindlimbs (Jones 1993). These "zamia staggers" happen mostly in cattle and sheep, but also in horses, goats, and pigs. There are reports, however, of some animals, namely emus, brush-tailed opossums, and chuditch, that can ingest the seeds of *Macrozamia riedlei,* which contain high levels of macrozamin, without ill effects (e.g., Ladd et al. 1993).

Despite the occurrence of these highly poisonous compounds, humans have long learned to neutralize the toxicity in cycad seeds by roasting or leaching in preparation for human consumption (Hegnauer 1962; Jones 1993). Australian aboriginals, for example, began utilizing *Cycas* and *Macrozamia* as a food source at least 13,000 years ago (Jones 1993). Both the starch-filled seeds and the large starch reserves in the pith of the trunks, known as sago or arrowroot, are harvested for making small loaves, cakes, or porridge (Jones 1993). Moreover, the roots of three species of *Zamia—Z. boliviana, Z. lecontei,* and *Z. ulei*—are used in Brazil for medicinal purposes as an antidote to snakebite (Hegnauer 1962).

It is not known whether the bennettitaleans and Mesozoic cycads contained the same abundance and range of toxins as the living cycads.

Co-occurrence with Major Late Jurassic Sauropod Faunas

There is a wealth of cycadophyte leaves throughout the Morrison Formation, from southern Utah to southern Canada (Tidwell 1990b; Ash & Tidwell 1998). Pollen, on the other hand, occurs in very small amounts—less than 1% of the entire sporomorph flora (Hotton & Baghai-Riding 2010)—and is difficult to distinguish from that of ginkgophytes. In Tendaguru, minor amounts of cycadophyte wood and cuticle have been reported (Kahlert et al. 1999).

In Patagonia, both cycads and bennettitaleans were integral parts of the Mesozoic flora. In the Triassic, for example, cycadophytes are so common and rigorously constrained in their stratigraphic occurrence that they can be used as biostratigraphic and chronostratigraphic markers (Spalletti et al. 1999; Stipanicic 2002). Cycadophytes occur in the Early Jurassic but are curiously absent in the Middle and Late Jurassic (Cúneo et al. 2010), unlike elsewhere in the world, for example in the Late Jurassic Morrison Formation, where they reach their climax. It is instead in the Cretaceous when the cycadophytes reach their peak of diversity and abundance in Patagonia, becoming major components of the flora (Cúneo et al. 2010). A recent study testing possible co-evolution between herbivorous nonavian dinosaurs and cycads in the Late Triassic and in the Cretaceous found no significant spatiotemporal correlation between these two groups; nor did it find any unequivocal support for co-evolutionary interactions (Butler et al. 2009, 2010).

Remarks and Rating

On the bases of the paucity of leaves borne by most cycads and bennettitaleans, as well as their slow growth and leaf regeneration, potentially high toxin content, and low energy yield, cycads and possibly also the bennettitaleans were a poor dietary option for sauropods. Given the common occurrence of cycadophytes in the Mesozoic landscape in both hemispheres, a cycadophyte may have been offered an occasional biteful of leaves when encountered by a herbivorous dinosaur, but it is doubtful that they were truly a staple of their diet.

MESOZOIC SEED FERNS AND OTHER ENIGMATIC GYMNOSPERMS

Seed ferns, also known as pteridosperms, are not a natural group of plants but are lumped together on the basis of their fern-like foliage and seed-bearing habit. Like *Ginkgo* and the conifers, they are considered gymnosperms because their seeds are unprotected by an ovary or fruit. Although seed ferns such as *Medullosa* and *Glossopteris* are well known from the Paleozoic, a completely different suite of seed ferns thrived during the Late Paleozoic–Mesozoic. This included the Caytoniales, Corystospermales, Petriellales, and Peltaspermales (Taylor et al. 2006). Other enigmatic Mesozoic gymnosperms are the Gigantopteridales, Vojnovskyales, Iraniales, Pentoxylales, Czekanowskiales, and Hermanophytales (Taylor et al. 2009). Because they are extinct, there is no information on seed ferns in regard to biomass production, regrowth, digestibility, and consumption by modern herbivores; however, four of these orders—the Caytoniales, Corystospermales, Czekanowskiales, and Hermanophytales—will be discussed briefly here as a reminder that Mesozoic seed ferns also likely formed a part of the sauropod diet.

Habit and Habitat

The *Caytonia* plant that bore the palmately compound leaves of *Sagenopteris* may have been part of swamp vegetation during Early Cretaceous times because it is associated with lignites in the Kootenai Formation in Montana, which overlies the Late Jurassic Morrison Formation (LaPasha & Miller 1984). The Corystospermales, which grew primarily in Gondwana, were probably small to large woody shrubs (Taylor et al. 2009), although some of them may have been medium-sized, unbranched, tree fern-like plants, or even had a scrambling habit (Taylor et al. 2006).

The existence of the Czekanowskiales is known through their leaves. Virtually nothing is known of growth forms in this group, although some information on habitat preference can be inferred from the fossil record. Czekanowskialean remains occur with coals and floras rich in hydrophilic plants, such as *Equisetum* and ferns, in the Morrison Formation and elsewhere in the world, and suggest that *Czekanowskia* grew under humid, temperate to tropical conditions (Ash 1994; Tidwell 1990b; Ash & Tidwell 1998; Tidwell et al. 1998 and references therein).

The Hermanophytales are an enigmatic group characterized by permineralized stems with internal wedges of secondary xylem (wood). *Hermanophyton* trunks range from 3 to 22.5 cm in diameter, measure over 18 m in length, and currently are known from only four sites in the Morrison Formation (Arnold 1962; Tidwell & Ash 1990; Tidwell 1990b, 2002; Ash & Tidwell 1998). In regard to habit, it is thought that *Hermatophyton* may represent a small- to medium-sized, narrow-stemmed tree with a crown of small leaves, or a long liana (Taylor et al. 2009).

Co-occurrence with Major Late Jurassic Sauropod Faunas

Three of the orders of seed ferns under discussion here have been found in the Morrison Formation. The foliage of *Caytonia*—represented by the form genus *Sagenopteris*—occurs in the northern part of the Morrison Formation, in Belt, Montana (Brown 1972), as well as in the Early Cretaceous Kootenai Formation in Montana (LaPasha & Miller 1985) and western Canada (Bell 1956) which overlies the Morrison Formation. Both *Czekanowskia* and *Hermatophyton* have been described from several sites from the Morrison Formation (Ash 1994; Tidwell 2002). The bisaccate pollen of the Caytoniales (*Vitreisporites* type) and Czekanowskiales (*Alisporites* type) are also represented in the Morrison Formation (Litwin et al. 1998; Hotton & Baghai-Riding 2010).

At Tendaguru, bisaccate pollen of *Alisporites* that could pertain to the Corystospermales has been reported (Schrank 1999, 2010). However, fossil remains of the Caytoniales, Czekanowskiales, and Hermanophytales have not been found.

Within the region consisting of southern South America and the nearby Antarctic Peninsula, *Sagenopteris* leaflets pertaining to the Caytoniales have been described from the Jurassic of Antarctica (Halle 1913; Gee 1989; Cantrill 2000), although they have not been reported from Patagonia. However, as was mentioned earlier, the corystosperms were a major element of the understory in Triassic forests of Patagonia, alongside the bennettitaleans (Artabe et al. 2001; Cúneo et al., 2010). Neither the Czekanowskiales nor the Hermanophytales are known from Patagonia.

Remarks and Rating

Because Mesozoic seed ferns and other enigmatic gymnosperms were a distinctive component of Mesozoic floras, they should not be overlooked in surveys of the Mesozoic flora. However, because little is known about their palatability, nutritional value, and ability to regrow after herbivory, it is difficult to evaluate their potential as a dietary option. They may have played at least a minor role in the diets of large sauropods, especially if they grew in more open areas.

Discussion

As summarized in this chapter, a number of different variables have been taken into account in the evaluation of plants as sauropod fodder. Central to this discussion is the availability of the individual plant groups to a large sauropod, which is reflected in part in the plant's habit (growth form) and habitat (environment in which it is found). Trees have the general advantage of being able to offer more leaf biomass on their spreading branches than smaller herbaceous plants such as ferns. One exception is *Equisetum*; although herbaceous and usually small in stature, horsetails have a fast growth rate and a dense, colony-forming habit that boost their productivity and biomass. In regard to habitat, plants that grew in more open areas and not in dense forests would have theoretically been more accessible to the larger sauropods, which would have had problems maneuvering in tight spaces. Continuing this line of thought, large conifers growing as solitary trees in open habitats, in lightly wooded areas, or at the edge of a dense forest would have been the most accessible to large sauropods.

Factors such as biomass production (growth rate) and potential for recovery (regrowth of eaten plant parts) are also important issues when assessing good long-term food sources. Plants with low biomass production and low potential for recovery, such as the cycads and most ferns, are less likely to serve as recurring food sources than plants with high biomass production and greater potential for recovery. Examples of the latter group include *Equisetum*, ginkgophytes, and conifers.

In this study, the potential for recovery not only means the simple replacement of browsed plant parts, but also includes the longevity of individual plants and taxa through geologic time. To this end, it should be noted that a number of nearest living relatives of the Mesozoic flora have dormant basal and/or aerial buds that will grow and replace roots, stems, and branches in the face of injury. This includes all of the living gymnosperms under study here: *Ginkgo*, the Araucariaceae, Podocarpaceae, Cupressaceae, and Pinaceae, and even the cycads.

Whether these dormant buds are called epicormic buds, coppice shoots, aerial chichi, basal chichi, lignotubers, or burls, woody plants with such dormant buds are known as *sprouters* in the ecological literature and are common in both temperate and tropical forests. The sprouting response leads to long-lived individuals with long generation times that will preserve genetic diversity, even in small populations (Bond & Midgley 2001). This is certainly the case in the living fossil *Ginkgo* and in most of the conifer taxa discussed in this chapter.

Sprouters are resistant to disturbances and catastrophes such as those caused by fire, hurricane damage, drought, flooding, landslides, and herbivory (Bond & Midgley 2001). Indeed, in modern ecosystems, it is thought that "sprouting behavior might also be important in determining tree species survival in the face of browsing by heavy mammals" (Bond & Midgley 2001, p. 49). In this case, reference was being made to the many species of savanna trees of a coppice in response to damage caused by the largest herbivores on earth today, the elephants. Assuming that sauropods fed on foliage, twigs, and branches, and that they may have damaged tree trunks, bark, and roots in their wake, being able to recover, regrow, and resprout would have been advantageous for woody plants during the Mesozoic.

Digestibility, which is quantified in the in vitro fermentation experiments of Hummel et al. (2008), and palatability, which is reflected in part in the consumption of plant groups by modern herbivores, are two more key factors. Although the fermentation experiments were designed to reflect herbivore hindgut digestion in general (Hummel & Clauss, this volume), the palatability of certain plants appears to be dependent on the herbivore, as evidenced by the survey presented in this chapter.

Most plant groups in this survey do co-occur with sauropods in the areal extent of the three major sauropod faunas of the Late Jurassic, which indicates that there were indeed opportunities for herbivore–plant interactions at many sites. On the basis of this brief review of co-occurrences, the most common food options available in the Western Interior of North America (Morrison Formation), eastern Africa (Tendaguru), and southern South America (Patagonia) would have been trees of the Cheirolepidiaceae, Araucariaceae, and Podocarpaceae.

Comparative Rating of Plant Groups

By taking an integrated botanical and paleobotanical approach, it is clear that several plant groups would have been good, accessible sources of food to the large sauropods in the Mesozoic. Leading would be plants such as *Equisetum, Araucaria,* and the Cheirolepidiaceae, which were common in Mesozoic ecosystems around the world and could offer much biomass as fodder. These plants grew in thick stands around freshwater (*Equisetum*) or brackish (Cheirolepidiaceae) bodies of water, or as large trees in open forests (*Araucaria,* Cheirolepidiaceae). Additional features attractive to sauropods may have been the high energy content of *Equisetum* shoots and rhizome tips, the unusual fermentation behavior of *Araucaria* that yields more energy when retained in the hindgut for longer periods of time (e.g., more than 30 hours), and the succulence of the leaves of the Cheirolepidiaceae.

Although limited to the Northern Hemisphere, ginkgoes may have been a good source of energy and especially protein. Similarly, conifers such as the Podocarpaceae and Cupressaceae—and the Pinaceae to a lesser extent, due to their Northern Hemisphere distribution—were convenient sources of fodder, providing much biomass but fewer kilojoules per mouthful than those plants mentioned above.

Because of their generally smaller stature and need for moist, humid conditions such as those found in dense, closed forests, most ferns would have been difficult for large sauropods to encounter and ingest regularly. This would have included *Angiopteris* and *Osmunda,* which are excellent sources of energy today. Fast-growing ferns that were better adapted to more open areas under mesic conditions would have had to occur in great abundance (the "fern prairies" of Taggart & Cross 1997, for example) for a sauropod to make a meal of low-biomass fronds of individual ferns.

Table 3.1. Likeliness of Mesozoic Plant Groups as a Preferred Food Source for the Sauropod Dinosaurs

Plant group	Rating	Likeliness as a food source
Araucaria, Equisetum, Cheirolepidiaceae	*****	Very likely
Ginkgo	****	Rather likely
Other conifer families (e.g., Podocarpaceae, Cupressaceae, Pinaceae)	***	Moderately likely
Ferns (e.g., *Angiopteris, Osmunda*)	**	Less likely
Cycads and bennettitaleans	*	Least likely
Seed ferns, other enigmatic gymnosperms	Unknown	Cannot be evaluated at this time

Cycads were probably last on the list of desirable food options as a result of their low energy yield, low biomass production, extremely slow growth rates, and high content of toxins. However, because cycads (and bennettitaleans) were common plants in open areas, sauropods may have occasionally included them in their diet as well.

These plant groups can be compared to one another in regard to their likelihood as a preferred dietary option or commonly eaten source of food for the large sauropod dinosaurs (Table 3.1). Hence, taxa with five stars in Table 3.1 were most likely to have been an accessible, dependable, plentiful, and energy-producing food, while those with one star were the least attractive as a food source. The relative ranking given in Table 3.1 is offered as a baseline for future refinement and discussion.

Conclusions

This survey of the Mesozoic flora, which took into account plant habit, habitat, biomass production, potential for recovery, digestibility, consumption by modern herbivores, and co-occurrence with sauropods in the Late Jurassic, revealed that some plant groups were likely a more accessible, sustaining, or preferred source of nutrition for the sauropod dinosaurs. This included *Araucaria, Equisetum,* the Cheirolepidiaceae, and *Ginkgo.* Sauropods were only moderately likely to browse other conifers, such as the Podocarpaceae, Cupressaceae, and Pinaceae, and less likely to feed on forest-dwelling ferns such as *Angiopteris* and *Osmunda*. The least likely eaten plants may have been the cycads and bennettitaleans.

Acknowledgments

Sincere thanks are due to my colleagues in the DFG Research Unit 533 for their helpful discussions, especially Martin Sander, Jürgen Hummel (both at the University of Bonn), and Marcus Clauss (University of Zurich), as well as to reviewers Bill Chaloner (University College London) and Jim Farlow (Purdue University, Fort Wayne, Indiana) for their cogent remarks and corrections. I am also indebted to Kerry Hanrahan (Atherton Forestry Office, Forestry Plantations Queensland) for an extensive, all-day tour through the *Araucaria* plantations and to David Warmington (Cairns Botanic Garden, Cairns) for organizing a trip to see the northernmost natural population of *Araucaria bidwillii* in northern Queensland, as well as to Hans Hagdorn (Muschelkalkmuseum Ingelfingen) and Ted Delevoryas (University of Texas at Austin) for the use of their photos of *Osmunda* and *Cycadeoidea*, respectively. This is contribution number 62 of the DFG Research Unit 533 "Biology of the Sauropod Dinosaurs: The Evolution of Gigantism."

References

Aberhan, M., Bussert, R., Heinrich, W.-D., Schrank, E., Schultka, S., Sames, B., Kriwet, J. & Kapilima, S. 2002. Palaeoecology and depositional environments of the Tendaguru Beds (Late Jurassic to Early Cretaceous, Tanzania).—*Mitteilungen aus dem Museum für Naturkunde in Berlin, Geowissenschaftliche Reihe* 5: 19–44.

Akashi, N. 2006. Height growth of young larch (*Larix kaempferi*) in relation to the frequency of deer browsing damage in Hokkaido, Japan.—*Journal of Forest Research* 11: 153–156.

Archangelsky, A. 1963. A new Mesozoic flora from Ticó, Santa Cruz Province, Argentina.—*Bulletin of the British Museum (Natural History), Geology* 8: 45–92.

Archangelsky, A., Andreis, R. R., Archangelsky, S. & Artabe, A. E. 1995. Cuticular characters adapted to volcanic stress in a new Cretaceous cycad leaf from Patagonia, Argentina.—*Review of Palaeobotany and Palynology* 89: 213–233.

Archangelsky, S. 1966. New gymnosperms from the Ticó flora, Santa Cruz Province, Argentina.—*Bulletin of the British Museum (Natural History), Geology* 13: 259–295.

Archangelsky, S. 1968. On the genus *Tomaxellia* (Coniferae) from the Lower Cretaceous of Patagonia (Argentina) and its males and female cones.—*Journal of the Linnean Society (Botany)* 61: 153–165.

Archangelsky, S. (ed.). 2003. *La Flora Cretácica del Grupo Baqueró, Santa Cruz, Argentina*. Museo Argentino de Ciencias Naturales, Buenos Aires.

Arnold, C. A. 1962. A *Rhenoxylon*-like stem from the Morrison Formation of Utah.—*American Journal of Botany* 49: 883–886.

Artabe, A., Morel, E. M. & Spalletti, L. A. 2001. Paleoecología de las floras triásicas argentines. *In* Artabe, A. E., Morel, E. M. & Zamuner, A. B. (eds.). *El Sistema Triásico en la Argentina*. Fundación Museo La Plata, La Plata: pp. 199–225.

Ash, S. 1994. First occurrence of *Czekanowskia* (Gymnospermae, Czekanowskiales) in the United States.—*Review of Palaeobotany and Palynology* 81: 129–140.

Ash, S. R. & Tidwell, W. D. 1998. Plant megafossils from the Brushy Basin Member of the Morrison Formation near Montezuma Creek Trading Post, southeastern Utah.—*Modern Geology* 22: 321–339.

Axsmith, B. J. & Jacobs, B. F. 2005. The conifer *Frenelopsis ramosissima* (Cheirolepidiaceae) in the Lower Cretaceous of Texas: systematic, biogeographical, and paleoecological implications.—*International Journal of Plant Sciences* 166: 327–337.

Beck, A. L. & Labandeira, C. C. 1998. Early Permian insect folivory on a gigantopterid-dominated riparian flora from north-central Texas.—*Palaeogeography, Palaeoclimatology, Palaeoecology* 142: 139–173.

Behrensmeyer, A. K., Damuth, J. D., DiMichele, W. A., Potts, R., Sues, H.-D. & Wing, S. L. (eds.). 1992. *Terrestrial Ecosystems through Time: Evolutionary Paleoecology of Terrestrial Plants and Animals*. University of Chicago Press, Chicago.

Bell, W. A. 1956. Lower Cretaceous floras of Western Canada.—*Geological Survey of Canada Memoir* 285: 1–331.

Bellingham, P. J., Wiser, S. K., Hall, G. M. J., Alley, J. C., Allen, R. B. & Suisted, P. A. 1999. Impacts of possum browsing on the long-term maintenance of forest biodiversity.—*Science for Conservation* 103: 1–59.

Bergström, R. & Bergqvist, G. 1997. Frequencies and patterns of browsing by large herbivores on conifer seedlings.—*Scandinavian Journal of Forest Research* 12: 288–294.

Bond, W. J. & Midgley, J. J. 2001. Ecology of sprouting in woody plants: the persistence niche.—*Trends in Ecology and Evolution* 16: 45–51.

Brown, J. T. 1972. *The Flora of the Morrison Formation (Upper Jurassic) of Central Montana*. Ph.D. Dissertation. University of Montana, Missoula.

Brown, J. T. 1975. Upper Jurassic and Lower Cretaceous ginkgophytes from Montana.—*Journal of Paleontology* 49: 724–730.

Bucyanayandi, J.-D., Bergeron, J.-M. & Menard, H. 1990. Preference of meadow voles (*Microtus pennsylvanicus*) for conifer seedlings: chemical components and nutritional quality of bark of damaged and undamaged trees.—*Journal of Chemical Ecology* 16: 2569–2579.

Burns, R. M. & Honkala, B. H. (eds.). 1990. *Silvics of North America. 1. Conifers*. Agriculture Handbook 654. USDA Forest Service, Washington, D.C.

Burrows, G. E. 1990. Axillary meristem ontogeny in *Araucaria cunninghamii* Aiton ex D. Don.—*Australian Journal of Botany* 34: 357–375.

Burrows, G. E., Offord, C. A., Meagher, P. F. & Ashton, K. 2003. Axillary meristems and the development of epicormic buds in Wollemi Pine (*Wollemia nobilis*).—*Annals of Botany* 92: 835–844.

Butler, R. J., Barrett, P. M., Kenrick, P. & Penn, M. G. 2009. Testing co-evolutionary hypotheses over geological timescales: interactions between Mesozoic non-avian dinosaurs and cycads.—*Biological Reviews* 84: 73–89.

Butler, R. J., Barrett. P. M., Penn, M. G. & Kenrick, P. 2010. Testing evolutionary hypotheses over geological time scales: interactions between Cretaceous dinosaurs and plants.—*Biological Journal of the Linnean Society* 100: 1–15.

Calder, M. G. 1953. A coniferous petrified forest in Patagonia.—*Bulletin of the British Museum (Natural History), Geology* 2: 99–138.

Cantrill, D. J. 2000. A new macroflora from the South Orkney Islands, Antarctica: evidence of an Early to Middle Jurassic age for the Powell Island conglomerate.—*Antarctic Science* 12: 185–195.

Cantrill, D. J. & Hunter, M. A. 2005. Macrofossil floras of the Latady Basin, Antarctic Peninsula, New Zealand.—*Journal of Geology and Geophysics* 48: 537–553.

Carter, K. K. & Snow, A. G., Jr. 1990. *Pinus virginiana* Mill., Virginia pine. *In* Burns, R. M. & Honkala, B. H. (eds.). *Silvics of North America. 1. Conifers.* USDA Forest Service, Washington, D.C.: pp. 513–519.

Chaw, S.-M., Walters, T. W., Chang, C.-C., Hu, S.-H. & Chen, S.-H. 2005. A phylogeny of cycads (Cycadales) inferred from chloroplast *matK* gene, *trnK* intron, and nuclear rDNA ITS region.—*Molecular Phylogenetics and Evolution* 37: 214–234.

Chen, D.-X., Li Y.-D., Luo, T- S., Lin, M.-X. & Sun, Y.-X. 2004. Study on biomass and net primary production of *Podocarpus imbricatus* plantation in Jianfengling, Hainan Island.—*Forest Research, Chinese Academy of Forestry* 17: 604–608.

Christenhusz, M. J. M. 2009. *Danaea.* January 23, 2009. Available at: http://tolweb.org/Danaea/56784. Accessed May 13, 2009.

Christenhusz, M. J. M., Tuomisto, H., Metgar, J. S. & Pryer, K. M. 2008. Evolutionary relationships within the Neotropical, eusporangiate fern genus *Danaea* (Marattiaceae).—*Molecular Phylogenetics and Evolution* 46: 34–48.

Coe, M. J., Dilcher, D. L., Farlow, J. O., Jarzen, D. H. & Russell, D. A. 1987. Dinosaurs and land plants. *In* Friis, E. M., Chaloner, W. G. & Crane, P. R. (eds.). *The Origins of Angiosperms and Their Biological Consequences.* Cambridge University Press, Cambridge: pp. 225–258.

Cúneo, N. R., Escapa, I., Villar de Seoane, L., Artabe, A. & Gnaedinger, S. 2010. Review of the fossil cycads and bennettitaleans of Argentina. *In* Gee, C. T. (ed.). *Plants in Mesozoic Time: Morphological Innovations, Phylogeny, Ecosystems.* Indiana University Press, Bloomington: pp. 187–212.

Currie, P. J., Koppelhus, E. B. & Muhammad, A. F. 1995. "Stomach" contents of a hadrosaur from the Dinosaur Park Formation (Campanian, Upper Cretaceous) of Alberta, Canada. *In* Sun, A. & Wang, Y. (eds.). *Sixth Symposium on Mesozoic Terrestrial Ecosystems and Biota, Short Papers.* China Ocean Press, Beijing: pp. 111–114.

Crepet, W. L. & Stevenson, D. W. 2010. The Bennettitales (Cycadeoidales): a preliminary perspective on this arguably engimatic group. *In* Gee, C. T. (ed.). *Plants in Mesozoic Time: Morphological Innovations, Phylogeny, Ecosystems.* Indiana University Press, Bloomington: pp. 215–244.

Daghlian, C. P. & Person, C. P. 1977. The cuticular anatomy of *Frenelopsis varians* from the Lower Cretaceous of central Texas.—*American Journal of Botany* 64: 564–569.

Delevoryas, T. & Hope, R. C. 1973. Fertile coniferophyte remains from the Late Triassic Deep River Basin, North Carolina.—*Americal Journal of Botany* 60: 810–818.

Delevoryas, T. & Hope, R. C. 1987. Further observations on the Late Triassic conifers *Compsostrobus neotericus* and *Voltzia andrewsii.*—*Review of Palaeobotany and Palynology* 51: 59–64.

Del Fueyo, G. 1996. Microsporogenesis and microgametogenesis of the Argentinian species of *Podocarpus* (Podocarpaceae).—*Botanical Journal of the Linnean Society* 122: 171–182.

Del Fueyo, G., Archangelsky, S., Llorens, M. & Cúneo, R. 2008. Coniferous ovulate cones from the Lower Cretaceous of Santa Cruz Province, Argentina.—*International Journal of Plant Sciences* 169: 799–813.

Del Tredici, P. 1991. Ginkgos and people: a thousand years of interaction.—*Arnoldia* 51: 3–15.

Del Tredici, P. 1992a. Natural regeneration of *Ginkgo biloba* from downward growing cotyledonary buds (basal chichi).—*American Journal of Botany* 79: 522–530.

Del Tredici, P. 1992b. Where the wild ginkgos grow.—*Arnoldia* 52: 2–11.

Del Tredici, P. 2004. *Ginkgo biloba* Linné. *In* Schütt, P., Weisgerber, H., Schuck, H. J., Lang, K. J., Stimm, B. & Roloff, A. (eds.). *Lexikon der Nadelbäume.* Nikol Verlagsgesellschaft, Hamburg: pp. 187–196.

Del Tredici, P. 2007. The phenology of sexual reproduction in *Ginkgo biloba:* ecological and evolutionary implications.—*Botanical Review* 73: 267–278.

Del Tredici, P. 2008. Wake up and smell the ginkgos.—*Arnoldia* 66: 11–21.

Donoso, C., Lara, A. & Alarcon, D. 2004. *Araucaria araucana* (Mol.) K. Koch, 1795. *In* Schütt, P., Weisgerber, H., Schuck, H. J., Lang, K. J., Stimm, B. & Roloff, A. (eds.). *Lexikon der Nadelbäume.* Nikol Verlagsgesellschaft, Hamburg: pp. 93–98.

Earle, C. J. 2009a. Pinophyta. The Gymnosperm Database. March 4, 2009. Available at: http://www.conifers.org/zz/pinales.htm. Accessed March 20, 2009.

Earle, C. J. 2009b. Cupressaceae. The Gymnosperm Database. January 3, 2009. Available at: http://www.conifers.org/cu/index.htm. Accessed March 20, 2009.

Earle, C. J. & Frankis, M. 2009. Pinaceae. The Gymnosperm Database. January 3, 2009. Available at: http://www.conifers.org/pi/index.htm. Accessed April 30, 2009.

Elias, T. S. & Dykeman, P. A. 1990. *Edible Wild Plants: A North American Field Guide.* Sterling Publishing, New York.

Enright, N. J. & Ogden, J. 1995. The southern conifers: a synthesis. *In* Enright, N. J. & Hill, R. S. (eds.). *Ecology of the Southern Conifers.* Smithsonian Institution Press, Washington, D.C.: pp. 271–321.

Escapa, I., Cúneo, R. & Axsmith, B. 2008. A new genus of the Cupressaceae (sensu lato) from the Jurassic of Patagonia: implications for conifer megasporangiate cone homologies.—*Review of Palaeobotany and Palynology* 151: 110–122.

Falaschi, P. 2009. *Sistemática, paleoecologia e indicaciones paleoclimáticas de la tafoflora Monumento Natural Bosques Petrificados, Jurásico Medio, Patagonia, República Argentinia*. Ph.D. Dissertation. Universidad Nacional de La Plata, Argentina.

Falaschi, P., Zamuner, A. B. & Foix, N. 2009. Una nueva Equisetal fértil de la Formación La Matilde, Jurásico Medio, Argentina.—*Ameghinana* 46: 263–272.

Falder, A. B., Rothwell, G. W., Mapes, G., Mapes, R. H. & Doguzhaeva, L. A. 1998. *Pityostrobus milleri* sp. nov., a pinaceous cone from the Lower Cretaceous (Aptian) of southwestern Russia.—*Review of Palaeobotany and Palynology* 103: 253–261.

Francis, J. E. 1983. The dominant conifer of the Jurassic Purbeck Formation, England.—*Palaeontology* 26: 277–294.

Gardarsson, A. & Sigurdsson, J. B. 1972. *Research on the Pink-footed Goose (Anser brachyrhynchus) in 1971. Other Studies in Thjorsarver in 1971.* Unpublished report, Icelandic National Energy Authority Reykjavik, Iceland.

Gee, C. T. 1989. Revision of the Late Jurassic/Early Cretaceous flora from Hope Bay, Antarctica.—*Palaeontographica Abt. B* 213: 149–214.

Gee, C. T. & Tidwell, W. D. 2010. A mosaic of characters in a new whole-plant *Araucaria, A. delevoryasii* Gee sp. nov., from the Late Jurassic Morrison Formation of Wyoming, USA. *In* Gee, C. T. (ed.). *Plants in Mesozoic Time: Morphological Innovations, Phylogeny, Ecosystems*. Indiana University Press, Bloomington: pp. 67–94.

Godman, R. M. & Lancaster, K. 1990. *Tsuga canadensis* (L.) Carr. Eastern hemlock. *In* Burns, R. M. & Honkala, B. H. (eds.). *Silvics of North America. 1. Conifers*. USDA Forest Service, Washington, D.C.: pp. 604–612.

Gomez, B., Martín-Closas, C., Barale, G., Solé de Porta, N., Thévenard, F. & Guignard, G. 2002. *Frenelopsis* (Coniferales: Cheirolepidiaceae) and related male organ genera from the Lower Cretaceous of Spain.—*Palaeontology* 45: 997–1036.

Gnaedinger, S. 2004. Estudio preliminar de la xilotafoflora de la Formación La Matilde (Jurásico Medio) del gran bajo de San Julián, Santa Cruz, Argentina.—*Comunicaciones Cientificas y Tecnológicas, Universidad Nacional del Nordeste, SGCCyT, UNNE*.

Gnaedinger, S. 2007. Podocarpaceae woods (Coniferales) from Middle Jurassic La Matilde Formation, Sant Cruz province, Argentina.—*Review of Palaeobotany and Palynology* 147: 77–93.

Grant, T. A., Henson, P. & Cooper, J. A. 1994. Feeding ecology of trumpeter swans breeding in south central Alaska.—*Journal of Wildlife Management* 58: 774–780.

Green, W. A. 2005. Were Mesozoic ginkgophytes shrubby? *In* Popp, M. & Bolhar-Nordenkampff, H. (eds.). *XVII International Botanical Congress Abstracts*, Vienna: p. 384.

Green, W. A. 2007. *Using Leaf Architectural Data for Phenetic Ecological Comparison of Modern and Fossil Forest Stands*. Ph.D. Dissertation. Yale University, New Haven.

Griebeler, E.-M. & Werner, J. This volume. The life cycle of sauropod dinosaurs. *In* Klein, N., Remes, K., Gee, C. T. & Sander, P. M. (eds.). *Biology of the Sauropod Dinosaurs: Understanding the Life of Giants*. Indiana University Press, Bloomington: pp. 263–275.

Halle, T. G. 1913. Some Mesozoic plant-bearing deposits in Patagonia and Tierra del Fuego and their floras.—*Kungliga Svenska Vetenskapsakademiens Handlingar* 51: 1–58.

Harris, T. M. 1931. The fossil flora of Scoresby Sound, East Greenland. Part 1: Cryptogams (exclusive of Lycopodiales).—*Meddelelsler om Grønland* 2: 1–102.

Harris, T. M. 1961. *The Yorkshire Jurassic Flora. I. Thallophyta-Pteridophyta*. British Museum (Natural History), London.

Hauke, R. L. 1969. The natural history of *Equisetum* in Costa Rica.—*Revista de Biología Tropical* 15: 269–281.

Hegnauer, R. 1962. *Chemotaxonomie der Pflanzen. Band 1, Thallophyten, Bryophyten, Pteridophyten und Gymnospermen*. Birkhäuser Verlag, Basel.

Herbst, R. 1971. Palaeophytología. III. 7. Revisión de las especies argentinas del género *Cladophlebis*.—*Ameghinana* 8: 265–281.

Hill, C. R. 1976. Coprolites of *Ptilophyllum* cuticles from the Middle Jurassic of North Yorkshire.—*Bulletin of the British Museum (Natural History), Geology* 27: 289–293.

Hill, C. R. 1987. Jurassic *Angiopteris* (Marattiales) from North Yorkshire.—*Review of Palaeobotany and Palynology* 51: 65–93.

Hill, R. S. 1995. Conifer origin, evolution and diversification in the Southern Hemisphere. *In* Enright, N. J. & Hill, R. S. (eds.). *Ecology of the Southern Conifers*. Melbourne University Press, Carlton: pp. 10–29.

Holzhüter, G., Narayanan, K. & Gerber, T. 2003. Structure of silica in *Equisetum arvense*.—*Analytical and Bioanalytical Chemistry* 376: 512–517.

Honda, H. 1997. *Ginkgo* and insects. *In* Hori, T., Ridge, R. W., Tulecke, W., Del Tredici, P., Trémouillaux-Guiller J. & Tobe, H. (eds.). *Ginkgo biloba, a Global Treasure: From Biology to Medicine*. Springer, Berlin: pp. 243–250.

Hotton, C. L. & Baghai-Riding, N. L. 2010. Palynological evidence for conifer dominance within a heterogeneous landscape in the Late Jurassic Morrison Formation, USA. *In* Gee, C. T. (ed.). *Plants in Mesozoic Time: Morphological Innovations, Phylogeny, Ecosystems*. Indiana University Press, Bloomington: pp. 295–328.

Hummel, J. & Clauss, M. This volume. Sauropod feeding and digestive physiology. *In* Klein, N., Remes, K., Gee, C. T. & Sander, P. M. (eds.). *Biology of the Sauropod Dinosaurs: Understanding the Life of Giants*. Indiana University Press, Bloomington: pp. 11–13.

Hummel, J., Gee, C. T., Südekum, K.-H., Sander, P. M., Nogge, G. & Clauss, M. 2008. In vitro digestibility of fern and gymnosperm foliage: implications for sauropod feeding ecology and diet selection.—*Proceedings of the Royal Society B: Biological Sciences* 275: 1015–1021.

Husby, C. 2003. How large are the giant horsetails? March 19, 2003. Available at: http://www.fiu.edu/chusb001/GiantEquisetum/HowLarge.html. Accessed May 13, 2009.

Jones, D. L. 1987. *Encyclopaedia of Ferns*. British Museum (Natural History), London.

Jones, D. L. 1993. *Cycads of the World: Ancient Plants in Today's Landscape*. Smithsonian Institution Press, Washington, D.C.

Jones, W. G., Hill, K. D. & Allen, J. M. 1995. *Wollemia nobilis*, a new living Australian genus and species in the Araucariaceae.—*Telopea* 6: 173–176.

Jud, N., Rothwell G. W. & Stockey, R. A. 2008. *Todea* from the Lower Cretaceous of western North America: implications for the phylogeny, systematics, and evolution of modern Osmundaceae.—*American Journal of Botany* 95: 330–339.

Jung, W. 1974. Die Konifere *Brachyphyllum nepos* Saporta aus den Solnhofener Plattenkalken (unteres Untertithon), ein Halophyt.—*Mitteilungen der Bayerischen Staatssammlung für Paläontologie und Historische Geologie* 14: 49–58.

Kahlert, E., Schultka, S. & Süß, H. 1999. Die mesophytische Flora der Saurierlagerstätte am Tendaguru (Tansania). Erste Ergebnisse.—*Mitteilungen aus dem Museum für Naturkunde in Berlin, Geowissenschaftliche Reihe* 2: 185–199.

Kimmins, J. P., Binkley, D., Chatarpaul, L. & Catanzaro, J. de. 1985. Biogeochemistry of temperate forest ecosystems: literature on inventories and dynamics of biomass and nutrients.—*Canadian Forestry Service, Information Report* PI-X-47E/F.

Kramer, K. U., Schneller, J. J. & Wollenweber, E. 1995. *Farne und Farnverwandte*. Georg Thieme Verlag, Stuttgart.

Krüssmann, G. 1972. *Handbuch der Nadelgehölze*. Paul Parey, Berlin.

Kunzmann, L., Mohr, B. A. R., Bernardes-de-Oliveira, M. E. C. & Wilde, V. 2006. Gymnosperms from the Early Cretaceous Crato Formation (Brazil). II. Cheirolepidiaceae.—*Fossil Record* 9: 213–225.

Kupferschmid, A. D. & Bugmann, H. 2005. Effect of microsites, logs and ungulate browsing on *Picea abies* regeneration in a mountain forest.—*Forest Ecology and Management* 205: 251–265.

Labandeira, C. C., Johnson, K. R. & Lang, P. 2002. Preliminary assessment of insect herbivory across the Cretaceous–Tertiary boundary: major extinction and minimum rebound.—*Geological Society of America Special Paper* 361: 297–327.

Ladd, P. G., Connell, S. W. & Harrison, B. 1993. Seed toxicity in *Macrozamia riedlei*. *In* Stevenson D. W. & Norstog, K. J. (eds.). *Proceedings of CYCAD 90, the Second International Conference on Cycad Biology*. Palm and Cycad Societies of Australia, Milton: pp. 37–44.

Lanner, R. M. 2002. *Conifers of California*. Cachuma Press, Los Olivos, Calif.

LaPasha, C. A. & Miller, C. N., Jr. 1984. Flora of the Early Cretaceous Kootenai Formation in Montana, paleoecology.—*Palaeontographica Abt. B* 194: 109–130.

LaPasha, C. A. & Miller, C. N., Jr. 1985. Flora of the Early Cretaceous Kootenai Formation in Montana, bryophytes and tracheophytes.—*Palaeontographica Abt. B* 196: 111–145.

Large, M. F. & Braggins, J. E. 2004. *Tree Ferns*. Timber Press, Portland, Ore.

Lieth, H. 1975. Primary production of the major vegetation units of the world. *In* Lieth, H. & Whittaker, R. H. (eds.). *Primary Productivity of the Biosphere*. Springer-Verlag, Berlin: pp. 203–215.

Little, S. & Garrett, P. W. 1990. *Pinus rigida*, pitch pine. *In* Burns, R. M. & Honkala, B. H. (eds.). *Silvics of North America. 1. Conifers*. Agriculture Handbook 654. USDA Forest Service, Washington, D.C.

Litwin, R. J., Turner, C. E. & Peterson, F. 1998. Palynological evidence on the age of the Morrison Formation, Western Interior U.S.—*Modern Geology* 22: 297–319.

Llorens, M. & Del Fueyo, G. 2003. Coniferales fértiles de la formación Kachaike, Cretácico medio de la provincia de Santa Cruz, Argentina.—*Revista del Museo Argentino de Ciencias Naturales* 5: 241–244.

Mabberly, D. J. 1993. *The Plant-Book. A Portable Dictionary of the Higher Plants*. Cambridge University Press, Cambridge.

Midgley, J. J., Midgley, G. & Bond, W. J. 2002. Why were the dinosaurs so large? A food quality hypothesis.—*Evolutionary Ecology Research* 4: 1093–1095.

Mohabey, D. M. 2005. Late Cretaceous (Maastrichtian) nests, eggs, and dung mass (coprolites) of sauropods (titanosaurs) from India. *In* Tidwell, V. & Carpenter, K. (eds.). *Thunder-Lizards. The Sauropodomorph Dinosaurs*. Indiana University Press, Bloomington: pp. 466–489.

NSW National Parks and Wildlife Service. 2001. *Recovery Plan for the Giant Fern (Angiopteris evecta)*. NSW National Parks and Wildlife Service, Hurstville.

Nugent, G., Fraser, K. W. & Sweetapple, P. J. 1997. Comparison of red deer and possum diets and impacts in podocarp–hardwood forest, Waihaha Catchment, Pureora Conservation Park.—*Science for Conservation* 50: 1–61.

Ogden, J. & Stewart, G. H. 1995. Community dynamics of the New Zealand conifers. *In* Enright, N. J. & Hill, R. S. (eds.). *Ecology of the Southern Conifers*. Smithsonian Institution Press, Washington, D.C.: pp. 81–119.

Ohmae, Y., Shibata, K. & Yamakura, T. 1996. Seasonal change in nagilactone contents in leaves in *Podocarpus nagi* forest.—*Journal of Chemical Ecology* 22: 477–489.

Palmer, L. J. 1944. Food requirements of some Alaska game mammals.—*Journal of Mammalogy* 2: 49–54.

Rees, P. M. & Cleal, C. J. 2004. Lower Jurassic floras from Hope Bay and Botany Bay, Antarctica.—*Special Papers in Palaeontology* 72: 1–90.

Relva, M. A. & Veblen, T. T. 1998. Impacts of introduced large herbivores on *Austrocedrus chilensis* forests in northern Patagonia, Argentina.—*Forest Ecology and Management* 108: 27–40.

Rigg, L. S., Enright, N. J. & Jaffre, T. 1998. Stand structure of the emergent *Araucaria laubenfelsii*, in maquis and rainforest, Mont Do, New Caledonia.—*Australian Journal of Ecology* 23: 528–538.

Rogers, G. 1997. Trends in the health of pahautea and Hall's totara in relation to possum control in central North Island.—*Science for Conservation* 52: 1–49.

Rothwell, G. W. & Holt, B. 1997. Fossils and phylogeny in the evolution of *Ginkgo biloba*. *In* Hori, T., Ridge, R. W., Tulecke, W., Del Tredici, P., Trémouillaux-Guiller, J. & Tobe, H. (eds.). *Ginkgo biloba, a Global Treasure: From Biology to Medicine*. Springer-Verlag, Berlin: pp. 223–230.

Sander, P. M. & Clauss, M. 2008. Sauropod gigantism.—*Science* 322: 200–201.

Sander, P. M., Christian, A. & Gee, C. T. 2009. Sauropods kept their heads down. Response.—*Science* 323: 1671–1672.

Sander, P. M., Gee, C. T., Hummel, J. & Clauss, M. 2010a. Mesozoic plants and dinosaur herbivory. *In* Gee, C. T. (ed.). *Plants in Mesozoic Time: Morphological Innovations, Phylogeny, Ecosystems.* Indiana University Press, Bloomington: pp. 331–359.

Sander, P. M., Christian, A., Clauss, M., Fechner, R., Gee, C. T., Griebeler, E. M., Gunga, H.-C., Hummel, J., Mallison, H., Perry, S., Preuschoft, H., Rauhut, O., Remes, K., Tütken, T., Wings, O. & Witzel, U. 2010b. Biology of the sauropod dinosaurs: the evolution of gigantism.—*Biological Reviews of the Cambridge Philosophical Society.* doi: 10.1111/j.1469-185X.2010.00137.x

Sander, P. M., Klein, N., Stein, K. & Wings, O. This volume. Sauropod bone histology and its implications for sauropod biology. *In* Klein, N., Remes, K., Gee, C. T. & Sander, P. M. (eds.). *Biology of the Sauropod Dinosaurs: Understanding the Life of Giants.* Indiana University Press, Bloomington: pp. 276–302.

Schrank, E. 1999. Palynology of the dinosaur beds of Tendaguru (Tanzania)—preliminary results.—*Mitteilungen aus dem Museum für Naturkunde in Berlin, Geowissenschaftliche Reihe* 2: 171–183.

Schrank, E. 2010. Pollen and spores from the Tendaguru Beds, Upper Jurassic and Lower Cretaceous of southeast Tanzania: palynological and paleoecological implications.—*Palynology* 43: 3–42.

Schütt, P., Weisgerber, H, Schuck, H.-J., Lang, K. J., Stimm, B. & Roloff, A. 2004. *Lexikon der Nadelbäume.* Hamburg, Nikol-Verlag.

Schwendemann, A. B., Taylor, T. N., Taylor, E. L., Krings, M. & Osborn, J. M. 2010. Modern traits in Early Mesozoic sphenophytes: the *Equisetum*-like cones of *Spaciinodum collinsonii* with in situ spores and elaters from the Middle Triassic of Antarctica. *In* Gee, C. T. (ed.). *Plants in Mesozoic Time: Morphological Innovations, Phylogeny, Ecosystems.* Indiana University Press, Bloomington: pp. 15–33.

Serbet, R. & Rothwell, G. W. 1999. *Osmunda cinnamomea* (Osmundaceae) in the Upper Cretaceous of western North America. Additional evidence for exceptional species longevity among filicalean ferns.—*International Journal of Plant Sciences* 160: 425–433.

Smith, A. G., Smith, D. G. & Funnell, B. M. 2004. *Atlas of Mesozoic and Cenozoic Coastlines.* Cambridge University Press, Cambridge.

Son, Y. & Kim, H.-W. 1998. Above-ground biomass and nutrient distribution in a 15-year-old ginkgo (*Ginkgo biloba*) plantation in central Korea.—*Bioresearch Technology* 63: 173–177.

Spalletti, L., Artabe, A., Morel, E. & Brea, M. 1999. Biozonación paleoflorística y cronoestratigrafía del Triásico argentino.—*Ameghiniana* 36: 419–451.

Spicer, R. A. & Parrish, J. T. 1986. Paleobotanical evidence for cool north polar climates in middle Cretaceous (Albian–Cenomanian) time.—*Geology* 14: 703–706.

Stewart, W. N. & Rothwell, G. W. 1993. *Paleobotany and the Evolution of Plants.* Cambridge University Press, Cambridge.

Stipanicic, P. N. 2002. Introducción.—*Revista de la Asociación Geológica Argentina, B* 26: 1–24.

Stockey, R. A. 1975. Seeds and embryos of *Araucaria mirabilis.*—*American Journal of Botany* 62: 856–868.

Stockey, R. A. 1978. Reproductive biology of Cerro Cuadrado fossil conifers: ontogeny and reproductive strategies in *Araucaria mirabilis* (Speg.) Windhausen.—*Palaeontographica, Abt.* B 166: 1–15.

Stockey, R. A. 2002. A reinterpretation of the Cerro Cuadrado fossil "seedings," Argentina. *In* Dernbach, U. & Tidwell, W. D. (eds.). *Secrets of Petrified Plants, Fascination from Millions of Years.* D'ORO Publishers, Heppenheim: pp. 164–171.

Süss, H. & Schultka, S. 2001. First record of *Glyptostroboxylon* from the Upper Jurassic of Tendaguru, Tanzania.—*Botanical Journal of the Linnean Society* 135: 421–429.

Sweetapple, P. J. & Nugent, G. 2007. Ship rat demography and diet following possum control in a mixed podocarp–hardwood forest.—*New Zealand Journal of Ecology* 31: 186–201.

Taggart, R. E. & Cross, A. T. 1997. The relationship between land plant diversity and productivity and patterns of dinosaur herbivory. *In* Wolberg, D. L., Stump, E. & Rosenberg, G. (eds.). *Dinofest.* International Academy of Natural Sciences, Philadelphia: pp. 403–416.

Taylor, E. L., Taylor, T. N., Kerp, H. & Hermsen, E. J. 2006. Mesozoic seed ferns: old paradigms, new discoveries.—*Journal of the Torrey Botanical Society* 133: 62–82.

Taylor, T. N., Taylor, E. L. & Krings, M. 2009. *Paleobotany: The Biology and Evolution of Fossil Plants.* Academic Press, San Diego.

Thomas, V. G. & Prevett, P. J. 1982. The role of horsetails (Equisetaceae) in the nutrition of northern-breeding geese.—*Oecologia* 53: 359–363.

Tian, N., Wang, Y.-D. & Jiang, Z.-K. 2008. Permineralized rhizomes of the Osmundaceae (Filicales): diversity and tempospatial distribution pattern.—*Palaeoworld* 17: 183–200.

Tidwell, W. D. 1990a. A new osmundaceous species (*Osmundacaulis lemonii* n. sp.) from the Upper Jurassic Morrison Formation, Utah.—*Hunteria* 2: 3–6.

Tidwell, W. D. 1990b. Preliminary report on the megafossil flora of the Upper Jurassic Morrison Formation.—*Hunteria* 2: 1–11.

Tidwell, W. D. 1994. *Ashicaulis,* a new genus for some species of *Millerocaulis* (Osmundaceae).—*Sida, Contributions to Botany* 16: 253–261.

Tidwell, W. D. 2002. The Osmundaceae: a very ancient group of ferns. *In* Dernbach, U. & Tidwell, W. D. (eds.). *Secrets of Petrified Plants, Fascination from Millions of Years.* D'ORO Publishers, Heppenheim: pp. 134–147.

Tidwell, W. D. & Ash, S. R. 1990. On the Upper Jurassic stem *Hermanophyton* and its species from Colorado and Utah, USA.—*Palaeontographica Abt. B* 218: 77–92.

Tidwell, W. D. & Rushforth, S. R. 1970. *Osmundacaulis wadei,* a new osmundaceous species from the Morrison Formation (Jurassic) of Utah.—*Bulletin of the Torrey Botanical Club* 97: 137–144.

Tidwell, W. D., Britt, B. B. & Ash, S. R. 1998. Preliminary floral analysis of the Mygatt-Moore Quarry in the Jurassic Morrison

Formation, west-central Colorado.—*Modern Geology* 22: 341–378.

Tidwell, W. D., Connely, M. & Britt, B. B. 2006. A flora from the base of the Upper Jurassic Morrison Formation near Como Bluff, Wyoming, USA.—*New Mexico Museum of Natural History and Science Bulletin* 36: 171–181.

Tiffney, B. H. 1997. Land plants as food and habitat in the age of dinosaurs. *In* Farlow, J. O. & Brett-Surman, M. K. (eds.). *The Complete Dinosaur.* Indiana University Press, Bloomington: pp. 352–370.

Torrano, L. & Valderrábano, J. 2005. Grazing ability of European black pine understorey vegetation by goats.—*Small Ruminant Research* 58: 253–263.

Troncoso, A., Gnaedinger, S. & Herbst, R. 2000. *Heidiphyllum, Rissikia* y *Desmiophyllum* (Pinophyta, Coniferales) en el Triasico del norte chico de Chile y sur de Argentina.—*Ameghiniana* 37: 119–125.

Van Soest, P. J. 1994. *Nutritional Ecology of the Ruminant.* Cornell University Press, Ithaca.

Villar de Seoane, L. 1998. Comparative study of extant and fossil conifer leaves from the Baqueró Formation (Lower Cretaceous), Santa Cruz Province, Argentina.—*Review of Palaeobotany and Palynology* 99: 247–263.

Villar de Seoane, L. 2005. *Equisetites pusillus* sp. nov. from the Aptian of Patagonia, Argentina.—*Revista del Museo Argentino de Ciencias Naturales* 7: 43–49.

Walsh, R. A. 1994. *Osmunda cinnamomea. In* Fire Sciences Laboratory (ed.). *Fire Effects Information System.* U.S. Department of Agriculture, Forest Service, Rocky Mountain Research Station.

Watson, J. 1988. The Cheirolepidiaceae. *In* Beck, C. B. (ed.). *Origin and Evolution of Gymnosperms.* Columbia University Press, New York: pp. 382–447.

WMAC(NS). 2007. Muskox fact sheet number 5—habitat and diet. Available at: http://www.wmacns.ca/pdfs/116_Muskox%20Fact%20Sheet%20Number%205.pdf. Accessed May 13, 2009.

Yang, S., Wang, J. & Pfefferkorn, H. W. 2008. *Marattia aganzhenensis* sp. nov. from the Lower Jurassic Daxigou Formation of Lanzhou, Gansu, China.—*International Journal of Plant Sciences* 169: 473–482.

Yao, X., Taylor, T. N. & Taylor, E. L. 1997. A taxodiaceous seed cone from the Triassic of Antarctica.—*American Journal of Botany* 84: 343–354.

Zamuner, A. B. & Falaschi, P. 2005. *Agathoxylon matildense* n. sp., leno araucariaceo del Bosques Petrificados del cerro Madre e Hija, Formación La Matilde (Jurásico medio), provincial de Santa Cruz, Argentinia.—*Ameghinana* 42: 339–346.

Zanoni, T. A. & Adams, R. P. 1973. Distribution and synonymy of *Juniperus californica* Carr. (Cupressaceae) in Baja California, Mexico.—*Bulletin of the Torrey Botanical Club* 100: 364–367.

Zhou, Z. 1983. *Stalagma samara,* a new podocarpaceous conifer with monocolpate pollen from the Upper Triassic of Hunan, China.—*Palaeontographica Abt. B* 185: 56–78.

Zhou, Z. & Zhang, F. 2002. A long-tailed, seed-eating bird from the Early Cretaceous of China.—*Nature* 418: 405–409.

4

The Diet of Sauropod Dinosaurs: Implications of Carbon Isotope Analysis on Teeth, Bones, and Plants

THOMAS TÜTKEN

SAUROPODS WERE MEGAHERBIVORES that fed predominantly on nonangiosperm vegetation such as gymnosperms, sphenophytes, and pteridophytes. In this chapter, the potential of carbon isotope ($\delta^{13}C$) analysis in skeletal apatite for inferring the diet and niche partitioning of sauropods was tested. The carbon isotope composition of food plants is transferred with a metabolic offset to higher trophic levels along the food chain, which suggests that differences in isotopic composition of sauropod food plants can be used to infer sauropod feeding behavior. For this purpose, the $\delta^{13}C$ values of sauropod bones and teeth, primarily from the Late Jurassic Morrison Formation, USA, and the Tendaguru Beds, Tanzania, East Africa, were analyzed, as were the leaves of extant and fossil potential sauropod food plants such as *Araucaria,* cycads, ferns, horsetails, and ginkgo. The metabolic carbon isotope fractionation between diet and enamel apatite estimated for sauropods is 16‰. By means of this fractionation, a diet based only on terrestrial C_3 plants can be reconstructed for sauropods. Therefore, sauropods did not ingest significant amounts of plants with high, C_4 plant-like $\delta^{13}C$ values such as marine algae or C_4 plants. However, plants that used crassulacean acid metabolism for biosynthesis and possibly freshwater aquatic plants may have contributed to the diet of sauropods. A more detailed discrimination of exactly which type of food plants was consumed by sauropods based on apatite $\delta^{13}C$ values alone is difficult because taxon-specific differences between C_3 plants are small and not well constrained. Mean enamel $\delta^{13}C$ values of sympatric sauropods differ by approximately 3‰, which may indicate a certain niche partitioning. Differences in mean $\delta^{13}C$ values for the living representatives of potential sauropod food plants suggest that a differentiation between low-browsing taxa feeding on ferns or horsetails with lower $\delta^{13}C$ values and high-browsing taxa feeding on conifers with higher $\delta^{13}C$ values might be possible.

Sauropod Feeding Behavior: What Do We Know?

Sauropod dinosaurs are one of the most successful groups of dinosaurs in terms of taxon longevity (Late Triassic to Late Cretaceous), taxonomic diversity, and geographic distribution (Dodson 1990; Wilson 2002; Rees et al. 2004; Upchurch et al. 2004; Barrett & Upchurch 2005; Sander et al. 2010a). They reached their highest abundance and diversity during the Jurassic, and as megaherbivores, they had an important influence on terrestrial ecosystems (Upchurch & Barrett 2000). Sauropods lived in a gymnosperm- and pteridophyte-dominated environment with a variety of conifers, as well as some ginkgoes, cycads, ferns, seed ferns, and horsetails (Plate 4.1) that potentially constituted the major sauropod food plants (Coe et al. 1987; Tiffney 1997; Rees et al. 2004; Hummel et al. 2008; Gee, this volume). In contrast, during the Late Cretaceous, sauropods lived in angiosperm-rich environments and probably also fed on angiosperm plants (Bakker 1978; Barrett & Willis 2001; Mohabey & Samant 2003) or even occasionally on grass (Prasad et al. 2005) or algae (Ghosh et al. 2003; Mohabey & Samant 2003).

As a result of their simple dentition, most sauropods had limited masticatory abilities for oral food processing (Sander et al. 2010a, 2010b; Hummel & Clauss, this volume). Therefore, sauropods (especially *Diplodocus*) were initially interpreted as having fed on soft aquatic plants or even invertebrates such as clams (Holland 1924; Haas 1963). Differences in dental morphology and wear, as well as reconstructed feeding envelopes and inferred browsing heights, suggest a certain degree of specialization and niche partitioning for sauropods and use of food resources at different heights above ground level (Bakker 1978; Stevens & Parrish 1999, 2005a, 2005b; Christian & Dzemski, this volume). Especially the long necks of sauropods could be a fundamental adaptation for increasing the vertical and horizontal feeding envelope (Bakker 1971, 1978; Stevens & Parrish 1999, 2005a, 2005b; Sander et al. 2009). Those taxa with a full battery of spatulate teeth presumably used some oral processing and likely were specialized for tougher forage, and those with a reduced battery of peg-like teeth may have fed predominantly on softer plants or possibly used their teeth to strip branches. Even an ontogenetic switching of diets has been postulated for certain taxa (e.g., Fiorillo 1991). However, it is difficult to constrain exactly which plants the different sauropod taxa used as food resources. Because sauropods were large megaherbivores, a more generalized feeding strategy is also likely, as known in living megaherbivores (Owen-Smith 1988).

Often several (up to five) different sauropod taxa co-occur at one site (Dodson 1990), which also supports the idea of a certain niche partitioning. The lack of well preserved definite sauropod gastrointestinal products such as coprolites makes it difficult to determine which plants and parts thereof formed the diet of individual sauropod taxa (see recent critical review by Sander et al. 2010b). The sole exceptions may be some Late Cretaceous putative sauropod coprolites from India with preserved silicified plant remains and phytoliths, indicating that a variety of different plants, including grasses, were ingested indiscriminately by titanosaurs (Mohabey & Samant 2003; Mohabey 2005; Prasad et al. 2005), if indeed these structures were coprolites and can be assigned to the titanosaurs (Sander et al. 2010b). Therefore, the reconstruction of sauropod diets and feeding behavior remains rather speculative.

A chemical proxy to infer diet, feeding behavior, and partitioning of food resources of extant and extinct vertebrates is the carbon isotope composition of bones and teeth (Cerling et al. 1997; Koch 1998, 2007; Kohn & Cerling 2002). Most studies have investigated the diet of mammals using carbon isotopes, but only a few have looked at the carbon isotopes of dinosaur skeletal remains (Bocherens et al. 1993; Ostrom et al. 1993; Stanton Thomas & Carlson 2004; Tütken et al. 2004, 2007; Fricke & Pearson 2008; Fricke et al. 2008, 2009), coprolites (Ghosh et al. 2003), or eggshells (Folinsbee et al. 1970; Erben et al. 1979; Sarkar et al. 1991; Cojan et al. 2003; Mohabey 2005) to infer their diet and feeding behavior. One major problem with analyzing the isotope composition of dinosaur remains is the diagenetic alteration of the ancient skeletal tissues. A recent detailed study of Fricke et al. (2008) showed that ecological information on diet and environment can indeed be preserved, at least in the enamel of dinosaur teeth. The preservation of original isotope compositions is, however, less likely for dinosaur bone (Kolodny et al. 1996; Trueman et al. 2003). In this study, carbon isotope analysis of sauropod bones and teeth as well as of their potential food plants will be used to explore sauropod feeding behavior and niche partitioning.

Carbon Isotopes in Plants and Dinosaur Fossils

CARBON ISOTOPES

Carbon is the most important building block of organic matter and life, and it is thus the major element in all living organisms. It can be either bound in soft tissues (e.g., muscles, skin, hair) or in biogenic hard parts (e.g., shell, eggshells, vertebrate skeletal tissues made of biogenic apatite such as bones and teeth). The element carbon consists of three different isotopes: two stable isotopes, ^{12}C and ^{13}C, and one radioactive isotope, ^{14}C, which disintegrates over time, with a constant half-life of 5,730 years and which is used for ^{14}C radiocarbon dating. The light carbon isotope ^{12}C is by far the most abundant (98.89%), followed by the heavy isotope ^{13}C (1.11%), and negligible amounts of the radioactive isotope ^{14}C (Hoefs 2008). The three isotopes, and the molecules containing them, react chemically in the same way but at different rates because their atoms have different atomic masses. Thus, different chemical and metabolic processes change the ratios of the isotopes in characteristic ways. Carbon isotope composition is expressed by the ratio of the heavy to light isotope ($^{13}C/^{12}C$). The carbon isotope composition of any substance is reported as a $\delta^{13}C$ value in the conventional δ notation in per mil (‰), relative to the known isotope reference standard V-PDB (Vienna Pee Dee Belemnite) (Coplen 1994). The δ value is defined according to equation (1), where R_{sample} and $R_{standard}$ are the $^{13}C/^{12}C$ ratios in the sample and the standard, respectively.

$$\delta^{13}C\ (‰) = [(R_{sample}/R_{standard}) - 1] \times 1{,}000. \quad (1)$$

By definition, the V-PDB standard itself has a δ value of 0‰. Samples enriched in the heavy isotope ^{13}C relative to the standard have positive $\delta^{13}C$ values, and those that are depleted in the heavy isotope ^{13}C have negative $\delta^{13}C$ values.

Carbon isotope composition can change significantly during the formation processes of inorganic and especially organic matter. Most organisms preferentially incorporate the light carbon isotope ^{12}C so that biogenic tissues are normally enriched in the light isotope relative to inorganic substances. The discrimination of isotopes against each other during formation processes is called fractionation. The fractionation of carbon isotopes during chemical and biological reactions and processes is responsible for the different isotope compositions of carbon-containing materials, which are in our case plant and animal tissues. The carbon isotope composition can be measured and allows inferences on the processes involved in tissue formation. The $\delta^{13}C$ value is usually measured in CO_2 gas generated by the phosphoric acid dissolution of carbonaceous or carbonate-containing minerals (in our case, the carbonate-containing skeletal bioapatite, Fig. 4.1), or in CO_2 gas liberated by thermal combustion for organic matter. The CO_2 gas is analyzed with a gas mass spectrometer with an analytical precision better than ±0.1‰.

CARBON ISOTOPES IN PLANTS

Plant groups show significant differences in $\delta^{13}C$ values as a result of different photosynthetic pathways. Additionally, environmental and climatic conditions during biosynthesis can greatly influence plant $\delta^{13}C$ values. These isotopic differences enable us to reconstruct the diet of herbivorous animals because they fractionate carbon isotopes taken up from the plants in a predictable way (DeNiro & Epstein 1978, 1981; Cerling & Harris 1999; Passey et al. 2005; Clementz et al. 2007).

The primary production of plants forms the basis of the food chain on land and in the ocean. Plants consist of biomass produced by the photosynthetic assimilation of atmospheric CO_2. During photosynthesis, plants strongly fractionate the carbon isotopes of this atmospheric CO_2 (Park & Epstein 1960; Farquhar et al. 1989). Therefore, plants are depleted in ^{13}C relative to atmospheric CO_2. The plant $\delta^{13}C$ values are mainly determined by the following: (1) the type of photosynthetic pathway, (2) the $\delta^{13}C$ value of atmospheric CO_2 (O'Leary 1988; Farquhar et al. 1989; Arens et al. 2000), and (3) local environmental factors (e.g., water stress, light, nutrients, temperature) (Heaton 1999). The three different photosynthetic pathways used by plants are referred to as C_3 pathway, C_4 pathway, and crassulacean acid metabolism (CAM).

Terrestrial Plants

During photosynthesis, most plants use the Calvin cycle, which utilizes the enzyme ribulose bisphosphate carboxylase–oxygenase (RuBisCo) to fix atmospheric CO_2, forming a three-carbon sugar, hence the term C_3. The majority of extant angiosperms, gymnosperms, pteridophytes, and sphenophytes, use the C_3 photosynthetic pathway. C_3 plants include trees, shrubs, cool-growing-season, high-altitude, and high-latitude grasses, and many aquatic plants. C_3 plants represent about 85% of the world's terrestrial plant biomass today. C_4 plants use the Hatch-Slack cycle during photosynthesis and a different enzyme to fix atmospheric CO_2, the phosphoenolpyruvate carboxylase, resulting in a four-carbon acid (as an intermediate product of photosynthesis), and hence the term C_4. Three fourths of all C_4 plants are tropical and temperate grasses and sedges growing in warm, dry habitats. C_4 plants are characterized by a special leaf anatomy known as Kranz anatomy that is documented in grass fossils as old as the late Miocene (Thomasson et al. 1986). C_4 plants account for 10% of terrestrial plant biomass. However, the C_4 photosynthetic pathway is restricted to angiosperms. Because this pathway presumably evolved in the mid Tertiary (20–30 million years ago) (Keeley & Rundel 2003), C_4 plants were not a component of Mesozoic floras (e.g., Bocherens et al. 1993) and accordingly were not available as food plants to dinosaurs.

The third photosynthetic pathway, CAM, is common in desert succulents (Cactaceae), tropical epiphytes, the fern families Polypodiaceae and Vittariaceae, and aquatic plants such as *Isoetes* (Keeley & Rundel 2003), but represents only 5% of terrestrial plant biomass (Ehleringer et al. 1991). CAM is an ancient photosynthetic pathway that likely has been present ever since the Paleozoic in plants from shallow-water palustrine habitats, and it is thought that CAM evolved convergently many times (Keeley & Rundel 2003). In terrestrial settings, CAM plants often grow in climatically stressful conditions (e.g., arid settings) too inhospitable for C_3 plants. On the basis of carbon isotope data of soil organic matter, it has been suggested that CAM plants may have existed in water-stressed environments in the Mesozoic (Decker & de Wit 2006).

Carbon Isotopes in Terrestrial C_3 and C_4 Plants

Because C_3 plants discriminate more against the heavy ^{13}C isotope during photosynthesis than C_4 plants, both plant groups have different carbon isotopic compositions (Farquhar et al. 1989). The average $\delta^{13}C$ value of extant C_3 plants is −27‰ (ranging from −36 to −20‰), while C_4 plants have an average $\delta^{13}C$ value of −13‰ (ranging from −17 to −9‰) (Deines 1980; O'Leary 1988) (Fig. 4.1). CAM plants exhibit a wide range of $\delta^{13}C$ values (−30 to −11‰) that can overlap with that of C_3 and C_4 plants (Bender 1971; Deines 1980; Farquhar et al. 1989; Fig. 4.1). $\delta^{13}C$ values for C_3 plants were mainly determined in angiosperms; however, gymnosperms, pteridophytes, and sphenophytes show similar values (see Fig. 4.4; Smith & Epstein 1971; Tütken et al. 2007). The distinctly different $\delta^{13}C$ values of C_3 and C_4 plants (mostly C_4 grasses) are often used in paleoecological and paleodietary studies to distinguish browsers from grazers in ecosystems where both types of plants were present (e.g., Cerling et al. 1997; MacFadden et al. 1999). The occurrence of a C_4 component in the diet of herbivorous mammals is interpreted as the intake of C_4 grass and thus as feeding in open grasslands. However, C_4 grasses became only globally widespread since the late Miocene (approximately 7 million years ago) (Cerling et al. 1993, 1997).

Recently, grass phytoliths have been found in putative sauropod coprolites from the Late Cretaceous of India, indicating the presence of some C_3 grasses by the end of the Mesozoic (Prasad et al. 2005). It has been suggested that CAM plants may have served as food for some dinosaurs (Bocherens et al. 1993; Stanton Thomas & Carlson 2004; Fricke et al. 2008), but this is questionable, given the extremely few non-angiosperm plants that use CAM today.

As noted above, C_3 plants can vary significantly in their $\delta^{13}C$ values because of environmental factors such as light, humidity, and the canopy effect (e.g., O'Leary 1988; Tieszen 1991; Heaton 1999; Arens et al. 2000). Therefore, plants that use the same photosynthetic pathway, or even plants from the same species that grow in different environments (xeric, mesic, aquatic, or canopy), may show a few per mil difference in $\delta^{13}C$ values (Ehleringer et al. 1986; Tieszen 1991; Fig. 4.1). The intraspecific variability of $\delta^{13}C$ values caused by environmental factors at a single locality is mostly less than 2–3‰ for living and fossil C_3 plants (O'Leary 1988; Bocherens et al. 1993; Nguyen Tu et al. 1999). $\delta^{13}C$ values of terrestrial plants vary seasonally and spatially with fluctuations in soil moisture and water use efficiency (Farquhar et al. 1989) as well as with the growth cycle, because some metabolic products are enriched in ^{13}C (e.g., sugars) or depleted in ^{13}C (e.g., lignin) (Park & Epstein 1960; Deines 1980; DeNiro & Hastorf 1985; Schleser et al. 1999). This is also the reason why different plant organs (e.g., leaves, fruits, seeds, wood) may have different $\delta^{13}C$ values (e.g., Tieszen 1991).

Another important local effect on plant $\delta^{13}C$ values is the canopy effect. In densely forested ecosystems, plants growing under a closed canopy with low irradiance and using ^{13}C-depleted CO_2 from biomass degradation near the forest floor for photosynthesis have $\delta^{13}C$ values as low as −36 to −32‰ (van der Merwe & Medina 1991). In such forest ecosystems, there are gradients of plant $\delta^{13}C$ values from the canopy toward the ground level. If such gradients are seen in herbivore communities in such habitats, they suggest that feeding occurs on different levels and on different food plants (Cerling et al. 2004).

Aquatic Plants

Aquatic plants in freshwater or marine ecosystems assimilate most of their carbon from the surrounding waters and have a broad range of $\delta^{13}C$ values from −30 to −8‰ (Fig. 4.1). Most fully marine plants (including sea grasses) use the C_3 photosynthetic pathway. However, the $\delta^{13}C$ values of marine C_3 plants are more enriched than those of terrestrial C_3 plants because the carbon in the oceans may be derived from bicarbonate (HCO_3^-), which is enriched in ^{13}C relative to atmospheric CO_2, as well as from dissolved atmospheric CO_2 (Boutton 1991; Keeley & Sandquist 1992; Hemminga & Mateo 1996). Most marine phytoplankton has $\delta^{13}C$ values similar to those of C_3 plants, whereas some intertidal kelps and seaweeds tend to resemble C_4 plants (Sackett et al. 1965; Clementz et al. 2006 and references therein). Aquatic plants have distinct mean $\delta^{13}C$ values of approximately −11‰ for sea grasses, approximately −18.5‰ for marine algae, and approximately −27‰ for freshwater plants (Clementz et al. 2007 and references therein). The $\delta^{13}C$ values of coastal aquatic plants are variable because of the multiple sources from which plants draw and fractionate carbon (Boutton 1991; Keeley & Sandquist 1992). Salt marsh C_4 plants have a mean $\delta^{13}C$ value of −14.9‰ (ranging from −17.7 to −12.8‰), while C_3 halophytes have similar $\delta^{13}C$ values as terrestrial C_3 plants. However, they vary (up to 10‰) with substrate salinity (Cloern et al. 2002). On average, marine C_3 plants and algae have higher $\delta^{13}C$ values than terrestrial C_3 plants, and animals feeding on marine food resources therefore have values about 7‰ higher than those feeding on terrestrial and/or freshwater plants (Chrisholm et al. 1982; Clementz et al. 2006, 2007). Thus, on the basis of carbon isotopes, herbivores that feed on marine and terrestrial food resources can be differentiated from one another.

In freshwater systems, as well as in the oceans, the carbon sources for photosynthesis include carbonate, bicarbonate, and dissolved carbon dioxide. $\delta^{13}C$ values of freshwater aquatic plants can be quite variable, but in general, they resemble those of terrestrial C_3 plants (Clementz et al. 2007), although they may sometimes have lower $\delta^{13}C$ values than terrestrial C_3 plants (Fig. 4.1). Low $\delta^{13}C$ values in aquatic freshwater plants occur because the assimilated carbon from the bicarbonate has low $\delta^{13}C$ values that originate from the degradation of organic matter with low $\delta^{13}C$ values. Thus, the tissue of herbivores that fed on freshwater plants show lower $\delta^{13}C$ values than herbivores feeding on terrestrial vegetation.

MESOZOIC PLANTS

The $\delta^{13}C$ value of atmospheric CO_2 has a globally constant mean value that shifts only as a result of changes in the global carbon cycle. For example, because of the burning of ^{12}C-rich fossil hydrocarbons, the current atmospheric CO_2 ($\delta^{13}C_{CO2}$ = −8‰) is 1.5‰ depleted in ^{13}C, compared to preindustrial CO_2, with a $\delta^{13}C$ value of −6.5‰ (Friedli et al. 1986). $\delta^{13}C$ values of carbonized plant fossils do not shift significantly during fossilization (DeNiro & Hastorf 1985) or experimental diagenesis and coalification (e.g., Schleser et al. 1999; Turney et al.

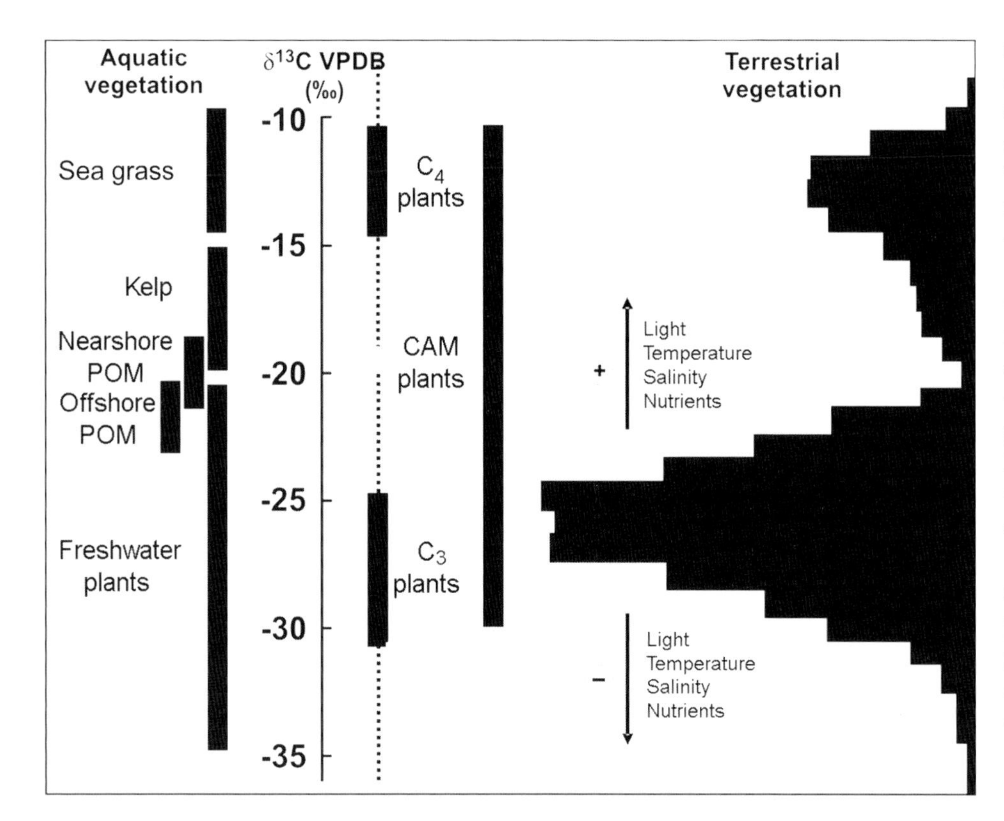

FIGURE 4.1. Carbon isotope composition of plants. Ranges of $\delta^{13}C$ values in extant terrestrial and aquatic plants are given. Thick vertical bars represent ranges in $\delta^{13}C$ values for terrestrial (Bender 1971; Deines 1980) and aquatic vegetation (Clementz et al. 2006 and references therein). The distribution of the $\delta^{13}C$ values of terrestrial C_3 and C_4 plants is also given in histogram form. Environmental influences and their effects on the $\delta^{13}C$ values of C_3 plants (Tieszen 1991; Heaton 1999; Gröcke 2002) are schematically shown. POM, particulate organic matter, a proxy for phytoplankton. *Modified after Clementz et al. (2006 and references therein).*

2006). For example, only small shifts (≤ 1.3‰), often toward slightly more negative $\delta^{13}C$ values due to the preferential loss of isotopically heavy ^{13}C-rich components, were observed during experimental coalification of wood (e.g., Schleser et al. 1999; Turney et al. 2006). Therefore, it can be assumed that carbonized plant remains from the Mesozoic have roughly retained their original carbon isotope composition.

Mesozoic plants have $\delta^{13}C$ values ranging from −28 to −20‰ (e.g., Bocherens et al. 1993; Nguyen Tu et al. 1999; Gröcke 2002). Values higher than −23‰ are rare and probably reflect water-stressed C_3 and/or CAM plants. The occurrence of CAM plants in arid Mesozoic terrestrial ecosystems is supported by $\delta^{13}C$ values of sedimentary organic matter lower than −20‰ in Upper Triassic to Jurassic strata of South Africa (Decker & de Wit 2006). However, CAM plants were not a major component of Mesozoic floras. Apart from a few global carbon isotope excursions, Mesozoic plant remains and wood have predominantly yielded $\delta^{13}C$ values typical for C_3 plants, with an average value of approximately −24‰ (Gröcke 2002). Late Cretaceous soil organic matter yielded similar mean $\delta^{13}C$ values—approximately −24.7 to −24.5‰ (Fricke et al. 2008). These $\delta^{13}C$ values are about 3‰ higher than for extant C_3 plants, probably as a result of higher $\delta^{13}C_{CO2}$ values of the Mesozoic atmosphere (Gröcke 2002). They likely represent the average carbon isotope composition of plants that dinosaurs fed on during most of the Mesozoic.

SKELETAL TISSUES: BONES, TEETH, AND EGGSHELLS

As noted, dietary isotopic signals are recorded in both soft tissues (muscles, skin, hair) and hard tissues (bones, teeth, eggshells) of vertebrates (Kohn & Cerling 2002; Koch 2007; Tütken 2010). Soft tissues are generally not preserved in dinosaur fossils, except for rare cases of protein preservation in dinosaur bones (Ostrom et al. 1993; Bocherens et al. 1993; Schweitzer et al. 2005, 2007). Therefore, mineralized hard tissues such as teeth, bones, and eggshells hold the best potential for recording the diet of dinosaurs.

Bones consist of a protein matrix of collagen (approximately 30 wt%) and a mineral phase of bioapatite (approximately 70 wt%). Bioapatite is a carbonate containing nonstoichiometric hydroxyapatite ($Ca_5(PO_4, CO_3)_3(OH, CO_3)$) and is also the major mineral phase of enamel and dentin. Carbonate replaces several percent (by weight) of the PO_4^{3-} ion and the subordinate OH group in the hydroxyapatite lattice (Fig. 4.2). Both bone collagen and bone mineral may record dietary information. However, the collagen usually degrades within a few thousand to 10,000 years, and only the mineral phase is preserved in fossil bones over millions of years. In addition, the nanometer-sized bioapatite crystals of bone recrystallize during fossilization and may chemically interact with the embedding sediment. Bones can record a long-term dietary signal because they form over several years and are remodeled throughout the lifetime of the animal. Fast-growing fibrolamellar bone of sauropods should register a record of several years of dietary intake over most of the life span of the individual sauropod (Curry 1999; Sander 1999; Klein & Sander 2008).

Dinosaur teeth were replaced continuously, and replacement teeth grew over the course of several months in most dinosaurs, but for up to two years in some theropods (Erickson 1996; Straight et al. 2004; D'Emic et al. 2009). Sauropods seem to have had very fast tooth replacement rates of only a few weeks to months. This was determined for *Diplodocus* (approximately 1 month) and *Camarasaurus* (approximately 2 months) by counts of von Ebner lines laid down in dentin (D'Emic et al. 2009). Thus, sauropod teeth represent the isotopic composition of the diet over a period of a few months. The third skeletal tissue that provides information on dinosaur diets is tooth enamel. Its formation time is about the same as for dentin. Enamel, with its low organic content (<1%), large apatite crystal size, and low porosity, shows the best preservation of original isotope compositions and dietary information (Stanton Thomas & Carlson 2004; Straight et al. 2004; Fricke et al. 2008).

Dinosaurs had calcareous eggshells, with a record extending back to the Late Triassic. In modern birds, eggshell forms within one day (e.g., 21 hours in a chicken; Folinsbee et al. 1970), and it is reasonable to assume a similar time frame for the formation of dinosaur eggs. Therefore, the calcitic eggshells likely record a short-term dietary signal of only a few days. The stable isotope composition of eggshell calcite seems to be preserved relatively well over millions of years, and it allows dietary and environmental inferences to be drawn for sauropods (Erben et al. 1979; Sarkar et al. 1991; Cojan et al. 2003; Mohabey 2005). Eggshell $\delta^{13}C$ data indicate that sauropods fed exclusively on C_3 plants (Erben et al. 1979; Sarkar et al. 1991; Cojan et al. 2003; Mohabey 2005).

CARBON ISOTOPES AS AN INDICATOR OF DIET AND ECOLOGY

The carbon isotope composition of fossil bioapatite of bones and teeth has been widely used as an archive for reconstructing the paleoecology and paleodiet of extinct vertebrates (Koch 1998, 2007; Kohn & Cerling 2002; Kohn et al. 2005). The carbon isotope composition of skeletal bioapatite is directly controlled by the $\delta^{13}C$ values of an animal's diet (Ambrose & Norr 1993; Tieszen & Fagre 1993). Plant carbon ingested by herbivores is metabolized and incorporated as carbonate (CO_3^{2-}) into the biogenic apatite and eggshell carbonate. The skeletal bioapatite forms in isotopic equilibrium with the blood.

Before dietary reconstructions can be made, the magnitude of the carbon isotope fractionation between skeletal apatite and diet, that is, the metabolic offset between apatite and diet

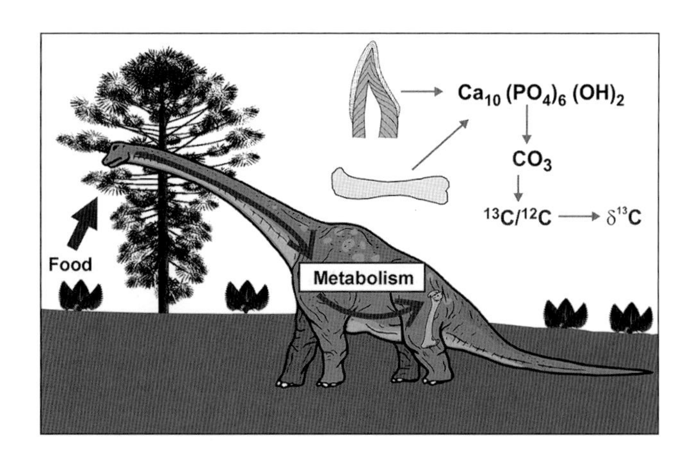

FIGURE 4.2. Schematic drawing of the incorporation of carbon from food into the bioapatite of bones and teeth of a sauropod by ingestion, metabolism, and tissue mineralization. *Plant drawings modified after Hinz (2010).*

($\Delta\delta^{13}C_{apatite\text{-}diet}$), must be known. However, it varies among different animals; in large mammalian herbivores, it ranges from approximately 12–15‰ (Koch 1998; Cerling & Harris 1999; Passey et al. 2005). The most extensive field study of wild African large herbivorous ungulates (which consists of grazers and browsers as well as foregut and hindgut fermenters) found an average $\Delta\delta^{13}C_{enamel\text{-}diet}$ enrichment factor of 14.1 ± 0.5‰ (Cerling & Harris 1999). In a controlled feeding study, Passey et al. (2005) found different $\Delta\delta^{13}C_{enamel\text{-}diet}$ enrichment factors of 13.3 ± 0.3‰ for nonruminant pigs and 14.6 ± 0.3‰ for ruminant cows raised on an isotopically identical diet. Therefore, digestive physiology (e.g., the rate of ^{12}C-rich methane production and its loss during digestion and rumination) can have an important influence on the $\Delta\delta^{13}C_{enamel\text{-}diet}$ enrichment factor (Passey et al. 2005). Thus, enamel $\delta^{13}C$ values mostly reflect the diet of the animal, but they may be additionally affected to some degree by its digestive physiology (Passey et al. 2005; Zanazzi & Kohn 2008). This may have applied to dinosaurs as well (Fricke et al. 2008). The $^{13}C_{eggshell\text{-}diet}$ enrichment factor between eggshell carbonate and bulk diet for large birds (ostriches) is 16.2‰ (von Schirnding et al. 1982; Johnson et al. 1998). Because the $\Delta\delta^{13}C_{apatite\text{-}diet}$ enrichment factor of extant reptiles and dinosaurs is unknown, the use of the ostrich enrichment factor for sauropods seems justified as a first approximation because these birds are surviving saurischian dinosaurs—the group to which the sauropods belong as well.

Recently, Fricke et al. (2008) found a $\Delta\delta^{13}C_{apatite\text{-}diet}$ value of 18‰ in ornithischian dinosaurs, based on the $\delta^{13}C$ difference between soil organic matter (assumed to represent the average $\delta^{13}C$ value of the plants that the dinosaurs fed on) and tooth enamel of Late Cretaceous herbivorous hadrosaurid dinosaurs. This value is even higher than that in extant herbivorous birds. This study attempts to derive a $\Delta\delta^{13}C_{apatite\text{-}diet}$

value for sauropods using enamel and food plant $\delta^{13}C$ data from the same locality. On the basis of this fractionation factor, $\delta^{13}C$ values of the food plants ingested by other sauropods can be calculated.

DIAGENESIS: PRESERVATION OF ORIGINAL CARBON ISOTOPE COMPOSITIONS?

For all geochemical, paleoecological, and paleodietary reconstructions based on fossil skeletal remains, it is critical to evaluate whether the original isotopic composition of the living animal tissue is still preserved. Chemical and biological processes during the fossilization of skeletal tissues can lead to significant changes of this in vivo isotope composition of the bioapatite (e.g., Kolodny et al. 1996; Tütken 2003; Tütken et al. 2008). Such changes hamper, or in the worst case even prevent, reconstruction of diet and environmental conditions. Because of its physical and chemical properties (see above), enamel is least affected by diagenesis (Ayliffe et al. 1994; Kohn et al. 1999; Tütken et al. 2008), and the original carbon isotope composition of fossil enamel can be preserved over millions of years (e.g., Lee-Thorp & van der Merwe 1987; Wang & Cerling 1994; Lee-Thorp & Sponheimer 2005), even in Mesozoic reptile and dinosaur teeth (Botha et al. 2005; Fricke et al. 2008; Fricke & Pearson 2008).

In fossil bone, however, the original carbon and oxygen isotope composition of the bone mineral can be partly or even completely altered (Schoeninger & DeNiro 1982; Nelson et al. 1986; Kolodny et al. 1996; Trueman et al. 2003; Kohn & Law 2006). In the latter case, they should equilibrate with soil carbonates (Kohn & Law 2006). Therefore, sauropod bones should at least reflect the long-term average $\delta^{13}C$ value of vegetation cover, but not the specific values of the preferred food plants ingested by the sauropods.

There are four different geochemical approaches to test for preservation of original isotopic composition and to monitor diagenetic alteration. First is comparison of skeletal tissues with different diagenetic resistance (e.g., enamel → dentin → bone) (Ayliffe et al. 1994; Tütken et al. 2008). Even though different skeletal tissues may form at different times in the life of the animal, they should have similar isotope compositions if no major dietary shifts occurred during ontogeny. Dentin and bone should more readily equilibrate with the diagenetic environment than enamel. Second is the comparison of the isotope composition of the fossil bioapatite and the carbonate from the embedding sediment (Fricke et al. 2008). If there are significant $\delta^{13}C$ differences between them, this confirms that no or only limited isotope exchange and alteration occurred and that at least a partial preservation of the original compositions is likely. Third is the preservation of expected differences in isotopic composition between ecologically and/or physiologically different taxa (Kolodny et al. 1996; Fricke & Rogers 2000; Trueman et al. 2003; Fricke & Pearson 2008). Finally, the mineralogical and chemical composition of the fossil skeletal tissues may be compared to that of fresh tissues by analyzing parameters such as apatite crystallinity, secondary mineral content, and the concentration of diagenetically incorporated trace elements such as rare earth elements (e.g., Hubert et al. 1996; Trueman 1999; Trueman & Tuross 2002; Pucéat et al. 2004; Trueman et al. 2004).

However, rare earth elements are not a good proxy for the diagenetic alteration of stable carbon and oxygen isotope signatures (e.g., Tütken et al. 2008). The same is true for apatite crystallinity (Pucéat et al. 2004). As long as the P-O and C-O bonds of the phosphate and carbonate, respectively, in the apatite are not broken, the isotope composition will not change (e.g., Fricke 2007). Nevertheless, the enamel of the sauropod teeth analyzed here has concentrations of rare earth elements one to two orders lower than sauropod bones from the same strata (Herwartz et al. 2010), indicating a less severe diagenetic alteration of the enamel than of the bones. This is further supported by the fact that the enamel has a $\delta^{18}O_{CO3}$ a few per mil higher (Tütken, pers. obs.) than the dentin from the same teeth and the bones from the same taxon. The enamel also has in most cases lower $\delta^{13}C$ values than the dentin and bone (see also Figs. 4.6, 4.8). The carbon isotope composition of enamel is generally less susceptible to diagenetic alteration than the oxygen isotope composition of enamel carbonate (Wang & Cerling 1994). Enamel $\delta^{13}C$ differences between teeth of different sauropod taxa from the same locality that underwent a similar diagenetic history suggest preservation of the original biogenic composition.

Carbon Isotope Analysis of Sauropod Skeletal Apatite, Potential Food Plants, and Living Relatives of Mesozoic Plants

SAMPLE OF SAUROPOD SKELETAL REMAINS

Long bones (mostly femora and humeri, n = 64) from 12 different sauropod taxa and teeth (n = 19) from five different sauropod taxa, mostly of Late Jurassic age, and teeth and bones of one prosauropod (*Plateosaurus*) of Late Triassic age were analyzed for their carbon isotope composition (Table 4.1). Most of the specimens come from different outcrops in either the Morrison Formation, USA, or the Tendaguru Beds, Tanzania, both of Late Jurassic age. These two rock formations have yielded a large number of sauropods and are among the richest and most famous dinosaur-bearing strata worldwide. In both formations, there were several sympatric sauropod taxa, at least five in the Tendaguru Beds and up to seven in the Morrison Formation (Dodson et al. 1980; Weishampel et al. 2004). The Tendaguru and Morrison sauropods lived in terrestrial floodplain environments characterized by a season-

Table 4.1. Sauropod Bones and Teeth Analyzed for Their $\delta^{13}C$ Values

Taxon	Skeletal tissue	n	Locality	Age
Ampelosaurus atacis	Bone	5	Espéraza, France	Late Cretaceous
Apatosaurus sp.	Bone	6	Howe-Stephens Quarry, Morrison Formation, USA	Late Jurassic
	Bone	3	Dinosaur National Monument, Morrison Formation, USA	Late Jurassic
	Bone	1	Kenton Quarry, Morrison Formation, USA	Late Jurassic
	Bone	1	Dry Mesa Quarry, Morrison Formation, USA	Late Jurassic
	Tooth	2	Howe-Stephens Quarry, Morrison Formation, USA	Late Jurassic
Barosaurus sp.	Bone	3	Tendaguru Beds, Tanzania	Late Jurassic
Brachiosaurus sp.	Bone	1	Howe-Stephens Quarry, Morrison Formation, USA	Late Jurassic
Brachiosaurus brancai	Bone	4	Tendaguru Beds, Tanzania	Late Jurassic
	Tooth	7	Tendaguru Beds, Tanzania	Late Jurassic
Camarasaurus sp.	Bone	2	Dry Mesa Quarry, Morrison Formation, USA	Late Jurassic
	Bone	2	Freezeout Hills, Morrison Formation, USA	Late Jurassic
	Bone	1	Dinosaur National Monument, Morrison Formation, USA	Late Jurassic
	Bone	1	Kenton Quarry, Morrison Formation, USA	Late Jurassic
	Tooth	2	Howe-Stephens Quarry, Morrison Formation, USA	Late Jurassic
Cetiosauriscus greppini	Bone	4	Moutier, Switzerland	Late Jurassic
Diplodocus sp.	Bone	8	Howe-Stephens Quarry, Morrison Formation, USA	Late Jurassic
	Bone	3	Dry Mesa Quarry, Morrison Formation, USA	Late Jurassic
	Tooth	2	Howe-Stephens Quarry, Morrison Formation, USA	Late Jurassic
Dicraeosaurus sp.	Bone	1	Tendaguru Beds, Tanzania	Late Jurassic
Europasaurus holgeri	Bone	2	Oker, Germany	Late Jurassic
Janenschia robusta	Bone	1	Tendaguru Beds, Tanzania	Late Jurassic
Mamenchisaurus sp.	Bone	3	Junggar Basin, China	Late Jurassic
	Tooth	4	Junggar Basin, China	Late Jurassic
Isanosaurus sp.	Bone	1	Nam Phong Formation, Thailand	Late Triassic
Plateosaurus engelhardti	Bone	4	Frick, Switzerland	Late Triassic
	Bone	7	Trossingen, Germany	Late Triassic
	Tooth	1	Frick, Switzerland	Late Triassic
	Tooth	1	Trossingen, Germany	Late Triassic

The bone specimens analyzed were taken from the cortex of sauropod long bone shafts, mostly femora and humeri. The specimens (n = number of specimens) originate mostly from different quarries in the Morrison Formation and from the Tendaguru Beds. Sauropod teeth from the Morrison Formation all come from the Howe-Stephens Quarry, whereas the *Brachiosaurus* teeth are from different microsites within the Tendaguru Beds.

ally dry climate (Aberhan et al. 2002; Engelmann et al. 2004; Rees et al. 2004; Foster 2007). However, Tendaguru was situated closer to the paleocoastline than the Morrison Formation environment. Within the Tendaguru Formation, the sauropod remains were deposited in fine-grained sediments of a coastal to tidal flat setting (Bussert et al. 2009). A few bones from *Cetiosauriscus* and *Europasaurus* from Late Jurassic marine near-shore settings of northern Switzerland and northern Germany, respectively, were also analyzed (Sander et al. 2006; Meyer & Thüring 2003). Furthermore, a few *Mamenchisaurus* bones and teeth from the terrestrial Late Jurassic sediments of the Junggar Basin in northwest China, *Ampelosaurus* bones from Late Cretaceous strata of the Pyrenean foreland in southwestern France, and one humerus of the Late Triassic early sauropod *Isanosaurus* from Thailand (Buffetaut et al. 2002) were analyzed.

METHODS

Carbon Isotope Measurement of Bones and Teeth

Bulk samples of bone, dentin, and enamel were taken with a handheld drill with diamond-studded drill bits. The sample powder (10 mg) was chemically pretreated with 2% NaOCl and 1 M calcium acetate acetic acid buffer solution, according to the methods described by Koch et al. (1997), to remove organics and diagenetic or non-lattice-bound carbonate, respectively. To measure the carbon isotope composition of the skeletal apatite, the structurally bound carbonate has to be extracted from the apatite lattice. This is achieved by the classic phosphoric acid (H_3PO_4) reaction used for carbonates following McCrea (1950), which liberates CO_2 from the sample. About 2 mg pretreated bone, dentin, or enamel powder was allowed to react with 100% H_3PO_4 for 90 minutes at 70°C with a ThermoFinnigan Gas Bench II (Spötl & Vennemann 2003). In this reaction, the acid fractionation factor between calcite and CO_2 applies. Carbon and oxygen isotope ratios of the generated CO_2 were measured in continuous flow mode in a Finnigan MAT 252 isotope ratio gas mass spectrometer at the University of Tübingen. The carbon and oxygen isotopic compositions measured were normalized to the in-house Carrara marble calcite standard that has been calibrated against the international NBS-19 calcite standard. The international phosphorite rock standard NBS 120c, treated and measured in the same way as the samples, yielded a $\delta^{13}C$ value of $-6.37 \pm 0.07‰$ (n = 27).

Carbon Isotope Measurement of Extant and Fossil Plants

The leaves of extant plants were air dried and then powdered and homogenized in a ball mill. About 150 μg of the powder was placed into tin capsules. The carbonized fossil plant specimens were treated with 1 M HCl for several hours to remove potential carbonate contamination, then rinsed several times with deionized water. A total of 50–80 μg of the pretreated material was placed into tin capsules. To measure the carbon isotope composition, the tin capsules were thermally combusted at 1,050°C in a Carlo Erba 1500 Elemental Analyzer. The released CO_2 was separated in a gas chromatographic column and then injected via a CONFLO II interface into a Finnigan MAT DeltaPlus XL continuous-flow isotope ratio gas mass spectrometer to measure the carbon isotopic composition. The reproducibility for $\delta^{13}C$ measurements was better than 0.1‰.

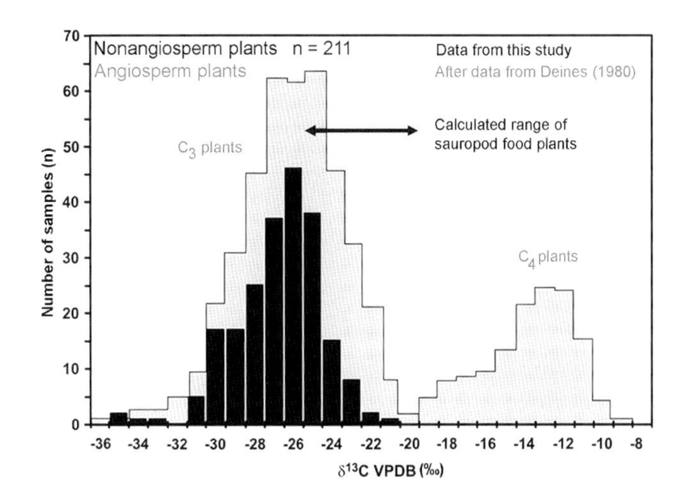

FIGURE 4.3. Histogram of $\delta^{13}C$ values of leaves in extant gymnosperms, pteridophytes, and horsetails (Table 4.2) collected from different botanical gardens in comparison to those of extant angiosperm C_3 and C_4 plants (Deines 1980). All the nonangiosperm plants analyzed have $\delta^{13}C$ values typical of C_3 plants. The $\delta^{13}C$ values calculated for the food plants ingested by the sauropods (see text for details) fall into the upper range of living C_3 plants.

Results

$\delta^{13}C$ VALUES OF EXTANT NONANGIOSPERM PLANTS

The carbon isotope composition of 240 leaf samples of extant gymnosperms (cycads, conifers, ginkgoes), pteridophytes (ferns, tree ferns), and sphenophytes (horsetails) were analyzed (Table 4.2). Plant samples were collected from different geographic settings, mostly from botanical gardens in Bonn, Germany, and Cape Town, South Africa. The $\delta^{13}C$ values of all nonangiosperm leaf samples range from −36.4 to −21.9‰ (n = 240) and cover the entire spectrum of values typical for terrestrial angiosperm C_3 plants with a mean $\delta^{13}C$ value of −27.5‰ (Tütken et al. 2007; Fig. 4.3). No $\delta^{13}C$ values characteristic of C_4 plants of approximately −12‰ were detected. However, CAM plants may be represented in this data set, especially because a few extant cycads and ferns are known to be CAM plants (Ong et al. 1986; Vovides et al. 2002).

Though all extant leaf samples have $\delta^{13}C$ values in the range of C_3 plants, differences in the carbon isotope composition exist (Fig. 4.3). The mean $\delta^{13}C$ values are highest for cycads (−26.0 ± 2.0‰) and *Araucaria* (−26.5 ± 1.6‰), other conifers have intermediate values (−27.3 ± 1.2‰), and the lowest values occur in horsetails (−28.9 ± 2.1‰), tree ferns (−29.2 ± 0.9‰), and ferns (−29.5 ± 2.4‰). The observed differences of 3.4‰ between the mean $\delta^{13}C$ values for these plant groups are much smaller than those between C_3 and C_4 plants (Fig. 4.3). Furthermore, the ranges of $\delta^{13}C$ values of the different groups overlap (Fig. 4.4). This makes a distinction between dinosaurs feeding on specific plant groups difficult because of the environmental factors discussed above that can also change plant $\delta^{13}C$ values by a few per mil (Tieszen 1991; Heaton 1999). However, there is a growing body of carbon isotope studies that have successfully determined niche partitioning and feeding behavior of herbivorous mammals in modern and ancient C_3 plant ecosystems (Cerling et al. 2004; MacFadden & Higgins 2004; Feranec & MacFadden 2006; Feranec 2007; Fricke et al. 2008; Zanazzi & Kohn 2008; Tütken & Vennemann 2009).

$\delta^{13}C$ VALUES OF CARBONIZED MESOZOIC PLANTS

In addition to the leaves of living, nonangiosperm plants, seven coalified fossil plant specimens (mostly leaves) from extinct groups such as bennettitaleans and pteridosperms from six different Jurassic and Early Cretaceous floras were analyzed for their carbon isotope composition. Furthermore, $\delta^{13}C$ values of four pieces of conifer charcoal from sauropod-bearing strata of the Howe-Stephens Quarry, Morrison Formation, Wyoming, were also analyzed (Table 4.3).

Table 4.2. Extant Gymnosperm, Pteridophyte Leaves, and Sphenophyte and Lycophyte Shoots Analyzed for Their $\delta^{13}C$ Values

Taxon	n	Type of plant
Equisetum arvense	1	Horsetail
Equisetum giganteum	1	Horsetail
Equisetum sp.	32	Horsetail
Cibotium schiedei	2	Tree fern
Cyathea brownii	2	Tree fern
Cyathea dealbata	1	Tree fern
Cyathea dregei	6	Tree fern
Cyathea sp.	1	Tree fern
Dicksonia antarctica	9	Tree fern
Dicksonia selowiana	1	Tree fern
Dicksonia squarrosa	1	Tree fern
Angiopteris evecta	1	Fern
Asplenium sp.	1	Fern
Blechnum brasiliense	1	Fern
Blechnum tabulare	3	Fern
Dacrydium cupressicum	1	Fern
Drynaria descensa	1	Fern
Drynaria quercifolia	1	Fern
Microlepia speluncae	2	Fern
Osmunda regalis	1	Fern
Phyllitis sp.	1	Fern
Platycerium sp.	1	Fern
Polypodium aureum	2	Fern
Polypodium crassifolium	1	Fern
Polypodium vulgare	1	Fern
Pteris tremula	1	Fern
Todea barbara	1	Fern
Araucaria araucana	8	Conifer
Araucaria angustofolia	35	Conifer
Araucaria bidwillii	2	Conifer
Araucaria columnaris	1	Conifer
Araucaria heterophylla	1	Conifer
Araucaria laubenfelsii	1	Conifer
Araucaria subulata	2	Conifer
Araucaria sp.	22	Conifer
Buja chilensis	1	Conifer
Cedrus sp.	2	Conifer
Chamaecyparis lawsoniana	1	Conifer
Metasequoia glyptostroboides	3	Conifer
Pinus sylvestris	1	Conifer
Podocarpus macrophyllus	1	Conifer
Pseudotsuga douglasii	1	Conifer
Sequoia sempervirens	1	Conifer
Sequoiadendron giganteum	2	Conifer
Taxodium distichum	1	Conifer
Taxus baccata	1	Conifer
Thuja plicata	1	Conifer
Torreya californica	1	Conifer
Ginkgo biloba	15	Ginkgoalean
Cycas circinalis	1	Cycad
Cycas revoluta	3	Cycad
Cycas thouarsii	1	Cycad
Cycas sp.	2	Cycad
Dioon edule	1	Cycad
Encephalartos altensteinii	4	Cycad
Encephalartos arenarius	2	Cycad
Encephalartos caffer	3	Cycad
Encephalartos friderici-guilielmi	6	Cycad
Encephalartos ghellinckii	1	Cycad
Encephalartos heenanii	1	Cycad
Encephalartos horridus	3	Cycad
Encephalartos inopinus	1	Cycad
Encephalartos lanatus	1	Cycad
Encephalartos lebomboensis	2	Cycad
Encephalartos lehmannii	2	Cycad
Encephalartos longifolius	3	Cycad
Encephalartos manikensis	1	Cycad
Encephalartos paucidentatus	1	Cycad
Encephalartos transvenosus	3	Cycad
Encephalartos trispinosus	2	Cycad
Encephalartos villosus	3	Cycad
Encephalartos sp.	1	Cycad
Stangeria eriopus	1	Cycad
Zamia pumila	1	Cycad
Ephedra distachya	1	Gnetalean
Gnetum gnemon	1	Gnetalean
Lycopodium squarrosum	1	Lycophyte
Selaginella causiana	1	Lycophyte
Selaginella pubescens	1	Lycophyte

The plant specimens analyzed were mostly collected from botanical gardens in Kirstenbosch, Cape Town, South Africa, and Bonn, Germany, during the spring.

The fossil plant samples have a mean $\delta^{13}C$ value of −23.4‰ (ranging from −25.7 to −21.4‰, n = 11). These values are similar to those measured for other coalified plant remains from the Jurassic with a mean $\delta^{13}C$ value of −24.6‰ (ranging from −26.4 to −23.4‰, n = 14; Bocherens et al. 1993). Thus, $\delta^{13}C$ values of Jurassic plants are enriched about 3‰ compared to the extant C_3 plants (Fig. 4.5). This difference between the Mesozoic and the extant plants is most likely due to changes in $\delta^{13}C$ values of atmospheric carbon dioxide (e.g., Gröcke 2002) because the experimental work described above indicates only minor changes during fossilization (e.g., DeNiro & Hastorf 1985; Schleser et al. 1999; Turney et al. 2006). Although the carbon isotope data for Mesozoic plants are still limited, I conclude that sauropods fed on plants that were enriched in ^{13}C relative to their modern counterparts, with an average $\delta^{13}C$ value of approximately −24‰ (Fig. 4.5).

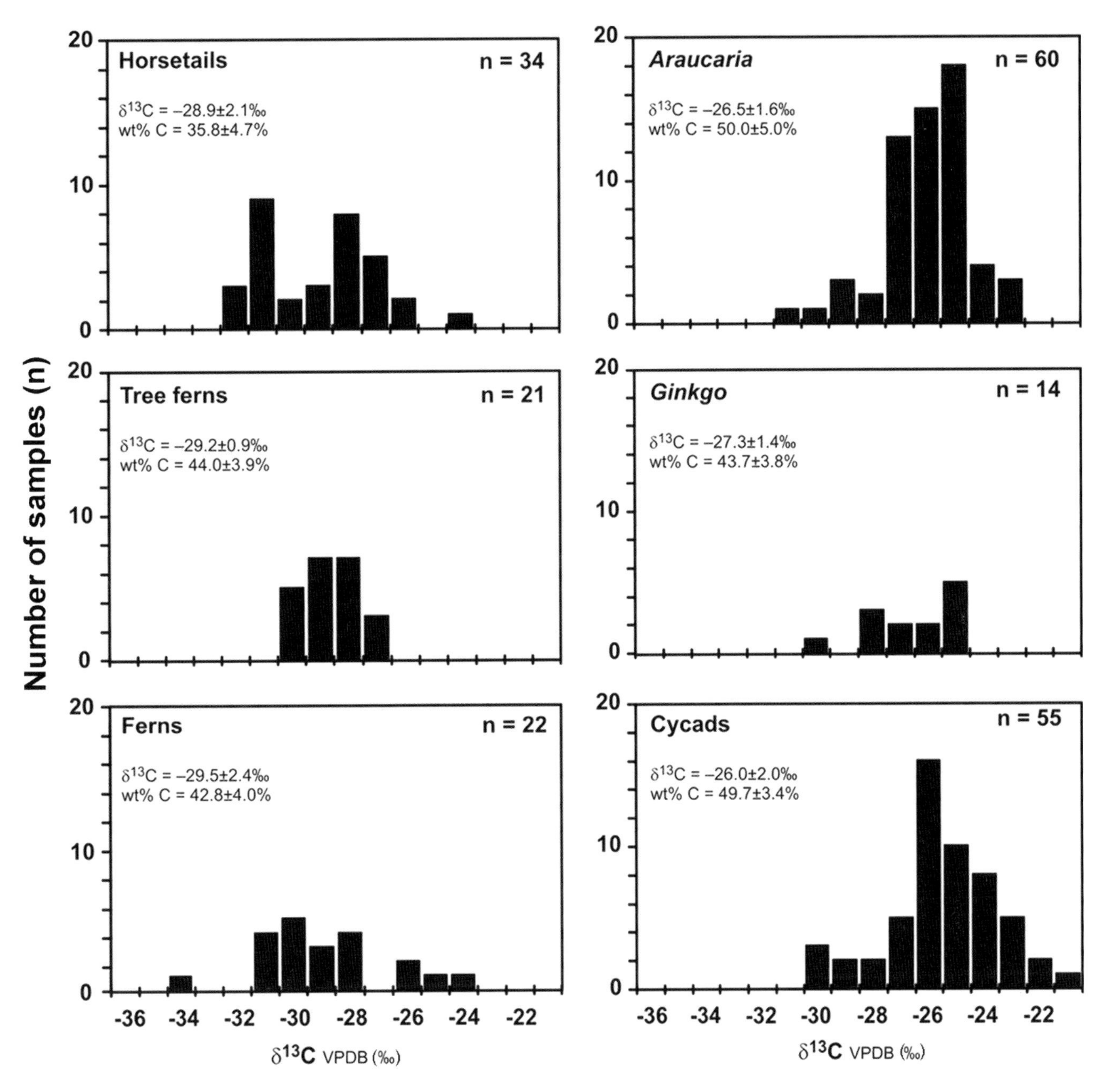

FIGURE 4.4. Histograms of the leaf $\delta^{13}C$ values of six different extant plant groups. Conifers such as *Araucaria* and *Ginkgo* as well as cycads have higher $\delta^{13}C$ values with mean values of approximately −27 to −26‰. Ferns, tree ferns, and horsetails, however, have lower $\delta^{13}C$ values with mean values of approximately −29‰. n = number of samples.

$\delta^{13}C$ VALUES OF SAUROPOD BONES AND TEETH

Bones and tooth enamel measured in this study (Table 4.1) have similar mean $\delta^{13}C$ values of −7.1 ± 1.4‰ (ranging from −10.9 to −4.7‰; n = 64) and −8.0 ± 1.2‰ (ranging from −9.1 to −4.1‰; n = 19), respectively (Fig. 4.5). All these $\delta^{13}C$ values fall within the range of terrestrial C_3 plant feeders; however, most enamel and bone $\delta^{13}C$ values are near the upper limit of other herbivores feeding on terrestrial C_3 plants (Fig. 4.6). Higher $\delta^{13}C$ apatite values characteristic of vertebrates feeding on C_4 and marine plants were not found.

In addition to the bulk bone samples, one femur bone of an adult *Apatosaurus* and one tooth of an adult *Camarasaurus* (Fig. 4.7), both from the Howe-Stephens Quarry of the Morrison Formation in Wyoming, as well as a tooth from a *Brachiosaurus* from the Tendaguru Beds, were serially sampled to determine the range of intraindividual $\delta^{13}C$ variation. The *Apatosaurus* bone yielded a mean $\delta^{13}C$ value of −7.4 ± 0.1‰ (ranging

Table 4.3. Mesozoic Plant Specimens Analyzed for Their Carbon Isotope Composition

Taxon	n	Type of plant	Locality	Age
Pinus sp.	1	Conifer	La Louvière, Belgium	Early Cretaceous
Cycadopteris cycadeoidea	1	Bennettitalean	Samogy, Hungary	Early Jurassic
undetermined conifer	4	Conifer	Howe-Stephens Quarry, USA	Late Jurassic
Podozamites feneonis	1	Conifer	Jura, France	Late Jurassic
Baiera gracilis	1	Ginkgoalean	Yorkshire, England	Middle Jurassic
Ginkgo huttonii	1	Ginkgoalean	Yorkshire, England	Middle Jurassic
Otozamites beanii	1	Bennettitalean	Yorkshire, England	Middle Jurassic
Williamsonia sp.	1	Bennettitalean	Yorkshire, England	Middle Jurassic

Coalified leaf specimens of Jurassic and Early Cretaceous age are from the collection of the Goldfuss Museum, University of Bonn. The charcoal of undetermined conifer wood comes from the sauropod-bearing strata of the Howe-Stephens Quarry, from which all the sauropod teeth of the Morrison Formation analyzed in this study originate.

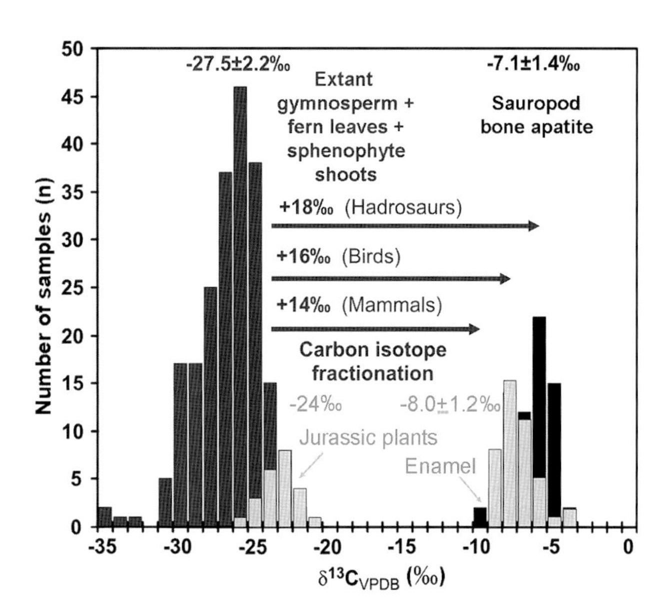

FIGURE 4.5. Histogram of $\delta^{13}C$ values of extant and Mesozoic nonangiosperms as well as sauropod bones and enamel samples. Mean $\delta^{13}C$ values of each histogram are also given. Carbon isotope fractionation between diet and hard tissues for different vertebrates is given: $\Delta\delta^{13}C_{enamel\text{-}diet}$ for large, herbivorous mammals approximately 14‰ (Cerling & Harris 1999), $\Delta\delta^{13}C_{eggshell\text{-}diet}$ for birds approximately 16‰ (von Schirnding et al. 1982), $\Delta\delta^{13}C_{enamel\text{-}diet}$ for hadrosaurs approximately 18‰ (Fricke et al. 2008). $\delta^{13}C$ data for carbonized Jurassic plant remains were generated in this study and were taken from Bocherens et al. (1993). The Jurassic plant remains have a mean $\delta^{13}C$ value about 3‰ higher than extant C_3 plants. The $\Delta\delta^{13}C_{carbonate\text{-}diet}$ fractionation between these plants remains and the enamel of sauropods is approximately 16‰, which is similar to the value for extant ostrich.

from −7.6 to −6.7‰; n = 40), and the *Camarasaurus* tooth has a $\delta^{13}C$ value of −6.8 ± 0.3‰ (ranging from −7.4 to −6.4‰; n = 25). Thus, the intratissue variability of $\delta^{13}C$ values is less than 1‰ in both sauropods. The same is true in the serially sampled *Brachiosaurus* tooth with a mean $\delta^{13}C$ value of −9.0 ± 0.1‰ (ranging from −9.2 to −8.8; n = 9). These limited data indicate only a small intraindividual $\delta^{13}C$ variability of 1‰ or less.

The enamel of the sauropod teeth analyzed has in most cases lower $\delta^{13}C$ values than dentin and the mean bone $\delta^{13}C$ value in the same taxon (Figs. 4.6, 4.8). Although a certain degree of diagenetic alteration cannot be excluded, enamel $\delta^{13}C$ values should most closely reflect dietary intake. *Brachiosaurus* from Tendaguru (−8.6 ± 0.8‰, n = 7) has the lowest enamel $\delta^{13}C$ value of all sauropods analyzed, similar to that of the Late Triassic *Plateosaurus* from Trossingen, Germany, and Frick, Switzerland (−8.5 ± 0.3‰, n = 2) (Fig. 4.8). The other sauropod taxa all have higher enamel $\delta^{13}C$ values, including *Mamenchisaurus* from the Junggar Basin in China (−7.2 ± 0.8‰, n = 4) and the sympatric sauropods from the Morrison Formation: *Diplodocus* (−7.9 ± 0.3‰, n = 2), *Camarasaurus* (−6.8 ± 0.1‰, n = 2), and *Apatosaurus* (−5.2 ± 0.5‰, n = 2), which has the highest enamel $\delta^{13}C$ value of all sauropod taxa investigated (Fig. 4.8). Because dentin is more prone to diagenesis than enamel, $\delta^{13}C$ values of dentin are always higher than enamel values (Fig. 4.8). This suggests at least a partial preservation of original enamel $\delta^{13}C$ values. Although the dentin of the teeth of different sauropods from the Morrison Formation has similar $\delta^{13}C$ values, the enamel values are different.

Implications of Carbon Isotopes for the Diet of Sauropods

Because of the small number of sauropod teeth analyzed here, the following discussion of the carbon isotope data and their implication for potential sauropod niche partitioning must be considered as preliminary. Additional enamel $\delta^{13}C$ data are needed to substantiate and refine the interpretations

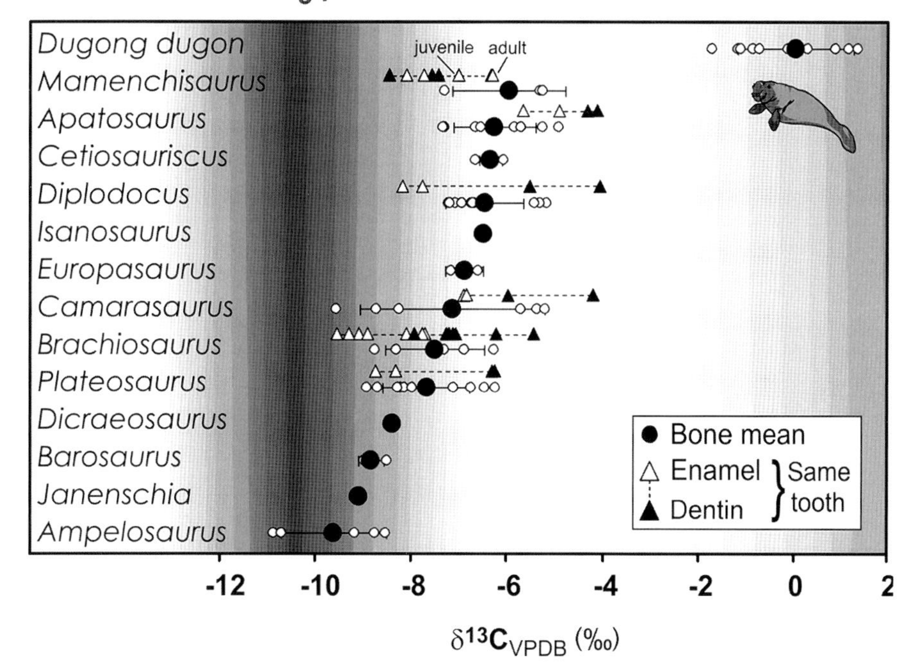

FIGURE 4.6. $\delta^{13}C$ values of apatite of bones and teeth from different sauropod and prosauropod taxa. For comparison, $\delta^{13}C$ values for rib bones of the extant sea cow (*Dugong dugon*), a marine herbivorous mammal feeding on aquatic plants, predominantly sea grass, are given. Most $\delta^{13}C$ values of the sauropod bones and teeth occur in the upper range of values expected for extant large herbivorous mammals consuming on C_3 plants. No values fall in the range of C_4-plant-like values, which suggests feeding on terrestrial C_3 plants and that marine food resources or C_4 plants were not used by sauropods. Note that enamel $\delta^{13}C$ values are always lower than the dentin values of the same tooth. Both dentin and bone $\delta^{13}C$ values are likely to be biased by diagenetic alteration toward more positive values.

about sauropod resource partitioning and feeding behavior given below.

To study the diet and feeding behavior of sauropods by analyzing the $\delta^{13}C$ values of their skeletal tissues, it is important to know the carbon isotope fractionation during metabolism of the food and incorporation into the bioapatite. The carbon isotope fractionation has been determined for extant herbivorous mammals (approximately 14‰; Cerling & Harris 1999; Passey et al. 2005) and birds (approximately 16‰; von Schirnding et al. 1982), but not for reptiles. However, a recent study of herbivorous ornithischian dinosaurs has estimated the $\Delta\delta^{13}C_{apatite\text{-}diet}$ to be approximately 18‰ (Fricke et al. 2008). To determine whether the $\Delta\delta^{13}C_{apatite\text{-}diet}$ values for ornithischian dinosaurs also applies to saurischian dinosaurs, a $\Delta\delta^{13}C_{apatite\text{-}diet}$ value for sauropods was estimated. This was done by using the mean $\delta^{13}C$ value of enamel (−8.0‰) and that of carbonized Jurassic plant remains (−24‰), yielding a $\Delta\delta^{13}C_{apatite\text{-}diet}$ value of approximately 16‰ (Fig. 4.5). This value is identical to the $\Delta\delta^{13}C_{calcite\text{-}diet}$ value found for ostrich eggshells (von Schirnding et al. 1982; Johnson et al. 1998). However, this value is 2‰ lower than the $\Delta\delta^{13}C_{apatite\text{-}diet}$ value of 18‰ (Fricke et al. 2008). The reason for this 2‰ difference is unclear. One explanation might be that the limited number of Jurassic plant remains used to calculate the $\Delta\delta^{13}C_{apatite\text{-}diet}$ value are not from the same strata as the sauropod teeth. Thus, their $\delta^{13}C$ values may not exactly represent those of the food plants ingested by the sauropods.

Although diagenesis might have altered the original $\delta^{13}C$ values of sauropod enamel, this likely does not apply to fossil plant tissue, as discussed above. If we assume that the 2‰ difference between the $\Delta\delta^{13}C_{apatite\text{-}diet}$ value of ornithischian and saurischian dinosaurs (sauropods) is real, it might reflect differences in biochemical processes during metabolism and digestion of the food plants, such as the rate of methane production, which have an important influence on the $\Delta\delta^{13}C_{apatite\text{-}diet}$ value (Passey et al. 2005). Fricke et al. (2008) speculated that the herbivorous hadrosaurs might have had higher rates of methane production and thus degassing of ^{12}C-rich methane from the gastrointestinal tract, which is responsible for their higher $\Delta\delta^{13}C_{apatite\text{-}diet}$ value compared to large herbivorous mammals. If this hypothesis is correct, then the lower, bird-like $\Delta\delta^{13}C_{apatite\text{-}diet}$ value of 16‰ determined for sauropods in this study might suggest that sauropods had intermediate rates of methane production compared to large herbivorous mammals and hadrosaurs. Assuming that a constant $\Delta\delta^{13}C_{apatite\text{-}diet}$ value of 16‰ applies to all sauropod taxa, the $\delta^{13}C$ values of their ingested food plants can be calculated as −25.1 to −20.1‰ and −26.7 to −20.7‰ on the basis of the enamel and bone samples, respectively. All sauropod enamel and bone $\delta^{13}C$ values are consistent with those expected for terrestrial herbivores purely feeding on C_3 plants (Fig. 4.6). The calculated $\delta^{13}C$ values of sauropod food plants fall within the range measured for extant gymnosperm and fern foliage and sphenophyte shoots but lie in the upper range of extant C_3 plants (Fig. 4.3). This may be due to higher $\delta^{13}C_{CO2}$ values in the Mesozoic atmosphere (e.g., Gröcke 2002) and/or water-stressed environments, which would agree with the seasonally dry settings currently being proposed for the Morrison Formation and Tendaguru Beds (Aberhan et al. 2002; Engelmann et al. 2004; Rees et al. 2004).

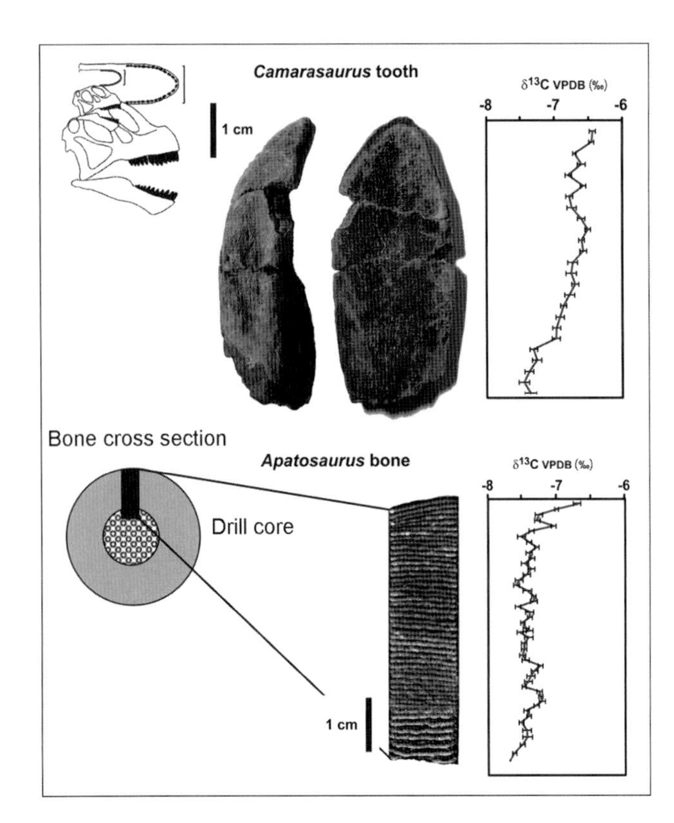

FIGURE 4.7. Intratooth and intrabone variation of the carbon isotope composition of a *Camarasaurus* tooth and an *Apatosaurus* femur bone, respectively. The grooves on the bone section are the sample marks. The enamel of the *Camarasaurus* tooth was serially sampled every 2 mm in a fashion similar to sampling of the bone. The top photo shows the tooth before sampling. Both skeletal elements are from the Late Jurassic Morrison Formation at the Howe-Stephens Quarry, Wyoming.

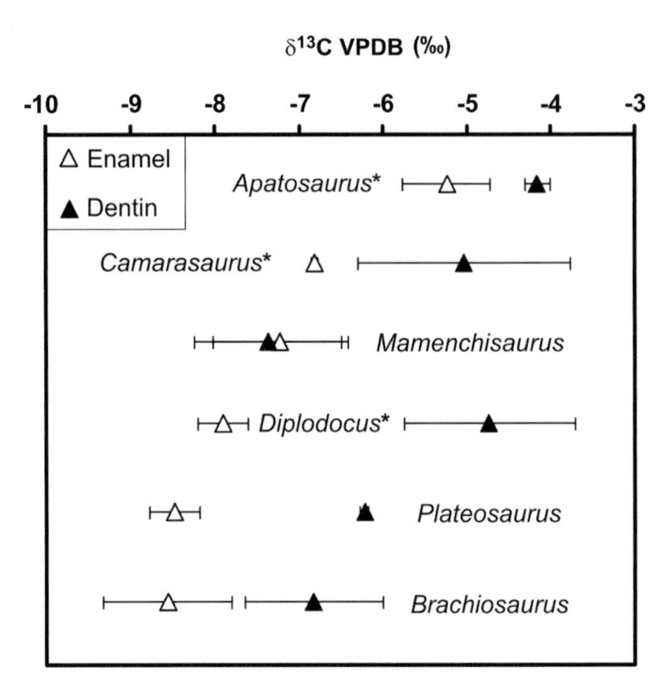

FIGURE. 4.8. Mean $\delta^{13}C$ values of sauropod and prosauropod enamel and dentin samples. Sauropods marked with an asterisk are from the Howe-Stephens Quarry, Wyoming (Morrison Formation). They were likely sympatric and thus lived and died in the same environmental setting. Differences in enamel $\delta^{13}C$ values between these taxa may therefore reflect the ingestion of isotopically distinct food plants and thus a certain niche partitioning. Note that the number of teeth (n = 2 to 4) per taxon analyzed is small, which means that these results are preliminary.

SAUROPOD DIET AND NICHE PARTITIONING

Sauropod diets so far have been inferred mostly from tooth morphology, dental wear, and neck posture and resulting feeding envelope, as well as digestibility and abundance of potential food plants (Bakker 1978; Weaver 1983; Fiorillo 1998; Stevens & Parrish 1999, 2005a, 2005b; Upchurch & Barrett 2000; Hummel et al. 2008; Christian & Dzemski, this volume; Gee, this volume; Hummel & Clauss, this volume). Despite different neck lengths and feeding envelopes, there is a considerable overlap in browse height among the different sauropod genera (Fig. 4.9). Nevertheless, high-browsing taxa such as *Brachiosaurus* and *Camarasaurus* (Fig. 4.9) were potentially able to feed on all arborescent gymnosperms that may have been out of reach of low-browsing taxa such as *Dicraeosaurus* and *Diplodocus*.

Because the differences of mean enamel $\delta^{13}C$ values between sauropod taxa (Fig. 4.8) are larger than intraspecific variability (Fig. 4.7), $\delta^{13}C$ values may be used to study preferential feeding on isotopically distinct food plants. Especially the comparison of enamel $\delta^{13}C$ data from sympatric sauropods can yield information about niche partitioning. All teeth analyzed from *Diplodocus, Apatosaurus,* and *Camarasaurus* (Fig. 4.8) come from the Howe-Stephens Quarry, Wyoming (cf. Ayer 1999). Thus, presumably these sauropods lived under the same environmental conditions. Assuming a similar physiology and carbon isotope fractionation, the differences in enamel $\delta^{13}C$ values (Fig. 4.8) likely reflect feeding on plants or parts thereof with different $\delta^{13}C$ values, and thus imply a certain niche partitioning between these sympatric sauropods.

Diplodocus

Diplodocus has the lowest mean enamel $\delta^{13}C$ value, indicating the ingestion of food plants with lower $\delta^{13}C$ values than the other two sympatric sauropods. On the basis of the differences in mean $\delta^{13}C$ values between modern gymnosperms, pteridophytes, and sphenophytes (Fig. 4.4), *Diplodocus* might have preferentially fed on ferns and/or horsetails, which have lower $\delta^{13}C$ values than that of conifers and cycads (Fig. 4.4). Such a low-browsing behavior of *Diplodocus* also accords with

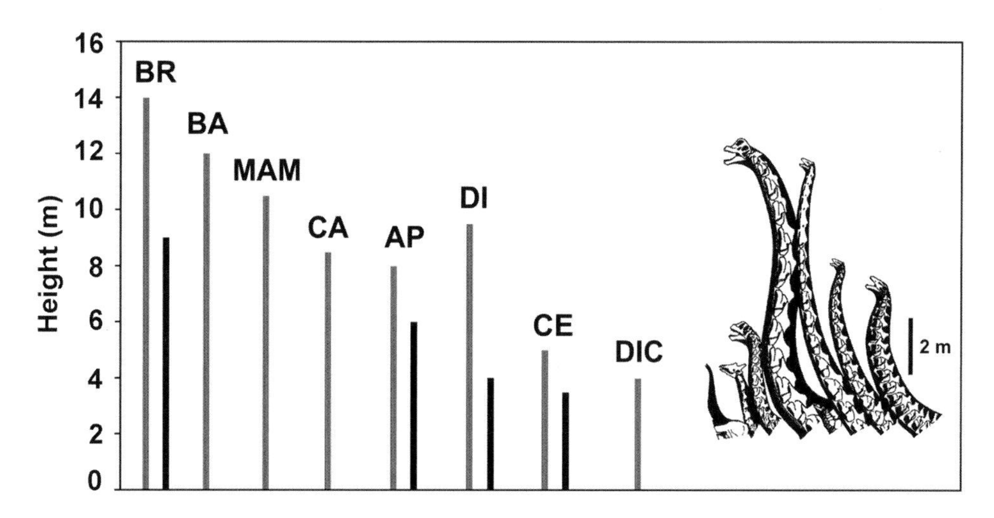

FIGURE 4.9. Vertical feeding ranges of different sauropod taxa in comparison to a giraffe and an elephant. BR, *Brachiosaurus;* BA, *Barosaurus;* MAM, *Mamenchisaurus;* CA, *Camarasaurus;* AP, *Apatosaurus;* DI, *Diplodocus;* CE, *Cetiosauriscus;* DIC, *Dicraeosaurus*. Sauropod necks modified after Paul (1998). Although the exact feeding heights and envelopes of the long-necked sauropods are still a matter of debate (Stevens & Parrish 1999, 2005a, 2005b; Christian & Dzemski, this volume), certain taxa such as *Brachiosaurus, Barosaurus,* and *Mamenchisaurus* were probably capable of high browsing on arborescent plants while other taxa were more likely low browsers, for example, *Dicraeosaurus. Modified after data from Upchurch & Barrett (2000) and Stevens & Parrish (2005a).*

the vertical feeding range of 4 m or less, the horizontal neck posture, and the ventral inclination of its skull as an adaptation for downward feeding, which allowed the animal to feed even below ground level (Stevens & Parrish 1999, 2005a). However, dorsoventral neck flexibility of *Diplodocus* was larger than suggested by Stevens & Parrish (1999) allowing it to raise the head well above the height of the shoulders and to utilize different neck postures during feeding (Dzemski & Christian 2007; Christian & Dzemski, this volume). Lacustrine feeding on water plants in *Diplodocus,* as suggested by Stevens & Parrish (2005a), does not seem to be supported by the limited $\delta^{13}C$ data. Feeding on ferns is more plausible because ferns may have been an abundant ground cover in the ecosystem of the Morrison Formation (e.g., Taggart & Cross 1997). The dental wear of *Diplodocus* suggests unilateral branch stripping as feeding mechanism, and thus may represent a more selective food-gathering strategy (Barrett & Upchurch 1994).

Camarasaurus

The mean enamel $\delta^{13}C$ value of *Camarasaurus* is intermediate between that of *Diplodocus* and *Apatosaurus* and might indicate a more mixed or generalistic feeding strategy and ingestion of a broader range of plants. Such an interpretation is in agreement with the large spatulate teeth of *Camarasaurus* (Fig. 4.7), which enable grasping of coarser food plants and limited oral food processing. This is also supported by the variable dental microwear patterns (Fiorillo 1991, 1998), which indicate a varied diet. Finally, on the basis of its neck length and posture, *Camarasaurus* is interpreted to be a high browser with a large vertical feeding range of approximately 8.5 m (Upchurch & Barrett 2000; Fig. 4.9). This allowed for feeding in the canopy of conifers, which have higher leaf $\delta^{13}C$ values than ferns and horsetails (Fig. 4.4). The intermediate enamel $\delta^{13}C$ values of *Camarasaurus* probably reflect both feeding on plants with low $\delta^{13}C$ values such as ferns and horsetails, and on plants with high $\delta^{13}C$ values such as *Araucaria,* other conifers, and cycads. Thus, *Camarasaurus* had a higher dietary flexibility compared to other sauropods.

Brachiosaurus

Brachiosaurus brancai from the Tendaguru Beds has the lowest mean enamel $\delta^{13}C$ value of all Late Jurassic sauropods analyzed (Fig. 4.8). The estimated vertical feeding range ($\geq$9 m) of *Brachiosaurus* is the largest for all sauropods (Fig. 4.9). Thus, *Brachiosaurus* is an undisputed high browser that probably fed on conifers such as Podocarpaceae, Cheirolepidoiaceae, and Araucariaceae (Fig. 4.10), which were abundant in the coastal plain depositional environment of the Tendaguru Beds (Schrank 2010). However, because *Araucaria* leaves have higher $\delta^{13}C$ values than other plant foliage (Fig. 4.4), increased enamel $\delta^{13}C$ values would be expected if *Araucaria* or other conifers were the main food plants. Unfortunately, so far, no enamel $\delta^{13}C$ data of other sympatric sauropods from the Tendaguru Beds—that is, from low-browsing *Dicraeosaurus*—could be measured for comparison. It may be possible that enamel $\delta^{13}C$ values of *Brachiosaurus* are low despite feeding on conifer foliage with lower $\delta^{13}C$ values as a result of the more mesic environment in the coastal plain of Tendaguru, resulting in lower plant $\delta^{13}C$ values compared to the more continental, in part seasonally dry setting of the Morrison Formation. The small enamel $\delta^{13}C$

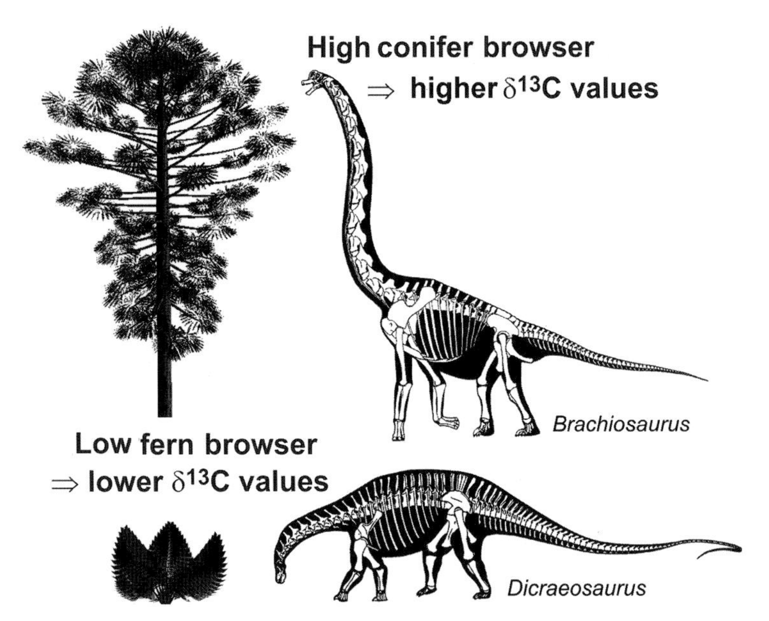

FIGURE 4.10. Schematic drawing of the influence of different vertical feeding ranges on the feeding strategy and niche partitioning in sauropods. On the basis of the differences in $\delta^{13}C$ values of extant ferns, sphenophytes, and gymnosperm foliage (Fig. 4.4) as well as on the so-called canopy effect (van der Merwe & Medina 1991), low (fern) browsing may result in low $\delta^{13}C$ values of skeletal apatite, while high (conifer) browsing may lead to higher $\delta^{13}C$ values. *Sauropod reconstructions are taken from Paul (1998), and plant drawings are from Hinz (2010).*

intratooth variation (<0.4‰) for one *Brachiosaurus* tooth is in accordance with feeding on plants with similar $\delta^{13}C$ values that possibly suffered little water stress. The alternative interpretation would be that *Brachiosaurus,* despite its high-browsing capability, fed on nonarborescent plants with low $\delta^{13}C$ values such as ferns and horsetails (Fig. 4.4), or even on freshwater plants (Fig. 4.1), although moisture-loving plants such as ferns and tree ferns growing in the understory of dense, closed forests would have been much less accessible to fully grown individuals of *Brachiosaurus* (Gee, this volume).

SEASONAL OR ONTOGENETIC SHIFTS IN THE DIET

The analysis of intrabone and intratooth $\delta^{13}C$ profiles (Fig. 4.1) does not reveal significant seasonal shifts in feeding strategy even though the investigated sauropods may have lived in seasonal environments (Aberhan et al. 2002; Engelmann et al. 2004; Rees et al. 2004). Dietary shifts during ontogeny are likely to have occurred because young sauropods had much smaller feeding envelopes than adult individuals (Fig. 4.9) and were thus restricted to low browsing. They probably had to compete for low-growing vegetation with other sympatric non-sauropod herbivorous dinosaurs and low-browsing adult sauropods before attaining their giant adult size and the ability, at least for some taxa, to exploit the high-browsing niche. Ontogenetic dietary changes can only be recorded in bones of sauropods that grew at high growth rates over several years to decades (Curry 1999; Sander & Tückmantel 2003; Klein & Sander 2008).

However, the carbon isotope composition of bone is prone to diagenetic alteration (e.g., Kohn & Law 2006) and may not reflect the original dietary composition. Therefore, the intrabone variation of the $\delta^{13}C$ values from the *Apatosaurus* bone (Fig. 4.7) must be interpreted with caution because original heterogeneities might have been reduced and/or absolute values may have shifted as a result of diagenetic alteration. Nevertheless, such a high-resolution intrabone $\delta^{13}C$ profile of a predominantly nonremodeled, fibrolamellar bone is the only way to look at a long-term, multiple year dietary record of an individual sauropod. Though likely altered, the small intrabone $\delta^{13}C$ variability seems to suggest no significant change of the $\delta^{13}C$ values of the ingested diet.

A better way to infer potential ontogenetic shifts in the diet of sauropods is the analysis of enamel from teeth of juvenile and adult individuals. However, dinosaurs constantly replace their teeth throughout their lifetime (Erickson 1996; D'Emic et al. 2009). Sauropods generally have fast tooth replacement rates of about one to two months, although tooth formation takes longer in *Camarasaurus*—several months to a year (D'Emic et al. 2009). Therefore, sauropod teeth only record short-term (from a few months to a year) dietary signals in their chemical composition and dental wear. Fiorillo (1998) found different microwear patterns for juvenile and adult *Camarasaurus* teeth and interpreted these as an indication of an ontogenetic shift in diet. Intratooth variability of enamel $\delta^{13}C$ values of a *Camarasaurus* tooth is ≤1‰ (Fig. 4.7). Thus, no major shift in the carbon isotope composition of the diet of this individual occurred during the tooth formation period of several months (D'Emic et al. 2009). Enamel $\delta^{13}C$ values of teeth from a juvenile and an adult individual of *Mamenchisaurus* differ by only 0.7‰ (Fig. 4.6). Thus, no significant shift in the carbon isotope composition of the diet occurred during the ontogeny of this very long-necked sauropod. However, on the basis of these very limited $\delta^{13}C$ data, it is not possible to infer whether sauropods changed their diets during ontogeny.

SAUROPODS FEEDING ON AQUATIC PLANTS?

Marine Feeding?

Sauropod tracks are often found in coastal environments and on tidal flats (e.g., Dodson 1990; Meyer 1993; Lockley 1994; Meyer & Thüring 2003; Wright 2005). Some sauropods, such as *Europasaurus* and *Magyarosaurus,* seem to have lived on islands (Sander et al. 2006; Stein et al. 2010). In the absence of sufficient terrestrial plants, these sauropods presumably had to, at least on a short-term basis, rely on marine food resources such as algae. Because marine C_3 plants and algae have higher $\delta^{13}C$ values than terrestrial C_3 plants (Fig. 4.1), animals feeding on marine food resources have about 7‰ higher $\delta^{13}C$ values than terrestrial animals (Chrisholm et al. 1982).

Consumption of marine food resources should therefore be recorded in the skeletal apatite by increased $\delta^{13}C$ values compared to terrestrial C_3 plant feeders. The extant herbivorous sea cow *Dugong dugon,* for example, has significantly higher $\delta^{13}C$ values than all sauropod skeletal remains (Fig. 4.6). Only the ingestion of three dietary resources could generate such high skeletal apatite $\delta^{13}C$ values: (1) terrestrial C_4 grasses, (2) sea grass, and (3) certain marine algae (Cloern et al. 2002; Clementz et al. 2007). C_4 plants evolved in the Tertiary (Thomasson et al. 1986; Cerling et al. 1997), and thus, no C_4 grasses were present in Mesozoic ecosystems; sea grasses did not evolve before the Late Cretaceous (van der Ham et al. 2007). Therefore, only marine algae with C_4-like $\delta^{13}C$ values (Fig. 4.1) remain as a possible near-coastal food resource for Jurassic sauropods.

To test the hypothesis of sauropods that used marine food resources, bones from Late Jurassic sauropods that lived in near-coastal settings, such as *Cetiosauriscus* from Moutier, Switzerland, and *Europasaurus holgeri* from Oker, Germany (Sander et al. 2006), were analyzed for their carbon isotope composition and compared to those of sauropods from terrestrial settings (e.g., the Morrison Formation). Despite being fossilized in marine near-coastal settings with high $\delta^{13}C$ values of the embedding sediments (Tütken, pers. obs.), the *Europasaurus* and *Cetiosauriscus* bone specimens do not have the increased $\delta^{13}C$ values typical of marine feeders, but fall well within the range of other sauropods from terrestrial settings (Fig. 4.6). Thus, they did not ingest significant amounts of marine plants, and the bone apatite did not equilibrate with the carbonate from the embedding marine sediment with high $\delta^{13}C$ values. Similarly, the low $\delta^{13}C$ values of the *Brachiosaurus* fossils indicate that this animal did not ingest marine food plants, although the Tendaguru locality was near the Jurassic coastline. On the basis of skeletal apatite $\delta^{13}C$ values, the consumption of significant amounts of marine plants by sauropods can thus be excluded.

Freshwater Feeding?

The diplodocids *Apatosaurus* and *Diplodocus* had sufficient neck flexibility to reach below ground level (Stevens & Parrish 1999, 2005a; Dzemski & Christian 2007; Christian & Dzemski, this volume) to feed on water plants. Because the $\delta^{13}C$ values of freshwater and terrestrial C_3 plants largely overlap (Fig. 4.1), a differentiation of feeding on terrestrial versus freshwater C_3 plants is difficult. Some lacustrine plants, however, have lower $\delta^{13}C$ values than terrestrial plants (Fig. 4.1). Therefore, sauropods feeding on freshwater plants may have lower apatite $\delta^{13}C$ values than terrestrial feeders. Freshwater plants were probably restricted to perennial rivers (Parrish et al. 2004) and may not have been an abundant food source in the Morrison Formation. Furthermore, the enamel $\delta^{13}C$ values of *Diplodocus* teeth are well within the range of terrestrial C_3 plant feeders (Fig. 4.6). Therefore, aquatic feeding does not seem to have been of major importance in *Diplodocus.* The same is true for *Apatosaurus,* which has even higher enamel $\delta^{13}C$ values than *Diplodocus* (Fig. 4.8).

Conclusions

The carbon isotope composition of living nonangiosperm plants (gymnosperms, pteridophytes, and horsetails), which represent the relatives of potential dinosaur food plants, is similar to angiosperm C_3 plants. Thus, the effects of the paleofloral change on sauropod feeding behavior and niche shifting toward an angiosperm-dominated ecosystem during the Cretaceous cannot be detected by means of carbon isotope analysis. However, the foliage of extant *Araucaria* and other conifers as well as cycads have, on average, about 3‰ higher $\delta^{13}C$ values than ferns and horsetails. On the basis of these leaf carbon isotope data, low (fern) browsing sauropods are expected to have lower apatite $\delta^{13}C$ values than high (conifer) browsing and/or low (cycad) browsing taxa.

Carbon isotope compositions of their bones and tooth enamel suggest that sauropods fed predominantly, if not exclusively, on terrestrial C_3 plants. A certain degree of niche partitioning is likely because the different sauropods display intertaxon differences in mean enamel and bone $\delta^{13}C$ values of about 3‰, while, in contrast, intraspecific and intraindividual differences are less than 1‰. Thus, presumably no significant ontogenetic shift of the diet occurred. Sauropods did not ingest a significant portion of plants with high C_4-plant-like $\delta^{13}C$ values, such as marine vascular plants and algae. However, the consumption of freshwater aquatic plants cannot be excluded. Sauropods fed on C_3 plants with $\delta^{13}C$ values increased a few per mil compared to modern C_3 plants. The higher plant $\delta^{13}C$ values are most likely due to increased atmospheric $\delta^{13}C_{CO2}$ values and/or feeding on water-stressed C_3

plants in drier settings. Beyond this, pinpointing exactly which food plants were consumed by sauropods by means of skeletal apatite $\delta^{13}C$ values is difficult.

On the basis of the difference between mean $\delta^{13}C$ values of carbonized Jurassic plant remains and sauropod enamel, a $\Delta\delta^{13}C_{apatite\text{-}diet}$ value of 16‰ was determined for sauropods. This value is higher than for large herbivorous mammals but similar to that of birds; however, it is lower than that of ornithischian dinosaurs (Fricke et al. 2008). This suggests that sauropods may have had a different digestive physiology and carbon isotope fractionation compared to large herbivorous mammals and hadrosaurs, possibly as a result of intermediate rates of methane production during digestion.

Acknowledgments

Special thanks must go to Hans-Jakob Siber (Sauriermuseum Aathal), Nicole Klein (then Museum für Naturkunde, Berlin), and Daniela Schwarz-Wings (Museum für Naturkunde, Berlin) for generously supplying skeletal remains and for permission to sample sauropod teeth for isotope analysis. Furthermore, I thank Martin Sander (University of Bonn) for the supply of all the sauropod bone samples from the Morrison Formation and the Tendaguru Beds, for the *Europasaurus* specimens, and for discussions about sauropod paleobiology; Oliver Wings (Museum für Naturkunde, Berlin) for the supply of *Mamenchisaurus* specimens; Christian Meyer (Natural History Museum, Basel) for access to *Cetiosauriscus* and *Plateosaurus* bone material; and Rainer Schoch (State Museum of Natural History, Stuttgart) for the sampling of a skull and postcranial material of *Plateosaurus*. The botanical gardens of Bonn and Cape Town as well as the Cologne Zoo are acknowledged for permission to sample plant leaves for carbon isotope analysis. Jürgen Hummel and Carole Gee (both University of Bonn) are acknowledged for kindly supplying leaf samples of extant gymnosperms, pteridophytes, and horsetails and for some Jurassic charcoal and fossil plant remains, and for discussions on Mesozoic plants. Further, I want to thank Sebastian Viehmann (University of Bonn) for sampling and chemical pretreatment of sauropod teeth and Bernd Steinhilber (University of Tübingen) for carbon isotope measurements. Finally, Juliane Hinz (University of Tübingen) is acknowledged for the graphic reconstructions of *Araucaria* and the fern *Osmunda*. This chapter benefited from the critical comments of an anonymous reviewer and Carole Gee. This research was funded by the Deutsche Forschungsgemeinschaft (grant TU 148/1-1) and the Emmy Noether Program (grant TU 148/2-1). This is contribution number 63 of the DFG Research Unit 533 "Biology of the Sauropod Dinosaurs: The Evolution of Gigantism."

References

Aberhan, M., Bussert, R., Heinrich, W.-D., Schrank, E., Schultka, S., Sames, B., Kriwet, J. & Kapilima, S. 2002. Palaeoecology and depositional environments of the Tendaguru Beds (Late Jurassic to Early Cretaceous, Tanzania).—*Mitteilungen aus dem Museum für Naturkunde in Berlin, Geowissenschaftliche Reihe* 5: 19–44.

Ambrose, S. H. & Norr, L. 1993. Experimental evidence for the relationship of the carbon isotope ratios of whole diet and dietary protein to those of bone collagen and carbonate. *In* Lambert, J. B. & Grupe, G. (eds.). *Prehistoric Human Bone—Archaeology at the Molecular Level*. Springer Verlag, Berlin: pp. 1–37.

Arens, N. C., Jahren A. H. & Amundson R. 2000. Can C_3 plants faithfully record the carbon isotopic composition of atmospheric carbon dioxide?—*Paleobiology* 26: 137–164.

Ayer, J. 1999. *The Howe Ranch Dinosaurs*. Sauriermuseum Aathal, Switzerland.

Ayliffe, L. K., Chivas, A. R. & Leakey, M. G. 1994. The retention of primary oxygen isotope compositions of fossil elephant skeletal phosphate.—*Geochimica et Cosmochimica Acta* 58: 5291–5298.

Bakker, R. T. 1971. Ecology of the brontosaurs.—*Nature* 229: 172–174.

Bakker, T. B. 1978. Dinosaur feeding behaviour and the origin of flowering plants.—*Nature* 274: 661–663.

Barrett, P. M. & Upchurch, P. 1994. Feeding mechanisms of *Diplodocus*.—*Gaia* 10: 195–204.

Barrett, P. M. & Upchurch, P. 2005. Sauropod diversity through time: macroevolutionary and paleoecological implications. *In* Curry Rogers, K. A. & Wilson, J. A. (eds.). *The Sauropods: Evolution and Paleobiology*. University of California Press, Berkeley: pp. 125–156.

Barrett, P. M. & Willis, K. J. 2001. Did dinosaurs invent flowers? Dinosaur–angiosperm coevolution revisited.—*Biological Reviews* 76: 411–447.

Bender, M. M. 1971. Variations in the $^{13}C/^{12}C$ ratios of plants in relation to the pathway of photosynthetic carbon dioxide fixation.—*Phytochemistry* 10: 1239–1244.

Bocherens, H., Friis, E. M., Mariotti, A. & Pedersen, K. R. 1993. Carbon isotopic abundances in Mesozoic and Cenozoic fossil plants: palaeoecological implications.—*Lethaia* 26: 347–358.

Botha, J., Lee-Thorp, J. & Chinsamy, A. 2005. The palaeoecology of the nonmammalian cynodonts *Diademodon* and *Cynognathus* from the Karoo Basin of South Africa, using stable light isotope analysis.—*Palaeogeography, Palaeoclimatology, Palaeoecology* 223: 303–316.

Buffetaut, E., Suteethorn, V., Le Loeuff, J., Cuny, C., Tong, H. & Khansubha, S. 2002. The first giant dinosaurs: a large sauropod from the Late Triassic of Thailand.—*Comptes Rendue Palevol* 1: 103–109.

Boutton, T. W. 1991. Stable carbon isotope ratios of natural materials: II. Atmospheric, terrestrial, marine, and freshwater environments. *In* Coleman, D. C. & Fry, B. (eds.). *Carbon Isotope Techniques*. Academic Press, New York: pp. 173–185.

Bussert, R., Heinrich, W.-D. & Aberhan, M. 2009. The Tendaguru Formation (Late Jurassic to Early Cretaceous, southern Tanzania): definition, palaeoenvironments, and sequence stratigraphy.—*Fossil Record* 12: 141–174.

Cerling, T. E. & Harris, J. M. 1999. Carbon isotope fractionation between diet and bioapatite in ungulate mammals and implications for ecological and paleoecological studies.—*Oecologica* 120: 347–363.

Cerling, T. E., Wang, Y. & Quade, J. 1993. Expansion of C_4 ecosystems as an indicator of global ecological change in the late Miocene.—*Nature* 361: 344–345.

Cerling, T. E., Harris, J. M., MacFadden, B. J., Leakey, M. G., Quade, J., Eisenmann, V. & Ehleringer, J. R. 1997. Global vegetation change through the Miocene/Pliocene boundary.—*Nature* 389: 153–158.

Cerling, T. E., Hart, J. A. & Hart, T. B. 2004. Stable isotope ecology in the Ituri Forest.—*Oecologia* 138: 5–12.

Chrisholm, B. S., Nelson D. E. & Schwarz, H. P. 1982. Stable-carbon isotope ratios as a measure of marine versus terrestrial protein in ancient diet.—*Science* 216: 1131–1132.

Christian, A. & Dzemski, G. This volume. Neck posture in sauropods. *In* Klein, N., Remes, K., Gee, C. T. & Sander, P. M. (eds.). *Biology of the Sauropod Dinosaurs: Understanding the Life of Giants*. Indiana University Press, Bloomington: pp. 251–260.

Clementz, M. T., Goswami, A., Gingerich, P. D. & Koch, P. L. 2006. Isotopic records from early whales and sea cows: contrasting patterns of ecological transition.—*Journal of Vertebrate Paleontology* 26: 355–370.

Clementz, M., Koch, P. L. & Beck, C. A. 2007. Diet induced differences in carbon isotope fractionation between sirenians and terrestrial ungulates.—*Marine Biology* 151: 1773–1784.

Cloern, J. E., Canuel, E. A. & Harris, D. 2002. Stable carbon and nitrogen isotope composition of aquatic and terrestrial plants of the San Francisco Bay estuarine system.—*Limnology and Oceanography* 47: 713–729.

Coe, M. J., Dilcher, D. L., Farlow, J. O., Jarzen, D. H. & Russell, D. A. 1987. Dinosaurs and land plants. *In* Friis, E. M., Chaloner, W. G. & Crane, P. R. (eds.). *The Origins of Angiosperms and Their Biological Consequences*. Cambridge University Press, Cambridge: pp. 225–258.

Cojan, I., Renard, M. & Emmanuel, L. 2003. Palaeoenvironmental reconstruction of dinosaur nesting sites based on a geochemical approach to eggshells and associated palaeosols (Maastrichtian, Provence Basin, France).—*Palaeogeography, Palaeoclimatology, Palaeoecology* 191: 111–138.

Coplen, T. B. 1994. Reporting of stable hydrogen, carbon, and oxygen isotopic abundances.—*Pure Applied Chemistry* 66: 273–276.

Curry, K. A. 1999. Ontogenetic histology of *Apatosaurus* (Dinosauria: Sauropoda): new insights on growth rates and longevity.—*Journal of Vertebrate Paleontology* 19: 654–665.

Decker, J. E. & de Wit, M. J. 2006. Carbon isotope evidence for CAM photosynthesis in the Mesozoic.—*Terra Nova* 18: 9–17.

Deines, P. 1980. The isotopic composition of reduced organic carbon. *In* Fritz, P. & Fontes, J. C. (eds.). *Handbook of Environmental Isotope Geochemistry*. Elsevier, New York: pp. 331–406.

DeNiro, M. J. & Epstein, S. 1978. Influence of diet on the distribution of carbon isotopes in animals.—*Geochimica et Cosmochimica Acta* 42: 495–506.

DeNiro, M. J. & Epstein, S. 1981. Influence of diet on the distribution of nitrogen isotopes in animals.—*Geochimica et Cosmochimica Acta* 45: 341–351.

DeNiro, M. J. & Hastorf, C. A. 1985. Alteration of $^{15}N/^{14}N$ and $^{13}C/^{12}C$ ratios of plant matter during the initial stages of diagenesis: studies utilising archaeological specimens from Peru.—*Geochimica et Cosmochimica Acta* 49: 97–115.

D'Emic, M., Wilson, J. & Fisher, D. 2009. The evolution of tooth replacement rates in sauropod dinosaurs.—*Journal of Vertebrate Paleontology* 29: 84A.

Dodson, P. 1990. Sauropod paleoecology. *In* Weishampel, D. B., Dodosn, P. & Osmólka, H. (eds.). *The Dinosauria*. University of California Press, Berkeley: pp. 402–407.

Dodson, P., Behrensmeyer, A. K., Bakker, R. T. & McIntosh, J. S. 1980. Taphonomy and paleoecology of the dinosaur beds of the Jurassic Morrison Formation.—*Paleobiology* 6: 208–232.

Dzemski, G. & Christian, A. 2007. Flexibility along the neck of the ostrich (*Struthio camelus*) and consequences for the reconstruction of dinosaurs with extreme neck length.—*Journal of Morphology* 268: 701–714.

Ehleringer, J. R., Field, C. B., Lin, Z. F. & Kuo, C. Y. 1986. Leaf carbon isotope and mineral-composition in subtropical plants along an irradiance cline.—*Oecologia* 70: 520–526.

Ehleringer, J. R., Sage, R. F., Flanagan, L. B. & Pearcy, R. W. 1991. Climate change and the evolution of C_4 photosynthesis.—*Trends in Ecology and Evolution* 6: 95–99.

Engelmann, G. F., Chure, D. J. & Fiorillo, A. R. 2004. The implications of a dry climate for the paleoecology of the fauna of the Upper Jurassic Morrison Formation.—*Sedimentary Geology* 167: 297–308.

Erben, H. K., Hoefs, J. & Wedepohl, K. H. 1979. Paleobiological and isotopic studies of eggshells from a declining dinosaur species.—*Paleobiology* 5: 380–414.

Erickson, G. M. 1996. Incremental lines of von Ebner in dinosaurs and the assessment of tooth replacement rates using growth line counts.—*Proceedings of the National Academy of Sciences of the United States of America* 93: 14623–14627.

Farquhar, G. D., Ehleringer, J. R. & Hubrick, K. T. 1989. Carbon isotope fractionation and photosynthesis.—*Annual Reviews of Plant Physiology and Molecular Biology* 44: 503–537.

Feranec, R. S. 2007. Stable carbon isotope values reveal evidence of resource partitioning among ungulates from modern C_3-dominated ecosystems in North America.—*Palaeogeography, Palaeoclimatology, Palaeoecology* 252: 575–585.

Feranec, R. S. & MacFadden, B. J. 2006. Isotopic discrimination of resource partitioning among ungulates in C_3-dominated communities from the Miocene of Florida and California.—*Paleobiology* 32: 190–205.

Fiorillo, A. R. 1991. Dental microwear on the teeth of *Camarasaurus* and *Diplodocus:* implications for sauropod paleoecology.

—*Contributions of the Palaeontological Museum, University of Oslo* 364: 23–24.

Fiorillo, A. R. 1998. Dental microwear patterns of the sauropod dinosaurs *Camarasaurus* and *Diplodocus:* evidence for resource partitioning in the Late Jurassic of North America.—*Historical Biology* 13: 1–16.

Folinsbee, R. E., Fritz, P., Krouse, H. R. & Robblee, A. R. 1970. Carbon-13 and oxygen-18 in dinosaur, crocodile, and bird eggshells indicate environmental conditions.—*Nature* 168: 1353–1356.

Foster, J. 2007. *Jurassic West: The Dinosaurs of the Morrison Formation and Their World.* Indiana University Press, Bloomington.

Fricke, H. C. 2007. Stable isotope geochemistry of bonebed fossils: reconstructing paleoenvironments, paleoecology, and paleobiology. *In* Rogers, R. R., Eberth, D. A. & Fiorillo, A. R. (eds.). *Bonebeds: Genesis, Analysis, and Paleobiological Significance.* University of Chicago Press, Chicago: pp. 437–490.

Fricke, H. C. & Pearson, D. A. 2008. Stable isotope evidence for changes in dietary niche partitioning among hadrosaurian and ceratopsian dinosaurs of the Hell Creek Formation.—*Paleobiology* 34: 534–552.

Fricke, H. C. & Rogers, R. R. 2000. Multiple taxon–multiple locality approach to providing oxygen isotope evidence for warm-blooded theropod dinosaurs.—*Geology* 28: 799–802.

Fricke, H. C., Rogers, R. R., Backlund, R., Dwyer, C. N. & Echt, S. 2008. Preservation of primary stable isotope signals in dinosaur remains, and environmental gradients of the Late Cretaceous of Montana and Alberta.—*Palaeogeography, Palaeoclimatology, Palaeoecology* 266: 13–27.

Fricke, H. C., Rogers, R. R. & Gates, T. A. 2009. Hadrosaurid migration: inferences based on stable isotope comparisons among Late Cretaceous dinosaur localities.—*Paleobiology* 35: 270–288.

Friedli, H., Lotscher, H., Oeschger, H., Siegenthaler, U. & Stauver, B. 1986. Ice core record of the $^{13}C/^{12}C$ ratio of atmospheric CO_2 in the past two centuries.—*Nature* 324: 237–238.

Gee, C. T. This volume. Dietary options for the sauropod dinosaurs from an integrated botanical and paleobotanical perspective. *In* Klein, N., Remes, K., Gee, C. T. & Sander, P. M. (eds.). *Biology of the Sauropod Dinosaurs: Understanding the Life of Giants.* Indiana University Press, Bloomington: pp. 34–56.

Ghosh, P., Bhattacharya, S. K., Sahni, A., Kar, R. K., Mohabey, D. M. & Ambwani, K. 2003. Dinosaur coprolites from the Late Cretaceous (Maastrichtian) Lameta Formation of India: isotopic and other markers suggesting a C_3 plant diet.—*Cretaceous Research* 24: 743–750.

Gröcke, D. R. 2002. The carbon isotope composition of ancient CO_2 based on higher-plant organic matter.—*Philosophical Transactions of the Royal Society A: Mathematical, Physical and Engineering Sciences* 360: 633–658.

Haas, G. 1963. A proposed reconstruction of the jaw musculature of *Diplodocus.*—*Annals of the Carnegie Museum of Natural History* 36: 139–157.

Heaton, T. H. E. 1999. Spatial, species, and temporal variations in the $^{13}C/^{12}C$ ratios of C_3 plants: implications for palaeodiet studies.–*Journal of Archaeological Science* 26: 637–649.

Hemminga, M. A. & Mateo, M. A. 1996. Stable carbon isotopes in seagrasses: variability in ratios and use in ecological studies.—*Marine Ecology Progress Series* 140: 285–298.

Herwartz, D., Tütken, T., Münker, C., Jochum, K.-P., Stoll, B. & Sander, P. M. 2010. Timescales and mechanisms of REE and Hf uptake in fossil bones.—*Geochimica et Cosmochimica Acta.* doi: 10.1016/j.gca.2010.09.036.

Hinz, J. K., Smith, I., Pfretzschner, H.-U., Wings, O. & Sun, G. 2010. A high-resolution three-dimensional reconstruction of a fossil forest (Upper Jurassic and Shishugou Formation, Junggar Basin, Northwest China).—*Palaeobiodiversity and Palaeoenvironments* 90: 215–240.

Hoefs, J. 2008. *Stable Isotope Geochemistry.* Springer, Berlin.

Holland, W. J. 1924. The skull of *Diplodocus.*—*Memoires of the Carnegie Museum* 9: 379–403.

Hubert, J. F., Panish, P. T., Chure, D. J. & Prostak, K. S. 1996. Chemistry, microstructure, petrology and diagenetic model of Jurassic dinosaur bones, Dinosaur National Monument, Utah.—*Journal of Sedimentary Research* 66: 531–547.

Hummel, J. & Clauss, M. This volume. Sauropod feeding and digestive physiology. *In* Klein, N., Remes, K., Gee, C. T. & Sander, P. M. (eds.). *Biology of the Sauropod Dinosaurs: Understanding the Life of Giants.* Indiana University Press, Bloomington: pp. 11–33.

Hummel, J., Gee, C., Südekum, K.-H., Sander, P. M., Nogge, G. & Clauss, M. 2008. In vitro digestibility of fern and gymnosperm foliage: implications for sauropod feeding ecology and diet selection.—*Proceedings of the Royal Society B: Biological Sciences* 275: 1015–1021.

Johnson, B. J., Fogel, M. L. & Miller, G. H. 1998. Stable isotopes in modern ostrich eggshell: a calibration for paleoenvironmental applications in semi-arid regions of southern Africa.—*Geochimica et Cosmochimica Acta* 62: 2451–2461.

Keeley, J. E. & Rundel, P. W. 2003. Evolution of CAM and C_4 carbon-concentration mechanisms.—*International Journal of Plant Sciences* 164: 55–77.

Keeley, J. E. & Sandquist, D. R. 1992. Carbon: freshwater plants [commissioned review].–*Plant, Cell and Environment* 15: 1021–1035.

Klein, N. & Sander, P. M. 2008. Ontogenetic stages in the long bone histology of sauropod dinosaurs.—*Paleobiology* 34: 247–263.

Koch, P. L. 1998. Isotopic reconstruction of past continental environments.—*Annual Reviews in Earth and Planetary Science* 26: 573–613.

Koch, P. L. 2007. Isotopic study of the biology of modern and fossil vertebrates. *In* Michener, R. & Lajtha, K. (eds.).—*Stable Isotopes in Ecology and Envrionmental Science.* Blackwell, Malden, Mass.: pp. 99–154.

Koch, P. L., Tuross, N. & Fogel, M. L. 1997. The effects of sample treatment and diagenesis on the isotopic integrity of carbonate in biogenic hydroxylapatite.—*Journal of Archaeological Science* 24: 417–429.

Kohn, M. J. & Law, J. M. 2006. Stable isotope chemistry of fossil bone as a new paleoclimate indicator.—*Geochimica et Cosmochimica Acta* 70: 931–946.

Kohn, M. J., Schoeninger, M. J. & Barker, W. B. 1999. Altered states: effects of diagenesis on fossil tooth chemistry.—*Geochimica et Cosmochimica Acta* 63: 2737–2747.

Kohn, M. J. & Cerling, T. E. 2002. Stable isotope compositions of biological apatite. *In* Kohn, M. J., Rakovan, J. & Hughes, J. M. (eds.).—*Phosphates: Geochemical, Geobiological, and Materials Importance. Reviews in Mineralogy and Geochemistry.* Mineralogical Society of America, Washington, D.C.: pp. 455–488.

Kohn, M. J., McKay, M. P. & Knight, J. L. 2005. Dining in the Pleistocene—who's on the menu?—*Geology* 33: 649–652.

Kolodny, Y., Luz, B., Sander, M. & Clemens, W. A. 1996. Dinosaur bones: fossils or pseudomorphs? The pitfalls of physiology reconstruction from apatitic fossils.—*Palaeogeography, Palaeoclimatology, Palaeoecology* 126: 161–171.

Lee-Thorp, J. A. & Sponheimer, M. 2005. Opportunities and constraints for reconstructing palaeoenvironments from stable light isotope ratios in fossils.—*Geological Quarterly* 49: 195–204.

Lee-Thorp, J. A. & van der Merwe, N. J. 1987. Carbon isotope analysis of fossil bone apatite.—*South African Journal of Science* 83: 712–715.

Lockley, M. G. 1994. Dinosaur ontogeny and population structure: interpretations and speculations based on fossil footprints. *In* Carpenter, K. & Currie, P. J. (eds.). *Dinosaur Systematics: Approaches and Perspectives*. Cambridge University Press, Cambridge: pp. 211–220.

MacFadden, B. J. & Higgins, P. 2004. Ancient ecology of 15-million-year-old browsing mammals within C_3 plant communities from Panama.—*Oecologia* 140: 169–182.

MacFadden, B., Solounias, N. & Cerling, T. E. 1999. Ancient diets, ecology, and extinction of 5-million-year-old horses from Florida.—*Science* 283: 824–827.

McCrea, J. M. 1950. On the isotopic chemistry of carbonates and a paleo-temperature scale.—*Journal of Chemical Physics* 18: 849–857.

Meyer, C. A. 1993. A sauropod dinosaur megatracksite from the Late Jurassic of northern Switzerland.—*Ichnos* 3: 29–38.

Meyer, C. A. & Thüring, B. 2003. Dinosaurs of Switzerland.—*Comptes Rendus Palevol* 2: 103–117.

Mohabey, D. M. 2005. Late Cretaceous (Maastrichtian) nests, eggs, and dung mass (coprolites) of sauropods (titanosaurs) from India. *In* Curry Rogers K. A. & Wilson, J. A. (eds.). *The Sauropods: Evolution and Paleobiology.* University of California Press, Berkeley: pp. 467–489.

Mohabey, D. M. & Samant, B. 2003. Floral remains from Late Cretaceous fecal mass of sauropods from central India: implications to their diet and habitat.—*Gondwana Geological Magazine, Special Volume* 6: 225–238.

Nelson, B. K., DeNiro, M. J., Schoeninger, M. J., DePaolo, D. J. & Hare, P. E. 1986. Effects of diagenesis on strontium, carbon, nitrogen and oxygen concentration and isotopic composition of bone.—*Geochimica et Cosmochimica Acta* 50: 1941–1949.

Nguyen Tu, T. T., Bocherens, H., Mariotti, A., Baudin, F., Pons, D., Broutin, J., Derenne, S. & Largeau, C. 1999. Ecological distribution of Cenomanian terrestrial plants based on $^{13}C/^{12}C$ ratios.—*Palaeogeography, Palaeoclimatology, Palaeoecology* 145: 79–93.

O'Leary, M. H. 1988. Carbon isotopes in photosynthesis.—*BioScience* 38: 328–336.

Ong, B. L., Kluge, M. & Friemert, V. 1986. Crassulacean acid metabolism in the epiphytic ferns *Drymoglossum piloselloides* and *Pyrrosia longifolia:* studies on responses to environmental signals.—*Plant, Cell and Environment* 9: 547–557.

Ostrom, P. H., Macko, S. A., Engel, M. H. & Russell, D. A. 1993. Assessment of trophic structure of Cretaceous communities based on stable nitrogen isotope analyses.—*Geology* 21: 491–494.

Owen-Smith, R. N. 1988. *Megaherbivores: The Influence of Large Body Size on Ecology.* Cambridge University Press, Cambridge.

Passey, B. J., Robinson, T. F., Ayliffe, L. K., Cerling, T. E., Sponheimer, M., Dearing, M. D., Roeder, B. L. & Ehleringer, J. R. 2005. Carbon isotope fractionation between diet, breath CO_2, and bioapatite in different mammals.—*Journal of Archaeological Science* 32: 1459–1470.

Park, R. & Epstein, S. 1960. Carbon isotope fractionation during photosynthesis.—*Geochimica et Cosmochimica Acta* 21: 110–126.

Parrish, J. T., Peterson, F. & Turner, C. E. 2004. Jurassic "savannah"-plant taphonomy and climate of the Morrison Formation (Upper Jurassic, Western USA).—*Sedimentary Geology* 167: 137–162.

Paul, G. S. 1998. Terramegathermy and Cope's rule in the land of titans.—*Modern Geology* 23: 179–217.

Prasad, V., Strömberg, C. A. E., Alimohammadian, H. & Sahni, A. 2005. Dinosaur coprolites and the early evolution of grasses and grazers.—*Science* 310: 1177–1180.

Pucéat, E., Reynard, B. & Lécuyer, C. 2004. Can crystallinity be used to determine the degree of chemical alteration of biogenic apatites.—*Chemical Geology* 205: 83–97.

Rees, P. M., Noto, C. R., Parrish, J. M. & Parrish, J. T. 2004. Late Jurassic climates, vegetation, and dinosaur distributions.—*Journal of Geology* 112: 643–653.

Sackett, W. M., Eckelmann, W. R., Bender, M. C. & Be, A. W. H. 1965. Temperature dependence of carbon isotope composition in marine plankton and sediment.—*Science* 148: 235–237.

Sander, P. M. 1999. Life history of the Tendaguru sauropods as inferred from long bone histology.—*Mitteilungen aus dem Museum für Naturkunde in Berlin, Geowissenschaftliche Reihe* 2: 103–112.

Sander, P. M. & Tückmantel, C. 2003. Bone lamina thickness, bone apposition rates, and age estimates in sauropod humeri and femora.—*Paläontologische Zeitschrift* 76: 161–172.

Sander, P. M., Mateus, O., Laven, T. & Knötschke, N. 2006. Bone histology indicates insular dwarfism in a new Late Jurassic sauropod dinosaur.—*Nature* 441: 739–741.

Sander, P. M., Christian, A. & Gee, C. T. 2009. Sauropods kept their heads down. Response.—*Science* 323: 1671–1672.

Sander, P. M., Christian, A., Clauss, M., Fechner, R., Gee, C. T., Griebeler, E. M., Gunga, H.-C., Hummel, J., Mallison, H., Perry, S., Preuschoft, H., Rauhut, O., Remes, K., Tütken, T., Wings, O. & Witzel, U. 2010a. Biology of the sauropod dinosaurs: the evolution of gigantism.—*Biological Reviews of the Cambridge Philosophical Society.* doi: 10.1111/j.1469=185X.2010.00137.x.

Sander, P. M., Gee, C. T., Hummel, J. & Clauss, M. 2010b. Mesozoic plants and dinosaur herbivory. *In* Gee, C. T. (ed.). *Plants in Mesozoic Time: Morphological Innovations, Phylogeny, Ecosystems.* Indiana University Press, Bloomington: 331–359.

Sarkar, A., Bhattacharya, S. K. & Mohabey, D. M. 1991. Stable isotope analyses of dinosaur egg shells: palaeoenvironmental implications.—*Geology* 19: 1068–1071.

Schleser, G. H., Frielingsdorf, J. & Blair, A. 1999. Carbon isotope behaviour in wood and cellulose during artificial aging.—*Chemical Geology* 158: 121–130.

Schoeninger, M. J. & DeNiro, M. J. 1982. Carbon isotope ratios of apatite from fossil bone cannot be used to reconstruct diets of animals.—*Nature* 297: 557–578.

Schrank, E. 2010. Pollen and spores from the Tendaguru Beds, Upper Jurassic and Lower Cretaceous of southeast Tanzania: palynological and paleoecological implications.—*Palynology* 43: 3–42.

Schweitzer, M. H., Wittmeyer, J. L., Horner, J. L. & Toporski, J. K. 2005. Soft-tissue vessels and cellular preservation in *Tyrannosaurus rex*.—*Science* 307: 1952–1954.

Schweitzer, M. H., Suo, Z., Avci, R., Asara, J. M., Allen, M. A., Arce, F. T. & Horner, J. R. 2007. Analyses of soft tissue from *Tyrannosaurus rex* suggest the presence of protein.—*Science* 316: 277–280.

Smith, B. N. & Epstein, S. 1971. Two categories of $^{13}C/^{12}C$ ratios for higher plants.—*Plant Physiology* 47: 380–384.

Spötl, C. & Vennemann, T. W. 2003. Continuous-flow IRMS analysis of carbonate minerals.—*Rapid Communications in Mass Spectrometry* 17: 1004–1006.

Stanton Thomas, K. J. & Carlson, S. J. 2004. Microscale $\delta^{18}O$ and $\delta^{13}C$ isotopic analysis of an ontogenetic series of the hadrosaurid dinosaur *Edmontosaurus:* implications for physiology and ecology.—*Palaeogeography, Palaeoclimatology, Palaeoecology* 206: 257–287.

Stein, K., Csikib, Z., Curry Rogers, K., Weishampel, D. B., Redelstorff, R., Carballidoa, J. L. & Sander, P. M. 2010. Small body size and extreme cortical bone remodelling indicate phyletic dwarfism in *Magyarosaurus dacus* (Sauropoda: Titanosauria).—*Proceedings of the National Academy of Sciences of the United States of America* 107: 9258–9263.

Stevens, K. A. & Parrish, J. M. 1999. Neck posture and feeding habits of two Jurassic sauropod dinosaurs.—*Science* 284: 798–800.

Stevens, K. A. & Parrish, J. M. 2005a. Neck posture, dentition, and feeding strategies in Jurassic sauropod dinosaurs. *In* Tidwell, V. & Carpenter, K. (eds.). *Thunder-Lizards: The Sauropodomorph Dinosaurs.* Indiana University Press, Bloomington: pp. 212–232.

Stevens, K. A. & Parrish, J. M. 2005b. Digital reconstructions of sauropod dinosaurs and implications for feeding. *In* Curry Rogers, K. A. & Wilson, J. A. (eds.). *The Sauropods: Evolution and Paleobiology.* University of California Press, Berkeley: pp. 178–200.

Straight, W. H., Barrick, R. E. & Eberth, D. A. 2004. Reflections of surface water, seasonality and climate in stable oxygen isotopes from tyrannosaurid tooth enamel.—*Palaeogeography, Palaeoclimatology, Palaeoecology* 206: 239–256.

Taggert, R. E. & Cross, A. T. 1997. The relationship between land plant diversity and productivity and patterns of dinosaur herbivory. *In* Wolberg, D. L., Stump, E. & Rosenberg, G. D. (eds.). *Dinofest International*. Academy of Natural Sciences, Philadelphia: pp. 403–416.

Tieszen, L. L. 1991. Natural variations in the carbon isotope values of plants: implications for archaeology, ecology and paleoecology.—*Journal of Archaeological Science* 18: 227–248.

Tieszen, L. L. & Fagre, T. 1993. Effect of diet quality and composition on the isotopic composition of respiratory CO_2, bone collagen, bioapatite and soft tissues. *In* Lambert, J. B. & Grupe, G. (eds.). *Prehistoric Human Bone: Archaeology at the Molecular Level*. Springer Verlag, Berlin: pp. 121–155.

Tiffney, B. H. 1997. Land plants as food and habitat in the Age of Dinosaurs. *In* Farlow, J. O. & Brett-Surman, M. K. (eds.). *The Complete Dinosaur.* Indiana University Press, Bloomington: pp. 352–370.

Trueman, C. N. 1999. Rare earth element geochemistry and taphonomy of terrestrial vertebrate assemblages.—*Palaios* 14: 555–568.

Trueman, C. N. G. & Tuross, N. 2002. Trace elements in recent and fossil bone apatite.—*Reviews in Mineralogy and Geochemistry* 48: 489–521.

Trueman, C. N. G., Chenery, C., Eberth, D. A. & Spiro, B. 2003. Diagenetic effects on the oxygen isotope composition of bones of dinosaurs and other vertebrates recovered from terrestrial and marine sediments.—*Journal of the Geological Society* 160: 895–901.

Trueman, C. N. G., Behrensmeyer, A. K., Tuross, N. & Weiner, S. 2004. Mineralogical and compositional changes in bones exposed on soil surfaces in Amboseli National Park, Kenya: diagenetic mechanisms and the role of sediment pore fluids.—*Journal of Archaeological Science* 31: 721–739.

Thomasson, J. R., Nelson, M. E. & Zakrzewski, R. J. 1986. A fossil grass (Gramineae: Chloridoideae) from the Miocene with Kranz anatomy.—*Science* 233: 876–878.

Tütken, T. 2003. *Die Bedeutung der Knochenfrühdiagenese für die Erhaltungsfähigkeit in vivo erworbener Element-und Isotopenzusammensetzungen in fossilen Knochen.* Ph.D. Dissertation. Eberhard-Karls Universität, Tübingen.

Tütken, T. & Vennemann, T. 2009. Stable isotope ecology of Miocene large mammals from Sandelzhausen, southern Germany.—*Paläontologische Zeitschrift* 83: 207–226.

Tütken, T., Pfretzschner, H.-U., Vennemann, T. W., Sun, G. & Wang, Y. D. 2004. Paleobiology and skeletochronology of Jurassic dinosaurs: implications from the histology and oxygen isotope compositions of bones.—*Palaeogeography, Palaeoclimatology, Palaeoecology* 206: 217–238.

Tütken, T., Sander, M., Hummel, J. & Gee, C. 2007. Ernährung und Mobilität von Sauropoden: Informationspotential der Isotopenzusammensetzung von Knochen und Zähnen.—*Hallesches Jahrbuch der Geowississenschaften, Beiheft* 23: 85–92.

Tütken, T., Vennemann, T. W. & Pfretzschner, H.-U. 2008. Early diagenesis of bone and tooth phosphate in fluvial and marine settings: constraints from combined oxygen isotope, nitrogen and REE analysis.—*Palaeogeography, Palaeoclimatology, Palaeoecology* 266: 254–268.

Tütken, T. 2010. Die Isotopenanalyse fossiler Skelettreste: Bestimmung der Herkunft and Mobilität von Menschen und Tieren. *In* Meller, H. & Alt, K. W. (eds.). Anthropologie, Isotopie und DNA: Biografische Annäherung an namenlose vorgeschichtliche Skelette. Tagungsband 2. Mitteldeutscher Archäologentag.—*Tagungen des Landesmuseums für Vorgeschichte Halle,* Band 3: 33–51.

Turney, C. S. M., Wheeler, D. & Chivas A. R. 2006. Carbon isotope fractionation in wood during carbonization.—*Geochimica et Cosmochimica Acta* 70: 960–964.

Upchurch, P. & Barrett, P. M. 2000. The evolution of sauropod feeding. *In* Sues, H.-D. (ed.). *Evolution of Herbivory in Terrestrial Vertebrates: Perspectives from the Fossil Record.* Cambridge University Press, Cambridge: pp. 79–122.

Upchurch, P., Barrett, P. M. & Dodson, P. 2004. Sauropoda. *In* Weishampel D. B., Dodson, P. & Osmólska, H. (eds.). *The Dinosauria. 2nd edition.* University of California Press, Berkeley: pp. 259–322.

van der Ham, R. W. J. M., van Konijnenburg-van Cittert, J. H. A. & Indeherberge, L. 2007. Seagrass foliage from the Maastrichtian type area (Maastrichtian, Danian; NE Belgium, SE Netherlands).—*Review of Palaeobotany and Palynology* 144: 301–321.

van der Merwe, N. J. & Medina, E. 1991. The canopy effect, carbon isotope ratios and foodwebs in Amazonia.—*Journal of Archaeological Science* 18: 249–259.

von Schirnding, Y., van der Merwe, N. J. & Vogel, J. C. 1982. Influence of diet and age on carbon isotope ratios in ostrich eggshell.—*Archaeometry* 24: 3–20.

Vovides, A. P., Etherington, J. R., Dresser, P. Q., Groenhof, A., Iglesias, C. & Ramirez, J. F. 2002. CAM-cycling in the cycad *Dioon edule* Lindl. in its natural tropical deciduous forest habitat in central Veracruz, Mexico.—*Botanical Journal of the Linnean Society* 138: 155–162.

Wang, Y. & Cerling, T. E. 1994. A model of fossil tooth and bone diagenesis: implications for paleodiet reconstruction from stable isotopes.—*Palaeogeography, Palaeoclimatology, Palaeoecology* 107: 281–289.

Weaver, J. C. 1983. The improbable endotherm: energetics of the sauropod dinosaur *Brachiosaurus.*—*Paleobiology* 9: 173–182.

Weishampel, D. B., Barrett, P. M., Coria, R. A., Le Loeuff, J., Xu, X., Zhao X., Sahni, A., Gomani, E. M. P. & Noto, C. R. 2004. Dinosaur distribution. *In* Weishampel, D. B., Dodson, P. & Osmólska H. (eds.). *The Dinosauria. 2nd edition.* University of California Press, Berkeley: pp. 517–607.

Wilson, J. A. 2002. Sauropod dinosaur phylogeny: critique and cladistic analysis.—*Zoological Journal of the Linnean Society of London* 136: 217–276.

Wright, J. L. 2005. Steps in understanding sauropod biology: the importance of sauropod tracks. *In* Curry Rogers, K. A. & Wilson, J. A. (eds.). *The Sauropods: Evolution and Paleobiology.* University of California Press, Berkeley: pp. 252–284.

Zanazzi, A. & Kohn, M. 2008. Ecology and physiology of White River mammals based on stable isotope ratios of teeth.—*Palaeogeography, Palaeoclimatology, Palaeoecology* 257: 22–37.

PART TWO

PHYSIOLOGY

5

Structure and Function of the Sauropod Respiratory System

STEVEN F. PERRY, THOMAS BREUER, AND NADINE PAJOR

BECAUSE DINOSAUR LUNGS do not fossilize, reconstruction of the sauropod respiratory system must rely on indirect evidence. We combine extant phylogenetic bracketing and functional morphological approximation to draw conclusions on the structure and function of the sauropod respiratory system. The combination of these techniques leads to strong evidence for the presence of lungs that consisted of two parts: a gas exchange region that was attached to the ribs and vertebrae, and a sac-like region below the exchange region, close to the liver and intestine. This respiratory system is similar to the efficient lung–air sac system of birds. It is highly adaptable and could have served to supply oxygen, remove carbon dioxide, and help with temperature control.

Introduction

Indirect evidence for the reconstruction of the respiratory system of extinct animals can come from the skeleton itself, mainly from the ribs and the rib cage. Most interesting here is the morphology of the uncinate processes, the course and density of Sharpey's fibers within the ribs, and the heads of the ribs and their articulation to the vertebrae.

Extant phylogenetic bracketing (Witmer 1995) is used to deduce the possible structural type of the respiratory system, whereas functional morphological approximation (Perry & Sander 2004) is applied in a second step to give information on the functional/physiological implications of these structures.

In this chapter, we first give an overview of general lung structure and function in amniotes. We then focus on differences in lung systems of the closest living relatives of dinosaurs: birds and reptiles. Finally, we discuss the indirect evidence for the presence of an avian-like structural type of lung in sauropod dinosaurs. This structure has functional implications, which in turn either limit or make possible certain metabolic strategies. It is capable of supporting a broad scope of metabolic demands, including the high demand of a rapidly growing and highly active juvenile and the relatively low demand of a resting adult. Furthermore, an avian-like lung system could have helped in thermoregulation by providing large heat-transfer surfaces. Finally, the presence of air sacs lowers the overall density and thus the mass of an animal of a given size, resulting in lower energy requirement and mechanical advances compared to models with nonavian lung structures. We conclude that the lung structure in sauropods contributed to making gigantism possible (see also Sander et al. 2010).

Reconstruction of the Sauropod Respiratory Tract

RESPIRATION AND RESPIRATORY ORGANS IN AMNIOTES

Lung Structure

Because every vertebrate depends on a continuous and adequate oxygen supply, respiratory and circulatory systems are necessary in order to transport oxygen to the cells and to remove the cellular waste gas, carbon dioxide, from the body (Torday et al. 2007). Respiratory systems of living amniotes, the group including mammals, reptiles, and birds, in general consist of a pump and an exchanger (Perry 1989). The pump mechanically ventilates the exchanger and supplies it with fresh, oxygen-rich air. Gas exchange takes place in the exchanger, typically a richly subdivided region of the lung with a large surface area and a thin barrier between air and blood (Torday et al. 2007). Air is taken into the lung through the trachea (windpipe) that in the lung splits into smaller branches called bronchi. Mammalian lungs (Fig. 5.1) are combined pumps and exchangers: the whole lung is inflated as the rib cage expands, sucking in air. Gas exchange takes place in the so-called alveoli, which are homogeneously distributed throughout almost the whole volume of the lung, excluding the space occupied by the bronchi. Lungs of living reptiles (we do not use the term *reptiles* in a phylogenetic sense here, but only to cover living amniotes that are neither birds nor mammals) such as turtles, lizards, snakes, and crocodiles (Fig. 5.1) feature a partial functional and structural separation of pump and exchanger (Perry 2010): the lungs consist of large chambers mainly serving as bellows and for air storage, and gas exchange takes place in dense exchange tissue in the chamber walls.

Birds show the most complete separation of pump and exchanger among living vertebrates (Torday et al. 2007): bellows-like structures, called air sacs, extending throughout the whole thorax, serve as pumps, and are expanded and compressed by rib movement during breathing (Figs. 5.2, 5.3). The air sacs of birds completely lack exchange tissue and thus have no direct gas exchange function at all. From anterior to posterior in the body, the air sacs include the cervical, interclavicular, anterior thoracic, posterior thoracic, and abdominal air sacs. The interclavicular and anterior thoracic sacs constitute the anterior functional group, the posterior thoracic and abdominal sacs make up the posterior functional group, and the cervical air sacs are encased in bone and immobile, and thus do not contribute to air movement. The exchangers in this system are the neopulmo and paleopulmo (Duncker 1971, 1978), each consisting of dozens of parallel air ducts (parabronchi) that are surrounded by a dense air–blood capillary network (Fig. 5.2B). The neopulmo and paleopulmo, as opposed to the air sacs, show almost no volume change during breathing. The bird lung is characterized by a thin air–blood diffusion barrier combined with a crosscurrent gas exchange mechanism; the average directions of air and blood flow are perpendicular to each other (Piiper & Scheid 1972, 1986). This system is highly efficient in extracting oxygen from the air, making it capable of satisfying the high energy requirements of the animal.

Respiratory Mechanics

In spite of different lung structures, the general mechanical processes for breathing are the same in all living amniotes (Brainerd 1999). Breathing is accomplished by expansion and compression of the rib cage, in some cases supported by accessory breathing structures such as the diaphragm in mammals and the m. diaphragmaticus in crocodiles (Brainerd & Owerkowicz 2006). During inspiration (breathing in), the external intercostal muscles (flat, sheet-like muscles between adjacent ribs) contract and result in a forward swinging movement of the ribs, expanding the rib cage (Brainerd 1999). Because of adhesion of the lungs to the body wall in mammals, or physical attachment of the air sacs to the body wall in birds, these hollow organs expand, and air is sucked in as a result of the resulting pressure drop (Perry 2004). In some animals, this inspiratory movement is supported by the accessory breathing structures, but expiration (breathing out) is often due to the contraction of the lung itself. However, it can be supported by the action of the internal intercostal muscles and the transversus muscle sheet, swinging the ribs backwards or pressing the internal organs against the lungs and air sacs, respectively.

Diffusion and Gas Transport

Once the air is inside the lungs, gas exchange takes place through extremely thin (less than 1/1,000 of a millimeter) barriers that separate the lung air from the blood, flowing in the tiny vessels (capillaries) of the exchange tissue. The movement of oxygen from the lung air to the blood and that of carbon dioxide from the blood to the lung air is a passive process known as diffusion, and it relies entirely on the molecular movement (Brownian motion) of particles. Diffusive processes that allow gas exchange result from unequal concentrations of the respiratory gases oxygen and carbon dioxide in blood and air. If separated by a membrane that allows passage of these gases, such as the lining layers of the lung and capillaries, then diffusion takes place through the membrane in the direction of the lower concentration (Cameron 1989).

Diffusion is made faster by the airflow generated by breathing and the rapid flow of blood, which causes large concentration differences that constantly replenish the air and blood. In addition, substances such as hemoglobin (a respiratory pigment in the red blood cells) bind oxygen temporarily and do not allow it to diffuse back out into the lung air. Bonded to hemoglobin, oxygen is then transported to the different

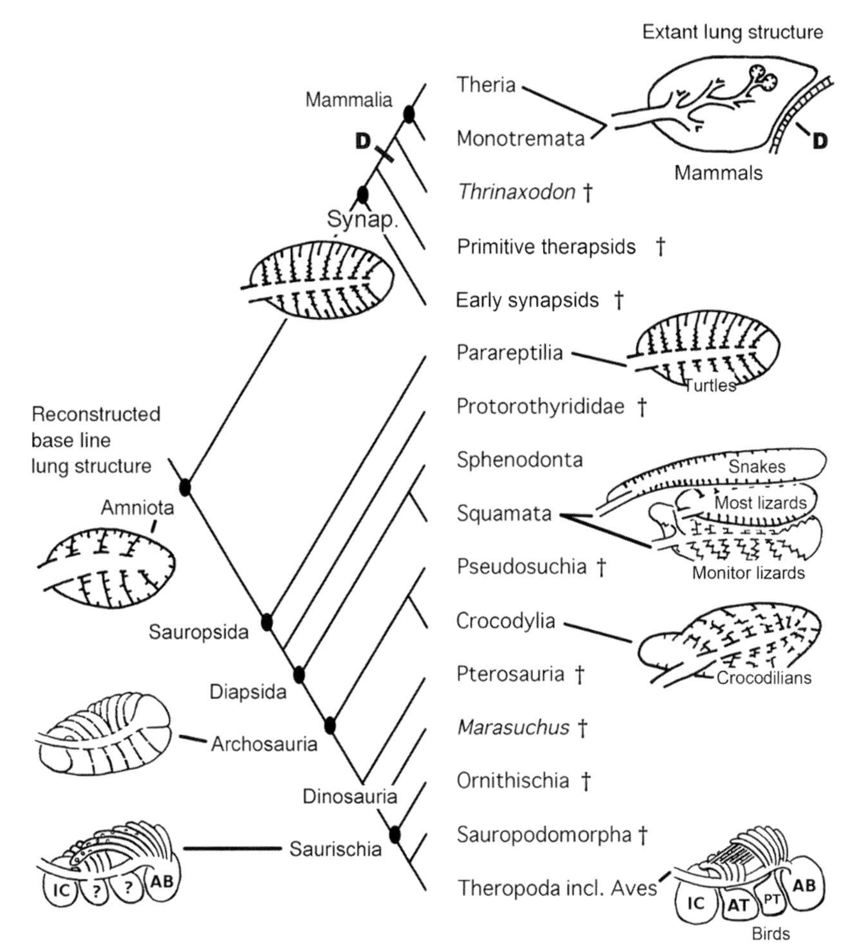

FIGURE 5.1. Phylogenetic tree depicting lung evolution in amniotes. Lung diagrams are idealized sagittal sections or lateral views (dinosaurs and birds). Note the location of sauropods (Sauropodomorpha; lungs not illustrated) between birds (Theropoda incl. Aves) and crocodiles (Crocodylia). Extant lung structure at right; reconstructed stem line lung structure at left. D, mammalian diaphragm; Synap., Synapsida; AB, abdominal air sacs; AT, anterior thoracic air sacs; IC, interclavicular air sac(s); PT, posterior thoracic air sacs.

tissues of the body by the circulatory system. Here it is set free to diffuse as a result of the relatively high acidity in the body tissue. The high acidity, caused by the production of carbon dioxide in metabolic processes of all the cells of the body, makes the hemoglobin less able to bind oxygen, which then diffuses into the tissue. Because of its high concentration in the tissues, carbon dioxide diffuses into the blood, where it is then transported in both red blood cells and plasma. Thus, in the tissues, oxygen diffuses from the blood into the cells and carbon dioxide diffuses from the cells to the blood, which transports it away to the heart and from there to the lungs, where it is breathed out.

LUNG SYSTEMS IN THE EXTANT PHYLOGENETIC BRACKET

Extant phylogenetic bracketing (Witmer 1995) is based on comparison of certain features (in the present case, the lungs) of the nearest extant relatives in order to reconstruct these features in extinct animals. It is based on the assumption that soft tissue features shared by the living descendent of a common ancestor were also present in extinct members located between them in the phylogenetic tree. This is not a black-and-white issue, but only puts certain limits on the possible bauplan (*bauplan* is a word of German origin indicating a structural type or blueprint) of the structure in question, and it helps us reconstruct the most likely (in anatomical and phylogenetic terms) bauplan of the feature in question in the extinct animal.

Sauropod dinosaurs are phylogenetically bracketed by birds and crocodiles (Fig. 5.1). Thus, their lungs must have been characterized by at least those structural traits common to both groups (Perry & Sander 2004). These are presented and discussed below.

Multichambered Lungs

Multichambered lungs are composed of a large number of saccular or tubular units (chambers) that emanate separately from a bronchus. Such multichambered lungs are seen in turtles, varanid lizards, and crocodilians, and it is assumed that the avian lung–air sac system also evolved from a multichambered reptilian lung (Perry 1992; Fig. 5.1).

Separation of the Pump and Exchanger

This condition is seen in crocodiles and other reptiles as well as in birds. Most reptilian lungs feature large sac-like portions

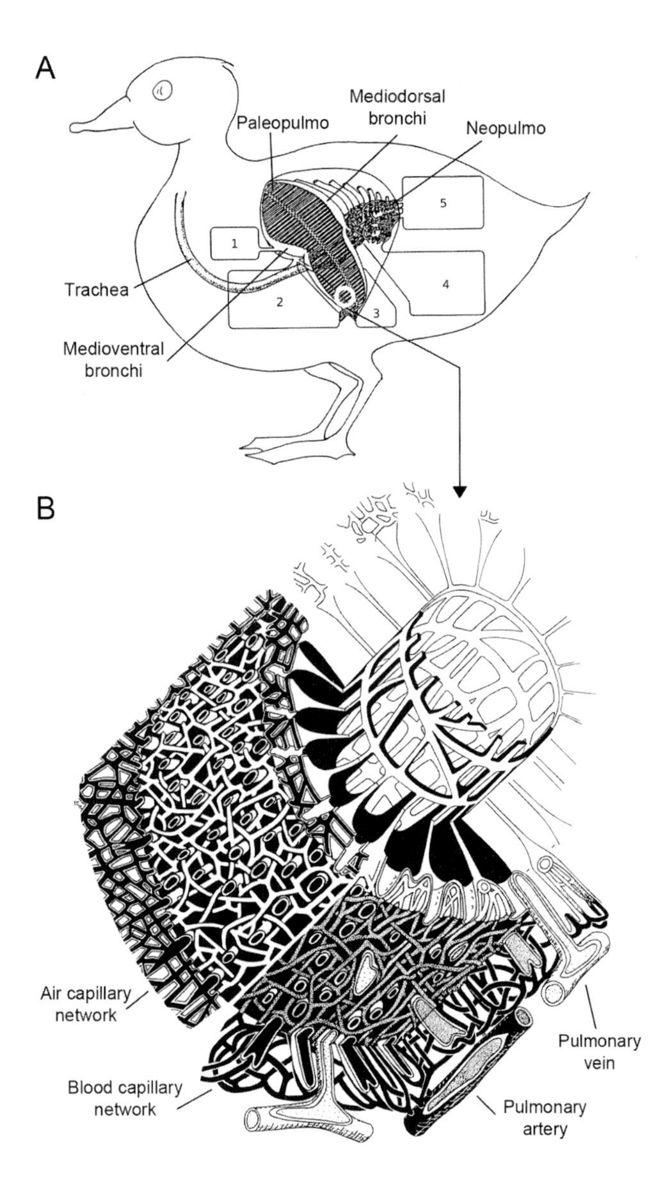

FIGURE 5.2. Structure of avian lung–air sac system. (A) Components of the respiratory system. The lung consists of a paleopulmo and neopulmo. The air sacs are numbered: (1) noninflatable cervical; (2) usually unpaired interclavicular; (3) paired anterior thoracic; (4) paired posterior thoracic; (5) paired abdominal. (B) Microanatomy of a parabronchus. The air capillary network (left) and blood capillary network (right) are depicted here separately, although in reality they are intermeshed. *Modified after Perry (2010).*

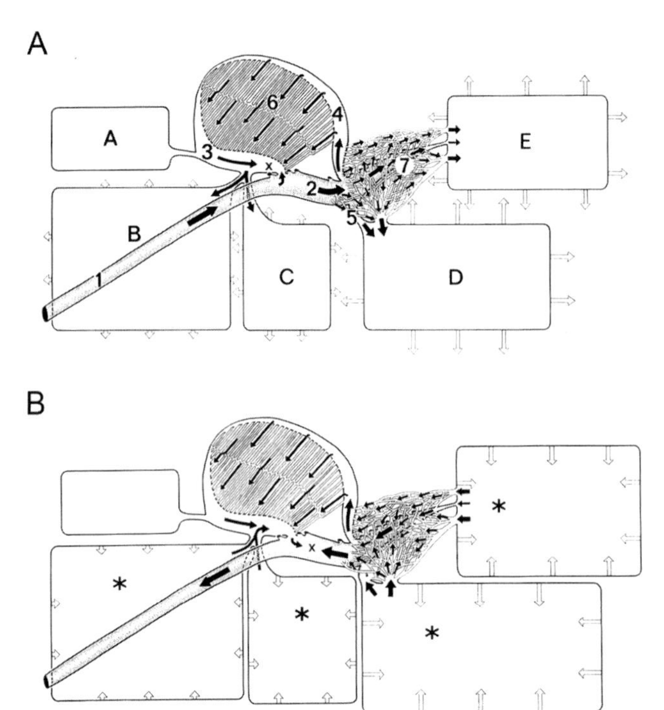

FIGURE 5.3. Airflow in the avian lung–air sac system. (A) During inspiration, all air sacs increase in volume. Air flows from the primary bronchus (2) through saccobronchi (5) or parabronchi of the neopulmo (7) into the posterior thoracic (D) or abdominal (E) air sacs. Alternatively, air bypasses the posterior air sac group, enters the mediodorsal secondary bronchi (4), flows through parabronchi of the paleopulmo (6), and continues via the mediodorsal secondary bronchi (3) to the anterior thoracic (C) and interclavicular air sacs. Air flowing in the mediodorsal bronchi is blocked (x) from reentering the primary bronchus by air being drawn into the same secondary bronchi by expansion of the anterior air sac group. The cervical air sacs (A) are not involved in ventilation of the lung. (B) During expiration air is moved from the air sacs (*) in the case of the posterior air sac group, back through the neopulmo, and then via the mediodorsal secondary bronchi through the paleopulmo, medioventral secondary bronchi, and out the lung. Air from the anterior air sac group exits the mediodorsal bronchi directly. Some of this air enters the primary bronchus but is prevented from recirculating (x) by air leaving the posterior air sac group. *Modified after Duncker (1971).*

that serve primarily in gas storage or as bellows for pumping air (Wolf 1933). In the American alligator (*Alligator mississippiensis*), these lung regions have been shown to play a major role in buoyancy control (Uriona & Farmer 2008), while gas exchange takes place in the parenchyma, which is the gas exchange tissue, and the enclosed air spaces.

In crocodilians, the parenchyma is composed of honeycomb-like structures on the chamber walls called ediculae. The parenchyma tends to be concentrated in the dorsal region of the lungs, whereas the ventral part is composed of sac-like structures. As noted above, in birds, the respiratory system shows a much greater separation of gas exchange and gas storage regions (Fig. 5.2A).

Finger-like Lung Chambers or Secondary Bronchi

In the crocodilian lung, sets of finger-like chambers extend from an unbranched airway, which is called the intrapulmonary bronchus. Classical embryological studies (Locy & Larsell

A

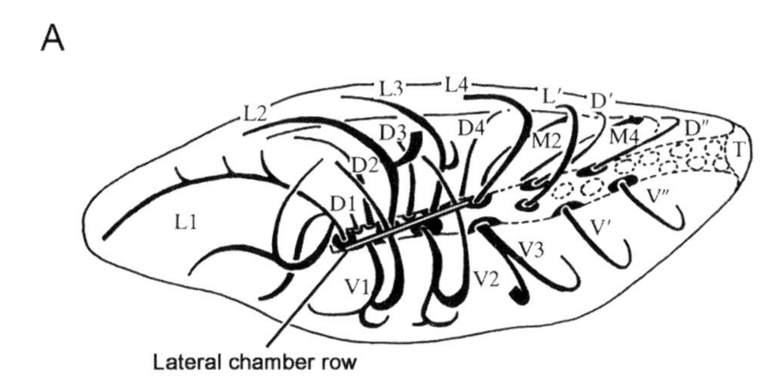

B

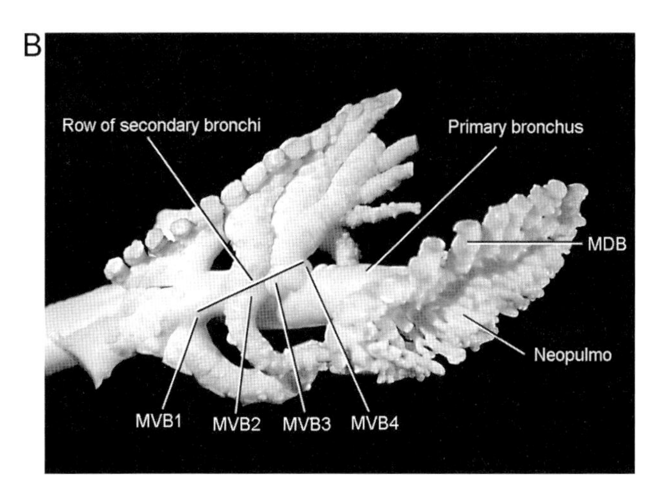

FIGURE 5.4. (A) Schematic drawing of the left lung of *Crocodylus niloticus* in lateral view, showing the entrances (ostia) to chambers lying in a left-hand spiral around the main bronchus. The lung chambers occur in four groups (1–4), each consisting of a dorsal (D), lateral (L), and ventral (V) rows. Medial chambers (M) also exist in the more posterior regions of the lung. The portion of the primary intrapulmonary bronchus not supported by cartilage rings and the numerous ostia located there are indicated by dotted lines. (B) Cast of the primary bronchus of the quail (*Coturnix coturnix*), showing the same slanted location of the ostia of the secondary bronchi as observed in the crocodile. The secondary bronchi are designated according the standard terminology for avian lungs: the anterior group (medioventral bronchi, MVB1-4) and the posterior group (numerous mediodorsal bronchi, MDB). Parabronchi of the neopulmo are also indicated. *Modified after Perry & Sander (2004).*

1916a, 1916b) have demonstrated that in early developmental stages, the bauplan of the avian lung closely resembles that of the adult crocodile (Fig. 5.4). In birds, though, the finger-like extensions are called secondary bronchi. During its further embryological development, the bird lung establishes contact between the secondary bronchi extending from the posterior and the anterior part of the lung. These contacts are in the form of tubes called parabronchi that contain the gas exchange tissue in the adult bird. Crocodiles also show contacts, in the form of holes called intercameral foramina (Perry 1998) that are very reminiscent of truncated avian parabronchi (Farmer & Sanders 2010). These holes are located at the base of the honeycomb-like ediculae that cover the chamber walls. Thus, the bird lung can be seen as a further development of the unbranched intrapulmonary bronchus and the perforated, finger-like chambers observed in crocodiles (Perry 1992).

Internal Subdivision of the Body Cavity and Formation of Air Sacs

In both crocodilians and birds, the lungs are stabilized by two membranes: the postpulmonary septum, which lacks muscle tissue but, like the mammalian diaphragm, lies between the liver and the lungs; and the posthepatic septum, which arises posterior to the liver and makes a compartment containing both the lungs and liver (Plate 5.1). In crocodilians, the posthepatic septum also serves as the origin for the m. diaphragmaticus, which inserts on the pelvic girdle. Contraction of this muscle pulls the liver–lung unit back, inflating the lungs if the larynx (voice box) is open. When the larynx is closed, this mechanism can shift the air within the lung, displacing the center of buoyancy posteriorly and allowing the animal to control its orientation in the water (Uriona & Farmer 2008). It thus appears that the m. diaphragmaticus may represent a special adaptation to aquatic life (Uriona & Farmer 2008).

Although birds also have both a postpulmonary and a posthepatic septum, they lack an m. diaphragmaticus. Also unlike crocodilians, the liver is not attached to the lungs but is separated from them by the thoracic air sacs, which invade the postpulmonary septum, splitting it into a medial part that is anchored to the liver and a lateral part that is attached to the body wall and moves with the ribs. The lungs maintain a constant volume, and air moves through them by movement of the body wall expanding and collapsing the air sacs.

Unidirectional airflow through avian parabronchi does not require the anterior and posterior thoracic air sacs (Brackenbury & Amaku 1990). Although the presence of air spaces in the middle thoracic vertebrae of highly derived sauropods indicates the contact of these vertebrae with the respiratory system (Wedel 2009), it cannot be concluded with certainty that thoracic air sacs homologous to those of birds were present. On the contrary, in our opinion, the unique developmental pathway of formation of the thoracic air sacs by invading the postpulmonary septum, as mentioned above, makes the separate origin of this mechanism in sauropods and theropods highly unlikely, and these structures are indicated by question marks on the hypothetical saurischian lung in Fig. 5.1. Nevertheless, their presumed position, should they have existed, is shown in Plate 5.1C, D. The interclavicular air sacs (Plate 5.1C–F), on the other hand, appear to be homologous to the anterior chambers of the crocodilian lung, and the abdominal air sacs are foreshadowed by the terminal chamber of the crocodilian lung, which in birds perforates the postpulmonary septum and expands into the body cavity.

Further Similarities between Crocodilian and Avian Lungs

The gross anatomy of the crocodilian lung bears further similarities to that of birds (Fig. 5.4). In both cases, the intrapulmonary bronchus is reinforced by cartilage only in the anterior half. Here, chambers (secondary bronchi in birds) that extend toward the body's midline are reduced or lacking, but in crocodilians, four or five groups of openings (ostia) to the chambers form dorsal, lateral, and ventral rows. In birds, only a single row of secondary bronchi is present in the anterior part of the intrapulmonary bronchus, but as in crocodilians, there are four or five of them (Fig. 5.3).

In both crocodilians and birds, the posterior part of the intrapulmonary bronchus, which does not have cartilage, has numerous ostia to the chambers/secondary bronchi, but they are not arranged in rows as they are in the anterior part. Furthermore, the intrapulmonary bronchus shows a monopodial branching pattern; that is, all major chambers emerge from one central stem, like the branches on a fir tree (Fig. 5.4).

As mentioned above, the ostia are arranged in rows in the anterior part of the intrapulmonary bronchus. These rows are not parallel to the long axis of the bronchus, but spiral around it counterclockwise in the left lung and clockwise in the right in both crocodiles and birds (marked by lines in Fig. 5.4).

In summary, extant phylogenetic bracketing tells us that a dinosaur lung probably had all of the following structures that are present in both crocodiles and birds. The lungs had an unbranched intrapulmonary bronchus containing cartilage in its anterior part. This intrapulmonary bronchus gave rise to tubular chambers (or secondary bronchi) that were connected with others through foramina, which could have evolved into parabronchi. There were maximally three rows of such chambers in the anterior part; medial ones were lacking. The ostia to the rows of chambers spiraled counterclockwise around the intrapulmonary bronchus in the left lung and clockwise in the right one. The posterior part of the lung lacked cartilage reinforcement of the intrapulmonary bronchus and gave rise to a large number of chambers that were not arranged in discrete rows. The ventral parts of the lung tended to form sac-like regions, which could have evolved into true air sacs.

FUNCTIONAL MORPHOLOGICAL APPROXIMATION

Functional morphological approximation is a technique by which, based on known analogs, a functionally plausible bauplan of an organ or organ system is reconstructed within the anatomical limits set by extant phylogenetic bracketing, and the physiological consequences of this bauplan are evaluated here (Perry & Sander 2004).

Crosscurrent Model in Sauropods

Because the average directions of blood and airflow in the tubular crocodilian lung chambers and in the avian parabronchi are perpendicular to each other (Farmer & Sanders, 2010), this gas exchange principle is called crosscurrent exchange (Fig. 5.5). Whereas the crocodilian lung appears overdesigned for the usual long-term metabolic state of extant crocodiles and alligators (Coulson & Hernandez 1983; Perry 1988), in birds, it results in a versatile and highly efficient respiratory organ both at rest and during exercise (Scheid 1979).

In resting alligators, posterior-to-anterior unidirectional air movement has been observed (Farmer & Sanders, 2010), thus maintaining a slow respiratory airflow in the lungs between breaths. The unidirectional flow characteristic of the paleopulmo part of the avian lung may have been conserved in nonavian and avian dinosaurs, even as different mechanisms for ventilating the archosaurian respiratory system evolved. This principle of conservation of a vital function despite the evolution of different mechanisms of achieving it is well illustrated by the presence of countercurrent exchange (countercurrent flow of blood and water in the gills, allowing high oxygen extraction from the water) in hagfish, lampreys, sharks, and bony fish, although each group has radically different breathing mechanisms (Perry 2010).

The specialized respiratory tract of birds (Fig. 5.3) achieves a unidirectional flow through the paleopulmo of the lung: expansion of the body cavity causes negative pressure in the air sacs. About half of the inspired air flows through the parabronchi of the neopulmo into the posterior air sac group (Fig. 5.3A). The other half is drawn into the paleopulmo by the negative pressure in the anterior air sac group. Upon compression of the body cavity, air from the posterior air sac group is pushed into the paleopulmo, passing through the neopulmo again (Figs. 5.3B, 5.5A). Thus the airflow in the neopulmo is bidirectional but the gas exchange model remains crosscurrent (Fig. 5.5B). This respiratory cycle results in an almost continuous airflow on the respiratory surfaces of the neopulmo and paleopulmo, allowing constant extraction of oxygen from the air (Fig. 5.5). Comparison of the structure of the ribs in maniraptoran dinosaurs and birds suggests that the same biomechanical ventilatory mechanisms existed in at least some theropods (Codd et al. 2007).

Thus, both the low metabolic rate during resting in an adult sauropod and the extremely high metabolic rate experienced by an active juvenile that is rapidly growing (Sander & Clauss 2008) could profit from an avian-like bauplan. In addition, although unidirectional flow may have already existed in the lungs of ancestral archosaurs, it does not have to be main-

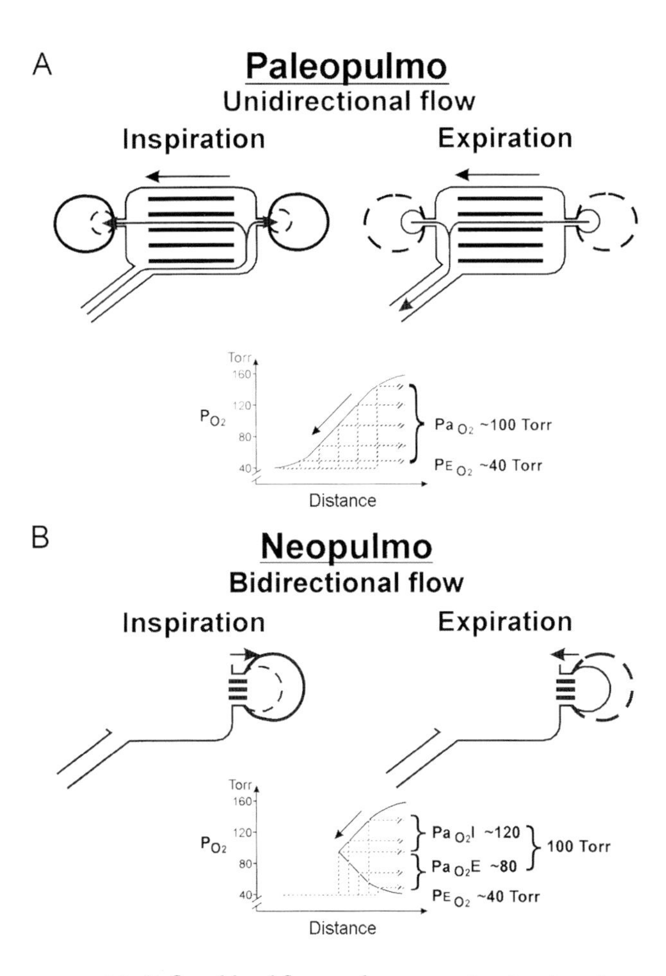

FIGURE 5.5. Airflow, blood flow, and oxygen extraction in avian lungs. (A) Unidirectional airflow through the paleopulmo during inspiration and expiration. (B) Bidirectional airflow in the neopulmo. Airflow is indicated by solid arrows, blood flow by dashed arrows. Because of the crosscurrent exchange model, the mean partial pressure of oxygen in blood leaving the lung (Pa_{O2}) can exceed that of the expired air (PE_{O2}), both in unidirectional and in bidirectional models. *Modified after Perry & Reuter (1999).*

tained in order for the crosscurrent model to apply. The release from this constraint makes the crosscurrent model even more plausible in sauropods.

Implications of Air Spaces in Bones

Pneumaticity (air spaces inside bones) is extensively developed in the vertebrae of sauropods (Wedel 2003a, 2003b, 2005, 2006, 2009) and theropods (O'Connor 2006), and in some sauropod species in the ribs as well (Janensch 1950; Wilson 2002; Upchurch et al. 2004; Lovelace et al. 2007). These air spaces are presumed to emanate from the respiratory system. This observation has fundamental functional implications for the lungs (O'Connor & Claessens 2005). If the air spaces come from the lung, then the lung must have been attached to the bones at that point. If the lung is attached to the ribs, which form the body wall, then a mammalian-type lung must be ruled out because mammalian lungs slide freely on a fluid film in the body cavity.

In sauropsids (all amniotes except mammals and their ancestral line), highly heterogeneous lungs are always fused to the body wall (Perry 1983). The reverse, however, is not always the case because sea turtles, soft-shelled turtles, and some snakes that have homogeneously distributed parenchyma also have lungs that are fused to the body wall. It is highly probable that sauropods had heterogeneous lungs in which the parts with the greatest concentration of respiratory surface area were located dorsally (Plate 5.1C, D). As in testudinid turtles and birds, the intercostal spaces between adjacent rib articulations to the vertebral column as well as the subjacent third of the body cavity must have contained dense lung parenchyma. The rest of the lung invaded the body cavity as air sacs or sac-like lung regions (see below).

It is probable that the pneumatic spaces evolved separately in sauropodomorphs and theropods and that they therefore are not strictly homologous in sauropods and birds (Wedel 2007, 2009). Basal sauropodomorphs (prosauropods), for example, lack pleurocoels (Wedel 2007; Rauhut et al., this volume). Nevertheless, their presence in both groups is consistent with common ancestry from a crocodilian-like lung structure and could reflect the same response to metabolic demands and structural constraints. The bottom line remains: the presence of pleurocoels strongly indicates heterogeneous lungs in sauropods. In addition, given the trend toward increased pneumaticity that progresses from cranial to caudal in dorsal vertebrae of more derived sauropods, especially among diplodocoids and titanosaurids (Britt 1993; Sanz et al. 1999; Wedel 2009), one can also postulate increased heterogeneity of the lung bauplan in more derived forms (Plate 5.1A, B, Fig. 5.4).

A consequence of large saccular regions or true air sacs is that because of their large volume (air sacs can make up a third of the volume of the body cavity in birds), a long trachea can be present without compromising respiratory efficiency resulting from its large dead space. The long neck of sauropods certainly housed a long trachea that would not have functioned without large saccular regions or true air sacs.

True Air Sacs and Pulmonary Edema

The greatest heterogeneity of tetrapod respiratory systems evolved independently in snakes and birds (Wolf 1933). The parenchyma of the lungs in these two groups has the greatest surface-to-volume ratio within the respective groups: squamates (snakes and lizards) and archosaurs (crocodilians and birds) (Stinner 1982; Duncker & Güntert 1985). The air sacs, on the other hand, are completely devoid of pulmonary vasculature and thus are incapable of gas exchange.

Snakes have elongate lungs that can pervade up to 80% of the body length. Given the relatively low blood pressure in snakes, such long lungs would result in blood flow problems and fluid accumulation in the lungs (pulmonary edema) during vertical movement if the exchange tissue were distributed throughout the whole lung. Concentration of the exchange tissue near the heart solves this problem because pulmonary blood pressure just has to serve the needs of the gas exchange tissue, not that of the nonrespiratory storage/pumping regions (Lillywhite 1987).

The multichambered lungs of turtles, varanid lizards, and crocodilians also display heterogeneity but show a gradient from densely subdivided to sparsely subdivided regions (Perry 1989). The same principle applies to the single-chambered lungs of lizards such as geckos (Perry et al. 1994). The sparsely subdivided sac-like regions in both cases are not devoid of gas exchange function but have mechanical properties similar to true air sacs (Perry & Duncker 1980).

We define true air sacs as being completely devoid of pulmonary vasculature and therefore lacking pulmonary gas exchange function. Sac-like lung regions, on the other hand, may lack parenchymal structures but still contain pulmonary capillaries with a thin air–blood barrier. They are easily ventilated and their diffusing capacity can make up a significant proportion of the total (Perry 1992; Perry et al. 1994).

Although the mechanical properties of sac-like lung regions may be similar to those of true air sacs, their hemodynamic constraints are the same as in other parts of the lungs: the maximum vertical distance of pulmonary vasculature to the heart must be reflected by pulmonary blood pressure. Because lung veins do not have valves, the blood cannot be pushed along from valve to valve by local external pressure on the vessel wall as in systemic veins. True air sacs are free of such gravitational constraints because they have no pulmonary vasculature.

A highly heterogeneous lung bauplan, if present in sauropods, would have allowed them to develop an effective, highly dynamic respiratory system, in which the gas exchange tissue is located near the level of the heart or above (dorsal to it) (Plate 5.1C, D). This means that those parts of the lung that lie more than 1 m below the heart were probably true air sacs. The pulmonary venous blood pressure therefore did not have to be so high (approximately 100 mm Hg) that return of oxygenated blood over a height of 1 m to the heart would be guaranteed. Here, the definition of "true" air sacs is purely anatomical and does not imply homology.

Mass Distribution

As stated by Mallison (Chapter 13 in this volume), the location of the lungs is crucial for determining the center of mass and the distribution of weight on the four legs of sauropods. Of particular importance is the location of the parts that are not flexible, that is, the pneumatic cavities in the bones and that part of the lung that is suspended in the concavity of the dorsal thorax because their positions and volumes stay fixed during breathing and thereby permanently influence the center of gravity of the animal. The air sacs, in turn, also influence the position of the center of gravity but may slightly shift it around the point defined by the rest of the body mass during breathing.

Air Sacs and Temperature Control

Air sacs could have played a role in temperature control as a site of heat transfer to respiratory air from the crop (if present), liver, and cecal fermentation sacs, the organs that account for the greatest percentage of the heat production of an animal at rest (Perry et al. 2009). This function, however, could have been accomplished by extensive interclavicular and abdominal air sacs alone and does not require anterior and posterior thoracic air sacs. Also, to function in heat transfer, the air sac walls do not need pulmonary vasculature because animal tissues mainly consist of water, which has a high thermal storage capacity and conductance. Heat from metabolic processes is thus rapidly distributed throughout the body (also with the help of the circulatory system) and can easily pass through the air sac walls without the need for high vascularization.

How Did Sauropods Breathe?

As noted above, costal breathing (breathing by movement of the ribs) is the plesiomorphic mechanism in amniotes (Brainerd & Owerkowicz 2006). Saurischian dinosaurs, like birds, have distinctly two-headed ribs. The upper head, or tuberculum, articulates on the transverse process of the vertebra, while the lower head, the capitulum, usually articulates on the body of the vertebra. In addition to being stable, this two-headed construction functions like a door hinge, allowing each rib to move in only a single, well defined arc: the swing plane. Because the articulation point of the tuberculum is usually located posterior to that of the capitulum as well as above it, the ribs tend to swing outward and forward at the same time, expanding the rib cage and allowing the animal to breathe in.

The location of the dual articulations of a given rib relative to each other changes as one moves down the spinal column of a sauropod dinosaur: in general, the first two ribs are nearly vertically oriented, with the tuberculum and capitulum placed one above the other, suggesting a supportive rather than respiratory function. In this region, the shoulder girdle is attached to the trunk via the mm. serrati, and the ribs have to support a large proportion of the body weight. As in mounted specimens of theropods (Hengst 1998), the more posterior ribs in sauropods appear to diverge at progressively greater angles, suggesting a respiratory function. Species differences among basal and derived groups are now being investigated to see whether there are any correlations with pneumaticities in the vertebrae. Taken together, these two sources of information could tell us more about the degree of heterogeneity of the lung.

DID SAUROPODS HAVE A CROCODILE-LIKE M. DIAPHRAGMATICUS?

Ruben et al. (1997, 1999) suggested that theropods had a topographic relationship of lung and liver similar to that of crocodiles, namely, that the lungs and heart filled the anterior thorax and the posterior surface of the lungs broadly contacted the anterior surface of the liver. They further postulated that a muscle connected the posthepatic septum with the pubis and, like the m. diaphragmaticus of crocodiles, affected inspiration by pulling the liver back. Finally, they proposed that this mechanism was not a specialization of crocodile-like animals (Crocodyliformes), but rather a feature of all archosaurs, including sauropods, and was later lost in birds. Ruben et al. (1997) further maintained that the pulling back of the liver in nonavian theropods would increase the abdominal pressure during inspiration, thereby making the formation of abdominal air sacs (and hence the theropod origin of birds) functionally impossible.

This side effect is avoided in crocodiles by a simultaneous retraction of the pubis, which articulates freely on the ischium (Farmer & Carrier 2000), providing space for the retraction of liver and digestive tract. Fossil evidence shows that dinosaurs lack a mobile pubis, and a mobile pubis is in fact a characteristic of crocodiliforms (Claessens 2004). Because it is more likely that the m. diaphragmaticus evolved together with the kinetic pubis rather than without it, the kinetic pubis is an indicator of the presence of this muscle. Thus, as pointed out by Hicks & Farmer (1998) and by Claessens (2004), sauropods, lacking a kinetic pubis, also must have lacked a m. diaphragmaticus. This conclusion is also supported by indirect evidence for the presence of abdominal air sacs provided by pneumaticities of the sacral and proximal caudal vertebrae of highly derived sauropods (Wedel 2003a, 2009). Pending some direct evidence to the contrary, it is therefore most likely that sauropods were rib breathers.

Conclusions

Extant phylogenetic bracketing and functional morphological approximation suggest a highly heterogeneous, bird-like respiratory system in sauropod dinosaurs. Functional morphological approximation suggests that a voluminous respiratory system with air sacs would solve hemodynamic problems associated with large lungs and also act as an active cooling mechanism, keeping the body temperature of the animal within realistic ranges. Furthermore, the high-performance, bird-like lung type with its crosscurrent exchange model would be best suited to serve a broad spectrum of metabolic rates most likely present over the life span of a sauropod dinosaur (Sander & Clauss 2008).

Breathing was accomplished by movement of the posterior ribs, which expanded the lateral wall of the air sacs on inspiration, the medial wall being stabilized against the large mass of liver, intestine, and cecal fermentation sacs. Because the respiratory cycle resulted in almost constant air movement along the respiratory surfaces, continuous gas exchange for highly effective oxygen supply became possible.

The implications of a bird-like respiratory system for sauropod dinosaurs are that the capacity of the respiratory system to deliver oxygen was certainly not a limiting factor in gigantism. On the contrary, the large air sacs allowed the animals to become extremely large without a proportional increase in mass. In fact, one probable key adaptation to gigantism in sauropods (Sander & Clauss 2008; Sander et al. 2010), the long neck, could not have evolved without true air sac or sacular regions having overcome the problem of tracheal dead space.

Acknowledgments

We thank Martin Sander (University of Bonn) and Mathew Wedel (Department of Anatomy and College of Podiatric Medicine, Pomona) for discussion. This chapter was greatly improved by the comments of an anonymous reviewer. We thank the Deutsche Forschungsgemeinschaft for funding. This is contribution number 64 of the DFG Research Unit 533 "Biology of the Sauropod Dinosaurs: The Evolution of Gigantism."

References

Brackenbury, J. H. & Amaku, J. A. 1990. Respiratory responses of domestic fowl to hyperthermia following selective air sac occlusions.—*Experimental Physiology* 75: 391–400.

Brainerd, E. L. 1999. New perspectives on the evolution of lung ventilation mechanisms in vertebrates.—*Experimental Biology* 4: 11–28.

Brainerd, E. L. & Owerkowicz, T. 2006. Functional morphology and evolution of aspiration breathing in tetrapods.—*Respiratory Physiology and Neurobiology* 154: 73–88.

Britt, B. B. 1993. *Pneumatic Postcranial Bones in Dinosaurs and Other Archosaurs*. Ph.D. Dissertation. University of Calgary, Calgary.

Cameron, J. N. 1989. *The Respiratory Physiology of Animals*. Oxford University Press, Oxford.

Claessens, L. 2004. Archosaurian respiration and the pelvic girdle aspiration breathing of crocodyliforms.—*Proceedings of the Royal Society B: Biological Sciences* 271: 1461–1465.

Codd, J. R., Manning, P. L., Norell, M. A. & Perry, S. F. 2007.

Avian-like breathing in maniraptoran dinosaurs.—*Proceedings of the Royal Society B: Biological Sciences* 275: 157–161.

Coulson, R. A. & Hernandez, T. 1983. Alligator metabolism. Studies on chemical reactions in vivo.—*Comparative Biochemistry and Physiology* 74: 1–18.

Duncker, H.-R. 1971. The lung air sac system of birds. *Advances in Anatomy, Embryology and Cell Biology* 45: 1–171.

Duncker, H.-R. 1978. General morphological principles of amniotic lungs. *In* Piiper, J. (ed.). *Respiratory Function in Birds, Adults and Embryonic.* Springer, Berlin: pp. 2–22.

Duncker, H.-R. & Güntert, M. 1985. The quantitative design of the avian respiratory system—from humming-bird to mute swan. *In* Nachtigall, W. (ed.). *BIONA-Report 3: Bird Flight: Aspects of Breathing and Energetics*. Gustav Fischer, Stuttgart: pp. 361–378.

Farmer, C. G. & Carrier, D. R. 2000. Pelvic aspiration in the American alligator (*Alligator mississippiensis*).—*Journal of Experimental Biology* 203: 1679–1687.

Farmer, C. G. & Sanders, K. 2010. Unidirectional airflow in the lungs of alligators.—*Science* 327: 338–340.

Hengst, R. 1998. Lung ventilation and gas exchange in theropod dinosaurs.—*Science* 281: 47.

Hicks, J. W. & Farmer, C. G. 1998. Lung ventilation and gas exchange in theropod dinosaurs.—*Science* 281: 45–46.

Janensch, W. 1950. Die Wirbelsäule von *Brachiosaurus brancai.*—*Palaeontographica S* 7 (3): 27–93.

Lillywhite, H. B. 1987. Circulatory adaptations of snakes to gravity.—*American Zoologist* 27: 81–95.

Locy, W. W. & Larsell, O. 1916a. The embryology of the bird's lung based of observations of the domestic fowl. Part 1.—*American Journal of Anatomy* 19: 447–504.

Locy, W. W. & Larsell, O. 1916b. The embryology of the bird's lung based of observations of the domestic fowl. Part 2.—*American Journal of Anatomy* 20: 1–44.

Lovelace, D. M., Hartman, S. A. & Wahl, W. R. 2007. Morphology of a specimen of *Supersaurus* (Dinosauria, Sauropoda) from the Morrison Formation of Wyoming, and a re-evaluation of diplodocoid phylogeny.—*Arquivos do Museu Nacional, Rio de Janeiro* 65: 527–544.

Mallison, H. This volume. *Plateosaurus* in 3D: how CAD models and kinetic–dynamic modeling bring an extinct animal to life. *In* Klein, N., Remes, K., Gee, C. T. & Sander, P. M. (eds.). *Biology of the Sauropod Dinosaurs: Understanding the Life of Giants.* Indiana University Press, Bloomington: pp. 219–236.

O'Connor, P. M. 2006. Postcranial pneumaticity: an evaluation of soft-tissue influences on the postcranial skeleton and the reconstruction of pulmonary anatomy in archosaurs.—*Journal of Morphology* 267: 1199–1226.

O'Connor, P. M. & Claessens, L. 2005. Basic avian pulmonary design and flow-through ventilation in non-avian theropod dinosaurs.—*Nature* 436: 253–256.

Perry, S. F. 1983. Reptilian lungs. Functional anatomy and evolution.—*Advances in Anatomy, Embryology and Cell Biology* 79: 1–81.

Perry, S. F. 1988. Functional morphology of the lungs of the Nile crocodile *Crocodylus niloticus:* non-respiratory parameters.—*Journal of Experimental Biology* 134: 99–117.

Perry, S. F. 1989. Morphometry of crocodilian lungs.—*Fortschritte der Zoologie* 35: 546–549.

Perry, S. F. 1992. Gas exchange strategies in reptiles and the origin of the avian lung.—*Lung Biology in Health and Disease* 56: 149–167.

Perry, S. F. 1998. Lungs: comparative anatomy, functional morphology and evolution. *In* Gans, C. & Gaunt, A. S. (eds.). *Biology of the Reptilia, Vol. 19: Functional Morphology and Evolution.* Society for the Study of Amphibians and Reptiles, Ithaca: pp. 1–92.

Perry, S. F. 2010. Atmungsorgane. *In* Westheide, W. & Rieger, R. (eds.). *Spezielle Zoologie Teil 2: Wirbel-oder Schädeltiere,* 2nd edition. Spektrum Akademischer Verlag, Heidelberg: pp. 127–141.

Perry, S. F. & Duncker, H.-R. 1980. Interrelationship of static mechanical factors and anatomical structure in lung evolution.—*Journal of Comparative Physiology Part B: Biochemical, Systems, and Environmental Physiology* 138: 321–334.

Perry, S. F. & Reuter, C. 1999. Hypothetical lung structure of *Brachiosaurus* (Dinosauria: Sauropoda) based on functional constraints. *Mitteilungen aus dem Museum für Naturkunde in Berlin, Geowissenschaftliche Reihe* 2: 75–79.

Perry, S. F. & Sander, P. M. 2004. Reconstructing the evolution of the respiratory apparatus in tetrapods.—*Respiratory Physiology and Neurobiology* 144: 125–139.

Perry, S. F., Hein, J. & van Dieken, E. 1994. Gas exchange and morphometry of the lungs of the tokay, *Gekko gekko* (Reptilia: Squamata: Gekkoniodae).—*Journal of Comparative Physiology Part B: Biochemical, Systems, and Environmental Physiology* 164: 206 -214.

Perry, S., Christian, A., Breuer, T. Pajor, N. & Codd, J. 2009. Implications of an avian-style respiratory system for gigantism in sauropod dinosaurs.—*Journal of Experimental Zoology Part A: Ecological Genetics and Physiology* 311A: 600–610.

Piiper, J. & Scheid, P. 1972. Maximum gas transfer efficiency of models for fish gills, avian lungs and mammalian lungs.—*Respiration Physiology* 14: 115–124.

Piiper, J. & Scheid, P. 1986. Models for comperative functional analysis of gas exchange organs in vertebrates.—*Journal of Applied Physiology* 53: 1321–1329.

Rauhut, O. W. M., Fechner, R., Remes, K. & Reis, K. This volume. How to get big in the Mesozoic: the evolution of the sauropodomorph body plan. *In* Klein, N., Remes, K., Gee, C. T. & Sander, P. M. (eds.). *Biology of the Sauropod Dinosaurs: Understanding the Life of Giants.* Indiana University Press, Bloomington: pp. 119–149.

Ruben, J. A., Jones, T. D., Geist, N. R. & Hillenius, W. J. 1997. Lung structure and ventilation in theropod dinosaurs and early birds.—*Science* 278: 1276–1270.

Ruben, J. A., Dal Sasso C., Geist, N. R., Hillenius, W. J., Jones, T. D. & Signore, M. 1999. Pulmonary function and metabolic physiology of theropod dinosaurs.—*Science* 283: 514–516.

Sander, P. M. & Clauss, M. 2008. Sauropod gigantism.—*Science* 322: 200–201.

Sander, P. M., Christian, A., Clauss, M., Fechner, R., Gee, C. T., Griebeler, E. M., Gunga, H.-C., Hummel, J., Mallison, H., Perry, S., Preuschoft, H., Rauhut, O., Remes, K., Tütken, T., Wings, O. & Witzel, U. 2010. Biology of the sauropod dinosaurs: the evolution of gigantism.—*Biological Reviews of the Cambridge Philosophical Society*. doi: 10.1111/j.1469=185X.2010.00137.x.

Sanz, J. L., Powell, J. E., LeLoeuff, J., Martinez, R. & Pereda Superbiola, X. 1999. Sauropod remains from the Upper Cretaceous of Laño (north central Spain). Titanosaur phylogenetic relationships.—*Estudios del Museo de Ciencias Naturales de Alava* 14: 235–255.

Scheid, P. 1979. Mechanisms of gas exchange in bird lungs.—*Respiration Physiology* 86: 138–186.

Stinner, J. N. 1982. Functional anatomy of the lung of the snake *Pituophis melanoleucus*.—*American Journal of Physiology: Regulatory, Integrative and Comparative Physiology* 243: 251–257.

Torday, J. S., Rehan, V. K., Hicks, J. W., Wang, T., Maina, J., Weibel, E. R., Hsia, C. C. W., Sommer, R. J. & Perry, S. F. 2007. Deconvoluting lung evolution: from phenotypes to gene regulatory networks.—*Integrative and Comparative Biology* 47: 601–609.

Upchurch, P., Barrett, P. M. & Dodson, P. 2004. Sauropoda. *In* Weishampel, D. B., Dodson, P. & Osmólska, H. (eds.). *The Dinosauria. 2nd edition*. University of California Press, Berkeley: 259–322.

Uriona, T. J. & Farmer, C. G. 2008. Recruitment of the diaphragmaticus, ischiopubis and other respiratory muscles to control pitch and roll in the American alligator (*Alligator mississippiensis*).—*Journal of Experimental Biology* 211: 1141–1147.

Wedel, M. J. 2003a. Vertebral pneumaticity, air sacs, and the physiology of sauropod dinosaurs.—*Paleobiology* 29: 243–255.

Wedel, M. J. 2003b. The evolution of vertebral pneumaticity in sauropod dinosaurs.—*Journal of Vertebrate Paleontology* 23: 344–357.

Wedel, M. J. 2005. Postcranial skeletal pneumaticity in sauropods and its implications for mass estimates. *In* Curry Rogers, K. & Wilson, J. A. (eds.). *The Sauropods: Evolution and Paleobiology*. University of California Press, Berkeley: pp. 201–228.

Wedel, M. J. 2006. Origin of postcranial skeletal pneumaticity in dinosaurs.—*Integrative Zoology* 2: 80–85.

Wedel, M. J. 2007. What pneumaticity tells us about "prosauropods," and vice versa. *Special Papers in Palaeontology* 77: 207–222.

Wedel, M. J. 2009. Evidence for bird-like air sacs in saurischian dinosaurs.—*Journal of Experimental Zoology A* 311: 1–18.

Wilson, J. A. 2002. Sauropod dinosaur phylogeny: critique and cladistic analysis.—*Zoological Journal of the Linnean Society of London* 136: 217–276.

Witmer, L. M. 1995. The extant phylogenetic bracket and the importance of reconstructing soft tissues in fossils. *In* Thomason, J. J. (ed.). *Functional Morphology in Vertebrate Paleontology*. Cambridge University Press, New York: pp. 19–33.

Wolf, S. 1933. Zur Kenntnis von Bau und Funktion der Reptilienlunge.—*Zoologische Jahrbücher, Abteilung Anatomie und Ontogenie der Tiere* 57: 139–190.

6

Reconstructing Body Volume and Surface Area of Dinosaurs Using Laser Scanning and Photogrammetry

STEFAN STOINSKI, TIM SUTHAU, AND HANNS-CHRISTIAN GUNGA

CRUCIAL BASELINE DATA in dinosaur paleobiology and for reconstructing dinosaurs as living animals are accurate measurements of one- to three-dimensional features such as length, surface area, and volume. Dinosaur skeletons mounted and on display in museums offer the opportunity to obtain such data and to create digital models of them. These models, in turn, serve as the basis for estimating physiological and other biological parameters. In this chapter, we provide an overview of data capture using laser scanning and photogrammetrical methods and describe the working steps from the captured point clouds of the skeleton to the final volume model. Because dinosaur skeletons are complex objects with irregular structures, laser scanning proved to be much more accurate for capturing their shape than previously used methods such as photogrammetry. The modeling of the body surface area and body volume with digital techniques is also more accurate than established methods that are based on scale models. Here, nonuniform rational B spline (NURBS) curves and CAD software are used to reconstruct the body surface and for surface area and volume calculations.

Introduction

Body size is one of the most fundamental attributes of any organism, and it has direct consequences for design, life history, and ecology (Clutton-Brock et al. 1980; Peters 1983; Schmidt-Nielsen 1984; Alexander 1998; Makarieva et al. 2005; Bonner 2006; Hunt & Roy 2006). This applies particularly to the largest terrestrial vertebrates that ever lived, the sauropod dinosaurs, and reliable estimates of their body mass are central for the understanding of their biology and gigantism. Using allometric functions (Schmidt-Nielsen 1984, 1997; Calder 1996), a multitude of physiological data can be estimated if body mass is known. However, highly disparate body mass estimates for sauropods are found in the literature (Peczkis 1994; Appendix).

Two types of methods are used for body mass estimates. First, various approaches are used to produce some estimate of body volume which is then converted into body mass using the specific density of the animal (Colbert 1962; Gunga et al. 1999). Second, some workers have produced estimates that are based on a biomechanical or scaling approach, for example using long bone circumferences (Anderson et al. 1985; formula corrected by Alexander 1989). The pros and cons of each method have already been extensively discussed (Lambert 1980; Schmidt-Nielsen 1984; Anderson et al. 1985; Haubold 1990; Paul 1997; Henderson 1999; Montani 2001; Seebacher 2001; Christiansen & Farina 2004; Mazzetta et al. 2004; Foster 2007).

In this chapter, we review our method for estimating body mass on the basis of reconstructed body volume, using either a laser scanner or classical photogrammetry (Gunga et al. 1995, 1999, 2002, 2007, 2008) with a particular focus on sauropod dinosaurs. We discuss the advantages of both methods, provide a step-by-step description for each method, and offer some results.

Data Acquisition: Digitizing Dinosaur Skeletons

A mounted dinosaur skeleton in a museum consists of many individual bones, the position and form of which must be determined, and the bones are held together by partially visible support elements. The purpose of both classical photogrammetry and laser scanning is to describe the skeleton as a number of points in space, which is the principle of geodetic measurements. However, geodetic measurements require a network of geodetically surveyed control or reference points. This applies to capture methods for dinosaur skeletons as well as to the impossibility of capturing the entire skeleton from a single camera or scanner position. Thus, several viewpoints will have to be used during a capture campaign, but the reference points will remain the same. The reference points are numbered and are captured from every scanner or camera position. Depending on the size of the object to be captured, in this study, 10 to 20 reference points were placed on and around each skeleton. The reference points consists of styrofoam spheres with a diameter of either 60 or 80 mm and hollow plastic spheres (Fig. 6.1) supplied with the Mensi S25 laser scanner we use. The plastic spheres have a diameter of 76.2 mm and are mounted on a magnetic base plate. The use of spheres of a known size as reference points is the optimal method because in a point cloud generated by a scanner, the center of the sphere can be computed from any viewpoint. In photographic images, circles or ellipses are easy to detect and thus can be matched easily. Further reference points for photogrammetric images that we used were reflective paper markers attached to the skeleton and background.

PHOTOGRAMMETRY

The goal of photogrammetric methods is to generate a 3D object from 2D data. Compared to laser surface scanning for capturing dinosaur skeletons, the advantages of photogrammetry are that it records color information, has an unlimited range, offers fast data acquisition, and can use historical imagery. The capture principle is stereophotogrammetry. Pictures of the same object are taken from two different viewpoints; the distance between the projection centers of the two pictures is called the *base*. The viewing direction is perpendicular to the base, as in the case with our eyes. If the image arrangement, which is the plane, orientation, and distance to the object, is in agreement for each eye, we are able to see a 3D image. This principle is used in analytical stereophotogrammetric instruments such as the Kern DSR 11 at the Technical University of Berlin, Germany. With this digital photogrammetric surveying instrument, it is possible to calculate 3D points from a pair of 2D images by means of special mathematical formulas known as colinearity equations. The images generated by the camera are used in digital form by the instrument.

The first step in stereophotogrammetry is to calculate the exterior orientation of the image arrangement during capturing. The exterior orientation is the position of the photo camera in space, the view directions, and the correlation between the camera positions during image capture. The second bit of information needed is the interior orientation of the cameras (the geometric parameters of the imaging process such as focal length). After determining the exterior orientation of the cameras, the reference points must be identified in both images. As a result of this work, the object becomes visible in 3D on the screen. Now prominent object points (such as the tip of the snout and the tail, and the tips of the toes) can be identified. These points are then saved to a file and connected to each other in CAD software such as AutoCAD. The final result is a wireframe model of the captured object. However, analytical stereophotographic instruments are rarely used today because of the rapid development of computer hardware and software that allow creating a 3D point cloud on a PC screen with a scanner and appropriate software.

In the 1990s, the Department of Photogrammetry and Remote Sensing (now Computer Vision and Remote Sensing) at the Technical University of Berlin, Germany, started capturing images of dinosaur skeletons with a Zeiss TMK 6 measurement photo camera on glass negatives. The advantage of using these over any type of flexible film is that they do not suffer from distortion during photographic exposure. Furthermore, a Réseau grid was used, which is a thin glass sheet covered with an etched grid of known dimensions. Because the grid was visible on every image, distortions that may result from deformations of the image during darkroom processing can be compensated for mathematically.

LASER SCANNING

Just like photogrammetry, laser scanning is a remote and noninvasive capture method that on a smaller scale is increasingly used in paleontology (Breithaupt et al. 2004; Bates et al. 2008, 2009; Petti et al. 2008). The advantages of laser scanning are the direct acquisition of 3D points and the enormous number of 3D points that can be captured to describe a surface. This leads to laser scanning being a prime technique for the capture of irregular surfaces such as reliefs, sculptures, and columns in architecture, or, in our case, dinosaur skeletons, and the results are available within a very short time. Another advantage is that the emerging point cloud is displayed during the capture campaign, that is, as soon as the scan is performed. This allows for an immediate check for the completeness of coverage and the subsequent selection of additional viewpoints.

Since 2003, the main focus of our team was on the survey of dinosaur skeletons using a laser scanner. However, we are still using a SLR Canon EOS 1D Mark II camera to photograph the dinosaur skeletons. As in the earlier photogrammetric work, we applied reflective paper markers to the skeletons for detecting the position of the camera in space. The use of photographic images is necessary for backup and to texture the point cloud resulting from the laser scan. This means that a photo of the object is projected onto the corresponding point cloud so that objects such as individual bones can be identified from the point cloud and separated from the background.

The scanner we are using is a model S25 laser scanner manufactured by Mensi SA corporation (now Trimble Inc.). The measurement range of the Mensi S25 is between 0.8 and 25 m. The speed is up to 100 points per second, and the accuracy perpendicular to range is about 0.8–3.8 mm vertically and 0.2–3.4 mm horizontally. The distance accuracy is about 0.2 mm at 4 m and 1.4 mm at 10 m (Boehler et al. 2003). Tall objects can be captured quickly as well because the field of view is about 320° × 46°.

As a triangulation laser scanner, the Mensi S25 uses the principle of triangulation, which relies on the proper positioning of its components, that is, the laser, acceptor optic, reflectance mirror, and a charge-coupled device (CCD) sensor, to each other (Fig. 6.1). The laser creates a spotlight on the surface to be scanned, the reflection of which is collected by the acceptor optic and received by the CCD sensor. The software of the scanner then computes the position of the spotlight on the surface in a local coordinate system. Before scanning, the object is surrounded by numbered reference spheres (Fig. 6.1; Plate 6.1), the location of which must not be changed during the scan. This is important because the scanner defines a new local coordinate system for each viewpoint (Plate 6.1).

Before scanning the whole object, only the reference spheres are scanned first. This way, the accuracy of the scan can be determined immediately because the diameter of the spheres is known and can be compared to the value obtained by the scanner. The scanner software then automatically selects the spacing of the scan lines and the distance of the scan points, but the spacing can also be set manually. Each viewpoint of the scanner captures a different point cloud of the object, with all point clouds having their own local coordinate system (Plate 6.1C). To obtain a single point cloud, all of the local systems have to be transformed into a common one. This procedure is called registration (Plate 6.1D) and requires a minimum of three reference spheres for each viewpoint. The better the scan of the reference spheres from each viewpoint, the better the registration accuracy.

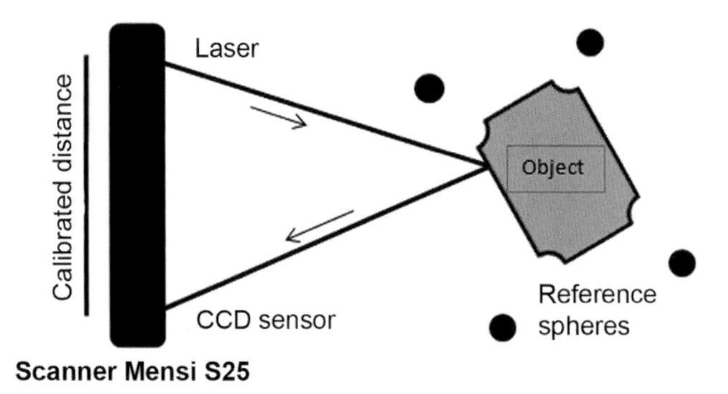

FIGURE 6.1. Principle of triangulation laser scanning. The purpose of laser triangulation is to measure the distance between the laser source and the object point as well as to determine the direction to this point. Using the distance between the CCD sensor and the laser source, the CCD sensor and scanner software locate the reflected laser beam in the digital image generated by the CCD and then calculate the 3D coordinates of the point in a local coordinate system. This process is repeated many times along scan lines, generating a point cloud image in the computer of the part of the object that is visible to the scanner. Reference spheres are placed around the object to facilitate the merger of point clouds obtained from different scanner positions (registration).

The next step is one of the most time consuming. This is the editing of the point cloud to remove all parts that are not of interest, such as the support elements of the skeleton, the floor and background, and the reference spheres. The result of this editing is a cleared point cloud representing only the skeleton (Plate 6.1E). The editing is done using a special function in the scanner software.

Modeling Body Volumes

The first step in obtaining the body volume of a dinosaur is the modeling of the body surface. In the 1990s, when we started modeling dinosaurs (Gunga et al. 1995), paper copies of scaled-down silhouettes of the captured skeletons were printed. Then the physiologist on our team drew the anatomically plausible outline of the animal on the paper. The body surface was obtained by fitting several geometrical parts, especially regular geometrical primitives such as cylinders, cones, and truncated cones, into the reconstructed body outline (Gunga et al. 1995; Fig. 6.2). The surface area and the volume of these geometric solids were then calculated and added up to obtain the total volume and surface area of the reconstructed animal.

With the data generated by the laser scanner, namely, the point clouds, we initially took a CAD-based approach. Again on the basis of general anatomical and physiological considerations, the body surface of the animal was reconstructed and then separated into rotational solids in AutoCAD

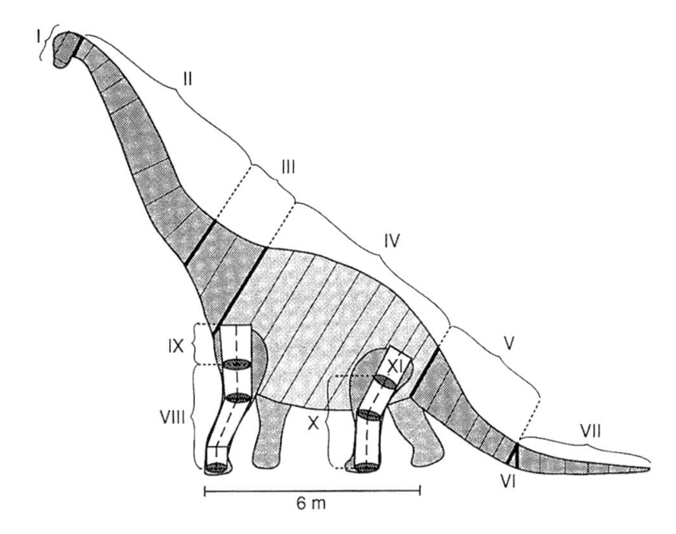

FIGURE 6.2. The first volume model of a dinosaur using photogrammetry. The model is based on the historic mount of *Brachiosaurus brancai* on exhibit at the Museum für Naturkunde, Berlin, Germany. To obtain the volume estimate, the body was divided into segments consisting of geometric primitives such as truncated cones and cylinders. The Roman numerals indicate the body regions. *From Gunga et al. (1995).*

(Fig. 6.2). The solids were then merged in the CAD program, and body volume and body surface area were calculated with ease. This method resulted in a better approximation of volume and surface area than for the initial models that are based on photogrammetry.

The advantage of this approach was that we did not have to depend on the use of geometrical primitives. However, the result still did not look realistic, as illustrated by our modeling of an Indian elephant (Fig. 6.3), leading us to search for software that would allow modeling dinosaurs in a more realistic way. The solution is the use of nonuniform rational B splines (NURBS). These are mathematical curves used for creating and displaying 2D lines, curves, and rectangles, but they also work in creating 3D freehand shapes and 3D volume solids. The outlines of the body shape, which originally had been drawn by hand, are now produced digitally. To obtain more accurate models, several outlines from different view angles and from cross sections must be drawn. Parts of the body, such as the neck or a leg, can be outlined separately. When all outlines are drawn, the 3D model is created with a few mouse clicks from these outlines. The outlines and models are relatively easy to deform in order to obtain a better approximation of the body surface.

After this initial reconstruction, the models are reviewed in collaboration with physiologists and paleontologists. If the model lives up to their expectations, the body volume and the body surface area are calculated. The body mass in kilograms is obtained by multiplying the body volume (expressed in liters) by the hypothesized specific density of the dinosaurs. In most land vertebrate groups, specific density is close to 1, which is the value used in our earlier studies (Gunga et al. 1995, 1999, 2002). However, the current consensus is that sauropod dinosaurs had a heavily pneumatized skeleton and large air sacs (Perry 2001; Henderson 1999; Wedel 2003a, 2003b, 2005; Schwarz & Fritsch 2006; Perry et al., this volume), resulting in a much lower specific density of only 0.8 to 0.9 kg/l. Thus, a value of 0.8 kg/l is used in our current studies (Gunga et al. 2007, 2008), following a suggestion by Wedel (2005).

The size of interior organs and organ systems can be estimated by allometric equations that are based on the initial values for body volume and body surface area (Ganse et al., this volume). These estimates can then be used for checking the accuracy of the initial model by fitting the organs and organ systems into the appropriate space in the body cavity of the model. If the fit is poor (e.g., the organs are too large or too small for the predicted ratio between body mass and organ size), the model must be adapted. For example, in our recent study of the basal sauropodomorph *Plateosaurus engelhardti* (Gunga et al. 2007), a robust model was initially created, but after the reconstruction of the organs and organ systems, this model turned out to be too bulky (Fig. 6.4, Plate 7.1). In response, a slimmer version of *Plateosaurus* was created that shows better agreement with organ size, which also resulted in a different center of gravity (Gunga et al. 2007; Fig. 6.4, Plate 7.1).

SCANNING EXTANT ANIMALS FOR MODEL VALIDATION

To test the accuracy of the volume estimates obtained by our CAD reconstruction method, we scanned a stuffed rhinoceros of known life body mass. The specimen is on display in the Senckenberg Naturhistorische Sammlung Dresden, Germany (formerly the Tierkundemuseum Dresden). It took us two days using 15 reference spheres and 16 scanner positions to scan this object, which fits into a space of about $3.00 \times 0.80 \times 1.50$ m. The resulting point cloud includes over 570,000 points, and the accuracy of the registration was better than 3 mm, which is also the overall accuracy of the scan (Fig. 6.5). According to museum records, the life mass of the rhinoceros was 1,050 kg. Because the computed volume enclosed by the scanned body surface is 0.914 m^3, a specific density of 1.15 kg/l for this specimen was obtained.

Next, to test the accuracy of volume reconstructions based solely on skeletons, we also scanned the skeleton of an Indian elephant of known life mass. The skeleton comes from a juvenile individual and is mounted in the Zoological Museum of the University of Copenhagen, Denmark (Fig. 6.3A). We also took two days to capture the skeleton and used 7 scanner positions and 15 reference points. The elephant skeleton fits into a space of about $1.70 \times 1.50 \times 0.70$ m. The registration accuracy of the over 920,000 points is better than 1 mm.

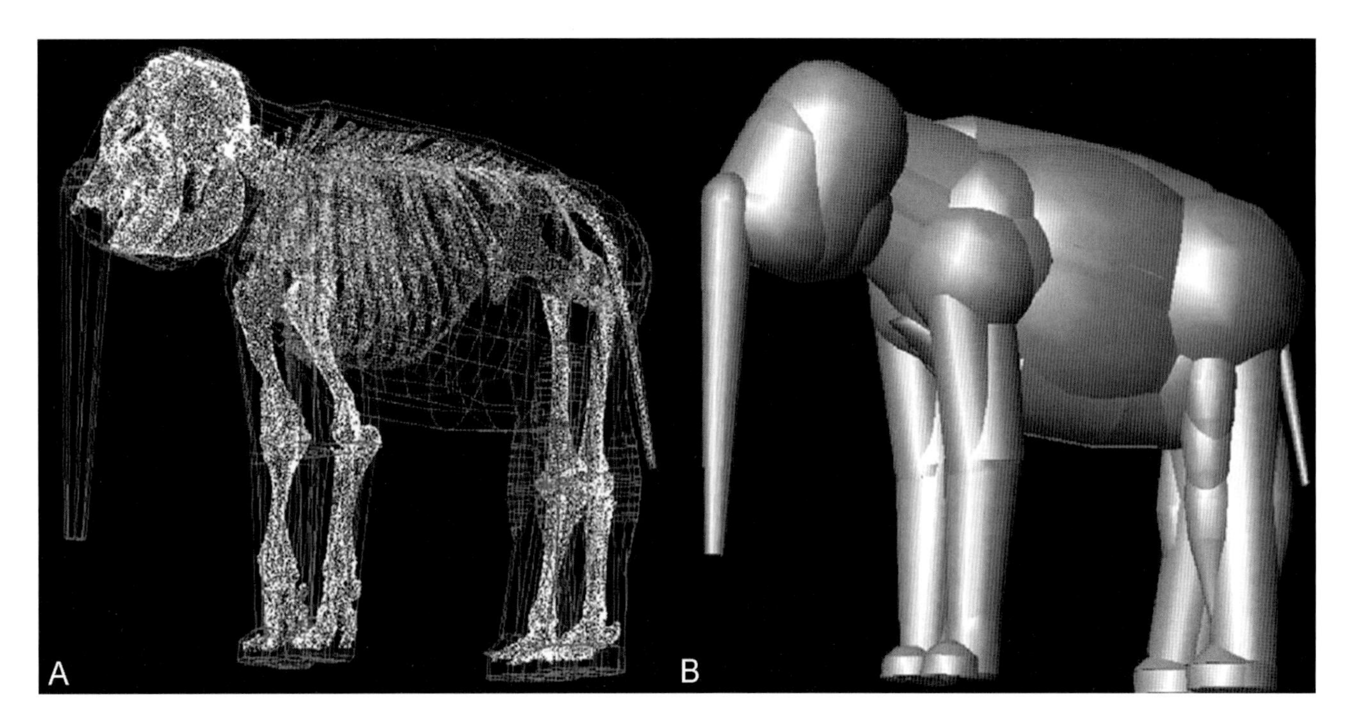

FIGURE 6.3. Volume model of a juvenile Indian elephant based on a laser scan of a skeleton mounted at the Zoological Museum of the University of Copenhagen. (A) Wireframe model reconstruction of the animal based on the point cloud resulting from the laser scan of the skeleton. (B) In the next step, the wireframe model was separated into rotational solids with the software AutoCAD. The volume was obtained by merging the rotational solids in AutoCAD.

When the body surface of the juvenile elephant was modeled (Fig. 6.3B), it was found to enclose a volume of 0.622 m^3, resulting in a mass estimate of 715 kg using the specific density of 1.15 kg/l obtained from the stuffed rhino specimen. This mass compares well to the known life mass of the elephant of 850 kg, as our method underestimated its true mass by only 16%. This underestimate could be the result of density differences, a volume underestimation, or a combination of both.

SCANNED DINOSAUR SKELETONS

Several sauropodomorph skeletons in museums in Europe, Africa, and China have been scanned by our group since 2004 (Table 6.1). The goal of our project is to obtain 3D data for taxa representative of sauropod diversity in space and time. In addition, several other dinosaur and one fossil mammal skeletons were scanned because this could be done with relatively little effort.

The first dinosaur to be scanned in 2004 was a skeleton of the basal sauropodomorph *Plateosaurus engelhardti* on display at the University of Tübingen (specimen GPIT 1). We used 10 scanner positions and 17 reference spheres. The skeleton fits into a space of about $4.30 \times 0.80 \times 1.90$ m^3, and it took three days to capture it, resulting in over 1,100,000 points. The accuracy was better than 3 mm.

The skeleton of the basal macronarian sauropod *Atlasaurus imelakei* from Morocco was scanned in 2004 as well. The skeleton is on display in the Moroccan Ministry of Energy and Mining in Rabat and nearly takes up the entire hall it resides in. This made it very difficult to find the right scanner positions. We needed four days to scan 660,000 points from 16 different scanner positions using 23 reference spheres.

In conjunction with a meeting of the Research Unit at the Sauriermuseum in Aathal, Switzerland, we took the opportunity to scan a wall-mounted *Camarasaurus* skeleton nicknamed ET and freestanding *Allosaurus* and *Stegosaurus* skeletons (Fig. 6.6). We used only one scanner position each for the *Camarasaurus* and the *Allosaurus* skeletons because of the nature of the mount of the former and the inaccessibility (it was mounted high up on a ledge) of the latter. The *Stegosaurus* skeleton was scanned from two scanner positions, but without reference spheres. This meant that the point clouds had to be registered using their own points, which is less accurate and more difficult to perform without specialized software (e.g., PolyWorks).

An extensive scanning campaign in China, at the Beijing Natural History Museum and at the Zigong Dinosaur Museum, was possible in 2005. In addition to the necessary permits to scan the specimens, we also needed authorization for importing the laser scanner and associated equipment. The cooperation between authorities and scientists was excel-

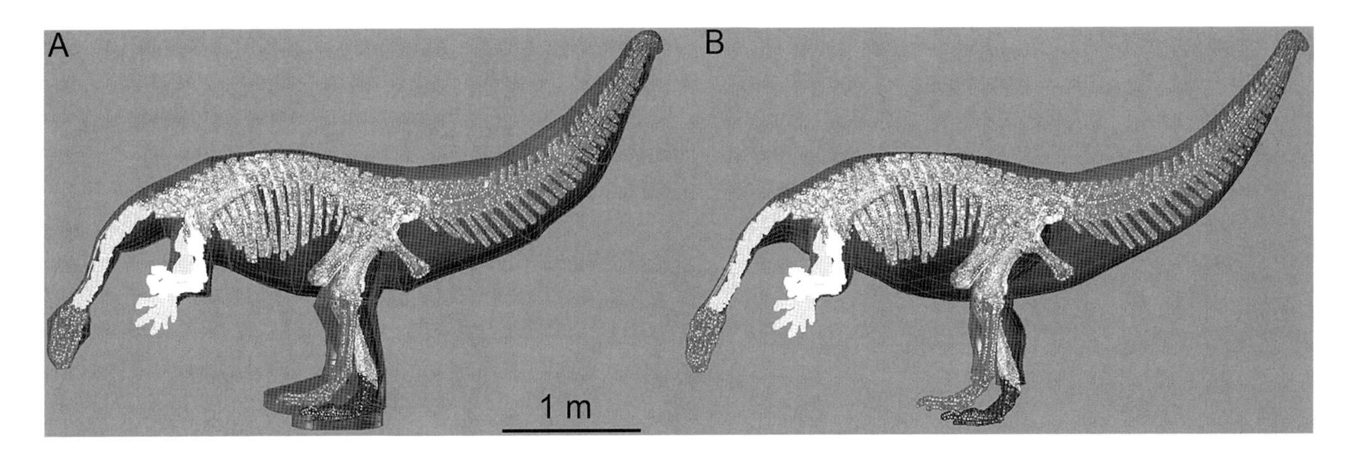

FIGURE 6.4. For a color version of this illustration see Plate 7.1. Alternative volume models of *Plateosaurus engelhardti* based on laser scans of the mounted skeleton GPIT 1 at the University of Tübingen, Germany. The volume models were reconstructed with NURBS curves. The point cloud generated by scanning the skeleton is visible inside the models. The colors denote the different body regions. A robust version (A) and a slim version (B) were generated to test which model would be more consistent with the space required by the internal organs. The thinner model turned out to be the preferred reconstruction. Note the much thinner hindlimbs and feet in the slim model.

lent, and we scanned more specimens than we had originally planned (Table 6.1). The sauropodomorphs scanned at the Beijing Museum included a skeleton of the basal sauropodomorph *Lufengosaurus huenei* and a skeleton of the extremely long-necked basal eusauropod *Mamenchisaurus jingyanensis*. At the Zigong Dinosaur Museum, we were able to scan a total of 10 skeletons (Table 6.1), five of which were sauropods. These include a juvenile and an adult of the basal sauropod *Shunosaurus lii*, another skeleton of *Mamenchisaurus* (labeled as *M. hochuanensis*), a skeleton of *Omeisaurus tianfuensis*, which is closely related to *Mamenchisaurus*, and finally *Datousaurus bashanensis*, which is of uncertain systematic affinities.

The skeletons in Beijing were scanned in seven days with more than 8 viewpoints and 15 reference spheres. The registration accuracy was better than 4 mm. In Zigong, we needed 42 reference spheres and 20 viewpoints, and the accuracy was better than 14 mm.

In the 1990s, the original skeletons of *Brachiosaurus brancai* and *Dicraeosaurus hansemanni* and the cast skeleton of *Diplodocus carnegii* mounted in the Museum für Naturkunde, Berlin, were captured via photogrammetric methods. The year 1997 was the first in which we used a laser scanner to capture the same skeletons of *Dicraeosaurus* and *Diplodocus*. The measurement accuracy of the scans was better than 5 mm, and more than 800,000 object points for *Diplodocus* and more than 500,000 points for *Dicraeosaurus* were captured from 21 and 8 viewpoints, respectively.

In 2007, the Museum für Naturkunde, Berlin, refurbished its famous dinosaur hall, an occasion for which the gigantic *Brachiosaurus brancai* skeleton and the *Diplodocus carnegii* and *Dicraeosaurus hansemanni* skeletons were remounted (Remes et al., this volume). For comparison with the old mount, the skeleton of *Brachiosaurus* was scanned again in 2008. It took us only one day and seven viewpoints to scan the skeleton with an accuracy of better than 5 mm. The accuracy of the scan of the long neck and the head are only better than 10 mm because the accuracy of the laser scanner decreases with distance, and we did not have a lift platform to capture the higher parts of the skeleton.

REVISED BODY MASS ESTIMATE FOR *BRACHIOSAURUS*

As described earlier, the modeling of the *Plateosaurus engelhardti* skeleton GPIT 1 had taught us that the reconstructed organs and organ systems seem to fit better with a thin model, although a considerable degree of uncertainty exists in regard to the numerous factors influencing the volume estimates (see Gunga et al. 2007 for discussion).

Thus, we applied the methodology established for *Plateosaurus* to create a revised model of the *Brachiosaurus brancai* skeleton in Berlin (Gunga et al. 2008). However, we based our work on the historic, pre-2005 mount because the data were already available in a processed form. Specifically, we used a wireframe model of this skeleton resulting from the photogrammetric capture in the 1990s (Gunga et al. 1995; Wiedemann et al. 1999) and described it with NURBS. The new model of *Brachiosaurus* has a volume of 47.9 m^3. This is much less than the original model based on geometric solids, which had a volume of 74.4 m^3 (Wiedemann et al. 1999) and had led to a mass estimate of 74.4 metric tons based on a specific density of 1 kg/l. In comparison, the new mass estimate is only 38 metric tons because it is based on a specific density of 0.8 kg/l (Gunga et al. 2008). The new estimate for *Brachiosaurus brancai* is thus over 21 metric tons lighter than the old estimate.

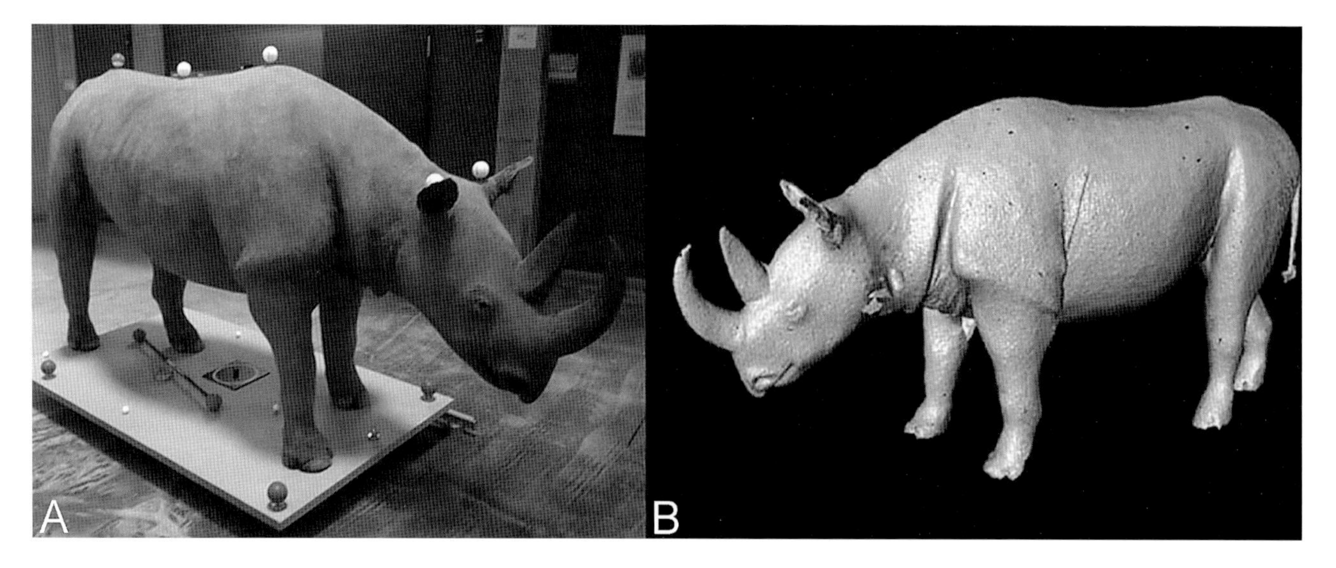

FIGURE 6.5. Laser scan and volume model of a stuffed rhinoceros exhibited in the Senckenberg Naturhistorische Sammlung Dresden, Germany (formerly the Tierkundemuseum Dresden). This mount was used as a reference object to calibrate the volume models based on skeletons and to estimate specific density of the animal because its life body mass was known. (A) The stuffed rhinoceros ready to be scanned. Note the reference spheres on the mount and surrounding it. (B) The point cloud generated by laser scanning after registration and editing.

Future work will involve determining the volume of the new mount (Remes et al., this volume) based on the recent laser scan and comparing it to the volume estimate for the old mount. This will be a prime test case for our method to understand the influence of mounting on mass estimates.

Comparison with Other Methods

Apart from the advantages and disadvantages of the different methods discussed above, the practical experience gained during our work raises some additional points. The survey and data capture of dinosaur skeletons via photogrammetrical methods is technically very easy. Only a few photos of the whole object have to be taken with a calibrated camera, and only a few specific distances have to be measured to obtain the scale of the images. In contrast, capture with a laser scanner requires a much greater effort. For scanning, the authorization and sometimes a formal invitation from the host museum (e.g., in China) is needed. The equipment for laser scanning is much bulkier than for photogrammetry and weighs in at more than 120 kg in our case. The camera used in photogrammetry only weighs 1 or 2 kg, to which a tripod is added if needed. In addition, scanning requires at least a team of two, whereas photogrammetry only requires a single person. These considerations are of particular relevance if overseas work is planned.

Like all methods for estimating body volumes and body masses of dinosaurs, the laser scanning method has certain sources of error that need to be understood in order to appreciate the difference between a mass estimate for a dinosaur and a value derived from measuring a living animal. The two major sources of error for the laser scanning method are the uncertainties involved in mounting the skeleton and the error introduced by digitally reconstructing the body surface.

Although at face value it may seem a good idea to use actual skeletons for body mass estimates, the authenticity of these skeletons needs to be carefully checked, and the anatomical quality of the mount needs to be evaluated. Most mounted dinosaur skeletons, and particularly those of sauropods, are not based on a single complete individual skeleton (e.g., Remes et al., this volume). Commonly, some parts of the skeleton have been reconstructed in plaster, and bones from other individuals have been incorporated into the mount. On the other hand, mounting a skeleton requires careful considerations of the possible mobility in the joints and the entire construction of the animal; skeletons mounted by scientific institutions are generally a good approximation of the skeleton of the living animal. As noted above, the Berlin *Brachiosaurus brancai* skeleton thus will represent a particularly interesting test case because we have data for two different mounts based on the same skeleton. A comparison of the two mounts will be made once the new volume model is finished.

The other source of error is the reconstruction of the body volume by means of NURBS or other digital techniques because it involves the reconstruction of soft parts that are not preserved with the skeleton. This source of error is com-

FIGURE 6.6. Laser scans of dinosaur skeletons pictured at the same scale, with repository of the mount and year the scan was made. (A) *Atlasaurus imelakei,* Moroccan Ministry of Energy and Mining, Morocco, 2004. (B) *Camarasaurus,* specimen "ET," Sauiermuseum Aathal, Switzerland, 2005. (C) *Stegosaurus,* specimen "Moritz," Sauiermuseum Aathal, Switzerland, 2005. (D) *Allosaurus,* specimen "Big Al II," Sauiermuseum Aathal, Switzerland, 2005. (E) *Plateosaurus engelhardti,* specimen GPIT 1, University of Tübingen, Germany, 2004.

mon to all methods that are based on the reconstruction of soft parts (Alexander 1989; Henderson 1999; Seebacher 2001) and is best minimized by rigorous reconstruction techniques (Paul 1987).

Our method for estimating dinosaur body masses and volumes has certain advantages over other methods. One is that it is closest to the actual specimen in that real skeletons are used and not idealized models and drawings as in, for instance, Alexander (1989), Henderson (1999), and Seebacher (2001). This advantage is shared by the method of Anderson et al. (1985), which is based on the circumference of the long bones of actual skeletons. The use of actual skeletons also precludes errors introduced by scaling calculations, where a small error in a scale model can result in a multiton error in the reconstruction of the living sauropod. A case in point is the classical study by Colbert (1962), in which a scale model with

Table 6.1. Sauropodomorph Skeletons Scanned by Our Group Since 2004

Object	Point cloud (side view)	Object	Point cloud (side view)
Plateosaurus engelhardti, Tübingen, Germany		*Lufengosaurus huenei*, Beijing, China	
Shunosaurus lii (adult), Zigong, China		*Shunosaurus lii* (juvenile), Zigong, China	
Mamenchisaurus hochuanensis, Zigong, China		*Mamenchisaurus jingyanensis*, Beijing, China	
Omeisaurus tianfuensis, Zigong, China		*Datousaurus bashanensis*, Zigong, China	
Atlasaurus imelakei, Rabat, Morocco		*Camarasaurus* sp., Aathal, Switzerland	
Yangchuanosaurus hepingensis, Zigong, China		*Yangchuanosaurus hepingensis*, Zigong, China	
Allosaurus sp., Aathal, Switzerland		*Gigantspinosaurus sichuanensis*, Zigong, China	
Stegosaurus sp., Aathal, Switzerland		*Bactrosaurus johnsoni*, Beijing, China	
Protoceraptos, Beijing, China		*Paraceratherium tianshanensis*, Beijing, China	
Elephas maximus, Copenhagen, Denmark	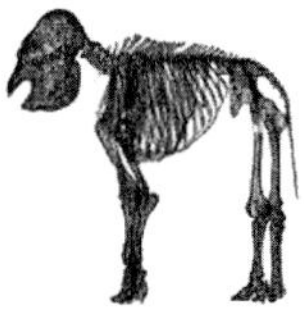	*Rhinoceros* sp., Dresden, Germany	

reconstructed muscles and skin was submerged into a water basin to measure the displaced volume. Here, small differences in sculpting of the model can have a large effect on the body mass estimate.

Conclusions

The main advantage of capturing dinosaur skeletons with a laser scanner is that the scans directly produce a 3D point cloud of the same (full) scale of the captured object and thus of a specific individual. To obtain volume data for the captured skeleton from the point cloud, it is necessary to model the body surface area of the living animal. Mass can then be calculated from volume if specific density of the animal is known (or estimated). Laser scanning is preferred over photogrammetry because the latter results in only a wireframe model of the captured object that does not provide the complete coverage of a point cloud. The best way to reconstruct the body surface from the point cloud generated by the scanner is NURBS. Older methods for modeling the body surface area (geometric primitives, rotational solids) of sauropod dinosaurs from wireframe models of the skeleton or from point clouds proved to be too inaccurate and resulted in excessive volume estimates.

Acknowledgments

We thank all the members of the research unit, especially Martin Sander (University of Bonn) and Oliver Rauhut (Bavarian State Collection for Paleontology and Geology), for their help in selecting the dinosaurs and organizing our journeys. Further thanks go to the staff of the institutions where we scanned the skeletons, especially for their warm response, patience, and support. In particular, we want to thank the following persons: Clara Stefen (Senckenberg Naturhistorische Sammlung Dresden), Heinrich Mallison (University of Tübingen and Museum für Naturkunde), Mogens Andersen and Per Christiansen (Zoological Museum of the University of Copenhagen), Najat Aquesbi and Mohammed Rochdy (Ministry of Energy and Mines, Rabat, Morocco), Hans-Jakob Siber and his team (Sauriermuseum Aathal), and Wolf-Dieter Heinrich (Museum für Naturkunde, Berlin). We also thank the Beijing Museum of Natural History and the Zigong Dinosaur Museum for the fruitful collaboration and the opportunity to carry out our scanning work in China. Last but not least, we thank the Deutsche Forschungsgemeinschaft (DFG) for funding. This paper is contribution number 65 of the DFG Research Unit 533 "Biology of the Sauropod Dinosaurs: The Evolution of Gigantism."

References

Alexander, R. M. 1989. Mechanics of fossil vertebrates.—*Journal of the Geological Society* 146: 41–52.

Alexander R. M. 1998. All-time giants: the largest animals and their problems.—*Palaeontology* 41: 1231–1245.

Anderson, J. F., Hall-Martin, A. & Russell, D. A. 1985. Long-bone circumference and weight in mammals, birds and dinosaurs.—*Journal of Zoology* 207: 53–61.

Bates, K. T., Manning, P. L. & Hodgetts, D. 2008. Three-dimensional modeling and analysis of dinosaur trackways.—*Palaeontology* 51: 999–1010.

Bates, K. T., Manning, P. L., Hodgetts, D. & Sellers, W. I. 2009. Estimating mass properties of dinosaurs using laser imaging and 3D computer modeling.—*PLoS One* 4: e4532.

Boehler, W., Bordas, M. & Marbs, A. 2003. Investigating laser scanner accuracy.—*In Proceedings of the International Committee for Architectural Photogrammetry, 14th Symposium*. Antalya, Turkey: pp. 696–702

Bonner, J. T. 2006. *Why Size Matters: From Bacteria to Blue Whales*. Princeton University Press, Princeton.

Breithaupt, B. H., Matthews, N. A. & Noble, T. A. 2004. An integrated approach to three-dimensional data collection at dinosaur tracksites in the Rocky Mountain West.—*Ichnos* 11: 11–26.

Calder, W. A. 1996. *Size, Function, and Life History*. Dover Publications, New York.

Christiansen, P. & Farina, R. A. 2004. Mass prediction in theropod dinosaurs.—*Historical Biology* 16: 85–92.

Clutton-Brock, T. H., Albon, S. D. & Harvey, P. H. 1980. Antlers, body size and breeding group size in the Cervidae.—*Nature* 285: 565–566.

Colbert, E. H. 1962. The weights of dinosaurs.—*American Museum Novitates* 2076: 1–16.

Foster, J. R. 2007. *Jurassic West: The Dinosaurs of the Morrison Formation and Their World*. Indiana University Press, Bloomington.

Ganse, B., Stahn, A., Stoinski, S., Suthau, T. & Gunga, H.-C. This volume. Body mass estimation, thermoregulation, and cardiovascular physiology of large sauropods. *In* Klein, N., Remes, K., Gee, C. T. & Sander, P. M. (eds.). *Biology of the Sauropod Dinosaurs: Understanding the Life of Giants*. Indiana University Press, Bloomington: pp. 105–115.

Gunga, H.-C., Kirsch, K., Baartz, F., Röcker, L., Heinrich, W.-D., Lisowski, W., Wiedemann, A. & Albertz, J. 1995. New data on dimensions of *Brachiosaurus brancai* and their physiological implications.—*Naturwissenschaften* 82: 190–192.

Gunga, H.-C., Kirsch, K., Rittweger, J., Röcker, L., Clarke, A., Albertz, J., Wiedermann, A., Mokry, S., Suthau, T., Wehr, A., Heinrich, W.-D. & Schultze, H. P. 1999. Body size and body volume in two sauropods from the Upper Jurassic of Tendaguru (Tanzania).—*Mitteilungen aus dem Museum für Naturkunde in Berlin, Geowissenschaftliche Reihe* 2: 91–102.

Gunga, H.-C., Kirsch, K., Rittweger, J., Clarke, J., Albertz, J., Wiedermann, A., Mokry, T., Wehr, A., Heinrich, W.-D. & Schultze, H. P. 2002. Dimensions of *Brachiosaurus brancai, Dicraeosaurus hansemanni* and *Diplodocus carnegii* and their implications for gravitational physiology. *In* Moravec, J., Takeda, N. & Singal, P. K. (eds.). *Adaptation, Biology and Medicine, Vol. 3*. Narosa Publishing House, New Delhi: pp. 156–169.

Gunga, H.-C., Suthau, T., Bellmann, A., Friedrich, A., Schwanebeck, T., Stoinski, S., Trippel, T., Kirsch, K. & Hellwich, O. 2007. Body mass estimations for *Plateosaurus engelhardti* using laser scanning and 3D reconstruction methods.—*Naturwissenschaften* 94: 623–630.

Gunga, H.-C., Suthau, T., Bellmann, A., Stoinski, S., Friedrich, A., Trippel, T., Kirsch, K. & Hellwich, O. 2008. A new body mass estimation of *Brachiosaurus brancai* Janensch, 1914 mounted and exhibited at the Museum of Natural History (Berlin, Germany).—*Fossil Record* 11: 33–38.

Haubold, H. 1990. *Dinosaurier, System-Evolution-Paläobiologie*. Neue Brehm-Bücherei Band 432. Ziemsen-Verlag, Wittenberg.

Henderson, D. M. 1999. Estimating the masses and centers of mass of extinct animals by 3-D mathematical slicing.—*Paleobiology* 25: 88–106.

Hunt, G. & Roy, K. 2006. Climate change, body size evolution and Cope's rule in deep-sea ostracodes.—*Proceedings of the National Academy of Sciences of the United States of America* 103: 1347–1352.

Lambert, D. 1980. *A Field Guide to Dinosaurs*. Avon, New York.

Makarieva, A. M., Gorshkov, V. G. & Li, B.-L. 2005. Temperature-associated upper limits to body size in terrestrial poikilotherms. —*Oikos* 111: 425–436.

Mazzetta, G. V., Christiansen, P. & Farina, R. A. 2004. Giants and bizarres: body size of some southern South American Cretaceous dinosaurs.—*Historical Biology:* 1–13.

Montani, R. 2001. Estimating body mass from silhouettes: testing the assumption of elliptical body cross-sections.—*Paleobiology* 27: 735–750.

Paul, G. S. 1987. The science and art of restoring the life appearance of dinosaurs and their relatives. *In* Czerkas, S. A. & Olson, E. C. (eds.). *Dinosaurs Past and Present, Vol. 2*. Natural History Museum of Los Angeles County, Los Angeles: pp. 4–49.

Paul, G. S. 1997. Dinosaur models: the good, the bad, and using them to estimate the mass of dinosaurs.—*In* Wollberg, D. L., Stump, E. & Rosenberg, G. D. (eds.). *Dinofest*. International Academy of Natural Sciences, Philadelphia: pp. 129–154.

Peczkis, J. 1994. Implications of body-mass estimates for dinosaurs.—*Journal of Vertebrate Paleontology* 14: 520–533.

Perry, S. F. 2001. Functional morphology of the reptilian and avian respiratory systems and its implications for theropod dinosaurs. *In* Gauthier, J. & Gall, L. F. (eds.). *New Perspectives on the Origin and Early Evolution of Birds*. Yale Peabody Museum, New Haven: pp. 429–441.

Perry, S. F., Breuer, T. & Pajor, N. This volume. Structure and function of the sauropod respiratory system. *In* Klein, N., Remes, K., Gee, C. T. & Sander, P. M. (eds.). *Biology of the Sauropod Dinosaurs: Understanding the Life of Giants*. Indiana University Press, Bloomington: pp. 57–79.

Peters, R. H. 1983. *The Ecological Implications of Body Size*. Cambridge University Press, New York.

Petti, F. M., Avanzini, M., Franceschi, M., Girardi, S., Remondino, F., Belvedere, M., Feretti, P. & Tommasoni, R. 2008. New approaches to study dinosaur tracks: 3D digital modeling by photogrammetry and active sensors.—*In Ichnia 2008. The Second International Congress on Ichnology. Abstract Book*. Jagiellonian University, Cracow.

Remes, K., Unwin, D. M., Klein, N., Heinrich, W.-D. & Hampe, O. This volume. Skeletal reconstruction of *Brachiosaurus brancai* in the Museum für Naturkunde, Berlin: summarizing 70 years of sauropod research. *In* Klein, N., Remes, K., Gee, C. T. & Sander, P. M. (eds.). *Biology of the Sauropod Dinosaurs: Understanding the Life of Giants*. Indiana University Press, Bloomington: pp. 305–316.

Schmidt-Nielsen, K. 1984. *Scaling: Why Is Animal Size So Important?* Cambridge University Press, Cambridge.

Schmidt-Nielsen, K. 1997. *Animal Physiology: Adaptation and Environment, Vol. 5*. Cambridge University Press, Cambridge.

Schwarz, D. & Fritsch, G. 2006. Pneumatic structures in the cervical vertebrae of the Late Jurassic (Kimmeridgian–Tithonian) Tendaguru sauropods *Brachiosaurus brancai* and *Dicraeosaurus*. —*Eclogae Geologicae Helvetiae* 99: 65–78.

Seebacher, F. 2001. A new method to calculate allometric length-mass relationships for dinosaurs.—*Journal of Vertebrate Paleontology* 21: 51–60.

Wedel, M. J. 2003a. Vertebral pneumaticity, air sacs, and the physiology of sauropod dinosaurs.—*Paleobiology* 29: 243–255.

Wedel, M. J. 2003b. The evolution of vertebral pneumaticity in sauropod dinosaurs.—*Journal of Vertebrate Paleontology* 23: 344–357.

Wedel, M. J. 2005. Postcranial skeletal pneumaticity in sauropods and its implications for mass estimates. *In* Curry Rogers, K. & Wilson, J. A. (eds.). *The Sauropods: Evolution and Paleobiology*. University of California Press, Berkeley: pp. 201–228.

Wiedemann, A., Suthau, T. & Alberts, J. 1999. Photogrammetric survey of dinosaur skeletons.—*Mitteilungen aus dem Museum für Naturkunde in Berlin, Geowissenschaftliche Reihe* 2: 113–119.

7

Body Mass Estimation, Thermoregulation, and Cardiovascular Physiology of Large Sauropods

BERGITA GANSE, ALEXANDER STAHN, STEFAN STOINSKI, TIM SUTHAU, AND HANNS-CHRISTIAN GUNGA

THIS CHAPTER PROVIDES AN OVERVIEW on thermoregulation and the cardiovascular physiology of sauropods on the basis of data obtained by laser scanning and surface modeling of the basal sauropodomorph *Plateosaurus engelhardti* and the basal macronarian sauropod *Brachiosaurus brancai*. Nonuniform rational B splines (NURBS) were used to obtain volume estimates of the thoracic cavity, and these estimates correspond well with vital organ masses as determined by allometric modeling. To reach body masses of about 50 metric tons, large sauropods might have had, at least partly during their life span, a high resting metabolic rate, and they might have been endothermic homeotherms to maintain thermoregulative control. Assuming a lack of sweat glands in sauropods, heat balance was likely to be regulated by processes of radiation, convection, and conduction. Heat transfer from the body surface via convection, especially during exercise (hyperthermia), was probably limited, and large bird-like air sacs as part of the lung structures might have served as "thermal windows" to help regulate the temperature. A four-chambered heart would have generated lower pressures in the pulmonary circulation and higher pressures in the systematic regulation. Additional physiological mechanisms such as high oxygen transport capacity, muscular venous pumps, tight skin layers, thick vessel walls, strong connective tissue, precapillary vasoconstriction, low permeability of capillaries to plasma proteins, and digital cushions in the feet were necessary to meet cardiovascular requirements by supporting fluid volume regulation and preventing edema in large sauropods. Thus, in regard to cardiovascular and thermoregulative control, sauropods were highly specialized animals.

Introduction

What were the average body masses and sizes of the sauropod dinosaurs? How did their hearts work, and how big were they? How did they regulate their body temperature? Were they endothermic or ectothermic?

These sorts of questions are asked by paleobiologists and physiologists working on biology of the largest terrestrial animals ever (Sander et al. 2010). Of particular interest are the issues of body mass, growth rates, thermoregulation, and cardiovascular constraints, and this chapter reviews the current knowledge in these areas of research in regard to sauropod biology. We begin with gigantism, body mass, and body mass estimations, and why these estimations are important for the reconstruction of extinct organisms. Specific examples for whole-body and segmental data obtained from *Plateosaurus engelhardti* and *Brachiosaurus brancai* are provided as a basis for a discussion on dinosaur physiology. General concepts of thermophysiology such as modes of thermoregulation are briefly described, then discussed in regard to aspects of heat production, heat dissipation, and metabolic rate in dinosaurs in general and in sauropods in particular. Next, we summarize the present hypotheses about the heart and circulation in these animals, with a specific emphasis on morphology and posture. Finally, this chapter concludes with an integrative discussion on the thermoregulative and cardiovascular limits of sauropods.

Gigantism, Body Mass, and Body Mass Estimations

Because sauropods were the largest land-living animals ever, gigantism obviously played a major role in their evolution. With increasing body size, the bauplan (blueprint) of an organism must change accordingly and will affect metabolism, heat balance, locomotion, and the cardiovascular system. And while it is quite easy to determine the body mass of a living organism, it is difficult to do so accurately for extinct animals because of variations in anatomy and function. The depiction of dinosaurs in the popular media is merely an interpretation of artists and scientists that is based on few facts. In this context, we share Henderson's (1997, p. 165) opinion: "The work that results from a collaborative effort between artists and scientists turns paleontology and related earth sciences into a rich and modern form of storytelling." The literature contains numerous methods for estimating body mass that differ considerably in their results (Peczkis 1994). These methods rely on the skeleton, which is often incompletely preserved, because fossilized remains of soft tissue are rare. However, because body mass and body mass distribution play such key roles in understanding the physiology of any organism, several attempts using various methods have been made to estimate the body mass of sauropods. The advantages and disadvantages of these methods have been discussed intensively in the literature (Colbert 1962; Anderson et al. 1985; Chapman 1997; Henderson 1999; Motani 2001; Seebacher 2001; Christiansen & Fariña 2004; Gunga et al. 2007; Bates et al. 2009; Franz et al. 2009; Sander et al. 2010).

Compared to most land vertebrates with a specific density of 1,000 kg/m^3, air sacs in the body cavity probably decreased the specific density of sauropods (Paul 1988). Furthermore, it was pointed out only recently that extensive postcranial skeletal pneumaticity in sauropods must be taken into account in body mass reconstructions (Wedel 2003, 2005), resulting in substantial reduction in the estimated mass for giant sauropods such as *Brachiosaurus brancai* (compare Gunga et al. 1995 vs. Gunga et al. 2008). These new studies assume a mean tissue density (specific density) of only 800 kg/m^3. The tissue density of birds lies in the range of 730 kg/m^3, and it is likely that the density of sauropod tissue was somewhere between 730 kg/m^3 and 800 kg/m^3. Therefore, Wedel (2005) suggested that the volume-based estimations of mass, which were published before the consensus that sauropod dinosaurs had a pneumatized axial postcranial skeleton and an avian-style respiratory system, should be reduced by about 10% to account for the reduced tissue density.

The two major approaches for mass estimation involve some measure of volume that is then converted into either body mass or biomechanical data, for instance, using long bone circumferences (Alexander 1989). One such volume-based technique includes digital 3D capture of skeletons that uses a laser scanner and CAD (computer-aided design) software (Gunga et al. 1995, 1999, 2007, 2008; Lovelace et al. 2007). As an example, Plates 7.1 and 7.2 show the steps in obtaining such a 3D model for the basal sauropodomorph *Plateosaurus engelhardti* based on a mount at the University of Tübingen (specimen GPIT 1) and for the basal macronarian sauropod *Brachiosaurus brancai,* based on the historic mount at the Museum für Naturkunde, Berlin. The skeleton of *P. engelhardti* was captured by a MENSI 25 laser scanner placed in different positions around the skeleton. The resulting point cloud describing the skeleton was stored, visualized, and given a body shape (Plate 7.1A, B; see also Stoinski et al., this volume). To provide realistic smooth body surfaces unrestricted by geometrical limitations, the modeling works with nonuniform rational B splines (NURBS), a mathematical model commonly used in computer graphics for generating and representing curves and surfaces.

On the basis of this computerized 3D skeleton, different body shapes can be tested, and calculations for each body part/segment can be made. The major advantage of the scanning of mounted skeletons over other techniques is that it limits measurement errors, if they occur, to body segments only, and thus does not affect the entire estimate, as happens when whole-body mass is extrapolated from a few long bone circumferences or from a scale model.

For the reconstruction of specific body parts, we used the elephant leg as an analog to design the shape, that is, the mass of the dinosaur's extremities. The shape of the thorax followed the anatomical limitations given mainly by the skeleton (ribs and vertebrae). By modeling the surface and approximating the body volume, it is possible to simulate robust (Plate 7.1A) and slim (Plate 7.1B) versions of an individual dinosaur based on the same mounted skeleton, in this case the *P. engelhardti* skeleton GPIT 1.

After determining body mass, we estimated the size of organs (heart, lungs, kidney, liver, spleen, gut, blood volume, skeleton, musculature, and fat) by means of allometric functions published by Anderson et al. (1979), Schmidt-Nielsen (1997), and Calder (1996). These are mostly based on mammals and not on birds or crocodiles, but this may not represent much of a problem, if a high metabolic rate in sauropods comparable to that in mammals is assumed. The organ masses for reptiles are generally lower than those of mammals and birds (Franz et al. 2009). We argue that allometric equations from mammals can be used because recent data from Sander (2000), Erickson et al. (2001), Sander & Tückmantel (2003), and Lehman & Woodward (2008) indicate that whole-organism growth rates for dinosaurs were much faster than those of living reptiles of equivalent size (see also Sander et al., this volume). We did not use allometric functions from birds because a database of specimens with a comparable body mass does not exist. In mammals, at least data from the megaherbivores such as the African elephant are available and have been analyzed intensively.

Finally, we tested whether the organs modeled by allometry would actually fit into the thoracic cavity delimited by the anatomical dimensions of the skeleton (Gunga et al. 2007, 2008). For a slim version of 630 kg body mass (total body surface and volume of 8.8 m^2 and 0.79 m^3, respectively), the main organ masses in the body cavity for *P. engelhardti* were as follows: skeleton approximately 21 kg, blood volume approximately 25 l, heart 3.4 kg, lung 6.5 kg, liver 9.0 kg, spleen 2.2 kg, kidney 1.7 kg, integument approximately 30 kg, and gut approximately 167 kg (Gunga et al. 2007). The muscle mass and fat mass of the thorax are calculated at approximately 95 kg and approximately 106 kg, respectively (Gunga et al. 2007). When all masses of the organs are added up, a total organ mass of approximately 454 kg results for the trunk region of *P. engelhardti*. This leads to a mass of approximately 176 kg for the remaining body parts. These results are in accordance with the anatomical limitations of the thorax represented by the skeleton of 0.530 m^3 as estimated according to scanning and NURBS modeling. For a robust reconstruction with a total body mass of 912 kg (total body volume of 1.14 m^3), the resulting thorax mass and volume of 597 kg and 0.746 m^3, respectively, seem to be at the upper limit of what could fit into the thoracic cavity as determined by 3D scanning (0.616 m^3) (Gunga et al. 2007).

The greatest uncertainties in this reconstruction concern the lungs and the gut mass. In particular, a key factor in determining the volume of the thoracic cavity is the modeling of the gastrointestinal tract. Although this organ has been shown to increase in direct proportion to body mass (Owen-Smith 1988; Clauss et al. 2005), we are just at the beginning of elucidating the nutrition and digestion of sauropodomorphs, and it might well be that an improved understanding will lead to a different size estimate of the gut. Furthermore, despite recent findings regarding the presence of some form of bird-like air sac system in sauropods (Wedel 2009), precise reconstruction of its dimensions is still difficult, further jeopardizing precise modeling of the thoracic cavity.

Finally, for the reconstruction of the complete organism, we assumed that the neck and tail as body parts to be elliptical in cross section, and we added their masses. For *Brachiosaurus brancai*, we calculated a body volume of 48 m^3 (Plate 7.2A, B), a body surface area of 119 m^2, and a body mass of 38 metric tons, assuming a tissue density of 800 kg/m^3, as for *P. engelhardti*. With a body mass of approximately 40 metric tons, the Berlin specimen of *B. brancai* is somewhat smaller than the giant brachiosaurid *Sauroposeidon* and the titanosaur *Argentinosaurus*, which, according to old estimates by Wedel et al. (2000) and Mazzetta et al. (2004), had an estimated body mass of approximately 50–70 metric tons. In terms of gigantism, body mass, and body mass estimations, it seems fair to say that most current estimates consistently place well known sauropods such as *Diplodocus*, *Apatosaurus*, and *Brachiosaurus* in the 15–50 metric ton category (Appendix).

Thermophysiology

TERMINOLOGY AND MODES OF THERMOREGULATION

Thermoregulation is the ability and the mechanisms by which an organism regulates its body temperature in relation to ambient temperatures. Body temperature greatly influences the speed of temperature-sensitive biochemical processes and enzymes acting in the body. In addition, the flow of information through the nerves runs fastest in tissue temperatures of approximately 37°C in most mammals (White & Seymour 2003). Several pairs of contrasting terms explain the mode of thermoregulation.

Organisms can be homeothermic or poikilothermic. Homeothermy describes the ability to maintain a stable internal body temperature (to a certain extent) independent of the environment. This temperature often exceeds ambient temperature. In mammals, normal core temperature is relatively constant, in most mammals between 36.4 and 37.4°C. Deviations can only be tolerated over a small range (stenothermy). Poikilothermy refers to organisms whose internal temperatures vary, often matching the temperature of the environment.

Furthermore, terms referring to the way heat is acquired, such as *endothermic* and *ectothermic*, deserve clarification. Endothermy describes the ability to maintain body temperature through internally produced heat, for example, through metabolic activity, muscle shivering, and heat production in the bowels. To avoid heat loss in endotherms in cold environments, especially in small organisms, an insulation of the body is necessary (subcutaneous fat tissue, hair, fur, feathers), thus preventing hypothermia; we will return to this point with respect to dinosaurs later on. Ectothermic means that the organism is dependent on external heat, and the temperature gradient to the environment is small (<5°C). The level of activity depends on ambient temperature. An advantage of ectothermy is the ability to live in a broad spectrum of temperatures (euthermy).

There are further terms used in thermophysiology that specifically describe the kind of metabolic rate of an organism: *tachymetabolic* and *bradymetabolic*. Tachymetabolic refers to animals with high resting metabolic rates, usually resulting in endothermy and homeothermy. Bradymetabolic describes animals with a low resting metabolic rate and a high metabolic rate during activity. In extreme environments, bradymetabolic animals are able to shut down their metabolism, reaching near-death states while waiting for better environmental conditions. Bradymetabolic animals tend to be ectothermic poikilotherms. A shortage of food can be tolerated for a longer time than in tachymetabolic animals.

The outdated terms *warm-blooded* and *cold-blooded* are subsumed today under the terms mentioned above. Warm-blooded generally refers to endothermy, homeothermy, and tachymetabolism, while cold-blooded implies ectothermy, poikilothermy, and bradymetabolism. Mammals and birds usually meet most of the criteria for being warm-blooded. Most animals originally thought to be cold-blooded, such as reptiles, insects, and fish, have now been shown to mostly incorporate different variations of the criteria mentioned above, without displaying the typical characteristics of cold-blooded animals in the classical sense. Furthermore, some bats and small birds are poikilothermic and bradymetabolic when they sleep, but homeothermic and tachymetabolic during active periods. For these creatures, another term was invented: *heterothermy*. Obviously—and this is one point we would like to stress—nature has provided a large variety of different strategies for an organism to cope with the environmental temperature constraints and the availability of food.

Concerning dinosaur thermophysiology, it may well be possible that some were homeothermic, some were heterothermic, and some possessed thermoregulatory mechanisms that cannot be clearly described by either category (Reid 1997). Previous views on the thermophysiology of dinosaurs were probably too narrow. Larger animals, because of their body mass, obviously take longer to change their temperature than smaller ones. Consequently, one hypothesis assumes that an ectotherm could become homeothermic if it grows large enough. This is known as mass homeothermy, inertial homeothermy, or gigantothermy (Paladino et al. 1997; Gillooly et al. 2006). According to this hypothesis, a large dinosaur would be as independent as an endothermic mammal in terms of its activity level during the day and at the same time, it would enjoy the advantages of a low metabolic rate. Colbert (1962) studied the thermoregulation of large alligators, extrapolated their data to dinosaurs, and concluded that a 10 metric ton "cold-blooded" dinosaur would take 86 hours to change its body temperature by 1°C. Many dinosaurs could have had "high" (30°C) and stable (daily amplitude 2°C) body temperatures without metabolic heat production, even in winter (Seebacher 2003). However, gigantothermy is a concept based on fully grown individuals, and it is important to remember that large dinosaurs started out as small juveniles (Seebacher 2003).

An argument for endothermic dinosaurs is provided by discoveries of dinosaurs inhabiting former polar regions (Rich et al. 2002). Because of the low ambient temperature in polar regions, it is likely that a kind of homeothermy was present in these dinosaurs. Alternatively, if polar dinosaurs were unable to remain active (warm) throughout the winter, they may have migrated (Dunham et al. 1989). We think, however, that the strongest argument for sauropods having been endothermic is the high growth rates recorded in the histology of their bones (Sander et al., this volume, and references therein). There seems to be no way for giant sauropods to reach a body mass of >50 metric tons in a reasonable lifetime without having—at least partly during their life span—a high resting metabolic rate comparable to or even higher than that in mammals. This is consistent with the evidence presented by Head et al. (2009) that even in the highest ambient temperature, maximum body size in terrestrial vertebrates is limited to slightly more than one metric ton.

HEAT PRODUCTION

As noted, several ways of heat production are known in animals, including shivering and nonshivering thermogenesis. In humans, the mechanical effectiveness of muscle work reaches only 20–30%, which means that 70–80% of the chemical energy is turned into heat (Gunga 2005). During hard physical work, energy conversion is increased, and additional heat accumulates. To maintain thermal equilibrium, the release of heat from the body must be increased as well. Furthermore, heat production through muscle shivering is what we do to purposely increase body core temperature in cold environments. In humans at rest, for example, approximately 80% of the heat is produced in the inner organs (nonshivering thermogenesis), whereas other body parts, including the extremities and muscles, contribute only about 20% (and the brain up to 18%). During physical work, this distribution changes rapidly. Up to 90% of the heat can now be generated by muscles, only up to 22% by inner organs, and up to 3% by the brain (Gunga 2005). Another source of heat production is digestion. The degradation of food and the storage of substances produced by the metabolism lead to an increase in body temperature.

Similar methods of heat production (metabolic processes, digestion, muscular activity) can be assumed in sauropods. The resource (food) availability probably limited the digestive heat production to a certain degree (Dunham et al. 1989), but especially during the period of fastest growth of sauropods during adolescence, considerable heat could also have been produced by digestive processes such as fermentation in the bowels. However, it should be noted that very large body sizes do not seem to necessarily imply a digestive advantage and thus substantially increased heat production (Clauss et al. 2007). More information on food intake and diet is available in Hummel & Clauss (this volume), Gee (this volume), and Tütken (this volume).

TEMPERATURES AND HEAT DISSIPATION

The body surface temperature in mammals varies according to local and environmental influences from approximately 28 to 40°C (White & Seymour 2003). The average core temperature —that in the head, chest, and abdomen—of mammals lies between 36 and 40°C; in birds it is higher and varies between 40 and 43°C (Alexander 1989). This core temperature is not constant but varies by ±0.5°C over 24 hours, following a circadian rhythm. On the basis of the scaling of body surface to body volume, sauropods with a body mass of >50 metric tons might have had difficulties getting rid of the heat they produced.

Before discussing thermal constraints and heat dissipation in large sauropods, we would like to briefly sum up the basic principles of thermoregulation. When describing the transfer of thermal energy physiologically, these four concepts are of special importance: conduction, convection, radiation, and evaporation (Gunga 2005). *Conduction* is the direct transfer of heat/energy by vibrations at the molecular level through a solid or fluid, equalizing the temperature differences. The thermal energy is transported as kinetic energy between neighboring molecules (heat conduction). Conduction is important within the body, that is, from the core to the skin surface.

Convection is the transfer of thermal energy through mass transport. There are two types: natural convective heat transfer and forced convective heat transfer. Natural convective heat transfer takes place by fluid or gas movement following temperature gradients. For instance, fluid or gas surrounding a warm body becomes less dense when it gets warmer and rises. The surrounding, cooler fluid or gas then moves up to replace it. In naked humans, this mass transport results in about 600 l of air per minute. Forced convective heat transfer (advective heat transfer) is present in moving media (wind, water, currents) or if the person/animal moves actively within the medium. The size and shape of the body play an important role as well. In smaller animals (e.g., mice), the heat loss is much greater in relation to the surface area than in large organisms (e.g., elephants).

Thermal radiation is electromagnetic radiation emitted from the surface of an object. It is released by all objects with a temperature above absolute zero (−273.15°C). The wavelength of the radiation depends on the temperature and usually lies in the infrared part of the spectrum.

The term *evaporation* implies the loss of heat by sweating. An organism constantly loses water not only by sweating, but also by diffusion through the skin (insensitive perspiration) and the mucosa of the respiratory tract. During the evaporation of sweat, a lot of thermal energy is necessary to bring this liquid from the fluid state into the gas state (2.4 kJ/min ≈ 1 g H_2O/min), making evaporation an effective heat-loss process.

In ambient conditions (25°C), about 50–60% of the total heat production in humans is emitted via infrared radiation. The remainder is released by conduction, convection, and evaporation in more or less equal parts. At higher ambient temperatures and/or increased physical activity, the only way for an animal to get rid of the heat is to increase the evaporation rate—if it has sweat glands—or to increase the heat loss via the respiratory tract by panting or conduction. In this context, recent findings of postcranial skeletal pneumaticity indicative of air sacs in dinosaurs are important (Wedel 2003, 2005, 2009). They suggest that these lung structures may have played a role in thermoregulation through panting and increased conductive heat loss via the lung. The latter is the case because the position of the air sacs at the highest point of the thorax along the vertebral column and their close anatomical proximity to the gastrointestinal tract would allow the use of large air sacs as a kind of "thermal window." The role of the air sacs in thermoregulation is still uncertain and the target of current research. Usually, within the body, heat transport mainly takes place through blood circulation by convection. The conveyance of heat from the vascularized subcutaneous tissue to the skin surface is conductive. By vasodilation or vasoconstriction, especially in the extremities, heat transport can be modified as required (Gunga 2005).

In sauropods, evaporation through the skin presumably did not play a role because they probably did not have sweat glands and a thick skin. Rare instances of fossilized skin were found to be similar to the skin of reptiles today, even in embryonic sauropods (Chiappe et al. 1998). In some dinosaurs, feathers or long, tubular, bristle-like structures were present (Mayr et al. 2002; Vinther et al. 2008), but there is no other indication that sauropods possessed any kind of insulating integumentary structures such as hair. Therefore, taken as a whole, only radiation, convection through the respiratory tract (panting), and conduction can be postulated for the large sauropods. Because of their enormous body mass, large sauropods presumably did not move very fast, thereby avoiding any excessive heat production resulting from physical exercise.

Metabolic Rate

By definition, basal metabolic rate (BMR) or standard metabolic rate is the amount of postabsorptive energy expended (in humans after 12 hours of fasting; the fasting time depends on the animal; White & Seymour 2003) at rest in an environment with a neutral temperature. In this state, energy is mainly expended by vital organs such as the heart, lungs, brain, kidneys, and liver (Gunga 2005). Any form of activity requires further energy. This is called the exercise metabolic rate or exercise energy expenditure. Endothermic animals have a much higher BMR than comparable ectotherms. Because of the endogenous metabolic heat production, the resting metabolic rate of an endotherm has to be three to four times higher than in ectotherms. This demands a high and continuous energy uptake. However, in some cases, active ectotherms have higher metabolic rates than resting endotherms (basal plus exercise metabolic rate). The BMR of endothermic creatures is not directly proportional to their body mass, meaning that mass-specific BMR (metabolic rate per kilogram body mass per day) is much lower in large animals, such as elephants, compared to small animals, such as mice. The scaling relation of body mass to BMR follows a power function: $BMR = 288.58 \times M^{0.72}$ (kJ/d), where M is body mass. Some scientists support a scaling exponent of 0.67 (most birds appear to adhere to this exponent), whereas others estimate it to be 0.75 (Clauss et al. 2008). McKechnie et al. (2006) showed that captive-raised birds had an exponent of 0.67, which is significantly lower than in wild-caught birds (0.744). According to White et al. (2006), the exponent for reptiles is significantly higher than that for birds and mammals. The overall exponent seems to be 0.72. Therefore, the application of allometric functions derived from mammals to large sauropods does not seem to be far-fetched. It is likely that different types of sauropods differed slightly in their exponent as well.

BMR depends on nutrition and varies with food composition (Hummel et al. 2008). Although great strides are being made in regard to the nutritional ecology of dinosaurian megaherbivores (Hummel & Clauss, this volume; Gee, this volume; Tütken, this volume), it is still difficult to determine their exact BMR. Other factors are interindividual fluctuations in BMR (Labra et al. 2007). Because of the negative scaling of BMR, sauropods presumably had a relatively low basal energy expenditure per kilogram per day. Heterothermic dinosaurs would have had a high exercise metabolic rate with extensive heat production but a low BMR, whereas endothermic dinosaurs would have had a relatively high BMR. Taking together all information on metabolic rate, growth rate, digestion, and the lack of sweat glands in the skin of dinosaurs, we conclude that thermoregulation was obviously a problem for multiton adult sauropods. Especially in fast-growing juvenile sauropods with a continuously high metabolic rate, thermoregulation seems to be a bottleneck per se, and it remains to be solved how they were able to avoid hypothermia during everyday life, even in the presumably higher ambient temperatures conditions of the Triassic, Jurassic, and Cretaceous.

The Cardiovascular System

HEART AND BLOOD PRESSURES

Present-day sauropsids—that is, all amniotes except the mammal lineage—have hearts that are either three-chambered (most reptiles, including sea snakes and lizards) or four-chambered (crocodiles, birds). Both types of hearts are theoretically capable of generating systolic blood pressures of over 100 mm Hg. The three-chambered heart consists of two atria and one variably partitioned ventricle, and the pulmonary circulation to the lungs and the systemic circulation to the body are connected, rather than separated as in four-chambered hearts (Reid 1997). This leads to relatively equal pressure levels in the pulmonary and systemic circulation of reptiles with three-chambered hearts. So, in fact, most reptiles actually have a five-chambered heart with two atria, a cavum arteriosum, a cavum venosum, and a cavum pulmonale, but the last three chambers are connected to one another without valves, and are commonly considered to make up a single chamber. The four-chambered heart permits lower blood pressures in the pulmonary circulation and higher pressures in the systemic circulation because the systems are totally disconnected. In smaller reptiles, three chambers are sufficient because the vertical hydrostatic forces due to the vertical distance between heart and lower and upper parts of the body are relatively low. Consequently, there is no need in these animals for the systemic arterial pressure to be much higher than the pulmonary arterial pressure.

In large sauropods, however, the vertical distance (hydrostatic column) from the heart to the brain was probably >8 m when the neck was maximally raised (Fig. 7.1; Plate 7.3). The systemic arterial pressure required to accommodate this would immediately damage the lung by creating leaks if applied to the pulmonary vessels. It is therefore most likely that dinosaurs had separate lung and body circulation, including a four-chambered heart, as has been suggested by Seymour (1976) and Seymour et al. (2004). This hypothesis is supported by the exceptional preservation of cardiac soft tissue of an ornithischian dinosaur that was analyzed by computed tomography by Fisher et al. (2000). Their study revealed structures that are suggestive of a four-chambered heart and a single systemic aorta. However, there is considerable controversy about the interpretations of these authors.

Besides the anatomical (hydrostatic) aspects, functional constraints such as neck posture must be taken into account. Recently, the discussion on the posture of the sauropod neck

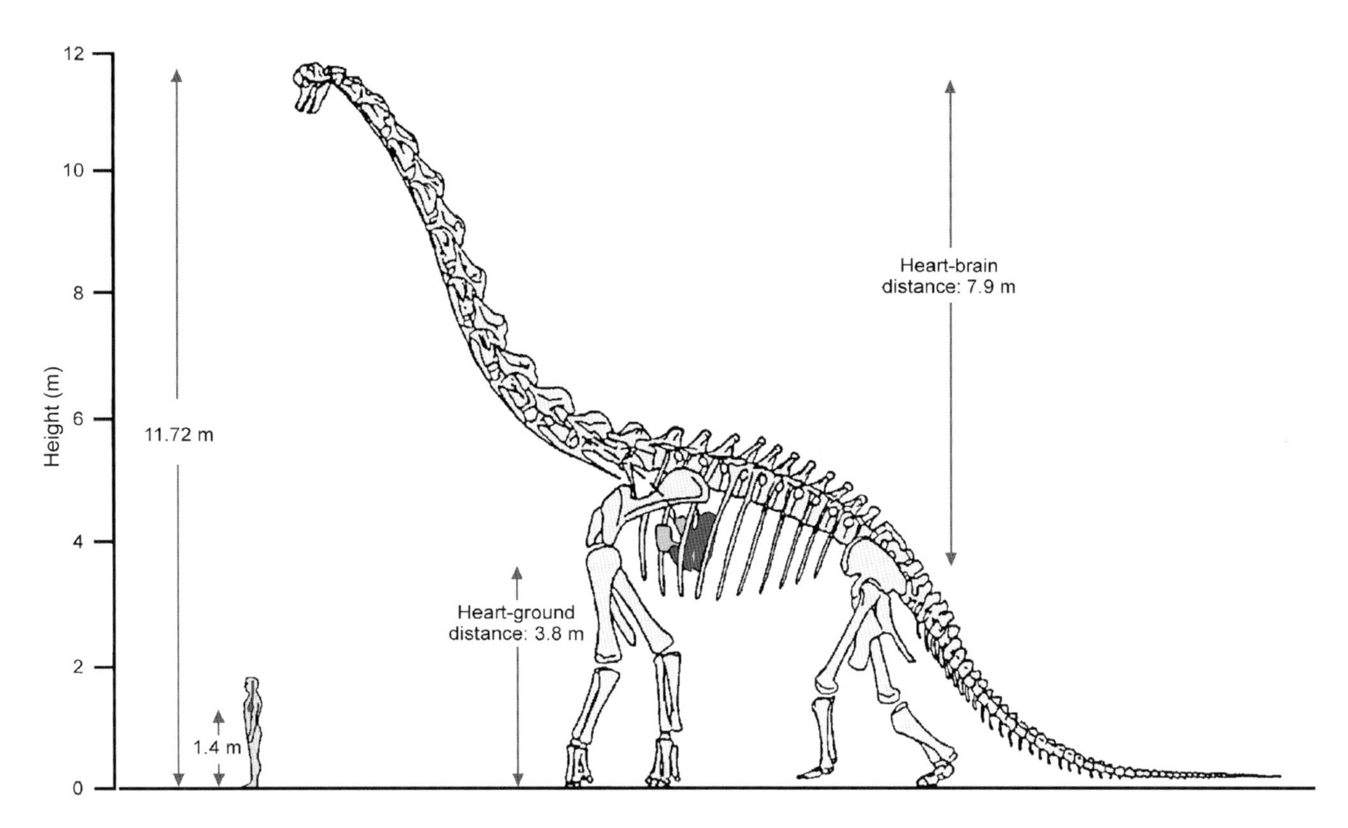

FIGURE 7.1. Comparison of the body, heart size, and vertical distances between a human and a sauropod. The sauropod skeleton is based on the side view of the historic mount of *Brachiosaurus brancai* on exhibit until 2005 at the Museum für Naturkunde, Berlin.

has received renewed attention. Cardiovascular evidence suggests that the neck could not have been raised high for any length of time (Seymour & Lillywhite 2000; Seymour 2009), while from anatomical evidence and comparative studies with living animals, we can infer that the neck was habitually raised high, at least in some sauropods (Dzemski & Christian 2007; Sander et al. 2009; Taylor et al. 2009; Christian & Dzemski, this volume).

If sauropods did hold their necks upright, systemic arterial blood pressures of over 700 mm Hg have to be assumed. To accommodate such high blood pressures, several different anatomical and/or physiological solutions have been hypothesized. Independent of the need to overcome the high hydrostatic pressure associated with neck posture, on the basis of the body mass alone, for the 38 metric ton Berlin specimen of the *Brachiosaurus brancai,* one would expect a heart mass of approximately 200 kg (Gunga et al. 2008). However, because the cardiac muscle adaptively adjusts its dimensions to normalize the cross-sectional stress in response to pressure and volume loading, a heart of this size is relatively unlikely. Problems in the blood supply of the walls during the diastole (relaxation of chambers) and mechanical problems would probably occur (Seymour & Lillywhite 2000). If the oxygen transport capacity of the blood had increased evolutionarily in sauropods, the stroke volume in relation to body mass could be lower. This feature would reduce the stroke volume, but because of the increased viscosity (more hemoglobin means increased hematocrit), the afterload (pressure required for ejecting blood from a chamber) would be higher, and the heart would have to pump harder against it. An increased oxygen transport capacity in sauropods is therefore unlikely.

Another suggestion made by several authors, including Bakker (1978) and Choy & Altman (1992), was that there were multiple hearts in series. The main heart would only have to pump the blood up to the next pump and so on, resulting in a reduced hydrostatic column. According to Choy & Altman (1992), the primary and secondary hearts might have been located in the thorax, and further pairs of hearts in series might have been located in the neck. They estimate the distances necessary between the hearts as approximately 2.4 m. Yet in our opinion, the appearance of accessory hearts is rather unlikely. First, there is no positive evidence, such as fossilized soft tissue, nor are any such structures known from extant vertebrates. Valves would have been necessary, and there are no valves in any artery of any extant vertebrate. Furthermore, a pulsatile contractile vessel wall would have had to evolve, which is extremely unlikely, considering the embryology of the arteries and the heart. Second, the neurovegetative coordination of such a multiple heart system is quite difficult and makes its existence in the past unlikely. These doubts have been expressed by other authors as well (Millard et al. 1992; Seymour & Lillywhite 2000).

Another way to overcome the constraints of high systemic systolic blood pressures is to postulate a siphon mechanism in the neck of sauropod dinosaurs (Hicks & Badeer 1989), facilitating uphill blood flow in arteries. However, this theory has been refuted by several authors, including Seymour & Johansen (1987) and Seymour (2000), and is no longer taken seriously.

At this point, it is useful to consider biological solutions for high blood pressure that currently exist in nature. The tall build of the giraffe *Giraffa camelopardalis* makes this animal quite an interesting model. The adult giraffe's heart weighs approximately 11 kg, and the left ventricular wall is thick compared to other animals (7.5 cm) in order to create systolic blood pressures exceeding 350 mm Hg (Goetz & Keen 1957; Goetz et al. 1960). The characteristics of giraffe blood have a viscosity and erythrocyte count higher than in humans. From a circulatory point of view, however, the adequate perfusion of a brain several meters above the heart, as in giraffes and especially in sauropods, is a problem.

Also problematic is fluid accumulation (edema) in any tissues below the heart, and mainly in the fore- and hindlimbs. Additional mechanisms in sauropods probably supported the return of the blood from the lower body parts to the heart. In horses, for example, a digital cushion between the bones in the foot and the sole collects blood. When the horse steps on the foot, it compresses the cushion, which moves the blood up through the veins (Plate 7.4). Hargens et al. (1987) have examined the anatomy of the giraffe and tried to determine in detail how giraffes avoid pooling of blood and tissue fluid in the interstitial space of their extremities. They found that blood and tissue fluid pressures, combined with a tight skin layer, move fluid upward against gravity. They also discovered that further mechanisms for edema prevention play a role in the giraffe, such as precapillary vasoconstriction and a low permeability of capillaries to plasma proteins. We suggest that similar solutions may have been found in the circulatory system of giant sauropods as well.

Discussion

On the basis of accurate body mass estimates, it is possible to analyze a great quantity of physiological data and parameters. Digital 3D capture of skeletons permits segmental body masses to be reconstructed, thus allowing calculations of the size of the inner organs. Specifically, it is possible to assess whether organ sizes regarded as physiologically necessary can fit into the anatomical dimensions of the animal given by the skeleton. Previously, the greatest uncertainties in the reconstruction of sauropod organs applied to the lungs and gut mass because little was known about these organs in dinosaurs. However, Wedel (2009) recently presented new evidence for bird-like air sacs in sauropods, and consequently an air sac-driven ventilation system. However, estimations of gut mass remain problematic. Furthermore, inferences on thermoregulation and the cardiovascular system of sauropods are still difficult to make. In regard to thermoregulation, metabolic rate and body surface area (heat exchange) are both essential factors in regulating body core temperature. Furthermore, skin thickness and composition are important factors to look at. From our body mass estimations, it seems obvious that large dinosaurs might have had problems with dissipating heat to the environment because they had a relatively small surface area in relation to body mass. Consequently, heat transfer from the body surface to the surrounding air, especially during exercise (hyperthermia), was limited. Thus, as hypothesized by Sander & Clauss (2008) and Sander et al. (2010), it is possible that the air sac structure of the lungs in combination with the cardiovascular system played a significant role for maintaining body temperature in sauropods.

The cardiovascular and microcirculatory system could have supported the heat dissipation from the body core to the shell. However, the cardiovascular system in large sauropods not only had to deal with the excessive heat transfer, but also had to cope with the extreme vertical dimensions of the animals that resulted in high hydrostatic forces. This leads to questions such as how big the heart was, and whether the blood vessels had anatomical features different from those of extant animals. From an anatomical point of view, the heart must have had four chambers (Seymour 1976; Reid 1997). A multiple heart system along the neck was suggested by Choy & Altman (1992) and discussed intensively in the literature, also in comparison with long-necked animals living today, such as the giraffe (Hargens et al. 1987; Withers 1992; Seymour & Lillywhite 2000). We believe that from a neurophysiological perspective, a multiple-heart system appears unlikely because the volumetric coordination of several hearts with this type of pump mechanism would have been extremely complicated.

It is rather likely that the characteristics of sauropod blood, such as its viscosity, differed from the blood of the animals we know today. Otherwise, blood pressures of 600 to 750 mm Hg would have been necessary to supply the brain properly; such cardiac pressure is unknown in today's fauna. However, in 3–4 m tall giraffes, blood pressures up to 380 mm Hg have been reported (Amoroso et al. 1947), and the giraffe's cerebral blood vessels can maintain a tissue perfusion pressure of 100 mm Hg during changing neck positions. Another anatomical feature controlling blood pressure in the brain is the rete mirabile (a complex network of fine arteries), which prevents blood pressure from increasing too much in the brain when the neck is lowered.

We speculate that especially *Brachiosaurus* and other sauropods that were able to raise their neck high might have possessed such a structure in the head area. The veins are a part of the circulatory low pressure system and contain by far the greatest portion of total blood volume, which is usually 5–10% of the body mass. In *Brachiosaurus brancai,* 80% of the blood volume (approximately 3,000 l) would have been located in the veins, the right side of the heart, and the vessels of the pulmonary circulation (Gunga et al. 2002). Because of the high hydrostatic load at the level of the extremities, further mechanisms of edema prevention must have been present, such as muscular venous pumps, precapillary vasoconstriction, low permeability of capillaries to plasma proteins, higher oxygen transport capacities of the blood, extremely strong connective tissue and digital cushions in the feet (Plate 7.4), and certain other mechanisms accompanied by thicker vessel walls (Hargens et al. 1987). Soft tissue discoveries may support these hypotheses in the future.

Conclusions

Many questions remain to be answered in the field of dinosaur physiology, particularly regarding the giant sauropods. Because soft tissue preservation is rare, many variables are difficult to determine, especially with regard to the cardiovascular system. However, on the basis of skeletal anatomy, it is possible to reconstruct not only the shape but also the anatomical and physiological features of the dinosaur's body. New methods allow 3D reconstructions and the approximation of organ sizes. In the field of thermophysiology, current key questions concern the heat transfer within the body and on the body surface, especially with regard to the problems of hyperthermia. Here, comparative studies in large animals such as giraffes and elephants may be helpful. The cardiovascular system is still a field of speculation. We conclude that sauropods were highly specialized with regard to their metabolism, circulation, and temperature regulation, and that they had unique morphological adaptations that allowed them to thrive despite their gigantic sizes.

Acknowledgments

We thank the Deutsche Forschungsgemeinschaft for funding and support, and Annette Sommer (Charité Graphics Department, Berlin) for her assistance. The chapter benefitted from constructive reviews by two anonymous reviewers. This is contribution number 66 of the DFG Research Unit 533 "The Biology of the Sauropod Dinosaurs: The Evolution of Gigantism."

References

Alexander, R. M. 1989. *Dynamics of Dinosaurs and Other Extinct Giants.* Columbia University Press, New York.

Amoroso, E. C., Edholm, O. G. & Rewell, R. E. 1947. Venous valves in the giraffe, okapi, camel and ostrich.—*Proceedings of the Zoological Society of London* 117: 435–440.

Anderson, J. F., Rahn, H. & Prange, H. D. 1979. Scaling of supportive tissue mass.—*Quarterly Review of Biology* 54: 139–148.

Anderson, J. F., Hall-Martin, A. & Russell, D. A. 1985. Long-bone circumference and weight in mammals, birds, and dinosaurs.—*Journal of Zoology* 207: 53–61.

Bakker, R. T. 1978. Dinosaur feeding behaviour and the origin of flowering plants.—*Nature* 274: 661–663.

Bates, K. T., Manning, P. L., Hodgetts, D. & Sellers, W. I. 2009. Estimating mass properties of dinosaurs using laser imaging and 3D computer modeling.—*PLoS ONE* 4: e4532.

Calder, W. A. 1996. *Size, Function, and Life History.* Harvard University Press, Cambridge.

Chapman, R. E. 1997. Technology and the study of dinosaurs. *In* Farlow, J. O. & Brett-Surman, M. K. (eds.). *The Complete Dinosaur.* Indiana University Press, Bloomington: pp. 112–135.

Chiappe, L. M., Coria, R. A., Dingus, L., Jackson, F., Chinsamy, A. & Fox, M. 1998. Sauropod dinosaur embryos from the Late Cretaceous of Patagonia.—*Nature* 396: 258–261.

Choy, D. S. J. & Altman, P. 1992. The cardiovascular system of *Barosaurus:* an educated guess.—*Lancet* 340: 534–536.

Christian, A. & Dzemski, G. This volume. Neck posture in sauropods. *In* Klein, N., Remes, K., Gee, C. T. & Sander, P. M. (eds.). *Biology of the Sauropod Dinosaurs: Understanding the Life of Giants.* Indiana University Press, Bloomington: pp. 251–260.

Christiansen, P. & Fariña, R. A. 2004. Mass prediction in theropod dinosaurs.—*Historical Biology* 16: 85–92.

Clauss, M., Robert, N., Walzer, C., Vitaud, C. & Hummel, J. 2005. Testing predictions on body mass and gut contents: dissection of an African elephant *Loxodonta africana* Blumenbach 1797.—*European Journal of Wildlife Research* 51: 291–294.

Clauss, M., Schwarm, A., Ortmann, S., Streich, W. J. & Hummel, J. 2007. A case of non-scaling in mammalian physiology? Body size, digestive capacity, food intake, and ingesta passage in mammalian herbivores.—*Comparative Biochemistry and Physiology Part A* 148: 249–265.

Clauss, M., Hummel, J., Streich, W. J. & Südekum, K.-H. 2008. Mammalian metabolic rate scaling to ⅔ or ¾ depends on the presence of gut contents.—*Evolutionary Ecology Research* 10: 153–154.

Colbert, E. H. 1962. The weights of dinosaurs.—*American Museum Novitates* 2076: 1–16.

Dunham, A. E., Overall, K. L., Porter, W. P. & Forster, C. A. 1989. Implications of ecological energetics and biophysical and developmental constraints for life-history variation in dinosaurs. —*Geological Society of America Special Papers* 238: 1–19.

Dzemski, G. & Christian, A. 2007. Flexibility along the neck of the ostrich (*Struthio camelus*) and consequences for the reconstruction of dinosaurs with extreme neck length.—*Journal of Morphology* 268: 701–714.

Erickson, G. M., Curry Rogers, K. A. & Yerby, S. A. 2001. Dinosaurian growth patterns and rapid avian growth rates.—*Nature* 412: 429–433.

Fisher, P. E., Russell, D. A., Stoskopf, M. K., Barrick, R. E., Hammer, M. & Kuzmitz, A. A. 2000. Cardiovascular evidence for an intermediate or higher metbolic rate in an ornithischian dinosaur.—*Science* 288: 503–505.

Franz, R., Hummel, J., Kienzle, E., Kölle, P., Gunga, H.-C. & Clauss, M. 2009. Allometry of visceral organs in living amniotes and its implications for sauropod dinosaurs.—*Proceedings of the Royal Society B: Biological Sciences* 276: 1731–1736.

Gee, C. T. This volume. Dietary options for the sauropod dinosaurs from an integrated botanical and paleobotanical perspective. *In* Klein, N., Remes, K., Gee, C. T. & Sander, P. M. (eds.). *Biology of the Sauropod Dinosaurs: Understanding the Life of Giants*. Indiana University Press, Bloomington: pp. 34–56.

Gillooly, J. F., Allen, A. P. & Charnov, E. L. 2006. Dinosaur fossils predict body temperatures.—*PLoS Biology* 4: e248.

Goetz, R. H. & Keen, E. N. 1957. Some aspects of the cardiovascular system in the giraffe.—*Angiology* 8: 542–564.

Goetz, R. H., Warren, J. V., Gauer, O. H., Patterson, J. L., Doyle, J. T., Keen, E. N. & McGregor, M. 1960. Circulation of the giraffe.—*Circulation Research* 1960: 1049–1058.

Gunga, H.-C. 2005. Wärmehaushalt und Temperaturregulation. *In* Deetjen, P., Speckmann, E. J. & Hescheler, J. (eds.). *Physiologie*. Urban & Fischer, München: pp. 669–698.

Gunga, H.-C., Kirsch, K., Baartz, F., Röcker, L., Heinrich, W.-D., Lisowski, W., Wiedemann, A. & Albertz, J. 1995. New data on the dimensions of *Brachiosaurus brancai* and their physiological implications.—*Naturwissenschaften* 82: 190–192.

Gunga, H.-C., Kirsch, K., Rittweger, J., Clarke, A., Albertz, J., Wiedemann, A., Mokry, S., Suthau, T., Wehr, A., Clarke, D., Heinrich, W.-D. & Schulze, H.-P. 1999. Body size and body volume distribution in two sauropods from the Upper Jurassic of Tendaguru/Tanzania (East Africa).—*Mitteilungen aus dem Museum für Naturkunde in Berlin, Geowissenschaftliche Reihe* 2: 91–102.

Gunga, H.-C., Kirsch, K., Rittweger, J., Clarke, A., Albertz, J., Wiedemann, A., Wehr, A., Heinrich, W.-D. & Schultze, H.-P. 2002. Dimensions of *Brachiosaurus brancai, Dicraeosaurus hansemanni* and *Diplodocus carnegii* and their implications for gravitational physiology. *In* Moravec, J., Takeda, N. & Singal, P. K. (eds.). *Adaptation Biology and Medicine: New Frontiers: 3*. Narosa Publishing House, New Delhi: pp. 156–169.

Gunga, H.-C., Suthau, T., Bellmann, A., Friedrich, A., Schwanebeck, T., Stoinski, S., Trippel, T., Kirsch, K. & Hellwich, O. 2007. Body mass estimations for *Plateosaurus engelhardti* using laser scanning and 3D reconstruction methods.—*Naturwissenschaften* 94: 623–630.

Gunga, H.-C., Suthau, T., Bellmann, A., Stoinski, S., Friedrich, A., Trippel, T., Kirsch, K. & Hellwich, O. 2008. A new body mass estimation of *Brachiosaurus brancai* Janensch, 1914 mounted and exhibited at the Museum of Natural History (Berlin, Germany).—*Fossil Record* 11: 33–38.

Hargens, A. R., Millard, R. W., Pettersson, K. & Johansen, K. 1987. Gravitational haemodynamics and oedema prevention in the giraffe.—*Nature* 329: 59–60.

Head, J. J., Bloch, J. I., Hastings, A. K., Bourque, J. R., Cadena, E. A., Herrera, F. A., Polly, P. D. & Jaramillo, C. A. 2009. Giant boid snake from the Palaeocene neotropics reveals hotter past equatorial temperatures.—*Nature* 457: 715–717.

Henderson, D. M. 1997. Restoring dinosaurs as living animals. *In* Farlow, J. O. & Brett-Surman, M. K. (eds.). *The Complete Dinosaur*. Indiana University Press, Bloomington: pp. 165–172.

Henderson, D. M. 1999. Estimating the masses and centers of masses of extinct animals by 3-D mathematical slicing.—*Paleobiology* 25: 88–106.

Hicks, J. W. & Badeer, H. S. 1989. Siphon mechanism in collapsible tubes: application to circulation of the giraffe head.—*American Journal of Physiology: Regulatory, Integrative and Comparative Physiology* 256: 567–571.

Hummel, J. & Clauss, M. This volume. Sauropod feeding and digestive physiology. *In* Klein, N., Remes, K., Gee, C. T. & Sander, P. M. (eds.). *Biology of the Sauropod Dinosaurs: Understanding the Life of Giants*. Indiana University Press, Bloomington: pp. 11–33.

Hummel, J., Gee, C. T., Südekum, K.-H., Sander, P. M., Nogge, G. & Clauss, M. 2008. In vitro digestibility of fern and gymnosperm foliage: implications for sauropod feeding ecology and diet selection.—*Proceedings of the Royal Society B: Biological Sciences* 275: 1015–21.

Labra, F. A., Marquet, P. A. & Bozinovic, F. 2007. Scaling metabolic rate fluctuations.—*Proceedings of the National Academy of Sciences of the United States of America* 104: 10900–10903.

Lehman, T. M. & Woodward, H. N. 2008. Modeling growth rates for sauropod dinosaurs.—*Paleobiology* 34: 264–281.

Lovelace, D. M., Hartman, S. A. & Wahl, W. R. 2007. Morphology of a specimen of *Supersaurus* (Dinosauria, Sauropoda) from the Morrison Formation of Wyoming, and a re-evaluation of diplodocid phylogeny.—*Arquivos do Museu Nacional, Rio de Janeiro* 65: 527–544.

Mayr, G., Peters, D. S., Plodowski, G. & Vogel, O. 2002. Bristle-like integumentary structures at the tail of the horned dinosaur *Psittacosaurus*.—*Naturwissenschaften* 89: 361–365.

Mazzetta, G. V., Christiansen, P. & Fariña, R. A. 2004. Giants and bizarres: body size of some southern South American Cretaceous dinosaurs.—*Historical Biology* 16: 71–83.

McKechnie, A. E., Freckleton, R. P. & Jetz, W. 2006. Phenotypic plasticity in the scaling of avian basal metabolic rate.—*Proceedings of the Royal Society B: Biological Sciences* 273: 931–937.

Millard, R. W., Lillywhite, H. B. & Hargens, A. R. 1992. Cardiovascular-system design and *Barosaurus*.—*Lancet* 340: 914.

Motani, R. 2001. Estimating body mass from silhouettes: testing the assumption of elliptical body cross-sections.—*Paleobiology* 27: 735–750.

Owen-Smith, R. N. 1988. *Megaherbivores: The Influence of Very Large Body Size on Ecology*. Cambridge University Press, Cambridge.

Paladino, F. V., Spotila, J. R. & Dodson, P. 1997. A blueprint for giants: modeling the physiology of large dinosaurs. *In* Farlow, J. O. & Brett-Surman, M. K. (eds.). *The Complete Dinosaur*. Indiana University Press, Bloomington: pp. 491–504.

Paul, G. S. 1988. The brachiosaur giants of the Morrison and Tendaguru Formations with a description of a new subgenus,

Giraffatitan, and a comparison of the world's largest dinosaurs. —*Hunteria* 2: 1–14.

Peczkis, J. 1994. Implications of body-mass estimates for dinosaurs.—*Journal of Vertebrate Paleontology* 24: 520–533.

Reid, R. E. H. 1997. Dinosaurian physiology: the case for "intermediate" dinosaurs. *In* Farlow, J. O. & Brett-Surman, M. K. (eds.). *The Complete Dinosaur.* Indiana University Press, Bloomington: pp. 449–473.

Rich, T., Vickers-Rich, P. & Gangloff, R. A. 2002. Polar dinosaurs. —*Science* 295: 979–980.

Sander, P. M. 2000. Long bone histology of the Tendaguru sauropods: implications for growth and biology.—*Paleobiology* 26: 466–488.

Sander, P. M. & Clauss, M. 2008. Sauropod gigantism. *Science* 322: 200–201.

Sander, P. M. & Tückmantel, C. 2003. Bone lamina thickness, bone apposition rates, and age estimates in sauropod humeri and femora.—*Paläontologische Zeitschrift* 77: 161–172.

Sander, P. M., Christian, A. & Gee, C. T. 2009. Sauropods kept their heads down. Response.—*Science* 323: 1671–1672.

Sander, P. M., Christian, A., Clauss, M., Fechner, R., Gee, C. T., Griebeler, E. M., Gunga, H.-C., Hummel, J., Mallison, H., Perry, S., Preuschoft, H., Rauhut, O., Remes, K., Tütken, T., Wings, O. & Witzel, U. 2010. Biology of the sauropod dinosaurs: the evolution of gigantism.—*Biological Reviews of the Cambridge Philosophical Society.* doi: 10.1111/j.1469=185X .2010.00137.x.

Sander, P. M., Klein, N., Stein, K. & Wings, O. This volume. Sauropod bone histology and its implications for sauropod biology. *In* Klein, N., Remes, K., Gee, C. T. & Sander, P. M. (eds.). *Biology of the Sauropod Dinosaurs: Understanding the Life of Giants.* Indiana University Press, Bloomington: pp. 276–302.

Schmidt-Nielsen, K. 1997. *Animal Physiology.* Cambridge University Press, Cambridge.

Seebacher, F. 2001. A new method to calculate allometric length–mass relationships of dinosaurs.—*Journal of Vertebrate Paleontology* 21: 51–60.

Seebacher, F. 2003. Dinosaur body temperatures: the occurrence of endothermy and ectothermy.—*Paleobiology* 29: 105–122.

Seymour, R. S. 1976. Dinosaurs, endothermy and blood pressure. —*Nature* 262: 207–208.

Seymour, R. S. 2000. Model analogues in the study of cephalic circulation.—*Comparative Biochemistry and Physiology Part A* 125: 517–524.

Seymour, R. S. 2009. Raising the sauropod neck: it costs more to get less.—*Biology Letters* 5: 317–319.

Seymour, R. S. & Johansen, K. 1987. Blood flow uphill and downhill: does a siphon facilitate circulation above the heart?—*Comparative Biochemistry and Physiology Part A* 88: 167–170.

Seymour, R. S. & Lillywhite, H. B. 2000. Hearts, neck posture and metabolic intensity of sauropod dinosaurs.—*Proceedings of the Royal Society B: Biological Sciences* 267: 1883–1887.

Seymour, R. S., Bennett-Stamper, C. L., Johnston, S. D., Carrier, D. R. & Grigg, G. C. 2004. Evidence for the endothermic ancestors of crocodiles at the stem of archosaur evolution. *Physiological and Biochemical Zoology* 77: 1051–1067.

Stoinski, S., Suthau, T. & Gunga, H.-C. This volume. Reconstructing body volume and surface area of dinosaurs using laser scanning and photogrammetry. *In* Klein, N., Remes, K., Gee, C. T. & Sander, P. M. (eds.). *Biology of the Sauropod Dinosaurs: Understanding the Life of Giants.* Indiana University Press, Bloomington: pp. 94–104.

Taylor, M. P., Wedel, M. J. & Naish, D. 2009. Head and neck posture in sauropod dinosaurs inferred from extant animals.—*Acta Palaeontologica Polonica* 54: 213–220.

Tütken, T. This volume. The diet of sauropod dinosaurs: implications of carbon isotope analysis on teeth, bones, and plants. *In* Klein, N., Remes, K., Gee, C. T. & Sander, P. M. (eds.). *Biology of the Sauropod Dinosaurs: Understanding the Life of Giants.* Indiana University Press, Bloomington: pp. 57–79.

Vinther, J, Briggs, D. E. G., Prum, R. O. & Saranathan, V. 2008. The colour of fossil feathers.—*Biology Letters* 4: 522–525.

Wedel, M. J. 2003. Vertebral pneumaticity, air sacs, and the physiology of sauropod dinosaurs.—*Paleobiology* 29: 243–255.

Wedel, M. J. 2005. Postcranial skeletal pneumaticity in sauropods and its implications for mass estimates. *In* Curry Rogers, K. A. & Wilson, J. A. (eds.). *The Sauropods: Evolution and Paleobiology.* University of California Press, Berkeley: pp. 201–228.

Wedel, M. J. 2009. Evidence for bird-like air sacs in Saurischian dinosaurs.—*Journal of Experimental Zoology* 311A: 1–18.

Wedel, M. J., Cifelli, R. L. & Sanders, R. K. 2000. Osteology, paleobiology, and relationships of the sauropod dinosaur *Sauroposeidon*.—*Acta Palaeontologica Polonica* 45: 343–388.

White, C. R. & Seymour, R. S. 2003. Mammalian basal metabolic rate is proportional to body mass.—*Proceedings of the National Academy of Sciences of the United States of America* 100: 4046–4049.

White, C. R., Phillips, N. F. & Seymour, R. S. 2006. The scaling and temperature dependence of vertebrate metabolism.—*Biology Letters* 2: 125–127.

Withers, P. C. 1992. *Comparative Animal Physiology.* Saunders College Publishing, Fort Worth.

A
B
C
D
E
F
G
H

previous page **PLATE 3.1.** Living and fossil *Araucaria.* (A, B, D, E) *Araucaria araucana.* (A) A young tree at the Botanical Garden of the University of Bonn in the spring. (B) Young pollen cones developing on branches in the upper half of the *A. araucana* tree in (A). (C) Fossil pollen cone of *Araucaria delevoryasii* from the Howe-Stephens Quarry (Gee & Tidwell 2010). Note the long, tapered form of the microsporophylls. (D, E) Pollen cones collected from the *Araucaria araucana* tree in (A). (D) Immature pollen cone, depicted at the same scale as the fossil cone in (C). Note the similarity between the two cones in their general shape, size, and long, tapered form of the microsporophylls. (E) Mature pollen cone, showing a lax arrangement of microsporophylls. (F–H) *Araucaria cunninghamii* on plantations on the Atherton Tableland, Queensland, Australia. (F) Habit as an approximately 20 m tall tree. Note the cluster of leafy twigs at the ends of the branches, which are mostly directed upward. (G, H) Epicormic shoots approximately 15 months after damage inflicted by Cyclone Larry in March 2006. (G) Regrowth of leafy branches at the base of branches snapped off by Cyclone Larry. Arrows indicate broken branches still attached to tree trunk. (H) Luxuriant regrowth of leafy branches at the top of a tree that had been snapped off completely by the cyclone.

opposite **PLATE 3.2.** (A–C) Living and fossil *Equisetum,* also known as horsetails or scouring rushes. (A) Living *Equisetum hyemale* var. *robustum* at the Botanical Garden in Bochum, Germany. The leaves are united in a sheath that wraps around the shoot and occurs at regular intervals. Writing pen is approximately 14 cm long. (B) Fossil *Equisetum laterale* from the Jurassic Hope Bay flora of the Antarctic Peninsula (Gee 1989). Notice the leaf sheaths at regular intervals around the shoot. Scale bar = 1 cm. (C) Fossil *Equisetum* sp. from the Late Jurassic Morrison Formation at the Howe-Stephens Quarry, Wyoming (Gee, unpublished data). Again, the leaves are fused into leaf sheaths that wrap around the stem and occur at regular intervals. Scale bar = 5 mm. (D, E) Characteristic features of the Cheirolepidiaceae. (D) The pollen grain *Classopollis* from the Late Jurassic Morrison Formation at the Howe Quarry, Wyoming (Gee, unpublished data). Scale bar = 20 μm. (E) Cuticle with papillae that overarch the stoma from the Late Jurassic Morrison Formation at the Howe-Stephens Quarry, Wyoming (Gee, unpublished data). Scale bar = 25 μm. (F, G). The living fossil *Ginkgo biloba* (maidenhair tree), at the start of its autumnal leaf fall, growing in front of the Steinmann Institute, University of Bonn. (F) Habit as an approximately 20 m high tree. (G) Its distinctive single-notched, fan-shaped leaves on the tree in Plate 3.1F, which turn a brilliant golden color in the fall. (H) Seeds and leaves of *Podocarpus chinensis* at the Botanical Garden of the University of Bonn. These seeds are covered at their base by a fleshy ovuliferous scale known as an epimatium, boosting their attractiveness to birds and other animals as a source of food.

following page **PLATE 3.3.** (A–C) The tree habit, varied habitats, and woody cones in the conifer families Cupressaceae and Pinaceae. (A) *Taxodium distichum* (bald cypress) of the Cupressaceae in a freshwater swamp at Jean La Fitte State Park near New Orleans, Louisiana, in the spring before its leaves have emerged. (B) *Pinus torreyana* (Torrey pine) of the Pinaceae growing on sandy cliffs above the Pacific Ocean and endemic to a small area near San Diego, California. (C) The seed cones and needle-like foliage of *Pinus torreyana* in Plate 3.2B. Unlike the fleshy, single-seeded reproductive structure of *Podocarpus* in Plate 3.2H, the seed cones of the Pinaceae are compact, woody, multiseeded structures. (D–F) Living and fossil ferns of the family Osmundaceae. (D) Sterile and fertile pinnae of *Osmunda claytoniana* (interrupted fern) in the understory of a *Cunninghamia lanceolata* forest in southwestern Guizhou Province, China. (E) A frond of *Todea barbara* (austral king fern) in the greenhouses at the Heidelberg Botanical Garden, Germany. (F) Frond of *Cladophlebis denticulata* from the Jurassic Hope Bay flora of the Antarctic Peninsula (Gee 1989). Note the similarity in the structure and robustness of the frond and in the shape of the pinnules to those of *Todea barbara* in Plate 3.1E. Scale bar = 4 cm. (G–H) Living and fossil cycadophytes. (G) *Cycas revoluta* (sago palm) at Huntington Botanical Gardens, California. (H) Fossilized stem of the bennettitalean *Cycadeoidea dacotensis* (MacBride) Ward from the Early Cretaceous of North America. Scale bar = 10 cm. D: *Photo courtesy of Hans Hagdorn.* H: *Original photo courtesy of Ted Delevoryas.*

A
B
C
D
E
F
G
H

B
C
A
D
E
F
G
H

above **PLATE 4.1.** Living relatives of potential Mesozoic dinosaur food plants. (A) Cycad *Encephalartos altensteinii* in the National Botanical Garden Kirstenbosch, Cape Town, South Africa. (B) Bunya pine *Araucaria bidwillii,* southern Australia. (C) Tree fern *Dicksonia antarctica* in the National Botanical Garden Kirstenbosch, Cape Town, South Africa. (D) Single leaf of the maidenhair tree *Ginkgo biloba*. (E) Horsetail *Equisetum* sp., Olympic Peninsula, Washington State, USA. (F) Fern, Olympic Peninsula, Washington State, USA.

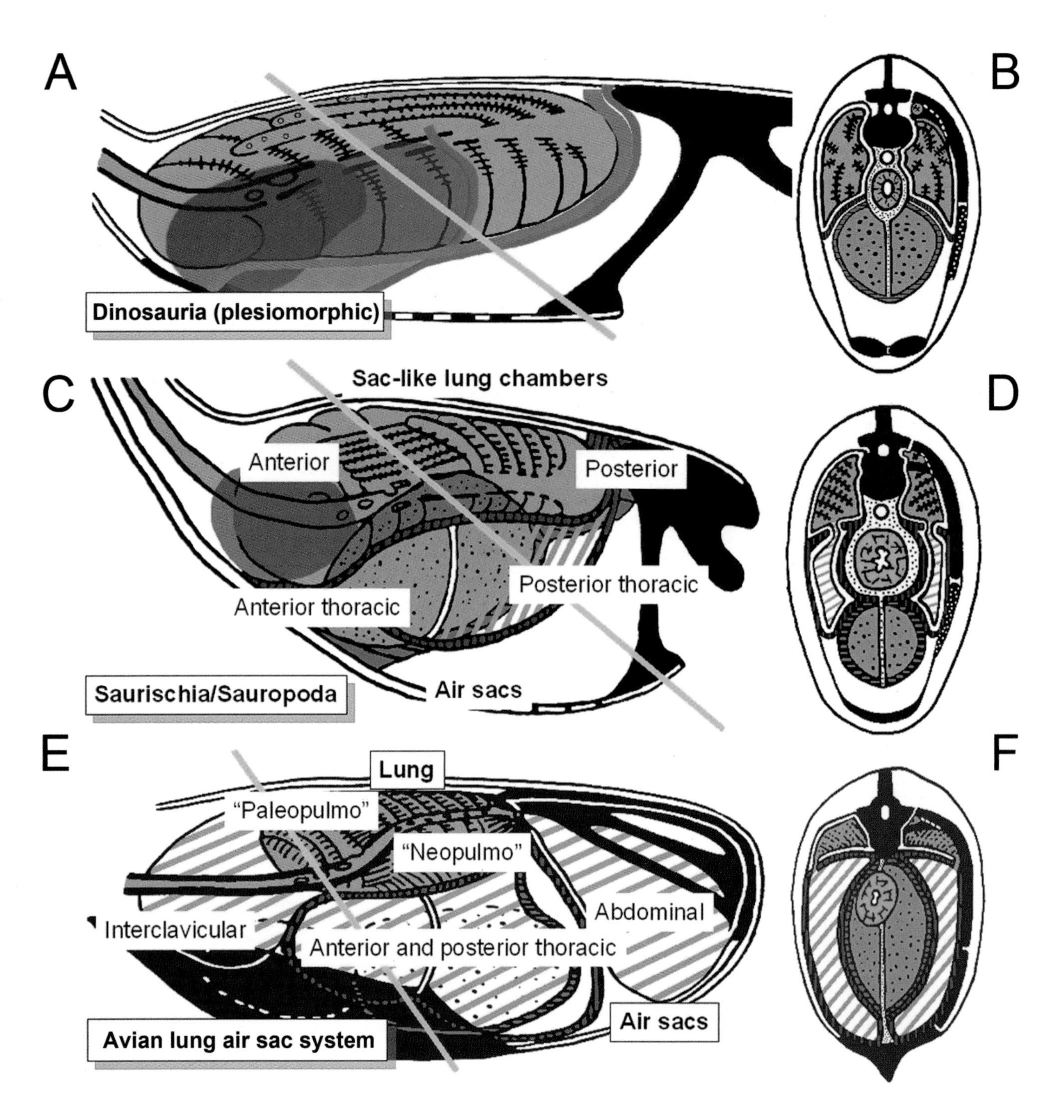

above **PLATE 5.1.** Comparison between plesiomorphic condition in Dinosauria (A, B), proposed sauropod lung structure (C, D), and avian lung–air sac system (E, F). The green line in (A), (C), and (E) indicates the level of the respective schematic cross sections in (B), (D), and (F). Postpulmonary and posthepatic septa are indicated in red and blue, respectively. The level of the pericardium (the sac that surrounds the heart) is indicated in brown, that of the liver in purple. Note long, arching posterior lung chambers directly contacting the more anterior chambers in (A). In the sauropod lung (C), foramina developing at these contact points are assumed to form parabronchus-like structures. Anterior and posterior thoracic air sacs are illustrated (C, D) invading the postpulmonary septum as in birds (E, F), although their presence is not necessary for lung function.

opposite **PLATE 6.1.** Scanning and digital reconstruction of the mounted skeleton of *Plateosaurus engelhardti* (GPIT 1) at the University of Tübingen, Germany, in 2004. (A) The laser scanner is ready to scan the skeleton. Note the red and white reference spheres on the skeleton and its surroundings, as well as the reference spheres on the tripod in the foreground. (B) The scanner has now been placed behind the skeleton to capture different viewpoints. (C) The point clouds captured from several viewpoints before registration. The colors identify the different point clouds, which each have their own local coordinate systems. (D) The point clouds after the registration, with the skeleton in red and all the unwanted background information in gray. (E) The skeleton after editing, with the contributions from the different viewpoints indicated by different colors.

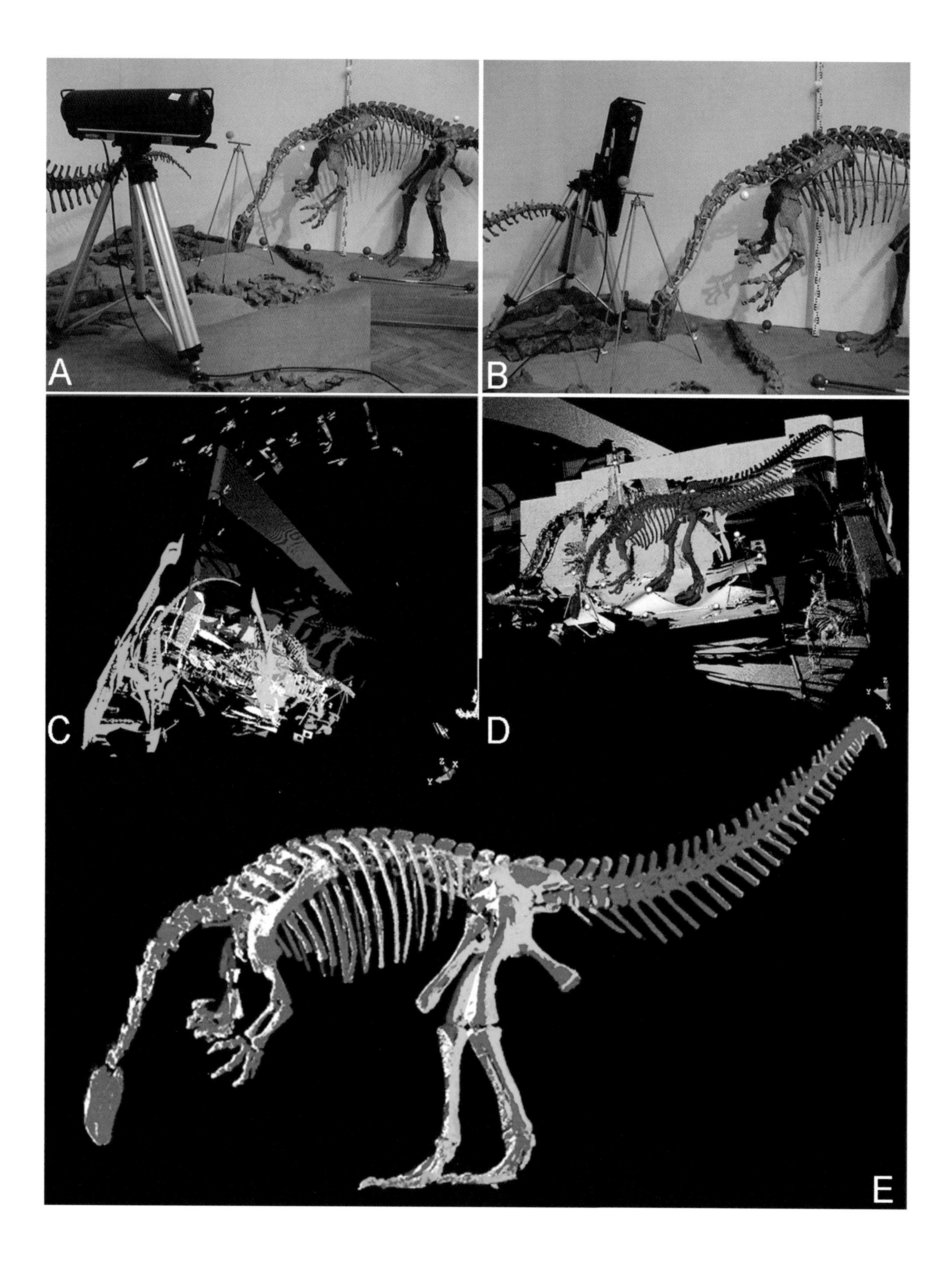

A
B
C
D
E

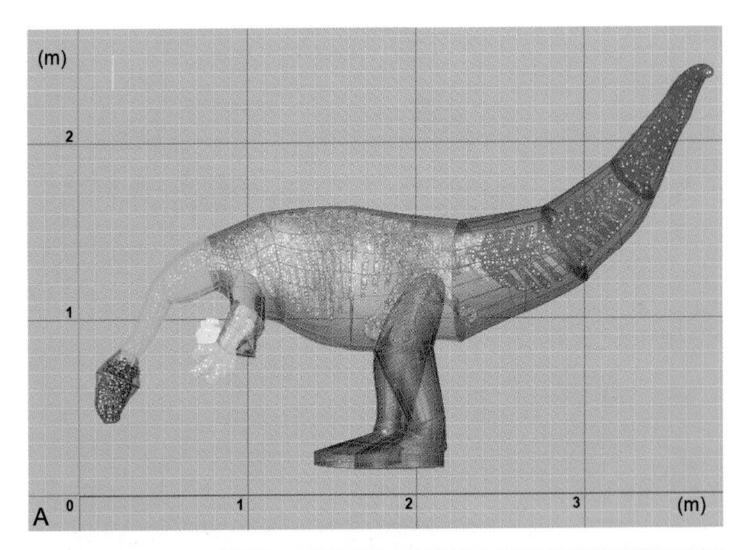

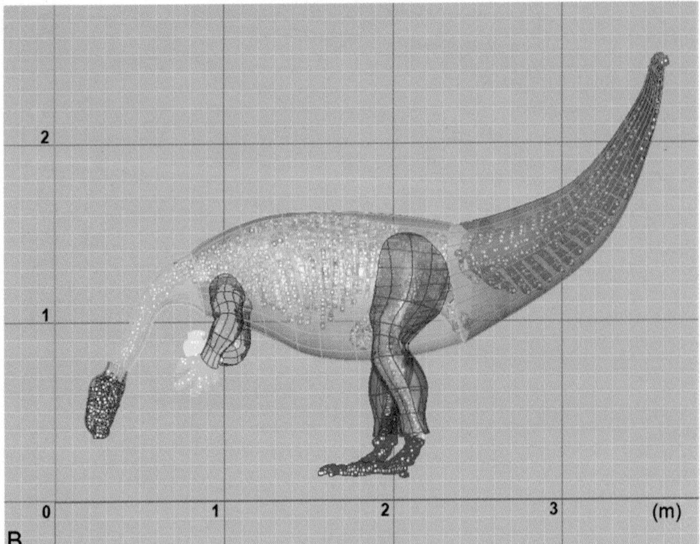

PLATE 7.1. 3D reconstruction of the soft tissue mass of *Plateosaurus engelhardti* based on a laser scan of a skeleton mounted at the University of Tübingen (GPIT 1). The different body compartments such as head, neck, and limbs are shown in different colors. By modeling body surface and body mass, a robust model (A) and a slim model (B) were reconstructed.

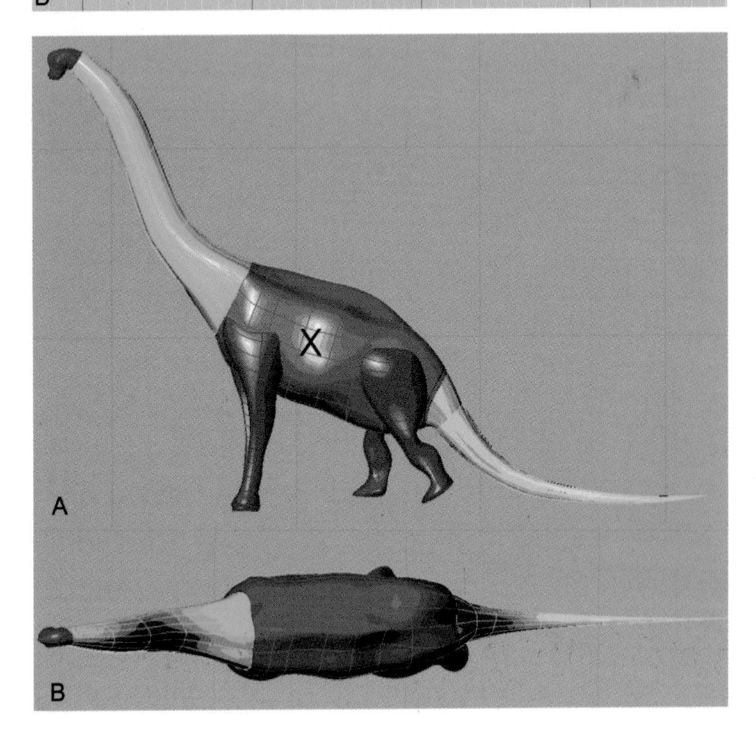

PLATE 7.2. 3D slim version of the reconstruction of *Brachiosaurus brancai* skeleton mounted and exhibited at the Museum für Naturkunde, Berlin, based on the historic mount of the skeleton on exhibit until 2005. (A) Side view. (B) Top view. The cross in (A) indicates the calculated center of mass.

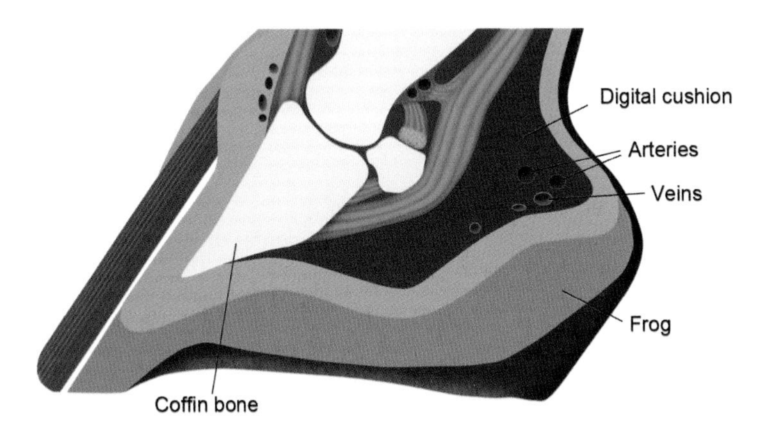

above **PLATE 7.3.** Schematic 3D model of *Brachiosaurus brancai* showing the skeleton (beige), arteries (red), veins (blue), the gastrointestinal tract (white and purple), and the respiratory system (green-yellow). A human figure is included as a scale.

left **PLATE 7.4.** Schematic drawing of a horse's foot, including the digital cushion which collects blood. By putting weight on the foot, the horse compresses the cushion, and the blood flows back up through the veins. A similar mechanism in the sauropod's lower extremities may also have supported the return of blood to the heart.

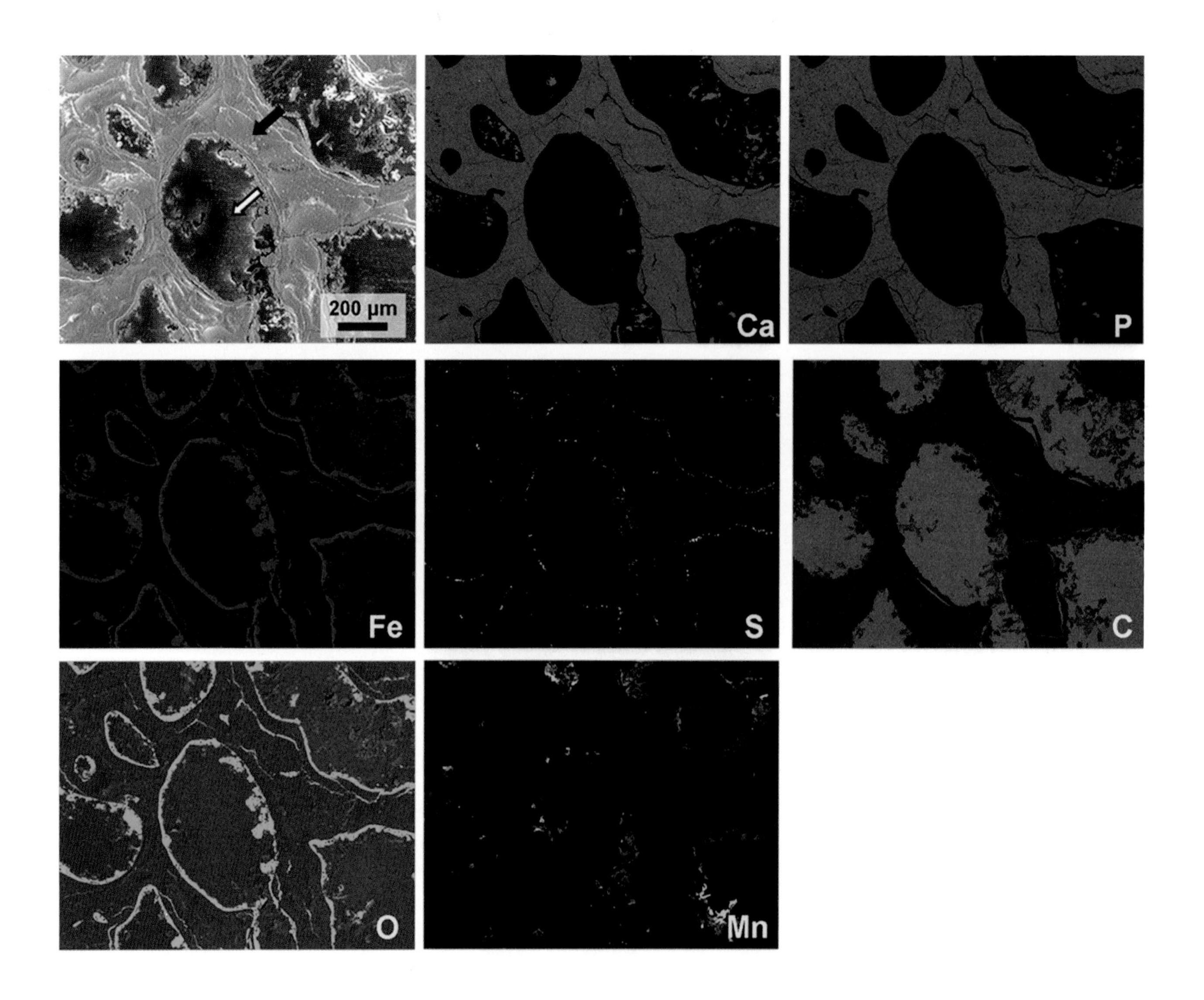

PLATE 9.1. SEM-EDS element maps of secondary trabecular bone in an *Apatosaurus* sp. cross-sectional sample (BYU 601-17328) from the Morrison Formation, USA (femur, 158.0 cm, histologic ontogenetic stage 12 of Klein & Sander 2008). The image in the upper left is an SEM image using the BSE detector. Black arrow, vascular spaces; white arrow, trabecular bone tissue. Letters indicate the elements mapped, each having been assigned an arbitrary color. The distribution of calcium and phosphorus reflects the distribution of bone tissue. The distribution of iron, sulfur, manganese, and oxygen indicates the first stage of the diagenetic infill of the vascular spaces, with Fe and S suggesting the presence of the mineral pyrite. Carbon is concentrated in the later diagenetic infill of the vascular spaces, suggesting that these consist of calcite or the resin used for sample preparation.

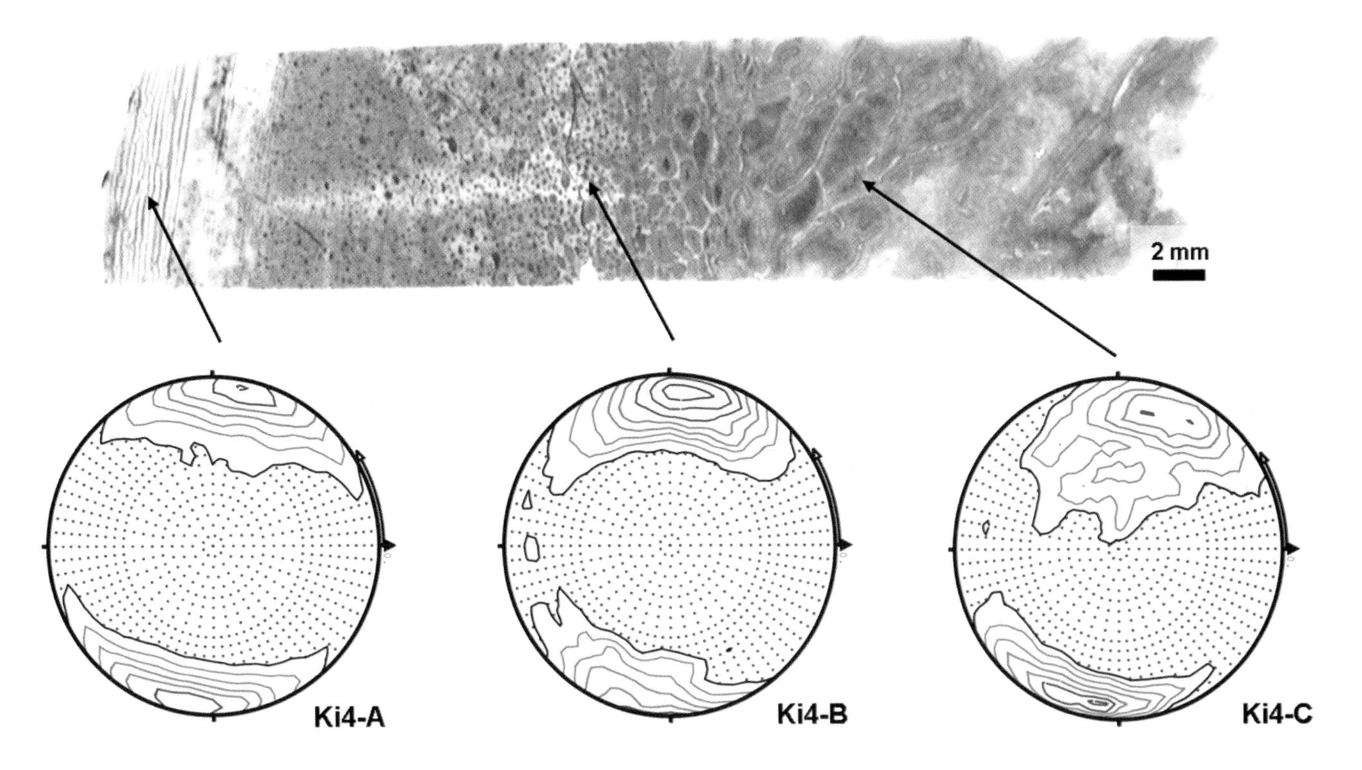

PLATE 9.2. 002 pole figures of apatite at three different positions along the bone cross section of a slowly growing indeterminate sauropod (MFN Ki 4a, left femur, 120.0 cm, periosteal margin of cortex on the left). The inner position (*right*) is in secondary trabecular bone, the middle one (*center*) is in Haversian bone, and the outer one (*left*) is in laminar FL bone. Level of the contour lines = 1.0, 1.2, . . . , 2.6 m.r.d. (multiple of random distribution). *Modified after Pyzalla et al. (2006).*

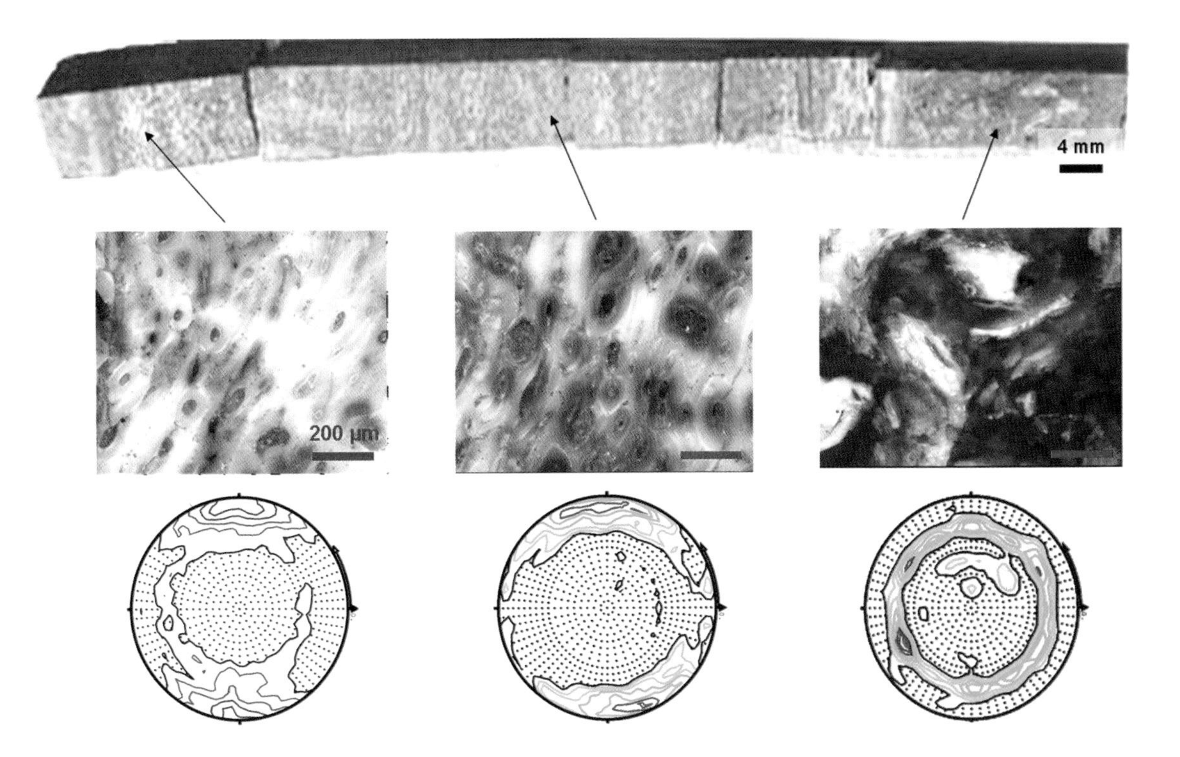

PLATE 9.3. 002 pole figures of apatite at three different positions along the bone cross section of a senescent individual of *Brachiosaurus brancai* (MFN XV, left femur, 219.0 cm; periosteal margin of cortex on the left) with insets showing the local histology at the positions. The inner position (*right*) is in secondary trabecular bone while the two outer ones (*center and left*) are in Haversian bone. Level of contour lines = 1.0, 1.2, . . . , 2.6 m.r.d. (multiple of random distribution). *Modified after Pyzalla et al. (2006).*

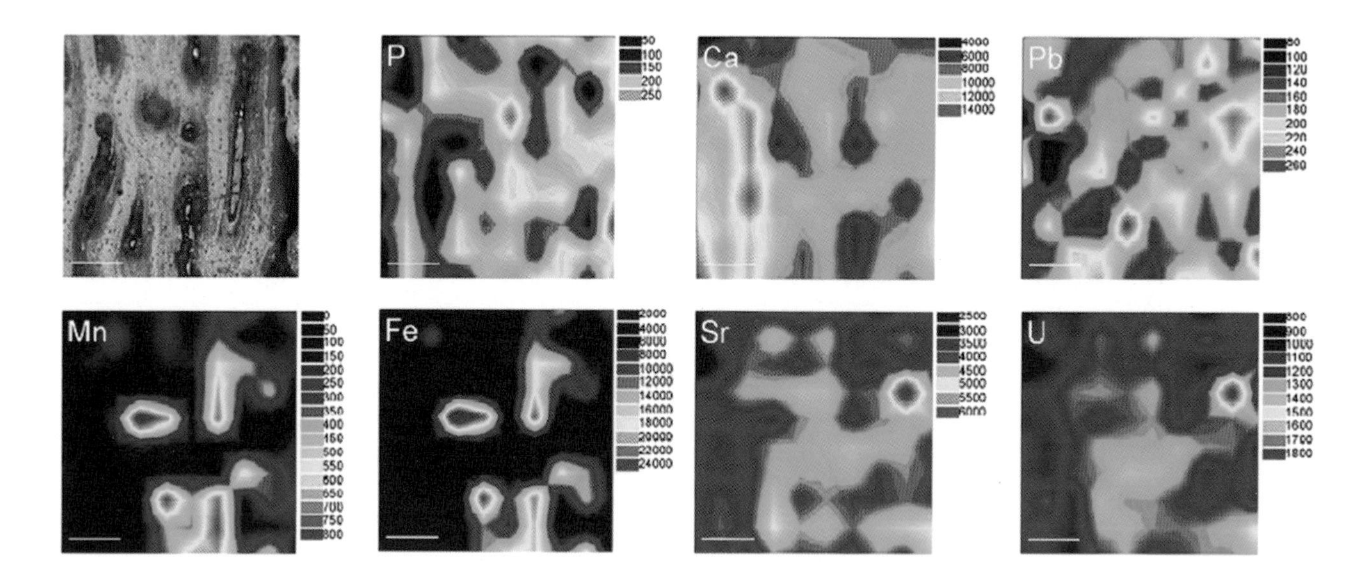

PLATE 9.4. Element maps obtained by µ-XFR in the FL bone of *Tornieria africana* (MFN A 1, right humerus, 99 cm; Sander 2000; Klein & Sander 2008). The image on the upper left is an incident light microscopic image of the area. Letters indicate the elements mapped; color spectrum indicates concentration, from lowest (blue) to highest (red). See text for details. *Modified after Ferreyro et al. (2006).*

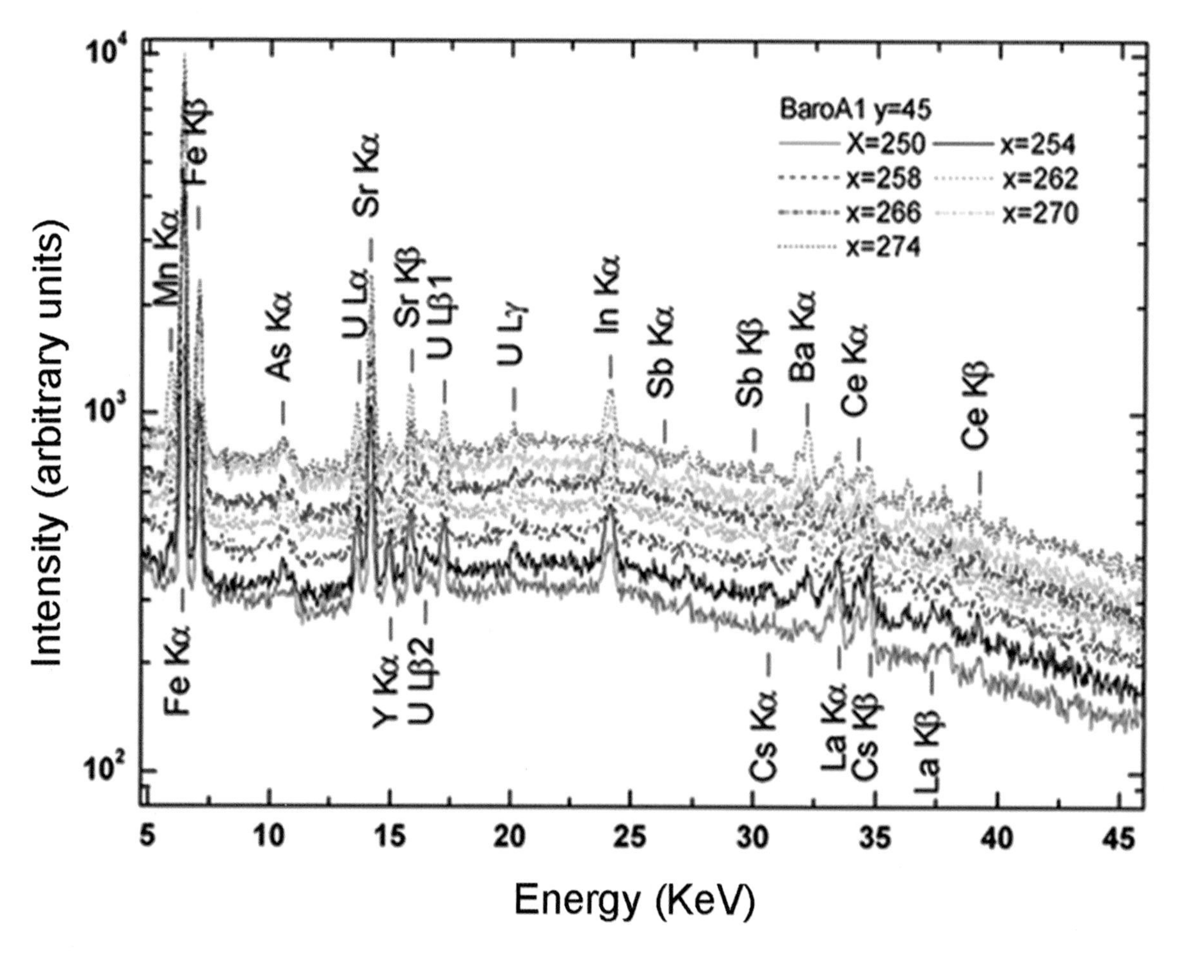

PLATE 9.5. PIXE spectrum across the cortex of *Tornieria africana* (MFN A 1, right humerus, 99 cm; Sander 2000; Klein & Sander 2008). The colored lines represent spectra measured at seven different locations ("x") across the bone cortex. *Modified after Stempniewicz & Pyzalla (2003).*

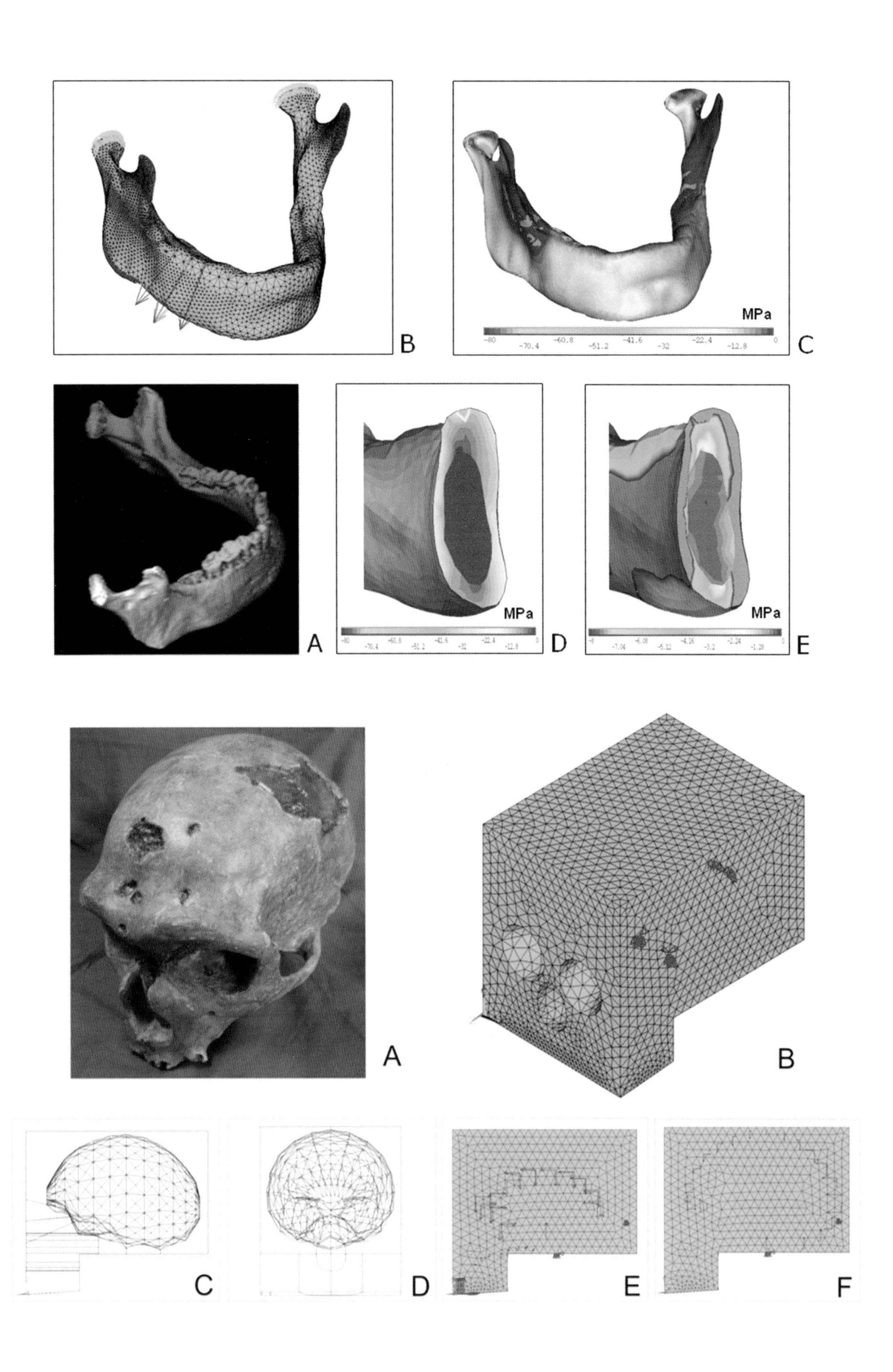

B
MPa
-80
-70.4
-60.8
-51.2
-41.6
-32
-22.4
-12.8
0
C
A
MPa
D
MPa
E
A
B
C
D
E
F

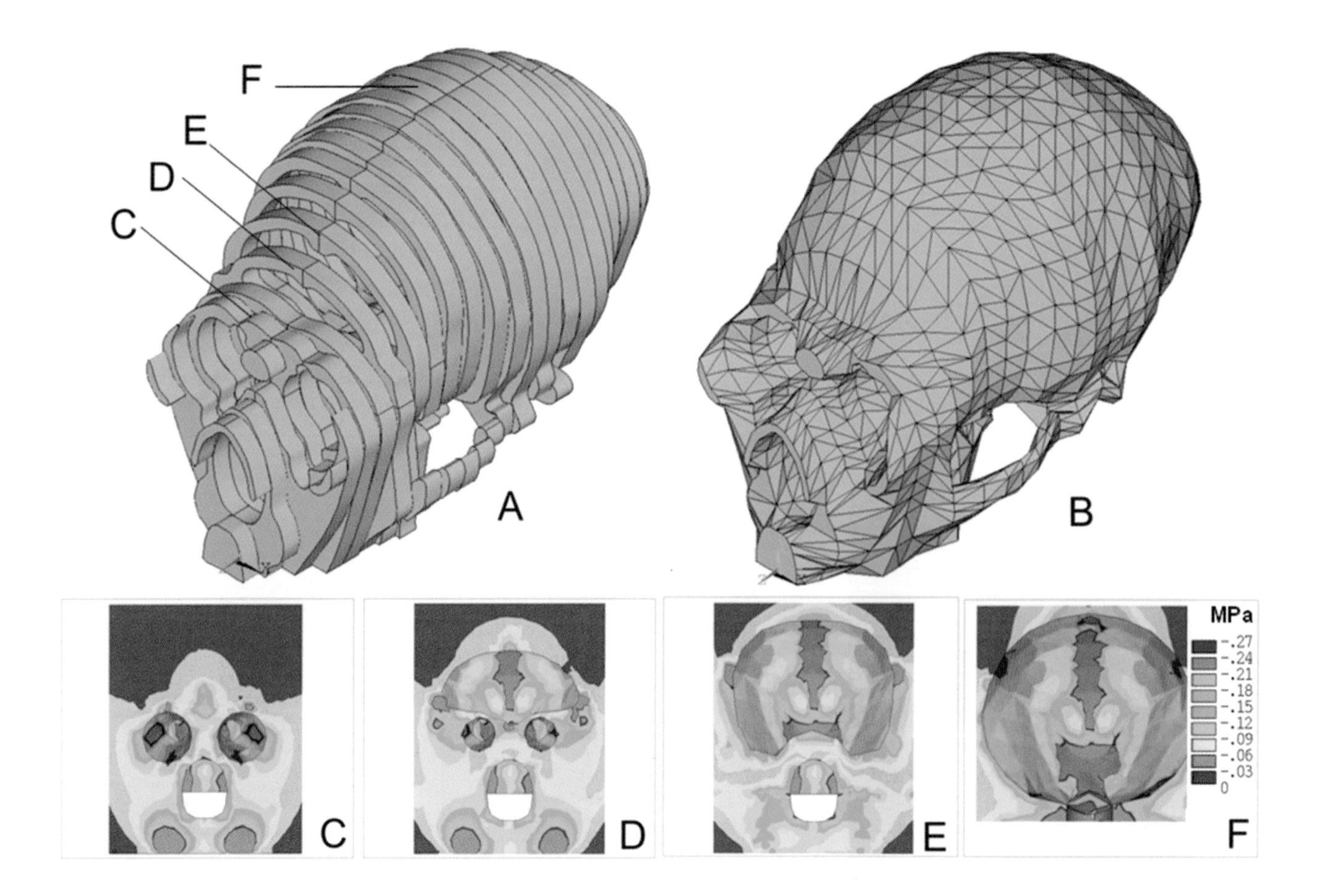

previous top **PLATE 10.1.** FESA of the human mandible. (A) Reconstruction of a human mandible based on CT scans. (B) Meshed 3D FE model with the finite elements, forces applied to the tooth positions (indicated by arrows). The supports of the model are located at the jaw joint. (C) Compressive stresses at the surface of the FE model. (D) Cross section through the mandible: compressive stress distribution in the cortical bone. Spreading scale of stresses as in (C). (E) Compressive stress distribution in the cancellous bone, using a spreading scale of stresses from 0 to −8 MPa. Gray areas are those that have higher stresses and thus represent cortical bone. Note that in engineering science, compressive stresses are denoted as negative values, in contrast to the geological sciences, where negative values indicate tension. Hence, red colors in this and the following plates indicate areas under low compressive stress, while yellow, green, and blue colors indicate areas exposed to higher compressive stress.

previous bottom **PLATE 10.2.** (A) Skull (cast) of *Homo neanderthalensis* from La Chapelle-aux-Saints. (B) Meshed bauraum for this skull. (C, D) Initial conditions for the first computation: functional spaces for the brain, eye sockets, and nose cavity inside the bauraum in (C) lateral and (D) posterior view. (E) Load case 1 (as an example for load cases in general): biting with incisors, with muscle forces indicated by arrows. (F) The stable three-point support, located at the two condyles, and a bearing at the insertion area of the neck muscles (blue markings). Forces for the falx membrane were generated by lateral acceleration of the brain (Witzel in press). A: *Photo courtesy of H. Preuschoft, Ruhr University of Bochum.*

above **PLATE 10.3.** Skull of *Homo neanderthalensis* after the first iteration of the FESS computation under the load case described in Plate 10.2E (biting with incisors). (A) First reduced stepped model. (B) Reduced FE model prepared for meshing and the next iteration. Note that in the older version of ANSYS used for this synthesis, the model surface had to be formed by flat triangles, while the current versions of ANSYS use spline surfaces. (C–F) Cross sections after the first iteration showing the distribution of compressive stresses. Location of cross sections is indicated in (A). For explanations of color codings, see Plate 10.1.

opposite top **PLATE 10.4.** First model of a *Diplodocus* skull: the initial bauraum. (A) The bauraum is shown with the meshed FEs, the applied muscle forces (red arrows), the bearings (turquoise triangles), and the location of the numbered cross sections shown in later illustrations. (B) Distribution of compressive stresses at the surface of the model. Areas under high load are the quadrato-mandibular joint (qu. m. j.), basioccipital (b. occ.), and supraoccipital (s. occ.). Dark blue indicates stresses of −2.24 to −2.52 MPa; see also the scale at lower right. Unloaded or little-stressed regions are the corners (unl. c.), with red indicating 0.0 to −0.28 MPa. Gray regions indicate high stress values off the scale.

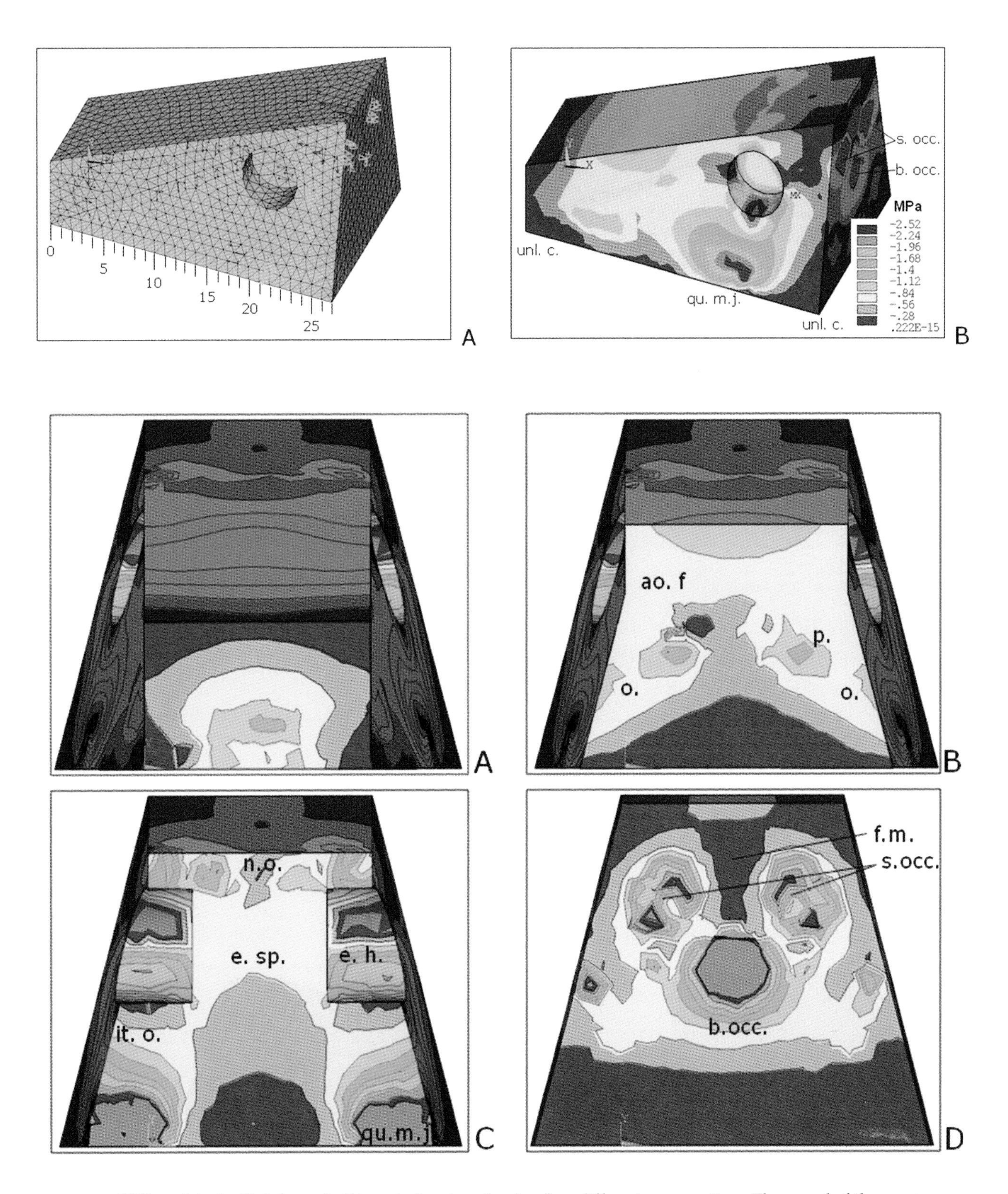

PLATE 10.5. FESS model of a *Diplodocus* skull in anterior view showing four different cross sections. The spread of the compressive stresses is from 0.0 to –2.52 MP. For explanations of color codings, see caption of Plate 10.1. For location of cross sections, see Plate 10.4. (A) Cross section 2 through the snout near the premaxillary teeth. (B) Cross section 14 through the palate (p.) with two openings (o.) and the central space, the dorsal bones of the snout, and the antorbital fenestrae (ao. f). (C) Cross section 20 through the eye sockets (e. h.) showing the quadratomandibular joint (qu. m. j.) as a high stress area, the infratemporal openings (it. o.), the central space (e. sp.), and the posterior end of the external nose opening (n. o.) bordering the frontal bone anteriorly. (D) Cross section 26 near the occipital end of the skull, showing the basioccipital (b. occ.) and supraoccipitals (s. occ.) as areas of high stress and the foramen magnum (f. m.) as an area of low stress. Gray regions in (C) and (D) indicate high stress values off the scale.

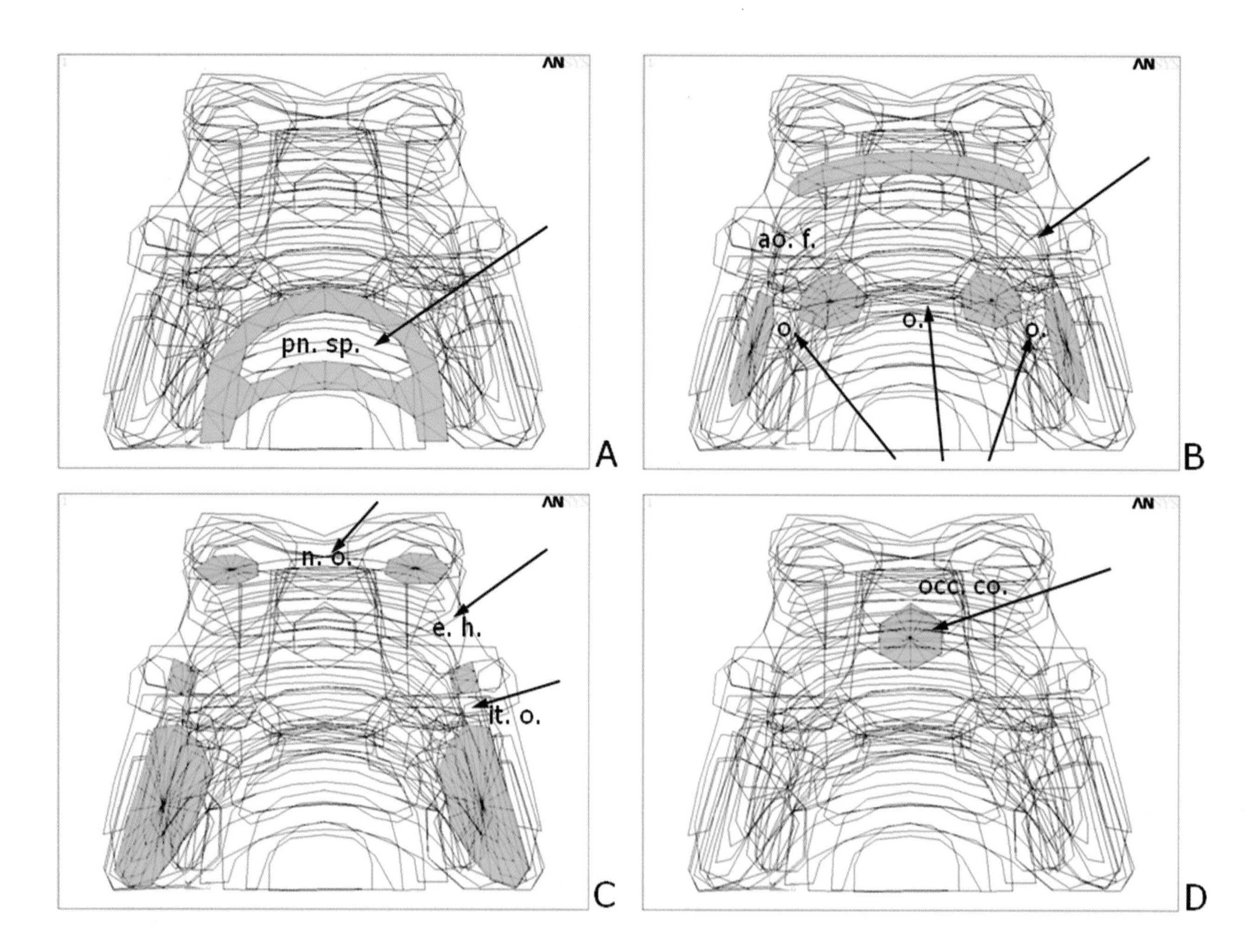

PLATE 10.6. Superimposed outlines of cross sections of FESS model of *Diplodocus* skull with stress-free areas from the first iteration deleted. Each panel shows a different cross section with high-stress areas (bone) shown in turquoise. For location of cross sections, see Plate 10.4. (A) Cross section 4 through the middle of the dental arcade shows the palate and the premaxilla, and a pneumatized space (pn. sp., arrow). (B) Cross section 14 through the three openings (o., arrows) in the palate and the antorbital fenestrae (ao. f., arrow). (C) Cross section 18 through the anterior end of the eye sockets (e. h., arrow), the middle of the nasal aperture (n. o., arrow), and the infratemporal fenestrae (it. o., arrow). (D) Last cross section through the occipital condyle (occ. co., arrow).

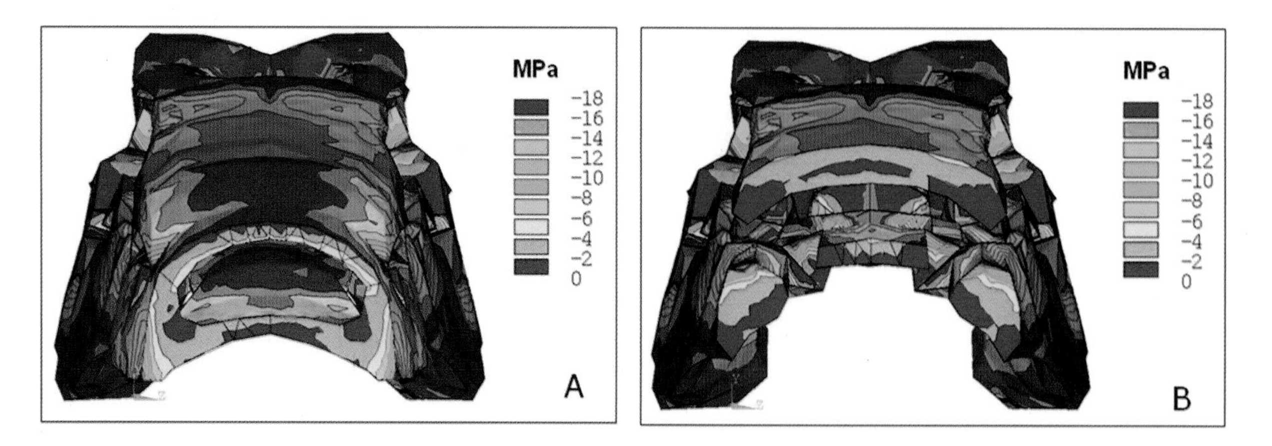

PLATE 10.7. FESS model of the *Diplodocus* skull after the second iteration of the FESS computation. The model is seen in anterior view cut along two different cross sections. Note the thinning effect on the cortical walls. Marginal areas of the synthesized structure are under low or no compressive stress (red) and can therefore be deleted in the next iteration. (A) Cross section 4 through the snout. (B) Cross section 9. Note the ventromedial thinning of the premaxilla. For location of cross sections, see Plate 10.4.

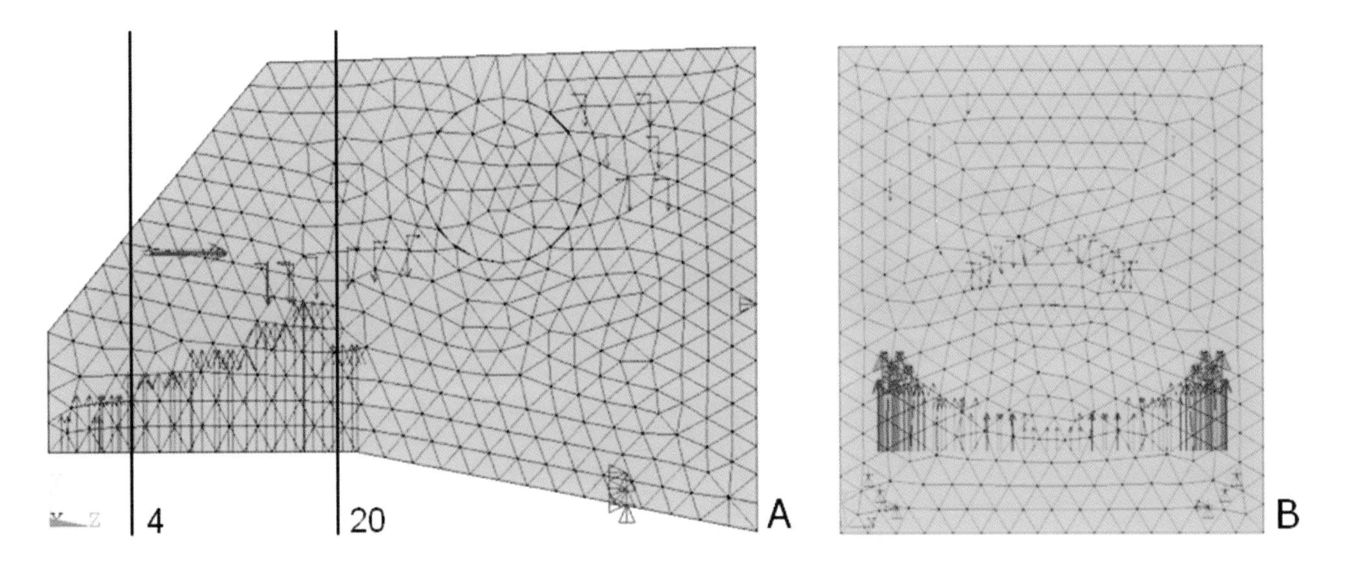

PLATE 10.8. Meshed 3D bauraum, the starting point for the FESS computation, of the skull of *Camarasaurus,* with bite and muscle forces chosen as a priori assumptions (red arrows). (A) Lateral view with positions of cross sections 4 and 20 indicated. (B) Posterior view. Compare to Plates 10.2 and 10.4.

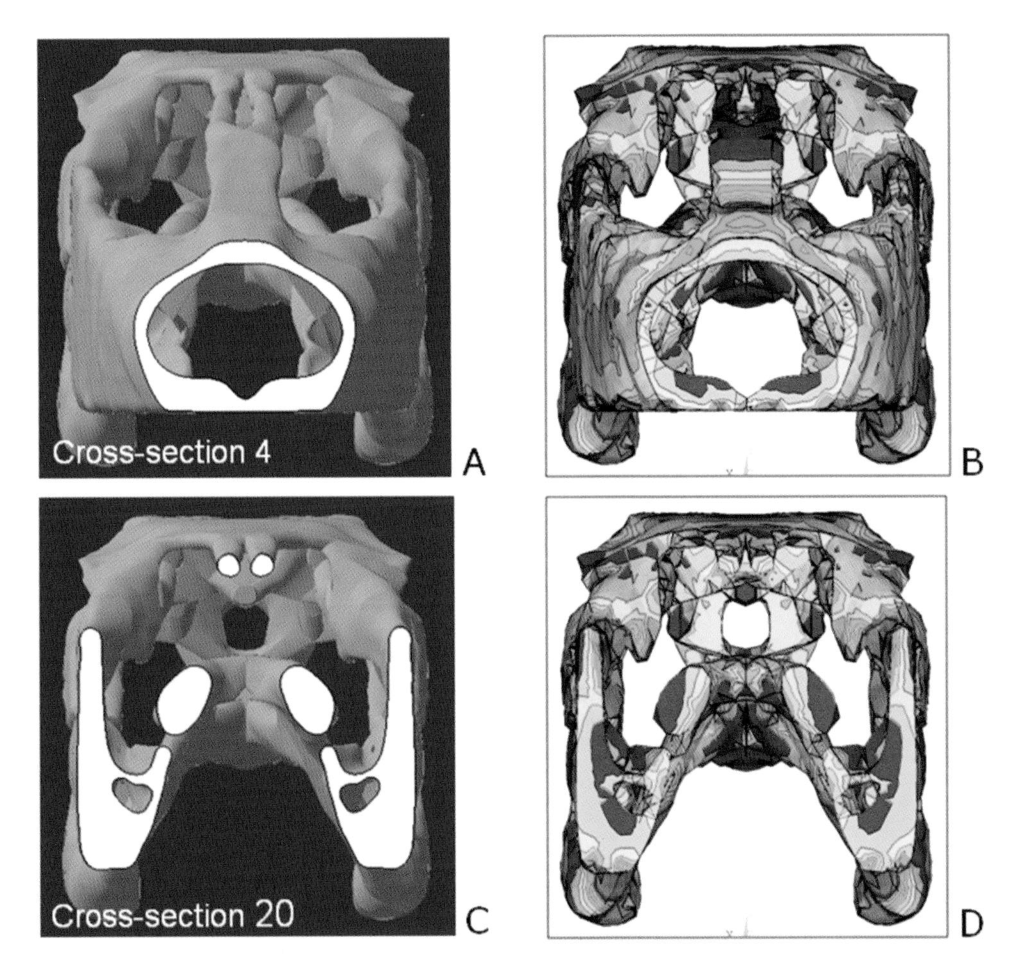

PLATE 10.9. Thinning effect of cortical walls as indicated by unstressed areas (red) with *Camarasaurus* skull FESS used as an example. The second and third iterations of the model are compared. (A) Anterior view cut along cross section 4 of the second iteration shown as CAD model. (B) Anterior view cut along cross section 4 of the third iteration showing compressive stresses and wall thinning, for example, on the dorsal side of the palate. (C) Posterior view cut along cross section 20 of the second iterations shown as CAD model. (D) Wall thinning after third iteration of the same cross section. Compare to Plate 10.7. For explanations of color coding, see Plate 10.1 caption. For location of cross sections, see Plate 10.8.

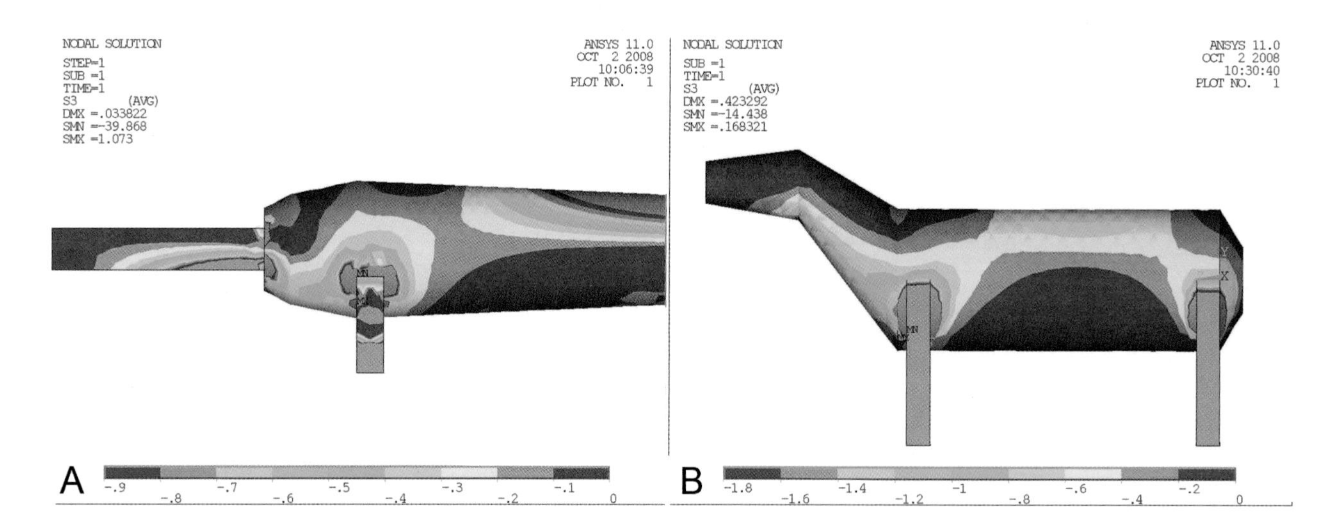

PLATE 11.1. Compressive stress distribution in 3D FE tetrapod models in lateral view. All limbs are in contact with the ground. (A) Sprawling model. (B) Erect model. High compressive stresses are indicated in blue; low compressive stresses are indicated in red; stresses significantly beyond the range indicated at the bottom of the frames are indicated in gray. Note the different distributions of compressive stresses along the body of the models.

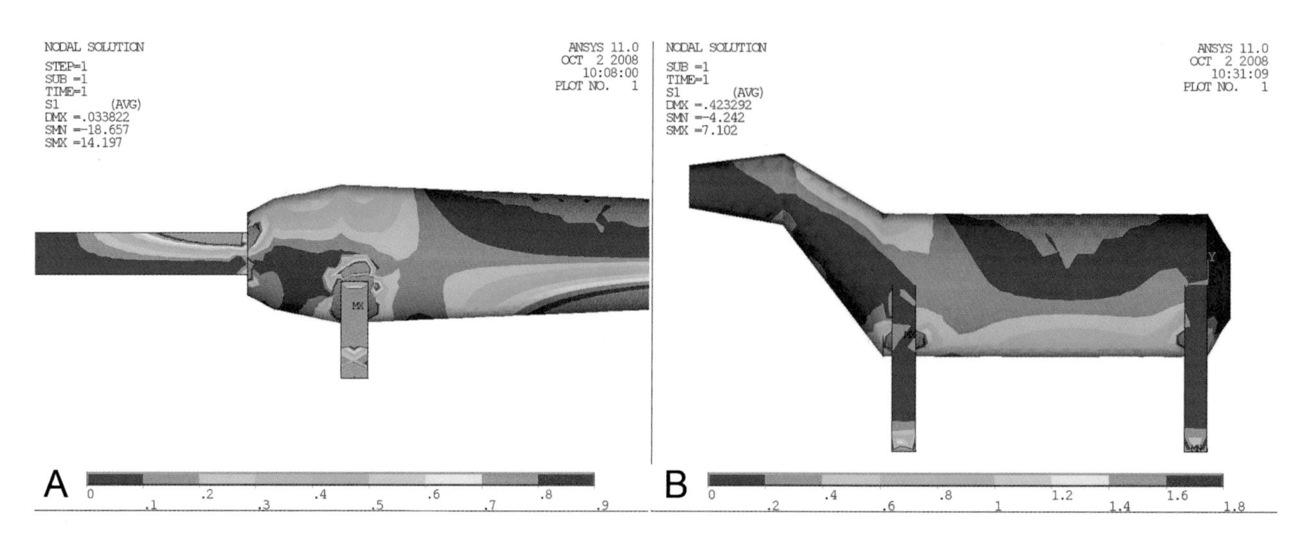

PLATE 11.2. Tensile stress distribution in 3D FE tetrapod models in lateral view. All limbs are in contact with the ground. (A) Sprawling model. (B) Erect model. High tensile stresses are indicated in red; low tensile stresses are indicated in blue; stresses significantly beyond the range indicated at the bottom of the frames are indicated in gray. Note the stress distributions on the ventral side and neck region.

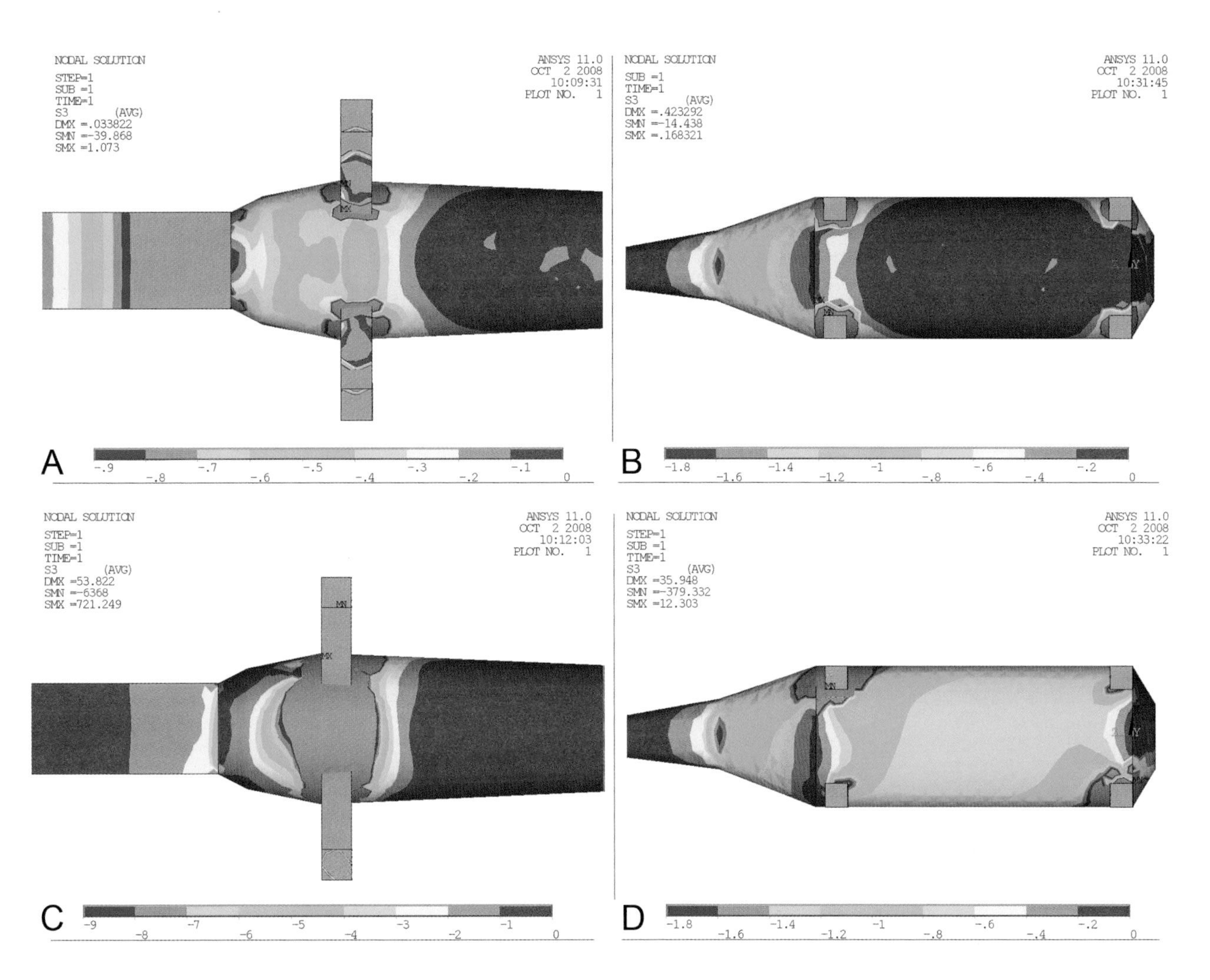

PLATE 11.3. Compressive stress distribution in 3D FE tetrapod models in ventral view (A, B) Model supported on all four limbs. (C, D) Model has only contralateral body support to simulate walking. Note the higher stress values and the increase in compressive stresses between the forelimbs of the sprawling model (A, C) compared to the erect model (B, D). Color coding as in Plate 11.1.

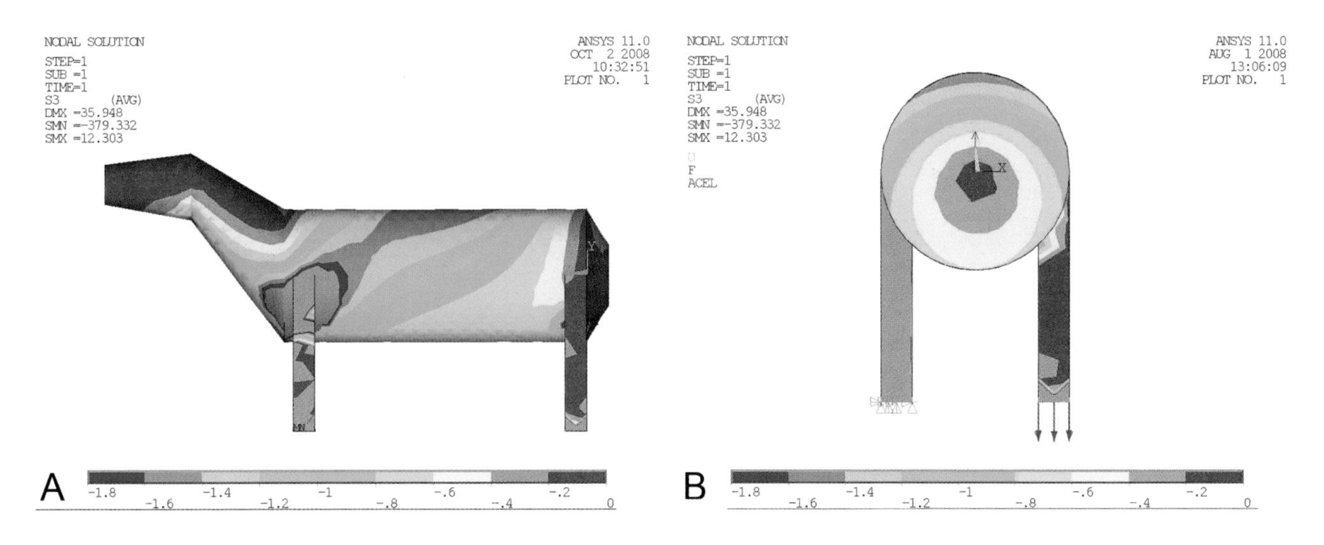

PLATE 11.4. (A) Erect 3D FE tetrapod model showing compressive stresses along the lateral side of the body. The left forelimb and right hindlimb are in contact with the ground. Note the compressive stresses oriented obliquely between the contralateral limbs with ground contact. (B) Cross section through the middle trunk region of the erect model. Compressive stresses are concentrated at the surface of the trunk, leaving the center unstressed. Color coding as in Plate 11.1.

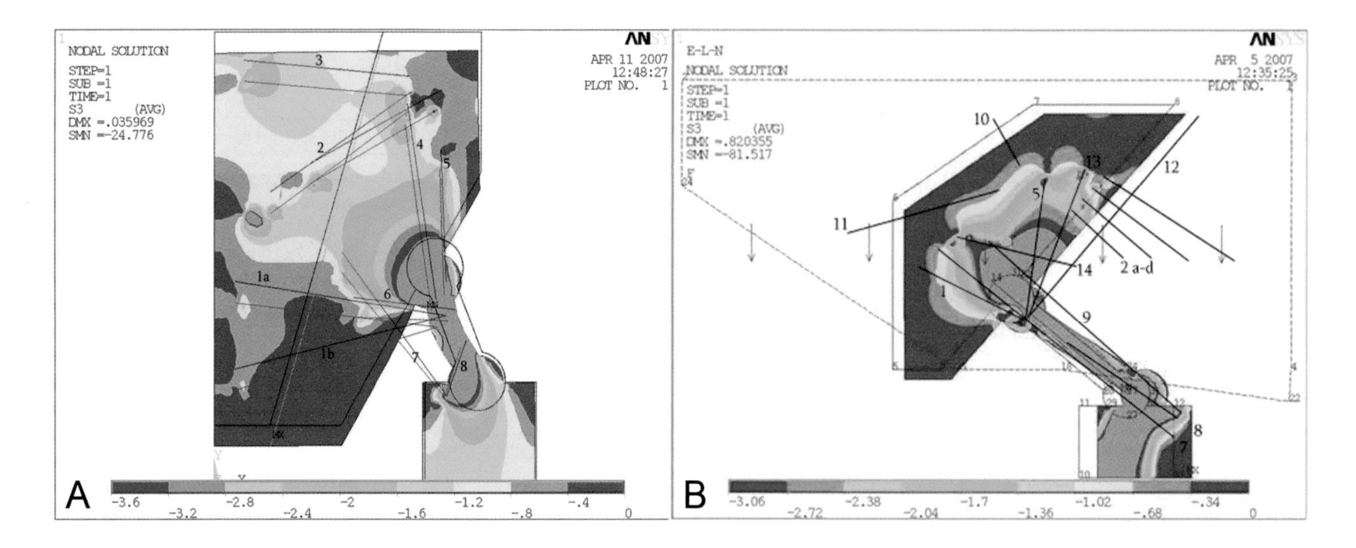

PLATE 11.5. Compressive stresses in the 2D FE crocodile shoulder model based on *Caiman crocodylus* in anterior (A) and lateral (B) views. The position of the four elements before the computation is shown by the thin black outlines. Color coding as in Plate 11.1. Muscles and their lines of action are indicated by numbered lines: 1, 1a, 1b, m. pectoralis; 2, 2a–d, m. serratus; 3, m. rhomboideus; 4, m. subscapularis; 5, m. dorsalis scapulae; 6, m. coracobrachialis brevis; 7, m. triceps brachii caput coracoideum; 8, m. triceps brachii caput humerale; 9, m. triceps brachii caput scapulare; 10, m. trapezius; 11, m. levator scapulae; 12, m. latissimus dorsi; 13, m. teres major; 14, m. costocoracoideus.

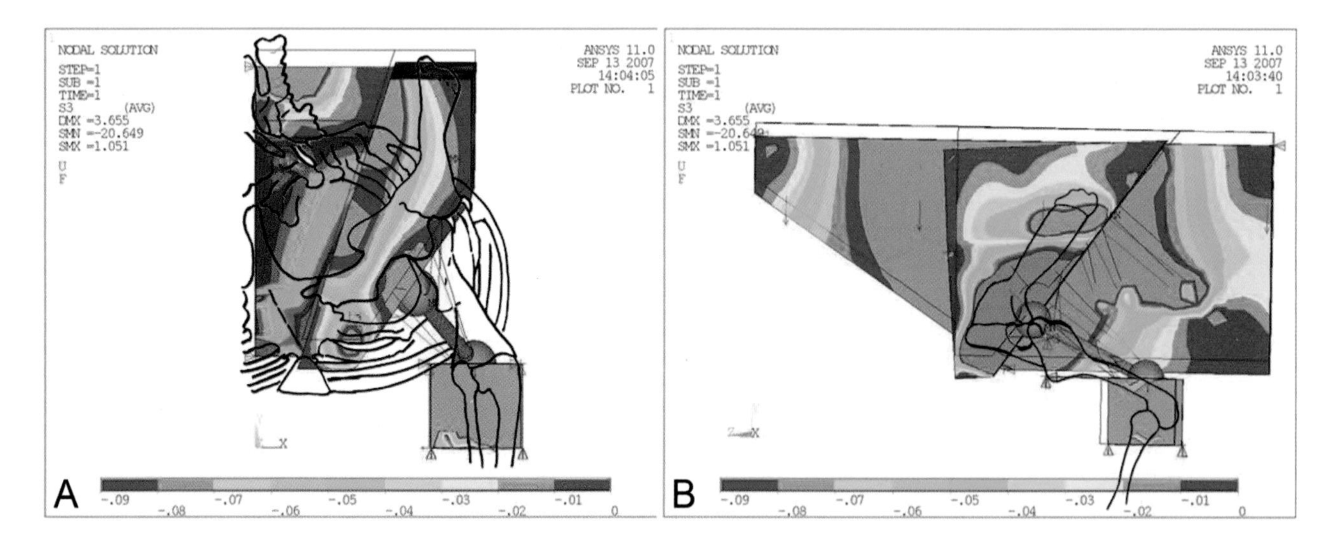

PLATE 11.6. 3D FE crocodile shoulder model based on *Caiman crocodylus*. Results of the first iteration in frontal (A) and lateral (B) views. A sketch of the original skeleton is placed as an overlay on the model. The high compressive stresses (yellow, green, and blue) are largely restricted to the area of the original scapulocoracoid. Color coding as in Plate 11.1.

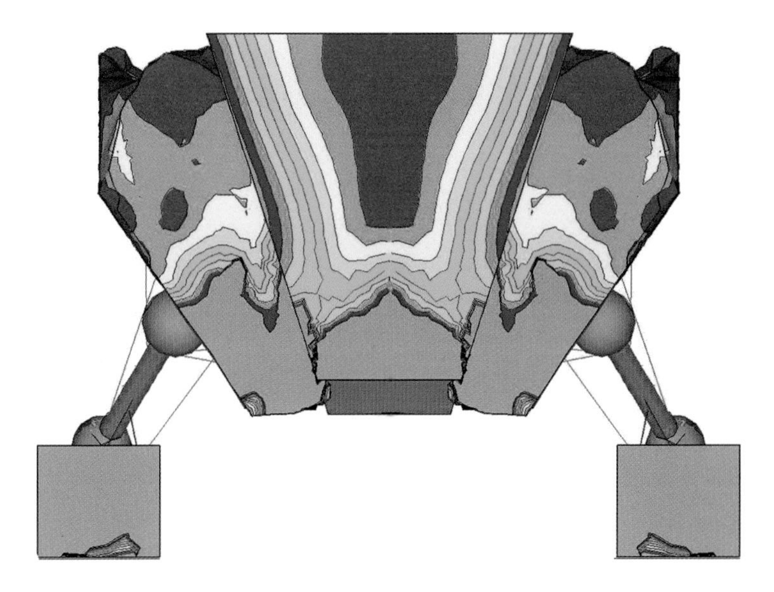

PLATE 11.7. Anterior view of the first iteration of the 3D FE crocodile shoulder model based *Caiman crocodylus,* which was paired with its mirror image in order to produce an entire shoulder region. This view shows the concentration of stresses at the margin of the trunk best, leaving the inner space unstressed. Note the stress distribution from the lower part of the scapula to the trunk, representing the connection between the coracoid plate and the sternum. Color coding as in Plate 11.1.

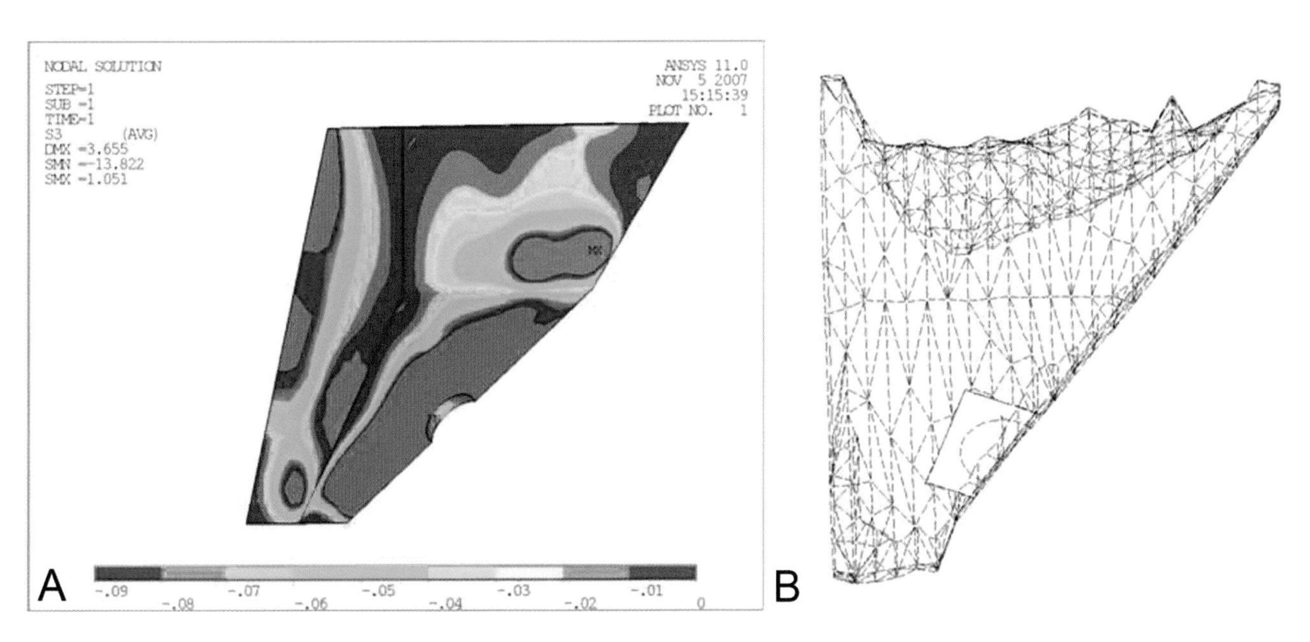

PLATE 11.8. Reduction of the bauraum of the scapula after the first iteration. Reduction involves eliminating the regions that show low compressive stress values. (A) Compressive stress distribution after the first iteration. (B) 3D grid model of the reduced bauraum after eliminating the low-stress areas. The reduced model provides the basis for the next iteration. Color coding as in Plate 11.1.

5 cm

Tortues et crocodiles

opposite top PLATE 16.1. Unhatched eggs of the oogenus *Megaloolithus* from the Late Cretaceous (Maastrichtian) of southern Europe. Eggs on the left are from Catalonia, northern Spain (top, Suterranya, Tremp Basin; bottom, Coll de Nargó). Eggs on the right are both from Rousset near Aix-en-Provence, southern France. Note the considerable size range in this material. All eggs figured are deposited at the Steinmann Institute, Division of Paleontology, University of Bonn, Germany.

opposite bottom PLATE 16.2. Mounted skeleton of *Ampelosaurus atacis*. This titanosaur species comes from the Late Cretaceous (early Maastrichtian) of southern France. It is a medium-sized species (length from snout to tail about 15 m) and is the probable producer of the eggs from southern France shown in Plate 16.1. The mounted skeleton and original bones of this titanosaur are exhibited in the Musée des Dinosaures, Espéraza, France. *Photograph courtesy of Jean Le Loeuff.*

above PLATE 16.3. A female American alligator (*Alligator mississippiensis*) in front of her nest. Females build a nest of vegetation, sticks, leaves, and mud in a sheltered place near the water. Eggs are covered by vegetation. The female remains close to the nest throughout the incubation period and defends her nest against potential predators. When the young begin to hatch, the female quickly digs them out and carries them to the water. Females guard their offspring during the first months of their life.

PART THREE

CONSTRUCTION

8

How to Get Big in the Mesozoic: The Evolution of the Sauropodomorph Body Plan

OLIVER W. M. RAUHUT, REGINA FECHNER, KRISTIAN REMES, AND KATRIN REIS

SAUROPOD (OR, MORE CORRECTLY, EUSAUROPOD) dinosaurs are highly distinctive, not only in their overall body form, but also in respect to many details of their anatomy. In comparison with basal dinosaurs, typical sauropods are characterized by small skulls, elongate necks, massive bodies, and an obligatory quadrupedal stance with elongate forelimbs and straight limbs in general. Tracing the anatomical changes that led to this distinctive body plan through sauropodomorph evolution is problematic as a result of the incompleteness of many basal taxa and phylogenetic uncertainty at the base of the clade. The decrease in skull size in sauropodomorphs seems to be abrupt at the base of the clade, but it is even more pronounced toward sauropods. Major changes in the sauropod skull are a relative shortening and broadening of the snout and an enlargement and retraction of the nares. Although the ultimate causes for these evolutionary changes are certainly manifold, most if not all of them seem to be related to the ecological and biomechanical requirements of the transition from a carnivorous to an herbivorous lifestyle, in which the skull is mainly used as a cropping device. A relatively elongate neck seems to be ancestral for sauropodomorphs, but the neck is further elongated on the lineage toward sauropods, especially by incorporation of two additional vertebrae at the base of Sauropoda. The relatively simple structure of the cervical vertebrae in basal sauropodomorphs might be a secondary reduction relative to basal saurischians as a result of changes in neck biomechanics in connection with the reduction of the size of the skull. Thus, the more complicated structure of sauropod cervicals probably reflects changing biomechanical requirements in connection with an elongation of the neck and an increase in body size, as does the opisthocoelous structure of the cervical vertebral centra. Limb evolution in sauropodomorphs is dominated by adaptations toward increasing body size and thus graviportality, with the limbs getting straighter and the distal limb segments relatively shorter. Body size increase in sauropodomorphs seems to have been rapid but even-paced, with the ancestral body size of the clade being in the 0–10 kg category, and the ancestral body size for sauropods probably being in the 1,000–10,000 kg category.

Introduction

Sauropod dinosaurs are certainly among the most recognizable fossil vertebrates known. With their massive bodies, tiny heads, long necks, and long tails, they are the stereotypical dinosaur and are found everywhere in popular culture, from cartoons to the advertising billboards of a big oil company. However, sauropods are not only popular with the general public. They represent the largest terrestrial animals in the history of life on earth, and hence they are of great scientific

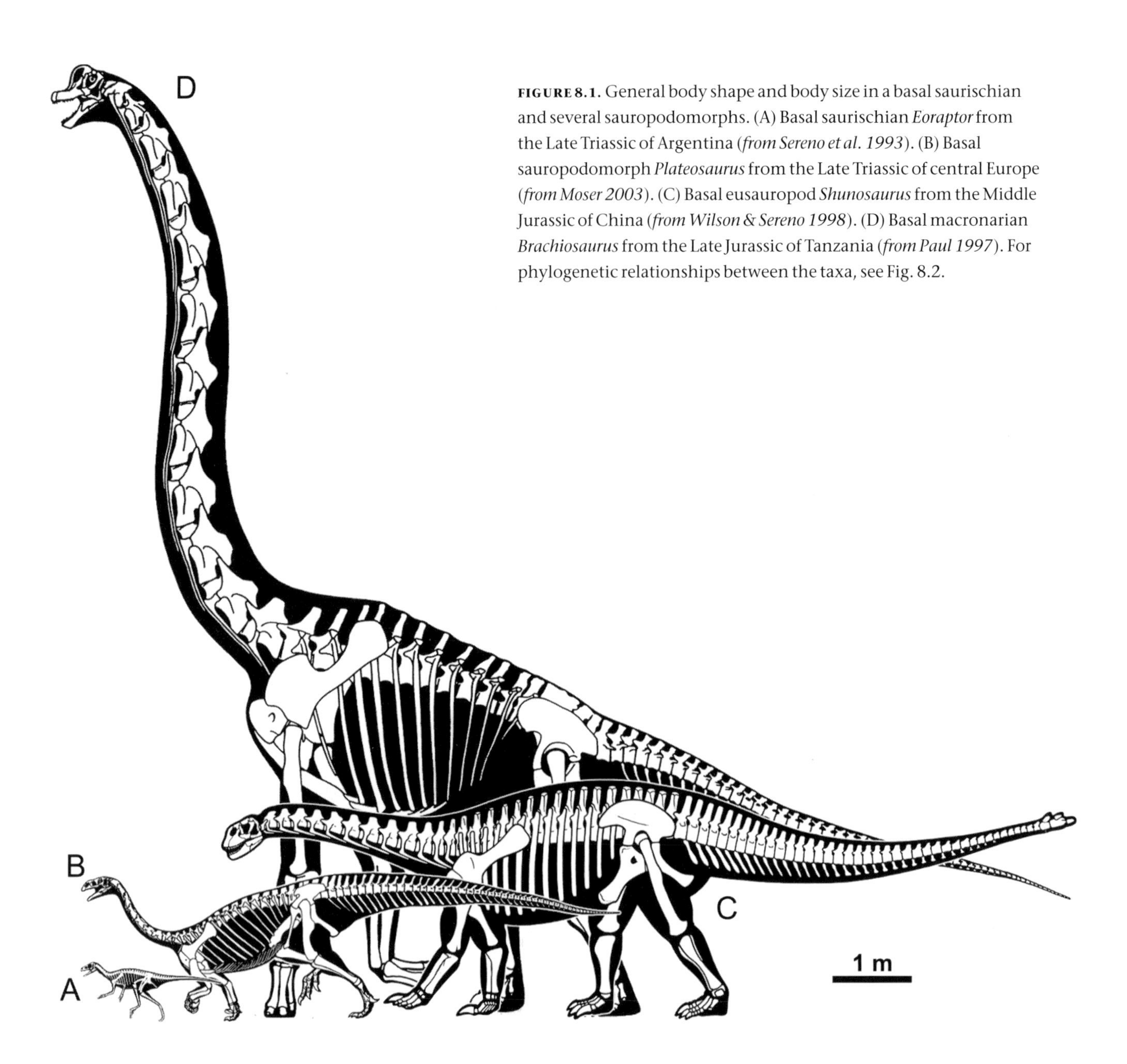

FIGURE 8.1. General body shape and body size in a basal saurischian and several sauropodomorphs. (A) Basal saurischian *Eoraptor* from the Late Triassic of Argentina (*from Sereno et al. 1993*). (B) Basal sauropodomorph *Plateosaurus* from the Late Triassic of central Europe (*from Moser 2003*). (C) Basal eusauropod *Shunosaurus* from the Middle Jurassic of China (*from Wilson & Sereno 1998*). (D) Basal macronarian *Brachiosaurus* from the Late Jurassic of Tanzania (*from Paul 1997*). For phylogenetic relationships between the taxa, see Fig. 8.2.

interest to students of natural history (Sander et al. 2010a). Scientists have long marveled at the giant size and anatomical specializations of these remarkable animals, and much has been published on their stance, gaits, physiology, and probable habits (see overviews in Coombs 1975; Dodson 1990; Upchurch et al. 2004a; Sander et al. 2010a). As noted by Wilson & Curry Rogers (2005), these animals are truly the monoliths of the Mesozoic!

However, two aspects of sauropod dinosaurs have long remained elusive: their origin and phylogenetic interrelationships. Although a number of proposals have been advanced since the early years of sauropod research (e.g., von Huene 1932), the notorious incompleteness of their skeletons has long severely hampered interpretations of their evolutionary history. Indeed, as recently as 1965, Charig et al. proposed an origin of sauropod dinosaurs different from that of other dinosaurs (see also Heerden 1997), and Romer (1968) noted, "It will be a long time, if ever, before we obtain a valid, comprehensive picture of sauropod classification and phylogeny." Although modern methods and an increased interest in sauropod interrelationships and origins have greatly improved our understanding of their phylogeny (e.g., Calvo & Salgado 1995; Upchurch 1995, 1998; Salgado et al. 1997; Wilson & Sereno 1998; Wilson 2002, 2005; Yates 2003, 2004, 2007a, 2007b; Yates & Kitching 2003; Upchurch et al. 2004a, 2007a; Curry Rogers 2005), many aspects of the anatomical, functional, and ecological transitions from small, basal dinosauriforms to the giants of the Mesozoic are still poorly understood. Several authors have explored aspects of the evolutionary history of sauropodomorphs, but most of these works focused on the evolutionary transitions within the Sauropoda (e.g., Upchurch 1995; Bonaparte 1999; Salgado 1999; Wilson 1999, 2005; Bonnan 2003; Wedel 2003a, 2005), and only few dealt with aspects of the evolutionary transitions from basal saurischians to sauropods (e.g., Yates & Kitching 2003; Barrett & Upchurch 2007; Wedel 2007). Thus, the purpose of the current chapter is to review our current knowledge of this evolutionary transition: What were the main anatomical changes that led from small basal saurischians to the giants of the Mesozoic (Fig. 8.1)?

MATERIALS AND METHODS

For this study, data on basal dinosaurs, basal sauropodomorphs, and sauropods were compiled from the literature and from personal observations of actual specimens. Table 8.1 provides a list of data and their respective sources.

Several people have noticed general trends in the evolution of sauropodomorph dinosaurs, such as a reduction in skull length and an increase in neck length (e.g., Benton 1990), or have commented on anatomical changes that occurred during the transition from basal dinosaurs to sauropods. However, most of the accounts regarding changes in overall body plan and more general changes (other than qualitative anatomical changes; e.g., Barrett & Upchurch 2007) from basal dinosaurs to sauropods are rather anecdotal and have not been backed up by quantitative data. Quantifying these changes is hampered mainly by the incomplete nature of many of the taxa involved. Data on well preserved skulls, for example, are only available in a few species, and for many of these species, the postcranial material necessary for the comparison of body proportions is not preserved, is incomplete, or has not been described in sufficient detail for a thorough survey of changes in these proportions. Because such a survey is beyond the scope of this work, in order to quantify these changes, we looked at several simple length ratios in basal dinosaurs, basal sauropodomorphs, and basal sauropods. These ratios were then plotted against the cladistic rank of the taxa measured to visualize changes on the evolutionary lineage toward advanced sauropods. Qualitative changes are commented on in several anatomical complexes that we looked at in more detail.

In this analysis of general body proportions, we measured and compared areas of the head, neck, body, limbs, and tail of several taxa from reconstructions of the animals in lateral view. Although these measurements are certainly not precise, they can be used as proxies for relative changes in body proportions because lateral area is correlated with the volume of the respective body parts (Seebacher 2001). Areas were measured using Auto Montage Pro software, version 5.02.

In a morphological transition as profound as that from basal dinosaurs to sauropods, it is difficult to find and define independent reference measurements, because all body proportions are affected by these changes. However, femur length is often used as a proxy for body size, and thus we used this measurement as a reference to look at changes in relative skull size. It should be noted, though, that there is a tendency to lengthen the femur in relation to the bones of the lower leg in sauropodomorphs, which may amplify the observed reduction in size of the skull. However, the general trend observed remains the same if skull size is compared to the length of femur plus tibia as well.

Looking at changes in skull shape and configuration, a simple shape analysis was carried out in addition to the calculation of ratios between different skull measurements. Furthermore, areas of skull openings (nares, antorbital fenestra, orbit, infratemporal fenestra) in lateral view were measured and compared to the overall area of the lateral skull profile in the few taxa for which good skulls are available (e.g., *Eoraptor, Plateosaurus, Shunosaurus, Mamenchisaurus*). For the shape analysis, skulls of different sauropodomorphs were scaled to the same skull length and oriented so that the tip of the snout and the quadrate condyle lay on a horizontal line. Landmarks marking the rostralmost, caudalmost, dorsalmost, and ventralmost points of the cranial openings were then traced through sauropodomorph phylogeny.

To analyze the evolution of body size in sauropodomorphs, we categorized sauropodomorphs in logarithmic size classes, which were then assigned character states (0: 0–10 kg; 1: 10–100 kg; 2: 100–1,000 kg; 3: 1,000–10,000 kg; 4: >10,000 kg). Data on body mass were mainly taken from Peczkis (1994), Seebacher (2001), Mazzetta et al. (2004), and Gunga et al. (2007), with the size class for a few taxa estimated on the basis of comparisons of overall body size. The size classes were then mapped onto the cladogram of Yates (2007b), with interrelationships of basal sauropods taken mainly from Allain & Aquesbi (2008), using MacClade 4.08 (Maddison & Maddison 2003). Size classes were treated as ordered characters, which results in a conservative interpretation of body size evolution in which smaller body size usually represents the plesiomorphic character state. The same criterion was applied to groups in which adult body size varied in included taxa (e.g., the genus *Plateosaurus,* in which *P. gracilis* is smaller than either *P. engelhardti* and *P. ingens*) or taxa in which mass estimates varied between two different classes.

The Origin and Interrelationships of Sauropodomorph Dinosaurs

The taxonomic name Sauropodomorpha was coined by von Huene (1932, pp. 17–18) to include the Sauropoda and their ancestors, the Prosauropoda (which was understood to be a

Table 8.1. Morphometric Data of Basal Saurischian and Sauropodomorph Dinosaurs

No.	Taxon	Specimen no.	Neck length	Trunk length	Humerus length	Radius length	Mc II length	Forelimb total	Femur length	Tibia length	Mt III length	Hindlimb total
1	*Anchisaurus*	YPM 1883		507	150	95	36	281	211	145	98	454
2	*Antetonitrus*	BP/I/4952			713	370	132	1215	794	512	197	1503
3	*Apatosaurus*	NSMT-PV 20375		1780	1033	616	208	1857	1470	943	210	2623
4	*Brachiosaurus*	MB t 1										
5	*Camarasaurus*	CM 11338	1020	952	426	292	140	858	580	350	88	1018
6	*Camarasaurus*	GMNH 101			1130		324		1485	930	223	2638
7	*Coelophysis*	AMNH 7224	403	460	122	82	39	243	205	213	121	539
8	?*Coloradisaurus*	PVL field Nr. 6								445	220	
9	*Dilophosaurus*	UCMP V4214			283	183	104	570	556	538	293	1387
10	*Diplodocus*	CM 11161										
11	*Efraasia*	SMNS 12667			170	90	50	310	230	225		
12	*Eoraptor*	PVSJ 512			84	61	19	164				
13	*Herrerasaurus*	PVSJ 373			175	153	58	386	364	315	165	844
14	*Herrerasaurus*	PVSJ 407			170	154						
15	*Heterodontosaurus*	SAM K 1332			83,5	58,5	23	165	112	145	68	325
16	*Jobaria*	MNN Tig 3, 4	4030		1360	1040	390	2790	1800	1080	300	3180
17	*Lufengosaurus*	IVPP V 15							560	350	212	1122
18	*Mamenchisaurus*	ZDM 0083	5959	1852	830	545	225	1600	1160	665		
19	*Marasuchus*	PVL 3870							42,4	50,1	28	120,5
20	*Massospondylus*	BP/I/4934										
21	*Massospondylus*	BP/I/5241										
22	*Melanorosaurus*	NM QR3314			435	230	80	745	560	365	145	1070
23	?*Patagosaurus*	MACN CH										
24	*Plateosaurus*	GPIT 1	948	1203	370	210	97	677	590	440	225	1255
25	*Plateosaurus*	SMNS 13200	1075	1440	400	240	100	740	680	490	240	1410
26	*Riojasaurus*	PVL 3808			470				600	510	200	1310
27	*Saturnalia*	MCP 3844-PV							157	158	84	399
28	*Scleromochlus*	BMNH R 3557			19,5	18	2,5	40	31,7	34,5	15,5	81,7
29	*Shunosaurus*	ZDM T 5402	2233	1890	720	482	160	1362	1200	682	175	2057
30	*Shunosaurus*	ZDM 65430										
31	*Shunosaurus*	ZDM T 5401										
32	*Yunnanosaurus*	IVPP V 20			417	219	94	730	435	350	170	955

Values are provided in millimeters. Mc II, metacarpal II; Mt III, metatarsal III; m, maxilla; pm, premaxilla.

paraphyletic group). It might be worth noting, though, that some taxa that are currently regarded as basal sauropodomorphs were considered then to be carnivorous, and thus regarded as basal carnosaurs by von Huene, in a supposedly monophyletic group called the Pachypodosauria that included sauropodomorphs and carnosaurs to the exclusion of coelurosaurs. It was not until 1985 that it was established that these allegedly basal carnosaurs were based on an erroneous association of basal sauropodomorph postcrania with cranial or dental remains of other carnivorous archosaurs (Galton 1985; see also Benton 1986).

With the growing acceptance of cladistic methodology in vertebrate paleontology, more rigorous analyses of the interrelationships of sauropods to other dinosaurs and archosaurs were published starting in the 1980s. In 1986, Gauthier established the monophyly of the Saurischia and gave the first, apomorphy-based diagnosis of Sauropodomorpha in the appendices of his work. He also offered several observations and comments on the possible interrelationships of basal sauropodomorphs and sauropods, noting that the classical "Prosauropoda" probably represented a paraphyletic clade.

Since then, a number of phylogenetic hypotheses have been

Table 8.1 *continued*

Nr.	Neck/ back ratio	Forelimb/ hindlimb ratio	Hindlimb/ trunk ratio	C 6 length/ height ratio	Skull length	Skull height	Skull width at level of orbitae	Skull width at level of pm-m	Reference
1		0.62			130				Galton (1976)
2		0.81							Own measurements (Remes); Yates & Kitching (2003)
3		0.71	1.47	2.20					Upchurch et al. (2004b)
4					733	379	370	247	Janensch (1935)
5	1.07	0.84	1.07		330				Gilmore (1925)
6				2,26					McIntosh et al. (1996)
7	0.88	0.45	1.17	3,27	215	65			Own measurements (O.W.M.R.)
8				4					Own measurements (O.W.M.R.)
9		0.41							Own measurements (O.W.M.R.)
10					515	243	182	157	Holland (1924)
11				2,75					Own measurements (O.W.M.R., R.F., Remes)
12					121	49	40	10	Own measurements (O.W.M.R., Remes)
13		0.46							Novas (1993)
14					300	97	72	32	Sereno & Novas (1993)
15		0.51							Santa Luca (1980)
16		0.88			730	346	241	142	Sereno et al. (1999)
17	0,84								Own measurements (R.F.); Barrett & Upchurch (2007)
18	3.22			3,76	510	220	178	87	Ouyang & Ye (2002)
19				2,25					Own measurements (O.W.M.R.); Mauersberger (2005)
20					215	89			Sues et al. (2004)
21					175	86	57	29	Sues et al. (2004)
22		0.70			279	105	115	61	Own measurements (Remes, R.F.); Yates (2007a)
23				2,40					Bonaparte (1986b)
24	0.79	0.54	1.04	2,05					Own measurements (Reis); Mauersberger (2005)
25	0.75	0.52	0.98		342	130	95	38	von Huene (1926)
26				2,70					Bonaparte (1972)
27			1,28						Langer (2003)
28		0.49							Own measurements (R.F.)
29	1.18	0.66	1.09	2,15	420				Zhang (1988)
30					495	212	127	57	Chatterjee & Zheng (2002)
31					320	159	92	57	Zhang (1988)
32		0.76		1,80					Own measurements (O.W.M.R., R.F., Remes)

published, both on the interrelationships of sauropods (e.g., Calvo & Salgado 1995; Upchurch 1995, 1998; Salgado et al. 1997; Wilson & Sereno 1998; Wilson 2002, 2005; Upchurch et al. 2004a; Curry Rogers 2005) and of basal sauropodomorphs (e.g., Sereno 1999; Benton et al. 2000; Yates 2003, 2004, 2007a, 2007b; Yates & Kitching 2003; Galton & Upchurch 2004; Upchurch et al. 2007a). However, whereas a consensus is starting to emerge on the interrelationships of at least advanced sauropods (Wilson 2005), no such consensus seems to be in sight on the interrelationships of basal sauropodomorphs and thus the origin of sauropods. Likewise, although the relative relationships of the basal eusauropods most commonly included in phylogenetic analyses are similar in most hypotheses, many taxa have never or only rarely been included in formal phylogenies, so that the evolution of basal eusauropods is also still poorly understood (Allain & Läng 2009).

Within basal sauropodomorphs, the extreme ideas range from a monophyletic Prosauropoda, which basically includes all taxa traditionally included in this group (e.g., Sereno 1999; Benton et al. 2000; Galton & Upchurch 2004), to a completely paraphyletic array of basal sauropodomorph taxa on the lineage to sauropods (e.g., Yates 2007a, 2007b). But recently

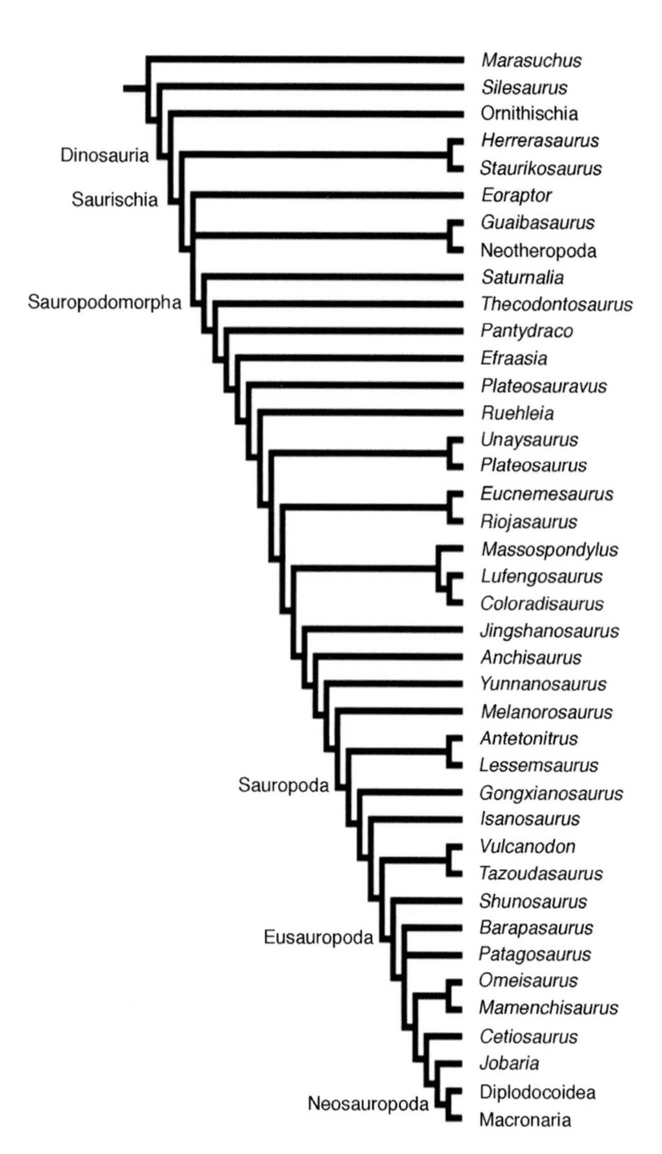

FIGURE 8.2. Composite consensus cladogram of basal sauropodomorph relationships. Basal, nonsauropodan sauropodomorphs based mainly on Yates (2007a, 2007b) and basal sauropods on Allain & Aquesbi (2008), with a few modifications.

there seems to be a growing consensus that basal sauropodomorphs represent a paraphyletic group, with some taxa being more closely related to sauropods than others (e.g., Smith & Pol 2007; Upchurch et al. 2007a; Yates 2007b). However, with the exception of a few taxa, the interrelationships of prosauropod taxa within this paraphyletic array still differ widely. Thus, concerning the interrelationships of basal sauropodomorphs, we mainly follow the phylogenetic hypothesis of Yates (2007b) here (Fig. 8.2) because this is the hypothesis that includes most characters and taxa. We also adopt the stem-based definition of Sauropodomorpha employed by Galton & Upchurch (2004) and Yates (2007b) because this encompasses all taxa that were traditionally regarded as sauropodomorphs.

In sauropods, the general consensus is that several basal taxa form successive outgroups to a large clade called Neosauropoda, which is subdivided into two lineages, the Diplodocoidea, which include rebbachisaurids, dicraeosaurids, and diplodocids, and the Macronaria, which include several basal taxa, such as *Camarasaurus*, and the Titanosauriformes (Fig. 8.2). Within the basal sauropods, however, interrelationships of the mainly Jurassic basal sauropods are much less certain, as is the definition of Sauropoda and thus the taxonomic content of this clade.

Although the definition of Sauropoda is a question of semantics rather than of phylogeny, it is nevertheless important to note which concept is followed in a given work to avoid confusion. Given a paraphyletic array of basal sauropodomorphs outside Sauropoda as proposed by Yates (2007b), the stem-based definition of this taxon used by Sereno (1998, 1999), which is based on *Plateosaurus* and *Saltasaurus*, leads to the inclusion of many classical basal sauropodomorph taxa, such as *Massospondylus*, *Riojasaurus*, or *Anchisaurus*, in the Sauropoda, and thus a considerable change in taxonomic content at the base of this clade (see also Upchurch et al. 2007a; Yates 2007a). On the other hand, the node-based definition given by Salgado et al. (1997), based on *Vulcanodon* and Eusauropoda, is also problematic because the definition of the latter clade is controversial (see discussion in Upchurch et al. 2007a, pp. 65–66). Therefore, Yates (2007a, p. 104) proposed a different definition, with sauropods being defined as all sauropodomorphs that are more closely related to *Saltasaurus loricatus* than to *Melanorosaurus readi*. This suggestion is followed here.

Concerning the interrelationships of basal sauropods, there is considerable controversy about the placement of several taxa, which may at least partially be due also to differential taxon sampling in different phylogenies (e.g., Upchurch 1998; Wilson 2002; Upchurch et al. 2004a; Allain & Aquesbi 2008). Nevertheless, the phylogenetic relationships between some key taxa are stable in all published phylogenies. Because these taxa seem to capture most of the known morphological diversity in basal sauropods, our study is mainly based on the consensus tree shown in Fig. 8.2.

The Evolution of General Body Shape in Sauropodomorphs

Sauropods are the quintessential dinosaurs: if asked to draw a dinosaur, most people would probably draw an animal with a massive body, small head, and long neck and tail. To this list of characteristics of typical sauropods, one might add columnar limbs and a large to gigantic size. It is interesting to note that despite the great taxonomic diversity of the group, there is surprisingly little deviation from this general body plan in sauropods as a whole (Upchurch et al. 2004a; Rauhut et al. 2005); only the most basal members of this clade, under the

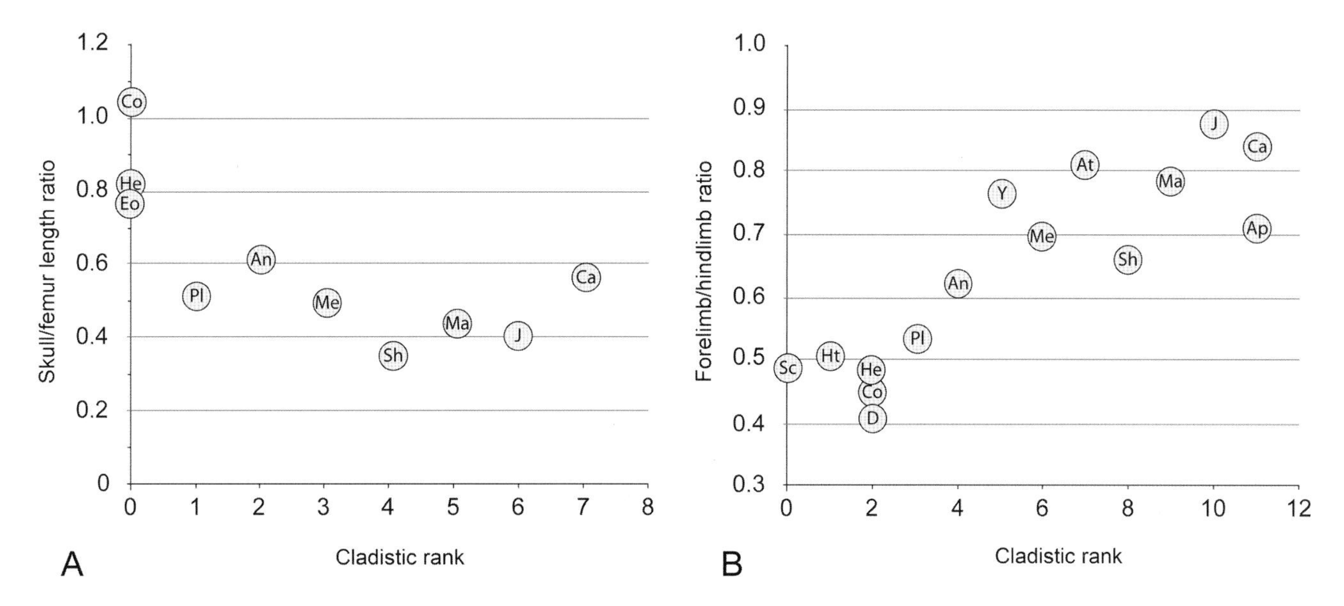

FIGURE 8.3. (A) Relative skull size in several taxa on the evolutionary lineage toward sauropods. The x-axis represents cladistic rank (=relative distance from the most basal node represented on the evolutionary lineage toward sauropods, based on the cladogram shown in Fig. 8.2); y-axis is the skull/femur length ratio. A trend of gradual reduction in relative skull size toward eusauropods becomes apparent; for explanations, see text. (B) Forelimb/hindlimb ratio of several taxa on the evolutionary lineage toward neosauropods. The x-axis represents cladistic rank; y-axis is forelimb/hindlimb ratio. With the notable exception of *Shunosaurus,* there is a gradual trend in the relative elongation of the forelimb toward neosauropods. An, *Anchisaurus;* Ap, *Apatosaurus;* At, *Antetonitrus;* Ca, *Camarasaurus;* Co, *Coelophysis;* D, *Dilophosaurus;* Eo, *Eoraptor;* He, *Herrerasaurus;* Ht, *Heterodontosaurus;* J, *Jobaria;* Ma, *Mamenchisaurus;* Me, *Melanorosaurus;* Pl, *Plateosaurus;* Sc, *Scleromochlus;* Sh, *Shunosaurus;* Y, *Yunnanosaurus.*

definition of Sauropoda adopted here, are still more basal sauropodomorph-like. However, ancestral dinosauriforms were quite different: they were small animals with greatly elongate hindlimbs and at least facultatively bipedal habits (Sereno & Arcucci 1994; Benton 2004; Langer 2004; Fechner 2006). So what were the major changes in general body plan that got these animals from one extreme to the other, and where did they occur?

Looking at relative skull size (Fig. 8.3A), we found that non-sauropodomorph saurischians and basal theropods have skulls that are only slightly shorter (*Herrerasaurus*) or even longer than the femur (*Coelophysis*). In basal sauropodomorphs, such as *Plateosaurus,* the skull is relatively much smaller, almost only half the length of the femur. The size is further reduced in sauropods, in which the skull is less than half of the length of the femur, although there seems to be some considerable variation in the relative size of the skull in sauropods (e.g., the relatively large skulls in *Anchisaurus* and *Camarasaurus*). However, in many vertebrates, relative skull size decreases during ontogeny, and thus some of this variation might be because animals of different ontogenetic stages were included in this study, such as the juvenile *Camarasaurus* specimen described by Gilmore (1925). Nevertheless, there seems to be a general tendency toward reduction of the relative size of the skull toward sauropods.

A similar picture appears when looking at limb ratios in the evolution of sauropodomorphs (Fig. 8.3B). Barrett & Upchurch (2007, p. 99) note that forelimb/hindlimb ratios in most basal sauropodomorphs lie between 0.5 and 0.6, but if the phylogenetic hypothesis of Yates (2007b) is accepted, a general trend in basal sauropodomorph evolution toward the elongation of the forelimbs can be observed (note that the limb lengths and ratios given by Barrett & Upchurch 2007 differ slightly from those given here because these authors did not take the metapodial elements into account that have been included here). Basal ornithodirans, such as *Scleromochlus,* have a forelimb/hindlimb ratio of approximately 0.5, and the same or a slightly lower value is also consistently found in basal dinosaurs, including the basal ornithischian *Heterodontosaurus* (0.51, Santa Luca 1980), the basal saurischians *Eoraptor* (0.42, PVSJ 512) and *Herrerasaurus* (0.47, PVSJ 373), and the basal theropods *Coelophysis* (0.45, AMNH 7224) and *Dilophosaurus* (0.41, UCMP V4214). In basal sauropodomorphs, such as *Plateosaurus,* the forelimbs are slightly elongate in relation to the hindlimbs (0.54, GPIT 1; 0.53, SMNS 13200), and there seems to be a tendency toward relative elongation of the forelimbs from these taxa via *Anchisaurus* (0.62, YPM 1883), *Yunnanosaurus* (0.76, IVPP V 20) and *Melanorosaurus* (0.7, NM QR3314) to *Antetonitrus* (0.81, BP/1/4952; Yates & Kitching 2003). Interestingly, the basal sauropod *Shunosaurus* has slightly shorter forelimbs (0.66, Zhang 1988), but the ratios in other basal sauropods are closely comparable to or slightly

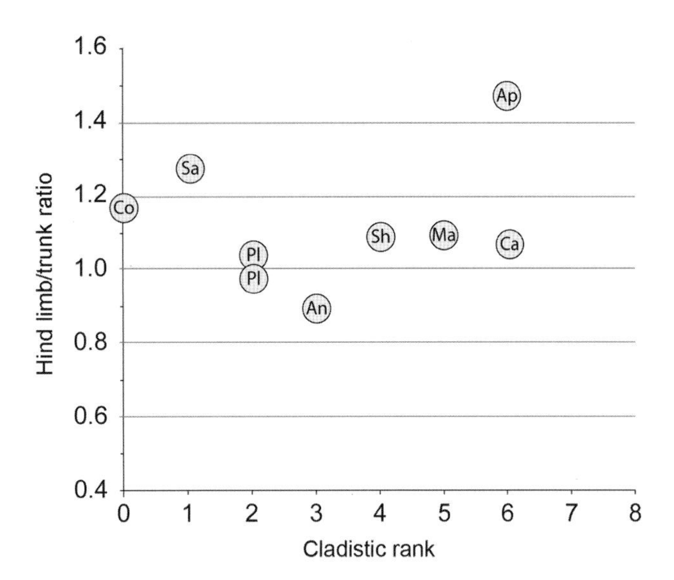

FIGURE 8.4. Hindlimb/trunk ratio of several taxa on the evolutionary lineage toward neosauropods. There is no clear evolutionary trend; the ratio appears more or less constant. Abbreviations as in Fig. 8.3, with the addition of Sa, *Saturnalia*.

higher than that of *Antetonitrus* (*Mamenchisaurus:* 0.79, Ouyang & Ye 2002; *Jobaria:* 0.88, Sereno et al. 1999; *Camarasaurus:* 0.84, Gilmore 1925). Within neosauropods, relative forelimb length further increases in at least some groups of macronarians (e.g., brachiosaurids; Janensch 1937), but decreases in diplodocoids (see also McIntosh 1990), with *Apatosaurus* showing a ratio of 0.71 (Upchurch et al. 2004b).

Another ratio that is often used in describing general body plan and especially in inferring body posture is the hindlimb to trunk length (Galton 1970, 1990; Galton & Upchurch 2004; Bonnan & Senter 2007). According to Galton (1970), a hindlimb/trunk ratio of over 1.2 is typical for obligatory bipedal animals, whereas obligate quadrupeds have ratios of 0.9 or lower; intermediate values stand for facultatively bipedal animals. Galton (1970) calculated the trunk length as the distance between the acetabulum and the glenoid cavity of the scapulocoracoid. However, we used the length of the dorsal vertebral column for our calculations (Fig. 8.4) because it largely coincides with the distance used by Galton but is more reliable (most acetabulum–glenoid distances seem to have been taken from skeletal reconstructions of often incomplete original specimens; see Langer 2003 for a discussion of the ratio given for *Thecodontosaurus*) and not influenced by problems such as the exact position and orientation of the shoulder girdle (e.g., Schwarz et al. 2007a).

The only outgroup taxa for which we had both a complete dorsal vertebral column and complete hindlimb was the basal theropod *Coelophysis* (AMNH 7224), which has a hindlimb/trunk ratio of 1.17. In the basal saurischians *Herrerasaurus* and *Guaibasaurus,* no complete dorsal vertebral column is available (Novas 1993; Bonaparte et al. 1999, 2007), and the complete skeleton of *Eoraptor* has not yet been described in sufficient detail, nor have we measured the dorsal column. However, when measured from the skeletal reconstruction given by Sereno et al. (1993), the hindlimb/trunk ratio of this taxon is approximately 1.34, but this value should be regarded with caution.

The most basal sauropodomorph, *Saturnalia,* has a hindlimb/trunk ratio of 1.28 (based on MCP 3845-PV, in which the missing caudalmost dorsal vertebrae and the metatarsal III reconstructed on the basis of MCP 3844-PV, which is closely comparable in size). This ratio is slightly higher than that given by Langer (2003), which may be due to differences in estimating the trunk length; unfortunately, Langer (2003) did not specify how he measured trunk length. The basal sauropodomorph *Plateosaurus* has a hindlimb/trunk ratio of about 1 (1.04 in GPIT 1 and 0.98 in SMNS 13200), which seems to be closely comparable to most other basal sauropodomorphs, for which Galton & Upchurch (2004, p. 258) gave a general range of 0.95 to 1.15. The advanced basal sauropodomorph *Anchisaurus* slightly deviates from this range, with a hindlimb/trunk ratio of 0.9 (based on YPM 1883; vertebral measurements taken from Galton 1976). However, this rather isolated value does not indicate that there might be a trend toward relative lengthening of the trunk on the lineage toward sauropods because the basal eusauropods *Shunosaurus* and *Mamenchisaurus* have hindlimb/trunk ratios of 1.09 (Zhang 1988; Ouyang & Ye 2002), which are well within the range of basal sauropodomorphs, and a similar value is found in the basal neosauropod *Camarasaurus* (1.07; Gilmore 1925). In *Apatosaurus,* the ratio is considerably higher, at 1.47 (Upchurch et al. 2004b). This is probably the result of a considerable shortening of the dorsal vertebral column in diplodocids in combination with the incorporation of additional vertebrae in the cervical series and does not indicate obligatory bipedal habits for this taxon.

Finally, looking at outline areas of different body sections as a proxy for their volume in one outgroup (*Herrerasaurus*), a basal sauropodomorph (*Plateosaurus*), and several sauropods (*Shunosaurus, Brachiosaurus, Diplodocus*) led to some interesting results (Fig. 8.5). First of all, the comparative values confirmed the trend toward a decrease in skull size. In *Herrerasaurus,* the skull accounts for approximately 10.5% of the total body area in lateral view (excluding limbs), whereas this value is 2.5% in *Plateosaurus* and further decreases in sauropods, from 2.3% in *Shunosaurus* to 1.6% in *Brachiosaurus* and 0.8% in *Diplodocus.* Interestingly, the relative neck area even slightly decreases from *Herrerasaurus* (9%) to *Plateosaurus* (7%), but then increases within sauropods (*Shunosaurus:* 14%; *Brachiosaurus:* 31%; *Diplodocus:* 23%). The values for the torso stay remarkably constant (55% in *Herrerasaurus, Shunosaurus,* and

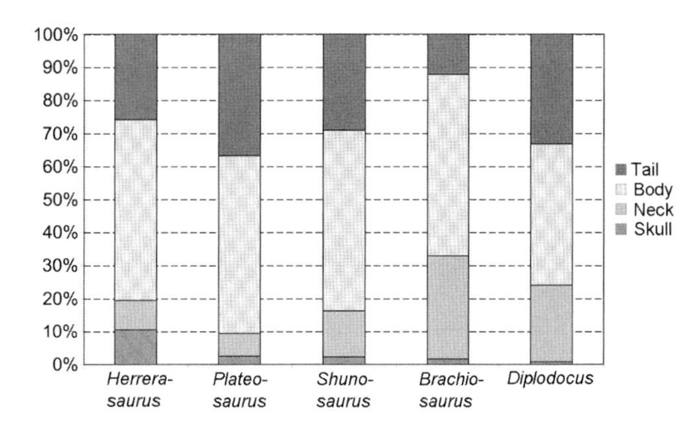

FIGURE 8.5. Relative percentages of the different body parts in one outgroup (*Herrerasaurus*), a basal sauropodomorph (*Plateosaurus*), a basal sauropod (*Shunosaurus*), and two neosauropods (*Brachiosaurus* and *Diplodocus*). Measurements were taken from body outlines in lateral view; limbs were not considered. See text for further details.

Brachiosaurus and 54% in *Plateosaurus*) and only show a notable decrease in *Diplodocus* (43%). The contribution of the tail varies strongly and seemingly more randomly, with values at 26% for *Herrerasaurus,* 37% for *Plateosaurus,* 29% for *Shunosaurus,* 12% for *Brachiosaurus,* and 33% for *Diplodocus.*

The Evolution of the Sauropodomorph Skull

Sauropod skulls (Fig. 8.6) are highly unusual. They are very small when compared to body size (see above); are short, broad, and high; have large eyes; and have strongly enlarged, dorsally placed nares. The interpretation of the evolution of the sauropodomorph skull is somewhat hampered by our lack of knowledge of the cranial anatomy of basal dinosauriforms (and thus of the plesiomorphic state for dinosaurs), and by the low number of good skulls available for basal sauropodomorphs and especially basal sauropods. Nevertheless, some general tendencies in the evolution of the sauropodomorph skull can be recognized.

The relative skull length in sauropodomorphs, measured as skull length divided by caudal skull height, decreases from 3 or more in the long-snouted basal saurischian or basal theropodan outgroups via basal sauropodomorphs, such as *Plateosaurus,* toward Sauropoda (Fig. 8.7). According to the reconstruction of the skull of *Melanorosaurus* by Yates (2007a, fig. 6), this direct outgroup taxon to sauropods had a rather low and long skull, which seems to contradict this gradual decrease in relative skull length on the lineage toward sauropods. However, the original specimen is strongly compacted dorsoventrally, and close inspection of the original specimen (NM QR3314) by one of us (O. W. M. R.) indicates that the reconstruction of Yates (2007a) considerably underestimates the height of the skull. Interestingly, relative skull length stays much the same, about 2, in basal sauropods and basal neosauropods. The same relative length occurs in derived basal sauropodomorphs, such as *Massospondylus,* although this may be because the only measured skull of this taxon represents a subadult individual (BPI/1/5241). Indeed, the largest skull of *Massospondylus* known (BPI/1/4934) is relatively longer and lies between basal sauropodomorphs, such as *Plateosaurus,* and basal sauropods.

A similar tendency is observed when looking at relative skull width (skull length divided by skull width, measured at the level of the orbitae), which, of course, is related to relative skull length (Fig. 8.7). Indeed, the relation of skull width to skull length closely follows that of relative skull length through basal sauropodomorph phylogeny. However, a marked increase in relative skull width is apparent in more derived macronarians, which is represented by *Brachiosaurus brancai* in the current sample.

An interesting tendency in the evolution of the sauropodomorph skull is relative muzzle width, which is measured as skull length divided by muzzle width at the premaxillary–maxillary suture (Fig. 8.7). Relative muzzle width increases only slightly from basal saurischians, such as *Herrerasaurus,* to basal sauropodomorphs, but then drastically increases toward Sauropoda. Again, *Massospondylus* already shows a close resemblance to basal sauropods in this respect, which may be a further indication that this taxon is indeed closely related to the latter clade, as argued by Yates (2007a, 2007b). In contrast to relative skull length, this resemblance is not due to the subadult status of the specimen because recent research in the ontogeny of sauropodomorph skulls indicates that these skulls increase in relative width during ontogeny (Whitlock et al. 2008). Within sauropods, muzzle width further increases toward neosauropods, in which relative muzzle width is more than double that of sauropodomorph outgroups.

Looking at shape changes in the skull on the evolutionary lineage toward sauropods, the change of the size and position of the external nares is striking (Fig. 8.8). In the evolution of sauropodomorph dinosaurs, the external nares were first enlarged and somewhat retracted in basal forms (basal sauropodomorphs) then strongly retracted at the base of Eusauropoda, and further enlarged and elevated toward neosauropods. These are by far the most conspicuous changes in overall skull design in sauropodomorphs, although a few other aspects might be noteworthy. One is the position of the rostral end of the infratemporal fenestra, which moves somewhat downward and foreward in sauropodomorph evolution, although much less markedly so than the change in the position of the nares. Nevertheless, in combination, these two changes result in the impression that these two openings describe a rotational movement around the orbit. This impression is supported by a slight rostral displacement of the distal

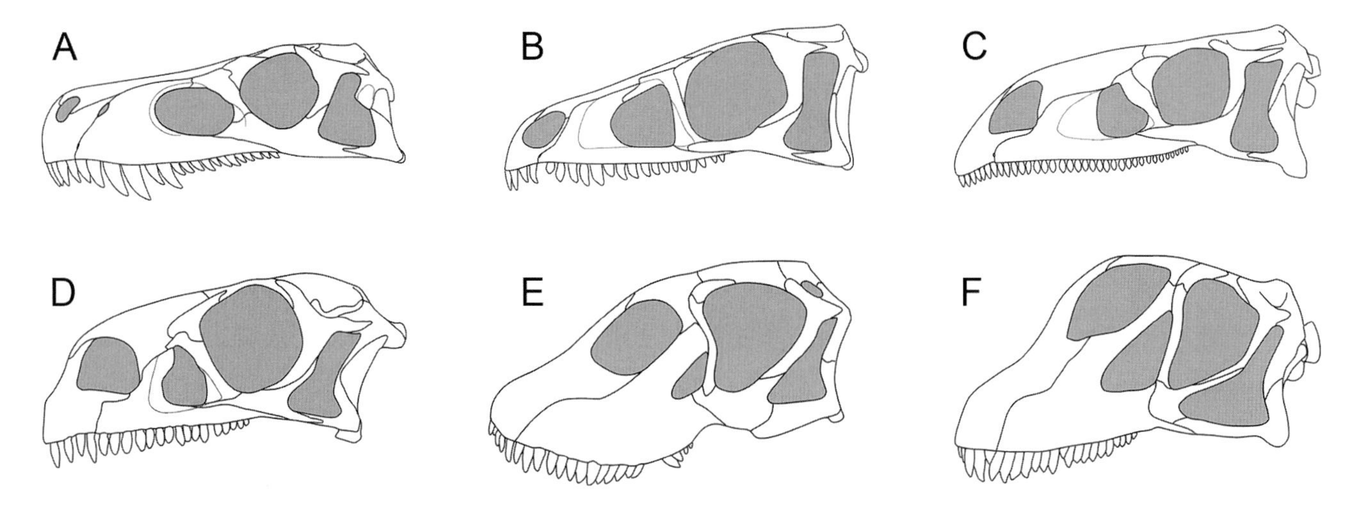

FIGURE 8.6. Skull reconstructions of two outgroups and several sauropodomorph taxa in lateral view. Note the enlargement of the external nares in basal sauropodomorphs and a caudodorsal retraction of the nares in sauropods. (A) Basal saurischian *Herrerasaurus* (based on Sereno & Novas 1993). (B) Basal saurischian *Eoraptor* (based on Sereno et al. 1993 and PVSJ 512). (C) Basal sauropodomorph *Plateosaurus* (based on MB.R.1937). (D) Advanced basal sauropodomorph *Massospondylus* (based on Sues et al. 2004 and BP/1/5241). (E) Basal sauropod *Shunosaurus* (based on Zhang 1988 and ZDM T 5403). (F) Advanced basal sauropod *Mamenchisaurus* (based on Ouyang & Ye 2002). Skulls not drawn to the same scale, but to the same length to better illustrate differences in proportions.

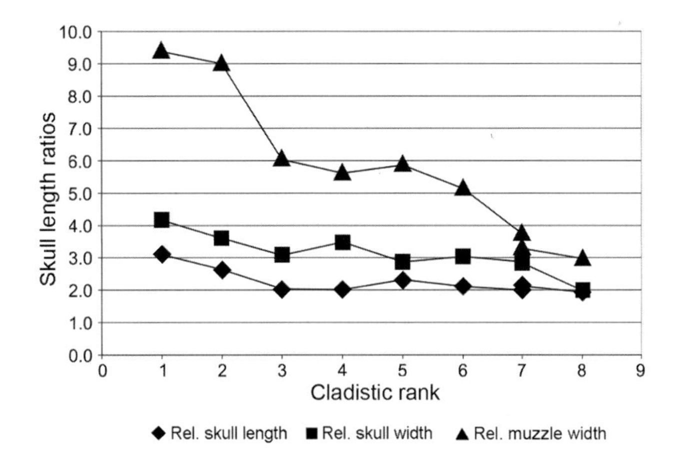

FIGURE 8.7. Skull proportions of several taxa on the evolutionary lineage toward sauropods. The x-axis represents cladistic rank. Relative skull length (diamonds) refers to the skull length/height ratio (height measured posteriorly); relative skull width (squares) refers to the posterior skull length/width ratio (width measured at the orbita); relative muzzle width (triangles) refers to the anterior skull length/width ratio (width measured at the premaxillary–maxillary suture). Note that in the latter two ratios lower values indicate greater relative width. (1) *Herrerasaurus;* (2) *Plateosaurus;* (3) *Massospondylus;* (4) *Shunosaurus;* (5) *Mamenchisaurus;* (6) *Jobaria;* (7) *Camarasaurus* and *Diplodocus;* (8) *Brachiosaurus*. Although there is a strong signal for gradual widening of the muzzle on the line toward neosauropods, relative skull length and width experience only a slight but constant decrease.

end of the tooth row, which is placed below the midsection of the orbit in sauropodomorh outgroups and basal sauropodomorphs, but rostral to the lacrimal in neosauropods. One further aspect of the skulls of sauropodomorh dinosaurs should be noted. The orbits of small and large sauropodomorh taxa are generally round to broadly oval in shape. On the other hand, in their sister group, the theropods, orbit shape seems to be strongly correlated with absolute skull size (and strength), with small taxa generally having round and large orbits and large taxa having narrowly oval to keyhole-shape orbits (Chure 2000; Henderson 2002).

Analysis of the surface area of the skull openings supports the conclusions drawn above (Fig. 8.9). Again, the most notable trend is an increase in area of the external nares, with a notable increase from basal saurischians to sauropodomorphs, but also a gradual increase within sauropodomorphs on the lineage to neosauropods. In contrast, the antorbital fenestra seems to decrease slightly in relative size in sauropodomorphs ancestrally, but then varies in area within this group, whereas no clear trend is seen in either the size of the orbit or the infratemporal fenestra.

In summary, the main evolutionary changes in the skull of sauropodomorphs are a reduction in skull length in relation to both overall body size and skull height, a marked increase in muzzle width, and an enlargement, retraction, and elevation of the external nares. As far as can be said on the basis of the limited data set, these changes are not completely correlated, but happened at different phases of sauropodomorph evolution. The first notable change was an enlargement of the nares, followed by an increase in muzzle width, which was

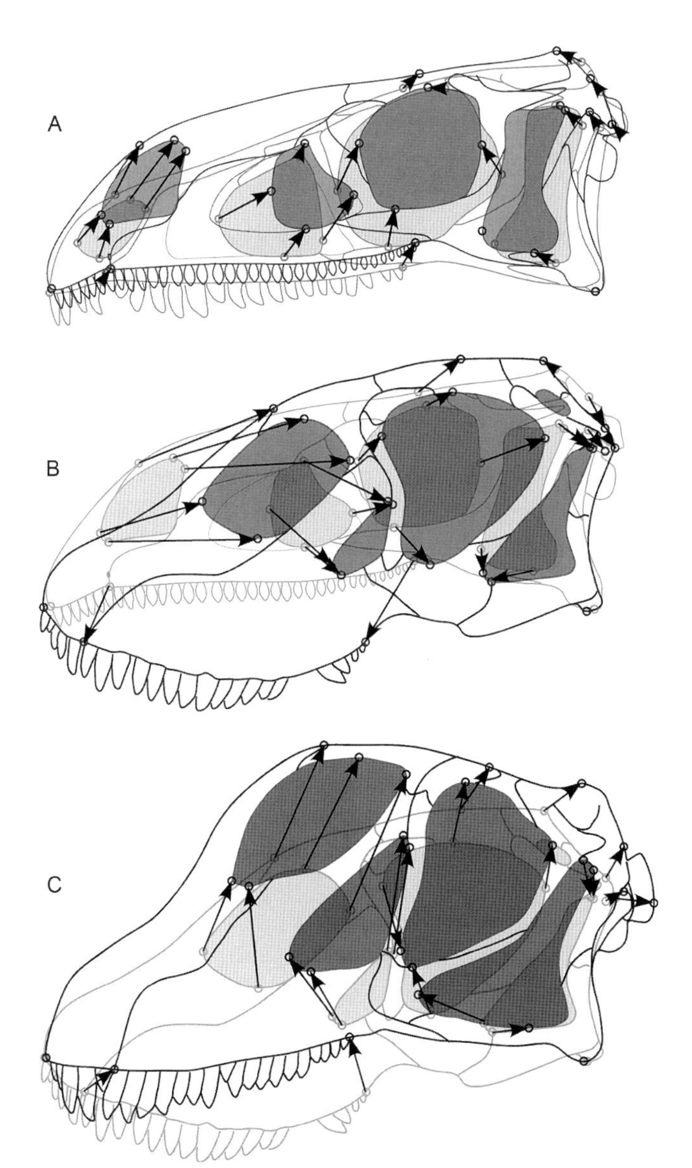

FIGURE 8.8. Major changes in the shape of the main skull openings and general skull shape from the basal saurischian *Eoraptor* to the advanced nonneosauropodan eusauropod *Mamenchisaurus*. (A) Changes from *Eoraptor* (light gray) to *Plateosaurus* (dark gray). Note the caudodorsal shift of both the external nares and antorbital fenestra. (B) *Plateosaurus* (light gray) to *Shunosaurus* (dark gray). Note the strong caudal shift and enlargement of the external nares and a clockwise rotation of the infratemporal fenestra. (C) *Shunosaurus* (light gray) to *Mamenchisaurus* (dark gray). The nares, and antorbital and infratemporal fenestrae are further enlarged and rotate clockwise around the orbit. Arrows represent the direction and amount of change of equivalent landmarks from one taxon to the next.

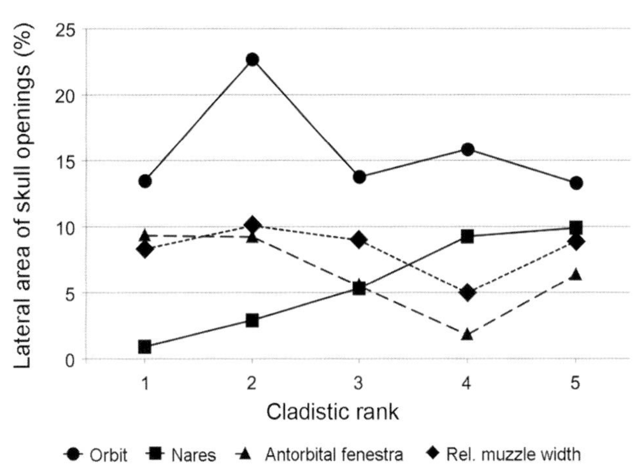

FIGURE 8.9. Changes in the relative size of cranial openings, measured as a percentage of lateral area of these openings in relation to overall skull area in lateral view from basal saurischians to sauropods. Taxa represented: (1) *Herrerasaurus*, (2) *Eoraptor*, (3) *Plateosaurus*, (4) *Shunosaurus*, (5) *Mamenchisaurus*. As the relative size of the nares gradually increases on the line toward neosauropods, no clear trend can be recognized in the remaining skull openings. For details, see text.

again followed by a further increase and retraction of the nares. The most notable of these changes apparently happened rather rapidly at the base of Sauropoda, between taxa such as *Melanorosaurus* and basal eusauropods.

The Evolution of the Vertebral Column

The long neck and tail and the elaborate structure of the presacral vertebrae are certainly some of the most conspicuous characters of sauropods, and more has been written on the evolution of these structures than on most other parts of the sauropod skeleton (e.g., Bonaparte 1986a, 1999; Wilson & Sereno 1998; Wilson 1999; Wedel 2003a, 2007; Parrish 2005). However, many aspects of the evolution of the sauropodomorph vertebral column are still poorly understood, such as the exact pattern of neck elongation in basal sauropodomorphs on the lineage toward sauropods or the sequence of vertebral incorporation in the neck. This may partially be because there are surprisingly few basal sauropodomorphs known in which complete vertebral columns are preserved (see also Barrett & Upchurch 2007).

Parrish (2005) presented an analysis of neck elongation on the lineage to sauropods. However, many of the measurements listed there could not be verified by our own observations or by a literature survey. For example, the neck length of *Marasuchus* (*Lagosuchus* in Parrish 2005) was given as 2.5 cm and the neck/back ratio as 0.58, but personal observations on the only specimen in which a complete neck is preserved (PVL 3870) indicates that the neck in this animal is approximately

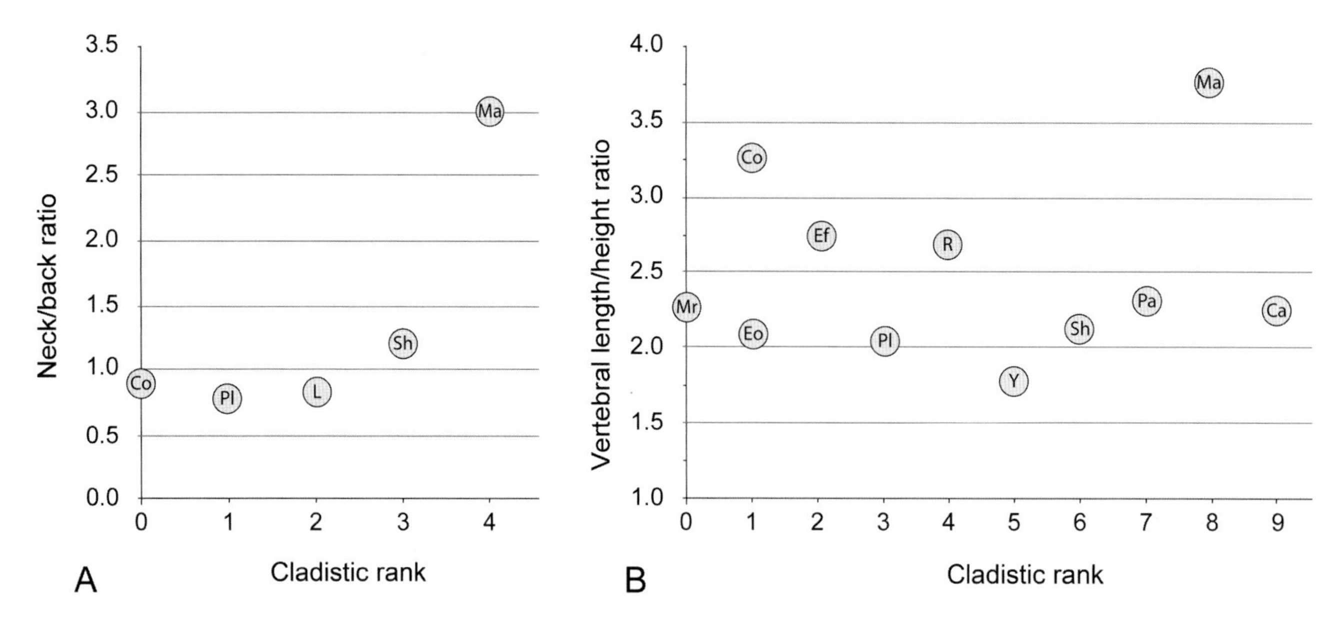

FIGURE 8.10. (A) Neck/back ratio of several taxa on the evolutionary lineage toward neosauropods. A gradual increase can be seen from basal sauropodomorphs to basal eusauropods; the exceptionally elongate neck of *Mamenchisaurus* is an autapomorphic specialization not found in most neosauropods. (B) Relative length of the sixth cervical vertebrae in several taxa on the evolutionary lineage toward neosauropods, measured as the ratio of the length of the vertebral body divided by its posterior height. Elongate midcervicals are also found in basal dinosauriforms, basal saurischians, basal theropods, and basal sauropodomorphs; they are not an evolutionary novelty of sauropods. Exceptionally elongate cervicals such as those in *Mamenchisaurus* are autapomorphic specializations. *Camarasaurus* and *Apatosaurus* have the same ratio. Abbreviations as in Fig. 8.3, with the addition of Ef, *Efraasia;* L, *Lufengosaurus;* Mr, *Marasuchus;* Pa, ?*Patagosaurus;* R, *Riojasaurus*.

4 cm in length, and the dorsal series is incomplete (only 10 of probably 15 or 16 dorsals are preserved). The same is true in *Shunosaurus,* for which neck length was given as 267 cm and Zhang (1988) cited as the source for these data. However, the approximate total neck lengths for the two specimens that were measured by Zhang (1988, p. 38) are actually 150 and 225 cm, respectively (both specimens are missing some vertebrae, the length of which had been estimated on the basis of their surrounding elements), and the neck/back ratio of these specimens is about 1.2, rather than 0.87, as stated by Parrish (2005). Furthermore, complete neck and dorsal vertebral column length were given for several taxa in which no complete columns are known, such as *Anchisaurus* (Galton 1976; Galton & Cluver 1976) and *Herrerasaurus* (Novas 1993; Sereno & Novas 1993). Because too few basal sauropodomorphs are complete enough to yield a neck/back ratio with any certainty, and even fewer have measurements cited in the literature, a formal analysis of relative neck length will not be attempted here, but some general comments on the evolution of this trait will be made.

In nonsaurischian dinosauriforms and basal saurischians, neck length seems to be about half or less than the length of the dorsal vertebral column (Parrish 2005; Barrett & Upchurch 2007). The neck is elongate in more derived saurischians (Fig. 8.10A). The basal theropod *Coelophysis* has a neck/back ratio of about 0.88 (measured on a cast of AMNH 7224), although Colbert (1989) noted some considerable variation for this ratio in this taxon, because some specimens have necks that are longer than their dorsal vertebral column. A similar value is found in the basal sauropodomorph *Plateosaurus,* for example a ratio of 0.75 in specimen SMNS 13200 (measurements taken from von Huene 1926) and 0.78 in GPIT 1 (O.W.M.R., pers. obs.). Thus, these data indicate that a relatively long neck (75% or more the length of the dorsal series) represents the plesiomorphic condition for eusaurischians. In eusauropods, the neck is even more elongate, as it is longer than the back (Fig. 8.10A). As noted above, in the basalmost eusauropod for which good vertebral material is known, *Shunosaurus,* the neck/back ratio is about 1.2.

Interestingly, neck/back ratios were given for *Lufengosaurus* and *Jingshanosaurus* by Barrett & Upchurch (2007) as 0.84 and 1.08, respectively. In the phylogenetic hypothesis of Yates (2007b), these taxa are subsequently closer outgroups to sauropods, indicating that neck length evolution in basal sauropodomorphs toward the sauropod condition might have been gradual. Such a gradual evolution of neck length in sauropods would be somewhat surprising, though, because all basal sauropodomorphs for which complete necks are known have 10 cervicals, whereas at least 12 vertebrae form the neck in all eusauropods for which a complete neck is known (Galton & Upchurch 2004; Upchurch et al. 2004a). No complete neck is known for the most basal sauropods, but the immediate outgroup, *Melanorosaurus,* has 10 cervical vertebrae (NM QR 3314). A gradual increase in neck length would thus signify a gradual increase in the relative length of individual vertebrae during the course of evolution of basal sauropodomorphs, followed by an increase in number of vertebrae in early sauropods, in which individual elements would have been relatively shorter than in advanced basal forms.

Looking at the relative length of individual cervical vertebrae, we divided the corporal length of the sixth cervical by caudal centrum height for a number of taxa in which these measurements are available (Fig. 8.10B). Although there is considerable variation in the length of individual elements, no general trend either in the elongation of cervical vertebrae in basal sauropodomorphs or in a marked decrease in length in basal sauropods is obvious. Indeed, most basal sauropodomorphs have relative cervical vertebral lengths that are roughly comparable to those of basal saurischians and even basal dinosauriforms such as *Marasuchus.* In the latter, the length/height ratio in the sixth cervical (C6) is approximately 2.25 (based on PVL 3870), and the same ratio is 2.1 in *Eoraptor* (PVSJ 512), 2.75 in *Efraasia* (von Huene 1932), 2.05 in *Plateosaurus* (von Huene 1926), 2.7 in *Riojasaurus* (Bonaparte 1972), and 1.8 in *Yunnanosaurus huangi* (Young 1951). Basal theropods, such as *Coelophysis,* even show higher ratios (3.25; AMNH 7224). In basal sauropods, *Shunosaurus* has a length/height ratio in C6 of 2.15 (Zhang 1988) and *Patagosaurus* of 2.4 (MACN CH 936), whereas the cervicals are, not surprisingly, considerably elongated in *Mamenchisaurus* (3.75; Ouyang & Ye 2002). The neosauropods *Camarasaurus* and *Apatosaurus* exhibit values of 2.25 (Gilmore 1925; Upchurch et al. 2004b). Thus, we could not confirm the assertion by Galton & Upchurch (2004, p. 244) that most basal sauropodomorphs have cervical vertebral centra that are more than three times longer than high. However, it should be noted that the current sample is incomplete and that the length/height ratio of midcervical vertebrae is certainly considerably higher in massospondylids (sensu Yates 2007b) than in other basal sauropodomorphs. Indeed, a specimen from the Los Colorados Formation (PVL field no. 6) that Yates (2007b, table 1) referred to the massospondylid *Coloradisaurus* has a length/height ratio of an unspecified midcervical vertebra of almost 4. In contrast, a cranial to midcervical vertebra of the basal sauropod *Lessemsaurus* has a considerably lower ratio of about 2 (Pol & Powell 2007), underlining the high variability of this value in sauropodomorphs in general.

Within Sauropoda, there is considerable variation not only in the relative length of individual elements, but also in the relative length of the complete neck (Upchurch & Barrett 2000), with the neck/back ratio ranging from more than 3 in *Mamenchisaurus* (Upchurch & Barrett 2000; Ouyang & Ye 2002; Parrish 2005) to less than 1 in *Brachytrachelopan* (Rauhut et al. 2005).

In summary, although more data on neck length in basal sauropodomorphs are certainly needed, especially if the hypothesis of a paraphyletic "Prosauropoda" is accepted, these data thus suggest that basal sauropodomorphs did not show a notable increase in relative neck length in comparison to close outgroups, neither in respect to overall neck length, nor in respect to the relative length of individual elements (contra Galton & Upchurch 2004). However, the neck is elongated at the base of sauropods or gradually on the lineage toward this clade. Current evidence indicates that the increase of neck length at the base of sauropods rather have been due to an increase in cervical number from 10–12 rather than to an increase in the length of individual elements.

With the increase in neck length and body size, the system of lateral laminae (bony plates and struts that allow for a light but sturdy construction) becomes more sophisticated in sauropods, and laminar architecture is one of the main taxonomic characters in these animals (Bonaparte 1999; Wilson 1999). In saurischians in general at least the dorsal vertebrae have well developed diapophyseal laminae, which connected this structure to the zygapophyses and the centrum (Wilson 1999). To which extent this vertebral lamination might be ancestral in archosaurs is debated: whereas such laminae are present in erythrosuchians (Gower 2001), *Euparkeria* (although less well developed; O. W. M. R., pers. obs.), several basal crurotarsans (e.g., *Rauisuchus*; Lautenschlager 2008), and saurischians, they are absent in many crurotarsans (including crocodiles), basal dinosauriforms (Sereno & Arcucci 1994), and ornithischians (e.g., Thulborn 1972; Santa Luca 1980). However that may be, these laminae are well developed in the dorsal vertebrae in basal sauropodomorphs, but less so in the cervical vertebrae. Basal sauropodomorphs, such as *Plateosaurus,* lack a diapophysis in the axis, in which the cervical rib is one-headed. In subsequent vertebrae, the diapophysis appears as a small lateral projection of the vertebral body, but is not supported by well developed vertebral laminae (Fig. 8.11B). The laminae first appear in the midcervical vertebrae (cervicals 5–7), and are developed here only as small, horizontal cranial and caudal extensions of the diapophysis, connecting this structure with the craniodorsal and dorsal rim of the middle part of the centrum, respectively. These laminae probably represent the cranial and caudal centrodiapophyseal laminae. In vertebrae 6 and 7, a further lamina appears, the prezygodiapophyseal lamina, and rapidly becomes more pronounced than the cranial centrodiapophyseal lamina, which it overhangs. However, the postzygodiapophyseal lamina, which is usually well developed in the cervicals of sauropods, only appears in the caudalmost cervicals in these basal sauropodomorphs. In contrast, all zygodiapophyseal and centrodiapophyseal laminae are typically well developed in all postaxial cervicals in sauropods (Fig. 8.11C). Unfortunately, no good neck material is known or has been described in taxa that are closest to the

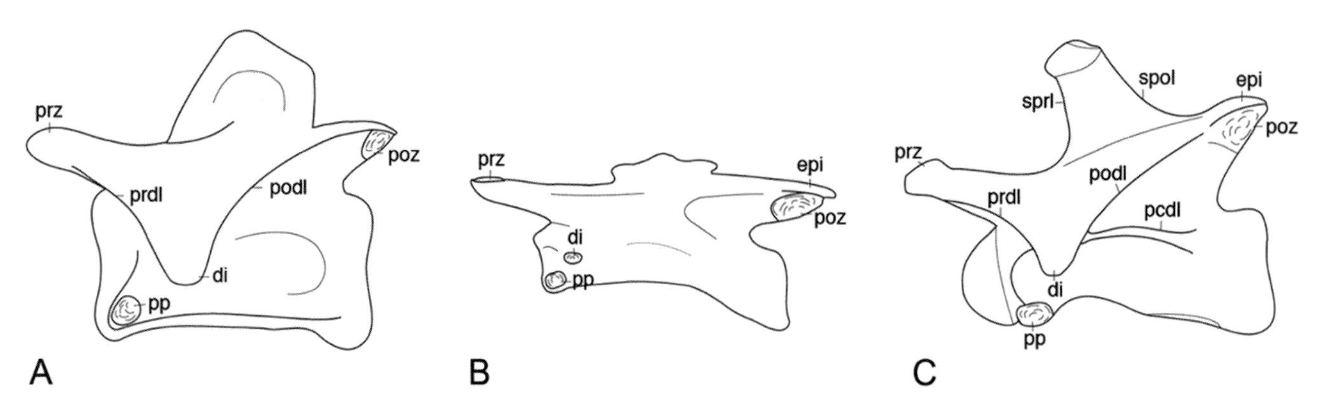

FIGURE 8.11. Anterior cervical vertebrae of the basal saurischian *Herrerasaurus* (A), the basal sauropodomorph *Plateosaurus* (B), and the basal eusauropod *Patagosaurus* (C) in lateral view. Although some diapophyseal laminae are already present in *Herrerasaurus* (and other, more basal archosaurs), they are missing in *Plateosaurus* but elaborate in sauropods. *(A) Based on Sereno & Novas (1993) and PVSJ 373; (B) based on GPIT 1; (C) based on MACN CH 936.* di, diapophysis; epi, epipophysis; pcdl, posterior centrodiapophyseal lamina; podl, postzygodiapophysel lamina; poz, postzygapophysis; pp, parapophysis; prdl, prezygodiapophyseal lamina; prz, prezygapophysis; spol, spinopostzygapophyseal lamina; sprl, spinoprezygapophyseal lamina.

origin of sauropods, so that the pattern of acquisition of all laminae in sauropodomorph cervicals cannot be followed. However, the cervical vertebrae in the immediate sauropod outgroup *Melanorosaurus* (NM QR 3314) and the possibly basal sauropod *Lamplughsaura* (Kutty et al. 2007) still seem to follow the basal sauropodomorph pattern. Immediate eusauropod outgroups, such as *Tazoudasaurus,* already seem to have all the typical lateral laminae in at least the midcervicals (Allain & Aquesbi 2008), indicating that the acquisition of these laminae in all cervicals happened rapidly at the base of the sauropods. The appearance of additional laminae in the dorsal and cervical vertebrae, such as pronounced, strongly expanded spinozygapophyseal laminae and spinodiapophyseal laminae, happened within sauropods and has been described by Bonaparte (1999) and Wilson (1999).

As in the vertebral laminae, pneumatic depressions and spaces in the presacral vertebrae seem to have appeared rather rapidly in basal eusauropods. Wedel (2007) reviewed the evidence for vertebral pneumaticity and thus the presence of birdlike air sacs in basal sauropodomorphs, and concluded that only the type specimen of *Pantydraco caducus* may show some evidence for the presence of cervical air sacs in basal sauropodomorphs. However, although no conclusive evidence for either cervical or abdominal air sacs exists in basal sauropodomorphs (as pointed out by Wedel 2007), some observations on the morphology of the presacral vertebrae in *Plateosaurus* may provide additional evidence for the presence of both cervical and abdominal air sacs in this animal and, by inference, in basal sauropodomorphs in general. Wedel (2003b, fig. 2) demonstrated that paravertebral air sacs are present in the cervical vertebrae in birds, which run along the lateral side of the vertebral body. The presence of similar structures is indicated in sauropods by the large pleurocoels on the lateral side of the vertebral centra (Schwarz et al. 2007b). In *Plateosaurus,* there is a shallow sulcus below the diapophysis in cervical vertebra that runs along the length of the vertebral centrum (Fig. 8.12). This sulcus thus may represent an osteological correlate of such paravertebral diverticula, although more data are needed to confirm this. Likewise, the last dorsal vertebrae in *Plateosaurus* have deep depressions on the lateral sides of their centra. Although such depressions could have been caused by other structures than air sacs (O'Connor 2006), the occurrence of these depressions in only the last dorsals is noteworthy and might indicate the presence of abdominal air sacs in these basal sauropodomorphs (Wedel 2003b, 2007). In sauropods, pneumatic depressions, recesses, and internal spaces are generally well developed, and their evolution has been outlined by Wedel (2003a).

A last notable change in the presacral vertebrae in sauropo-

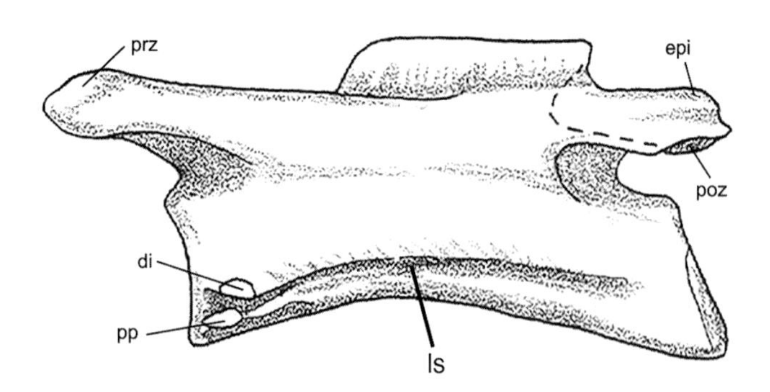

FIGURE 8.12. Anterior cervical vertebra of *Plateosaurus* in lateral view, illustrating the lateral sulcus (ls) on the vertebral body, a potential homolog of the pleurocentral depression of sauropods and therefore possibly indicating the presence of air sacs (*from Bonaparte 1999*). Other abbreviations as in Fig. 8.11.

domorphs is the change from amphicoelous articular surfaces to an opisthocoelous condition. Again, this change obviously occurred rapidly because the immediate sauropod outgroup *Melanorosaurus* (Galton et al. 2005) and some of the most basal sauropods, *Antetonitrus* (Yates & Kitching 2003) and *Lessemsaurus* (Pol & Powell 2007), still exhibit the amphicoelous condition, whereas the cervical vertebrae are opisthocoelous in the close eusauropodan outgroups *Isanosaurus* (Buffetaut et al. 2000) and *Tazoudasaurus* (Allain & Aquesbi 2008). In *Tazoudasaurus,* the rostral dorsal vertebrae are platy-amphicoelous to slightly opisthocoelous (Allain & Aquesbi 2008), but these vertebrae are strongly opisthocoelous in basal eusauropods, such as *Patagosaurus* (MACN CH 936). Finally, in macronarians, all dorsal vertebrae are opisthocoelous (Upchurch 1998; Wilson & Sereno 1998).

The Evolution of the Pectoral Girdle and Forelimb

As expected for animals as large as eusauropods, both fore- and hindlimbs are graviportal, meaning that the limb elements are oriented vertically and are placed directly under the body. In most eusauropods, the forelimb is shorter than the hindlimb but is clearly adapted to one single purpose, namely locomotion. Nevertheless, the case is not as clear in the most basal sauropods, basal sauropodomorphs, or more basal dinosaurs close to the origin of the Sauropodomorpha.

Early archosaurs, relatively small animals of about the size of a cat, inherited a massive and robust pectoral girdle from their ancestors among basal Diapsida. This plesiomorphic structure is characterized by a short but wide scapular blade, a large and wide coracoid on the ventral side of the body, and massive, bracing clavicles. A cartilaginous suprascapula is found in all extant nonavian diapsids (including crocodilians), and was therefore probably also present at the base of the Archosauria. The glenoid of these early forms is laterally to caudolaterally directed, indicating a sprawling to semierect posture of the forelimb. *Euparkeria capensis* from the Anisian of South Africa, a well known form closely related to the base of the Archosauria, illustrates this ancestral bauplan (Fig. 8.13A). In this form, the humerus has a large triangular deltopectoral crest and strongly expanded proximal and distal ends. The transverse axis of the distal end is twisted relative to that of the proximal axis at about 45°. The medial tuberosity at the proximal end of the humerus is prominent, and the ulnar condyle is expanded distally relative to the radial condyle. Radius and ulna are shorter than the humerus, and moderately slender (Ewer 1965).

Unfortunately, the evolution of the pectoral girdle in early Ornithodira before the advent of the Dinosauria is only poorly represented in the fossil record. The earliest representatives of the sister group of the Dinosauromorpha, the Pterosauria, are

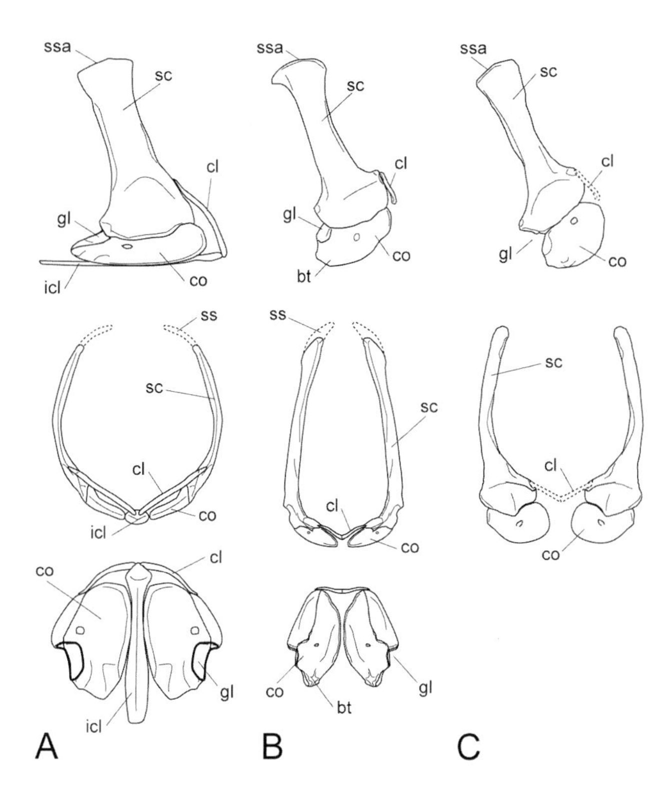

FIGURE 8.13. Reconstructions of the pectoral girdle of (A) *Euparkeria,* (B) *Massospondylus,* and (C) *Patagosaurus.* First row, lateral view; second row, cranial view; third row, ventral view. Note the relative reduction in size of the ventral part of the girdle (coracoid and clavicle), and a counterclockwise rotation relative to the scapular blade in sauropods. Pectoral girdles not drawn to scale, but to the same height for a better comparison of proportions. bt, biceps tubercle on coracoid; cl, clavicle; co, coracoid; gl, glenoid; icl, interclavicle; sc, scapula; ss, suprascapula; ssa, suprascapular attachment.

already characterized by highly modified forelimbs. The basalmost ornithodiran currently known (Benton 1999, 2004), *Scleromochlus taylori* from the late Carnian of Scotland, occurs some 15 million years after the split of the pterosaurian and dinosaurian branches of the ornithodiran tree and therefore may represent its own specialized branch of the Ornithodira, one not necessarily indicative of the ancestral ornithodiran forelimb bauplan. Although the forelimb is rather short in comparison to the hindlimbs (see above; Fig. 8.3B), all forelimb elements are elongate and slender.

Early representatives of the lineage leading to dinosaurs in most cases lack forelimb material. A partial forelimb has been described in the basal dinosauriform *Marasuchus lilloensis* (Bonaparte 1975; Sereno & Arcucci 1994), but these elements were added to the type material after the original description of Romer (1971) and most likely belong to a sphenosuchian crocodylomorph (Remes 2008). Only two species of basal dinosauriforms preserve good, unambiguously associated forelimb material, including the type specimen of *Lewisuchus ad-*

mixtus from the Ladinian of Argentina and *Silesaurus opolensis* from the late Carnian of southern Poland. In *Lewisuchus*, only the pectoral girdle and humerus are preserved (Romer 1972). In both forms, the coracoid is reduced in size relative to the scapular head, and the scapular blade is long and slender. This indicates a high oval body profile in the shoulder region, in contrast to the circular cross section in basal archosaurs (Fig. 8.13A). Clavicles have not been found in any basal dinosaur, but their presence in more derived forms including theropods, ornithischians (*Psittacosaurus*), and sauropodomorphs (see below) demonstrates that they must have been present. Their weak connection to the rest of the skeleton and their position embedded in the shoulder muscles (which are attractive for predators and scavengers) leads to a low preservation potential of these elements. Interclavicles are not known from any ornithodiran, and therefore they were probably lost early in the evolution of this group, in combination with the reconfiguration of the pectoral girdle.

The humerus retains a well developed deltopectoral crest in *Lewisuchus*, while this feature is strongly reduced in *Silesaurus* (Dzik 2003). Both forms possess a markedly more elongate humerus than *Euparkeria*, *Silesaurus* much more so than *Lewisuchus*. However, both also preserve a well developed medial tuberosity and a 45° twist of the shaft. Although the extremely elongate antebrachial elements in concert with overall body proportions indicate an obligate quadrupedal habit of *Silesaurus* (Dzik 2003), *Lewisuchus* is too incomplete to draw definite conclusions about its preferred mode of locomotion.

The earliest saurischian dinosaurs, like *Eoraptor* and *Saturnalia* (the latter already regarded as a basal sauropodomorph; Langer et al. 1999; Langer 2004), preserve the same general morphology of the pectoral girdle. The humerus of *Eoraptor* remains as elongate as in *Lewisuchus*. It retains a prominent deltopectoral crest and a well developed medial tuberosity, but the grade of axial twist is unknown as a result of compression (Remes, pers. obs.). Relative to the humerus, the antebrachium is considerably shortened. The manus is incomplete because the lateral part of the slab that bears the hand elements is broken off. Nevertheless, the manus exhibits large, robust proximal and distal carpalia, five well developed metacarpals (with only a slight reduction of metacarpal V), and block-like phalanges ending in short unguals in the inner three digits (see Fig. 8.15A). In *Saturnalia*, the basalmost sauropodomorph known, the humerus is much shorter and more robust than in other basal dinosaurs, with strongly expanded proximal and distal ends, as well as an enlarged and elongate deltopectoral crest that extends almost the length of the entire proximal half of the element (Langer 2007). As in later sauropodomorphs, radius and ulna are short and robust, and the ulna is characterized by a well developed olecranon process, a plesiomorphic trait lost by more derived forms (Langer 2007). Other early saurischians, that is, herrerasaurids and basal theropods, show autapomorphic modifications of the forelimb related to predatory behavior and therefore cannot serve as a good model for the forelimb anatomy at the beginning of the evolutionary lineage leading to sauropods.

In early sauropodomorphs more derived than *Saturnalia* (e.g., *Thecodontosaurus*, *Efraasia*), initially the plesiomorphic configuration of the pectoral girdle is retained without significant modifications. However, with increasing size, the humerus develops a prominent medial tuberosity, and most forms preserve the plesiomorphic axial twist of about 45°. The antebrachial elements are elongate and slender in some forms (e.g., *Thecodontosaurus*), but short and robust in others (*Efraasia*). Similarly, the metacarpals of *Thecodontosaurus* (YPM 2195) are slender (except metacarpal I), and considerably reduced in the outer two digits, while all metacarpals are well developed and robust in *Efraasia*. In the latter form, phalanges are not preserved, but at least those of *Thecodontosaurus* are considerably elongate relative to those of *Eoraptor*. The unguals of *Thecodontosaurus* became strongly curved and transversely slender, and ungual I is enlarged relative to those of digits II and III. *Thecodontosaurus*, resembling *Eoraptor*, possesses three distal carpals, but the medial one is enlarged and caps metacarpals I and II (Benton et al. 2000).

Basal sauropodomorphs more derived than *Efraasia* and *Thecodontosaurus* develop a more pronounced acromial facet on the craniodorsal edge of the scapular head and an extremely large, protruding biceps tubercle caudolaterally on the coracoid. The former serves for articulation with small clavicles (arranged in a V shape in cranial view), which until now have only been identified in *Plateosaurus* (von Huene 1926) and *Massospondylus* (Fig. 8.13B; Yates & Vasconcelos 2005). The enlargement of both the acromial facet and the biceps tubercle may be the result of scaling effects because the general retention of the plesiomorphic dinosauriform bauplan makes disproportionally larger shoulder muscles necessary, which probably led to the observed reinforcement of their osteological correlates (Remes 2008). In *Massospondylus* and more derived sauropodomorphs, oval or subrectangular sternal ossifications develop, which articulate with the caudomedial corners of the coracoids via thickened, rugose craniolateral processes (Fig. 8.13B).

The bauplan of the humerus in basal sauropodomorphs is subject to only minor variations, the most notable exception being *Plateosaurus*, with a long, slender humeral shaft and reduced proximal and distal expansions (see also Mallison, Chapter 13 of this volume). All other basal sauropodomorphs are characterized by heavy, robust humeri with extremely expanded deltopectoral crest and proximal and distal ends, a large, protruding medial tuberosity, and a plesiomorphic 45° twist in the shaft axis (Fig. 8.14). Moreover, all basal sauropodomorphs exhibit a large subquadrangular facet on the distomedial corner of the humerus, which is probably related to

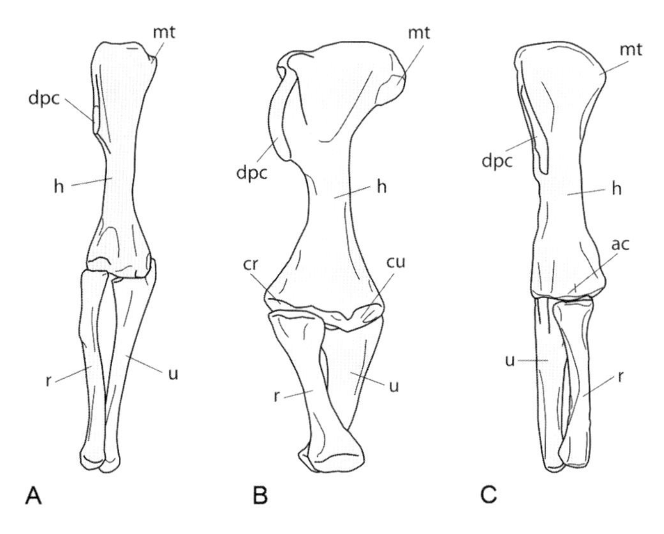

FIGURE 8.14. The right forelimb of the basal saurischian *Herrerasaurus* (A), the basal sauropodomorph *Lufengosaurus* (B), and the basal sauropod *Patagosaurus* (C). Note the robustness and enlarged muscular attachment sites in basal sauropodomorphs, and the counterclockwise rotation of ulna and radius in sauropods. Forelimbs not drawn to the same scale, but to the same height for a better comparison of proportions. ac, accessory condyles; cr, condylus radialis; cu, condylus ulnaris; dpc, deltopectoral crest; h, humerus; mt, medial tuberosity; r, radius; u, ulna.

specialized manual flexor musculature (Remes 2008). Radius and ulna are generally short and stout elements in basal sauropodomorphs. The shaft of the ulna is usually curved cranially toward the distal end, while the proximal end is characterized by its subtriangular shape with a blunt, rounded, and short lateral process. The radius is a straight bone with slightly expanded proximal and distal ends; in some forms, it exhibits a prominent tubercle at about midshaft on its lateral side; a similar structure in theropod dinosaurs is often referred to as biceps tubercle (Weishampel et al. 2004). The wide separation of the radial and ulnar condyles of the distal end of the humerus would have allowed for a certain degree of crossing of the radius cranial to the ulna (and therefore for the manus pronation necessary for an effective quadrupedal walk, Bonnan & Yates 2007) in most forms (Fig. 8.14B) except for *Plateosaurus* (Mallison 2007, Chapter 13 of this volume).

Most basal sauropodomorph hands known lack ossified proximal carpals, although they have been described in *Massospondylus* (Broom 1911; see also Läng & Goussard 2007). They usually retain a large distal carpal 1 that extends laterally and caps the proximal face of distal carpal 2 (at least in *Lufengosaurus* and *Massospondylus,* where three distal carpals are preserved; Fig. 8.15B). As in *Efraasia,* metacarpals remain rather short and robust elements in most basal sauropodomorphs where the manus is known; the most notable exception is *Anchisaurus* with a slender hand similar to that of *Thecodontosaurus* (Yates 2004). Likewise, the manual phalanges preserve a short and robust form, which is about as long as wide. Again, *Anchisaurus* has elongate phalanges obviously adapted for grasping, while *Plateosaurus* is intermediate between both extremes. All basal sauropodomorphs are characterized by a hypertrophied first digit with an enormous, strongly recurved ungual (Fig. 8.15B; Galton & Upchurch 2004). Interestingly, the size discrepancy between the first and second digit is largest in those basal sauropodomorphs currently regarded as most closely related to the origin of sauropods, for example *Lufengosaurus* and *Yunnanosaurus* (Yates 2007a, 2007b). Overall, most basal sauropodomorph hands seem to be adapted to multiple purposes, including both grasping and locomotion, and grasping adaptations are stronger in the smallest forms.

Currently, there are a number of transitional forms regarded as early sauropods or immediate outgroups of the Sauropoda that preserve forelimb material, including *Melanorosaurus* (Bonnan & Yates 2007), *Antetonitrus* (Yates & Kitching 2003), and *Gongxianosaurus* (He et al. 1998). These have in common a relatively shorter but craniocaudally wider scapular blade, combined with enlarged coracoids and sternal plates; the humerus of these forms is slightly more elongate and has a slightly reduced axial twist (approximately 25°). Radius and ulna remain short and stout, but the ulna bears an enlarged lateral process (Bonnan & Yates 2007) that develops a secondary articulation between humerus and ulna on the radial condyle (Remes 2006). Carpals are not preserved in these forms; their presence in basal sauropods (see below), however, might indicate that this is only a taphonomic effect. As can be seen in *Melanorosaurus,* the distal phalanges (except of digit I) become reduced, because with increasing size, the locomotor function of the forelimb began to dominate over the grasping function, although both functions were most probably still in use.

Until recently, forelimb material of "true" basal sauropods was only known from fragments (Cooper 1984, Buffetaut et al. 2000, 2002), but the recent description of *Tazoudasaurus* from the Early Jurassic of Morocco (Allain & Aquesbi 2008) filled a substantial gap in the fossil record of sauropod forelimb evolution. Complete forelimbs of other basal sauropods are only known from Middle Jurassic sediments in mainly China and Argentina, and comprise derived eusauropods only. In *Tazoudasaurus,* the coracoid still preserves a rugose caudolateral eminence (homologous to the biceps tubercle of basal sauropodomorphs) which is completely reduced in later sauropods. The scapula is not preserved in *Tazoudasaurus,* but other basal sauropods are characterized by a slender scapular blade without a marked distal expansion (Fig. 8.13C). As in basal sauropodomorphs, the rugose structure of the distal edge of the blade indicates the presence of a cartilaginous suprascapula. The angle between scapular head and shaft is smaller than in basal sauropodomorphs, which together with

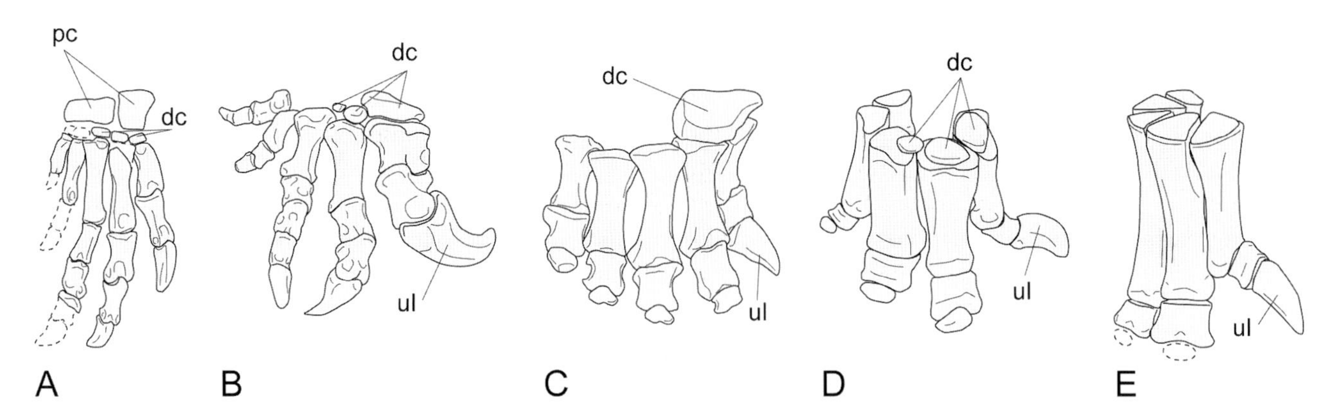

FIGURE 8.15. Comparison of the right manus in the basal saurischian *Eoraptor* (A), the basal sauropodomorph *Massospondylus* (B), the basal sauropod *Tazoudasaurus* (C), the basal eusauropod *Shunosaurus* (D), and the neosauropod *Camarasaurus* (E). Note the increasingly circular, pillar-like arrangement of the metacarpus and the shortening of the proximal and reduction of the distal phalanges. Proximal carpals are lost at the base of the Sauropodomorpha; distal carpals are reduced in neosauropods. Manus not drawn to scale, but to roughly the same height for a better comparison of proportions. dc, distal carpals; pc, proximal carpals (radiale, ulnare); uI, ungual of digit I.

a cranially rotated coracoid allows for a ventral orientation of the glenoid (Fig. 8.13C; Remes 2006, 2008). Clavicles are rarely preserved in sauropods, but those of basal forms are heavy, spear-shaped elements (He et al. 1988).

In concert with the fully vertical orientation of the forelimb, sauropods modified the shape of the humerus toward a straight element without shaft curvature or axial twist; however, in the basal form *Gongxianosaurus* and material assigned to *Kotasaurus* (Yadagiri 2001; the material referred to *Kotasaurus* represents more than one taxon; see Rauhut & López-Arbarello 2008), a slightly sigmoidal form of the shaft is preserved. These forms and *Tazoudasaurus* also retain a twist in the shaft, leading to a slight angle between the axes of the proximal and distal humeral ends of about 10–15°. In all basal sauropods known (including *Tazoudasaurus*), the proximal and distal ends of the humerus are much less expanded than in basal sauropodomorphs (Fig. 8.14C), although early forms such as *Barapasaurus, Tazoudasaurus,* and a humerus referred to *Kotasaurus* preserve a marked medial tuberosity. In the former genera, the humeral shaft has a subcircular cross section, while it is more compressed craniocaudally in more derived sauropods. The deltopectoral crest is reduced in basal sauropods as compared to "prosauropods", but it is still more prominent than in neosauropods. In addition, all known sauropods except for *Tazoudasaurus* and the humerus referred to *Kotasaurus* are characterized by small, paired tubercles on the cranial edge of the distal end ("accessory condyles"). Taking these details together, the basalmost sauropods show a number of transitional characters, indicating a gradual optimization of the humerus during the Early Jurassic toward the shape known in all higher eusauropods.

In the antebrachium, differences between the basal sauropodomorph grade and the sauropod condition are distinct and appear abruptly in the current fossil record. In the basalmost sauropods known, *Vulcanodon* and *Tazoudasaurus,* radius and ulna are already markedly more elongate and slender, and entirely straight. Moreover, the proximal end of the ulna already has the V-like or triradiate shape characteristic for all sauropods, which resulted from an enlargement and elongation of the lateral process. As Bonnan (2003) has shown, this modification enables the reconfiguration of the antebrachium in bringing the radius from the craniolateral to the craniomedial side in front of the ulna, therefore allowing for a permanent pronation of the manus during locomotion (Fig. 8.14).

The early evolution of the sauropod manus is only poorly represented in the fossil record, the sole complete hands known from *Tazoudasaurus* and the basal eusauropod *Shunosaurus*. Both taxa have short, robust metacarpals similar to those of advanced basal sauropodomorphs such as *Yunnanosaurus,* but with relatively enlarged metacarpals IV and V, and a metacarpal I that is not much wider than the other metacarpals (Fig. 8.15C, D). Most importantly, the metacarpus is arranged in a semicircle, a stage transitional to the bound, tubular metacarpus found in neosauropods (Upchurch et al. 2004a). As in basal sauropodomorphs, the inner three distal carpals are ossified in *Shunosaurus*. In *Tazoudasaurus* and neosauropods, one large distal carpal caps metacarpals I and II; however, a small second distal carpal may be present. In all nonneosauropod sauropods more derived than *Tazoudasaurus,* the phalangeal formular of the manus is reduced to 2-2-2-2-2, with block-like proximal and strongly reduced distal phalanges (Fig. 8.15). *Tazoudasaurus* has a similar phalangal morphology but retains three phalanges on the second digit (Allain & Aquesbi 2008). Only digit I retains an ungual in sauropods, which is usually hypertrophied (Upchurch 1994), but is relatively small in *Tazoudasaurus*. With the current incompleteness of the fossil record, it cannot be decided whether this represents an autapomorphic trait of this taxon or a secondary reenlargement of the claw in eusauropods.

In forms like *Patagosaurus,* the fundamental bauplan of the sauropod forelimb was complete. In most neosauropods, only minor variations occur (e.g., degree of distal expansion in the scapular blade, relative size of the scapular head, prominence of the acromial crest, varying shapes of the coracoid and sternal plates, long bone elongation in basal titanosauriforms, and further reduction of the phalanges). In titanosaurs, however, again, a number of structural modifications of the forelimb evolved that will not be addressed in this chapter (for overviews, see Upchurch et al. 2004a; Curry Rogers 2005; Wilson 2005).

The Evolution of the Pelvic Girdle and Hindlimb

In contrast to the pectoral girdle, the pelvic girdle of tetrapods has a fixed connection to the axial skeleton via the sacral vertebrae. Two sacral vertebrae are present in nondinosaurian dinosauromorphs, with the exception of *Silesaurus,* in which three sacrals are counted (Dzik & Sulej 2007). The number of sacral vertebrae in basal dinosaurs is uncertain. Although the presence of five sacrals is the primitive condition in ornithischians (e.g., Weishampel 2004), the basal saurischian *Herrerasaurus* has two sacrals (Novas 1993; although a third, dorsal vertebra has a thin connection to the ilium in at least one specimen, PVSJ 461), and three sacrals are present in *Eoraptor* (PVSJ 512) and *Guaibasaurus* (MCN-PV 2355; Bonaparte et al. 2007).

With the exception of *Herrerasaurus* (PVL 2655; Novas 1993), the ilium of basal saurischians is craniocaudally longer than dorsoventrally high (Fig. 8.16). The preacetabular process is of subtriangular shape and small. The postacetabular process is also small, but it is more robust than the preacetabular process. The acetabulum of nondinosaurian dinosauromorphs is initially closed. In herrerasaurids, the acetabulum is fully open, whereas the acetabulum of *Guaibasaurus* (MCN-PV 2355; Bonaparte et al. 1999, 2007) is only semiperforated. A prominent supraacetabular crest overhangs the acetabulum of basal saurischians craniodorsally. A brevis fossa on the caudal margin of the ilium first occurs in the nondinosaurian dinosauromorph *Silesaurus* (Dzik 2003) but is absent in *Guaibasaurus* (MCN-PV 2355; Bonaparte et al. 1999, 2007). Novas (1993) and other authors argued that a brevis fossa is present in *Herrerasaurus,* and Langer (2004) noted a brevis fossa-like structure in *Staurikosaurus,* which has already been depicted by Colbert (1970). However, direct examination of the material revealed that the "brevis fossa" of *Herrerasaurus* is actually a ridge on the lateral surface of the caudal ilium. In *Staurikosaurus,* the "brevis fossa" is a misidentified transverse process of a cranial caudal vertebra fused to the medial surface of the caudal margin of the ilium (R.F., pers. obs.). Thus, it is concluded here that a brevis fossa is not present in herrerasaurids.

The pubis of basal saurischians has a mesopubic orientation in which the pubis is directed roughly ventrally (Fig. 8.16A). The pubis of basal dinosauriforms and basal saurischians forms a broad pubic apron, which is oriented in the transverse plane. The ischium of saurischians can roughly be subdividedinto an obturator plate cranioproximally and a strap-like ischial shaft. The obturator blade is prominent in *Eoraptor* (PVSJ 512) and *Herrerasaurus* (PVL 2655; Novas 1993) but reduced in *Staurikosaurus* (MCZ 1699; Colbert 1970; Galton 1977) and *Guaibasaurus* (MCN-PV 2355; Bonaparte et al. 1999, 2007).

The femur of basal saurischians is straight in cranial and sigmoidally curved in lateral view. The femoral head is deflected from the transverse plane to 50–70° in cranial direction. On the craniolateral aspect of the proximal femur, a lesser trochanter and trochanteric shelf are present. In *Guaibasaurus,* however, the trochanteric shelf is reduced (MCN-PV 2355; Bonaparte et al. 1999, 2007). In basal saurischian dinosaurs, the fourth trochanter is located on the caudal side of the femoral shaft relatively proximally on the femur.

Tibia and fibula of basal saurischian dinosaurs are slender elements of subequal length. The cranial aspect of the proximal tibia displays a cranially projecting process, the cnemial crest. The distal end of the tibia in saurischians is rectangular (*Herrerasaurus,* PVL 2655; Novas 1993), oval (*Staurikosaurus;* Colbert 1970; Galton 1977), or subtriangular (*Guaibasaurus,* MCN-PV 2355; Bonaparte et al. 1999, 2007) in outline. A groove on the laterodistal aspect of the tibia accommodates the ascending process of the astragalus. The fibula is a strap-like bone with expanded proximal and distal ends. A prominent tuberculum m. iliofibularis fibulae is present on the craniolateral or lateral side of the fibula in all basal saurischians (e.g., *Herrerasaurus,* PVL 2655; Novas 1993; *Staurikosaurus,* MCZ 1699; Colbert 1970; Galton 1977). The astragalus is robust and rectangular in shape. The ascending process of basal saurischians is located in the medial half on its dorsal side (e.g., *Herrerasaurus,* PVSJ 2655; Novas 1993). The calcaneum of basal saurischians is a flat disc of subtriangular shape in medial view. A keel on the medial aspect of the calcaneum slides into a groove of the dorsolateral aspect of the astragalus to form a keel-and-groove articulation (e.g., *Herrerasaurus,* PVL 2655; Novas 1993; *Guaibasaurus,* MCN-PV 2355; Bonaparte et al. 1999, 2007). The distal tarsus of basal saurischian dinosaurs consists of two disc-like elements, distal tarsal III and IV. Distal tarsal III caps the proximal end of metatarsal III, and distal tarsal IV caps the proximal end of metatarsal IV (e.g., *Herrerasaurus,* PVL 2655; Novas 1993). The pes of basal saurischians is mesaxonic with a bundled metatarsus. Metatarsal III is the longest element of the metatarsus, and metatarsals II and IV are of subequal length. Metatarsal I is considerably

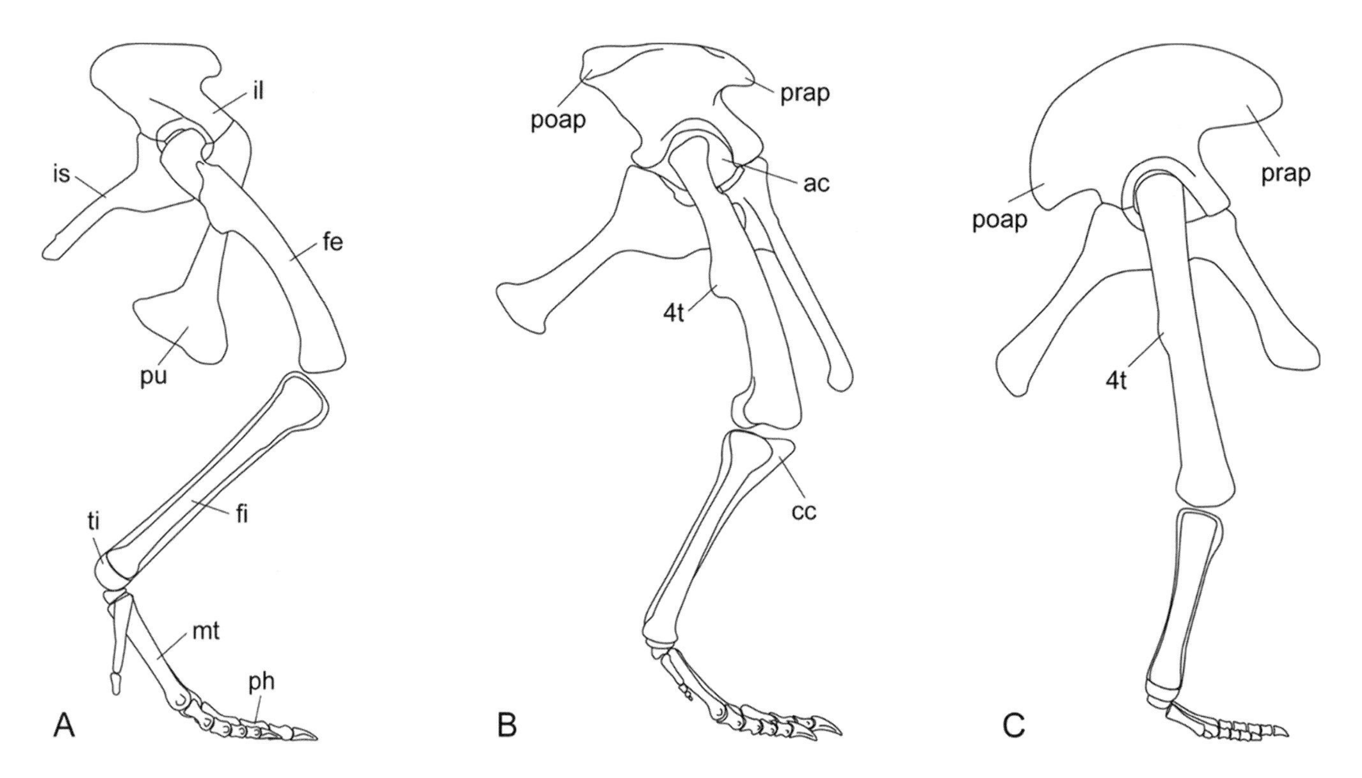

FIGURE 8.16. Pelvis and right hindlimb in lateral view of the basal saurischian *Herrerasaurus* (A), the basal sauropodomorph *Plateosaurus* (B), and the basal sauropod *Shunosaurus* (C). Note the distal placement of the fourth trochanter and the relative shortening of the distal limb elements on the line to the sauropods. In sauropods, the preacetabular process of the iliac blade is considerably enlarged. Pelves and hindlimbs drawn to the same height for better illustration of differences in proportions. ac, acetabulum; cc, cnemial crest; fe, femur; fi, fibula; il, ilium; is, ischium; mt, metatarsus; ph, phalanges; poap, postacetabular process; prap, preacetabular process; pu, pubis; ti, tibia; 4t, fourth trochanter.

reduced in length and less robust compared to metatarsals II–IV; metatarsal V is reduced to a bone splint. In *Herrerasaurus,* a penultimate phalanx is preserved. The phalangeal formula of basal saurischians is 2-3-4-5-0 (Langer 2004) and in *Herrerasaurus* 2-3-4-5-1, respectively (PVL 2655; Novas 1993). In basal saurischians, the distal limb is markedly elongate compared to the femur (Fig. 8.16). The distal limb/femur ratio ranges between 1.38 and 1.54.

Modification of the osteology of the pelvic girdle and hindlimb in nonsauropodan sauropodomorphs is located mainly in the pelvis and proximal limb. As in basal saurischian dinosaurs, two sacral vertebrae are present in the basal sauropodomorphs *Saturnalia* (MCN 3844-PV, MCN 3845-PV; Langer 2003) and *Thecodontosaurus* (Benton et al. 2000). In all other nonsauropodan sauropodomorphs (sensu Yates 2007b), with the exception of *Melanorosaurus,* the sacrum consists of three sacral vertebrae. In *Melanorosaurus* (NM QR1551; Galton et al. 2005), the number of sacrals is increased to four elements. However, there is some confusion as to how the sacrum is formed, whether by incorporation of a caudosacral or a dorsosacral; both possibilities seem to exist in basal sauropodomorphs (Galton 2001). Although the number of sacral vertebrae increases, the iliac blade of nonsauropodan sauropodomorphs retains the form of basal saurischian dinosaurs. In *Anchisaurus* (YPM 1883; Yates 2004), however, the preacetabulum is cranially expanded to form a spine-like process. With exception of *Saturnalia,* the acetabulum of Sauropodomorpha is fully open. In *Saturnalia* (MCN 3844-PV, MCN 3845-PV; Langer 2003), the acetabulum is semiperforated, as seen in *Guaibasaurus.* The perforation of the acetabulum in sauropodomorphs is associated with the elongation of the pubic and ischial peduncle. Whereas the ischial peduncle remains small in basal sauropodomorphs and is even reduced to a knob-like structure in derived sauropodomorphs, a consistent trend in the elongation of the pubic peduncle is observed on the evolutionary line toward sauropods (e.g., *Saturnalia,* MCN 3844-PV, MCN 3845-PV; Langer 2003; *Plateosaurus,* SMNS 12100; von Huene 1926; *Melanorosaurus,* NM QR1551; Galton et al. 2005). The prominent supraacetabular crest of basal saurischian dinosaurs is reduced in basal sauropodomorphs and moved to a more cranial position along the dorsal margin of the acetabulum (e.g., *Plateosaurus,* SMNS 13200; von Huene 1926). Nonsauropodan sauropodomorphs retain the brevis fossa on the caudal margin of the ilium (e.g., *Saturnalia,* MCN 3844-PV, MCN 3845-PV; Langer 2003; *Melanorosaurus,* NM QR1551; Galton et al. 2005). Contrary to basal saurischians, which are characterized by a mesopubic orientation of the pubis, in all sauropodomorphs the pubis is directed cranioventrally, corresponding to a propubic condition (e.g., *Saturnalia,* MCN 3844-PV, MCN 3845-PV; Langer 2003; *Melanorosaurus,* NM QR1551; Galton et al. 2005). The high deflection of the femoral head from the transverse plane of basal saurischians is reduced to around 45° in cranial direction in basal nonsauropodan sauropodomorphs (e.g., *Plateosaurus,* SMNS 13200; von Huene 1926). In more derived sauropodomorphs, such as *Lufengosaurus* (Young 1941), the femoral head is oriented in the transverse plane. Whereas *Saturnalia* (MCN 3844-PV, MCN 3845-PV; Langer 2003) retains the strongly sigmoidal femur in lateral view in the basal saurischians, the consistent trend in the straightening of the femoral shaft is observed in more derived sauropodomorphs (e.g., *Saturnalia,* MCN 3844-PV, MCN 3845-PV; Langer 2003; *Melanorosaurus,* NM QR1551; Galton et al. 2005). A trochanteric shelf is present in *Saturnalia* (MCN 3844-PV, MCN 3845-PV; Langer 2003) but absent in all more derived sauropodomorphs (Yates 2007b). The lesser trochanter is reduced to a low ridge but occurs in all basal sauropodomorphs. As in basal saurischians, the fourth trochanter in *Saturnalia* is located relatively proximally on the caudal side of the femur (MCN 3844-PV, MCN 3845-PV; Langer 2003). Toward more derived nonsauropodan sauropodomorphs, however, the fourth trochanter moves to a more distal position on the caudal femoral shaft (e.g., *Plateosaurus,* SMNS 12100; von Huene 1926; *Melanorosaurus,* NM QR1551; Galton et al. 2005). The distal limb of nonsauropodan sauropodomorphs retains the condition observed in basal saurischian dinosaurs in most aspects. A consistent trend in the increase of the robustness of the elements, such as tibia and fibula, and metatarsals and phalanges is evident. However, the cnemial crest on the proximocranial tibia becomes more prominent in sauropodomorphs, rendering the distal end of the tibia subtriangular in outline (e.g., *Plateosaurus,* SMNS 12100; von Huene 1926). The distal limb/femur ratio in nonsauropodan sauropodomorphs ranges between 1.07 and 1.45, indicating that the femur is elongated in relation to the distal limb as compared to basal saurischian dinosaurs.

As in *Melanorosaurus,* the number of sacrals is increased to four elements in most basal sauropods (e.g., *Shunosaurus,* IVPP V 9065; Dong et al. 1983), and five in more advanced basal eusauropods and neosauropods. With the exception of basal sauropods such as *Lessemsaurus* (Pol & Powell 2007), the preacetabular process is strongly modified toward gravisaurians (eusauropods + vulcanodontids; Allain & Aquesbi 2008). First, it is cranially expanded in basal gravisaurians, such as *Tazoudasaurus* (Allain & Aquesbi 2008), the type ilium of *Kotasaurus* (Yadagiri 2001), and *Shunosaurus* (IVPP V 9065; Dong et al. 1983), and then dorsally expanded to form a lobe-shaped ilium in neosauropods. This development is accompanied by a reduction of the postacetabular process (Fig. 8.16). The acetabulum of sauropod dinosaurs more derived than *Lessemsaurus* is relatively large compared to the femoral head (e.g., *Vulcanodon;* Raath 1972; Cooper 1984). The supraacetabular crest is lost in gravisaurians but still present in basal sauropods, such as *Lessemsaurus* (Pol & Powell 2007). The brevis fossa is present in basal sauropods sensu Yates (2007b), but is lost in gravisaurians (Allain & Aquesbi 2008). In eusauropods, the shaft of the pubis is less apron-like, twisted to face more caudomedially, and has an enlarged pubic foramen proximally. Toward sauropods, a consistent trend to increase the robustness of the femur and to reduce the sigmoid curvature in lateral view is observed (Fig. 8.16), but is present in *Lessemsaurus* and reduced in *Antetonitrus*. In *Isanosaurus* (Buffetaut et al. 2000) and more derived sauropods, the femur is completely straight in lateral view. The lesser trochanter is reduced to a low ridge but present in basal sauropods, with the exception of *Gongxianosaurus* (He et al. 1998) and *Isanosaurus* (Buffetaut et al. 2000). In *Shunosaurus* (IVPP V 9065; Dong et al. 1983) and more derived sauropods, the lesser trochanter is lost. Toward sauropods, the fourth trochanter moves to a more distal position on the caudal side of the femoral shaft and is also placed more medially on the shaft in many taxa (*Isanosaurus,* Vulcanodontidae). In most gravisaurians, the fourth trochanter is located at approximately the midshaft of the femur. Furthermore, the trochanter is developed as a stout flange in basal sauropodomorphs, but reduced to a stout but low ridge in sauropods. In the latter, a prominent groove is usually present medial to the trochanter on the medial side of the femoral shaft. The tibia becomes more robust and somewhat flattened craniocaudally in eusauropods. The cnemial crest becomes more prominent, but also more slender toward sauropods. In *Shunosaurus* (IVPP V 9065; Dong et al. 1983) and more derived eusauropods, the cnemial crest projects craniolaterally. In mamenchisaurids, the orientation of the cnemial crest is almost entirely lateral. As in the case of the tibia, the fibula becomes more robust on the lineage toward sauropods. With the reduction of the calcaneum within gravisaurians, the fibula becomes slightly elongate compared to the tibia, and the distal end is expanded medially to articulate with the astragalus. Both astragalus and calcaneum retain their form and function in basal sauropodomorphs and sauropods. In gravisaurians, the form of the astragalus becomes wedge-shaped (e.g., *Mamenchisaurus*).

With the exception of basal sauropods, both distal tarsals III and IV are reduced or are not ossified in sauropods. Toward the sauropods, the metatarsals and phalanges increase in robustness. The relative length of metatarsal III and IV is reduced, so that metatarsals I–V are subequal in length. With the exception of basal eusauropods, such as *Shunosaurus* (IVPP V 9065; Dong et al. 1983), the mesaxonic pes is modified to an ectaxonic pes, with metatarsal II being the longest element of the metatarsus. The phalanges of the lateral digits are reduced,

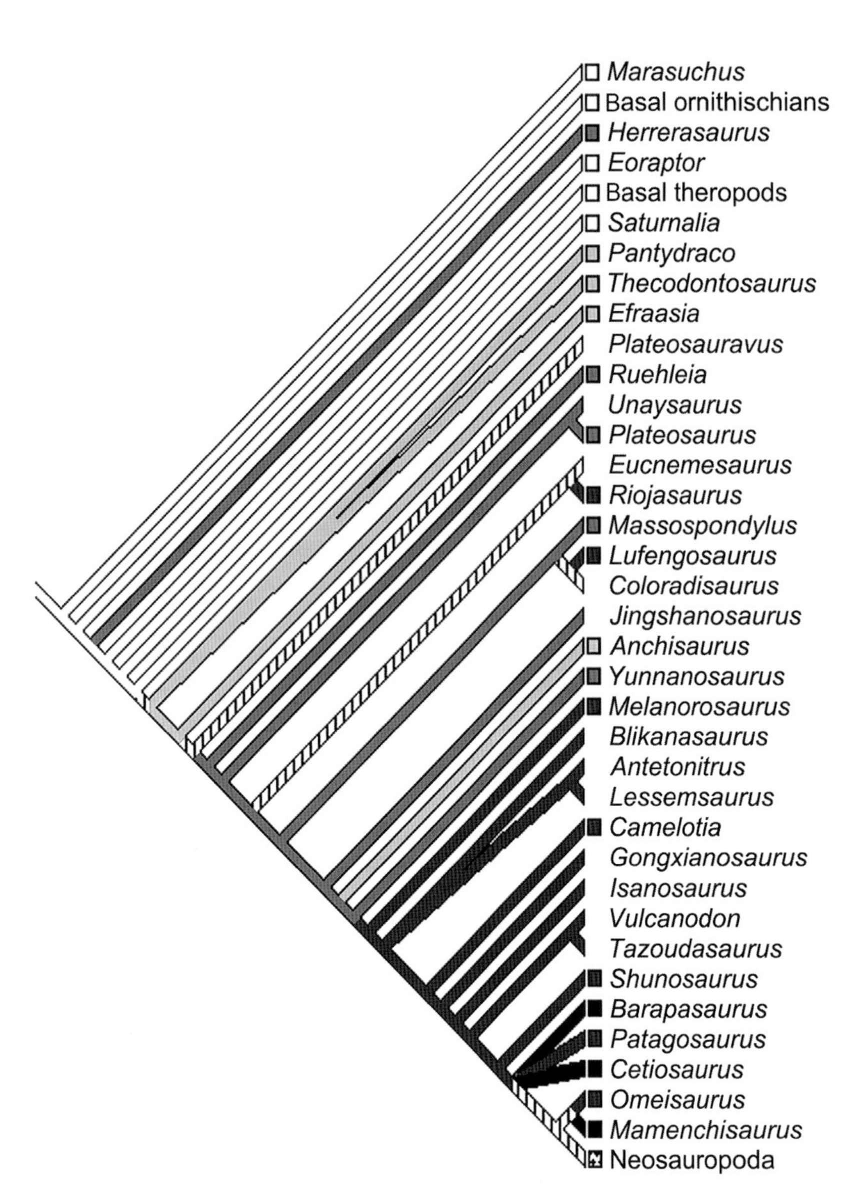

FIGURE 8.17. Logarithmic body size categories of sauropodomorphs mapped onto the cladogram shown in Fig. 8.2. White branches, 0–10 kg size class; light gray, 10–100 kg; medium gray, 100–1,000 kg; dark gray, 1,000–10,000 kg; black, >10,000 kg. Taxa without squares indicate missing data. See text for further explanations.

so that the phalangeal formula of sauropods is 2-3-4-2-1. In some basal sauropods, such as *Shunosaurus* (IVPP V 9065; Dong et al. 1983) and *Mamenchisaurus* (Allain & Aquesbi 2008), the phalangeal formula is 2-3-4-3-1. The metatarsus of sauropods is not bundled but broad. In basal sauropods, the distal limb/femur ratio is around 0.9, and less than 0.75 in basal eusauropods.

Body Size Evolution in Sauropodomorph Dinosaurs

Their enormous to gigantic size is, of course, one of the hallmarks of sauropod dinosaurs. However, dinosaur ancestors were rather small animals (e.g., Sereno & Arcucci 1993, 1994; Novas 1996; Benton 2004), and the first sauropodomorphs were not much bigger (Langer 2004). On the other hand, giant size seems to have been achieved early in sauropod evolution (Buffetaut et al. 2002) and relatively rapidly as well, with the first Late Triassic multiton giants being only some 30 million years younger than their Middle Triassic predecessors. How did this increase in size happen? Was it a gradual increase throughout sauropodomorph evolution, or was it a sudden event at the base of sauropods?

If the phylogenetic hypothesis of Yates (2007b) is accepted, the analysis of body size evolution in sauropodomorphs indicates that a gradual increase was the case (Fig. 8.17). Basal dinosauriforms, basal ornithischians, such as *Pisanosaurus, Heterodontosaurus,* and *Thecodontosaurus,* and basal saurischians, such as *Eoraptor* and *Guaibasaurus,* were small animals and fall within the 0–10 kg size class, which can thus be regarded as the ancestral size for dinosaurs. The same is probably also true for the first sauropodomorph *Saturnalia,* for which no formal estimations of body mass are available, but which, on the basis of comparisons of overall body size (Langer et al.

1999; Langer 2003), is an animal of about 10 kg body weight or slightly more. That body size is highly plastic in even basal dinosaurs is shown by herrerasaurids, the largest of which, *Frenguellisaurus* (Novas 1986), represents an animal of probably close to 1 metric ton. At the base of sauropodomorphs, body size increased into the 10–100 kg size class. However, whereas *Thecodontosaurus* and *Pantydraco* were probably still low within this size class, with estimated masses below 50 kg, *Efraasia* was already somewhat larger, and might have reached 100 kg in body weight (based on material at the SMNS). With the slightly more derived *Ruehleia,* sauropodomorphs enter the 100–1,000 kg size class. Although some taxa, such as large *Plateosaurus, Riojasaurus,* or *Lufengosaurus,* probably reached masses of more than 1,000 kg, and even larger taxa might have existed (O. W. M. R., pers. obs.), our analysis indicates that this size class represents the plesiomorphic state for much of the classic basal sauropodomorphs. An average body mass of more than 1 metric ton was then reached just below the node of Sauropoda as defined by Yates (2007a; see above) with *Melanorosaurus.* Thus, the plesiomorphic body size for sauropods lies beyond 1,000 kg, with masses of more than 10,000 kg having been achieved independently in several lineages of basal sauropods. Within later sauropods, body size is plastic, and macronarians even seem to show a tendency toward reduced size (Carrano 2005a).

In summary, our analysis indicates that body size evolution in sauropodomorphs was rapid but steady, and the ancestral body size of sauropods was more than 1,000 kg, a body mass reached by only few terrestrial mammals today (Peczkis 1988).

Discussion

In the following discussion, we will try to interpret the significance of the findings presented above for the evolution of the sauropodomorph body plan and the success of sauropodomorphs as gigantic herbivores from an intrinsic point of view. Thus, if we talk about the success of the group, this does not refer to their possible competition with other Mesozoic herbivores, but to the possible benefits that certain anatomical changes might have had for the evolution of the special herbivory employed by sauropods. Other aspects, such as changing vegetation, competition with other herbivores, or extinction events, certainly also had a profound impact on the evolution of sauropodomorphs. Considering the poor fossil record of the early stages of sauropodomorph evolution, we refrain from dealing with these aspects here.

Although the most conspicuous change from basal dinosaurs to sauropods might have been the enormous increase in body size (by a factor of more than 40,000 in body mass, from basal dinosauriforms such as *Marasuchus* to animals like *Cetiosaurus*), the evolutionary changes on the lineage toward sauropods deeply affected all aspects of the anatomy of the animals. Two main agents seem to have driven these changes: the ecological transition from a carnivorous to a herbivorous diet, and the biomechanical requirements of increasing size. Because herbivores tend to have larger body sizes than carnivores, and large body size might have been essential for the special type of herbivory employed by sauropods (Sander & Clauss 2008; Sander et al. 2010a, 2010b; Hummel & Clauss, this volume), both of these agents are undoubtedly strongly intrinsically connected.

Changes in the general body plan viewed have to be seen in the light of both of these aspects. Although the results of the calculations of the areas of different body sections are certainly in no way precise, the general tendencies observed in these data seems to be genuine and generally to fit well with the other results presented above. Both the results of this comparison and the skull length/femur length ratios demonstrate that the skull is indeed considerably reduced in size already at or close to the base of Sauropodomorpha and is further reduced in sauropods, although the decrease is far less marked. Area calculations and calculations of the neck/trunk ratio furthermore indicate that relative neck length seemingly did not notably increase at the origin or the early radiation of sauropodomorphs; the impression of a notably longer neck in these animals is instead due to the relative reduction in size of the skull in these animals. However, this skull size reduction might have been an important prerequisite to the later achievement of large body sizes, retention of a long neck, or even elongation of the neck. In theropods, larger taxa typically tend to have relatively shorter necks than smaller taxa, which is probably due to the biomechanical constraints posed on the neck by a large and powerful head. Thus, in sauropodomorphs, the reduction of the size of the skull allowed even large taxa to retain or even elongate the neck (Sander & Clauss 2008; Sander et al. 2009, 2010a). Relative skull size, in turn, was probably affected by the change from a carnivorous diet in basal dinosaurs to a herbivorous diet in sauropods. Barrett (2000) and Barrett & Upchurch (2007) made a strong case for omnivory in basal sauropodomorphs, so that this transition probably passed through an omnivorous phase. A small skull on a long neck would have been advantageous for larger-bodied omnivores, which might have used the neck for rapid snatching movements to secure small prey items, so an initial step in the development of a small skull might have been a specialization of carnivorous animals toward smaller prey. Although a limited amount of oral processing was probably present in basal sauropodomorphs and maybe even in basal sauropods (Upchurch et al. 2007b), adopting a herbivorous diet further allows the head to become relatively smaller because the skull is, apart from its function of housing the brain and sensory organs, mainly needed for cropping. As noted by Parrish (2005; see also Sander et al. 2009 and Hummel & Clauss, this volume; Christian & Dzemski, this volume), a long neck is also of further advantage for herbivorous animals in increasing the feeding envelope (Preuschoft et al., this volume). Thus, the plesiomorphically long neck in sauropodomorphs could be viewed as a preadaptation that facilitated herbivory and greatly aided in the success of these animals as herbivores. This effect of the long neck was improved with increasing neck length and body size on the lineage toward sauropods. Within sauropods, relative neck length is one of the most variable traits and might have been important in niche partitioning, as noted by Upchurch & Barrett (2000).

Interestingly, area comparisons indicate that the relative size of the torso did not change much from carnivorous sauropodomorph outgroups to even the certainly obligatory herbivorous neosauropods, although some variation certainly existed, as in the case of the especially long-necked diplodocids and mamenchisaurids—although in these animals, the relative size of the torso is even decreased. This is somewhat surprising because it has been suggested that the special herbivory of sauropods was only possible if especially large digestive tracts were present in these animals in which the plant fodder could be digested (Farlow 1987; Hummel et al. 2008; Sander & Clauss 2008; Sander et al. 2010a; Hummel & Clauss, this volume; see also Barrett & Upchurch 2007). If the relative size of the torso could not be increased in sauropodomorphs, the only way to assure the necessary gut capacity was thus an increase in overall body size. Thus, increasing specialization toward a herbivorous diet along the evolutionary lineage leading toward sauropods might have driven the generally gradual increase in body size in these animals.

Major changes in the skull of sauropodomorphs can also be explained by the change-over to the special herbivory employed by sauropods. If the skull is only used as a cropping device (as was probably the case in eusauropods; Upchurch et al. 2007b), snout length can be reduced and the tooth row can be restricted to the rostral parts of the jaw because only these are effectively used for cropping. As argued by Christiansen (1999), food intake rate is mainly dependent on skull width, not on skull length. Thus, this could account for the shortening of the skull and especially for the notable broadening of the muzzle that we see on the lineage toward sauropods. The latter aspect might thus both reflect a change from an omnivorous diet in taxa such as *Plateosaurus* to a strictly herbivorous diet in sauropods, and from more selective feeders at smaller body size to bulk feeders in larger herbivores. Interestingly, a relative broadening of the snout during ontogeny was noted in *Diplodocus* by Whitlock et al. (2008) and was also attributed to changing feeding habits during ontogeny in this taxon.

The major shape changes in the skulls of sauropodomorphs, namely the enlargement and retraction of the nares, may also have to do with the change in skull function. At the base of sauropodomorphs, the change from a carnivorous to an omnivorous diet, and thus to probably small prey items, led to a reduction of bite forces on the skull, which in turn allowed cranial openings to get relatively larger, especially in areas that are under high stress in carnivorous animals, such as the rostral part of the snout. This probable reduction in bite force probably also accounts for the thin internasal sutures in sauropodomorphs, which are frequently found broken in fossil specimens (e.g., Yates 2007b). This area of obviously low forces already roughly coincides with the position of the nares in more derived sauropods. Interestingly, Witzel & Preuschoft (2005) and Witzel et al. (this volume) found the position of the extremely retracted nares in *Diplodocus* coincides with an area of the skull that is not under stress during biting in this animal. Thus, the retraction of the nares in sauropodomorphs seems to have biomechanical reasons; the change to repeated cropping bites with the rostral teeth thus resulted in rather robust jaws, which were subjected to relatively lower, but frequent, forces, whereas the dorsal parts of the skull were under less stress than in the powerful bites of carnivorous animals. This hypothesis is also in general accordance with the observation that relative orbit size and shape did not change significantly during sauropodomorph evolution: as demonstrated by Henderson (2002), theropod taxa with higher bite forces tend to have relatively smaller and more elliptical orbits than taxa with lower bite forces.

Increased lamination of cervical vertebrae in sauropods most probably reflects the biomechanical requirements of additional neck elongation at a large body size in these animals (i.e., a compensation for increased forces caused by scaling effects). This elongation of the neck is also combined with an increase in pneumatization of its bony structures (assuming that all sauropodomorphs had cervical air sacs, see above; Wedel 2007; Perry et al., this volume), which also helped to define cervical laminae. Lamination and pneumatization may have thus been important aspects in retaining and even increasing a long neck in a large-sized animal. The observation that sauropods with shorter necks, such as dicraeosaurids, often show reduced pneumaticity and that some smaller sauropods, most notably small titanosaurs (e.g., Curry Rogers & Forster 2001), have reduced cervical lamination, may support this hypothesis. Interestingly, both increased lamination and pneumatization are also found in the necks of theropod dinosaurs, which are relatively shorter than in sauropods. However, in these animals, the skull is relatively larger than in sauropodomorphs, adding further weight and thus increasing mechanical forces acting on the neck, so that the elaboration of these structures may have to do with the same mechanical constraints as in sauropods. Given the comparable relative neck lengths in basal theropods and sauropodomorphs, the poor development of cervical laminae in the latter may thus represent a reduction of these structures in combination with a reduction of relative skull size at the base of Sauropodomorpha rather than the plesiomorphic condition. That this may have been the case is also indicated by cervical structure in the basal saurischians *Eoraptor* (PVSJ 512) and *Herrerasaurus* (Sereno & Novas 1993), which have well developed diapophyseal laminae (Fig. 8.11).

As in the case of vertebral lamination, the change from amphicoelous to opisthocoelous cervicals most likely has biomechanical reasons. Salisbury & Frey (2000) argue that the change from amphicoelous to procoelous vertebrae in crocodiles was due to increased shear stresses acting on the intervertebral articulations. The increased length of the neck, together with the increased overall body size in basal sauropods, certainly also resulted in higher stresses, including shear stresses, on the vertebral articulations, which means that the same probably applies to the formation of opisthocoelous cervicals in sauropods. As noted by Nopcsa (1930), the concave articular surface of vertebrae with one semispherical articular end tends to face toward the relatively more immobile part of the body, which applies to both the opisthocoelous cervicals in sauropods in general and the procoelous cranial caudals in many neosauropods. Nopcsa (1930) suggested that this may also be due to the mode of transmission of forces across the intervertebral articulations, but more research is needed to clarify the function and biomechanics of these articulations in sauropodomorphs. That Nopcsa's observation is not universally true in sauropods is shown by forms such as *Opisthocoelicaudia* in which the caudal vertebrae are opisthocoelous (Borsuk-Bialynicka 1977).

Interestingly, both the occurrence of well developed cervical vertebral lamination and opisthocoelous intervertebral articulations seem to have appeared rapidly at the base of Sauropoda and thus coincides with the increase of the number of cervical vertebrae in these animals from 10 to 12. It is tempting to suggest that the increase not only in neck length but also in joints within this structure resulting from the incorporation of two additional elements might have been responsible for the increase of forces acting on the vertebral column, which in turn led to these anatomical changes. However, further finds of basal sauropods are necessary to test this hypothesis.

The evolution of sauropodomorph limbs seems to have been dominated by the transition from cursorial to graviportal limbs (Coombs 1978; Carrano 2005b). In accordance with the more or less gradual increase in size, graviportal adaptations in the limbs of sauropodomorph dinosaurs also evolved gradually. The adoption of a graviportal limb also meant a reduction of the locomotor capacities in sauropods (Carrano 2005b). Straightening of the limbs and the relative shortening of the distal limb segments are typical adaptations toward increasing body size and are also observed in large mammals, ornithischians, and theropods (Coombs 1978; Carrano 1999, 2001, 2005b). Likewise, the relative elongation of the forelimbs if compared to the hindlimbs is probably also an adaptation to the graviportal posture rather than to improved quadrupedal locomotion, although it was certainly also beneficial for the latter aspect.

Thus, many changes in the evolution of the forelimbs became apparent in the context of both adaptations toward a graviportal posture and improved quadrupedal locomotion (Bonnan 2003; Remes 2006, 2008; Bonnan & Yates 2007). The rotation of the coracoid in the pectoral girdle allowed a more upright forelimb position, the rearrangement of the elbow joint allowed constant pronation of the manus, and a more digitigrade, tubular manus is clearly adapted as a weight-carrying device. The relative elongation of the forelimbs in comparison with the hindlimbs must also certainly be viewed in this context. However, the question of whether basal sauropodomorphs were obligate (e.g., Bonnan & Senter 2007) or only facultative bipeds, and whether there might have been an obligate bipedal phase in basal dinosaurian and sauropodomorph evolution (Fechner 2006, 2007), cannot be answered here but will be looked at in more detail in future studies. The often-used hindlimb/trunk ratio seems to be of limited information value in sauropodomorphs, given that some neosauropods, such as *Apatosaurus,* have ratios of "typical" obligate bipeds.

Most of the changes in the pelvic girdle and hindlimbs can also be attributed to the biomechanical necessities of increasing body size. An increase in the number of sacral vertebrae increases the strength of the connection between the pectoral girdle and the vertebral column, and the apparent gradual increase of sacral number in sauropodomorphs (from probably two at the base of Sauropodomorpha, to three in basal taxa, four in advanced basal taxa, and finally five in advanced eusauropods) nicely reflects the gradual increase in body size on the evolutionary lineage toward sauropods. In the pelvic girdle and hindlimbs, basal sauropodomorphs (e.g., *Saturnalia*) still largely resemble basal saurischian outgroups (e.g., *Eoraptor*), and basal sauropods (e.g., *Lessemsaurus, Antetonitrus*) resemble advanced basal sauropodomorphs. Especially in the pelvic girdle, the most important changes happened within basal sauropods. An increase in the size of the iliac blade offered more space for the attachment of hindlimb musculature, and a general reduction of obvious muscle attachment sites on the hindlimbs (e.g., lesser trochanter, fourth trochanter) is consistent with the observation that there was a tendency toward more fleshy muscle attachments, which left fewer osteological correlates, with increasing body size. A distal displacement of the fourth trochanter, the attachment site of the main propulsion muscle in reptiles, the m. caudofemoralis longus (Gatesy 1990), can also be viewed in this context: in this novel position, the muscle has better leverage and can thus generate more overall force, but the angular momentum of the rotation of the femur in the acetabulum is decreased, resulting in a slower but more powerful stride of the hindlimb.

Further adaptations toward increased body size are a semi-plantigrad foot posture, a short and broad metatarsus, and reduced phalanges in the foot.

In summary, the evolutionary history of the sauropodomorph bauplan is a complex process, in which ecological and interrelated biomechanical aspects led to a cascade of changes in all parts of the skeleton. Many of these changes might indeed have formed positive feedback loops, as suggested by Barrett & Upchurch (2007). Such correlated progression would also account for parallelisms evident in several stages in the evolutionary history of basal sauropodomorphs—for example, the increase in overall body size in some side lineages (e.g., plateosaurids), or the increase in neck length by increasing single-element length in massospondylids.

Conclusions

The evolutionary transition from small, carnivorous, basal dinosauriforms to gigantic sauropods deeply affected all parts of the anatomy of the animals. The most notable general tendencies in the evolution of sauropodomorphs were a pronounced decrease in relative skull size, a relative shortening and broadening of the skull, an increase in the relative size of the external nares, a lengthening of the neck, a straightening of the limbs, a relative reduction of the distal limb segments, and a relative lengthening of the forelimbs in comparison to the hindlimbs. However, all these changes happened during different phases of sauropodomorph phylogeny. Whereas some changes seem to have been rather gradual, marked, rapid changes can also be recognized, for example in the reduction of skull size at the base of the Sauropodomorpha, or in the increase in cervical vertebral lamination at the base of Sauropoda.

Furthermore, although body size is highly plastic in saurischians in general as well as in basal sauropodomorphs, a general tendency toward increasing overall body size is also apparent in basal sauropodomorphs up to eusauropods. Within eusauropods, absolute body size is again very plastic, although generally much greater than in basal saurischians. In contrast, and somewhat counterintuitively, there is no noticeable tendency either toward increasing trunk length in comparison to hindlimb length, or toward a notable increase in the size of the torso in comparison with other parts of the body. Most changes of the skull and neck can probably be accounted for by changing feeding habits, from carnivorous in basal saurischians via omnivorous in basal sauropodomorphs to bulk-feeding herbivory in sauropods (see also Barrett & Upchurch 2007), and by the associated changes in function and biomechanics of the respective structures. Changes in overall body shape and the limbs seem to have been driven by the biomechanical requirements of increasing body size.

Last, and maybe most importantly, this brief overview highlights our still incomplete knowledge of the anatomy and diversity of basal sauropodomorphs and basal sauropods, as well as the problems associated with uncertain phylogenetic relationships in this part of the dinosaur family tree. Many of the observations and interpretations presented here are strongly influenced by the choice of the phylogeny used and may change considerably under different phylogenetic hypotheses. Thus, new discoveries, but also more detailed studies of known taxa, are needed to better understand the evolutionary transition from basal saurischians to sauropods. Furthermore, it is to be hoped that a consensus of basal sauropodomorph relationships will soon be reached, similar to that in other parts of dinosaur phylogeny (e.g., theropods or eusauropods).

Acknowledgments

We thank all the members of the DFG Research Unit 533 for fruitful discussions, especially Martin Sander (University of Bonn) and Ulrich Witzel (Ruhr University Bochum). Furthermore, all the participants of the First International Workshop on Sauropod Biology and Gigantism in Bonn in November 2008 are thanked for fresh insights and exchange of ideas on sauropodomorph dinosaurs. Ronan Allain (Muséum National d'Histoire Naturelle, Paris) critically read this chapter and helped with many thoughtful comments. Too many colleagues to mention here have helped with providing access to collections and discussions during the years of this project, but all their help and input is greatly appreciated. This paper is a result of DFG project RA 1012/2 within the framework of the DFG Research Unit 533, and has also benefited from input by project RA 1012/5-1. This is contribution number 67 of the DFG Research Unit 533 "Biology of the Sauropod Dinosaurs: The Evolution of Gigantism."

References

Allain, R. & Aquesbi, N. 2008. Anatomy and phylogenetic relationships of *Tazoudasaurus naimi* (Dinosauria, Sauropoda) from the late Early Jurassic of Morocco.—*Geodiversitas* 30: 345–424.

Allain, R. & Läng, E. 2009. Origine et evolution des saurischiens.—*Comptes Rendus Palevol* 8: 243–256.

Barrett, P. M. 2000. Prosauropod dinosaurs and iguanas: speculations on the diets of extinct reptiles. *In* Sues, H.-D. (ed.). *Evolution of Herbivory in Terrestrial Vertebrates*. Cambridge University Press, Cambridge: pp. 42–78.

Barrett, P. M. & Upchurch, P. 2007. The evolution of feeding mechanisms in early sauropodomorph dinosaurs.—*Special Papers in Palaeontology* 77: 91–112.

Benton, M. J. 1986. The Late Triassic reptile *Teratosaurus*—a rauischian, not a dinosaur.—*Palaeontology* 29: 293–301.

Benton, M. J. 1990. Origin and interrelationships of dinosaurs. *In* Weishampel, D. B., Dodson, P. & Osmólska, H. (eds.). *The Dinosauria*. University of California Press, Berkeley: pp. 11–30.

Benton, M. J. 1999. *Scleromochlus taylori* and the origin of dinosaurs and pterosaurs.—*Philosophical Transactions of the Royal Society B: Biological Sciences* 354: 1423–1446.

Benton, M. J. 2004. Origin and relationships of Dinosauria. *In* Weishampel, D. B., Dodson, P. & Osmólska, H. (eds.). *The Dinosauria. 2nd edition.* University of California Press, Berkeley: pp. 7–19.

Benton, M. J., Juul, L., Storrs, G. W. & Galton, P. M. 2000. Anatomy and systematics of the prosauropod dinosaur *Thecodontosaurus antiquus* from the Upper Triassic of southwest England.—*Journal of Vertebrate Paleontology* 20: 77–108.

Bonaparte, J. F. 1972. Los tetrapodos del sector superior de la Formación Los Colorados, La Rioja, Argentina (Triásico Superior). I Parte.—*Opera Lilloana* 22: 1–183.

Bonaparte, J. F. 1975. Nuevos materiales de *Lagosuchus talampayensis* Romer (Thecodontia—Pseudosuchia) y su significado en el origen de los Saurischia. Chañarense inferior, Triásico Medio de Argentina.—*Acta Geologica Lilloana* 13: 5–90.

Bonaparte, J. F. 1986a. The early radiation and phylogenetic relationships of the Jurassic sauropod dinosaurs, based on vertebral anatomy. *In* Padian, K. (ed.). *The Beginning of the Age of Dinosaurs.* Cambridge University Press, Cambridge: pp. 245–258.

Bonaparte, J. F. 1986b. Les Dinosaures (Carnosaures, Allosauridés, Sauropodes, Cétiosauridés) du Jurassique moyen de Cerro Cóndor (Chubut, Argentine).—*Annales de Paléontologie* 72: 326–386.

Bonaparte, J. F. 1999. Evolución de las vértebras presacras en Sauropodomorpha.—*Ameghiniana* 36: 115–187.

Bonaparte, J. F., Ferigolo, J. & Riberio, A. M. 1999. A new early Late Triassic saurischian dinosaur from Rio Grande do Sul State, Brazil. *In* Tomida, Y., Rich, T. H. & Vickers Rich, P. (eds.). *Proceedings of the Second Gondwanan Dinosaur Symposium.* National Science Museum Monographs, Tokyo: pp. 89–109.

Bonaparte, J. F., Brea, G., Schultz, C. L. & Martinelli, A. G. 2007. A new specimen of *Guaibasaurus candelariensis* (basal Saurischia) from the Late Triassic Caturrita Formation of southern Brazil.—*Historical Biology* 19: 73–82.

Bonnan, M. F. 2003. The evolution of manus shape in sauropod dinosaurs: implications for functional morphology, forelimb orientation, and phylogeny.—*Journal of Vertebrate Paleontology* 23: 595–613.

Bonnan, M. F. & Senter, P. 2007. Were the basal sauropodomorph dinosaurs *Plateosaurus* and *Massospondylus* habitual quadrupeds?—*Special Papers in Palaeontology* 77: 139–155.

Bonnan, M. F. & Yates, A. M. 2007. A new description of the forelimb of the basal sauropodomorph *Melanorosaurus:* implications for the evolution of pronation, manus shape and quadrupedalism in sauropod dinosaurs.—*Special Papers in Palaeontology* 77: 157–168.

Borsuk-Bialynicka, M. 1977. A new camarasaurid sauropod *Opisthocoelicaudia skarzynskii* gen. n., sp. n. from the Upper Cretaceous of Mongolia.—*Palaeontologia Polonica* 37: 5–64.

Broom, R. 1911. On the dinosaurs of the Stromberg, South Africa.—*Annals of the South Africa Museum* 7: 291–308.

Buffetaut, E., Suteethorn, V., Cuny, G., Tong, H., Le Loeuff, J., Khansubha, S. & Jongautcharlyakul, S. 2000. The earliest known sauropod dinosaur.—*Nature* 407: 72–74.

Buffetaut, E., Suteethorn, V., Le Loeuff, J., Cuny, G., Tong, H. & Khansubha, S. 2002. The first giant dinosaurs: a large sauropod form the Late Triassic of Thailand.—*Comptes Rendus Palevol* 1: 103–109.

Calvo, J. O. & Salgado, L. 1995. *Rebbachisaurus tessonei* sp. nov. a new Sauropoda from the Albian–Cenomanian of Argentina; new evidence on the origin of the Diplodocidae.—*Gaia* 11: 13–33.

Carrano, M. T. 1999. What, if anything, is a cursor? Categories versus continua for determining locomotor habit in mammals and dinosaurs.—*Journal of Zoology* 247: 29–42.

Carrano, M. T. 2001. Implications of limb bone scaling, curvature and eccentricity in mammals and non-avian dinosaurs.—*Journal of Zoology* 254: 41–55.

Carrano, M. T. 2005a. Body-size evolution in the Dinosauria. *In* Carrano, M. T., Gaudin, T. J., Blob, R. W. & Wible, J. R. (eds.). *Amniote Paleobiology: Perspectives on the Evolution of Mammals, Birds, and Reptiles.* University of Chicago Press, Chicago: pp. 225–268.

Carrano, M. T. 2005b. The evolution of sauropod locomotion: morphological diversity of a secondarily quadrupedal radiation. *In* Curry Rogers, K. & Wilson, J. A. (eds.). *The Sauropods: Evolution and Paleobiology.* University of California Press, Berkeley: pp. 229–251.

Charig, A. J., Attridge, J. & Crompton, A. W. 1965. On the origin of the sauropods and the classification of the Saurischia.—*Proceedings of the Linnean Society of London* 176: 197–221.

Christian, A. & Dzemski, G. This volume. Neck posture in sauropods. *In* Klein, N., Remes, K., Gee, C. T. & Sander, P. M. (eds.). *Biology of the Sauropod Dinosaurs: Understanding the Life of Giants.* Indiana University Press, Bloomington: pp. 251–260.

Christiansen, P. 1999. On the head size of sauropodomorph dinosaurs: implications for ecology and physiology.—*Historical Biology* 13: 269–297.

Chure, D. J. 2000. On the orbit of theropod dinosaurs. *Gaia* 15: 233–240.

Colbert, E. H. 1970. A saurischian dinosaur from the Triassic of Brazil.—*American Museum of Arizona Bulletin* 57: 1–60.

Colbert, E. H. 1989. The Triassic dinosaur *Coelophysis.*—*Museum of Northern Arizona Bulletin* 57: 1–160.

Coombs, W. P. J. 1975. Sauropod habits and habitats.—*Palaeogeography, Palaeoclimatology, Palaeoecology* 17: 1–33.

Coombs, W. P. J. 1978. Theoretical aspects of cursorial adaptation in dinosaurs.—*Quarterly Review of Biology* 53: 393–418.

Cooper, M. R. 1984. A reassessment of *Vulcanodon karibaensis* Raath (Dinosauria: Saurischia) and the origin of the Sauropoda.—*Palaeontologia Africana* 25: 203–231.

Curry Rogers, K. 2005. Titanosauria: a phylogenetic overview. *In* Curry Rogers, K. & Wilson, J. A. (eds.). *The Sauropods: Evolution and Paleobiology.* University of California Press, Berkeley: pp. 50–103.

Curry Rogers, K. & Forster, C. A. 2001. The last of the dinosaur titans: a new sauropod from Madagascar.—*Nature* 412: 530–534.

Dodson, P. 1990. Sauropod paleoecology. *In* Weishampel, D. B., Dodson, P. & Osmólska, H. (eds.). University of California Press, Berkeley: pp. 402–407.

Dong, Z., Zhou, S. & Zhang, Y. 1983. The dinosaurian remains from Sichuan Basin, China.—*Palaeontologica Sinica* 162: 1–145.

Dzik, J. 2003. A beaked herbivorous archosaur with dinosaur affinities from the early Late Triassic of Poland.—*Journal of Vertebrate Paleontology* 23: 556–574.

Dzik, J. & Sulej, T. 2007. A review of the early Late Triassic Krasiejów biota from Silesia, Poland.—*Palaeontologia Polonica* 64: 3–27.

Ewer, R. F. 1965. The anatomy of the thecodont reptile *Euparkeria capensis* Broom.—*Philosophical Transactions of the Royal Society B: Biological Sciences* 248: 379–435.

Farlow, J. O. 1987. Speculations about the diet and digestive physiology of herbivorous dinosaurs.—*Paleobiology* 13: 60–72.

Fechner, R. 2006. Evolution of bipedality in dinosaurs.—*Journal of Vertebrate Paleontology* 26: 60A.

Fechner, R. 2007. Die frühe Entwicklung der Saurischia: Eine integrative Studie.—*Hallesches Jahrbuch für Geowissenschaften, Beiheft* 23: 55–56.

Galton, P. M. 1970. The posture of hadrosaurian dinosaurs.—*Journal of Paleontology* 44: 464–473.

Galton, P. M. 1976. Prosauropod dinosaurs (Reptilia: Saurischia) of North America.—*Postilla* 169: 1–98.

Galton, P. M. 1977. On *Staurikosaurus pricei,* an early saurischian dinosaur from the Triassic of Brazil, with notes on the Herrerasauridae and Poposauridae.—*Paläontologische Zeitschrift* 51: 234–245.

Galton, P. M. 1985. Diet of prosauropod dinosaurs from the late Triassic and early Jurassic.—*Lethaia* 18: 105–123.

Galton, P. M. 1990. Basal Sauropodomorpha—Prosauropoda. *In* Weishampel, D. B., Dodson, P. & Osmólska, H. *The Dinosauria.* University of California Press, Berkeley: pp. 320–344.

Galton, P. M. 2001. Prosauropod dinosaur *Sellosaurus gracilis* (Upper Triassic, Germany): third sacral vertebra as either a dorsosacral or a caudosacral.—*Neues Jahrbuch für Geologie und Paläontologie, Monatshefte* 2001: 688–704.

Galton, P. M. & Cluver, M. A. 1976. *Anchisaurus capensis* (Broom) and a revision of the Anchisauridae (Reptilia: Saurischia).—*Annals of the South African Museum* 69: 121–159.

Galton, P. M. & Upchurch, P. 2004. Prosauropoda. *In* Weishampel, D. B., Dodson, P. & Osmólska, H. (eds.). *The Dinosauria. 2nd edition.* University of California Press, Berkeley: pp. 232–258.

Galton, P. M., Heerden, J. v. & Yates, A. M. 2005. Postcranial anatomy of referred specimens of the sauropodomorph dinosaur *Melanorosaurus* from the Upper Triassic of South Africa. *In* Tidwell, V. & Carpenter, K. (eds.). *Thunder-Lizards: The Sauropodomorph Dinosaurs.* Indiana University Press, Bloomington: pp. 1–37.

Gatesy, S. M. 1990. Caudofemoral musculature and the evolution of theropod locomotion.—*Paleobiology* 16: 170–186.

Gauthier, J. A. 1986. Saurischian monophyly and the origin of birds.—*Memoirs of the Californian Academy of Science* 8: 1–55.

Gilmore, C. W. 1925. A nearly complete articulated skeleton of *Camarasaurus,* a saurischian dinosaur from the Dinosaur National Monument, Utah.—*Memoirs of the Carnegie Museum* 10: 347–384.

Gower, D. J. 2001. Possible postcranial pneumaticity in the last common ancestor of birds and crocodilians: evidence from *Erythrosuchus* and other Mesozoic archosaurs.—*Naturwissenschaften* 88: 119–122.

Gunga, H.-C., Suthau, T., Bellmann, A., Friedrich, A., Schwanebeck, T., Stoinski, S., Trippel, T., Kirsch, K. & Hellwich, O. 2007. Body mass estimations for *Plateosaurus engelhardti* using laser scanning and 3D reconstruction methods.—*Naturwissenschaften* 94: 623–630.

He, X.-L., Wang, C.-S., Liu, S.-Z., Zhou, F.-Y., Liu, T.-Q., Cai, K.-J. & Dai, B. 1998. A new species of sauropod dinosaur from the Early Jurassic of Gongxian County, Sichuan.—*Acta Geologica Sichuan* 18: 1–6.

Heerden, J. v. 1997. Prosauropods. *In* Farlow, J. O. & Brett-Surman, M. K. (eds.). *The Complete Dinosaur.* Indiana University Press, Bloomington: pp. 242–263.

Henderson, D. M. 2002. The eyes have it: the sizes, shapes, and orientations of theropod orbits as indicators of skull strength and bite force.—*Journal of Vertebrate Paleontology* 22: 766–778.

Holland, W. J., 1924. The skull of *Diplodocus.*—*Memoirs of the Carnegie Museum* 9(3): 379–403.

Hummel, J. & Clauss, M. This volume. Sauropod feeding and digestive physiology. *In* Klein, N., Remes, K., Gee, C. T. & Sander, P. M. (eds.). *Biology of the Sauropod Dinosaurs: Understanding the Life of Giants.* Indiana University Press, Bloomington: pp. 11–33.

Hummel, J., Gee, C. T., Südekum, K.-H., Sander, P. M., Nogge, G. & Clauss, M. 2008. In vitro digestibility of fern and gymnosperm foliage: implications for sauropod feeding ecology and diet selection.—*Proceedings of the Royal Society B: Biological Sciences* 275: 1015–1021.

Janensch, W. 1937. Skelettrekonstruktion von *Brachiosaurus brancai* aus den Tendaguru-Schichten Deutsch-Ostafrikas.—*Zeitschrift der Deutschen Geologischen Gesellschaft* 89: 550–552.

Kutty, T. S., Chatterjee, S., Galton, P. M. & Upchurch, P. 2007. Basal sauropodomorphs (Dinosauria: Saurischia) from the Lower Jurassic of India: their anatomy and relationships.—*Journal of Paleontology* 81: 1552–1574.

Läng, E. & Goussard, F. 2007. Redescription of the wrist and manus of *Bothryospondylus madagascariensis:* new data on carpus morphology in Sauropoda.—*Geodiversitas* 29 (4): 549–560.

Langer, M. C. 2003. The pelvic and hindlimb anatomy of the stem-sauropodomorph *Saturnalia tupiniquim* (Late Triassic, Brazil).—*PaleoBios* 23: 1–40.

Langer, M. C. 2004. Basal Saurischia. *In* Weishampel, D. B., Dodson, P. & Osmólska, H. (eds.). *The Dinosauria. 2nd edition.* University of California Press, Berkeley: pp. 25–46.

Langer, M. C. 2007. The pectoral girdle and forelimb anatomy of the stem-sauropodomorph *Saturnalia tupiniquim* (Upper Triassic, Brazil).—*Special Papers in Palaeontology* 77: 113–137.

Langer, M. C., Abdala, F., Richter, M. & Benton, M. J. 1999. A sauropodomorph dinosaur from the Upper Triassic (Carnian) of southern Brazil.—*Comptes Rendus de l'Académie de Sciences Série IIa: Sciences de la Terre et des Planetes* 329: 511–517.

Lautenschlager, S. 2008. *Revision of* Rauisuchus tiradentes *from the Late Triassic Santa Maria Formation of Brazil*. Diploma thesis. Ludwig-Maximilians-Universität, Munich.

Maddison, D. R. & Maddison, P. M. W. 2003. *MacClade 4.06*. Sinauer Associates, Sunderland, Mass.

Mallison, H. 2007. *Virtual Dinosaurs—Developing Computer Aided Design and Computer Aided Engineering Modeling Methods for Vertebrate Paleontology*. Ph.D. Dissertation. Geowissenschaftliche Fakultät, Eberhard-Karls-Universität Tübingen, Tübingen.

Mallison, H. This volume. *Plateosaurus* in 3D: how CAD models and kinetic–dynamic modeling bring an extinct animal to life. *In* Klein, N., Remes, K., Gee, C. T. & Sander, P. M. (eds.). *Biology of the Sauropod Dinosaurs: Understanding the Life of Giants*. Indiana University Press, Bloomington: pp. 219–236.

Mauersberger, M. 2005. *Morphometrische Untersuchungen der Vorder- und Hinterextremitäten der basalen Sauropodomorpha*. Diploma thesis. Freie Universität Berlin, Berlin.

Mazzetta, G. V., Christiansen, P. & Farina, R. A. 2004. Giants and bizarres: body size of some southern South American Cretaceous dinosaurs.—*Historical Biology* 2004: 1–13.

McIntosh, J. S. 1990. Sauropoda. *In* Weishampel, D. B., Dodson, P. & Osmólska, H. (eds.). *The Dinosauria*. University of California Press, Berkeley: pp. 345–401.

McIntosh, J. S., Miles, C. A., Cloward, K. C. & Parker, J. R. 1996. A new nearly complete skeleton of *Camarasaurus*.—*Bulletin of the Gunma Museum of Natural History* 1: 1–87.

Moser, M. 2003. *Plateosaurus engelhardti* Meyer, 1837 (Dinosauria: Sauropodomorpha) aus dem Feuerletten (Mittelkeuper; Obertrias) von Bayern.—*Zitteliana B* 24: 1–188.

Nopcsa, F. 1930. Über prozöle und opisthozöle Wirbel.—*Anatomischer Anzeiger* 69: 19–25.

Novas, F. E. 1986. Un probable terópodo (Saurischia) de la Formación Ischigualasto (Triásico superior), San Juan, Argentina.—*IV Congreso Argentino de Paleontología y Estratigrafía* 2: 1–6.

Novas, F. E. 1993. New information on the systematics and postcranial skeleton of *Herrerasaurus ischigualastensis* (Theropoda: Herrerasauridae) from the Ischigualasto Formation (Upper Triassic) of Argentina.—*Journal of Vertebrate Paleontology* 13: 400–423.

Novas, F. E. 1996. Dinosaur monophyly.—*Journal of Vertebrate Paleontology* 16: 723–741.

O'Connor, P. M. 2006. Postcranial pneumaticity: an evaluation of soft-tissue influences on the postcranial skeleton and the reconstruction of pulmonary anatomy in archosaurs.—*Journal of Morphology* 267: 1199–1226.

Ouyang, H. & Ye, Y. 2002. *The First Mamenchisaurian Skeleton with Complete Skull,* Mamenchisaurus youngi. Sichuan Science and Technology Press, Chengdu.

Parrish, J. M. 2005. The origins of high browsing and the effects of phylogeny and scaling on neck length in sauropodomorphs. *In* Carrano, M. T., Gaudin, T. J., Blob, R. W. & Wible, J. R. (eds.). *Amniote Paleobiology: Perspectives on the Evolution of Mammals, Birds, and Reptiles*. University of Chicago Press, Chicago: pp. 201–223.

Paul, G. S. 1997. Dinosaur models: the good, the bad, and using them to estimate the mass of dinosaurs. *In* Wolberg, D. L., Stump, E. & Rosenberg, G. D. (eds.). *Dinofest International*. Academy of Natural Sciences, Philadelphia: pp. 129–154.

Peczkis, J. 1988. Predicting body-weight distribution of mammalian genera in families and orders.—*Journal of Theoretical Biology* 132: 509–510.

Peczkis, J. 1994. Implications of body mass estimates for dinosaurs.—*Journal of Vertebrate Paleontology* 14: 520–533.

Perry, S. F., Breuer, T. & Pajor, N. This volume. Structure and function of the sauropod respiratory system. *In* Klein, N., Remes, K., Gee, C. T. & Sander, P. M. (eds.). *Biology of the Sauropod Dinosaurs: Understanding the Life of Giants*. Indiana University Press, Bloomington: pp. 83–93.

Pol, D. & Powell, J. E. 2007. New information on *Lessemsaurus sauropoides* (Dinosauria: Sauropodomorpha) from the Upper Triassic of Argentina.—*Special Papers in Palaeontology* 77: 223–243.

Preuschoft, H., Hohn, B., Stoinski, S. & Witzel, U. This volume. Why So Huge? Biomechanical reasons for the acquistion of large size in sauropod and theropod dinosaurs. *In* Klein, N., Remes, K., Gee, C. T. & Sander, P. M. (eds.). *Biology of the Sauropod Dinosaurs: Understanding the Life of Giants*. Indiana University Press, Bloomington: pp. 197–218.

Raath, M. A. 1972. Fossil vertebrate studies in Rhodesia: a new dinosaur (Reptilia: Saurischia) from near the Triassic–Jurassic boundary.—*Arnoldia* 5: 1–37.

Rauhut, O. W. M. & López-Arbarello, A. 2008. Archosaur evolution during the Jurassic: a southern perspective.—*Revista de la Asociación Geológica Argentina* 63: 557–585.

Rauhut, O. W. M., Remes, K., Fechner, R., Cladera, G. & Puerta, P. 2005. Discovery of a short-necked sauropod dinosaur from the Late Jurassic of Patagonia.—*Nature* 435: 670–672.

Remes, K. 2006. Evolution of forelimb functional morphology in sauropodomorph dinosaurs.—*Journal of Vertebrate Paleontology* 26: 115A.

Remes, K. 2008. *Evolution of the Pectoral Girdle and Forelimb in Sauropodomorpha (Dinosauria, Saurischia): Osteology, Myology, and Function*. Ph.D. Dissertation. Ludwig-Maximilians-Universität, Munich.

Romer, A. S. 1968. *Notes and Comments on Vertebrate Paleontology*. University of Chicago Press, Chicago.

Romer, A. S. 1971. The Chañares (Argentina) Triassic reptile fauna. X. Two new but incompletely known long-limbed pseudosuchians.—*Breviora* 394: 107.

Romer, A. S. 1972. The Chañares (Argentina) Triassic reptile fauna. XIV. *Lewisuchus admixtus* gen. et sp. nov., a further thecodont from the Chañares beds.—*Breviora* 390: 1–13.

Salgado, L. 1999. The macroevolution of the Diplodocimorpha (Dinosauria; Sauropoda): a developmental model.—*Ameghiniana* 36: 203–216.

Salgado, L., Coria, R. A. & Calvo, J. O. 1997. Evolution of

titanosaurid sauropods. I: phylogenetic analysis based on the postcranial evidence.—*Ameghiniana* 34: 3–32.

Salisbury, S. & Frey, E. 2000. A biomechanical transformation model for the evolution of semi-spheroidal articulations between adjoining vertebral bodies in crocodilians. *In* Grigg, G. C., Seebacher, F. & Franklin, C. E. (eds.). *Crocodilian Biology and Evolution*. Surrey Beatty & Sons, Chipping Norton: pp. 85–134.

Sander, P. M. & Clauss, M. 2008. Sauropod gigantism.—*Science* 322: 200–201.

Sander, P. M., Christian, A. & Gee, C. T., 2009. Sauropods kept their heads down. Response.—*Science* 323: 1671–1672.

Sander, P. M., Christian, A., Clauss, M., Fechner, R., Gee, C. T., Griebeler, E. M., Gunga, H.-C., Hummel, J., Mallison, H., Perry, S., Preuschoft, H. Rauhut, O., Remes, K. Tütken, T., Wings, O. & Witzel, U. 2010a. Biology of the sauropod dinosaurs: the evolution of gigantism. *Biological Reviews of the Cambridge Philosophical Society*. doi: 10.1111/j.1469=185X.2010.00137.x.

Sander, P. M., Gee, C. T., Hummel, J. & Clauss, M. 2010b. Mesozoic plants and dinosaur herbivory. *In* Gee, C. T. (ed.). *Plants in Mesozoic Time: Morphological Innovations, Phylogeny, Ecosystems*. Indiana University Press, Bloomington: 331–359.

Santa Luca, A. P. 1980. The postcranial skeleton of *Heterodontosaurus tucki* (Reptilia, Ornithischia) from the Stormberg of South Africa.—*Annals of the South Africa Museum* 79: 159–211.

Schwarz, D., Frey, E. & Meyer, C. A. 2007a. Novel reconstruction of the orientation of the pectoral girdle in sauropods.—*Anatomical Record* 290: 32–47.

Schwarz, D., Frey, E. & Meyer, C. A. 2007b. Pneumaticity and soft-tissue reconstructions in the neck of diplodocid and dicraeosaurid sauropods.—*Acta Palaeontologica Polonica* 52: 167–188.

Seebacher, F. 2001. A new method to calculate allometric length–mass relationships of dinosaurs.—*Journal of Vertebrate Paleontology* 21: 51–60.

Sereno, P. C. 1998. A rationale for phylogenetic definitions, with application to the higher-level taxonomy of Dinosauria.—*Neues Jahrbuch für Geologie und Paläontologie, Abhandlungen* 210: 41–83.

Sereno, P. C. 1999. The evolution of dinosaurs.—*Science* 284: 2137–2147.

Sereno, P. C. & Arcucci, A. B. 1993. Dinosaurian precursors from the Middle Triassic of Argentina: *Lagerpeton chanarensis*.—*Journal of Vertebrate Paleontology* 13: 385–399.

Sereno, P. C. & Arcucci, A. B. 1994. Dinosaur precursors from the Middle Triassic of Argentina: *Marasuchus lilloensis* gen. nov.—*Journal of Vertebrate Paleontology* 14: 53–73.

Sereno, P. C. & Novas, F. E. 1993. The skull and neck of the basal theropod *Herrerasaurus ischigualastensis*.—*Journal of Vertebrate Paleontology* 13: 451–476.

Sereno, P. C., Forster, C. A., Rogers, R. R. & Monetta, A. M. 1993. Primitive dinosaur skeleton from Argentina and the early evolution of Dinosauria.—*Nature* 361: 64–66.

Sereno, P. C., Beck, A. L., Dutheil, D. B., Larsson, H. C. E., Lyon, G. H., Moussa, B., Sadleir, R. W., Sidor, C. A., Varricchio, D. J., Wilson, G. P. & Wilson, J. A. 1999. Cretaceous sauropods from the Sahara and the uneven rate of skeletal evolution among dinosaurs.—*Science* 286: 1342–1347.

Smith, N. D. & Pol, D. 2007. Anatomy of a basal sauropodomorph dinosaur from the Early Jurassic Hanson Formation of Antarctica.—*Acta Palaeontologica Polonica* 52: 657–674.

Sues, H.-D., Reisz, R., Hinic, S. & Raath, M. A. 2004. On the skull of *Massospondylus carinatus* Owen, 1854 (Dinosauria: Sauropodomorpha) from the Elliot and Clarens formations (Lower Jurassic) of South Africa.—*Annals of the Carnegie Museum* 73: 239–257.

Thulborn, R. A. 1972. The postcranial skeleton of the Triassic ornithischian dinosaur *Fabrosaurus australis*.—*Palaeontology* 15: 29–60.

Upchurch, P. 1994. Manus claw function in sauropod dinosaurs.—*Gaia* 10: 161–171.

Upchurch, P. 1995. The evolutionary history of sauropod dinosaurs.—*Philosophical Transactions of the Royal Society B: Biological Sciences* 349: 365–390.

Upchurch, P. 1998. The phylogenetic relationships of sauropod dinosaurs.—*Zoological Journal of the Linnean Society of London* 124: 43–103.

Upchurch, P. & Barrett, P. M. 2000. The evolution of sauropod feeding mechanisms. *In* Sues, H.-D. (ed.). *Evolution of Herbivory in Terrestrial Vertebrates*. Cambridge University Press, Cambridge: pp. 79–122.

Upchurch, P., Barrett, P. M. & Dodson, P. 2004a. Sauropoda. *In* Weishampel, D. B., Dodson, P. & Osmólska, H. (eds.). *The Dinosauria. 2nd edition*. University of California Press, Berkeley: pp. 259–322.

Upchurch, P., Tomida, Y. & Barrett, P. M. 2004b. A new specimen of *Apatosaurus ajax* (Sauropoda: Diplodocidae) from the Morrison Formation (Upper Jurassic) of Wyoming, USA.—*National Science Museum Monographs* 26: 1–118.

Upchurch, P., Barrett, P. M. & Galton, P. M. 2007a. A phylogenetic analysis of basal sauropodomorph relationships: implications for the origin of sauropod dinosaurs.—*Special Papers in Palaeontology* 77: 57–90.

Upchurch, P., Barrett, P. M., Zhao, X. & Xu, X. 2007b. A re-evaluation of *Chinshakiangosaurus chunghoensis* Ye vide Dong 1992 (Dinosauria, Sauropodomorpha): implications for cranial evolution in basal sauropod dinosaurs.—*Geological Magazine* 144: 247–262.

von Huene, F. 1926. Vollständige Osteologie eines Plateosauriden aus dem Schwäbischen Keuper.—*Abhandlungen zur Geologie und Palaeontologie* 15: 1–43.

von Huene, F. 1932. Die fossile Reptil-Ordnung Saurischia, ihre Entwicklung und Geschichte.—*Monographien zur Geologie und Palaeontologie (Serie 1)* 4: 1–361.

Wedel, M. J. 2003a. The evolution of vertebral pneumaticity in sauropod dinosaurs.—*Journal of Vertebrate Paleontology* 23: 344–357.

Wedel, M. J. 2003b. Vertebral pneumaticity, air sacs, and the physiology of sauropod dinosaurs.—*Paleobiology* 29: 243–255.

Wedel, M. J. 2005. Postcranial pneumaticity in sauropods and its implications for mass estimates. *In* Curry Rogers, K. & Wilson, J. A. *The Sauropods: Evolution and Paleobiology.* University of California Press, Berkeley: pp. 201–228.

Wedel, M. J. 2007. What pneumaticity tells us about "prosauropods," and vice versa.—*Special Papers in Palaeontology* 77: 207–222.

Weishampel, D. B. 2004. Ornithischia. *In* Weishampel, D. B., Dodson, P. & Osmólska, H. (eds.). *The Dinosauria. 2nd edition.* University of California Press, Berkeley: pp. 323–324.

Weishampel, D. B., Dodson, P. & Osmólska, H. (eds.) 2004. *The Dinosauria. 2nd edition.* University of California Press, Berkeley.

Whitlock, J., Wilson, J. A. & Lamanna, M. C. 2008. Evidence for ontogenetic shape change in a juvenile skull of *Diplodocus.*—*Journal of Vertebrate Paleontology* 28: 160A.

Wilson, J. A. 1999. A nomenclature for vertebral laminae in sauropods and other saurischian dinosaurs.—*Journal of Vertebrate Paleontology* 19: 639–653.

Wilson, J. A. 2002. Sauropod dinosaur phylogeny: critique and cladistic analysis.—*Zoological Journal of the Linnean Society of London* 136: 217–276.

Wilson, J. A. 2005. Overview of sauropod phylogeny and evolution. *In* Curry Rogers, K. & Wilson, J. A. (eds.). *The Sauropods: Evolution and Paleobiology.* University of California Press, Berkeley: pp. 15–49.

Wilson, J. A. & Curry Rogers, K. 2005. Monoliths of the Mesozoic. *In* Curry Rogers, K. & Wilson, J. A. (eds.). *The Sauropods: Evolution and Paleobiology.* University of California Press, Berkeley: pp. 1–14.

Wilson, J. A. & Sereno, P. C. 1998. Early evolution and higher-level phylogeny of sauropod dinosaurs.—*Society of Vertebrate Paleontology Memoir* 5: 1–68.

Witzel, U. & Preuschoft, H. 2005. Finite-element model construction for the virtual synthesis of the skull in vertebrates: case study of *Diplodocus.*—*Anatomical Record Part A* 283A: 391–401.

Witzel, U., Mannhardt, J., Goessling, R., de Micheli, P. & Preuschoft, H. This volume. Finite element analyses and virtual syntheses of biological structures and their application to sauropod skulls. *In* Klein, N., Remes, K., Gee, C. T. & Sander, P. M. (eds). *Biology of the Sauropod Dinosaurs: Understanding the Life of Giants.* Indiana University Press, Bloomington: pp. 171–181.

Yadagiri, P. 2001. The osteology of *Kotasaurus yamanpalliensis,* a sauropod dinosaur from the Early Jurassic Kota Formation of India.—*Journal of Vertebrate Paleontology* 21: 242–252.

Yates, A. M. 2003. A new species of the primitive dinosaur *Thecodontosaurus* (Saurischia: Sauropodomorpha) and its implications for the systematics of early dinosaurs.—*Journal of Systematic Palaeontology* 1: 1–42.

Yates, A. M. 2004. *Anchisaurus polyzelus* (Hitchcock): the smallest known sauropod dinosaur and the evolution of gigantism among sauropodomorph dinosaurs.—*Postilla* 230: 1–58.

Yates, A. M. 2007a. The first complete skull of the Triassic dinosaur *Melanorosaurus* Haughton (Sauropodomorpha: Anchisauria).—*Special Papers in Palaeontology* 77: 9–55.

Yates, A. M. 2007b. Solving a dinosaurian puzzle: the identity of *Aliwalia rex* Galton.—*Historical Biology* 19: 92–123.

Yates, A. M. & Kitching, J. W. 2003. The earliest known sauropod dinosaur and the first steps towards sauropod locomotion.—*Proceedings of the Royal Society B: Biological Sciences* 270: 1753–1758.

Yates, A. M. & Vasconcelos, C. C. 2005. Furcula-like clavicles in the prosauropod dinosaur *Massospondylus.*—*Journal of Vertebrate Paleontology* 25: 466–468.

Young, C.-C. 1941. A complete osteology of *Lufengosaurus huenei* (gen. et sp. nov.) from Lufeng, Yunnan, China.—*Palaeontologica Sinica C* 7: 1–53.

Young, C.-C. 1951. The Lufeng saurischian fauna in China.—*Palaeontologica Sinica C* 13: 1–96.

Zhang, Y. 1988. *The Middle Jurassic Dinosaur Fauna from Dashanpu, Zigong, Sichuan: Sauropod Dinosaurs. Vol. 1,* Shunosaurus. Sichuan Publishing House of Science and Technology, Chengdu.

9

Characterization of Sauropod Bone Structure

MAÏTENA DUMONT, ANDRAS BORBÉLY, ALEKSANDER KOSTKA,
P. MARTIN SANDER, AND ANKE KAYSSER-PYZALLA

THIS CHAPTER DESCRIBES THE APPLICATIONS of some well established methods of material science in the examination of sauropod bone microstructure. Fossilized bone is characterized here at different levels of hierarchy, from the macro level (at which bone can be separated into cortical and cancellous bone) to the nano level (at which the bone is composed of an assemblage of collagen and mineral particles), and then compared to bone of extant animals. X-ray diffraction and fluorescence analysis in combination with electron microscopy permit the quantification of the influence of diagenetic processes on fossilized bone. The chapter emphasizes that there are a multitude of investigative techniques well suited for bone analysis at the different structural levels. For an in-depth understanding of dinosaur bone structure and its global preservation state, however, a combination of the methods is necessary.

Introduction

THE MASS OF SAUROPODS

Sauropod dinosaurs were the heaviest animals that ever inhabited the land, culminating in gigantic forms in at least three lineages: the Diplodocidae, Brachiosauridae, and Titanosauridae (Upchurch 1998; Wilson 2002). They are unique in exceeding the body mass of all other terrestrial tetrapods (i.e., mammals and other dinosaurs) by one order of magnitude. Body mass can be determined, for example, by photogrammetrical measurements (Gunga et al. 1999, 2008) or on the basis of graphical reconstructions of body shape (Seebacher 2001; Appendix). These estimates consistently place common sauropods in certain weight categories, such as 30–50 metric tons for *Brachiosaurus* and *Apatosaurus,* and 20 metric tons for *Barosaurus* (Appendix). The sheer size of these huge animals led to scale effects in their biology and physiology that are still poorly understood. The relative gracility of the sauropod bones indicates that midshaft circumference is an incomplete descriptor of bone strength (Anderson et al. 1985). It also suggests that sauropod bone strength may be attributed to other mechanisms of accommodation of the bone structure to the high loads caused by their large mass. The adaptation of bone to load was suggested as early as 1892 by Wolff (Wolff's law), and today it is generally accepted that mechanical loading plays an important role in the development and maintenance of the skeleton (Currey 1968, 1984; Tanck et al. 2001). However, if sauropod bone tissue shows no essential differences but instead a similarity with the hierarchical structure of mammals, this could be interpreted as the avoidance of high stress levels in bones, sauropod activities being less strenuous than those of other animals. This issue piqued our interest: we wondered whether there is an essential difference in the bone microstructure between sauropods and large herbivorous mammals such as cows, elephants, and rhinos.

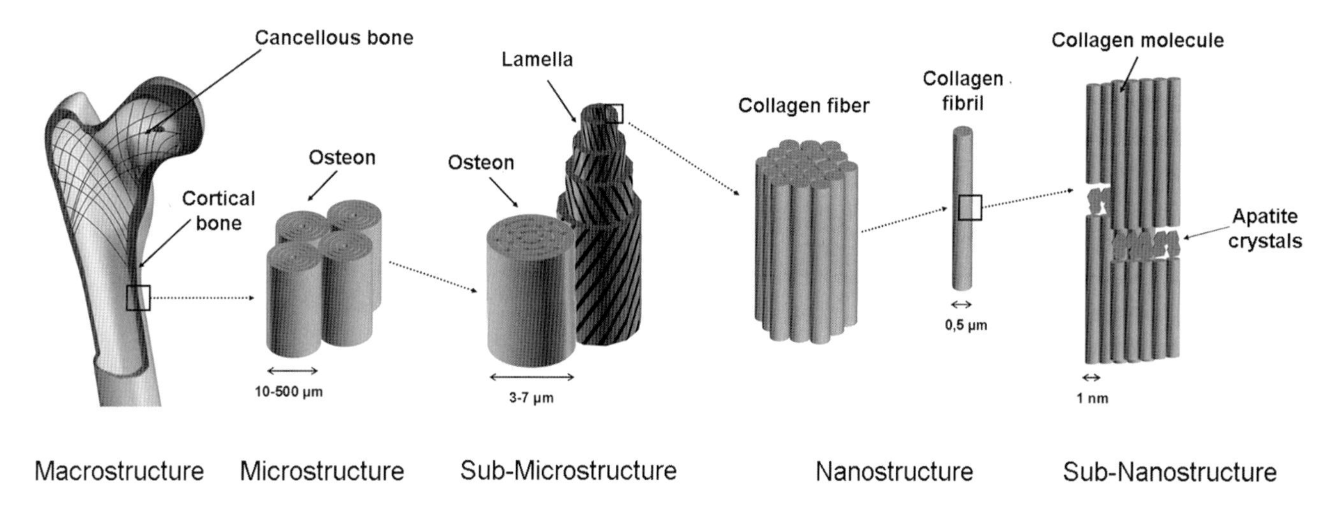

FIGURE 9.1. The hierarchical levels of bone structure. The six levels represent the bone structure from the nano scale (molecular level) to the macrostructure (macroscopic level). *Modified after Rho et al. (1998).*

BONE HIERARCHY

It is generally difficult to predict the mechanical behavior of bone because the bone material itself is a complex hierarchical structure (see Rho et al. 1998; Weiner & Wagner 1998; Fratzl & Weinkamer 2007). Figure 9.1 shows an arrangement of bone structures (of mammal bone) at several different hierarchical levels. The basic constituents of bone (level 1) are collagen and mineral particles (crystals of carbonated hydroxyapatite). The mineralized collagen fibrils (level 2) are assembled into fibers (level 3). The fibers arrays can also be arranged in different patterns of organization (level 4): parallel, woven (packed fibrils without any preferred spatial arrangement), and lamellar (sets of parallel fibers forming layers, each layer having a different orientation of fibers) (for more details, see Francillon-Vieillot et al. 1990; Pritchard 1956; Weiner et al. 1997). At the next hierarchical level, lamellae can be in some cases (i.e., mammalian and sauropod bone) arranged in concentric layers around a central canal to form a primary or secondary osteon (level 5). The osteon resembles a cylinder, in the case of secondary osteons often running parallel to the long axis of the cortical bone. The macroscopic level considers the bone as a whole entity, distinguishing only between typically solid cortical bone and cancellous bone (level 6), which are composed of the same histological elements but are organized differently. The cortical bone is relatively dense, while cancellous bone has a spongy structure composed of a network of bone trabeculae.

This hierarchically organized structure of bone displays a specific arrangement, with the orientation of the components making it a heterogeneous and anisotropic material. This means that the mechanical properties of bone, such as strength, toughness, and stiffness, depend on the relative position of the applied force with regard to the direction of the bone axis (Rho et al. 1998). In order to understand the mechanical properties of long bones of giant dinosaurs, it is important to characterize the structure of each phase at every hierarchical level (Weiner & Wagner 1998). To attain this goal, modern investigation techniques used in materials science are currently being applied. The use of different microscopic methods, such as scanning and transmission electron microscopy, as well as diffraction techniques, makes possible the understanding of bone structure and crystallographic texture in these huge animals. Nevertheless, the study of fossil bone is not straightforward, and attention must be paid to changes that bone has undergone during the fossilization process. Sauropod bones have sustained burial for more than 65 million years, with those in our study being 150 million years old, and thus have experienced significant diagenesis. These diagenetic changes often do not affect bone preservation at the histological level, but they lead to marked alterations in bone structure below the histological level.

In principle, investigations of bone structure (in both fossilized and recent bones) must include all three aspects of microstructural characterization. First, the morphology of the phases present in the bone (their size, shape, and spatial distribution) is characterized. Second, the crystallographic affiliation of the bone is identified, and finally its local chemical composition is determined.

ABBREVIATIONS FOR COLLECTIONS

BYU = Museum of Paleontology of Brigham Young University, Provo, Utah, USA; IPB = Paleontology collection, Steinmann Institute, University of Bonn, Germany; MFN = Museum für Naturkunde, Berlin, Germany; OMNH = Sam

Noble Oklahoma Museum of Natural History, Norman, Oklahoma, USA,

Microstructural Characterization

BASICS OF ELECTRON MICROSCOPY

The following section summarizes the electron microscopy and diffraction techniques used in the study of different bone levels, so that we might develop a concise idea of the functional adaptation of bone structure. The microstructure of fossilized bone is readily observed by traditional light microscopy techniques. In fact, important paleontological information concerning the growth, physiology, and evolution of dinosaurs has been derived from observations of the microstructure of bones (Enlow & Brown 1956; de Ricqlès 1976, 1980; Reid 1984; Sander et al., this volume). After it became clear that image resolution is limited not only by the construction of the microscope lenses but also by the wavelength of light, microscopy that uses electrons instead of light was applied to achieve higher magnifications. Modern electron microscopes, as a result of dramatic improvements in resolution and a wide range of detectors, allow for the ultimate microstructural characterization at the atomic level.

The basic components of the light microscope and the electron microscope are the same: there is always an illumination system containing condenser lenses, the objective lenses (the most important component determining the quality of the image and its resolution), and the imaging system, which enables image observation and recording. The main difference, apart from the use of electrons, is that electron microscopes use electromagnetic lenses and require a high vacuum, which allows the electron beam to travel through the instrument.

The capacity of electron microscopy to distinguish smaller details than light microscopy resides in the small wavelength of the electron beam (0.004 nm for electrons accelerated by 100 kV compared to 550 nm for green light, for example). The resolution of newest scanning electron microscopes (SEM) is about 1–5 nm, while transmission electron microscopes (TEM) can resolve 0.1 nm. This is 3–4 orders of magnitude greater than the light microscope, which cannot resolve details smaller than 1 μm. In TEM, the image is formed by the electron beam, which is transmitted through a thin sample. In SEM, the electron beam is scanned across the sample, and the detectors build up an image by mapping the detected signals with respect to the beam position. SEM imaging relies on surface processes rather than transmission; it is able to image large samples and has a much wider field of view than TEM.

Proper sample preparation is an essential process of the electron microscopy examination and is often the most difficult stage. Usually, sample preparation consists of several different steps, depending on the object of interest (polished or fractured surface, cross section, etc.) and the type of microscopy (SEM or TEM) (Goldstein et al. 2003; Egerton 2005). One important point to bear in mind is that the technique used for sample preparation must not affect the microstructure that is being investigated. Additionally, in case of nonconducting biological materials (such as bones of sauropods or of recent animals), the specimen must be coated with a thin layer of gold or carbon in order to prevent electrical charging during illumination by the electron beam.

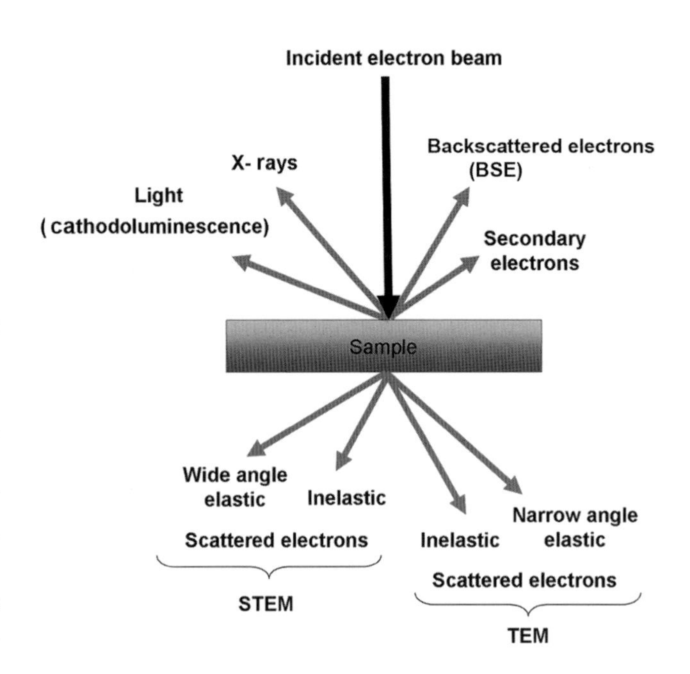

FIGURE 9.2. Signals generated during interaction of an electron beam with a sample. Most of these signals can be detected in SEM and TEM. The directions shown for each signal do not always represent the physical direction of the signal; they only indicate in a relative manner where the signal is the strongest or where it is measured.

Secondary Signals in Electron Microscopy

A major advantage of working with electrons is the wide range of secondary signals that arise from the interaction of the sample with the electron beam (Fig. 9.2). Many of these signals are used in analytical electron microscopy, providing chemical information and many other details about the material being studied. Signals generated during the interaction between the probe (the electron beam) and the specimen can be classified in two categories because they are the result of elastic or inelastic scattering of the electrons. In the first case, the energy of the incident electrons is conserved. In contrast, during the inelastic scattering process, the energy of the incident electrons is lost. Elastic scattering leads to diffraction contrast in TEM, and it is directly related to the nature of

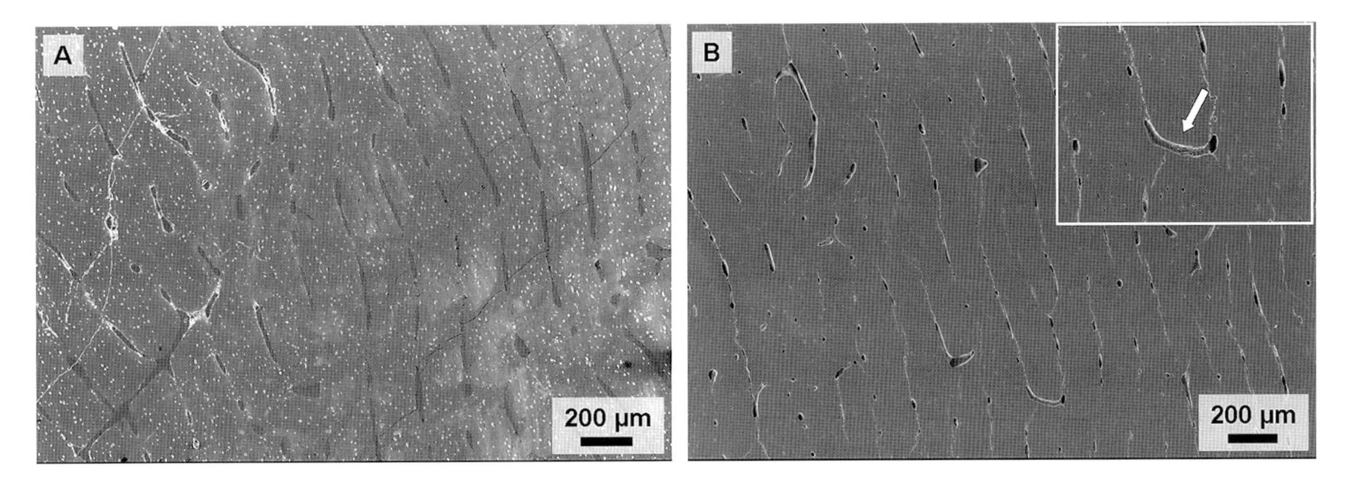

FIGURE 9.3. SEM micrographs using secondary electrons of FLB in cross section of (A) an *Apatosaurus* sample (BYU 725-17014) from the Morrison Formation, USA (femur, 97.0 cm; Klein & Sander 2008), and (B) a juvenile cow with a closer view of a radial vascular canal (white arrow). Note the laminar arrangement of the vascular canals resulting from their circumferential arrangement, which is parallel to the periosteal surface.

crystal lattice defects present in the sample (grain boundaries and other microstructural features). Secondary electrons are produced when the incident electrons knock out loosely bound conduction electrons. Because of their low energy, these secondary electrons (<50 eV) can only escape from a thin subsurface layer with a thickness of about 10 nm. The intensity of the detected signal depends on the angle between the beam and the specimen. These two factors mean that the secondary electron signal provides the highest topographic resolution. Secondary electrons are usually used in SEM to visualize the specimen. The backscattered electron (BSE) signal is produced in SEM by elastically scattered electrons, deflected through angles of 0° to 180° by atoms within the specimen. Those scattered by angles greater than 90° can reemerge from the specimen surface and still retain a high level of energy. Under similar operating conditions, the BSE signal will be produced from a larger volume of the sample than the secondary electron signal and has a lower topographical resolution. The BSE signal can also be used to give qualitative compositional information in heterogeneous samples. Phases consisting of heavy elements (high atomic number) will backscatter electrons more strongly than those with light elements (low atomic number), thus emitting a more intense signal and appearing brighter.

Electrons that lose energy during the interaction with the sample may eject lower-energy secondary electrons and X-rays. The X-rays emitted will have a characteristic energy that is unique for the element from which they originate, and their intensity is proportional to the concentration of the element. This analytical technique, called energy dispersive X-ray spectroscopy (EDS or EDX), is commonly used for determining qualitative element composition analysis in a specific material. It is a technique complementary to observations made with BSE. Whereas BSE imaging displays compositional contrast (resulting from a difference in atomic numbers), EDS allows for the identification of particular elements and their relative proportion in the sample.

LARGE-SCALE APPROACHES POSSIBLE WITH SEM

In the following examples of characterizing bone structure, investigations are presented from the micro scale (hierarchical levels 5 and 4) to the nano scale (levels 3 to 1). The microstructure of large mammal and dinosaur bones has been extensively studied in the past. At the level of the bone tissue, the lamellar structure of the bones can be well resolved by means of light and scanning microscopy techniques. Different sample preparation methods, such as mechanical polishing or chemical etching of surfaces, can provide specific information on the bone structure (Boyde et al. 1972, Pawlicki 1976). The most common sample preparation technique applied in the current study was mechanical grinding followed by polishing. This type of preparation turns out to be quite suitable for the comparison of microstructure in dinosaur and cow bones.

Bone Tissue Structures in SEM Micrographs

Figure 9.3 shows the structure of cortical bone in sauropod and cow bones tissues, respectively. Both micrographs reveal a typical fibrolamellar bone (FLB) organization (Enlow & Brown 1956, 1957; Francillon-Vieillot et al. 1990). The FLB is a 3D network that consists of two main components: woven bone tissue and primary osteons (de Ricqlès 1980; Klein & Sander 2008; Sander et al., this volume). This bone tissue is formed by the rapid deposition of a framework of woven bone matrix

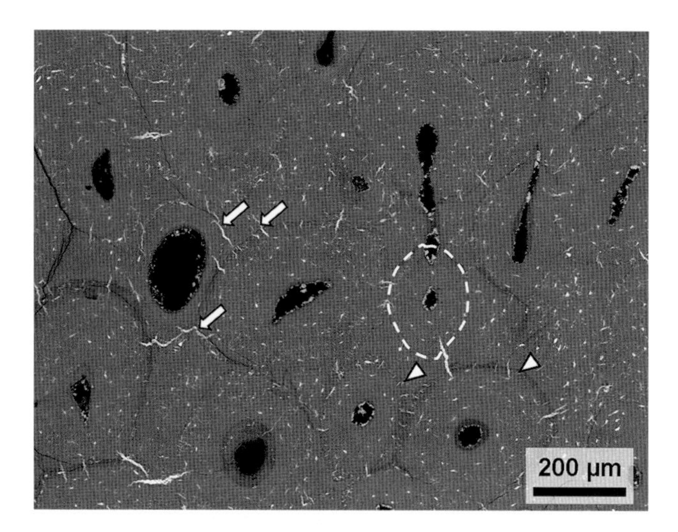

FIGURE 9.4. SEM micrograph using BSE of secondary osteons in a cross section of an *Apatosaurus* sp. sample (BYU 601-17328) from the Morrison Formation, USA (femur, 158.0 cm; Klein & Sander 2008). Secondary osteons (one of which is outlined by the dashed line) are delimited by cementing or resorption lines (arrowheads). Characteristic radial cracks (arrows) indicate that this bone was subjected to strain and/or fossilization.

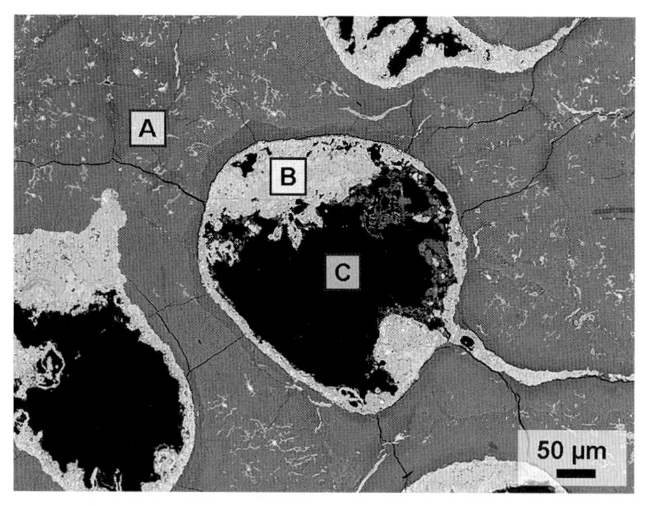

FIGURE 9.5. SEM micrograph using BSE of trabecular bone in a femur cross section of *Apatosaurus* sp. (BYU 601-17328) from the Morrison Formation, USA (femur, 158.0 cm; Klein & Sander 2008). Medium gray is bone tissue; bright and dark areas are diagenetic infill of the vascular spaces. (A–C) Area analysis by energy dispersive X-ray (see Table 9.1).

around blood vessels on the bone surface. These blood vessels then become incorporated into the framework of woven bone, and the resulting vascular canals are gradually filled in by lamellar bone matrix, forming primary osteons (Currey 1960). FLB is generally densely vascularized with various arrangements of vascular canals.

Both the sauropod bone and the cow bone in Fig. 9.3 are dominated by circumferential vascular canals, resulting in a laminar appearance of the bone tissue. However, in the cow bone (Fig. 9.3B), these circumferential canals are additionally connected by radial canals, forming a vascularization pattern called plexiform (see Francillon-Vieillot et al. 1990 for more details). This microstructure has already been noted in other large herbivorous mammals (e.g., Sander & Andrassy 2006) and was previously mentioned by Enlow & Brown (1956, 1958).

This bone tissue is interpreted as resulting from the rapid but normal deposition of new compact bone. Indeed, as a result of the high mass of large mammals and sauropods, their bones must increase in circumference quickly. As in large mammals (de Ricqlès et al. 1991), Haversian bone is also present in long bones of sauropods (Fig. 9.4; Sander 2000; Sander et al., this volume). Haversian bone results from the primary FLB being replaced during remodeling by secondary osteons. These form by erosion of the primary bone tissue and secondary infill by lamellar bone matrix. In long bones, secondary osteons are generally oriented longitudinally.

BSE Micrograph of Secondary Osteons

Figure 9.4 shows a micrograph of dense secondary osteons in sauropod bone. The BSE image reveals five different components in this bone tissue: vascular canals, primary and secondary bone, cementing lines, and osteocyte lacunae (bone cell spaces). Because the brightness of BSE images increases with increasing atomic number (Z), areas with different chemical composition can be recognized. The dark area corresponds to the open vascular canal of the secondary osteon (Fig. 9.4). The dark color is due to higher local amounts of carbon and oxygen (Bell 1990), which show up as the lighter elements. A detailed chemical composition analysis is presented later in this chapter (Fig. 9.5 and corresponding description). The gray shades correspond to primary and secondary bone, which are separated from each other by cementing lines (marked by arrowheads). These cementing lines surrounding secondary osteons (highlighted by the dashed line in Fig. 9.4) are formed by successive resorption and deposition during remodeling of the bone. They represent to regions of mechanical weakness (Francillon-Vieillot et al. 1990). When a bone is subjected to strain and/or fossilization, characteristic radial cracks appear in these regions (arrows in Fig. 9.4).

EDS for Chemical Analysis

An example of chemical composition analysis that uses the EDS technique is presented in Fig. 9.5 and Table 9.1. The gray area (marked with the letter A) corresponds to the bone tissue.

Table 9.1. Analysis of Three Different Areas in Trabecular Bone of *Apatosaurus* sp. (BYU 601-17328) from the Morrison Formation, USA[a]

	Element (wt%)						
Area[b]	Ca	P	F	C	O	Fe	Mn
A	38.4	17.92	2.57	14.26	25.94		
B				5.85	23.34	69.75	1.07
C				84.45	15.55		

[a]Femur, 158.0 cm; Klein & Sander (2008).
[b]A, B, and C are the areas marked in Fig. 9.5. A corresponds to the bone tissue, B to the first phase of diagenetic infill of the vascular space, and C to the final diagenetic infill.

EDS spectra from this region (Table 9.1) show a high concentration of calcium, phosphorus, and oxygen, indicating a typical apatite composition.

In the brightest areas (marked by the letter B, at the margins of the vascular spaces, in the cracks and in the osteocyte lacunae), an element with a higher atomic number is usually present. It is a component of minerals such as pyrite containing iron and manganese, which is in accordance with data presented in the literature (Pfretzschner 2000, 2001). This indicates the formation of iron and manganese oxides sulfides along the wall of the vascular space as the first diagenetic mineral deposit (Hubert et al. 1996; Farlow & Argast 2006; Turner-Walker & Jans 2008). The dark area corresponding to the later diagenetic infill of the vascular space (area indicated by the letter C) shows a lower BSE intensity, indicating the presence of lighter elements. In our sample, these are carbon (C) and oxygen (O) (Table 9.1), which may also derive from the resin filling up open pores during sample preparation (Turner-Walker & Jans 2008).

An additional possibility offered by SEM-EDS analysis is qualitative element mapping. The advantage of this technique is the range of magnifications and the information associated with the micro- and submicrostructure of the bone.

Plate 9.1 illustrates the iron oxides deposited on the margins of the vascular spaces in the trabecular tissue of the same *Apatosaurus* sample. The actual bone is depicted by areas high in calcium (Ca) and phosphorus (P). The embedding resin is composed of carbon, as previously demonstrated (Table 9.1). The intertrabecular spaces that were filled with soft tissue during the life of the animal are now filled with many elements incorporated during diagenesis. Indeed, these spaces are generally filled with trace elements. Chemical composition analy-

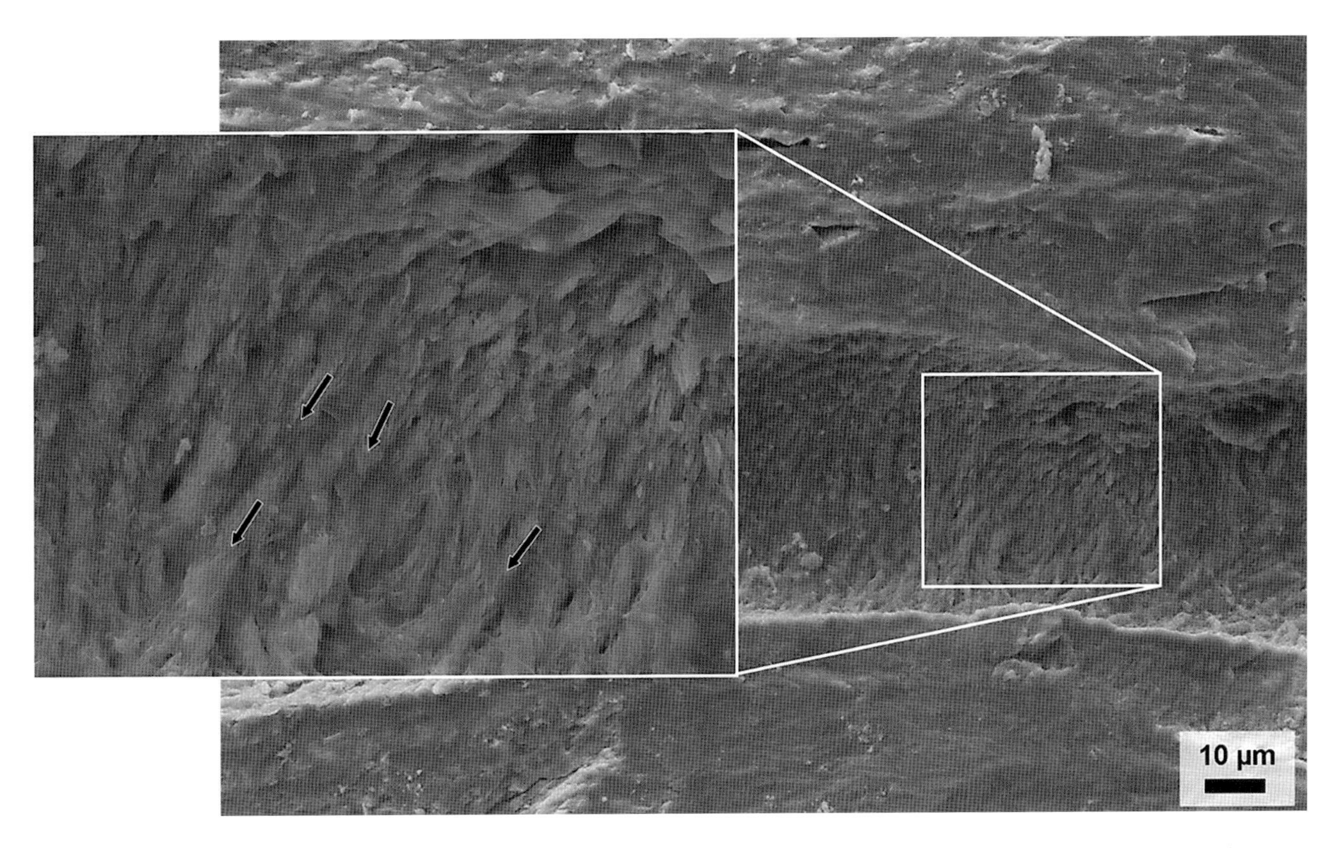

FIGURE 9.6. SEM micrograph of longitudinal fracture surface exposing a secondary osteon canal and showing bundle patterns of bone collagen matrix of the same orientation lining the canal (black arrows) of a tibia of *Astrapotherium* sp. (IPB M4468) from the Tertiary of Chubut, Argentina.

sis (Fig. 9.5; Plate 9.1; Table 9.1) shows that different elements were deposited: iron (Fe), manganese (Mn), and sulfur (S). Trace elements were found only in the cavities of the trabecular bone, suggesting that the original bone structure of the dinosaur is intact.

Fractography

Understanding bone structure in terms of mechanical adaptations requires understanding the submicrostructure and nanostructure of its components. Fractography is a method commonly used in materials science to examine the causes of failure (Wise et al. 2007). In the case of fractured samples, SEM shows its superiority over light microscopy in being able to distinguish between morphological features. Because of the large depth of focus of SEM, the fractured surface has a 3D appearance. It reveals the micro- and submicrostructure of the secondary osteons, the structure of the inner wall of the vascular canals, and of the osteocyte lacunae. Figure 9.6 shows the collagen fiber bundles in the osteon of an *Astrapotherium* (a South American fossil ungulate mammal). Here, fiber bundles show some degree of preferred arrangement, while narrow canals (completed osteons) observed in the same specimen, as well as in both the cow and dinosaur bone, show branching and a more random arrangement. The fractured bone surface can emphasize the manner in which collagen fibers are organized (inset of Fig. 9.6).

Figure 9.7 displays a different arrangement of fibers that are aligned in sheets oriented roughly along the bone long axis. The white arrowheads indicate a smooth, compact surface, while another distinct area around the canal (Fig. 9.7, black arrows) presents a rougher surface showing different orientations of the collagen fibers. A detailed description of diversity in fiber array organization (hierarchy level 4) was presented by Pritchard (1956) and Francillon-Vieillot et al. (1990). This observed pattern is common in both large mammals and sauropods. Their bones are indeed composed mostly of FLB with its two bone matrix types, woven and lamellar. Little information is known about the mechanical properties of woven bone matrix. It seems to be less stiff than normal bone matrix (Christel et al. 1981), with the flexibility to bend. Because of its highly mineralized nature, it is, however, more brittle, and when compared to lamellar bone matrix, it certainly has inferior mechanical properties.

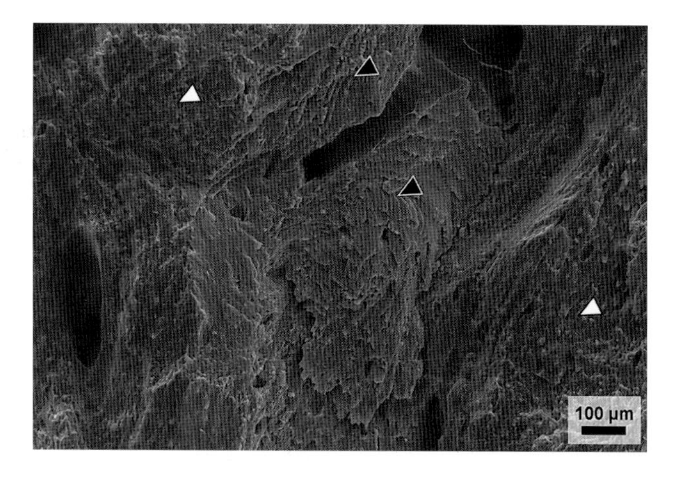

FIGURE 9.7. SEM micrograph of transversely fractured cow bone (same plane as the cross sections in Figs. 9.3–9.5). The fracture reveals a layered arrangement of different sheets of aligned fibrils: lamellar bone constituting a secondary osteon (black arrowheads) and primary bone (white arrowheads).

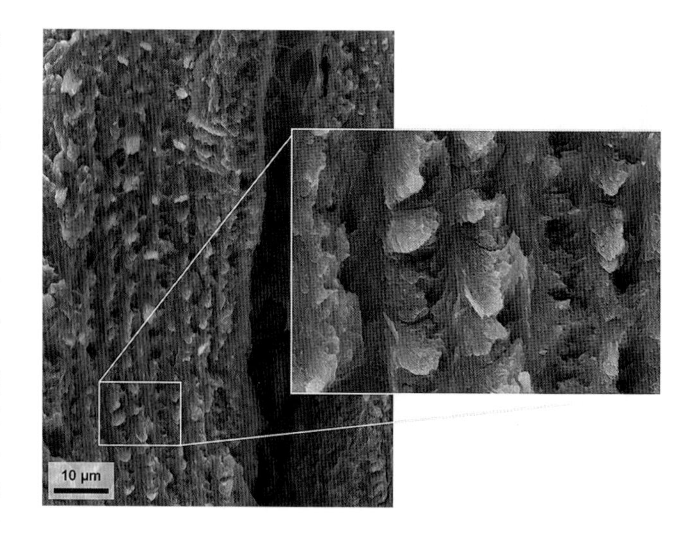

FIGURE 9.8. Transverse fracture surface (same plane as the cross sections in Figs. 9.3–9.5) of cow bone around a secondary osteon showing a succession of lamellae. The enlarged view of the lamellar bone shows an alternating structure of thinner and thicker lamellae as described by Ziv et al. (1996b).

The lamellar structure forming the secondary osteons or Haversian canals is more complex. As observed in Fig. 9.8, the lamellar bone of the cow is composed of alternating thinner and thicker lamellae. This was previously observed and described in human bone (Reid 1986, Ascenzi & Benvenuti 1986), in the calvaria (the roof of the skull) of a rat (Ziv et al. 1996a), and in a bovine femur (Ziv et al. 1996b). The thicker lamellae are characterized by a rough surface and the thinner ones by a smoother surface (Ziv et al. 1996a, 1996b). The packing of fibers in the thick lamella shows a well developed layered structure. These layers appear to change in their orientation from one lamella to the next, consistent with the "rotated plywood" model of Weiner et al. (1991), who have described this structure in great detail. It appears that this lamellar unit is composed of five sublayers decussating at an angle of roughly 30° (Weiner & Wagner 1998; Weiner et al. 1999). This specific decussation produces a structure with a tendency for isotropic properties (Ziv et al. 1996a). To understand the orga-

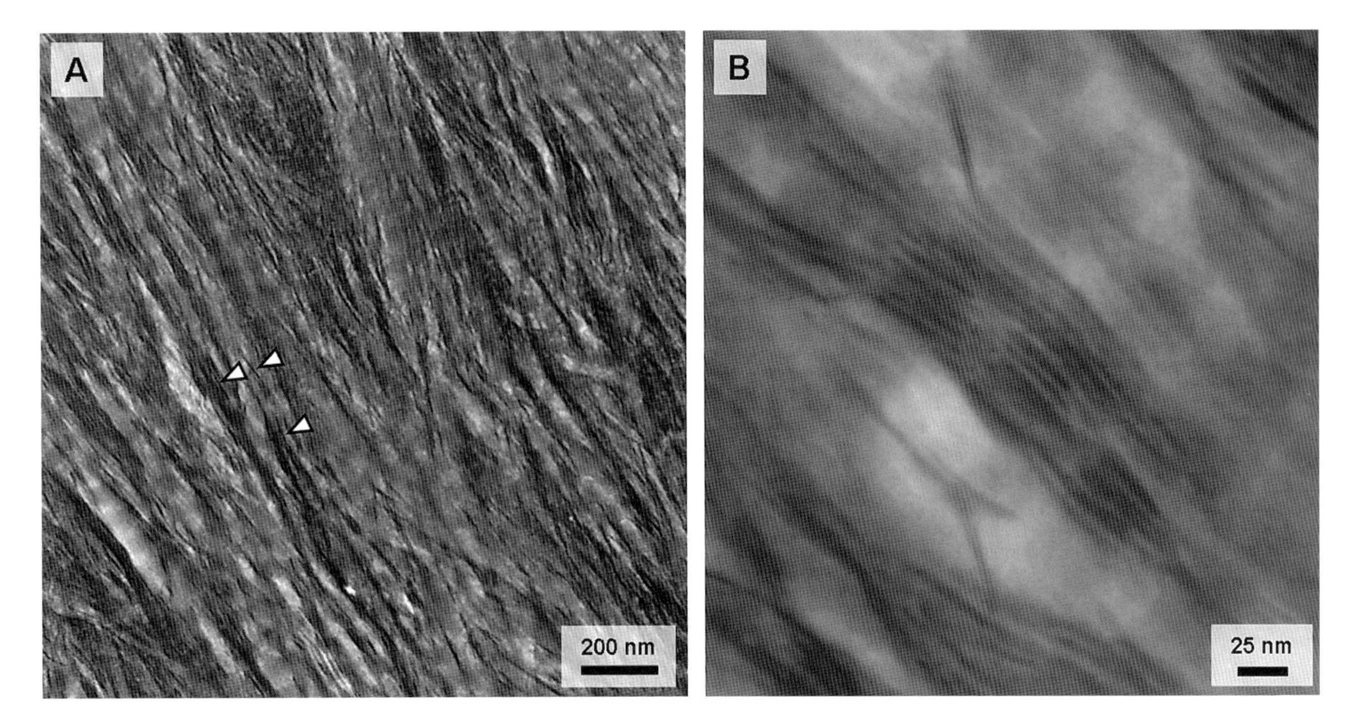

FIGURE 9.9. STEM of cow cortical bone. (A) Bright-field image showing mineralized collagen fibrils. The striated appearance results from the hydroxyapatite platelets. (B) Enlarged image of isolated collagen fibrils (bundles of type I collagen).

nization of bone structure at lower levels, the size and the organization of crystals in the nanometer and subnanometer size range can be studied with TEM.

TEM CHARACTERIZATION OF SAUROPOD BONES AT THE NANO SCALE

In TEM, the image is formed by an electron beam that is transmitted through a thin, transparent sample that is less than 100 nm thick. The most important characteristic of TEM is that it combines information from the real space of an object —at excellent resolution—with information from the reciprocal space of the object (electron diffraction pattern). TEM is a local area analysis technique and is an essential tool for characterizing fine bone structure—for example, the morphology of bone apatite crystals and their local preferred orientations as detected by a specific area electron diffraction pattern.

In our study, all TEM specimens were extracted and further thinned to the desired thickness by means of a focused ion beam system (Sugiyama & Sisegato 2004). TEM has rarely been used in studies of fossil bone structure (e.g., Zocco & Schwartz 1994; Hubert et al. 1996), despite its routine use in biology and materials science. Characterization of the extremely small apatite crystals (nanometer range) requires an instrument with ultimate resolution. The study of fossil bones by TEM permits for the characterization of the size, shape, and arrangement of bone apatite particles, as well as their relationship to one another. Moreover, phase identification also becomes possible through electron diffraction images from specific regions. The results, however, should be interpreted carefully because the chemical composition of the bone may have been modified by diagenesis, which influences the interatomic distances.

Bone Mineral Crystals in TEM

Living bone at hierarchical level 1 is composed of three major components: crystals of carbonated hydroxyapatite, water, and collagen fibrils. The manner in which the crystals and the collagen fibrils are organized is important for understanding the nearly axis-symmetrical properties of the structure. Moreover, it is this nanostructural level that is the foundation of all types of bone tissue (woven, lamellar, parallel-fibered). Therefore, in order to truly understand bone properties, it is necessary to examine its most basic level of organization: the nanostructured array of the bone mineral particles in the collagen matrix (level 1 of Fig. 9.1). Figure 9.9 shows two TEM micrographs of the mineralized collagen fibrils in the cortical bone of a cow. The fibrils are oriented in mostly one direction. The apatite crystals are arranged in parallel layers (banding patterns) across the fibrils (Traub et al. 1989; Weiner & Traub 1989; Weiner et al. 1991). Some areas (marked by white arrows) display small groups of crystals that are shaped like nee-

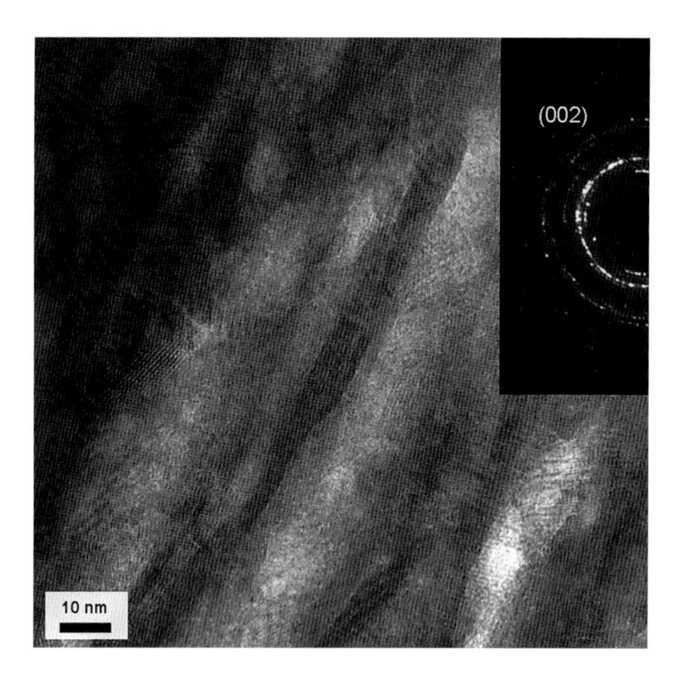

FIGURE 9.10. High-resolution TEM micrograph of mineralized collagen fibrils of cow bone and an electron diffraction pattern from a 10-μm-diameter area. The diffraction pattern shows contributions from both phases constituting the fibrils.

dles. According to Traub et al. (1989), these are caused by sectioning artifacts. Normally, these crystals are plate shaped. The magnified view of the mineralized collagen in Fig. 9.9B shows that the apatite crystals are aligned with both the crystal layers and the fibril axes (Weiner & Wagner 1998). The crystallographic c-axes of the crystals in the bone sample in Fig. 9.9 are preferentially oriented parallel to the collagen fibers which form bundles of type A collagen.

Figure 9.10 shows a TEM image at higher resolution, with the hydroxyapatite crystals that make up the mineralized fibril. This figure also shows an electron diffraction pattern of the bone, which was made with a specific area aperture of 10 μm larger than the size of the TEM image. The intensity distribution of the reflections pertaining to the apatite crystals (inner rings) appears discontinuous (circular arcs), indicating their preferential orientation (crystallographic texture). The direction of the longest edge of the particles lies mainly along the 001 crystallographic direction (the spots of the innermost arcs correspond to the 002 diffraction vectors).

Hydroxyapatite crystals of the sauropod *Apatosaurus* sp. (inset of Fig. 9.11) appear to have platelet shapes (Weiner 1986a, 1986b; Ziv & Weiner 1994; Reiche et al. 2002; Trueman et al. 2004); however, Wess et al. (2001) pointed out that as a result of diagenesis, samples of fossil bones may contain needle-shaped crystallites, too. On the basis of our TEM investigations, it was not possible to discern the true shape of the crystallites. The length of sauropod apatite crystals is in the range of 30–170 nm, averaging 80 nm. The width of the crystals ranges from 10–45 nm, averaging about 30 nm. A high-resolution TEM image of a single apatite crystal in the sauropod bone is shown in Fig. 9.12. The size of apatite crystals in sauropod bone is larger than in recent mammal bone, which average about 50 × 25 nm (Robinson 1952; Weiner 1986a, 1986b). These larger sizes were interpreted in the literature (Hubert et al. 1996; Trueman et al. 2004) as resulting from diagenetic growth. During fossilization, the addition of authigenic apatite to the original bone apatite crystals permits the growth and the infilling of intercrystalline porosity while preserving the original crystal alignment (Kolodny et al. 1996; Hubert et al. 1996).

Under higher magnification, the crystals appear to locally show preferential orientation, as has been noted previously in the large sauropod *Diplodocus hallorum* (formerly known as *Seismosaurus*), a mammoth, a recent crocodile (Zocco & Schwartz 1994), and in recent mammals such as a cow (Fig. 9.10). However, a look at the crystals at lower magnification indicates a large spread of orientations. The sauropod sample does not display any remaining traces of collagen, which is not always the case in fossil samples (Pawlicki et al. 1966; Doberenz & Wickoff 1967). However, as noted above, and according to Hubert et al. (1996) and Trueman & Tuross (2002), the orientation of the crystals with regard to bone axis is still preserved in fossil bone. We believe that the spread of crystallite orientations observed in *Apatosaurus* sp. is related to the area examined by TEM, namely, near the vascular canal filled with diagenetic material. Local crystallite orientation is probably influenced by nearby areas that have undergone large amounts of diagenesis. For this reason, investigation methods that cover larger areas, such as crystallographic texture or small-angle scattering measurements, are better suited for the analysis of the global preservation of the orientation of crystallites.

Crystal Orientation and Diagenesis

The diffraction pattern in Fig. 9.11 displays a partial ring diffracted by the crystal, which corresponds to the 002 planes and shows that the 001 crystallographic direction is aligned nearly parallel to the longest edge of the crystallites. In this case, it is probable that diagenesis did not affect the relationship between the particle's shape and its crystallography in this sauropod sample. Thus, the c-axis of the hexagonal crystal remains parallel to its longest edge. To confirm this situation, however, more and statistically relevant results must be obtained. It was shown by Zocco & Schwartz (1994) and Reiche et al. (2002) that TEM can be an excellent tool for microstructure characterization of lower hierarchical levels of

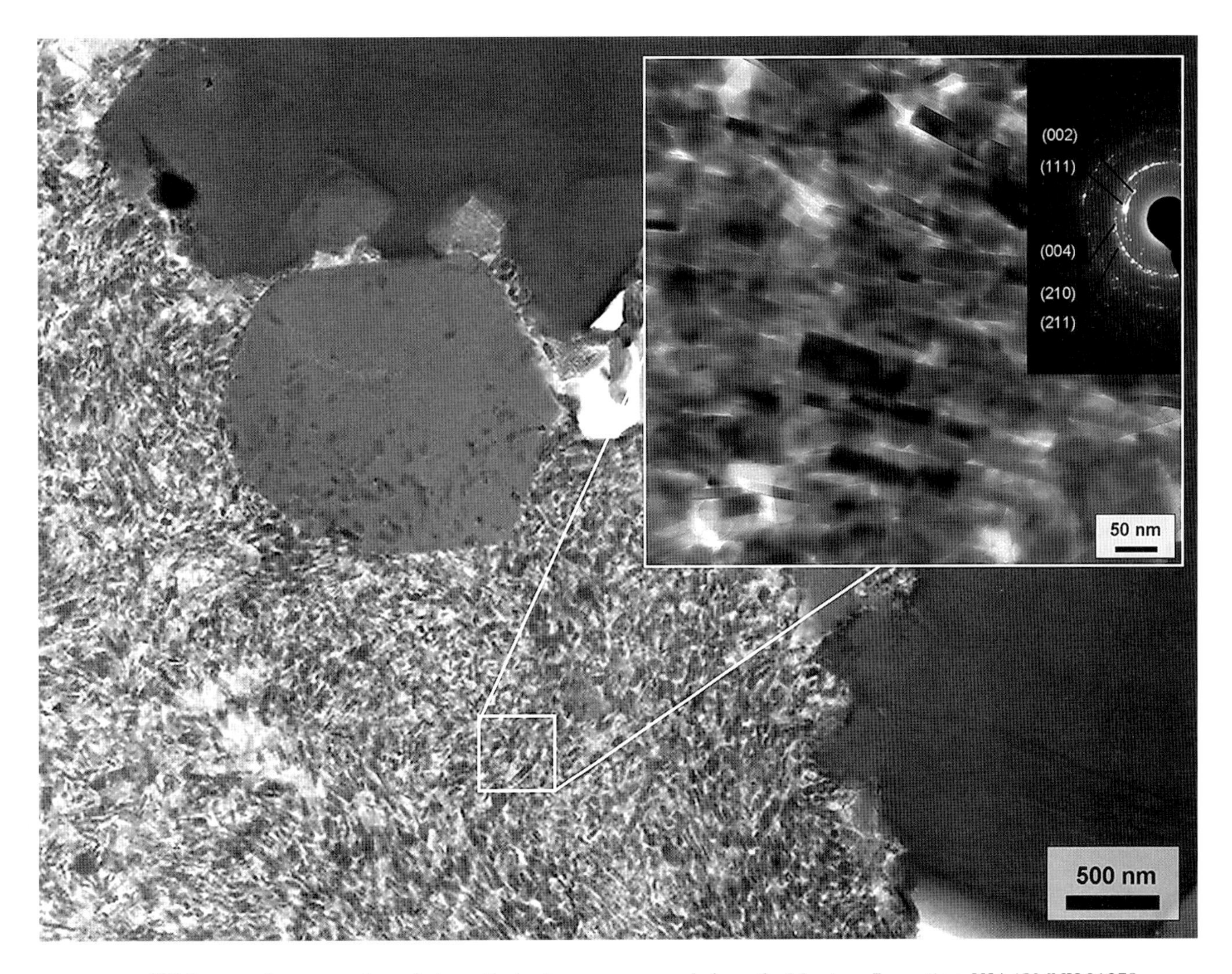

FIGURE 9.11. TEM image of a cross section of a juvenile *Apatosaurus* sp. sample from the Morrison Formation, USA (OMNH 01278, humerus, 25.8 cm; histologic ontogenetic stage 4 of Klein & Sander 2008) showing the submicrometer structure around a vascular canal. Diagenetic minerals filling the vascular canal and the primary bone are visible in the upper right corner. The inset shows the nano-size hydroxyapatite crystals of the bone and the associated diffraction pattern.

dinosaur bone, which are poorly documented in the scientific literature. More investigations are needed to better understand fibril structure and diagenetic effects.

Characterization of the Texture of Sauropod Bones

CRYSTALLOGRAPHIC TEXTURE

As extant bone, that of sauropod dinosaurs may be considered a composite material that, at the nanometer scale (level 1 in Fig. 9.1), is made up of at least two different constituent phases with significantly different physical and chemical properties (Fig. 9.13). As shown above, in life it consisted of a matrix of nearly contiguous collagen fibrils (first component) reinforced with apatite mineral particles (second component) (Landis 1995; Sasaki & Sudoh 1997; Lonardelli et al. 2005). The flexible matrix of collagen fibrils compensates for the brittleness of the stiff mineral particles. Texture, the distribution of orientations of the crystallites that compose a given material, greatly influences material properties, and this needs to be studied to understand sauropod bone from a materials science point of view.

At the atomic level, texture refers to the way in which the atomic planes that form the crystal lattice are positioned relative to a fixed reference (Randle & Engler 2000). In an isotropic material with no preferred crystallographic orientation, the distribution of mechanical properties, such as strength, toughness, and Young's modulus of elasticity (or stiffness), are independent of the direction in which these properties are

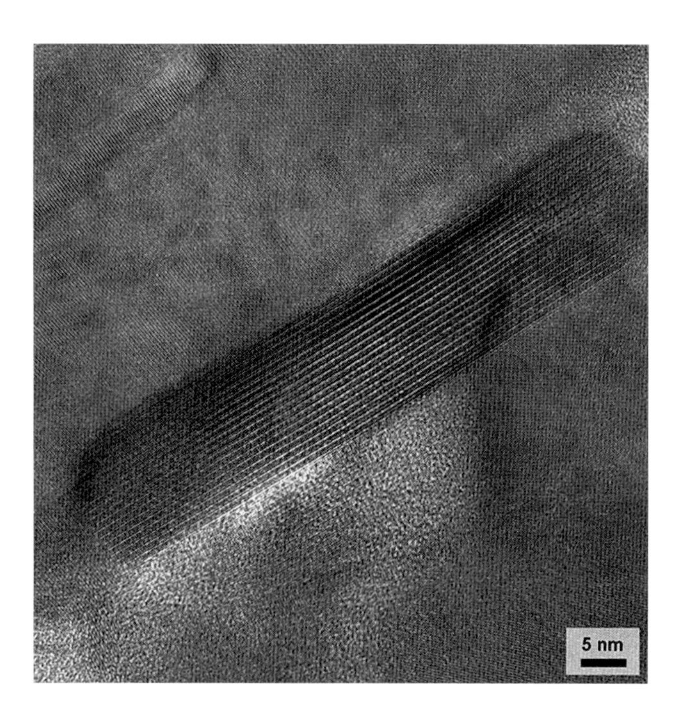

FIGURE 9.12. High-resolution TEM image of a single apatite crystal from a juvenile *Apatosaurus* sp. sample from the Morrison Formation, USA (OMNH 01278, humerus, 25.8 cm; histologic ontogenetic stage 4 of Klein & Sander 2008). The striations in the image correspond to the atomic planes that make up the crystal.

measured. However, mechanical properties are frequently direction dependent (Currey 1984, 2002). This is usually the consequence of crystallographic texture, which means that crystallographic directions show a preferential arrangement in the sample. Considering the bone structure as a result of functional adaptation, the existence of a specific geometrical arrangement among apatite crystallites, collagen matrix, and macroscopic bone shape can be expected. Indeed, the orientation distribution of apatite crystallites, for instance, is closely related to the mechanical function of the bone as a structural component that sustains body weight. In addition, lamellar bone presents a high level of spatial organization: closely packed collagen fibrils forming one lamella are mutually parallel to each other. The orientations of the lamellae, however, change in the bone (Francillon-Vieillot et al. 1990), making it more resistant to multiaxial tensile loads. The orientation of apatite crystals with regard to bone axis can be detected by texture measurements, usually performed with laboratory X-ray sources, with a synchrotron, or with neutron beams. In the following, we describe the principle of X-ray diffraction (for more details, see Warren 1990), which is the basis for texture characterization either by X-rays or neutrons, and a procedure that can be used to evaluate the crystallographic orientation of apatite crystals with regard to a specific direction in the macroscopic bone.

THE PRINCIPLE OF BRAGG DIFFRACTION

A monochromatic beam incident on a crystal is diffracted by the crystallographic lattice planes (defined by Miller indices, *hkl*) if they satisfy Bragg's law:

$$2d \sin \theta = \lambda, \quad (1)$$

where d is the distance between diffracting crystallographic planes, λ is the wavelength of the incident radiation, and θ is the diffraction angle measured between the incoming beam and the reflecting crystallite planes. The diffraction angle is also the angle formed between the detector (which registers the amount of reflected radiation as the number of incoming photons or neutrons) and the reflecting planes. A diffraction spectrum is obtained by recording the reflected intensities for gradually increasing angles θ. An example of a diffraction spectrum is shown in Fig. 9.14, in which the intensity of the reflected radiation is plotted as a function of twice the diffraction angle ($2 \times \theta$). When Bragg's law is satisfied—or, in other words, when a suitable combination of d and θ satisfying equation (1) occurs—a peak of intensity is generated, indicating that a greater amount of radiation is diffracted by the sample. The peaks observed in a diffraction spectrum are called *hkl* (or Bragg) reflections because each one of them corresponds to crystallographic lattice planes of different (*hkl*) indices. On the basis of the equation above, the interplanar distance d between two planes of atoms aligned in the same direction can be evaluated by measuring the diffraction angle θ. Different mineral phases are usually characterized by different interplanar distances. Therefore, by measuring the diffraction angle θ, different mineral phases present in a material can be identified, as exemplified in the case of apatite and calcite in Fig. 9.14. Assuming that the crystallographic structures of the constituent minerals or other crystalline materials are known (hexagonal for apatite and orthorhombic for calcite), the individual peaks can be assigned to their corresponding crystallographic planes (for more details, see Randle & Engler 2000).

THE PRINCIPLE OF POLE FIGURE MEASUREMENT

The crystallographic orientation of apatite (or any other crystallographic phase) with respect to specific directions in a bone can be obtained by performing pole figure measurements. A pole figure is a graphical representation of the distribution of a specific group of crystal planes in a material with respect to a reference coordinate system. Therefore, a single pole figure exists for each set of (*hkl*) planes of a certain mineral phase. If we consider, for example, the spectrum in Fig. 9.14, a pole figure can be obtained for the (002) and (310) apatite planes. However, it is important to start by defining

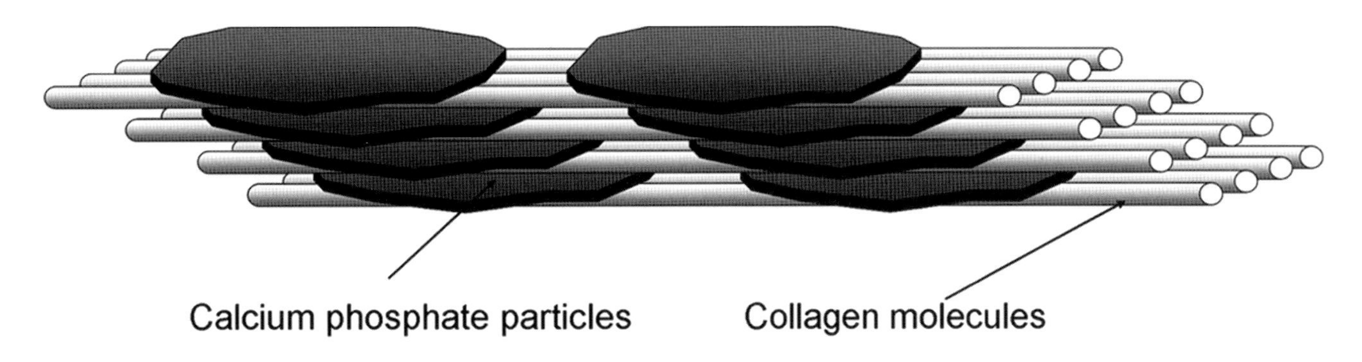

FIGURE 9.13. Reconstruction of the structure of sauropod bone tissue. The mineral crystals are platelet shaped (sensu Traub et al. 1989 and Sasaki & Sudoh 1997) and arranged parallel to each other and to the collagen fibrils (tubes) in the bone composite material. *Modified from Fratzl & Weinkamer (2007).*

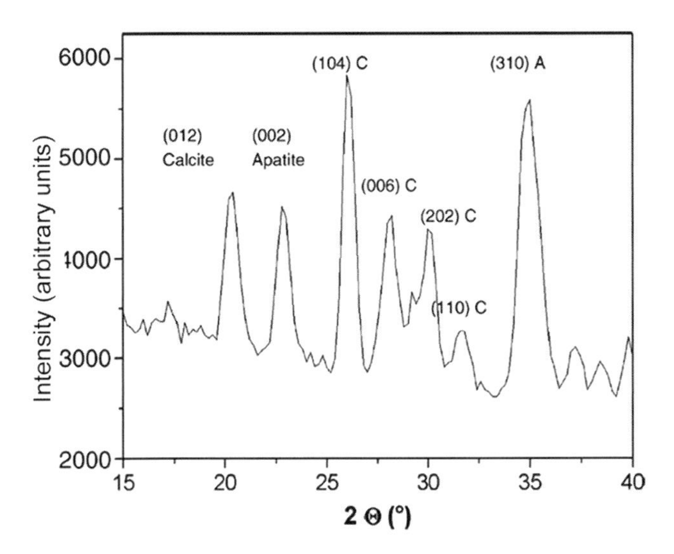

FIGURE 9.14. Neutron diffraction spectrum of a sauropod long bone (*Brachiosaurus brancai*, MFN J 12, right humerus, 170.0 cm; Pyzalla et al. 2006). The diffraction spectrum shows the different peaks corresponding to apatite (A) and calcite (C) with corresponding (*hkl*) reflections.

the reference coordinate system in which the orientation of crystallographic planes is described.

An example of the definition of a reference coordinate system is given in Fig. 9.15A. Here, the coordinate system is defined by axis LD (longitudinal direction) along the long axis of the bone, axis TD (transverse direction), and axis ND (normal direction), all of which are perpendicular to each other. The orientation of a given family of crystallographic planes in the reference coordinate system can be determined by measuring the intensity of the corresponding Bragg peak in several different directions in space. This is easily achieved if the detector is positioned at the expected θ angle, for example $2\theta \approx 23°$ in Fig. 9.14 in the case of (002) planes of the hexagonal apatite. By means of a monochromatic beam, the intensity recorded by the detector (number of incoming photons or neutrons) will be proportional to the number of crystal planes fulfilling the Bragg condition. By tilting the bone by the angle α (with regard to the diffraction plane determined by the incoming and scattered beams) and rotating it around the ND axis by the angle β, a different intensity is recorded.

The visualization of the results is usually done by plotting the intensities as a function of tilt angles α (radial) and β (azimuthal) in a so-called stereographic projection (Fig. 9.15B). The principle of stereographic projection is explained in Fig. 9.16, based on a hexagonal bone apatite crystal. The (001) crystallographic direction of the crystal in the bone system is described by the angles α and β, and it intersects the surface of the sphere at the pole P′. Because the representation of intensity in a 3D space is difficult, the equatorial plane of the sphere is used for this by projecting the pole P′ onto this plane. The stereographic projection P of the pole (001) is given by the intersection of the equatorial plane and the line connecting pole P′ with the south pole of the sphere. Given that the LD–ND axis plane is the diffraction plane, the apatite crystallite can be brought into diffracting position by rotating it by the angles $-\alpha$ and $-\beta$. In a similar way, we can obtain the pole figures for (101) poles of the apatite crystallites in a bone matrix. If the crystallites' orientation is completely random, the intensity will be uniformly distributed all over the equatorial plane (each point has the same value). If a preferred orientation of (*hkl*) planes is present, then the intensity will be concentrated in a certain area, permitting easy recognition of the orientation of diffracting (*hkl*) planes in the sample. The degree of orientation is quantified by the texture strength, which usually is given as a multiple of the intensity of the random distribution (m.r.d., multiple of random distribution). A higher m.r.d. number means a stronger texture and a higher degree of parallel arrangement in the crystallites (for more information, see Randle & Engler 2000).

Interesting results on texture were obtained by Sasaki &

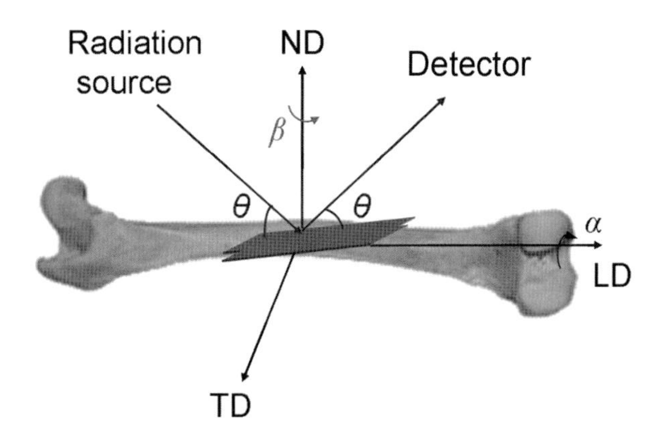

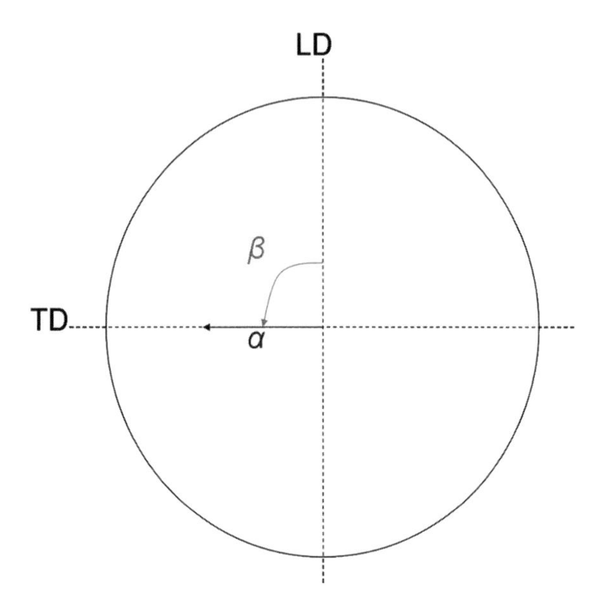

FIGURE 9.15. (*top*) The principle of pole figure measurement. The scattered intensity corresponding to a specific (*hkl*) family plane is recorded for each position of the sample defined by the tilt angle α (with regard to the diffraction plane) and rotation angle β (around sample normal axis ND). (*bottom*) In the stereographic projection is $\alpha = 0°$ at the center and $\alpha = 90°$ at the circle's perimeter. LD is the longitudinal direction, TD the transverse direction.

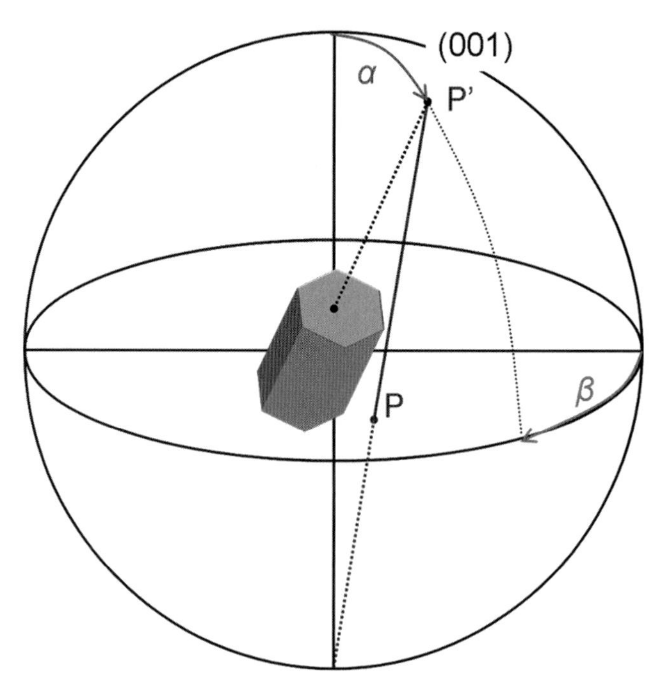

FIGURE 9.16. The principle of stereographic projection. P denotes the stereographic projection of the (001) pole P′ on the sphere. It is obtained as the intersection of the line connecting point P′ and the south pole of the sphere with the equatorial plane of the sphere.

Sudoh (1997) and Meneghini et al. (2003) on cow and human bones. The measured pole figures show that the c-axis of the apatite crystallites has an appreciable pole density peak in both the radial and tangential directions of the bone, whereas collagen molecules are oriented almost parallel to the direction of the long bone axis. Both working groups concluded that there are more than two morphologies in the apatite crystallites of bone mineral: one with the c-axis almost parallel to the bone axis, and another in which the c-axis is oriented nearly perpendicular to the bone axis (Sasaki & Sudoh 1997). On the other hand, observations by Bacon et al. (1977) and Dahms & Bunge (1989) indicate that the c-axis of apatite crystallites in ox bone appears to be oriented along the long bone axis.

TEXTURE OF SAUROPOD DINOSAUR BONE

Texture analysis was applied to three sauropod bones: *Tornieria africana* (Remes 2006) (sample MFN A 1, right humerus, 99.0 cm), an indeterminate sauropod (Remes 2009) previously assigned to *Barosaurus africanus* (MfN Ki 4a, left femur, 120.0 cm) (Sander 2000), and *Brachiosaurus brancai* (sample MFN XV, left femur, 219.0 cm) from Tendaguru (Tanzania, Africa) (Sander 2000; Pyzalla et al. 2003, 2006). *Tornieria africana* and the indeterminate sauropod differ in their growth strategies as observed in their microstructure (Sander 2000). The histology of *Tornieria africana* (MFN A 1), called type A by Sander (2000), is characterized by a thick cortex of primary bone of the laminar fibrolamellar type and few secondary osteons, which reflects rapid and continuous growth. In contrast, the histology of the indeterminate sauropod (MFN Ki 4a), called type B by Sander (2000), shows an inner cortex with heavily remodeled Haversian bone, secondary osteons scattered in the outer cortex, and several regularly spaced lines of arrested growth (LAGs) (Sander 2000). This histology can be interpreted as slower and interrupted growth. The femur of *Brachiosaurus brancai* (MFN XV) is from a fully grown, senescent individual, and its cortex is almost entirely remodeled by secondary osteons (Klein & Sander 2008, Sander et al., this volume).

These two growth strategies observed in *Tornieria africana* and the indeterminate sauropod (Remes 2006, 2009) not only affected bone histology, but also the orientation and distribution of the bone mineral. The slow-growing bone (type B histology of Sander 2000) displays bone apatite crystallites with a preferential orientation along the bone axis (Plate 9.2). This orientation seems to be constant along the cortical bone, whereas the fast-growing bone MFN A 1 (type A histology of Sander 2000) shows this preferential orientation only in the remodeled part. The texture index is lower for type A, which means that in this sample, the variations in the orientation of crystallites do not change as much as in type B. These results can be explained by ontogenetic differences: the sample of type A histology is mostly primary bone, while the sample of type B is heavily remodeled (Klein & Sander 2008; Sander et al., this volume).

Ontogenetic Variation in Sauropod Bones

Another consideration regarding the crystallographic texture of sauropod bones concerns the difference between ontogenetic stages. Further investigations have been conducted in two sauropod taxa, the indeterminate sauropod from Tendaguru (MFN Ki 4a) and *Brachiosaurus brancai* (Pyzalla et al. 2006). Plate 9.2 shows the three (002) pole figures for the indeterminate sauropod together with the corresponding measurement locations: at the primary FLB structure (left side), at primary and secondary osteons (center), and at the trabecular bone (right side). The (002) pole figure analysis of MFN Ki 4a shows that apatite crystallite texture varies only slightly along the cross section.

Plate 9.3 displays (002) pole figures measured at three locations in the senescent sauropod sample, that of *Brachiosaurus brancai*. In the fully remodeled cortex consisting of secondary osteons, the apatite crystallites again show a broad, nearly (001) fiber texture, as observed previously in the slowly growing MFN Ki 4a (Plate 9.2). The average c-axis of the bone mineral thus appears to be oriented along the direction of the long bone axis. This principal orientation is present in both the primary fibrolamellar bone of the cortex and in the secondary bone. The main difference between the slowly growing and senescent sauropods is observed in their trabecular bone. In case of the slowly growing individual, the axial orientation of the bone mineral suggests that the trabeculae actually originate from the resorption of compact bone, while in the case of the senescent individual, the orientation is lost as a result of the process of bone remodeling. The mineral is reoriented to meet the direction of the trabeculae, which can be interpreted as an adaptation of the apatite distribution to the mechanical load (long bones of large animals carry mostly axial loads).

Texture Strength

Fiber texture strengths determined on all sauropod samples were about 2.6 m.r.d. This value is similar to those determined in dinosaur tendons by Lonardelli et al. (2005), who have also shown that the c-axis of apatite in pachycephalosaurid dinosaur tendons is preferentially aligned to the tendon axis and has texture strength of 3.25 m.r.d. Apatite in the bone of living animals usually has much higher texture strengths as a result of particle alignment facilitated by the presence of collagen. For example, Wenk & Heidelbach (1999) measured a texture strength of 6 m.r.d. in ossified turkey tendon. Assuming that the apatite crystallites in the living dinosaur bone were aligned along the bone axis, a fiber strength of 2.6 m.r.d. suggests that there is considerable spread in the crystallographic orientation of apatite crystals, but their average alignment along the bone axis is preserved (the maximum of 2.6 m.r.d. being observed in the direction of the bone axis). This may indicate that diagenesis does indeed affect particle alignment, which is expected to be more pronounced in the neighborhood of the diagenetic material (see Fig. 9.11 as an example). A similar conclusion that diagenesis affects reorientation and growth of crystallites was obtained earlier, based on small-angle X-ray scattering data (Wess et al. 2001).

Trace Element Distribution in Sauropod Bones: A Window to Diagenesis

The preservation of structural features in dinosaur bones, such as primary and secondary osteons and osteocyte lacunae, has been known since the 19th century. De Ricqlès (1980), who performed studies on thin sections, first suggested the concept of molecule-by-molecule replacement of bone tissue without changes in the bone tissue structure. The prevalent view today is that the original bone mineral remains in the dinosaur bone (Zocco & Schwartz 1994). An intriguing question is related to the preservation state of bone at different levels of its hierarchical organization. A partial answer may be obtained by applying investigative methods sensitive to the chemical elements making up the bone.

After death, dinosaur bones were subjected to a multitude of physical, chemical, and biological processes, all of which are poorly understood. During the burial and fossilization of sauropod bones, their organic parts degenerated. This and other processes influencing the alteration of bone are usually referred to as diagenesis (Karkanas et al. 2000). Diagenesis has many consequences: it can affect bone histology, bone porosity, protein content, the crystallinity of the apatite, and the bone's content of chemical elements and compounds in general (Reiche et al. 2003). Organic substances, such as collagen, decompose after burial, but the inorganic (mineral) substances, mainly carbonated hydroxyapatite ($Ca_5(PO_4)_3(OH)$), are more resistant and generally preserved with only slight changes. Substitution commonly takes place, and the hydroxyapatite becomes fluorapatite ($Ca_5(PO_4)_3F$), with the fluorine replacing the hydroxyl group.

Each level of bone organization (Fig. 9.1) carries specific information concerning taphonomy, morphology, and paleobiology (see, e.g., isotopic studies on diet by Lambert et al. 1985; Tütken, this volume; and on thermophysiology by Tütken et al. 2004; Amiot et al. 2006).

ENERGY DISPERSIVE TECHNIQUES

Element maps with different spatial and concentration resolutions can be obtained by different techniques, three of which are presented here: (1) EDS in SEM (already described in the first part of this chapter), (2) X-ray fluorescence (XRF) spectroscopy, and (3) proton-induced X-ray emission (PIXE) spectroscopy. The techniques are exemplified by evaluating diagenetic effects in sauropod long bones at both the macroscopic and microscopic scale. An advantage of all energy dispersive techniques is that they do not require special sample preparation and are well adapted for fossil samples. Before presenting the different techniques and the results obtained, the physical principles of the methods are briefly reviewed here.

When an energy-rich beam of photons or particles (such as electrons or protons) strikes a sample, two effects may be observed: the beam can be absorbed by the atoms or scattered through the material. The first process is called *ionization* because a neutral atom or a molecule is transformed into an ion by adding or removing charged particles. During ionization, electrons from a lower orbital become excited (having higher energies), and a vacancy is created in the energetic structure of the atom. This state of the electronic structure is, however, unstable, and electrons with higher energies transfer to the lower orbital to fill the vacancy. During this transfer, a characteristic X-ray is emitted with an energy that is equal to the difference between the two binding energies of the corresponding orbitals. The electron energy levels in the atom are characteristic for each element, which is why the emitted radiation is also called *characteristic radiation*. The major difference between the spectroscopy techniques mentioned above is only the type of the ionizing radiation, which is a high-energy X-ray beam in XRF, an electron beam in standard EDS, and a proton beam in PIXE. By analyzing and quantifying the energy of the characteristic radiation, the elements contained in the sample can be identified (qualitative analysis), but element ratios for both mass and element percentages can be also calculated (quantitative analysis). The spectroscopy methods also allow for the identification of specific minerals; however, some care must be taken because of the lack of adequate standards and reduced energy resolution of available detectors.

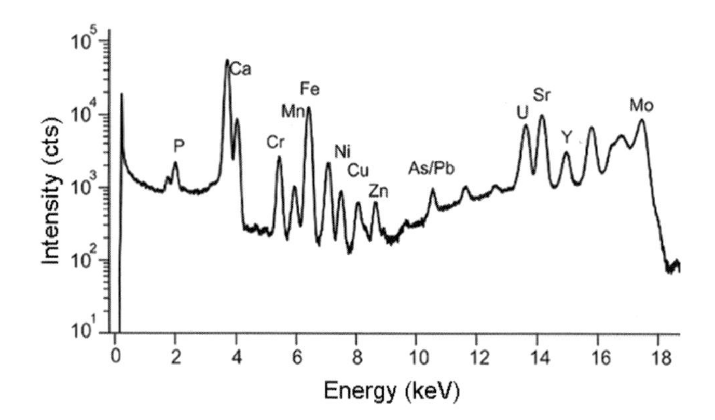

FIGURE 9.17. μ-XRF spectrum obtained for a *Tornieria africana* sample (MFN A 1, right humerus, 99.0 cm; Sander 2000; Klein & Sander 2008). The scanned area is 900 × 900 μm. cts, counts per second. *Modified after Ferreyro et al. (2006).*

X-RAY FLUORESCENCE SPECTROSCOPY

XRF has become an important tool to study the distribution of trace elements and is a common, nondestructive method (Carvalho et al. 2004; Reiche et al. 1999). In a first approximation, it provides a qualitative determination of the element distribution (with particular emphasis on area mapping and image features; however, linear scans can also be made). The fluorescent sample can then be analyzed by measuring the energy spectrum of photons emitted (see, e.g., Fig. 9.17). The intensity of each characteristic radiation is directly correlated with the amount of each element in the bone. XRF is a precise analytical method that can be used to analyze elements with atomic numbers (Z) larger or equal to that of beryllium ($Z \geq 5$) and that can complement SEM-EDS analyses. For light elements, the X-ray signal is weak, and the distinction among elements is difficult as a result of the proximity of the peaks in the spectrum. The concentrations are given in the form of pure elements, as in SEM-EDS analyses, but with a much more accurate relative detection limit in the 0.1–10 ppm range (Janssens & Van Grieken 2004).

Figure 9.18 shows the XRF spectrum across the cortex of a sauropod long bone from Tendaguru (Ferreyro et al. 2006). Synchrotron radiation–based micro-X-ray fluorescence (μ-XRF) with a spatial resolution of about 100 μm was used by Pyzalla et al. (2006) to analyze element concentration distributions in sauropod bones. The μ-XRF line scan in Fig. 9.18 shows that besides the elements expected in the hydroxyapatite, manganese (Mn), iron (Fe), copper (Cu), zinc (Zn), arsenic (As), lead (Pb), uranium (U), strontium (Sr), and yttrium (Y) are present in the sauropod bones. The spatial distribution of these elements shows maxima of calcium (Ca), manganese (Mn), iron (Fe), and copper (Cu) concentrations in the vascular canals. A continuously high Ca concentration occurs across the cortex of the bone. X-ray and neutron diffracto-

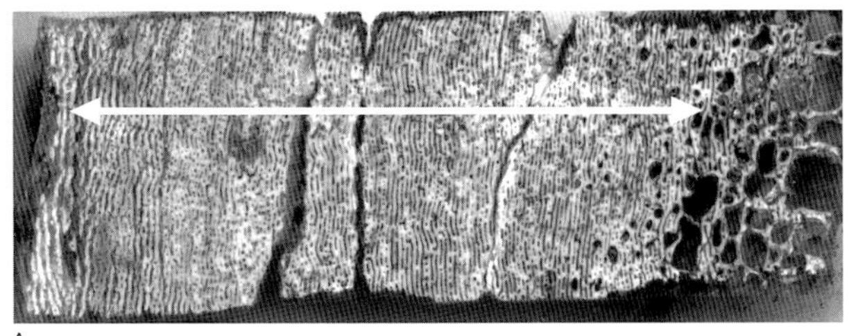

A

Periosteal   Endosteal

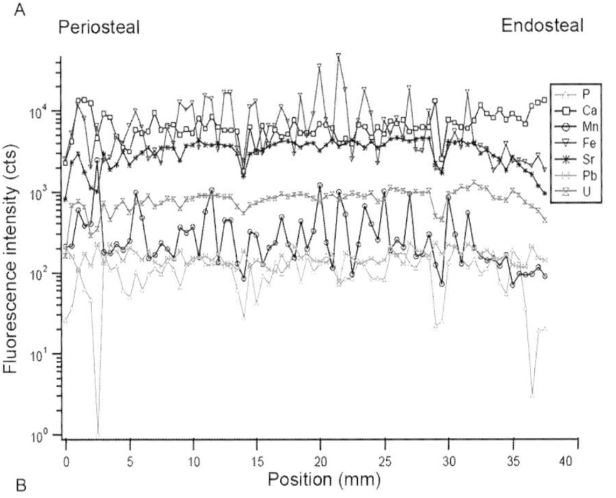

FIGURE 9.18. (A) Photographic image of the sample of *Tornieria africana* (MFN A 1, right humerus, 99.0 cm; Sander 2000; Klein & Sander 2008) consisting of a polished and bisected drill core. The sample thus represents a segment of the bone wall. The white line marks the position of the μ-XRF scan, and the arrows indicate the directions of the periosteal and endosteal margins of the cortex. (B) Fluorescent intensity distribution of elements P, Ca, Mn, Fe, Sr, Pb, and U along the scan line in (A). The scanned distance is 37.5 mm with a step size of 500 μm. The measurement time was 40 seconds per scan position.

grams of bones from the same fossil bed showed that Ca is mainly incorporated as calcite into the bone and can be observed as a deposit within the vascular canals (Carvalho et al. 2004). The carbonate hydroxyapatite that has been modified into fluorapatite (as evidenced by X-ray diffraction; Stempniewicz et al. 2004) are clearly bound to phosphorus (P). Unlike previous EDS analyses on other samples (as described above), Mn and Fe display similar peaks, indicating the presence of iron oxides in the former vascular canals (Parker & Toots 1970; Pfretzschner 2001).

Figure 9.18 also shows that the distribution of iron oxides is irregular. This may have resulted from diagenesis after the bone was buried, because Fe and Mn are not incorporated into the tissue by biological processes during the life of the animal. These elements are largely considered as contaminating elements (Lambert et al. 1983). A similar, correlated distribution is observed for the elements Sr, U, and P. Sr and U are more concentrated in the bone tissue than in the vascular canals (Plate 9.4); however, they are less correlated with vascular canals than Fe and Mn, indicating good preservation, as was previously observed by Carvalho et al. (2004). The concentration of Sr, U, and Pb are controlled by the phosphate in the bone (Romer 2001) and are usually used for determining age (U, Pb) or inferring diet (Sr). However, Romer (2001) has already shown for *Brachiosaurus brancai* that uncertainties that result from different lead concentrations (initial composition of the bone, uptake during diagenesis, and in situ grown radiogenic lead) may give erroneous ages.

Plate 9.4 shows element maps at a higher spatial resolution in the outer surface of *Tornieria africana*. The measurements were made with a beam size of 60 μm. The similarity of Fe and Mn distributions in this cortical region is remarkable. A correlation between Sr and U distribution can be observed as well; however, it is less pronounced. XRF also indicates which elements are incorporated during diagenesis. All elements mentioned previously as occurring in the inorganic part of the bone are present in many geological environments. However, their presence in the apatite is not exclusively due to diagenetic processes. During the life of the animal, it is probable that Na and Sr enter in the composition of the apatite. The variability of their concentration is lower than that of the other elements that are incorporated postmortem (e.g., Fe). The elements present in cracks or cavities seem to be clearly resulting from diagenesis. The elements Mn and Fe, as well as Ca and Pb, fall into this category. These results show the complexity of the exchange between the bone and environment that occur during diagenesis.

PROTON INDUCED X-RAY EMISSION (PIXE)

Compared to SEM-EDS and the standard XRF method, PIXE offers a superior peak-to-background ratio (which means that the peaks can be more clearly identified). This enables improving spatial resolution (Goodwin et al. 2007) and relative detection sensitivity in the 0.5–5 ppm range (Janssens & Van Grieken 2004). However, only elements heavier than fluorine ($Z = 9$) can be detected. The upper limit of detection is given by the ionization probability of the K electron shell. The usual proton energies are in the range of 1–4 MeV, which are well suited for the analysis of thin layers—for example, ink on parchment, or a painting composed of thin layers of paint (Neelmeijer et al. 2000). One advantage of PIXE is the ability to focus the proton beam at the micrometer scale (Roberts et al. 1999, Goodwin et al. 2007). Stempniewicz & Pyzalla (2003) have already reported PIXE results on Tendaguru sauropods from their analysis on the concentrations and the distributions of various elements across the bone cortex.

Plate 9.5 shows the PIXE spectrum at different points across the cortex of the humerus of *Tornieria africana* from Tendanguru (MFN A 1). The concentration of Fe appears to be higher in the intermediate zone of the bone cortex than toward the periosteal and endosteal margins (cf. Table 9.2). Barium (Ba),

Table 9.2. Quantitative Element Analyses by Proton-Induced X-ray Emission at Different Points across the Cortex of the Humerus of *Tornieria africana* from Tendanguru (MFN A 1)

Point[a]	Element concentration (ppm)								
	Mn	Fe	Zn	Sr	Y	In	Cs	Ba	U
1	271	3,547	0	755	162	216	42	181	1,353
2	245	3,532	39	727	126	205	26	176	1,314
3	64	12,934	52	772	29	227	25	179	816
4	276	13,838	22	776	14	217	0	146	828
5	162	4,407	22	732	35	188	0	117	196

[a]Points 1 and 2 are located toward the periosteal margin, points 3 and 4 are in the middle of the cortex, and point 5 is located toward the endosteal margin.

U, and Y, as well as cesium (Cs), show a gradient across the bone cortex, decreasing in concentration from the periosteal margin to the endosteal margin. These elements can enter bone tissue by diagenetic processes. No clear trend is visible for strontium (Sr), and its occurrence in bone close to the periosteal surface (point 1 and point 2 in Table 9.2) indicates that inhomogeneities may even occur on a rather local level. The flat concentration profile of Sr indicates an equilibrium concentration of this element or a fast exchange of this element between the bone and the surrounding environment (Reiche et al. 2003). The slow decrease could result from absorption and transport in hydroxyapatite and from the fact that Sr is diagenetically averaged, as was suggested earlier by Lambert et al. (1991).

The decrease of U toward the endosteal margin may indicate that some of it had leached out during diagenesis. Because U is used for dating bone by the uranium series method (Millard & Hedges 1995), an increasing interest in this element and the direct implication of its uptake in the bone have made uranium a popular topic in the literature on diagenesis. The problem arises that the uptake of uranium is common because bone is an open system. A well preserved bone will exhibit U-shaped concentration profiles across the bone cortex (Williams & Marlow 1987; Millard & Hedges 1995). The existence of diffusive or recharge mechanism is supported by an increase in U concentration from a U-shaped profile to a uniform one. Uranium is located in the fossil bone tissue as indicated by the uranyl ion showing a generally high affinity to carbonated hydroxyapatite (Plate 9.4) and is generally not present in accessory minerals in cavities or pore spaces (e.g., Fe or Mn). Quantitative results (Table 9.2) show a distinct decrease in U concentration from the periosteal to the endosteal margin of the bone. The enrichment of U in the outer cortex is the result of the greater surface area of bone in contact with groundwater. The large decrease toward the endosteal margin is an effect of porosity (Hedges & Millard 1995; Millard & Hedges 1995). Consequently, areas of high porosity will show low U content. Some recent models (Pike 2002, 2005) have allowed for an uptake history based on U concentrations profiles.

In conclusion, PIXE can quantitatively determine the concentrations of elements in fossil dinosaur bone. A combination of the different analysis techniques shows that although the bone seems to be unaffected at the histological level, sauropod fossils have endured strong diagenetic changes. The problem of diagenetic changes at the lower levels of bone structure is not yet solved.

Future Research

With regard to future research perspectives, there is also a type of transmission electron microscope, the scanning transmission electron microscope (STEM), that combines features of SEM and TEM (Fig. 9.9). In STEM, the electrons pass through the sample (as in TEM), but, as in SEM, the beam is scanned over the sample. This technique permits atomic-resolution images. An analytical electron microscopy study in the STEM complementary to the techniques discussed above could provide important information about the elemental composition of sauropod bone and the diagenetic processes at lower hierarchical levels.

With regard to the question posed at the beginning of this chapter, that is, whether specifics of the bone tissue of sauropod dinosaurs made their gigantism possible, we conclude that this was not the case. Instead, sauropod bone tissue does not appear to differ in its hierarchical structure and textural characteristics from that of extant large mammals. This can be stated with confidence despite the considerable diagenetic alterations experienced by sauropod dinosaur bone.

Acknowledgments

Our foremost thanks go to the numerous curators who made sampling of sauropodomorph long bones possible. These are Wolf-Dieter Heinrich and Hans-Peter Schultze (Museum für Naturkunde, Berlin), Richard Cifelli and Kyle Davies (Oklahoma Museum, Norman), and Ken Stadtman (Brigham Young University, Provo). We also would like to thank Dr. Andrea Denker and Dr. Magdalena Stempniewicz for providing us the results of the PIXE. We are grateful to Clive Trueman (University of Southampton) and one anonymous reviewer for their constructive suggestions, which substantially improved this chapter. Our research was founded by the Deutsche Forschungsgemeinschaft (DFG). This paper is contribution number 68 of the DFG Research Unit 533 "Biology of the Sauropod Dinosaurs: The Evolution of Gigantism."

References

Amiot, R., Lécuyer, C., Buffetaut, E., Escarguel, G., Fluteau, F. & Martineau, F. 2006. Oxygen isotopes from biogenic apatites suggest widespread endothermy in Cretaceous dinosaurs.—*Earth and Planetary Science Letters* 246: 41–54.

Anderson, J. F., Hall-Martin, A. & Russell, D. A. 1985. Long-bone circumference and weight in mammals, birds and dinosaurs.—*Journal of Zoology* 207: 53–61.

Ascenzi, A. & Benvenuti, A. 1986. Orientation of collagen fibers at the boundary between two successive osteonic lamellae and its mechanical interpretation.—*Journal of Biomechanics* 19: 455–463.

Bacon, G. E., Bacon, P. J. & Griffiths, R. K. 1977. The study of bones by neutron diffraction.—*Journal of Applied Crystallography* 19: 124–126.

Bell, L. S. 1990. Paleopathology and diagenesis: a SEM evaluation of structural changes using backscattered electron imaging.—*Journal of Archaeological Science* 17: 85–102.

Boyde, A., Vesely, P., MacKenzie, A. P., Thalmann, R., Comegys, T. H., Arenberg, I. K., Bessis, M., Weed, R. I., Nemanic, M. K., Nowell, J. A., Pangborn, J., Tyler, W. S., LoBuglio, A. F., Rinehart, J. J. & Balcerzak, S. P. 1972. Scanning electron microscopy studies of bone. *In* Bourne, G. H. (ed.). *The Biochemistry and Physiology of Bone.* 2nd edition. Academic Press, New York—Proceedings I of the 5th Annual Scanning Electron Microscope Symposium, II Workshop on Biological System Preparation for Scanning Electron Microscopy.

Carvalho, M. L., Marques, A. F., Lima, M. T. & Reuss, U. 2004. Trace elements distribution and post-mortem intake in human bones from Middle Age by total reflection X-ray fluorescence.—*Spectrochimica Acta Part B* 59: 1251–1257.

Christel, P., Cerf, C. & Pilla, A. 1981. Time evolution of the mechanical properties of the callus of fresh fractures.—*Annals of Biomedical Engineering* 9: 383–391.

Currey, J. D. 1960. Differences in the blood-supply of bone of different histological types.—*Quarterly Journal of Microscopical Science* 101: 351–370.

Currey, J. D. 1968. The adaptation of bones to stress.—*Journal of Theoretical Biology* 20: 91–106.

Currey, J. D. 1984. *The Mechanical Adaptation of Bones.*—Princeton University Press, Princeton.

Currey, J. D. 2002. *Bones: Structure and Mechanics.*—Princeton University Press, Princeton.

Dahms, M. & Bunge, H. J. 1989. The iterative series-expansion method for quantitative texture analysis. I. General outline.—*Journal of Applied Crystallography* 22: 439–447.

de Ricqlès, A. 1976. On bone histology of fossil and living reptiles, with comments on its functional and evolutionary significance. *In* Bellairs, A. d'A. & Cox, C. B. (eds.). *Morphology and Biology of Reptiles.* Linnean Society, London: pp. 123–150.

de Ricqlès, A. 1980. Tissue structures of dinosaur bone. Functional significance and possible relation to dinosaur physiology. *In* Thomas, D. K. & Olson, E. C. (eds.). *A Cold Look at the Warm-Blooded Dinosaurs.* Westview Press, Boulder: pp. 103–140.

de Ricqlès, A., Meunier, F.-J., Castanet, J. & Francillon-Vieillot, H. 1991. Comparative microstructure of bone. *In* Hall, B. K. (ed.). *Bone, Vol. 3: Bone Matrix and Bone Specific Products.* CRC Press, Boca Raton: pp. 1–78.

Doberenz, A. R. & Wyckoff, R. W. G. 1967. Fine structure in fossil collagen.—*Proceedings of the National Academy of Sciences of the United States of America* 57: 539–541.

Egerton, R. F. 2005. *Physical Principles of Electron Microscopy: An Introduction to TEM, SEM, and AEM.* Springer, Berlin.

Enlow, D. H. & Brown, S. O. 1956. A comparative histological study of fossil and recent bone tissues, part 1.—*Texas Journal of Science* 8: 405–443.

Enlow, D. H. & Brown, S. O. 1957. A comparative histological study of fossil and recent bone tissues, part 2.—*Texas Journal of Science* 9: 186–214.

Enlow, D. H. & Brown, S. O. 1958. A comparative histological study of fossil and recent bone tissues, part 3.—*Texas Journal of Science* 10: 187–230.

Farlow, J. O. & Argast, A. 2006. Preservation of fossil bone from the Pipe Creek Sinkhole (late Neogene, Grand County, Indiana, USA).—*Journal of the Paleontological Society of* Korea. 2: 51–75.

Ferreyro, R., Zoeger, N., Cernohlawek, N., Jokubonis, C., Koch, A., Streli, C., Wobrauschek, P., Sander, P. M. & Pyzalla, A. R. 2006. Determination of the element distribution in sauropod long bones by micro-XRF.—*Advances in X-ray Analysis* 49: 230–235.

Francillon-Vieillot, H., de Buffrénil, V., Castanet, J., Géraudie, J., Meunier, F. J., Sire, J. Y., Zylberberg, L. & de Ricqlès, A. 1990. Microstructure and mineralization of vertebrate skeletal tissues. *In* Carter, J. G. (ed.). *Skeletal Biomineralization: Patterns, Processes and Evolutionary Trends, Vol. 1.* Van Nostrand Reinhold, New York: pp. 471–530.

Fratzl, P. & Weinkamer, R. 2007. Nature's hierarchical materials.—*Progress in Materials Science* 52: 1263–1334.

Goldstein, J., Newbury, D., Joy, D., Lyman, C. & Echlin, P. 2003. *Scanning Electron Microscopy and X-ray Microanalysis.* Kluwer Academic Publishers, New York.

Goodwin, M. B., Grant, P. G., Bench, G. & Holroyd, P. A. 2007. Elemental composition and diagenetic alteration of dinosaur bone: distinguishing micro-scale spatial and compositional heterogeneity using PIXE.—*Palaeogeography, Palaeoclimatology, Palaeoecology* 253: 458–476.

Gunga, H.-C., Kirsch, K., Rittweger, J., Clarke, A., Albertz, J.,

Wiedemann, A., Mokry, S., Suthau, T., Wehr, A., Clarke, D., Heinrich, W.-D. & Schultze, H.-P. 1999. Body size and body volume distribution in two sauropods from the Upper Jurassic of Tendaguru, Tanzania (East Africa).—*Mitteilungen aus dem Museum für Naturkunde in Berlin, Geowissenschaftliche Reihe* 2: 91–102.

Gunga, H.-C., Suthau, T., Bellmann, A., Stoinski, S. & Friedrich, A. 2008. A new body mass estimation of *Brachiosaurus brancai* Janensch, 1914 mounted and exhibited at the Museum of Natural History (Berlin, Germany).—*Fossil Record* 11: 33–38.

Hedges, R. E. M. & Millard, A. R. 1995. Bones and groundwater: towards the modelling of diagenetic processes.—*Journal of Archaeological Science* 22: 155–164.

Hubert, J. F., Panish, P. T., Chure, D. J. & Prostak, K. S. 1996. Chemistry, microstructure, petrology, and diagenetic model of Jurassic dinosaur bones, Dinosaur National Monument, Utah.—*Journal of Sedimentary Research* 66: 531–547.

Janssens, K. & Van Grieken, R. 2004. *Non-Destructive Micro-analysis of Cultural Heritage Materials*. Elsevier, Amsterdam.

Karkanas, P., Bar-Yosef, O., Goldberg, P. & Weiner, S. 2000. Diagenesis in prehistoric caves: the use of minerals that form in situ to assess the completeness of the archaeological record.—*Journal of Archaeological Science* 27: 915–929.

Klein, N. & Sander, P. M. 2008. Ontogenetic stages in the long bone histology of sauropod dinosaurs.—*Paleobiology* 34: 247–263.

Kolodny, Y., Luz, B., Sander, P. M. & Clemens, W. A. 1996. Dinosaur bones: fossils or pseudomorphs? The pitfalls of physiology reconstruction from apatit fossils.—*Palaeogeography, Palaeoclimatology, Palaeoecology* 126: 161–171.

Lambert, J. B., Simpson, S. V., Buikstra, J. E. & Hanson, D. 1983. Electron microprobe analysis of elemental distribution in excavated human femurs.—*American Journal of Physical Anthropology* 62: 409–423.

Lambert, J. B., Simpson, S. V., Szpunar, C. B. & Buikstra, J. E. 1985. Bone diagenesis and dietary analysis.—*Journal of Human Evolution* 14: 477–482.

Lambert, J. B., Xue, L. & Buikstra, J. E. 1991. Inorganic analysis of excavated human bone after surface removal.—*Journal of Archaeological Science* 18: 363–383.

Landis, W. J. 1995. The strength of a calcified tissue depends in part on the molecular structure and organization of its constituent mineral crystals in their organic matrix.—*Bone* 16: 533–544.

Lonardelli, L., Wenk, H.-R., Luterotti, L. & Goodwin, M. 2005. Texture analysis from synchrotron diffraction images with the Rietveld method: dinosaur tendon and salmon scale.—*Journal of Synchrotron Radiation* 12: 354–360.

Meneghini, C., Dalconi, M. C., Nuzzo, S., Mobilio, and S. & Wenk, R. H. 2003. Rietveld refinement on X-ray diffraction patterns of bioapatite in human fetal bones. *Biophysical Journal* 84: 2021–2029.

Millard, A. R. & Hedges, R. E. M. 1995. The role of the environment in uranium uptake by buried bone.—*Journal of Archaeological Science* 22: 239–250.

Neelmeijer, C., Brissaud, I., Calligaro, T., Demortier, G., Hautojärvi, A., Mäder, M., Martinot, L., Schreiner, M., Tuurnala, T. & Weber, G. 2000. Painting—a challenge for XRF and PIXE analysis.—*X-ray Spectrometry* 29: 101–110.

Parker, R. B. & Toots, H. 1970. Minor elements in fossil bone.—*Geological Society of America Bulletin* 81: 925–932.

Pawlicki, R. 1976. Preparation of fossil bone specimens for scanning electron microscopy.—*Stain Technology* 51: 147–152.

Pawlicki, R., Korbel, A. & Kubiak, H. 1966. Cells, collagen fibrils and vessels in dinosaur bone.—*Nature* 211: 655–657.

Pfretzschner, H.-U. 2000. Microcracks and fossilization of Haversian bone.—*Neues Jahrbuch für Geologie und Paläontologie, Abhandlungen* 216: 413–432.

Pfretzschner, H.-U. 2001. Pyrite in fossil bone.—*Neues Jahrbuch für Geologie und Paläontologie, Abhandlungen* 220: 1–23.

Pike, A. W. G., Hedges R. E. M. & Van Calsteren, P. 2002. U-series dating of bone using the diffusion–adsorption model.—*Geochimica et Cosmochimica Acta* 66: 4273–4286.

Pike, A. W. G., Eggins, S., Gru, R., Hedges, R. E. M. & Jacobi, M. 2005. U-series dating of the Late Pleistocene mammalian fauna from Wood quarry (Steeley), Nottinghamshire, UK.—*Journal of Quaternary Science* 20: 59–65.

Pritchard, J. J. 1956. General histology of bone. *In* Bourne, G. H. (ed.). *The Biochemistry and Physiology of Bone*. Academic Press, New York: pp 1–25.

Pyzalla, A. R. & Stempniewicz, M. & Gunther, A. 2003. Growth strategy of sauropod dinosaurs studied by pole figure analysis.—*GeNF Experimental Report* 2003: 189–191.

Pyzalla, A. R., Sander, P. M., Hansen, A., Ferreyro, R., Yi, S.-B., Stempniewicz, M. & Brokmeier, H.-G. 2006. Texture analyses of sauropod dinosaur bones from Tendaguru.—*Materials Science and Engineering A* 437: 2–9.

Randle, V. & Engler, O. 2000. *Introduction to Texture Analysis: Macrotexture, Microtexture and Orientation Mapping*. Gordon and Breach Science Publishers, London.

Reiche, I., Quattropani, L.-F., Calligaro, C., Salomon, J., Bocherens, H., Charlet, L. & Menu, M. 1999. Trace element composition of archaeological bones and postmortem alteration in the burial environment.—*Nuclear Instruments and Methods in Physics Research* 150: 656–662.

Reiche, I., Vignaud, C. & Menu, M. 2002. The crystallinity of ancient bone and dentine: new insights by transmission electron microscopy.—*Archaeometry* 44: 447–459.

Reiche, I., Quattropani, L.-F., Vignaud, C., Bocherens, H., Charlet, L. & Menu, M. 2003. A multi-analytical study of bone diagenesis: the neolithic site of Bercy (Paris, France).—*Measurements, Science and Technology* 14: 1608–1619.

Reid, R. E. H. 1984. The histology of dinosaurian bone and its possible bearing of dinosaurian physiology.—*Symposium of the Zoological Society of London* 52: 629–663.

Reid, S. A. 1986. A study of lamellar organization in juvenile and adult human bone.—*Anatomy and Embryology* 174: 329–338.

Remes, K. 2006. Revision of the Tendaguru sauropod dinosaur *Tornieria africana* Fraas and its relevance for sauropod paleobiogeography.—*Journal of Vertebrate Paleontology* 26: 651–669.

Remes, K. 2009. Taxonomy of Late Jurassic diplodocid sauropods from Tendaguru (Tanzania).—*Fossil Record* 12: 23–46.

Rho, J.-Y., Kuhn-Spearing, L. & Ziuopos, P. 1998. Mechanical properties and the hierarchical structure of bone.—*Medical Engineering and Physics* 20: 92–102.

Roberts, M. L., Grant, P. G., Bench, G. S., Brown, T. A., Frantz, B. R., Morse, D. H. & Antolak, A. J. 1999. The stand-alone microprobe at Livermore.—*Nuclear Instruments and Methods in Physics Research B* 158: 24–30.

Robinson, R. A. 1952. An electron microscopic study of the crystalline inorganic component of bone and its relationship to the organic matrix.—*Journal of Bone and Joint Surgery* 34: 389–434.

Romer, R. L. 2001. Isotopically heterogeneous initial Pb and continuous ^{222}Rn loss in fossils: the U-Pb systematics of *Brachiosaurus brancai*.—*Geochimica et Cosmochimica Acta* 65: 4201–4213.

Sander, P. M. 2000. Longbone histology of the Tendaguru sauropods: implications for growth and biology.—*Paleobiology* 26: 466–488.

Sander, P. M. & Andrassy, P. 2006. Lines of arrested growth and long bone histology in Pleistocene large mammals from Germany: what do they tell us about dinosaur physiology?—*Palaeontographica Abt. A* 277: 143–159.

Sander, P. M., Klein, N., Stein, K. & Wings, O. This volume. Sauropod bone histology and its implications for sauropod biology. *In* Klein, N., Remes, K., Gee, C. T. & Sander, P. M. (eds.). *Biology of the Sauropod Dinosaurs: Understanding the Life of Giants*. Indiana University Press, Bloomington: pp. 276–302.

Sasaki, N. & Sudoh, Y. 1997. X-ray pole figure analysis of apatite crystals and collagen molecules in bone.—*Calcified Tissue International* 60: 361–367.

Seebacher, F. 2001. A new method to calculate allometric length–mass relationships of dinosaurs.—*Journal of Vertebrate Paleontology* 21: 51–60.

Stempniewicz, M. & Pyzalla, A. R. 2003. Tendaguru sauropod dinosaurs. Characterization of diagenetic alterations in fossil bone. *In Annual Report of Structural Research Abstracts*. Hahn Meitner Institut, Berlin: pp 42–43.

Stempniewicz, M., Hinkel, P., Weiss, S., Denker, A., Sander, P. M. & Pyzalla, A. R. 2004. Anwendung präparativer Methoden aus der Werkstoffkunde zur Untersuchung von Sauropodenknochen.—*Praktische Metallographie, Sonderband* 36: 217–221.

Sugiyama, M. & G. Sisegato. 2004. A review of focused ion beam technology and its applications in transmission electron microscopy.—*Journal of Electron Microscopy* 53: 527–536.

Tanck, E., Homminga, J., van Lenthe, G. H. & Huiskes, R. 2001. Increase in bone volume fraction precedes architectural adaptation in growing bone.—*Bone* 28: 650–654.

Traub, W., Arad, T. & Weiner, S. 1989. Three dimensional ordered distribution of crystals in turkey tendon collagen fibers.—*Proceedings of the National Academy of Sciences of the United States of America* 86: 9822–9826.

Trueman, C. N. & Tuross, N. 2002. Trace elements in recent and fossil bone apatite. *In* Kohn, M. J., Rakovan, J. & Hughes, J.-M. (eds.). *Phosphates: Geochemical, Geobiological, and Materials Importance*. Mineralogical Society of America: pp. 489–521.

Trueman, C. N., Behrensmeyer, A. K., Tuross, N. & Weiner, S. 2004. Mineralogical and compositional changes in bones exposed on soil surfaces in Amboseli National Park, Kenya: diagenetic mechanisms and the role of sediment pore fluids.—*Journal of Archaeological Science* 31: 721–739.

Tütken, T. This volume. The diet of sauropod dinosaurs: implications of carbon isotope analysis on teeth, bones, and plants. *In* Klein, N., Remes, K., Gee, C. T. & Sander, P. M. (eds.). *Biology of the Sauropod Dinosaurs: Understanding the Life of Giants*. Indiana University Press, Bloomington: pp. 57–79.

Tütken, T., Pfretzschner, H.-U., Vennemann, T. W., Sun, G. & Wang, Y. D. 2004. Paleobiology and skeletochronology of Jurassic dinosaurs: implications from the histology and oxygen isotope compositions of bones.—*Palaeogeography, Palaeoclimatology, Palaeoecology* 206: 217–238.

Turner-Walker, G. & Jans, M. 2008. Reconstructing taphonomic histories using histological analysis.—*Palaeogeography, Palaeoclimatology, Palaeoecology* 266: 227–235

Upchurch, P. 1998. The phylogenetic relationships of sauropod dinosaurs.—*Zoological Journal of the Linnean Society of London* 124: 43–103.

Warren, B. E. 1990. *X-Ray Diffraction*. Dover Publications, New York.

Weiner, S. 1986a. Disaggregation of bone into crystals.—*Calcified Tissue International* 39: 365–375.

Weiner, S. 1986b. Organization of hydroxyapatite crystals within collagen fibrils.—*FEBS Letters* 206: 262–266.

Weiner, S. & Traub, W. 1989. Crystal size and organization in bone.—*Connective Tissue Research* 21: 259–265.

Weiner, S. & Wagner, H. D. 1998. The material bone: structure–mechanical function–relations.—*Annual Review of Materials Science* 28: 271–98.

Weiner, S. Arad, T. & Traub, W. 1991. Crystal organization in rat bone lamellae.—*FEBS Letters* 285: 49–54.

Weiner, S., Arad, T., Sabanay, I. & Traub, W. 1997. Rotated plywood structure of primary lamellar bone in the rat: orientations of the collagen fibril arrays.—*Bone* 20: 509–514.

Weiner, S., Traub, W. & Wagner, D. 1999. Lamellar bone. structure–function-relations.—*Journal of Structural Biology* 126: 241–255.

Wenk, H.-R. & Heidelbach, F. 1999. Crystal alignment of carbonated apatite in bone and calcified tendon: results from quantitative texture analysis.—*Bone* 24: 361–369.

Wess, T., Alberts, I., Hiller, J., Drakopoulos, M., Chamberlain, A. T. & Collins, M. 2001. Microfocused small angle X-ray scattering reveals structural features in archaeological bone samples: detection of changes in bone mineral habit and size.—*Calcified Tissue International* 70: 103–110.

Williams, C. T. & Marlow, C. A. 1987. Uranium and thorium distributions in fossil bones from Olduvai gorge, Tanzania and Knam, Kenya.—*Journal of Archaeological Science* 14: 297–309.

Wilson, J. A. 2002. Sauropod dinosaur phylogeny: critique and cladistic analysis.—*Zoological Journal of the Linnean Society of London* 136: 217–276.

Wise, L. M., Wang, Z. & Grynpas, M. D. 2007. The use of fractogaphy to supplement analysis of bone mechanical properties in different strains of mice.—*Bone* 41: 620–630.

Wolff, J. 1892. *Das Gesetz der Transformation der Knochen.* Hirschwald, Berlin.

Ziv, V. & Weiner, S. 1994. Bone crystal sizes: a comparison of transmission electron microscopic and X-ray diffraction line with broadening techniques.—*Connective Tissue Research* 30: 165–175.

Ziv, V., Wagner, H. D. & Weiner, S. 1996a. Microstructure–microhardness relations in parallel-fibered and lamellar bone.—*Bone* 18: 417–428.

Ziv, V., Sabanay, I., Arad, T., Traub, W. & Weiner, S. 1996b. Transitional structures in lamellar bone.—*Microscopy Research and Technique* 33: 203–213.

Zocco, T. G. & Schwartz, H. L. 1994. Microstructural analysis of bone of the sauropod dinosaur *Seismosaurus* by transmission electron microscopy.—*Palaeontology* 37: 493–503.

10

Finite Element Analyses and Virtual Syntheses of Biological Structures and Their Application to Sauropod Skulls

ULRICH WITZEL, JULIA MANNHARDT, RAINER GOESSLING, PASCAL DE MICHELI, AND HOLGER PREUSCHOFT

IN MORPHOLOGY AND PALEONTOLOGY, the analysis of bony structures began with the art of drawing and the technique of photography. The first analytical calculations were possible by using simplified models, and quantitative measurements of strains on bone surfaces provided important opportunities for interpreting bony structures in recent animals. The development of finite element structure analysis (FESA) was a decisive step in obtaining spatial information about strain and stress distribution in models of both extinct and extant creatures. However, the inductive approach of FESA does not provide precise explanations for the existence of bone tissue in a specific position of a given finite element model. In contrast to FESA, the deductive technique of finite element structure synthesis (FESS) was developed for deducing a biological structure from a few initial conditions and boundary conditions. This makes FESS ideal for discovering which morphological structures can be explained in terms of mechanics and which cannot. Three examples of the applications of FESS illustrate its power: the virtual synthesis of the skull of a Neanderthal (*Homo neanderthalensis*) and of the skulls of the sauropods *Diplodocus* and *Camarasaurus*. These studies demonstrate the utility of FESS for the virtual synthesis of bony structures to test assumptions and hypotheses regarding the relationship between function and structure. By obtaining a high degree of conformity between the virtual model and the real object, the method is satisfyingly validated.

Introduction

For many fossilized bony structures that result from biological evolution, we know their form precisely and in much detail, and sometimes we understand something about their function. However, the exact biomechanical reasons for their shape and structure are mostly unknown. Thus paleontological research up to now has used the technique of photography and the art of drawing to analyze fossil finds (Fig. 10.1). Paleontologists have long attempted to identify mechanical functions by analyzing simplified models of animals or their parts. The strain gauge technique was used in recent animals as a tool in this research. It allows quantitative measurements of strain in a small area on the surface of models or bony structures during functional loading. However, spatial information about the strain distribution and insights into the 3D load behavior cannot be obtained this way. Although such 3D information is essential for an effective analysis, it cannot be determined by analytical calculations because vertebrate animals as biological structures are too complex. Strategies reducing complex forms and structures into simple bars and plates provide a helpful start into the analysis of skeletal elements. However, they cannot provide detailed biomechanical insights into these forms and structures.

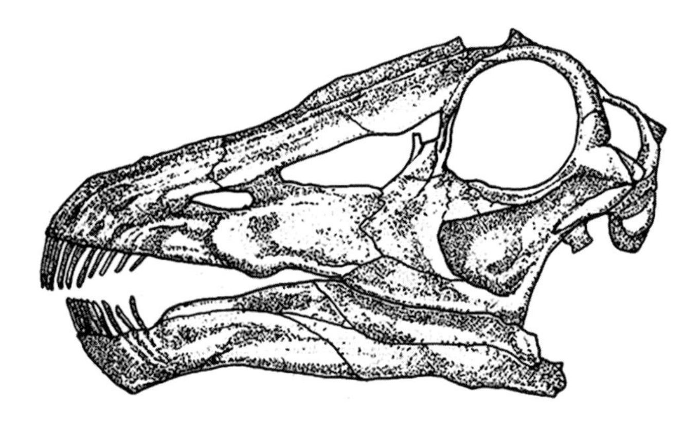

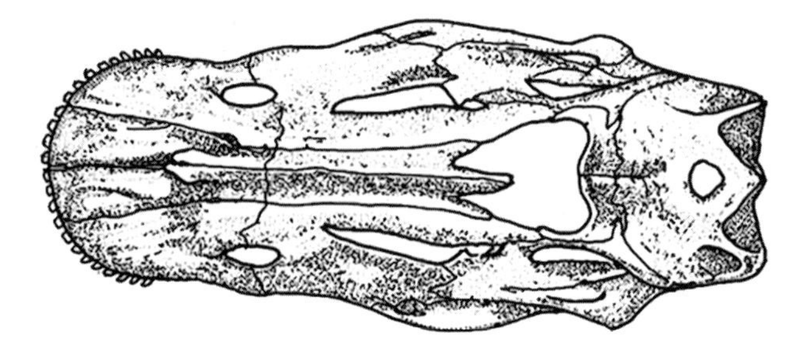

FIGURE 10.1. The skull of *Diplodocus* in lateral and dorsal views as an example of a paleontological drawing. The 3D shape is reduced to 2D projections, helping our understanding of the form as a whole. *Modified after Norman (1998).*

In approximately 1970, a new numerical method for calculating 3D strain–stress distribution in complex objects in engineering became established (Zienkiewicz 1971) that was called finite element analysis (FEA). FEA has since revolutionized mechanical engineering and has become the preferred method for optimizing engineering products with regard to strength and stability (Witzel 1980). Metal is the most important material in mechanical engineering, with its equal resistance to tensile and compressive stresses, and load cases can be calculated using FEA with axial and nonaxial loads, including bending moments. The numerical test of resistance calculates the tolerable equivalent stress, which is called the von Mises stress:

$$\sigma_e = \sqrt{0.5\,\{(\sigma_1 - \sigma_2)^2 + (\sigma_2 - \sigma_3)^2 + (\sigma_3 - \sigma_1)^2\}} = \sigma_F$$

according to the maximum energy of distortion theory

σ_i = principal stresses
σ_F = yield stress = e.g., 355 MPa in steel for medical implants.

The von Mises equation is a helpful calculation tool for technical structures.

FEA as an inductive method has been used in the field of orthopedic biomechanics for more than 20 years (Vasu et al. 1982; Huiskes 1987; Witzel 1996, 1998, 2000; Witzel et al. 2004a). In paleontology, the work of Rayfield (1998) yielded new insights into the structures of fossils some 10 years ago. Application of FEA also verified the hypothesis of evolutionary optimization—namely, that mass and energy are minimized—by a numerical method.

Finite element structure analysis (FESA) is now used in paleontology worldwide (Fastnacht et al. 2002; Rayfield 2004; Dumont et al. 2005; Ross et al. 2005; McHenry et al. 2006; Rayfield et al. 2007). However, this inductive approach is subordinate to the theory of morphogenesis and does not provide precise explanations for the existence of bone tissue in a specific position of a priori given finite element (FE) model (Fig. 10.2). In FESA, usually the design of the FE model is based on computed tomographic (CT) scans of existing skeletal structures. The absolute stresses after the application of bite and muscle forces remain uncertain because their values are not reliably known and because it is not known whether all force vectors acting in the living animal have been considered. FESA therefore does not lead to insights into features of bone shape and microanatomy that are not currently understood. FESA thus does not provide an explanation of shape, for example, of the skull, as an absolute necessity and an essential biological answer to mechanical loading (Wolff 1892; Witzel & Preuschoft 2005; Witzel 2007). Although FESA is useful for determining and testing assumed functions of a fossil skeletal structure, it is of limited value for answering questions concerning the evolutionary origin of this skeletal structure. Such questions cannot be solved by measuring the stress in models of existing structure or by calculating the stresses that occur in extinct forms. This is because the result of the evolutionary adaptation—for example, the shape of a skull or a vertebra—is already the basis of the investigation that is trying to explain this shape. The result is that the line of reasoning becomes circular. The foregoing considerations assume that skeletal structures evolved exclusively to sustain mechanical stresses and to fulfill biomechanical functions. However, we know that elements such as social display and developmental history also determine the final form.

The example in Plate 10.1 shows the FE structural analysis of a human mandible loaded on one premolar and two molars on the right side. The loads applied in this model are so-called physiological forces—that is, forces that are in the realm of what has been measured in real organisms. We might then want to examine the unknown influence of the bite and muscle forces on the contralateral side in this model (Plate 10.1B) designed on the basis of a CT reconstruction (Plate 10.1A) of an actual human mandible. The stress distribution on the surface of the mandible is shown in Plate 10.1C, and the internal stress distribution in the cortical bone and cancellous bone is represented in Plate 10.1D, E. This important information is not available in publications of other authors because their analyses show stress distributions exclusively on the bone surfaces. This can easily lead to erroneous conclusions because

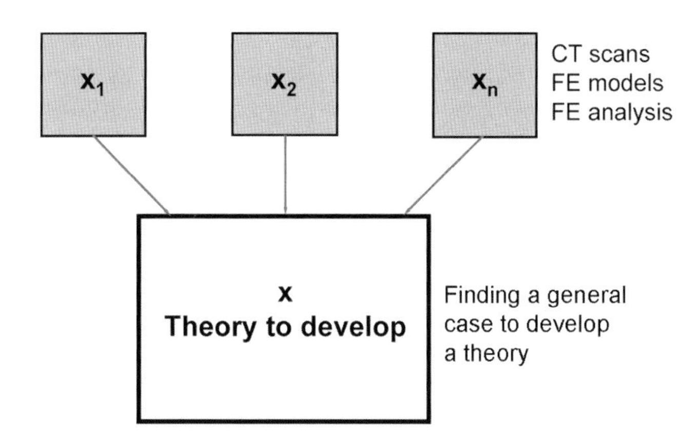

FIGURE 10.2. FESA as an inductive approach. FE computations of individual cases x_i (representing individual species) derived from CT-based models are used to develop a general theory of the evolutionary process.

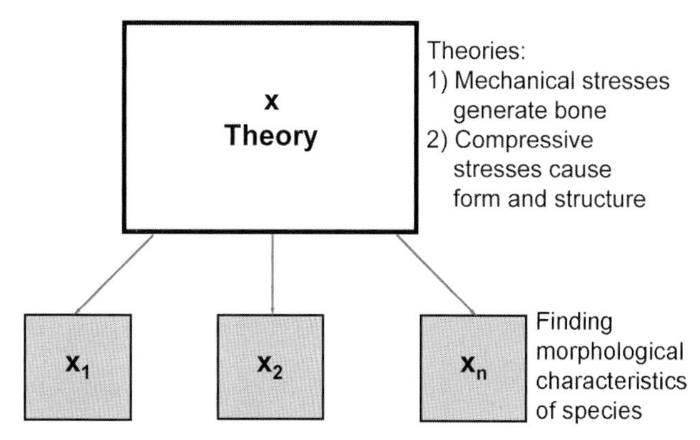

FIGURE 10.3. FESS as a deductive approach. Starting with well known theories (i.e., Wolff's law and related theories), FE computations are used to identify morphological characteristics of bony structures as a function of loading.

the stress distribution inside the bone is not uniform and there is usually a stress gradient. The 3D FESA method has the potential to show the complete stress flow in the entire volume of the structure of interest.

The opinion expressed in the literature (e.g., Rayfield et al. 2007) that very low or extremely high stresses in a FESA model loaded by physiological forces indicate an "overdesigned" or "suboptimally constructed" biological structure is not tenable because it contradicts Wolff's law (Wolff 1892). Stress extremes in such a model more likely indicate that the assumed way of loading, that is, the direction and strength of loading, is insufficient or incomplete because muscle activities have been ignored or the load application was not adequate.

For any FE model, a stable three-point support based on correct mechanical understanding and without contradiction to biological conditions must be defined. Beyond that, it has to be kept in mind that active tension chords (contracting muscles) and passive tension chords (ligaments) must be integrated into the biological structure. Tension chords are crucial because they reduce or cancel out bending stress. If bending is thus removed from a bony structure, it is loaded only by compressive forces. We posit that mainly compressive stresses occur in bony skeletal elements, and that tensile stresses are taken up by collagenous fibers in fasciae, Sharpey's fibers, ligaments, tendons, membranes, and so forth. This same principle serves to reduce material and weight in engineering design and is called *lightweight construction*. Examples of such lightweight constructions are reinforced concrete structures and suspension bridges, in which the steel reinforcements or steel ropes take up the tensile forces, the concrete itself being exposed to compression only. This makes sense because brittle materials such as concrete withstand compressive forces well but fail easily when tensile forces are applied, such as those caused by bending. We thus posit that bone is well adapted to compressive forces but poorly adapted to tensile forces, and that steel is a poor material analog for bone. The combination of bony structures with tension chords in vertebrates results in optimized lightweight construction (Rossmann et al. 2005; Witzel & Preuschoft 2005; Sverdlova & Witzel 2010).

In contrast to FESA, the deductive technique of finite element structure synthesis (FESS) (Fig. 10.3) is based on the well known theory that mechanical stresses lead to the deposition of bone tissue (within the genetic limits) and lack of stresses leads to atrophy of bone—notions identical to Wolff's law (Wolff 1892) and also recognized by Darwin (1859), Pauwels (1980), and Kummer (2005). FESS as a method was established in 1985 by synthesizing a cross section of the diaphysis of a human femur (Witzel 1985). A similar approach was developed later by Carter & Beaupre (2001). FESS starts from an unspecific homogeneous body of material, known as the *bauraum*, that offers the stresses ample volume for spreading between points of force application and supports. First, external forces are applied to the model. Then low-stress areas are iteratively removed. The resulting shapes are compared with those observed in nature. In this way, the method can be used to deduce a biological structure from a few initial conditions and boundary conditions. This makes FESS ideal for differentiating between morphological structures that can be explained in terms of mechanics and those that cannot.

FESS has been used to elucidate the mechanical reasons for the prominent bony nose in humans and other primates (Witzel & Preuschoft 1999), for the pneumatized spaces in the skulls of primates (Preuschoft et al. 2002), and for a virtual synthesis of the facial part of human and gorilla skulls (Witzel & Preuschoft 2002; Preuschoft & Witzel 2004). In the first example given here, the virtual synthesis of a typical Neanderthal skull obtained by the FESS technique demonstrates the direct correlation between functional loading and the biological structure and can therefore be used to test hypotheses regarding the relationship between structure and function during evolution. Changes in the form of the dental arcade, its position relative to the braincase, the origins of muscles, and in the volume of the brain lead to models that clearly resemble morphological differences between species or genera (Witzel, in press).

If we obtain a high degree of conformity between the model and the real object of interest, the FESS method has passed the strictest validation imaginable. FESS itself is definitely a more precise method of validation compared with strain gauge measurements, for example, which only assesses the surface of the skull at discrete points.

FESS, with its deductive approach, is normally based on the general premise of Wolff's law, but in the case of the Neanderthal skull, building the model using two specific premises offers further biomechanical insights. The first premise is that the structure of a skull is determined by powerful coupling elements consisting of lightweight constructions that connect functional spaces and loading regions such as the dental arcade, membrane fastenings, and muscle insertions. The relevant functional spaces contain the brain, the eyes, and the olfactory organ (Preuschoft et al. 2002; Witzel & Preuschoft 2002).

The second premise focuses on the relationship between skull function and shape. As noted, bone is mechanically determined (Wolff's law). This means that the morphology of bone is known to be heavily influenced by mechanical loading (Darwin 1859; Wolff 1892; Jansen 1920; Pauwels 1965; Frost 1988; Witzel 1993, 1994; Witzel & Hofmann 1993; Witzel & Preuschoft 2002, 2005). Loading due to neuromuscular activity plays a key role in bone formation and subsequent development of form in ontogeny, and phylogenetically, bone form adapts according to loading (Wong & Carter 1990).

An important mechanism for signaling loading on bone seems to be hydraulic pressure and fluid flow, which take place in Haversian and Volkmann canals (Imai et al. 1990; Qin et al. 1999, 2003). To increase the resistance of osteons to compression forces, they are surrounded by twisted collagenous fibers in several layers with different pitches (Ascenzi & Bonucci 1968; Amtmann & Doden 1981; Pidaparti & Burr 1992; Doden 1993). As a result of these fiber arrangements, which invest osteons with passive tension chords, we found by means of FE calculations that the compressive stiffness of these osteons was 24% higher than in osteons modeled to lack such fiber arrangements. With hydraulic filling of the Havesian canals, compressive stiffness increases to up to 47% (de Micheli & Witzel, unpublished data).

The considerations above indicate that bone represents an optimized compression structure. Accordingly, it is reasonable to attempt a virtual synthesis of bone form by modeling functionally relevant compression forces (Witzel & Preuschoft 1999, 2002, 2005; Preuschoft & Witzel 2004; Witzel 2007).

In 2006, we finished the first virtual synthesis of an entire *Homo neanderthalensis* skull (Plate 10.2A) (Witzel, in press). In the rest of this chapter, we provide a description of the FESS method which we are using successfully in orthopedics, anatomy, paleoanthropology, and paleontology, as well as specifically sauropod dinosaur skulls and vertebrae.

FESS Methodology as Applied to Neanderthal Skull

First, the bauraum is set up and modeled (Plate 10.2B) by more than 100,000 10-noded tetrahedral FEs (Rossmann et al. 2005; Witzel & Preuschoft 2005). The bauraum is then loaded by realistic functional, that is, physiological, forces. The Young's modulus of the bauraum is 17 GPa, and its Poisson ratio is 0.3. For each FE deformation, the mechanical stresses are calculated with the FE software ANSYS (ANSYS Inc., Canonsburg, PA). Mechanical stresses in the FEs show a specific load-relating force per area (expressed in N/mm^2 or MPa). These stresses can be transformed into fields of different bone densities: compressive stresses of -12 to -20 MPa correspond to densities of 1.3–2 g/cm^3 for cortical bone (Zioupos et al. 2008). Note that in engineering science, compressive stresses are denoted as negative values, in contrast to geological sciences, where negative values indicate tensional stresses.

The stable three-point support, which is a basic necessity for ANSYS FE calculations, is achieved by constraining the position of both the occipital condyles and the position of neck muscle insertions. Here reaction forces were applied: compression forces in the area of condyles and tensile forces in the attachment points of the neck muscles.

The insertions of the m. masseter, the m. temporalis, the temporal fascia as the active tension chord of the zygomatic arch (Witzel et al. 2004b), and the falx cerebri (Witzel & Hofmann 1993; Witzel 1994) were the locations of application of the muscle and membrane forces.

Before starting the FE computation, bite and chewing forces, muscle forces, and membrane forces were brought into equilibrium for each load case according to the following equation:

$\Sigma F_{i\ x,y,z} = 0$, with $F_{i\ x}$ = all forces in the direction of the x-coordinate, and so forth; and

$\Sigma M_{i\ x,y,z} = 0$, with $M_{i\ x}$ = all moments around the x-coordinate, and so forth.

The models were constructed on a 2.4 GHz portal and storage computer with 2 GB of RAM and a storage capacity of 240 GB. All calculations were performed on the central computer of Ruhr University of Bochum, an HP Superdome with

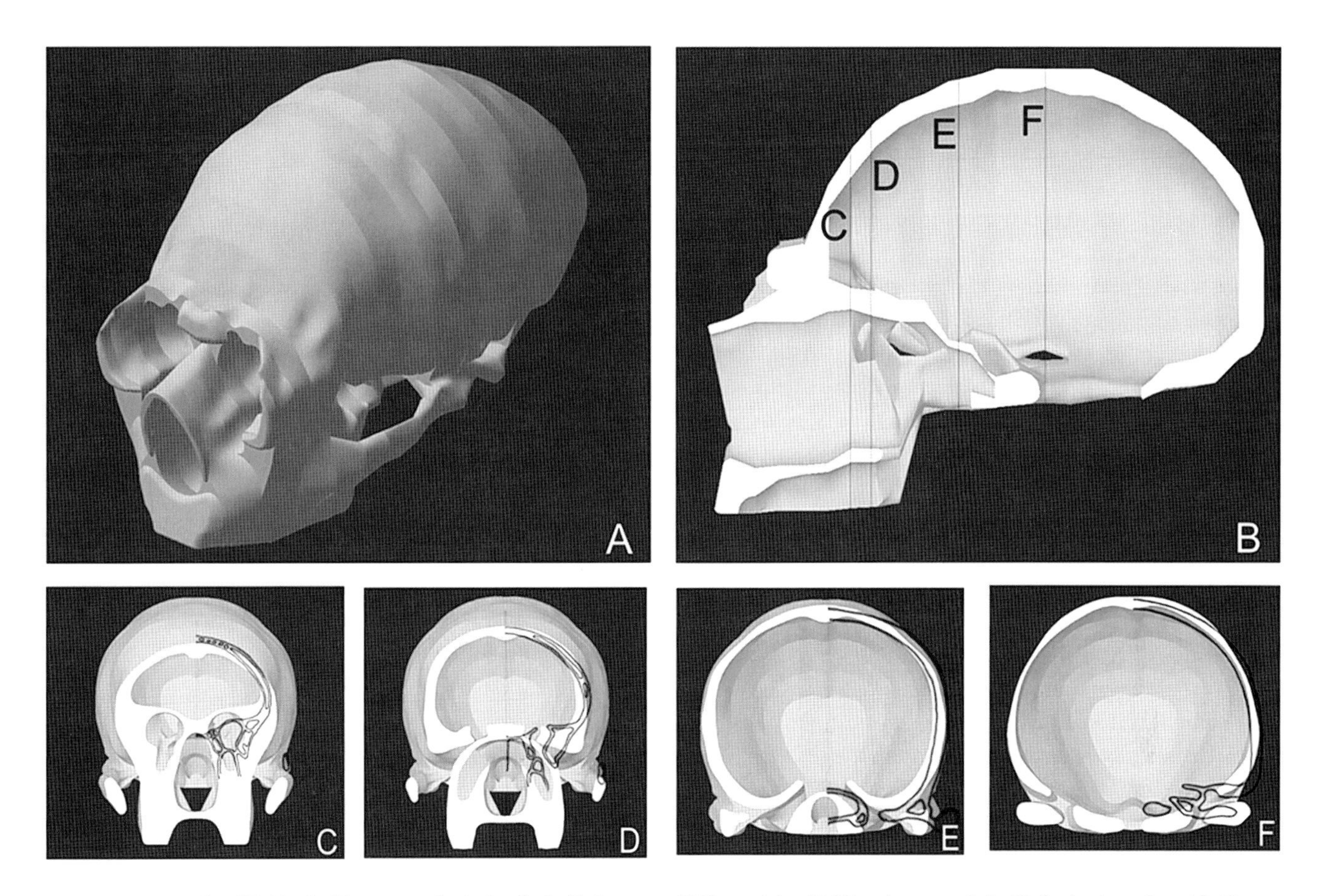

FIGURE 10.4. Result of FESS of a *Homo neanderthalensis* skull shown as CAD models. (A) View from top left. (B) Sagittal section. (C–F) Cross sections through skull, marked in (B), with superposition of the outline of bone tissue derived from CT scans of *Homo neanderthalensis* from Guattari. *Courtesy of L. Bondioli, Museo Nazionale Preistorico Etnografico, Rome.*

27 processors and 56 GB of RAM. After the models were loaded, the postprocessor of the FE program was used to extract all principal stresses. Wolff's law and Pauwels's theory of functional adaptation (Pauwels 1965, 1973, 1980) emphasize compressive stress (see above and below, second hypothesis). This was also the case in all of our studies because tensile stresses are usually taken up by muscle forces, and in all realistic physiological load cases, the bending moments are zero (Witzel & Preuschoft 2005).

The initial conditions employed in constructing the bauraum of the models are primarily the functional spaces for the brain, the eye openings, and the nasal cavity (Plate 10.2C, D). This seems justified because the brain and the sense organs precede the skeletal elements in ontogeny. They are already present before the development of the skeletal parts starts under the influence of mechanical stress (see below, first hypothesis). Further initial conditions are the muscle forces and the placement of the dental arcade, including assumed bite and chewing forces, and the spatial relationships of these (Plate 10.2E, F). By maintaining an equilibrium of forces, the primary 3D stress distribution in each load case (load cases are based on the same bauraum but employ different biting positions and movements) are added up by physiological superposition, a newly developed load case technique in biomechanical calculations. In physiological superposition, each FE accumulates the highest value of compressive stress from each load case. It is not logical to load the bauraum simultaneously with all possible external forces because this would be nonphysiological and would result in unrealistically high moments and stresses.

If the stress-free parts in all cross sections (e.g., Plate 10.3C–F) are eliminated and the summative stress flows are maintained, a reduced model results that is similar to the real skull (Plate 10.3A, B). This reduction of shape can be repeated iteratively and step by step leads to a more exact form. The final iteration is reached at a rather invariable physiological compressive stress distribution between −2 and −20 MPa (Witzel 2007). The first value is that of cancellous bone, and the second is that of cortical bone. Frost (1988) only published a single value of 11 MPa (±25%) for compact bone. The final FE model in Fig. 10.4A, B was created with the CAD software CATIA V5 (Dassault Systèmes, Paris, France), and the resulting cross sections were then compared with CT scans (Fig. 10.4C–F) of a real Neanderthal skull.

Application of FESS to Sauropods

Two case studies that used FESS based on the two premises explained above were carried out on sauropods within Research Unit 533. The first study (Witzel & Preuschoft 2005) consisted of the FE model construction and virtual synthesis of the skull of the diplodocoid sauropod *Diplodocus*, and the second resulted in the virtual synthesis of the skull and lower jaw of the basal macronarian sauropod *Camarasaurus* (Witzel 2007). In both cases, FESS helped us to identify muscle forces that played a role in shaping these bony structures. As a result, it will be possible to search for these muscles in future studies, their insertion areas in the fossils, and their forces.

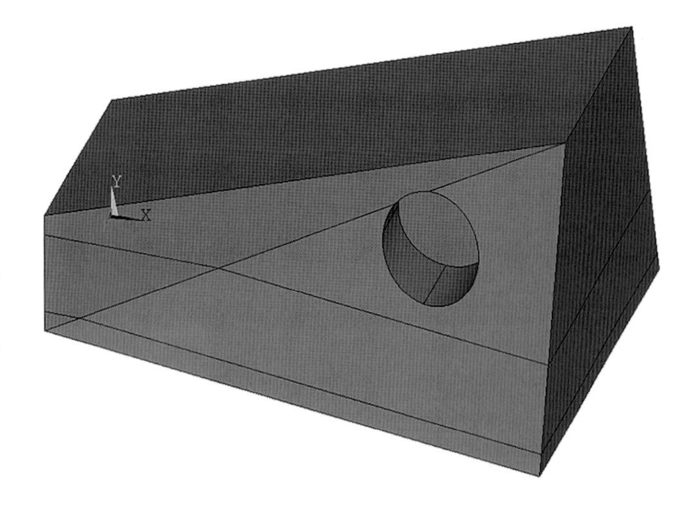

FIGURE 10.5. Wedge-shaped bauraum solid of the *Diplodocus* skull in caudolateral view. This 3D bauraum served as the starting point for the FESS. Only the position of the orbitae and the shape and position of the dental arcade (not shown here) were used as a priori constraints on the form.

VIRTUAL SYNTHESIS OF A *DIPLODOCUS* SKULL

To synthesize the skull of the sauropod dinosaur *Diplodocus* from the Upper Jurassic Morrison Formation of North America, the cast skull of the *Diplodocus* skeleton in the Senckenberg Museum, Frankfurt, Germany, was studied. The geometry of the bauraum (Fig. 10.5) was derived from the outer shape, the dental arcade, and the positions of the jaw joints of the fossil skull (Fig. 10.6). The insertions of the m. intramandibularis and m. pterygoideus anterior (which provide active tension chords, taking their origins on the mandible, maxilla, and premaxilla), the m. pseudotemporalis, the m. depressor mandibulae, and the longitudinal musculature (dorsal neck muscles) were used to determine the sites of application of the muscle forces in the model. The Young's modulus was assumed to be 17 GPa, and the Poisson ratio was set to 0.3 in the model.

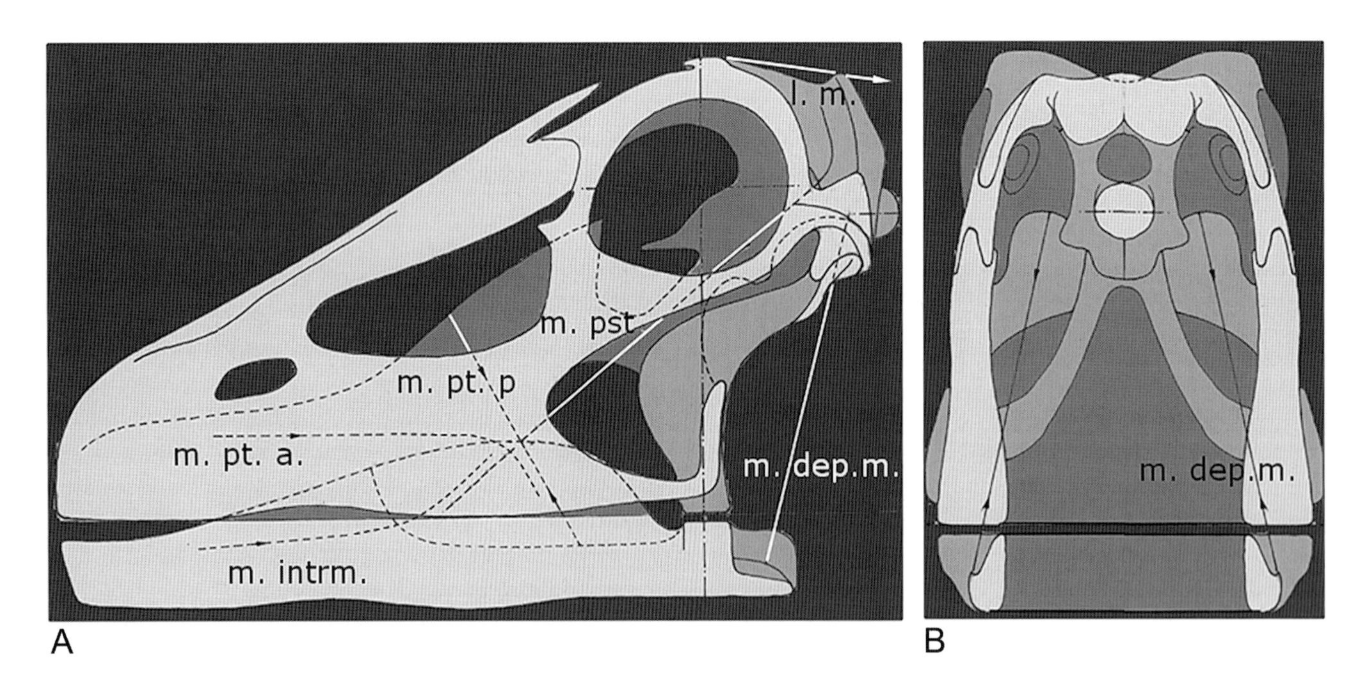

FIGURE 10.6. Analysis of the fossil skull of *Diplodocus* based on the cast skull of the mount on display at the Senckenberg Museum (Frankfurt, Germany). (A) Lateral view. (B) Caudal view. Assumptions are shown about the courses of muscles, serving as a priori constraints for the FESS. l. m., longitudinal musculature (dorsal neck muscles); m. dep. m., m. depressor mandibulae; m. intrm., m. intramandibularis; m. pst., m. pseudotemporalis; m. pt. a., m. pterygoideus anterior; m. pt. p., m. pterygoideus posterior.

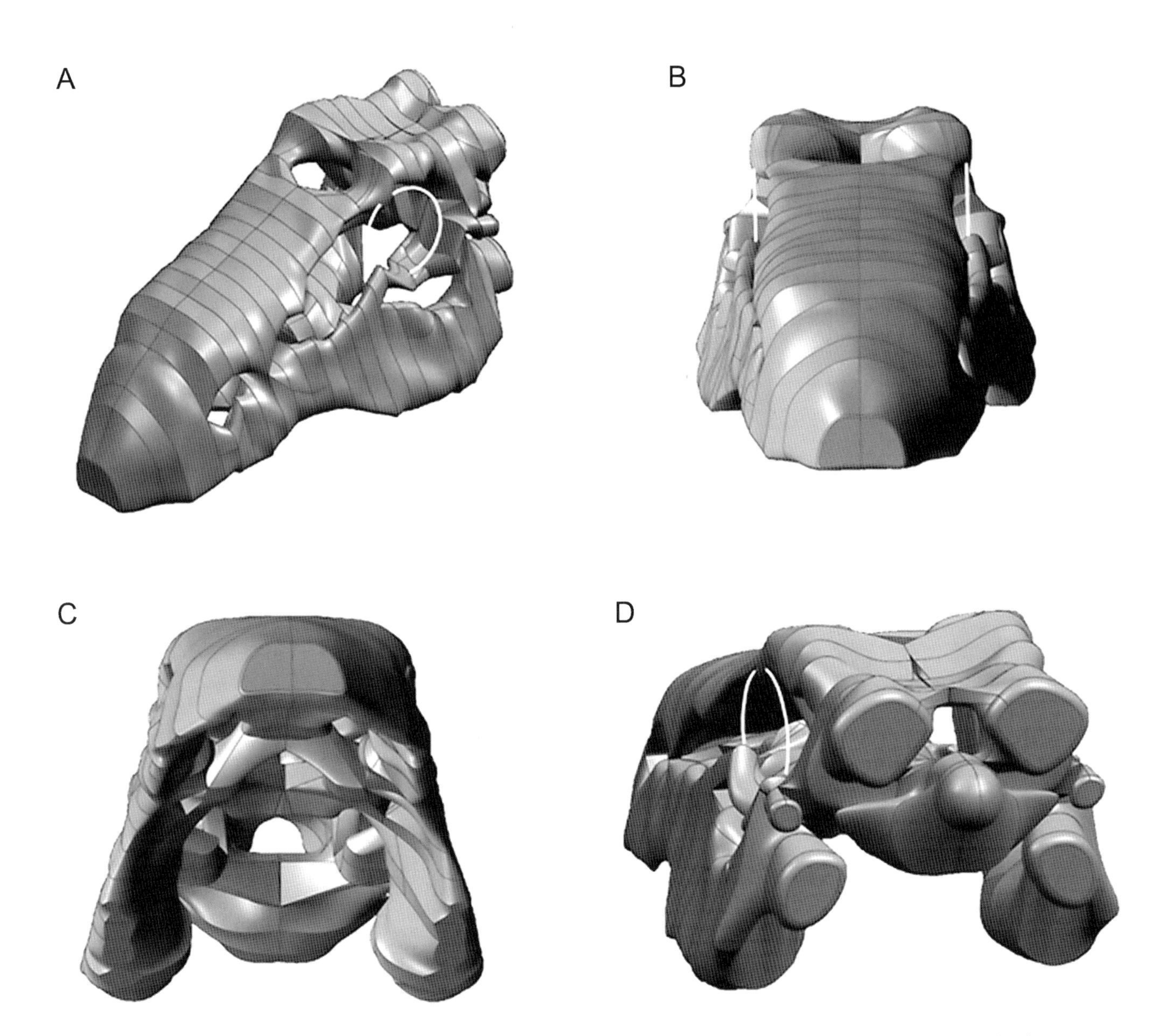

FIGURE 10.7. Complete reduced model of the *Diplodocus* skull in a CAD presentation created with CATIA V5 software. (A) Left dorsolateral view. (B) Cranial view. (C) Cranioventral view. (D) Caudolateral view. Note that FESS reproduces the position of the external nares, the antorbital and preantorbital fenestrae, and the foramen magnum. The white circle indicates the position of orbit as entered into the model as an a priori assumption.

To synthesize the skull of *Diplodocus,* only five a priori assumptions were made: the position of the two eye openings in the bauraum, the shape and position of the dental arcade, the relative bite forces, the muscle insertions, and the estimated activity of the muscles under the condition of equilibrium of the entire system. The stable three-point support was achieved by constraining the positions of the two quadratomandibular joints and of the occipital condyle. Load cases were biting, lateral pulling to either side, and opening of the mouth, as well as gravity. The primary 3D stress distributions of each load case were summed up by physiological superposition.

After FE calculation and the elimination of stress-free FEs, a reduced model resulted that was similar to the real skull. This reduction of shape was repeated iteratively three times and led to a more exact form, in which a physiological stress distribution in the total structure was reached. More specifically, in the different structural regions of the skull, the stresses assume rather uniformly moderate physiological values.

Plate 10.4B shows the distribution of compressive stress on the surface of the bauraum. Areas under high load are the quadratomandibular joint, the basioccipital, and the supraoccipital; the unloaded or little-stressed regions are the corners of the bauraum. An important result of this 3D calculation is that we were able to visualize the stress distribution in the interior of the model volume in cross section. For example, cross section 2 (Plate 10.5A; see Plate 10.4A for section

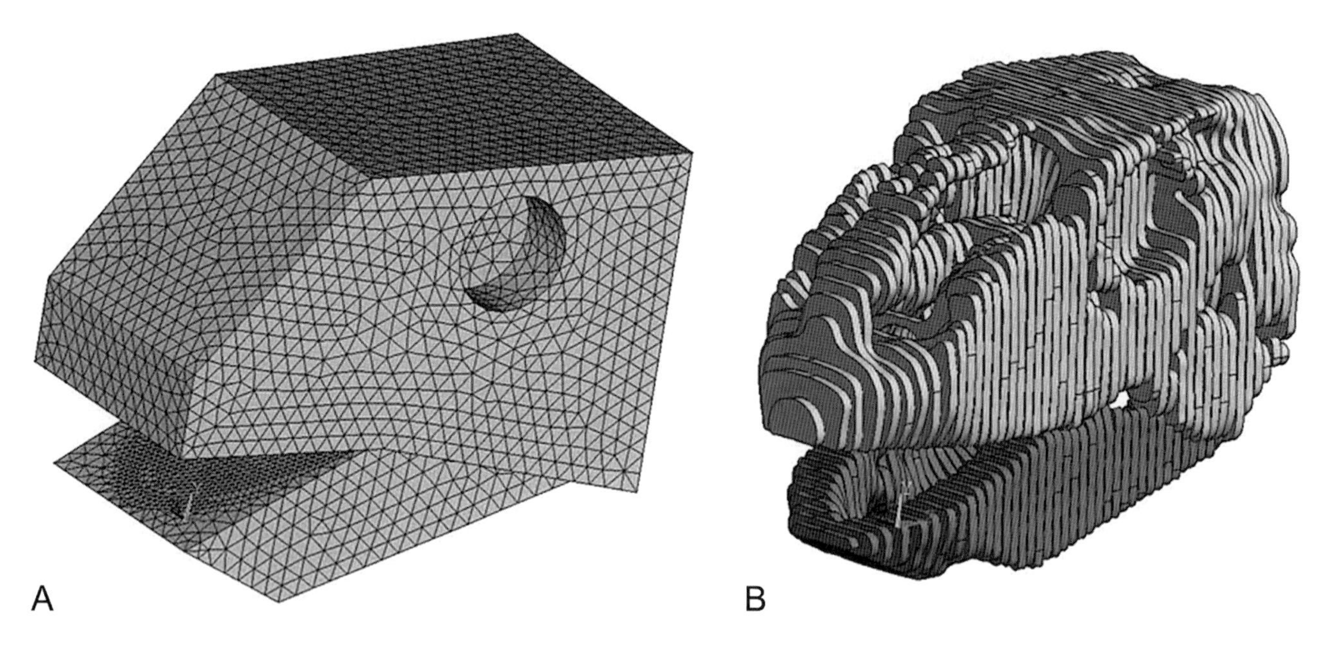

FIGURE 10.8. *Camarasaurus* skull model before and after the first iteration in the FESS computation. (A) Model meshed with FE. (B) Stepped model resulting from the first iteration. Unloaded or little-stressed regions were deleted. The positions of the external nares and infratemporal fenestra are already visible.

numbers) shows the snout close to the anterior teeth. Cross section 14 provides information on stress distributions in the palate (Plate 10.5B), showing three openings (low-stress areas corresponding to the positions of openings in the real skull) and the antorbital fenestra. Cross section 20 at the level of the orbits (Plate 10.5C) shows the quadratomandibular joint, the lower temporal opening, the unossified space below the skull roof, and the end of the nasal opening in the frontal. The basioccipital, supraoccipital, and foramen magnum are represented by different regions of low values of compressive stresses in cross section 26 (Plate 10.5D).

After the lightly stressed parts in the cross sections were deleted by applying an arbitrarily selected threshold of -0.8 N/mm^2, the equivalents of the bony elements remain (Plate 10.6). Cross section 4 (Plate 10.6A) through the middle of the dental arcade shows a low-stressed region between palate and premaxilla, corresponding to an open space in the *Diplodocus* skull, which contains the m. pterygoideus anterior as an active tension cord to prevent upward bending of the snout during the bite (Fig. 10.6A). The force in this active tension chord is considered in our analysis and is an important factor in shaping the bony structure. Cross section 14 shows the three openings in the palate and the antorbital fenestrae (Plate 10.6B). Cross section 18 cuts through the beginning of the eye openings (Plate 10.6C) and the middle of the nasal opening in the bones of the forehead. An opening in the skull roof was not expected at this location but exists in diplodocoid sauropods because the external nares have moved up onto the skull roof. Below the orbits, an intratemporal fenestra extends downward and forward to separate the postorbital from the quadratojugal and the quadrate, which forms a column bracing the mandibular joint. The occipital segment of the model (cross section 27, Plate 10.6D) shows the equivalent of the occipital condyle.

The merging of all cross sections leads to a 3D reconstruction of the reduced model as an embodiment of compressive stresses (Preuschoft & Witzel 2004). The result, rendered in the CAD software Solid Edge, is presented in Fig. 10.7. In Fig. 10.7A, the reduced model is seen from the top left with its characteristic silhouette of the snout. In front of the orbit (white circle) are the antorbital and preantorbital fenestrae. Below the orbit lies the infratemporal fenestra above the quadratomandibular joint. The region of the forehead and skull table is characterized by a wide opening, the nasal opening, and a massive plate for the insertions of the strong neck muscles. Figure 10.7B and C shows the anterior view from above and from below; in the latter view, there is a free passage through the braincase and the foramen magnum. Above the occipital condyle and on both sides of the foramen magnum, the skull model of *Diplodocus* shows two smooth surfaces (Fig. 10.7D) that could represent a joint-like contact with the anteriormost cervical vertebrae.

For the next step of reduction, this 3D model was again meshed with 10-noded tetrahedral FEs. The subsequent computation with the same loading regime (load case technique) as in the first iteration yields the surprising results shown in Plate 10.7. All cortical regions change to a more differentiated, walled construction. If the lightly stressed parts in all cross

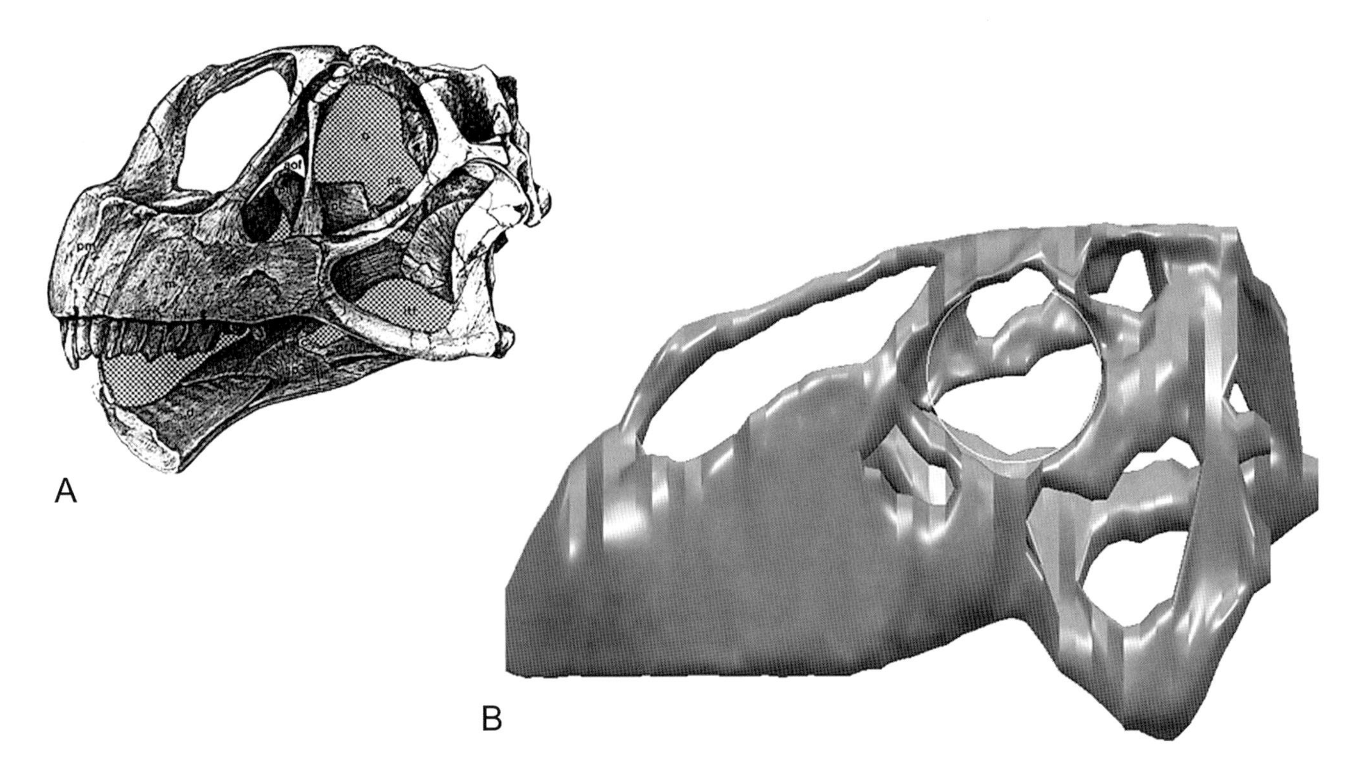

FIGURE 10.9. Skull of *Camarasaurus* in lateral view. (A) *Fossil skull as illustrated by Madsen et al. 1995.* (B) Result of the second iteration of the FESS computation (based on the first iteration model shown in Fig. 10.8B), visualized with the CAD software CATIA V5. Note that FESS clearly reproduces the position and relative size of the external nares, the antorbital fenestra, and both temporal openings. The white circle indicates the position of the orbit as entered into the model as an a priori assumption.

sections are eliminated with a threshold of -2 N/mm^2 (1% of the ultimate strength of Harversian bone: 172–220 N/mm^2) or are assigned a lower stiffness, the synthesized structure closely resembles the skull of *Diplodocus*. In Plate 10.7A, the wall-thinning effect is demonstrated in the snout. Other thick walls, such as the wall of the premaxilla (Plate 10.7B), are transformed into a sandwich structure consisting of a nearly empty space representing spongy bone between thin cortical layers. This is a principle that engineers also use in technical lightweight constructions.

VIRTUAL SYNTHESIS OF A *CAMARASAURUS* SKULL

The second example of a virtual synthesis of a sauropod skull is that of the basal macronarian *Camarasaurus* (Witzel 2007). The starting point was the bauraum of the skull that we designed based on specimen "ET" in the Sauriermuseum Aathal near Zurich, Switzerland. The dimensions of the bauraum solid are as follows: length 580 mm, height 380 mm, width 120 mm. The computation started with the same standard materials values of the bony material and the same five a priori assumptions about the shape of the bauraum that we used in the case study of *Diplodocus*. As in the *Diplodocus* synthesis, each of the 15 load cases (LC) consisted of the application of forces to two consecutively more posterior teeth, starting with the middle of the tooth row, that is, the symphysial teeth, and continuing down either side of the jaw. Thus, LC1 consists of 2×200 N applied to the symphysial teeth; LC2r consists of 2×320 N applied to the right side of the jaw; and LC2l is the same as LC2r but with the force applied to the left side. The forces for LC3r and LC3l were 2×444 N; those for LC4l and LC4r were 2×570 N; those for LC5l and LC5r were 2×720 N; those for LC6r and LC6l were 2×840 N; and finally, those for LC7r and LC7l were 2×602 N applied to the posteriormost teeth. On one hand, the force values for each load case were derived from a preliminary computation of possible bite forces with different lever arms and a constant biting moment, and on the other hand resulted from an analysis of mechanical tooth stability.

As mentioned above, it is essential for a successful FESS to locate the insertions of the muscles correctly, to bring the skull model into equilibrium and to include a stable three-point support. Plate 10.8 shows the meshed bauraum of the skull of *Camarasaurus* with bite and muscle forces included. Note that the m. pterygoideus anterior takes over the function of an active tension chord as in *Diplodocus* to eliminate the bending stress produced by biting. To test the effects of including the mandible, we modified the model as shown in Fig. 10.8A. The first iteration led to a stepped model (Fig. 10.8B). The structure synthesized in this first approach already resembled the fossil.

Further iterations led to an ever-closer approximation of the actual shape of the skull. This is demonstrated in Fig. 10.9, which compares the results of the FESS with anatomical drawings by Madsen et al. (1995) of another *Camarasaurus* skull.

The best result of this synthesis was produced by the third iteration, which is shown in Plate 10.9 in comparison with the second iteration. The distribution of compressive stresses (proportional to bone density; see above) represents a wall-thinning effect (Plate 10.9B, D) in the case of *Diplodocus*. The bone reduction takes place in internal FEs. The outer surfaces are exposed to rather high stresses. We call this the *skin effect* in FESS.

In our FESS studies, minor structures such as very thin shells and thin bars were not synthesized because of the practical limits of the FE software that is available at our university.

Conclusions

The success of our virtual syntheses, which has lead to shapes closely resembling the real shape of the skulls that we attempted to synthesize, confirms our assumptions and the theories underlying our deductive approach. Our results show the strong correlation between functional loading and biological structure and shape. The syntheses suggest that the evolution of skull form reflects natural selection for "optimal" skull construction, where the optimal shape is defined as maximal strength with minimal material.

We are convinced that these syntheses will provide access to new strategies for investigating the physiological loading of skulls as biological lightweight constructions. Our new biomechanical approach can be used for syntheses of fossils as well as extant vertebrates, and it makes it possible to obtain insight into morphological structures and the biomechanical constraints on the evolution of their shape.

Acknowledgments

We thank Gerhard Plodowski (Senckenberg Museum, Frankfurt) and Hans-Jakob Siber (Sauriermuseum Aathal) for access to the skulls of *Diplodocus* and *Camarasaurus*, respectively. Our special thanks go to Don Henderson (Royal Tyrell Museum, Drumheller) and one anonymous reviewer for careful reading and constructive criticism. We also thank the editors of this volume for all their support. This study is contribution number 69 of the DFG Research Unit 533 "Biology of the Sauropod Dinosaurs: The Evolution of Gigantism."

References

Amtmann, E. & Doden, E. 1981. Anpassung der Knochenstruktur an mechanische Beanspruchung.—*Zeitschrift für Morphologie und Anthropologie* 72: 1–21.

Ascenzi, A. & Bonucci, E. 1968. The compressive properties of single osteons.—*Anatomical Record* 161: 377–392.

Carter, D. R. & Beaupre, G. S. 2001. *Skeletal Function and Form: Mechanobiology of Skeletal Development, Aging and Regeneration.* Cambridge University Press, Cambridge.

Darwin, C. 1859. *On the Origin of Species by Means of Natural Selection.* John Murray, London.

Doden, E. 1993. The relationship between the function and the inner cortical structure of metacarpal and phalangeal bones. *In* Preuschoft, H. & Chivers, D. J. (eds.). *Hands of Primates.* Springer, Vienna: pp. 271–284.

Dumont, E. R., Piccirillo, J. & Grosse, I. R. 2005. Finite element analysis of biting behavior and bone stress in the facial skeletons of bats.—*Anatomical Record Part A* 283A: 319–330.

Fastnacht, M., Hess, N., Frey, E. & Weiser, H. P. 2002. Finite element analysis in vertebrate palaeontology.—*Senckenbergiana Lethaea* 82: 195–206.

Frost, H. M. 1988. Vital biomechanics: proposed general concepts for skeletal adaptations to mechanical usage.—*Calcified Tissue International* 42: 145–156.

Huiskes, R. 1987. Finite element analysis of acetabular reconstruction. Noncemented threaded cups.—*Acta Orthopaedica* 58: 620–625.

Imai, M., Shibata, T., Moriguchi, K. & Takada, Y. 1990. Fluid-path in the mandible and maxilla.—*Okajimas Folia Anatomica Japonica* 67: 243–247.

Jansen, M. 1920. *On Bone Formation, Its Relation to Tension and Pressure.* Longmans, Green, London.

Kummer, B. 2005. *Biomechanik, Form und Funktion des Bewegungsapparates.* Deutscher Ärzte-Verlag, Cologne.

Madsen, J. H. J., McIntosh, J. S. & Berman, D. S. 1995. Skull and atlas–axis complex of the Upper Jurassic sauropod *Camarasaurus* Cope (Reptilia: Saurischia).—*Bulletin of the Carnegie Museum of Natural History* 31: 1–115.

McHenry, C. R., Clausen, P. D., Daniel, W. J. T., Meers, M. B. & Pendharkar, A. 2006. Biomechanics of the rostrum in crocodilians: a comparative analysis using finite element analysis.—*Anatomical Record Part A* 288: 827–849.

Norman, D. B. 1998. *Dinosaurs.* Barnes & Noble, New York.

Pauwels, F. 1965. *Gesammelte Abhandlungen zur funktionellen Anatomie des Bewegungsapparates.* Springer, Berlin.

Pauwels, F. 1973. *Atlas zur Biomechanik der gesunden und kranken Hüfte.* Springer, Berlin.

Pauwels, F. 1980. *Biomechanics of the Locomotor Apparatus.* Springer, Berlin.

Pidaparti, R. M. V. & Burr, D. B. 1992. Collagen fiber orientation and geometry effects on the mechanical properties of secondary osteons.—*Journal of Biomechanics* 25: 869–880.

Preuschoft, H., Witte, H. & Witzel, U. 2002. Pneumatized spaces, sinuses and spongy bones in the skulls of primates.—*Anthropologischer Anzeiger* 60: 67–79.

Preuschoft, H. & Witzel, U. 2004. Functional structure of the skull in Hominoidea.—*Folia Primatologica* 75: 219–252.

Qin, L., Mak, A. T., Cheng, C. W., Hung, L. K. & Chan, K. M. 1999. Histomorphological study on pattern of fluid movement in cortical bone in goats.—*Anatomical Record* 255: 380–387.

Qin, Y. X., Kaplan, T., Saldanha, A. & Rubin, C. 2003. Fluid pressure gradients, arising from oscillations in intramedullary pressure, is correlated with the formation of bone and inhibition of intracortical porosity.—*Journal of Biomechanics* 36: 1427–1437.

Rayfield, E. J. 1998. Finite element analysis of the snout of *Megalosaurus bucklandi*.—*Journal of Vertebrate Paleontology* 18: 71A.

Rayfield, E. J. 2004. Cranial mechanics and feeding in *Tyrannosaurus rex*.—*Proceedings of the Royal Society B: Biological Sciences* 271: 1451–1459.

Rayfield, E. J., Milner, A. C., Xuan, V. B. & Young, P. G. 2007. Functional morphology of spinosaur "crocodile-mimic" dinosaurs.—*Journal of Vertebrate Paleontology* 27: 892–901.

Ross, C. F., Patel, B. A., Slice, D. E., Strait, D. S. & Dechow, P. C. 2005. Modeling masticatory muscle force in finite element analysis: sensitivity analysis using principal coordinates analysis.—*Anatomical Record Part A* 283A: 288–299.

Rossmann, T., Witzel, U. & Preuschoft, H. 2005. Mechanical stress as the main factor in skull design of the fossil reptile *Proterosuchus* (Archosauria). *In* Rossmann, T. & Tropea, C. (eds.). *Bionik: aktuelle Forschungsergebnisse in Natur-, Ingenieur- und Geisteswissenschaft*. Springer, Berlin: pp. 517–528.

Sverdlova, N. & Witzel, U. 2010. Principles of determination and verification of muscle forces in the human musculoskeletal system: muscle forces to minimise bending stress.—*Journal of Biomechanics* 43: 387–396.

Vasu, R., Carter, D. R. & Harris, W. H. 1982. Stress distributions in the acetabular region before and after total joint replacement.—*Journal of Biomechanics* 15: 155–164.

Witzel, U. 1980. Untersuchungen über die temperaturabhängige dynamische Tragfähigkeit von Seilendverbindungen mit Aluminium-Pressklemmen.—*Ruhr-Universität Bochum, Schriftenreihe des Instituts für Konstruktionstechnik* 80 (3): pp. 1–140.

Witzel, U. 1985. *Zur Biomechanik der ossären Schafteinbettung von Hüftendoprothesen. 5. Symposium über isoelastische RM-Hüftprothesen*. Mathys, Bettlach.

Witzel, U. 1993. Biomechanische Untersuchungen am Verbundsystem des Binde- und Stützgewebes mit der Methode der finiten Elemente. *In* Pesch, H. J., Stöss, H. & Kummer, B. (eds.). *Osteologie aktuell VII*. Springer-Verlag, Berlin: pp. 91–97.

Witzel, U. 1994. Über die Spannungsverteilung in der Calvaria aufgrund physiologischer und traumatischer Beschleunigungen des Schädels.—*Osteologie* 3 (Supplement 1): 50.

Witzel, U. 1996. *Mechanische Integration von Schraubpfannen*. G. Thieme Verlag, Stuttgart.

Witzel, U. 1998. Distribution of stress in a hemispherical RM cup and its bony bed. *In* Bergmann, E. G. (ed.). *The RM Cup-Monograph of a Coated Acetabular Implant*. Einhorn Presse Verlag, Hamburg: pp. 29–41.

Witzel, U. 2000. Biomechanische und tribologische Aspekte der Kniegelenkendoprothetik. *In* Eulert, J. & Hassenpflug, J. (eds.). *Praxis der Knieendoprothetik*. Springer, Berlin: pp. 19–32.

Witzel U. 2007. A case study of *Camarasaurus* as an illustrative example for a new method for virtual synthesis of skulls in vertebrates.—*Hallesches Jahrbuch für Geowissenschaften, Beiheft* 23: 73–78.

Witzel, U. In press. Virtual synthesis of the skull in Neanderthals by FESS. *In* Condemi, S., Schrenk, F. & Weniger, G. C. (eds.). *Neanderthals: Their Ancestors and Contemporaries*. Springer, Berlin.

Witzel, U. & Hofmann, P. 1993. Über den Einfluss der durch Beschleunigungskräfte belasteten dura mater enzephali auf die Spannungsverteilung in der Calvaria.—*Osteologie* 2 (Supplement 1): 37.

Witzel, U. & Preuschoft, H. 1999. The bony roof of the nose in humans and other primates.—*Zoologischer Anzeiger* 60: 103–115.

Witzel, U. & Preuschoft, H. 2002. Function-dependent shape characteristics of the human skull.—*Anthropologischer Anzeiger* 60: 113–135.

Witzel, U. & Preuschoft, H. 2005. Finite element model construction for the virtual synthesis of the skulls in vertebrates: case study of *Diplodocus*.—*Anatomical Record Part A* 283A: 391–401.

Witzel, U., Rieger, W. & Effenberger, H. 2004a. Dreidimensionale Spannungsanalyse bei Schraubpfannen und ihren knöchernen Lagern. *In* Effenberger, H. (ed.). *Schraubpfannen*. MCU, Austria: pp. 77–86.

Witzel, U., Preuschoft, H. & Sick, H. 2004b. The role of the zygomatic arch for the statics of the skull and its adaptive shape.—*Folia Primatolgica* 75: 202–218.

Wolff, J. 1892. *Das Gesetz der Transformation der Knochen*. Hirschwald, Berlin.

Wong, M. & Carter, D. R. 1990. A theoretical model of endochondral ossification and bone architectural construction in long bone ontogeny.—*Anatomy and Embryology* 181: 523–532.

Zienkiewicz, O. C. 1971. *The Finite Element Method in Engineering Science*. McGraw-Hill, London.

Zioupos, P., Cook, R. B. & Hutchinson, J. R. 2008. Some basic relationships between density values in cancellous and cortical bone.—*Journal of Biomechanics* 41: 1961–1968.

11

Walking with the Shoulder of Giants: Biomechanical Conditions in the Tetrapod Shoulder Girdle as a Basis for Sauropod Shoulder Reconstruction

BIANCA HOHN

MOST EXTANT STUDIES of dinosaur locomotor systems have concentrated on the hindlimbs and pelvic girdle. As a result, the functional morphology of the shoulder girdle and forelimbs is poorly understood. In this chapter, the biomechanics of the tetrapod shoulder girdle are investigated to provide a basis for understanding locomotion in sauropod dinosaurs. For this purpose, the finite element method is used in two different approaches. First, the basic static conditions of force transmission between the trunk and shoulder girdle in tetrapods are analyzed by means of rather simple finite element models. Second, a 3D finite element structure synthesis (FESS) of the scapulocoracoid in an extant crocodile was conducted. Because FESS is mainly based on Wolff's law and Pauwels's causal morphogenesis, both of which predict the relation between form and function in bones, this study examines the conditions at the shoulder girdle of a crocodilian (*Caiman crocodylus*) by synthesizing the scapulocoracoid. In doing so, the muscles that are necessary to keep the shoulder joints in equilibrium under static conditions were determined. Finally, a plausible reconstruction of the shoulder girdle in a sauropod dinosaur (*Diplodocus longus*) is presented. The reconstruction modeled on these results is discussed in detail, especially in view of their biomechanical implications for the statics of the shoulder girdle.

Introduction

Sauropod dinosaurs were the largest animals that ever walked on earth. The upper limit for body size in sauropod dinosaurs, based on rather conservative estimations, is currently assumed to be between 50 and 80 metric tons (Alexander 1998; Sander et al. 2010). In contrast, the maximum body size in theropod dinosaurs is between 7 and 14 metric tons (Anderson et al. 1985; Alexander 1989; Farlow et al. 1995; Therrien & Henderson 2007). The lighter weights of the theropods can be explained by their bipedal locomotion, by which their entire body weight is supported exclusively by the two hindlimbs (Therrien & Henderson 2007), and by the limitation of available food, which in fact holds true for all carnivores (Burness et al. 2001).

Most investigations concerning body size in sauropod dinosaurs refer to ecological and/or physiological determinants as the most influential factors (Janis & Carrano 1992; Farlow 1993; Farlow et al. 1995; Burness et al. 2001; Sander et al. 2010). In addition to these aspects, my approach considers the locomotion as it directly influences the life of each animal, that is, in regard to foraging or avoiding predators. Different types of locomotion require different functional adaptations of the musculoskeletal system to fulfill specific mechanical requirements using a minimum amount of energy. There is a direct correlation between the posture of the limbs, their biomechanics, energy-saving strategies, and costs of locomotion (Reilly et al. 2007). Because of this, each skeletal element must be adapted to the specific biomechanical and energetic requirements acting on the system. During locomotion on land, the most critical point is the transmission of body weight from the (heavy) trunk to the (supporting) limbs, which is accomplished by the limb girdles.

All sauropod dinosaurs were quadrupedal animals, but they differed from one another in the proportion of the weight carried by the forelimbs. On the basis of the position of the center of mass, the majority of the body weight in most sauropods was supported by the hindlimbs (Alexander 1989; Henderson 1999, 2006). In contrast, *Brachiosaurus* and titanosaurids, such as *Saltasaurus* and *Opisthocoelicaudia,* supposedly increased the amount of weight that could be borne by the forelimbs (Henderson 2006).

This study investigates the biomechanics of the shoulder girdle–forelimb system as a basis for the reconstruction of the shoulder girdle in sauropods and for further studies that may determine whether there are functional adaptations in the shoulder girdle that facilitate an increase in body weight. This approach may prove to be important in future reconstructions of the lifestyle and biology of sauropod dinosaurs.

HISTORICAL REVIEW

There is still little known about the functional morphology of the shoulder girdle and forelimbs of sauropods. Body posture has mainly been reconstructed on the basis of comparisons with living reptiles such as crocodiles, or even mammals. In the early days of dinosaur research, a sprawling posture of the limbs, similar to that of modern crocodiles, was assumed by some researchers (Hay 1908; Tornier 1909). Others doubted such reconstructions because of mechanical difficulties and instead hypothesized a semiaquatic lifestyle in which buoyancy would support the animals' enormous body weight (Marsh 1878; Colbert 1962; Romer 1966; Swinton 1970). In these early studies, some authors already maintained that the animals walked with straight limbs in a quadrupedal manner (Osborn 1899; Hatcher 1901; Matthew 1910; Abel 1910; Gilmore 1932), but it was not until the 1970s that the current consensus of sauropods as graviportal, fully terrestrial animals arose (Bakker 1971; Coombs 1975, 1978; Alexander 1976). More recent investigations have estimated gaits and speeds of sauropod dinosaurs (Alexander 1976, 1985, 1989; Thulborn 1990) and addressed the posture and evolution of the forelimbs, but the shoulder girdle itself has not yet been considered in detail (cf. Christian & Preuschoft 1996; Christiansen 1997; Christian et al. 1999a, 1999b; Wilson & Carrano 1999; Christian 2002; Bonnan 2003, 2004; Carrano 2005; Bonnan & Yates 2007).

RECONSTRUCTION OF THE SHOULDER GIRDLE

Most recent reconstructions of the shoulder girdle are based mainly on phylogenetic and morphological comparisons, especially those involving the position of the scapula, the orientation of the glenoid, and the placement of the sternal plates (Bakker 1971; Borsuk-Bialynicka 1977; McIntosh et al. 1997; Wilson & Sereno 1998; Wilhite 2003; Bonnan et al. 2005; Schwarz et al. 2007; Remes 2008; Rauhut et al., this volume).

The reconstruction of the musculature of the shoulder girdle in sauropods still lacks a consistent system, and only a few reconstructions of the forelimb musculature, for example, for *Opisthocoelicaudia* (Borsuk-Bialynicka 1977) and *Apatosaurus* (Filla & Redman 1994), exist. In a more recent and comprehensive study, Remes (2008) investigated the evolution of the shoulder girdle musculature from basal archosaurs to the Sauropoda by using extant phylogenetic bracketing. Schwarz et al. (2007) were the first to present an approach in which morphological comparisons as well as functional considerations were used to reconstruct the shoulder girdle elements and some of the adjacent musculature.

FUNCTIONAL ASPECTS OF THE SHOULDER GIRDLE

These previous studies contribute valuable information, but to determine the possible functional adaptations of the pectoral girdle, skeletal elements need to be reconstructed with respect to the biomechanical requirements for maintaining any given posture. This must include static as well as kinetic aspects. In the following, I will focus on statics, because standing is a precondition for walking, and walking simply consists of alternating stance phases on one forelimb. In contrast to the pelvic girdle, the pectoral girdle is connected to the trunk exclusively by muscles and tendons, which leave few indications for their reconstruction. Thus, the pectoral girdle elements first need to be arranged so that they support the trunk and transmit the body weight to the forelimbs. Second, these elements must be connected to one another and to the trunk by the appropriate muscles. Finally, the direction and magnitude of the forces acting to maintain equilibrium in the joints of the shoulder and forelimbs must be evaluated.

This approach begins by investigating the shoulder in mammals and living reptiles as a means of gaining basic information on mechanical conditions occurring in different types of shoulder girdle constructions in relation to limb posture. If sauropod dinosaurs had walked with extended limbs, they would have needed functional adaptations in the shoulder girdle skeleton similar to those observed in large extant mammals like elephants. The presence and distribution of muscles can be derived from living archosaurs, such as crocodilians, as a result of their close phylogenetic affinity to extinct dinosaurs, as well as from other extant reptiles, even though they may have different postural and locomotor habits. The biomechanical situation in mammals with extended limbs differs fundamentally from that in reptiles with sprawling limbs. Thus, two different situations are under investigation: the cursorial situation occurring in all mammals lacking a clavicle, and the sprawling situation observed in crocodiles, other reptiles, and anurans. On the basis of this information, conclusions on the limb posture of sauropod dinosaurs will be drawn.

THE FINITE ELEMENT METHOD AND FINITE ELEMENT STRUCTURE SYNTHESIS

A second biomechanical approach taken here uses the finite element (FE) method, a technique that indicates stress, strain, and deformation in solid and liquid structures (Zienkiewicz et al. 2005). It is commonly used in engineering sciences and orthopedics (Witzel 1985, 1996, 2000; Effenberger et al. 2001), but it has recently been applied in paleontology and zoology as well (Carter & Beaupré 2001). Applications of FE methods in biology and paleontology were recently surveyed by Rayfield (2007). The approach of previous studies mainly consisted of the analysis of in vivo stresses and strains in the skull on the basis of computed tomographic (CT) scans (Daniel & McHenry 2001; Rayfield 2004, 2005; Dumont et al. 2005; Ross et al. 2005; McHenry et al. 2006). These studies clearly show that the skull is highly adapted to resisting the bite forces acting on bony structures. However, although they do reveal the stresses occurring in a priori existing structures after the application of bite forces, they do not explain bone shape as a response to mechanical loading (Witzel & Preuschoft 2005), and therefore they do not offer any information on the functional adaptations of bony structures. A more deductive approach is finite element structure synthesis (FESS), which consists of the virtual synthesis of skeletal structures from a homogenous solid by applying muscle forces and computation in iterative steps. This analytical technique has been successfully applied to hominid skulls and femora, as well as to the skull of sauropod dinosaurs (Preuschoft & Witzel 2004, 2005; Witzel & Preuschoft 2005, Witzel et al., this volume). The method is based on Wolff's law (1892) and the "causal morphogenesis" hypothesis of Pauwels (1965), which predicts that bone material only is maintained where a considerable physiological amount of mechanical stresses exist. To test the biomechanical predictions, the FE method is used in two ways. First, the stress flow in simplified, 3D, solid tetrapod bodies is analyzed to determine the influence of gravity on the stress distribution on the body stem for two different limb postures.

Second, the novel method of FESS is here applied to the shoulder girdle and forelimbs (see Witzel & Preuschoft 2005; Witzel et al., this volume). To validate the use of FESS for noncranial skeletal elements and to evaluate the mechanical principles of the shoulder girdle in general, FESS is applied to an extant archosaur, *Caiman crocodylus*. These results are then incorporated into the study of the shoulder girdle in sauropod dinosaurs, which permits a mechanically plausible reconstruction of the shoulder girdle in a natural arrangement. It will also help to reveal mechanically based functional adaptations for locomotion in these extinct giants.

Materials and Methods

MATERIALS

Information on mammalian anatomy comes from either the literature cited or from personal observations made in the anatomical collection of the Institute of Anatomy at the Université Louis Pasteur in Strasbourg, France. Photographs of a mounted skeleton of *Caiman crocodylus* were taken in the anatomical collection of the Institute of Zoology and Neurobiology at the Ruhr University of Bochum. CT scans of a *Caiman* specimen were produced at the Radiological Institute of the Knappschaftskrankenhaus in Bochum. Dissection of an *Alligator* specimen was carried out at the School for Anatomical Preparation in Bochum. The caimans, which we used to study locomotion, are privately owned. No animal was harmed or killed for this study.

The reconstruction of *Diplodocus* is primarily based on the skeleton of *Diplodocus longus* housed in the Senckenberg Museum in Frankfurt, Germany (Fig. 11.1), with exception of the sternal plates, which are missing in this specimen. Instead, the sternal plates are based on skeleton HQ1 at the Sauriermuseum Aathal, Switzerland. The incompleteness of these skeletons and the fact that parts of different individuals were used in the skeletal mounts were taken into account. However, these slight inconsistencies are not deletrious for this study because the goal is not morphological reconstruction per se, but rather the functional understanding of shoulder construction in sauropods.

METHODS

All theoretical biomechanical considerations are founded on basic mechanical statics. The body of any animal must deal with different kinds of postural equilibria. First, the center of mass must be kept within the area of support to maintain stability. If weight forces pull toward the ground, ground reaction forces of the equal strength must act on the body in the opposite direction. If they do not run directly through a joint, mechanical torques will occur, which are defined as the product of the forces and lever arm length. The moments produced by the ground reaction forces tend to move the extremities

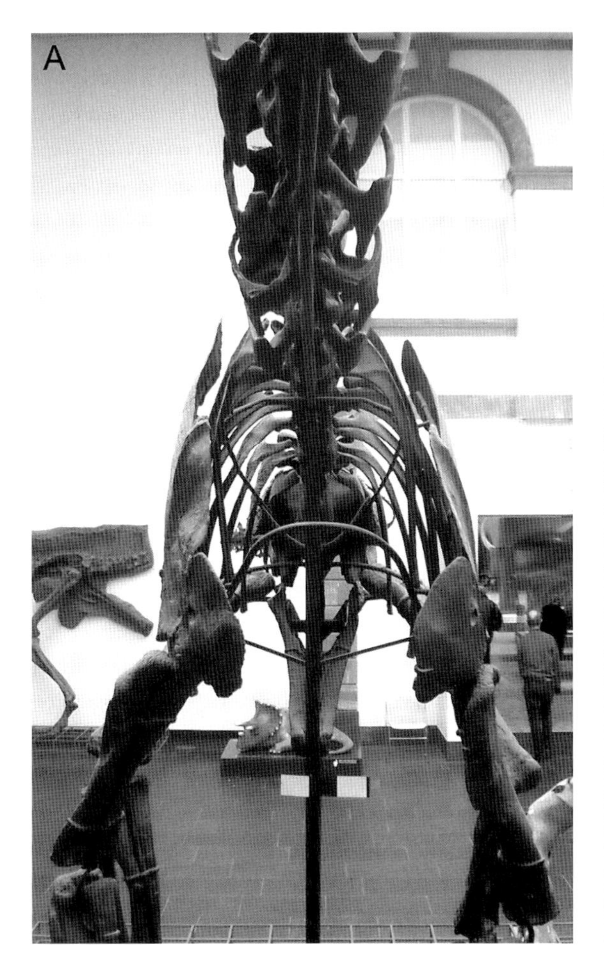

FIGURE 11.1. Shoulder girdle region of the mounted skeleton of *Diplodocus longus* at the Senckenberg Museum, Frankfurt, in frontal view (A), which also shows the base of the neck, the proximal forelimbs, and the rib cage, and in lateral view of the left forelimb (B).

around the joint's center of rotation. To keep the joints in position, these moments must be countered by active muscle forces that produce torques of equal strength. In a balanced static system, the sum of all moments produced by ground reaction forces is equal to the sum of the moments produced by the muscles, thus leading to a closed circle of forces (Kummer 1959; Preuschoft et al. 1994; Alexander 2002; Hildebrandt & Goslow 2001, 2004).

FE ANALYSIS OF SOLID TETRAPOD BODIES

All FE analyses were conducted with the FE analysis software ANSYS 11.0 (ANSYS Inc., Canonsburg, PA). Two different positions of the forelimbs were modeled. The overall anatomy and relative mass distribution of the first model correspond to those of extant crocodilians, lizards, and anurans (sprawling posture) (Fig. 11.2), while the second model reflects the situation in extant cursorial and graviportal mammals (erect posture) (Fig. 11.2B). The static condition is characterized by a symmetrical limb support at the shoulder region. During locomotion, limb support is asymmetrical during consecutive phases of a movement cycle, which results in a one-limb support at the shoulder region.

The predefined volumes of the model are divided into a finite number of 10-node tetrahedral elements. All parts of the modeled bodies possess equal material properties (Young's modulus, Poisson's ratio) to ensure that the stress flow is undisturbed and not bundled in areas with higher stiffness. The weight force is defined by the model's volume, density, and gravity. To position the model in 3D space, it is constrained by the application of bearings. In the current model, limbs with ground contact have bearings that restrict movements along the y-axis. To simulate locomotion, forces are applied that keep two limbs off the ground. Tensile and compressive stresses are labeled by convention with either a positive sign (tensile stress) or negative sign (compressive stress). The distribution of the stresses in Plates 11.1–11.8 is coded as follows. Areas with high compressive stresses are blue, areas of low compressive stresses are red, and tensile stresses are coded in complementary colors. A gray color indicates a region in which stresses are significantly beyond the spectrum of the selected color code.

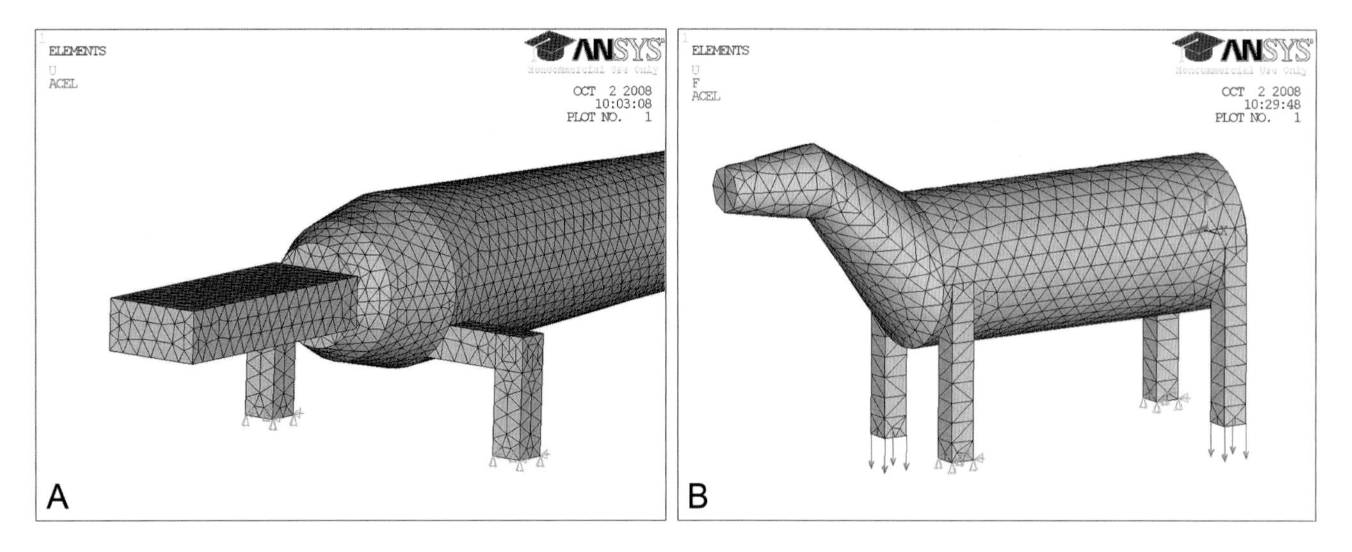

FIGURE 11.2. 3D FE models. Sprawling (A) and erect models (B) before computation. Triangles below the feet mark the bearings of the model and indicate limbs with ground contact. Arrows mark forces and indicate limbs held off the ground.

FESS OF THE SCAPULOCORACOID IN *CAIMAN CROCODYLUS*

Two-dimensional and 3D models were created on the basis of the original anatomy of the animal (see Witzel et al., this volume, for details of the FESS method). The aim was to model elements in the correct anatomical position, giving each position ample space for the stress to spread within the so-called bauraum. For this purpose, the virtual shoulder girdle elements of one caiman skeleton were arranged in accordance with information gained from CT scans, dissections of one formalin-preserved specimen, and observations made on living caimans. Photos in frontal and lateral view were taken to define the dimensions of the bauraum. These images were copied on graph paper, and the coordinates were transferred to the FE program. In this program, the bauraum in the caiman is defined by the dimensions of each part of the shoulder–forelimb system, that is, trunk, scapulocoracoid, humerus, and lower limb. The origins and insertions of the relevant muscles, including their lines of action, were based on the literature (Brinkmann 2000; Meers 2003) as well as from observations of a dissected specimen of an American alligator (*Alligator mississippiensis*).

In the first step, separate 2D models in lateral and frontal view were computed, which consisted of the four separate elements previously described as the bauraum. Only half of the body was modeled to reduce computation time. The position of the individual elements was stabilized by the application of bearings. The trunk was allowed to move vertically, whereas the lower limb was allowed to move sideways. At the interfaces between trunk and scapulocoracoid, as well as in the shoulder and elbow joints, contact elements were introduced to enable relative movements and the transmission of me-

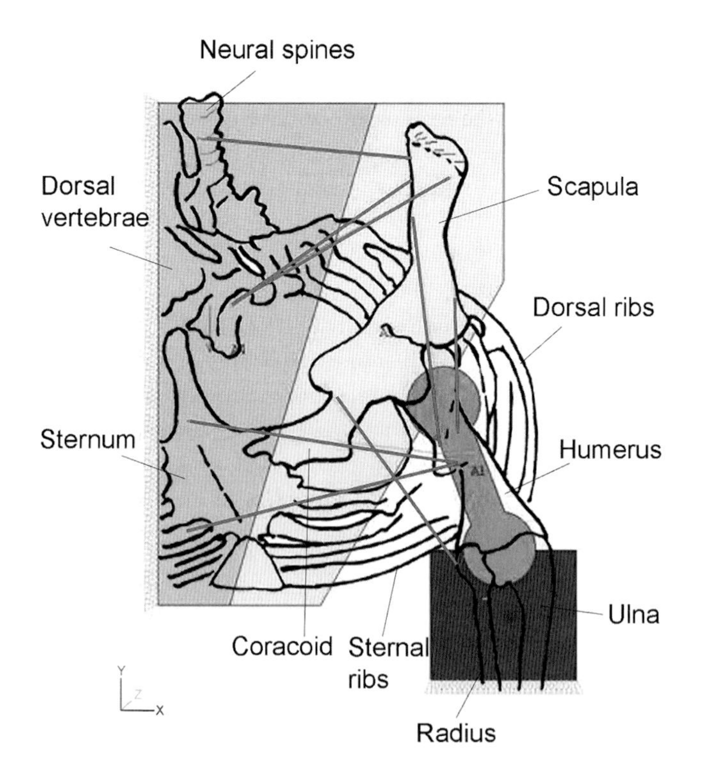

FIGURE 11.3. Drawing of the skeleton of the shoulder girdle region of *Caiman crocodylus* with an overlay of the bauraum of a 2D FE model in anterior view. Each skeletal element is represented in the bauraum by a different shading (medium light gray, trunk; light gray, scapulocoracoid; medium dark gray, humerus; dark gray, ulna and radius). The thick gray lines indicate the muscles.

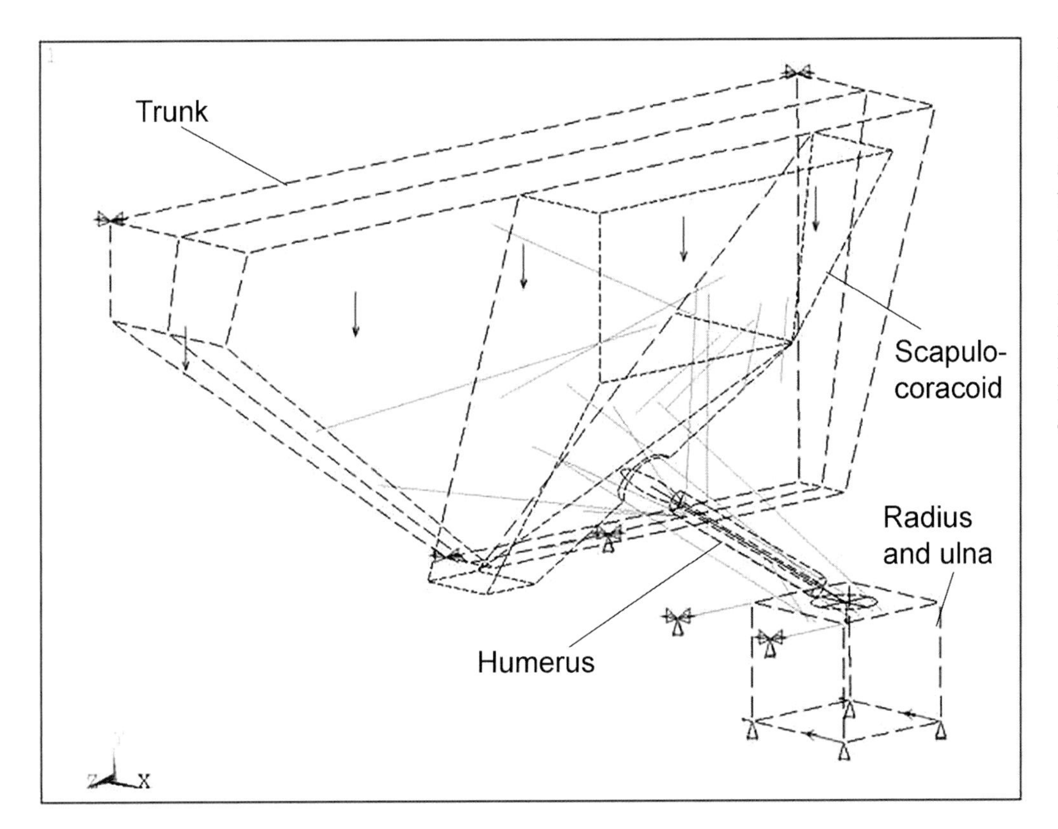

FIGURE 11.4. Four-part FE bauraum of the 3D crocodile model based on *Caiman crocodylus*. All applied loads (arrows), bearings (triangles), and muscles modeled as links (light gray lines) are indicated. Note that all four parts are able to move independently and are held together only by link elements.

chanical stress. Sufficient weight forces were applied to the trunk of the model for the parts to move apart (Fig. 11.3).

In contrast to the FESS of the skull (Witzel et al., this volume), link elements were applied instead of directed forces. These link elements have a given diameter (0.1 mm) and a set pretension (Fig. 11.3). These prestressed elements technically operate as forces. However, the absolute force values can only be derived from the pretension value by elaborate computation for each link at each iteration. Link elements function as active strings that hold each part in its relative position, thus avoiding marked deformations during computation. Because reaching equilibrium is the most critical point during the computation process by ANSYS 11.0, I used the qualitative approach of link elements instead of single force vectors to ascertain whether FESS could actually be applied to the modeling of the *Caiman* shoulder girdle.

For a more detailed investigation and the synthesis of the shoulder girdle region, the basic parameters of the frontal and lateral 2D models were transferred to a 3D model (Fig. 11.4). This combination included the coordinates (x, y, z) in space, the previous muscle arrangement, the forces applied to the trunk, and the restriction of movements of the model in the 3D space. Again, equilibrium in the shoulder and the elbow joint was maintained by applying certain link elements representing the musculature. In the first two iterative steps, the bauraum of the scapula was reduced by retaining only the parts that showed a certain minimal compressive stress in order to define a new, reduced bauraum.

Results

THEORETICAL BIOMECHANICAL ASPECTS

In animals with an erect stance, such as cursorial and graviportal mammals, the weight forces of the anterior part of the body are carried by the forelimb, scapula, ribs, and sternum. The reaction forces are transmitted as follows: (1) ground–forelimb–scapula via the m. serratus to the ribs and the vertebral column, and (2) forelimb and scapula via the m. pectoralis profundus to the sternum (Fig. 11.5A; Preuschoft et al. 2007; Preuschoft, unpublished data). Therefore, the first ribs are straight and strong, and the thorax has a narrow appearance. Cursorial mammals usually do not abduct the forelimbs appreciably to prevent horizontal forces, which cannot be countered by the existing structures. The narrowness of the trunk is also helpful in minimizing the rotational moments in an one-limb stance, which is essential for saving muscle energy (Preuschoft et al. 2007; Preuschoft, unpublished data).

A different situation occurs in living reptiles and amphibians as well as in the early tetrapods, which usually held their limbs in a sprawling position (Fig. 11.5B). The weight forces are carried by the humerus, scapulocoracoid, and ribs (based on my own observations on living and dissected animals). The sprawling posture requires horizontal tensional forces, which can be produced by the m. pectoralis (Fig. 11.5B). The resultant of ground reaction force and pull of m. pectoralis passes through the shoulder joint. These joint forces can be separated into a horizontal vector and a vertical vector. The hori-

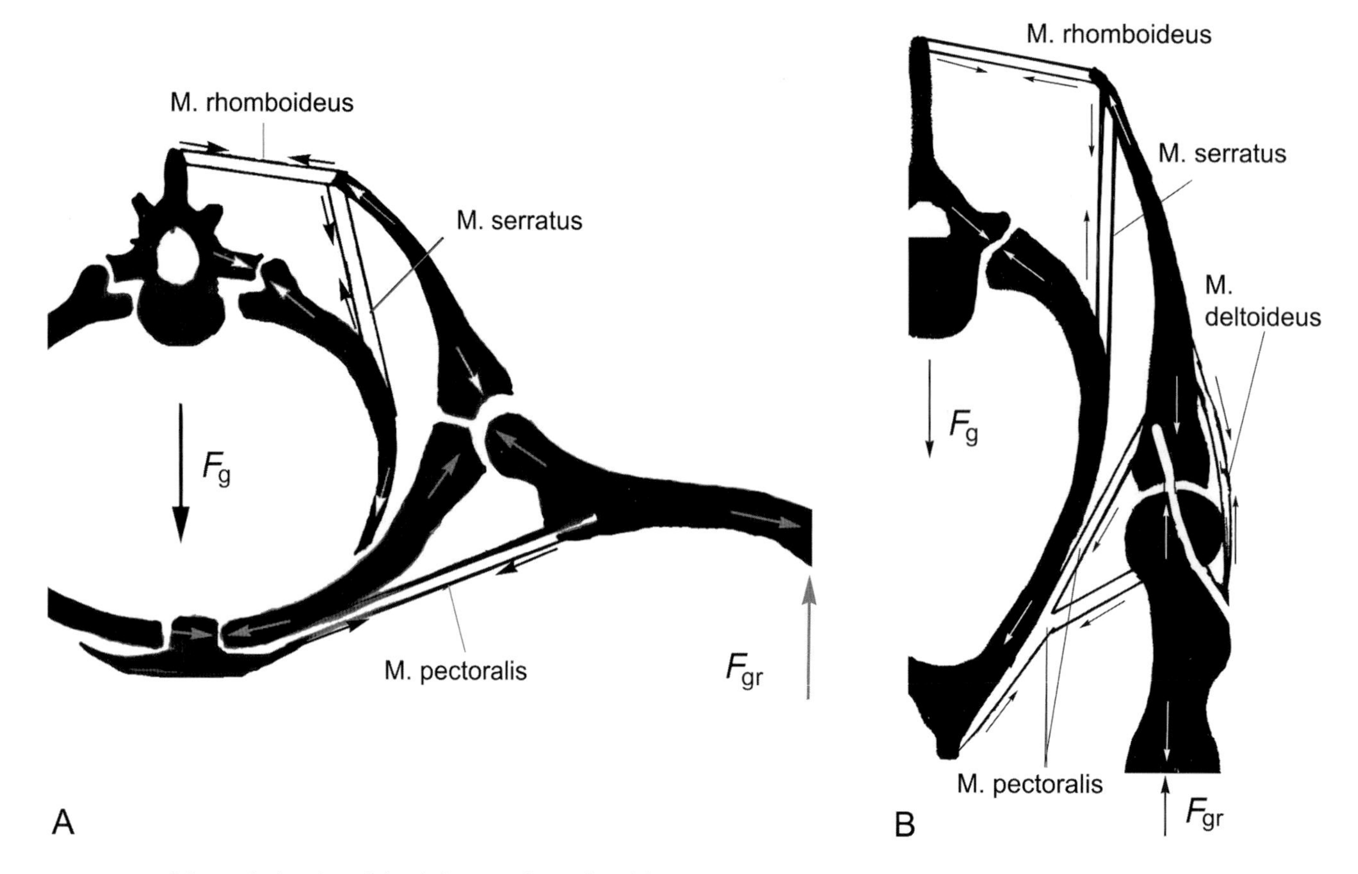

FIGURE 11.5. Schematic drawing of the skeleton and muscles of the shoulder region of a crocodile (A) and a cursorial mammal (B). The arrows indicate the direction of compressive stresses (arrows pointing toward each other) and tensile stresses (arrows pointing away from each other) induced by body weight (F_g) and ground reaction force (F_{gr}) through the structures involved.

zontal vector is transmitted through the coracoid plate to the sternal apparatus, whereas the vertical vector is transmitted via the scapula and the posterior part of the m. serratus into the ribs to the vertebral column (Preuschoft & Schmidt 2003; Preuschoft & Gudo 2005).

The transmission of body weight from the trunk to the forelimbs in reptiles and mammals is carried out in two different ways. In reptiles, the circle of forces is completed via the sternocoracoid joint, while in mammals, force transmission is maintained only by the muscles of the shoulder girdle. These loading conditions are reflected in mammals in the strong development of the anteriormost ribs, which are firmly connected to the sternum. In contrast, crocodilians possess short but stout anterior dorsal ribs and rather strong middle dorsal ribs, in which only the latter are connected to the sternum. Sauropods have strong middle dorsal ribs, but their anterior dorsal ribs are longer than those of crocodiles and likely reached the sternal elements.

STRESS DISTRIBUTION IN SOLID TETRAPOD BODIES

The FE models reveal stress patterns in regions representing the limb girdles and trunk. These stress patterns clearly depend on mass distribution in the head, neck, and trunk as well as on the position and posture of the supporting limbs. The compressive stresses observed in the sprawling model (Plate 11.1A; Preuschoft et al. 2007) as well as in the erect model (Plate 11.1B; Preuschoft & Schmidt 2003) are in remarkable accordance with the major skeletal elements. If an animal supports its body on both forelimbs, its trunk in lateral view is subject to bending moments that compress the dorsal and stretch the ventral side of the trunk (Plate 11.1). In living forms, the dorsally located vertebral column takes up these compressive stresses. On the ventral side, the tensile stresses are taken up by the musculature such as m. rectus abdominis. Concurrently, tensile forces occur on the dorsal side of the anterior trunk and posterior neck, which can be taken up by the neck and shoulder muscles (Plate 11.2).

The highest compressive stresses were found at the insertion sites of the limbs at the trunk, that is, at the glenoid joints. In both the sprawling and the erect models, compressive stresses spread dorsally to the side of the models with a slight caudalward inclination (Fig. 11.5). This inclination is more pronounced in the erect than in the sprawling model, which is indeed consistent with observations of living crocodiles. In the sprawling model, another part of the stress flow extends from the glenoid ventrally to the midline of the trunk, which is absent in the model with erect limbs (Plate 11.3A). In extant nonmammalian tetrapods, the stresses are taken up by the coracoid plate, which along with the proximal part of the scapula forms the glenoid joint. In the erect model, the medially directed stresses at this location are less pronounced and spread mainly from the erect humerus to the scapula (Plate 11.3B).

As soon as the animal lifts one forelimb off the ground (asymmetrical support) and begins to walk, the stress distributions and values start to shift. Tension occurs dorsally in the scapula of the standing limb, while compression increases ventrally. This increase is significantly higher in the sprawling model than in the erect model (Plate 11.3C, D). Thus, the alternating symmetrical and asymmetrical support during locomotion requires a continuous, compression-resistant structure located medially between the limbs. In living reptiles, the coracoids and the interclavicle are connected to the sternal elements in this regions. Mammals have a sternal element as well (see above).

Furthermore, in one-limb support, torsional stresses occur in both models on the surface of the trunk. These are higher and spread over larger areas than bending stresses in the symmetrical stance (Plate 11.4A). The torsional stresses extend obliquely from the scapula of the weight-bearing limb to the contralateral support (hindlimb). These stresses have tensile as well as compressive components. The former are taken up by the oblique muscles of the ventrolateral body wall; the latter necessitate segmental bony elements, that is, the ribs (Preuschoft et al. 2007). The compressive stresses are concentrated in the periphery, therefore leaving the center of the trunk devoid of stress (Plate 11.4B). Thus, the body cavity is less important for absorbing mechanical stresses resulting from locomotion and therefore can provide space for the internal organs. In comparison to sprawling with erect limbs, torsional stresses are greater in the former but do not show obvious difference in their distribution.

2D FE MODEL OF *CAIMAN CROCODYLUS*

After determining the static conditions in the shoulder girdle, two models—one in frontal view and the other in lateral view—of a caiman were calculated. The muscles involved in maintaining balance were determined for every possible movement of each part (Plate 11.5). Each of the link elements corresponds to muscles or muscle groups that were identified anatomically and that produce forces resulting in mechanical stresses in the skeletal elements. The computation was considered complete when equilibrium in the glenohumeral as well as in the elbow joint was established. The compressive stress flow in the part representing the scapulocoracoid provides a first approximation of the actual morphological structures (Plate 11.5).

3D FESS OF *CAIMAN CROCODYLUS*

On the basis of the 2D frontal and lateral model, a 3D model was designed. The weight forces applied and the muscular structures (link elements) are shown in Fig. 11.4. Weight forces of the trunk are carried by the mm. serrati. In lateral view, the m. trapezius and m. levator scapulae take up body weight while rotating the scapula cranially. These computations confirm the role of the m. pectoralis in preventing the lateral slip of the humerus. The m. teres major and m. latissimus dorsi are not involved in taking up the weight forces, but retract the humerus and thus exert forces, resulting in compressive stresses. An important role in maintaining equilibrium in the shoulder joint and avoiding the rotation of the coracoid relative to the sternum is fulfilled by the m. costocoracoideus, which pulls the coracoid caudally.

After computation of the first iteration, the stress patterns in the 3D model are already in accordance with the skeletal elements observed in the real animal (Plate 11.6). Deformations and movements of the model (black outlines in Plate 11.6) resulted from the application of forces but stay within tolerable limits, which is a precondition for computation. In a mirror image of the cross section, compressive stresses are concentrated at the periphery of the trunk, leaving the internal room unstressed, similar to the situation in the basic 3D FE models (Plate 11.7).

To reduce the bauraum to the shape seen in the actual animal, the forces are accumulated in iterative steps. Stress areas above a threshold value of compressive stress are then used to design the new bauraum (Plate 11.8). This reduced model is again computed under predefined conditions. After the second iteration and the last reduction step, the overall shape of the model of the scapulocoracoid closely resembles the original (Fig. 11.6). The slender scapula, its inclined orientation, and the connection of the coracoid to the trunk are clearly recognizable.

Discussion

FORM AND FUNCTION

The investigation of stress distribution in the 3D FE models reveals the necessity of skeletal elements within the tetrapod body where mechanical stresses occur. In the shoulder girdle, stress distribution and therefore bone arrangement are influenced by limb posture as well as locomotion. Basic morphological features such as the inclination of the scapula in mammals and the occurrence of ventral coracoid plates in reptiles can be explained by the mechanical forces such as gravity acting on the body.

With the aid of FESS, the main features of the bony elements in the scapulocoracoid of a caiman could be synthesized. The

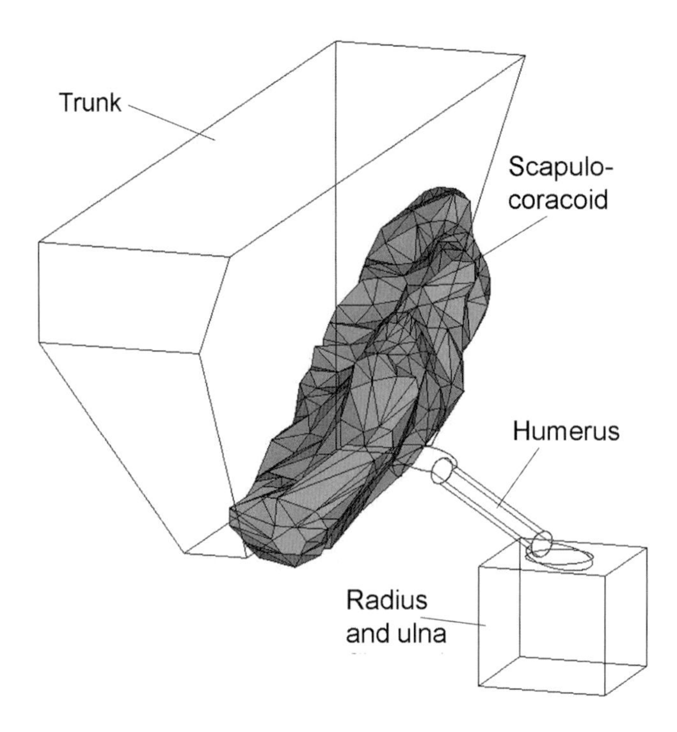

FIGURE 11.6. Reduced FE model of the scapulocoracoid after the second iteration. Note the overall reduction of the structure compared to the initial bauraum. The reduced bauraum has a slender, inclined part posterior to the glenoid joint (representing the scapula) and an anterior part medially connected to the trunk (representing the coracoid).

function of the shoulder muscles is to maintain equilibrium in the shoulder as well as in the elbow joint, which is a precondition for sprawling posture.

Because similar physical conditions lead to similar functional adaptations in morphological structures (Coombs 1978; Kummer 1985), functional adaptations of bones are a common topic in different fields of research. Examples include the reconstruction of life history in hominids (Wollfson 1950; Runestad et al. 1993; Trinkaus et al. 1994; Larsen 1997), applications in orthopedics and biomechanical engineering (Witzel 1985, 1996; Carter & Beaupré 2001), and especially the analysis of locomotion in animals and hominids (Christian & Preuschoft 1996; Demes et al. 2001; Patel & Carlson 2007; Preuschoft et al. 2007). Movement-dependent mechanical stimulation is responsible for cartilage and bone remodeling during embryogenesis and postembryonic life (Carter et al. 1987). If muscular activity is absent in very early ontogenetic stages, the formation and shape development of bony elements of the skull and the postcranial skeleton are inhibited (Lightfoot & German 1998; Gomez et al. 2007; Jones et al. 2007). The importance of mechanical influence on the skeletogenesis and remodeling as an epigenetic factor was recently pointed out by Newman & Muller (2005), who emphasized the influence of these factors as the probable source of the sudden phylogenetic appearance of the prominent fibular crest in birds (Müller & Streicher 1989). Thus, the structure of bone actually seems to depend on its mechanical function, and its shape directly results from mechanical stress acting within the locomotor system. Indeed, my investigations reveal that the location and shapes of bony structures and associated muscles depend on the occurrence and distribution of mechanical forces. This holds true not only for skulls and basic tetrapod body shapes, but also for freely moving structures such as the shoulder girdle.

In the present study, link elements instead of single forces were used for technical reasons. An estimation of absolute muscle forces was therefore not possible. In the application of FESS to an extinct animal, forces instead of links should be applied. This modification will allow for determining the actual force values working in the musculoskeletal system (see Witzel et al., this volume).

RECONSTRUCTION OF THE SAUROPOD SHOULDER GIRDLE

The FESS described above indicate how the pectoral parts must be arranged for the successful transmission of body weight to the forelimbs. The subsequent reconstruction of the shoulder girdle of *Diplodocus longus* as one example among the sauropods is based on FE model results and data from the literature.

The transmission of body weight depends on the assumed posture of the limbs and the direction in which they articulate in the glenoid joint. In general, an increase in body weight leads to an erect posture of the limbs, thus avoiding more muscle work (Alexander 1985; Biewener 1989). Considering the enormous body weights in sauropod dinosaurs, a vertical force transmission from the humerus to the scapula was postulated (McIntosh et al. 1997; Wilson & Sereno 1998; Bonnan 2003; Upchurch et al. 2004).

However, the angles between glenoid joint and humerus as well as that of the elbow joint, especially in lateral view, are controversial. If the humerus was positioned vertical to the glenoid joint, the scapula would have been inclined posteriorly at an angle of about 40° or less (McIntosh et al. 1997; Wilson & Sereno 1998; Bonnan 2003; Wilhite 2003; Upchurch et al. 2004). According to my simple FE models, an inclined position of the scapula is visible in both the sprawling as well as in the erect posture, but it never is inclined less than 50°. The analyses, however, also show that optimal force transmission is not limited to a strictly vertical humerus, but is also guaranteed by the humerus deviating less than 20° from the vertical (Fig. 11.7). According to Schwarz et al. (2007), a posterior inclination of the scapulocoracoid of greater than 45° results in a vertical position of the coracoid plates. Because the sternal plates were in direct contact with the coracoid plates for suc-

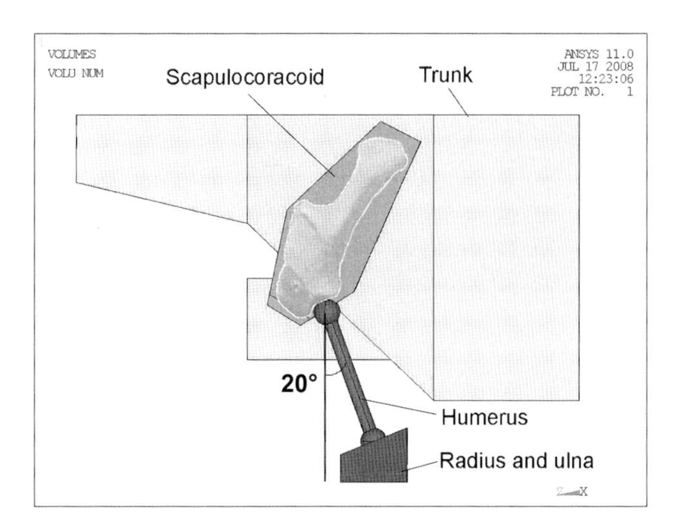

FIGURE 11.7. Bauraum of the FE model of *Diplodocus longus* consisting of four parts that represent the trunk (light gray), scapulocoracoid (medium gray, with actual bone included), humerus, and antebrachium (both dark gray). The humerus is modeled at an anteroposterior an angle of about 20° to the vertical.

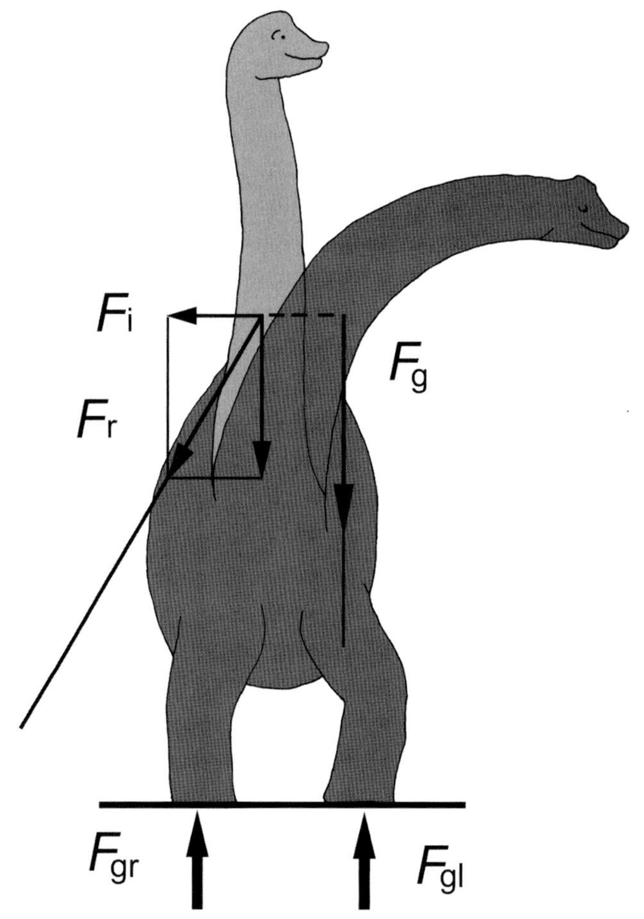

FIGURE 11.8. Long-necked sauropod dinosaur turning its neck to its left. The resultant (F_r) of the vertical weight force (F_g) and the horizontally inertial force (F_i) of the neck meets the ground to the side of the right foot of the dinosaur. Without an abductional movement of the right foot to the side, the animal would lose its balance. Therefore, the ability of forelimb abduction in the shoulder joint is clearly required in sauropod locomotion.

cessful force transmission, as I have suggested above, a vertical position of the coracoid would result in vertical sternal plates. Consequently, they would be prevented from contacting the cranialmost sternocostal rib segments (Schwarz et al. 2007), which would prevent force transmission from the limb to the trunk via the shoulder girdle impossible. Furthermore, the inclination of the scapula has a direct influence on the lines of action in the relevant musculature, which must be investigated thoroughly and is part of the ongoing research program described here.

On the basis of the results of the 3D FE models and the FESS of the caiman shoulder region, a low angle of about 20° between the glenoid joint and the humerus in lateral view is assumed (Fig. 11.7).

According to Christian et al. (1999a), the compressive stresses acting on the shoulder joint would be too high without any flexion in the elbow joint. This view was adopted in the current reconstruction. In addition, some abductional movements in the forelimbs are necessary because the enormous length and weight of the neck causes moments on the anterior part of the trunk that can only be compensated for by widely abducted forelimbs (Fig. 11.8; Preuschoft & Gudo 2005). The saddle-shaped glenoid facet and the flattened humeral head of sauropods indeed enable abduction and adduction as well as extension and flexion. Therefore, an erect posture of the forelimbs, with slight flexion in the shoulder as well as in the elbow joint in lateral view and a moderate ability for abductional movements in frontal view, is proposed here.

In sauropods, the anterior trunk is narrow, as in mammals, and the first trunk ribs are less robust than the middle ones, similar to the situation in crocodiles (Fig. 11.9). However, the curvature of the middle trunk ribs is similar to that in cursorial mammals. This suggests that the vertical part of the weight forces were transmitted through the humerus via the scapula to the posterior part of the m. serratus and to the middle ribs (Fig. 11.9) as in modern quadrupedal archosaurs such as crocodiles.

In contrast to most living reptiles, the coracoid plate in sauropods is reduced but still present. The sternal elements are of great interest, but there is still some confusion regarding their exact anatomical position. With the limb posture proposed here, a medial component of the forces transmitted through the glenoid extends to the midline of the trunk (Fig. 11.9). Therefore, the circle of forces in the horizontal plane can only be completed if the forces flow from the humerus via the coracoid to the bony sternum and back to the humerus through

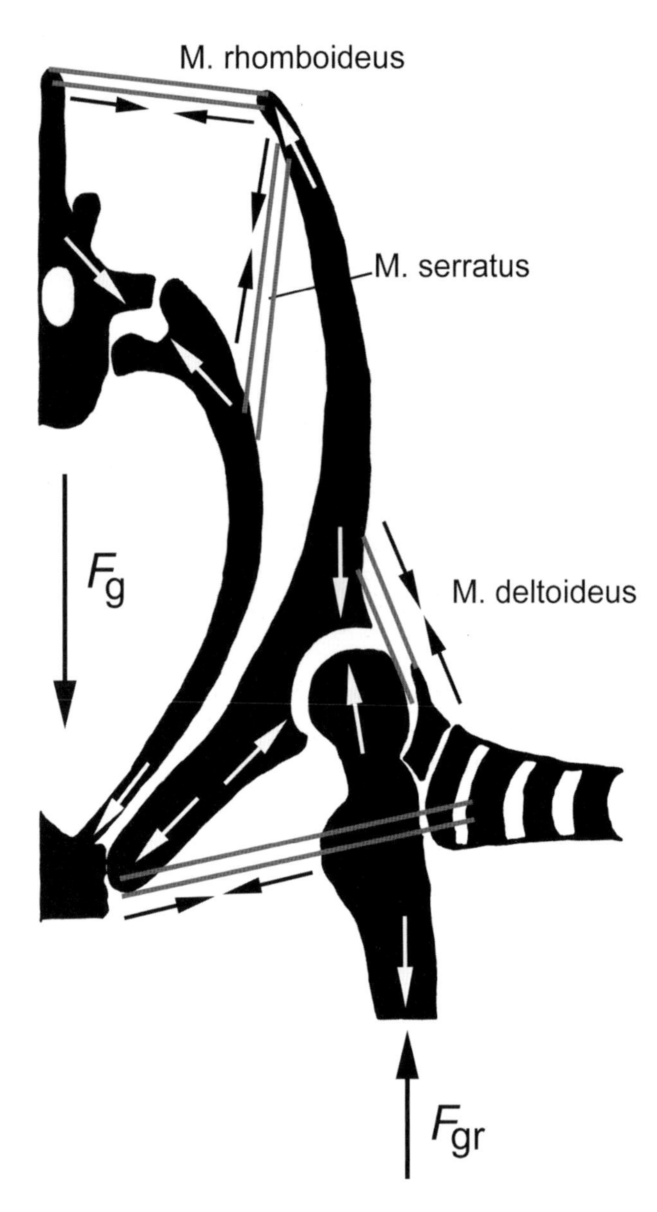

FIGURE 11.9. Loop of forces in a sauropod shoulder girdle in frontal view. Compressive forces (white arrows) are taken up by bony elements (black); tension forces (black arrows) are taken up by muscles. The weight force of the body (F_g) is transmitted by the m. serratus from the trunk to the scapula and from there on to the humerus via the glenoid joint. The medial component runs from the scapulocoracoid via the sternal plates back to the trunk, closing the loop of forces. In an abducted position of the humerus (banded humerus), medially directed tensile stresses are taken up by the m. pectoralis, which serves as the humeral adductor.

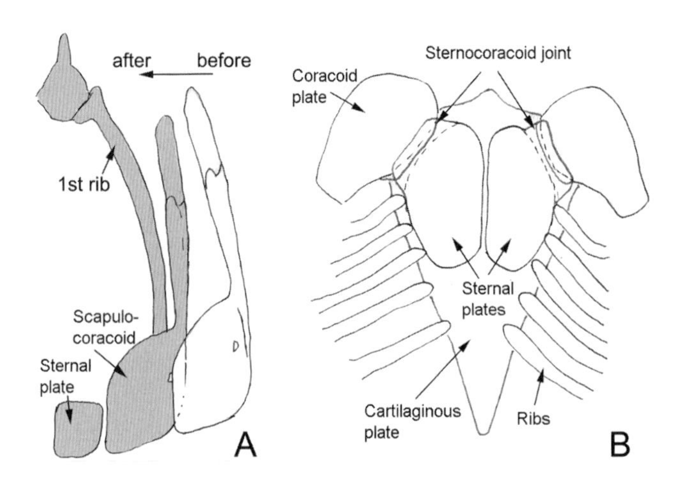

FIGURE 11.10. (A) Position of shoulder girdle of *Diplodocus longus* in frontal view (white) as mounted at the Senckenberg Museum (see Fig. 11.1A) compared to the position as suggested by this study (gray). In the reconstruction favored here, the scapulocoracoid is placed more parallel to the ribs, therefore reducing the distance between the two coracoid plates. Note the articulation of the coracoids with the sternal plates. (B) Sternal apparatus in ventral view as reconstructed in this study.

the m. pectoralis. The coracoid therefore resists the compressive stresses that occur between the shoulder joint and the sternum, which must have been located in close mediocaudal contact with the coracoid plate in order to transfer the forces to the ribs (Figs. 11.9, 11.10B).

According to Schwarz et al. (2007), the sternal plates laid in the same plane as the contour of the distal cartilaginous ends of the ribs, but their true position will also depend on their contact with the coracoid plates. In the current reconstruction, the scapula was positioned parallel to the surface of the rib cage and covered the first three ribs. This arrangement of the scapulocoracoid is significantly narrower than the original mount (Fig. 11.10A). A nearly parallel position of the scapula to the rib cage is derived from observations on living tetrapods. Nevertheless, the space between the two coracoid is sufficient for the sternal plates.

During quadrupedal locomotion, translational movements of the shoulder girdle occur. In crocodilians, these movements are made possible by rotation of the elongated coracoid relative to the sternum. The coracoid and sternum exhibit a tongue-and-groove structure made of cartilaginous tissue, where the sternum acts as the groove in which the coracoid slides (Fig. 11.11). In lizards (varanids and chameleons), the coracoid moves translational relative to the sternal apparatus (Peterson 1973; Jenkins & Goslow 1983; Lilje 2005). The junction between coracoid and sternum in these reptiles also exhibits a tongue-and-groove structure, but it consists of bone rather than cartilage. In both cases, the junction is surrounded by a joint capsule embedded in connective tissue. The mo-

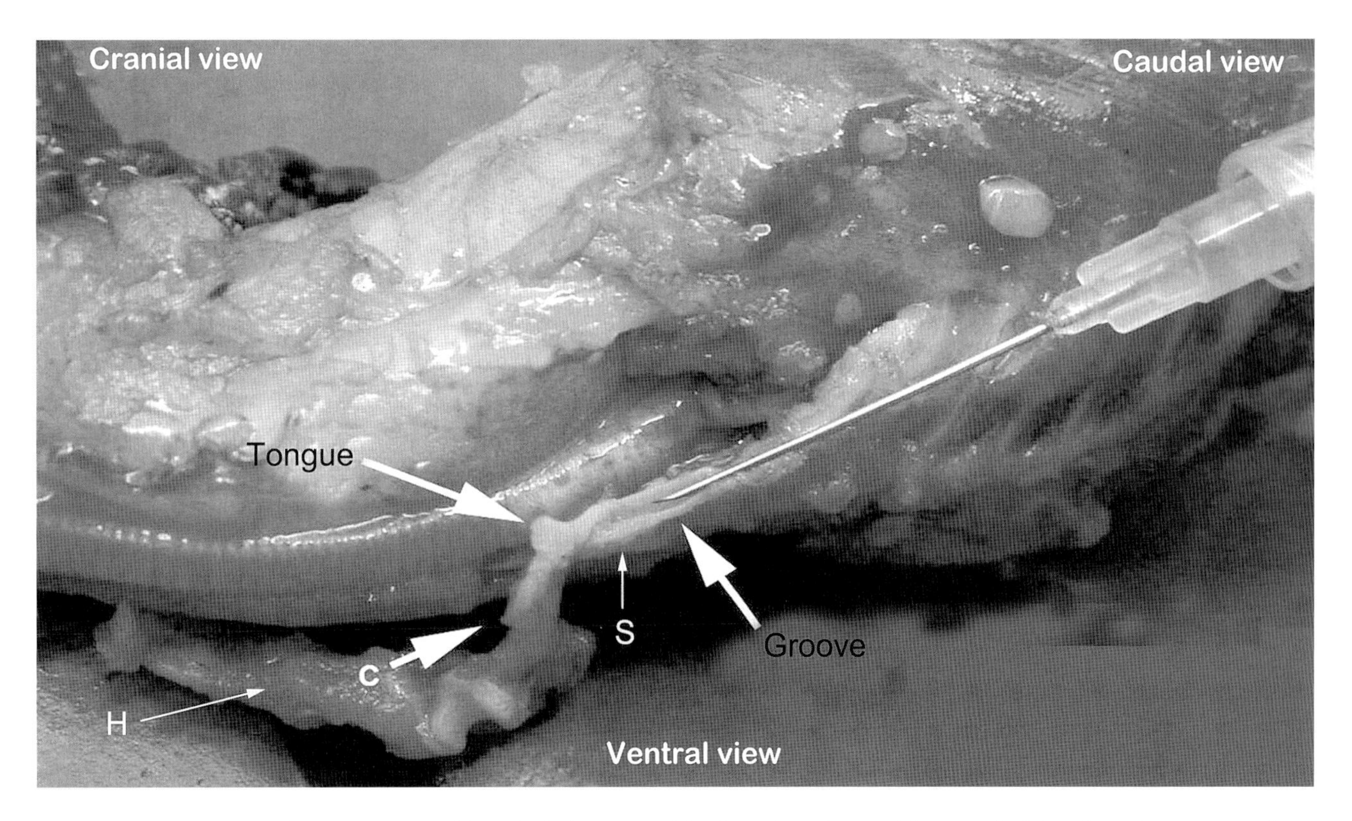

FIGURE 11.11. Dissection of the coracosternal joint in *Caiman crocodylus*. Scapula and coracoid were pulled dorsolaterally to expose the coracosternal joint which consists of a cartilaginous tongue-and-groove articulation. C, coracoid; H, scapula; S, sternum.

bility of this joint, although restricted, is indispensable for the function of the shoulder girdle because it makes force transmission from the coracoid to the sternum possible and at the same time leaves a certain degree of mobility in the shoulder region.

In sauropods, there is still no consensus whether and to what degree mobility in the shoulder girdle occurred. Some authors assume that the shoulder girdle was immobile (Bakker 1987; Paul 1987; Christiansen 1997; Wilhite 2003; Henderson 2006). However, my results and the observations discussed above suggest that certain mobility was present in the scapulocoracoid–sternal system, which must have been a prerequisite for force transmission and maintaining equilibrium, especially as body size increased evolutionarily. In addition, as body weight increases, equivalent muscle forces must evolve to counter the increasing weight forces and to keep the shoulder girdle in equilibrium. Therefore, a similar construction as in crocodiles is proposed in which the sternal plates were embedded in a cartilaginous plate, forming a joint with the coracoids and surrounded by thick connective tissue (Fig. 11.10B). This connection would have provided stability by allowing force transmission between the coracoid and the sternum. At the same time, it would have provided some mobility of the shoulder girdle relative to the trunk during locomotion.

Summary and Future Research

The shoulder girdle of sauropod dinosaurs must have been subjected to the same mechanical constraints that apply to all extant tetrapods. In a future, more advanced analysis of the biomechanics of the sauropod shoulder girdle, FESS will be reapplied. I expect to find mechanical explanations for a functional distribution and activity of the relevant musculature, which, when intergrated in synthesis with the former approach, will reveal a more fundamental understanding of the function of the shoulder girdle of sauropods in detail and sauropod locomotion in general.

Acknowledgments

I thank all members of the DFG Research Unit 533 for fruitful discussions. For their extensive support throughout the entire investigation, I am indebted to Holger Preuschoft and Ulrich Witzel (both Ruhr University of Bochum). For review of this chapter and constructive criticism, many thanks go to Ray Wilhite (Louisiana State University, Baton Rouge) and Daniela Schwarz-Wings (Museum für Naturkunde, Berlin). Further thanks go to Claudia Distler (Ruhr University of Bochum) for carefully proofreading this article and her overall support. Finally, Rainer Goessling and Nina Sverdlova are acknowledged for providing help with technical questions. The project was financially supported by the Deutsche Forschungsgemeinschaft. This is contribution number 70 of the DFG Research Unit 533 "Biology of the Sauropod Dinosaurs: The Evolution of Gigantism."

References

Abel, O. 1910. Die Rekonstruktion des *Diplodocus*.—*Abhandlungen der kaiserlichen und königlichen Zoologisch-Botanischen Gesellschaft in Wien* 5: 1–60.

Alexander, R. M. 1976. Estimates of speeds of dinosaurs.—*Nature* 261: 129–130.

Alexander, R. M. 1985. Mechanics of posture and gait of some large dinosaurs.—*Zoological Journal of the Linnean Society of London* 83: 1–25.

Alexander, R. M. 1989. *Dynamics of Dinosaurs and Other Extinct Giants*. Columbia University Press, New York.

Alexander, R. M. 1998. All time giants: the largest animals and their problems.—*Palaeontology* 41: 1231–1245.

Alexander, R. M. 2002. *Principles of Animal Locomotion*. Princeton University Press, New Jersey.

Anderson, J. F., Hall-Martin, A. & Russell, D. A. 1985. Long bone circumference and weight in mammals, birds and dinosaurs.—*Journal of Zoology A* 207: 53–61.

Bakker, R. L. 1971. Ecology of the brontosaurs.—*Nature* 229: 172–174.

Bakker, R. L. 1987. The return of the dancing dinosaurs. *In* Czerkas, S. A. & Olson, E. C. (eds.). *Dinosaurs Past and Present*. University of Washington Press, Seattle: pp. 38–69.

Biewener, A. A. 1989. Scaling body support in mammals: limb posture and muscle mechanics.—*Science* 245: 45–48.

Bonnan, M. F. 2003. The evolution of manus shape in sauropod dinosaurs: implications for functional morphology, forelimb orientation, and phylogeny.—*Journal of Vertebrate Paleontology* 23: 595–613.

Bonnan, M. F. 2004. Morphometric analysis of humerus and femur shape in Morrison sauropods: implications for functional morphology and paleobiology.—*Paleobiology* 30: 444–470.

Bonnan, M. F., Parrish, M. J., Graba, J. & Senter, P. 2005. Scapular position and function in the Sauropodomorpha (Reptilia: Saurischia).—*Journal of Vertebrate Paleontology* 25: 38A.

Bonnan, M. F. & Yates, A. M. 2007. A new description of the forelimb of the basal sauropodomorph *Melanorosaurus:* implications for the evolution of pronation, manus shape and quadrupedalism in sauropod dinosaurs.—*Special Papers in Palaeontology* 77: 157–168.

Borsuk-Bialynicka, M. 1977. A new camarasaurid sauropod *Opisthocoelicaudia skarzynskii* gen. n., sp. n. from the Upper Cretaceous of Mongolia.—*Acta Palaeontologia Polonica* 37: 5–64.

Brinkmann, J. 2000. Die Muskulatur der Vorderextremität von *Alligator mississippiensis* (Daudin, 1802). Diploma Thesis. Universität Tübingen.

Burness, G. P., Diamond, J. & Flannery, T. 2001. Dinosaurs, dragons, and dwarfs: the evolution of maximal body size.—*Proceedings of the National Academy of Sciences of the United States of America* 98: 14518–14523.

Carrano, M. T. 2005. The evolution of sauropod locomotion: morphological diversity of a secondarily quadrupedal radiation. *In* Curry Rogers, K. & Wilson, J. A. (eds.). *The Sauropods: Evolution and Paleobiology*. University of California Press, Berkeley: pp. 229–251.

Carter, D. R. & Beaupré, G. S. 2001. *Skeletal Function and Form: Mechanobiology of Skeletal Development, Aging and Regeneration*. Cambridge University Press, Cambridge.

Carter, D. R., Orr, T. E., Fyhrie, D. P. & Schurmann, D. J. 1987. Influences of mechanical stress on prenatal and postnatal skeletal development.—*Clinical Orthopaedics* 219: 237–250.

Christian, A. 2002. Neck posture and overall body design in sauropods.—*Mitteilungen aus dem Museum für Naturkunde in Berlin, Geowissenschaftliche Reihe* 5: 271–281.

Christian, A. & Preuschoft, H. 1996. Deducing the body posture of extinct large vertebrates from the shape of the vertebral column.—*Palaeontology* 39: 801–812.

Christian, A., Heinrich, W.-D. & Golder, W. 1999a. Posture and mechanics of the forelimbs of *Brachiosaurus brancai* (Dinosauria: Sauropoda).—*Mitteilungen aus dem Museum für Naturkunde in Berlin, Geowissenschaftliche Reihe* 2: 63–73.

Christian, A., Müller, R. H. G., Christian, G. & Preuschoft, H. 1999b. Limb swinging in elephants and giraffes and implications for the reconstruction of limb movements and speed estimates in large dinosaurs.—*Mitteilungen aus dem Museum für Naturkunde in Berlin, Geowissenschaftliche Reihe* 2: 81–90.

Christiansen, P. 1997. Locomotion in sauropod dinosaurs.—*GAIA* 14: 45–75.

Colbert, E. H. 1962. The weights of dinosaurs.—*American Museum Novitates* 2076: 1–16.

Coombs, W. P. J. 1975. Sauropod habits and habitats.—*Palaeogeography, Palaeoclimatology, Palaeoecology* 17: 1–33.

Coombs, W. P. J. 1978. Theoretical aspects of cursorial adaptation in dinosaurs.—*Quarterly Review of Biology* 53: 393–418.

Daniel, W. & McHenry, C. 2001. Bite force to skull stress correlation—modeling the skull of *Alligator mississippiensis*. *In* Grigg, G. C., Seebacher, F. & Franklin, C. E. (eds.). *Crocodilian Biology and Evolution*. Surrey Beatty and Sons, Chipping Norton: pp. 263–284.

Demes, B., Qin, J. X., Stern, J. T. J., Larson, S. G. & Rubin, C. T. 2001. Patterns of strain in the macaque tibia during functional activity.—*American Journal of Physical Anthropology* 116: 257–265.

Dumont, E. R., Piccirillo, J. & Grosse, I. R. 2005. Finite element analysis of biting behaviour and bone stress in the facial skeletons of bats.—*Anatomical Record* 283: 319–330.

Effenberger, H., Witzel, U., Lintner, F. & Rieger, W. 2001. Stress analysis of threaded cups.—*International Orthopaedics* 25: 228–235.

Farlow, J. O. 1993. On the rareness of big, fierce animals: speculations about the body sizes, population densities, and geographic ranges of predatory mammals and large carnivorous dinosaurs.—*American Journal of Science* 293: 167–199.

Farlow, J. O., Smith, M. B. & Robinson, J. M. 1995. Body mass, bone "strength indicator," and cursorial potential of *Tyrannosaurus rex*.—*Journal of Vertebrate Paleontology* 15: 713–725.

Filla, J. & Redman, P. D. 1994. *Apatosaurus yahnahpin:* preliminary description of a new species of diplodocid sauropod from the Late Jurassic Morrison Formation of southern Wyoming, the first sauropod dinosaur found with a complete set of "belly ribs."—*Wyoming Geological Association Guidebook* 4: 159–178.

Gilmore, C. W. 1932. On a newly mounted skeleton of *Diplodocus* in the United States National Museum.—*Proceedings of the United States National Museum* 81: 1–21.

Gomez, C., David, V., Peet, N. M., Vico, L., Chenu, C., Malaval, L. & Timothy, M. S. 2007. Absence of mechanical loading in utero influences bone mass and architecture but not innervation in Myod-Myf5-deficient mice.—*Journal of Anatomy* 210: 259–271.

Hatcher, J. B. 1901. *Diplodocus* (Marsh): its osteology, taxonomy, and probable habits, with a restoration of the skeleton.—*Memoirs of the Carnegie Museum* 1: 1–63.

Hay, O. P. 1908. On the habits and the pose of the sauropodous dinosaurs, especially of *Diplodocus*.—*American Naturalist* 42: 672–681.

Henderson, D. M. 1999. Estimating the masses and centers of mass of extinct animals by 3-D mathematical slicing.—*Paleobiology* 25: 88–106.

Henderson, D. M. 2006. Burly gaits: centres of mass, stability, and the trackways of sauropod dinosaurs.—*Journal of Vertebrate Paleontology* 26: 907–921.

Hildebrand, M. & Goslow, G. 2001. *Analysis of Vertebrate Structure*. 5th edition. John Wiley & Sons, New York.

Hildebrand, M. & Goslow, G. E. 2004. *Vergleichende und funktionelle Anatomie der Wirbeltiere*. Springer, Berlin.

Janis, C. M. & Carrano, M. T. 1992. Scaling of reproductive turnover in archosaurs and mammals: why are large terrestrial mammals so rare?—*Annales Zoologica Fennici* 28: 201–216.

Jenkins, F. A. & Goslow, G. E., Jr. 1983. The functional anatomy of the shoulder of the savannah monitor lizard (*Varanus exanthematicus*).—*Journal of Morphology* 175: 195–216.

Jones, D. C., Zelditch, M. L., Lightfoot, P. & German, R. Z. 2007. The effects of muscular distrophy on the craniofacial shape of *Mus musculus*.—*Journal of Anatomy* 210: 723–730.

Kummer, B. 1959. *Bauprinzipien des Säugetierskeletts*. Thieme, Stuttgart.

Kummer, B. 1985. Die funktionelle Anpassung des Bewegungsapparates in der Phylogenese der Wirbeltiere.—*Verhandlungen der Deutschen Zoologischen Gesellschaft* 78: 23–44.

Larsen, C. S. 1997. *Bioarchaeology: Interpreting Behavior from the Human Skeleton*. Cambridge University Press, Cambridge.

Lightfoot, P. & German, R. Z. 1998. The effects of muscular dystrophy on craniofacial growth in mice: a study of heterochrony and ontogenetic allometry.—*Journal of Morphology* 235: 1–16.

Lilje, K. E. 2005. Arboreale Lokomotion beim Chamäleon (*Chameleo calyptratus*). Ph.D. Dissertation. Friedrich-Schiller-Universität, Jena.

Marsh, O. C. 1878. Principal characters of American Jurassic dinosaurs. Part I.—*American Journal of Science (3)* 16: 411–416.

Matthew, W. D. 1910. The pose of sauropodous dinosaurs.—*American Naturalist* 44: 547–560.

McHenry, C. R., Clausen, P. D., Daniel, W. J. T., Meers, M. B. & Pendharkar, A. 2006. Biomechanics of the rostrum in crocodilians: a comparative analysis using finite element analysis.—*Anatomical Record* 288: 827–849.

McIntosh, J. S., Brett-Surman, M. K. & Farlow, J. O. 1997. Sauropods. *In* Farlow, J. O. & Brett-Surman, M. K. (eds.). *The Complete Dinosaur.* Indiana University Press, Bloomington: pp. 264–290.

Meers, M. B. 2003. Crocodylian forelimb musculature and its relevance to Archosauria.—*Anatomical Record Part A* 274A: 891–916.

Müller, G. B. & Streicher, J. 1989. Ontogeny of the syndesmosis tibiofibularis and the evolution of the bird hindlimb: a caenogenetic feature triggers phenotypic novelty.—*Anatomy and Embryology* 179: 327–339.

Newman, S. A. & Muller, G. B. 2005. Origination and innovation in the vertebrate limb skeleton: an epigenetic perspective.—*Journal of Experimental Zoology Part B: Molecular and Developmental Evolution* 304B: 593–609.

Osborn, H. F. 1899. A skeleton of *Diplodocus*.—*Memoirs of the American Museum of Natural History* 1: 191–214.

Patel, B. A. & Carlson, K. J. 2007. Bone density spatial patterns in the distal radius reflect habitual hand postures adopted by quadrupedal primates.—*Journal of Human Evolution* 52: 130–141.

Paul, G. S. 1987. The science and art of restoring the life appearance of dinosaurs and their relatives. *In* Czerkas, S. A. & Olson, E. C. (eds.). *Dinosaurs Past and Present. Vol. 2.* Natural History Museum of Los Angeles County, Los Angeles: pp. 4–49.

Pauwels, F. 1965. *Gesammelte Abhandlungen zur funktionellen Anatomie des Bewegungsapparates*. Springer Verlag, Berlin.

Peterson, J. A. 1973. *Adaptation for Arboreal Locomotion in the Shoulder Region of Lizards*. Ph.D. Dissertation. University of Chicago.

Preuschoft, H. & Gudo, M. 2005. Der Schultergürtel von Wirbeltieren: Biomechanische Überlegungen zu Bauprinzipien des Wirbeltierkörpers und zur Fortbewegung von Tetrapoden.—*Zentralblatt für Geologie und Paläontologie* 2: 339–361.

Preuschoft, H. & Schmidt, M. 2003. The influence of three dimensional movements of the forelimb on the shape of the thorax and its importance to erect body posture.—*Courier Forschungsinstitut Senckenberg* 243: 9–24.

Preuschoft, H. & Witzel, U. 2004. Functional structures of the skull in hominoidea.—*Folia Primatologica* 75: 219–259.

Preuschoft, H. & Witzel, U. 2005. The functional shape of the skull in vertebrates: which forces determine morphology? With a comparison of lower primates and ancestral synapsids.—*Anatomical Record Part A* 283: 402–413.

Preuschoft, H., Witte, H., Christian, A. & Recknagel, S. T. 1994. Körpergestalt und Lokomotion bei großen Säugetieren.—*Verhandlungen der Deutschen Zoologischen Gesellschaft* 87: 147–163.

Preuschoft, H., Witzel, U., Hohn, B., Schulte, D. & Distler, C. 2007. Biomechanics of locomotion and body structure in varanids with special emphasis on the forelimb.—*Advances in Monitor Research III, Mertensiella* 16: 59–78.

Rauhut, O. W. M., Fechner, R., Remes, K. & Reis, K. This volume. How to get big in the Mesozoic: the evolution of the sauropodomorph body plan. *In* Klein, N., Remes, K., Gee, C. T. & Sander, P. M. (eds.). *Biology of the Sauropod Dinosaurs: Understanding the Life of Giants*. Indiana University Press, Bloomington: pp. 119–149.

Rayfield, E. J. 2004. Cranial mechanics and feeding in *Tyrannosaurus rex*.—*Proceedings of the Royal Society B: Biological Sciences* 271: 1451–1459.

Rayfield, E. J. 2005. Aspects of comparative cranial mechanics in the theropod dinosaurs *Coelophysis, Allosaurus* and *Tyrannosaurus*.—*Zoological Journal of the Linnean Society of London* 144: 309–316.

Rayfield, E. J. 2007. Finite element analysis and understanding the biomechanics and evolution of living and fossil organisms.—*Annual Review of Earth Planetary Science* 35: 541–576.

Reilly, S. M., McElroy, E. J. & Biknevicius, A. R. 2007. Posture, gait and the ecological relevance of locomotor costs and energy-saving mechanisms in tetrapods.—*Zoology* 110: 271–289.

Remes, K. 2008. *Evolution of the Pectoral Girdle and Forelimb in Sauropodomorpha (Dinosauria, Saurischia): Osteology, Myology, and Function*. Ph.D. Dissertation. Ludwig-Maximilians-Universität, Munich.

Romer, A. S. 1966. *Vertebrate Paleontology*. University of Chicago Press, Chicago.

Ross, C. F., Patel, B. A., Slice, D. E., Strait, D. S. & Dechow, P. C. 2005. Modelling masticatory muscle force in finite element analysis: sensitivity analysis using principal coordinates analysis.—*Anatomical Record Part A* 283: 288–299.

Runestad, J. A., Ruff, C. B., Nieh, J. C., Thorington, J. R. & Teaford, M. F. 1993. Radiographic estimation of long bone cross-sectional geometric properties.—*American Journal of Physical Anthropology* 90: 207–213.

Sander, P. M., Christian, A., Clauss, M., Fechner, R., Gee, C. T., Griebeler, E. M., Gunga, H.-C., Hummel, J., Mallison, H., Perry, S., Preuschoft, H., Rauhut, O., Remes, K., Tütken, T., Wings, O. & Witzel, U. 2010. Biology of the sauropod dinosaurs: the evolution of gigantism.—*Biological Reviews of the Cambridge Philosophical Society*. doi: 10.1111/j.1469=185X.2010.00137.x.

Schwarz, D., Frey, E. & Meyer, C. A. 2007. Novel reconstruction of the orientation of the pectoral girdle in sauropods.—*Anatomical Record* 290: 32–47.

Swinton, W. E. 1970. *The Dinosaurs*. Wiley, New York.

Therrien, F. & Henderson, D. M. 2007. My theropod is bigger than yours . . . or not: estimating body size from skull length in theropods.—*Journal of Vertebrate Paleontology* 27: 108–115.

Thulborn, R. A. 1990. *Dinosaur Tracks*. Chapman and Hall, London.

Tornier, G. 1909. Wie war der *Diplodocus carnegii* wirklich gebaut?—*Sitzungsberichte der Gesellschaft naturforschender Freunde zu Berlin* 1909: 193–209.

Trinkaus, E., Churchill, S. E. & Ruff, C. B. 1994. Postcranial robusticity in *Homo*. Humeral bilateral asymmetry and bone plasticity.—*American Journal of Physical Anthropology* 93: 1–34.

Upchurch, P., Barrett, P. M. & Dodson, P. 2004. Sauropoda. *In* Weishampel, D. B., Dodson, P. & Osmólska, H. (eds.). *The Dinosauria. 2nd edition*. University of California Press, Berkeley: pp. 259–322.

Wilhite, D. R. 2003. *Biomechanical Reconstruction of the Appendicular Skeleton in Three North American Jurassic Sauropods*. Ph.D. Dissertation. Louisiana State University, Louisiana.

Wilson, J. A. & Carrano, M. T. 1999. Titanosaurs and the origin of "wide-gauge" trackways: a biomechanical and systematic perspective on sauropod locomotion.—*Paleobiology* 25: 252–267.

Wilson, J. A. & Sereno, P. C. 1998. Early evolution and higher-level phylogeny of sauropod dinosaurs.—*Society of Vertebrate Paleontology Memoir* 5: 1–68.

Witzel, U. 1985. *Zur Biomechanik der ossären Schafteinbettung von Hüftendoprothesen*. Mathys AG, Bettlach, Switzerland.

Witzel, U. 1996. *Mechanische Integration von Schraubpfannen*. Thieme Verlag, Stuttgart.

Witzel, U. 2000. Die knöcherne Integration von Schraubpfannen. *In* Perka, C. & Zippel, H. (eds.). *Pfannenrevisionseingriffe nach Hüft-TEP. Standards und Alternativen*. Einhorn Verlag, Reinbeck: pp. 11–16.

Witzel, U. & Preuschoft, H. 2005. Finite Element model construction for the virtual synthesis of the skulls in vertebrates: case study of *Diplodocus*.—*Anatomical Record Part A* 283: 391–401.

Witzel, U., Mannhardt, J., Goessling, R,. de Micheli, P. & Preuschoft, H. This volume. Finite element analyses and virtual syntheses of biological structures and their application to sauropod skulls. *In* Klein, N., Remes, K., Gee, C. T. & Sander, P. M. (eds.). *Biology of the Sauropod Dinosaurs: Understanding the Life of Giants*. Indiana University Press, Bloomington: pp.171–181.

Wolff, J. 1892. *Das Gesetz der Transformation der Knochen*. Hirschwald, Berlin.

Wollfson, D. M. 1950. Scapula shape and muscle function, with special reference to the vertebral border.—*American Journal of Physical Anthropology* 8: 331–341.

Zienkiewicz, O. C., Taylor, R. L. & Zhu, J. Z. 2005. *The Finite Element Method: Its Basis and Fundamentals*. Elsevier Butterworth-Heinemann, Amsterdam.

12

Why So Huge? Biomechanical Reasons for the Acquisition of Large Size in Sauropod and Theropod Dinosaurs

HOLGER PREUSCHOFT, BIANCA HOHN, STEFAN STOINSKI, AND ULRICH WITZEL

TO UNDERSTAND GIGANTISM, the pros and cons of large size must be clearly recognized. Although the *disadvantages* connected with extraordinary size are dealt with in the other chapters in this book, a better understanding of the biomechanical *advantages* of large body size is needed. We therefore focus on the question of which immediate, proximate advantages are connected with gigantic body size, and we analyze the biomechanical advantages and limitations of several size parameters. We discuss the neck length required for harvesting large volumes of food, which is limited by the muscle and skeletal mass necessary to maneuver a long neck. We also look at the limb length needed for increasing locomotor speed and reducing energy consumption per unit distance covered, although this is limited by reduced step frequency. Finally, in agonistic encounters, the decisive factors are the kinetic energy contained in the colliding bodies, and the forces and impulses exchanged between the animals. All factors depend on speed, so a deficiency of mass can be made up by greater speed. Great mass and length are equivalent to slowness, especially in the defensive and evasive movements of limbs and neck. Volume alone is a protective trait against bite attacks. Thickness of skin, and skeletal and muscular cover of the most vulnerable organs increase linearly with size. In short, an increase of body dimensions and body mass offers quantitative biomechanical advantages. These parameters, however, follow a linear or cube root function—that is, they are not very impressive, and in some cases, they reach asymptotes at larger sizes, so that their advantages become smaller with increasing size. Body mass and dimensions of body segments set limits to the quickness of evasive and defensive movements. After quantitatively defining the advantages and limitations on the basis of various biomechanical laws, we argue that these numerical advantages can be understood as selection pressures that have led to gigantism.

Introduction

One of the first questions raised when talking about dinosaurs is why, but not how, they became so huge. The same question can be put more precisely: what advantages resulted from their enormous size? Because the problems, preconditions, and necessities connected with extraordinary size are dealt with in other chapters of this book, we will focus on the question of which biomechanical advantages are connected with gigantic body size. A side effect of addressing this question is the insight gained into the problems connected with and limitations set by size increase—in short, the disadvantages of large size. We are aware of the wealth of literature, often within the framework of Cope's rule, on the phylogenetic increase of body size. No doubt these arguments are sound and convincing (Blankenhorn 2000; Rauhut et al., this volume), but aside from the conviction that "greater fitness" results from larger size, there is a deplorable lack of detailed biomechanical explanations for the most proximate, immediate advantages that an individual obtains by a larger size that help it to survive or to exist with a lesser input of energy.

As will be shown below, such advantages can indeed be defined quantitatively on the basis of relatively simple and generally known biomechanical principles. With the help of these principles, the advantages or disadvantages can be quantified in the form of mathematical functions. Interestingly, these biomechanical functions describe the exact selective pressures that may have acted on individuals and therefore led to gigantism. A similar approach for size differences in primates was presented by Preuschoft (2010); important results are in Sander et al. (2010). These sorts of ideas are not entirely new; they have been pursued explicitly in Preuschoft & Demes (1984, 1985) and especially in Demes (1991), who used the term *biomechanical allometry* for this line of reasoning. To avoid misunderstanding, it should be emphasized that *allometry* is commonly used as a purely descriptive method for analyzing the variation of measurements during ontogeny or among taxonomically related animals. It is also used to assess relations between functions and characteristics of shape.

We investigate here the effect of size and mass on the chance of survival under various clearly defined external conditions. These selected conditions are essential for survival: food acquisition, locomotion, and fighting with conspecifics or predators. A major factor for survival seems to be a limitation of energy expenditure. The relationships between the individual and other animals and its environment are described in the form of mathematical equations. Because our aim is to explain obvious traits and their interrelationships, we provide details on the laws of physics and equations within the context of our results. This chapter is divided into four sections: feeding, locomotion, agonistic activities, and the coevolution of predators. In each section, we first describe the physical laws that govern the activity, and second, we investigate the influence of mass or a selected dimension. The physical laws concern either statics—that is, the motionless state—or kinetics—that is, the changes in the situation as time continues. Because the laws of physics are timeless, to improve the readability of the text, we have opted to use the present tense in our discussion of sauropod mechanics even though the last sauropod lived some 65 million years ago. Evolution as a historical process was obviously constrained by the laws of physics, as much during the lifetime of dinosaurs as during the later evolutionary body size increase of mammals, and our conclusions concerning sauropods are therefore written in the past tense.

Table 12.1. Masses and Lever Arm Lengths of Forelimb and Neck Segments of *Brachiosaurus brancai*[a]

Site	Mass (kg)	Lever length (m)
Forelimb		
Segment "shoulder + upper arm"	1,000	0.61
Segment "forearm"	150	2.53
Segment "foot"	110	3.71
Neck (segments from proximal to distal)		
Segment 1	2,050.2	0.81
Segment 2	737.0	1.03
Segment 3	522.6	3.11
Segment 4	194.3	4.79
Segment "head"	290.0	5.80

For details about the generation of these data, see Stoinski et al. (this volume). The data were kindly provided by S. Stoinski.

[a]Total length of the forelimb is 3.96 m. The total mass moment of inertia of the forelimb (J) is 2,846.2 kgm^2. Total length of the neck plus head (=reaching distance) is 8.6 m. Total mass moment of inertia of the neck (J) is 20,050.163 kgm^2.

METHODS

The concept of size includes different aspects. In dinosaurs and quadrupedal mammals, for examples, the length of the entire body describes size well. In humans, body height is used to characterize size; similarly, we are impressed when we learn that the length of a basking shark is 9 m, or when we view a European bison or an Asiatic gaur from the side as "large"—but we are disappointed when we note their narrowness in anterior view. The only way to measure size exactly is to determine body mass. However, in the following, we will deal with all aspects of size because they all influence fitness.

Characterizing size can be difficult in the case of fossils, and we commonly have to find approximations for mass. Dimensions, or length measurements, can be obtained fairly exactly from preserved skeletons. Areas and volumes of skeletal structures can also be determined exactly by the multiplication of two or three measurements. For our purposes here, areas and volumes of the entire body or of body segments are needed. These parameters can only be found on the basis of estimates of the soft parts, which are usually not preserved, within some degree of plausibility. In the following, the terms *dimension, area,* and *volume* will be strictly differentiated from one another. Volume is proportional to mass, depending on the specific density of the body segment under consideration. Dimensions, meaning lengths and diameters, follow a linear function as they increase or decrease. Areas follow a square function (that is, with an exponent of 2); volume, and thus mass, follows a cube function (that is, with an exponent of 3).

The relationship between energy expenditure and speed, or mass, which the equations describe, is well known. All this can easily be found in textbooks of mathematics or mechanics. The following analyses are based on the handbook of mathematics by Engesser (1996), Lehmann's (1974–77) textbook of mechanics, and Dubbel's (1981) handbook of mechanical engineering, but others may do just as well.

For the approach employed here, it is essential to use absolute values as starting points. For this purpose, we have used the data obtained by Gunga et al. (1995, 1999, 2008), Christian et al. (1999a, 1999b), Christian (2002), Christian & Dzemski (2007), and Stoinski et al. (this volume) on the famous Berlin *Brachiosaurus brancai,* as given in Table 12.1. This individual had a total mass of 38 metric tons, according to the most recent estimates of Gunga et al. (2008).

Neck Length and Feeding

It is a matter of taste whether sauropods are described in this initial approach as animals of absolutely outstanding bulk (an order of size larger than other dinosaur groups), or as animals with the longest necks that ever existed. In the context of this chapter, we will follow the latter approach. More information on the statics of the enormous neck of *Brachiosaurus* is given in Christian & Heinrich (1998) and Christian & Dzemski (2007, this volume). Also relevant for our discussion are data on how modern giraffes, camels, and ostriches use their long necks, which can be found in Dzemski (2006) and Christian & Dzemski (2007).

ADVANTAGES OF GREAT NECK LENGTH

There is general agreement that the long neck of sauropods is related to the acquisition of food. Some divergence of opinion exists on whether neck length allows the animal to select rather rare, dispersed food sources high up in the trees (or below the water surface of ponds while the animal's body remains on firm ground; Stevens & Parrish 1999), or whether it is a means to nonselectively harvest a large volume of vegetation without needing to move the mass of the entire body. On the basis of experimental work by Hummel (pers. comm.; Hummel et al. 2006, 2008, Hummel & Clauss this volume), large sauropods had to consume between 60 and 200 kg of their most probable staple foods, namely *Araucaria* twigs with leaves and *Equisetum,* every day. In our calculations, the range of these values is of minor importance because the essential point in both cases is the reaching distance. The reaching distance r increases linearly with neck + skull length (Table 12.1, Fig. 12.1). The maximal volume within which the head can be moved to collect food (the so-called feeding envelope) is one quarter of a sphere, $V_g = 4\pi r^3/3$, which means that the accessible volume is

$$V_a = 4\pi r^3/12.$$

This holds true only if the trunk is resting on the ground and if the neck is quite flexible, as illustrated in Fig. 12.1. This latter condition, however, may not be met because Stevens & Parrish (1999, 2005a, 2005b) have provided arguments that the mobility of the sauropod neck was restricted. In fact, the necks of modern long-necked mammals such as camels and giraffes are rather stiff (Dzemski 2006; Christian & Dzemski 2007) and move through restricted angles. On the other hand, the flexible necks of birds give then access to any point within their feeding envelope. In living animals, the degree of mobility is often much more extensive than their skeleton seems to permit (Putz 1976, 1983; Preuschoft et al. 1988; Dzemski 2006; Christian & Dzemski 2007). In our model calculations, any restriction assumed for neck mobility would not fundamentally influence the increase of the feeding envelope with increasing neck length, because only a smaller part of the feeding envelope would be accessible, for example, an eighth or a tenth versus a quarter of a sphere, whereas the increase of the feeding volume would still scale with the cube of neck length.

The pressure exerted by the neck base onto the ground when the cranial segments are lifted seems to be a good reason to develop strong cervical ribs on the proximal neck segments. Strong cervical ribs would provide a means of shifting the pressure caused by neck movements away from the trachea and the esophagus and distribute the reaction force over several vertebrae.

If the center of the feeding sphere (=the base of the neck) is lifted above the ground by the forelimbs, the volume of a half cylinder ($V_c = \pi r^2 h$, where h is forelimb length) between the sphere and the ground must be added to the feeding envelope. This means that the total volume V_t (see above) that can be reached increases by

$$V_t = \pi r^2 h/2 + 4\pi r^3/12.$$

In Fig. 12.2, the increase of accessible volume with increasing neck and forelimb length is plotted up to a value of 5 m for the forelimb, because this is the extreme forelimb length found in *Brachiosaurus*. From here on, limb length is assumed to remain constant. As can be readily seen, the distance between the two curves changes from proportionally increasing to parallel. According to Christian & Dzemski (2007) and Dzemski (2005, 2006), living species do not necessarily make use of this volume, but instead confine themselves to a smaller sector to feed in. This means that only a fraction of the entire volume is exploited, and the curve shown in Fig. 12.2 remains at a lower level. If sauropods had confined their neck movements to only part of the accessible volume, as shown in Fig. 12.1 (as modern giraffes do; Dzemski 2005, 2006), then only a part of this accessible volume was exploited. However, in any case, the volume of accessible vegetation grows by the cube of neck length and linearly with forelimb length. The advantage of increasing neck length is obviously the greater reaching distance and thus an increase in the feeding envelope, within which the head can be moved without moving the trunk.

To obtain the necessary amount of food required for its subsistence, an animal needs to exploit the vegetation covering a

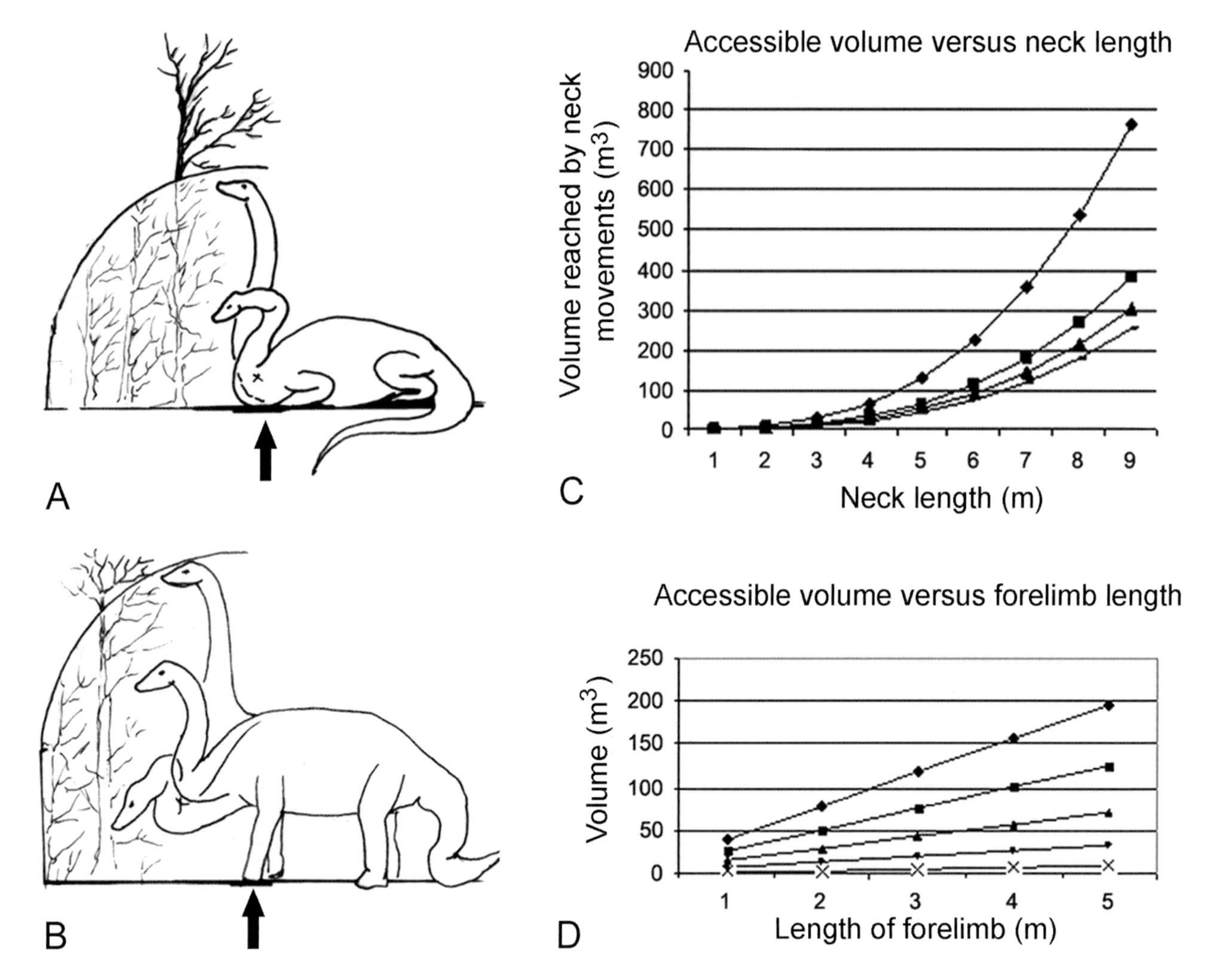

FIGURE 12.1. (A) Schematic illustration of a sauropod resting lazily on the ground while collecting food by neck movements alone, potentially covering a feeding envelope of a quarter sphere. Note that the length and mass of the neck leads to a reaction force between ground and basis of the neck, as indicated by the arrows. (B) The same sauropod standing while feeding. The maximal volume of vegetation that can be reached by neck movements alone, that is, the feeding envelope, is outlined. (C) The volume that can be reached by movements of the neck alone increases with neck length raised to the third power if the neck is able to cover the entire quarter sphere (diamonds). If the movements of the neck are restricted (as observed in giraffes because of their preferred angles of inclination of the neck), or if a specific layer of the canopy is exclusively exploited, only a smaller part out of the feeding envelope is used, and the accessible volume is reduced proportionally. The three lower curves show the results of the calculation if only one eighth (squares), one tenth (triangles), and one twelfth (small dots) of the sphere is accessible. (D) The feeding envelope also increases linearly with leg length. The individual curves are drawn for short (bottom) and increasingly long (top) necks.

certain area (Burness et al. 2001). The ability of the animal to harvest the resources in the area (Fig. 12.3) depends on the length of its neck. As a starting point, we arbitrarily imagine an area of 1 ha, that is, an area of 100 × 100 m (this area provides roughly the grass required per year by one horse under the climatic conditions of Central Europe). In our example, the hectare is thought to be covered with vegetation consisting of plants more than 15 m in height (Gee pers. comm.; this volume) and available to large, ground-living animals with necks of variable lengths (1–10 m). If the neck has a length of 1 m, the animal must change its place, the "feeding station" in the sense of Shipley et al. (1996) (Fig. 12.3A, B) as soon as the volume of a quarter of a sphere plus a semicylinder (as in Fig. 12.2) with a 1 m radius is used up. To harvest the resources of our basic hectare, the short-necked animal has to change its feeding station roughly 5,000 times. If its neck has a length of 2 m, only 1,250 changes are necessary, as is illustrated in Fig. 12.3. The animal with a 10 m neck requires less than 100 changes. This has major consequences for the energy required for foraging.

Each change of feeding station involves an acceleration and a deceleration. Both are probably of the same magnitude (see also Shipley et al. 1996), although in land animals, the metabolic cost of accelerating is greater than of decelerating, that is, braking is cheaper. This is mainly the result of energy lost to

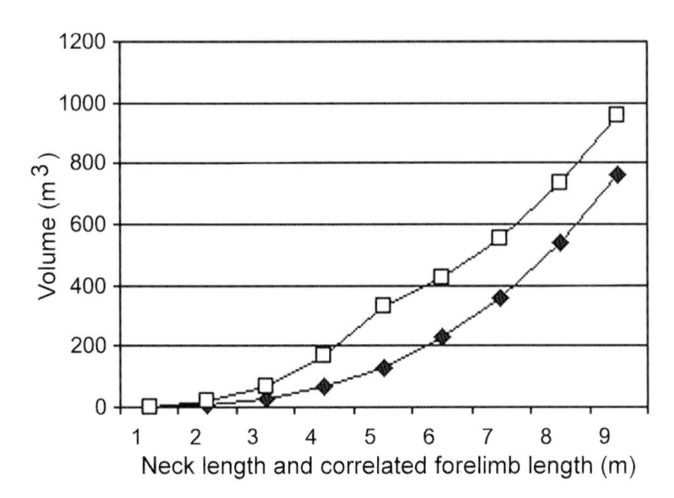

FIGURE 12.2. Influence of increasing neck length on the accessible volume of vegetation, that is, the feeding envelope size. The lower curve (diamonds) is equivalent to the top curve in Fig. 12.1C. The upper curve (open squares) includes the effects of both increasing neck length and increasing forelimb length, but only to a limb length of 5 m. While forelimb length increases with neck length, the two curves are proportional; after forelimb length has reached its maximum, the curves are parallel.

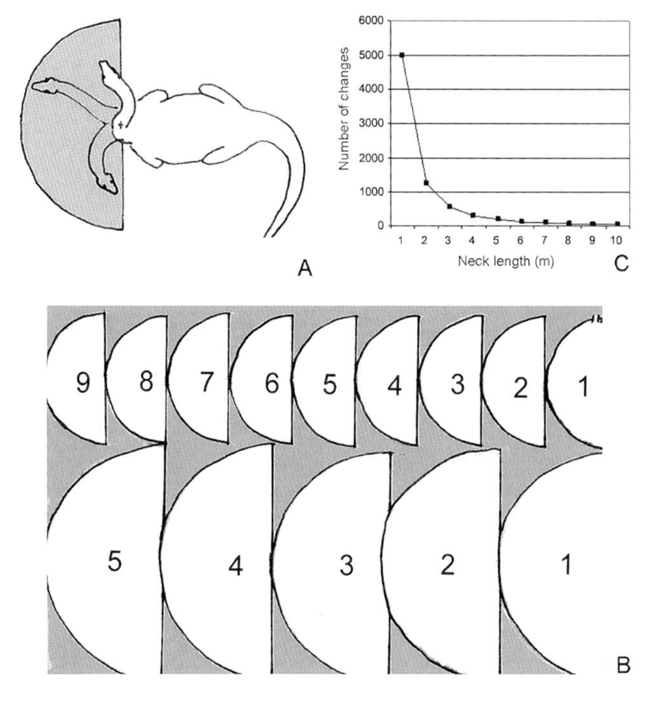

FIGURE 12.3. Influence of neck length on the number of feeding stations per area to be harvested. (A) At a given feeding station, the animal can harvest the vegetation within the reaching distance of its neck. A more rigid neck reduces the volume that can be exploited. (B) Sketch illustrating the relationship between neck length and number of feeding stations. The animal with the shorter neck (top) needs nine feeding stations, while the animal with the longer neck (bottom) needs only five feeding stations to cover the same distance, but about twice the area. (C) Influence of neck length on the number of feeding stations required to harvest an area of one hectare (100 m by 100 m). The curve starts off with neck lengths >1 m and under the assumption of great flexibility of the neck. Obviously, the longer the neck, the smaller the number of changes required to cover the same area. Note that the point of diminishing return is approached after a neck length of about 6–8 m.

internal friction, that is, damping in viscoelastic tissues that contribute to reducing the speed of movement. Propulsion, on the other hand, requires active muscle power. For the sake of simplicity, we have not considered this complication but instead multiplied the frequency of changes by 2. As anybody who has become tired from working knows, each change of speed requires power; that is, energy must be expended. There is no need to imagine sauropods moving forward in sudden bursts of speed from one feeding station to the next. We should instead imagine trucks moving slowly on a road in stop-and-go traffic. Although their accelerations are slow, they use much more fuel than in traffic flowing at constant speed. The force needed for acceleration and deceleration is basically proportional to mass and acceleration:

$$F = m\,a.$$

Hence, the dependence of mass and force is linear at any level of acceleration. In Fig. 12.4, the increase in forces required for acceleration are plotted for increasing body mass at a constant acceleration. Even if acceleration is as low as $a = 0.2$ g (where 1 g is the acceleration due to Earth's gravity, 1 g also being the accelaration of an average motor car), the huge mass of a sauropod implies considerable force requirements and energy expenditure. The figures in our example (but not in principle!) hold true only if the neck is flexible, so that the entire volume within reaching distance can be exploited. If the neck is rather immobile, as assumed by Stevens & Parrish (1999, 2005a, 2005b) and others, much less food can be harvested at each feeding station. Therefore, feeding stations must be changed more often. From the viewpoint of energy requirements, it is clear that the stiffness of the neck is a disadvantage.

It must be emphasized that a great number of changes from one feeding station to the next takes considerable time, especially if acceleration is low. It may well be that the time a large, short-necked animal would need for acceleration and deceleration is too long to leave sufficient time for the intake of food. This may become a critical factor for the chances of survival. If acceleration is greater, energy expenditure assumes greater values, but more food can be collected. The trick is that force requirements increase much more slowly if neck length increases (Fig. 12.5). The advantage of neck length, however, reaches the point of diminishing return at neck lengths exceeding 6–8 m.

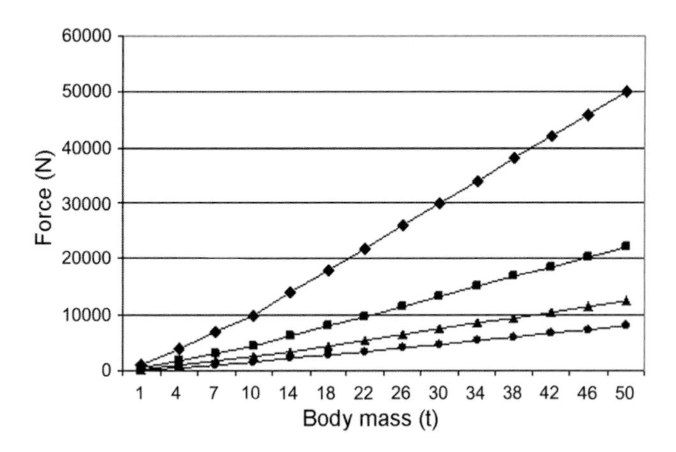

FIGURE 12.4. Increase of forces required for moving from one feeding station to the next with increasing body mass but constant acceleration. The curves were plotted for the following frequencies of feeding station change: 5,000 (diamonds), 2,218 (squares), 1,250 (triangles), and 800 (dots). Note that the forces increase linearly with mass but exponentially with the number of feeding stations. Compared to an animal that requires 800 changes, an animal of the same mass that requires 5,000 changes to cover the same area requires 2.25 times the force. For 2.218 changes, the factor is 1.78, and for 1,100 changes, the factor is 1.56.

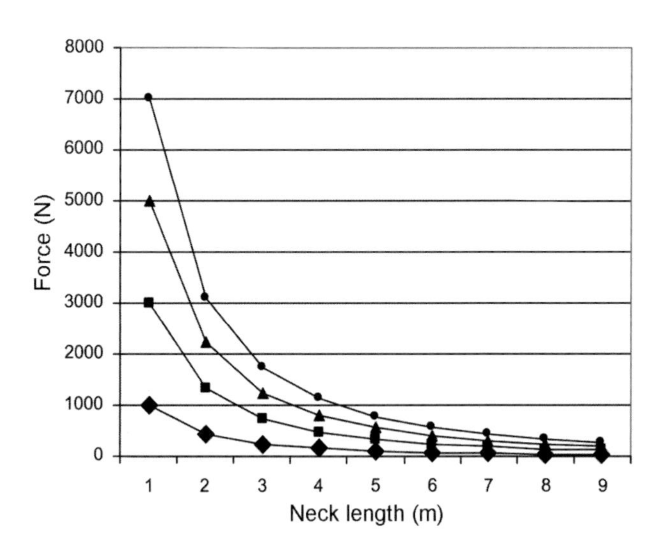

FIGURE 12.5. Forces required for moving constant masses at increasing acceleration *a* from one feeding station to the next. The curves show the force requirement depending on the speed of acceleration at neck lengths (x-axis) increasing from 1 to 9 m. The lowest acceleration value is 0.2 m/s^2 (diamonds) followed by 0.6 m/s^2 (squares), 1.0 m/s^2 (triangles), and finally 1.4 m/s^2 (dots). Note that great neck lengths reduce the force necessary for changing the feeding stations at any assumed acceleration. This effect is particularly pronounced at high accelaration speeds.

LIMITATIONS TO NECK LENGTH

The increased neck length, however, also entails disadvantages: the linear increase of lever arms (lengths of head and neck) requires an increase of either lever arms of muscles (as in giraffe, horse, and cattle), or of muscle forces to control the neck (as in tortoises, monitor lizards, and snakes). Both changes, of lever arms as well as of muscular cross sections (which are proportional to muscle force), lead to an increase in proximal neck diameter. This is equivalent to an increase in proximal neck mass to the third power (see also below; Fig. 12.7; Table 12.1). To illustrate this, we used the neck of *Brachiosaurus* (Fig. 12.6), the dimensions and volume of which were published by Gunga et al. (2008).

This entire load of the neck and head is balanced by muscles that have slightly increasing lever arms. As long as neck diameter does not grow in parallel with bending moments (which seems improbable from the view of virtually all available reconstructions), the necessary muscle forces become greater with increasing neck length, and the strongest muscle contractions must occur at the base of the neck, of course. Because the neck's mass moment of inertia J increases with mass m times length l to the second power,

$$J = m\,l^2,$$

and because mass scales with the third power of length l^3, J scales with the fifth power of length l^5 (Preuschoft et al. 1998). Therefore, quick movements of a very long neck will rapidly exceed the abilities of musculature. Even if quick movements can be avoided, neck elongation is not without limits (Fig. 12.7) as a result of the well known Galileo rule: muscle and bone strength depend on cross-sectional area and thus scale with a square function, whereas mass increases with the third power.

If muscular force is increased (Fig. 12.7), the curve illustrating the available force can be multiplied by factor of, say, 100. It is evident that neck lengths of more than 9 m are unlikely to evolve from a biomechanical point of view—unless the lever arms of muscles are increased.

To provide the necessary muscular forces with increasing neck length, and consequently the increasing strength of the vertebral column, the mass of the neck needs to increase further. The evolution of a very long neck is critically dependent on the balance between an increase in muscle energy required for maneuvering the heavy neck and an increase in energy intake because of the larger volume of food that can be harvested. To work out the influence of neck length, the diameter of the neck was kept constant in our calculations, and only the horizontal position was considered. Both assumptions are not realistic (as detailed in Christian & Dzemski 2007), but they exclude the influence of changing values of force and load.

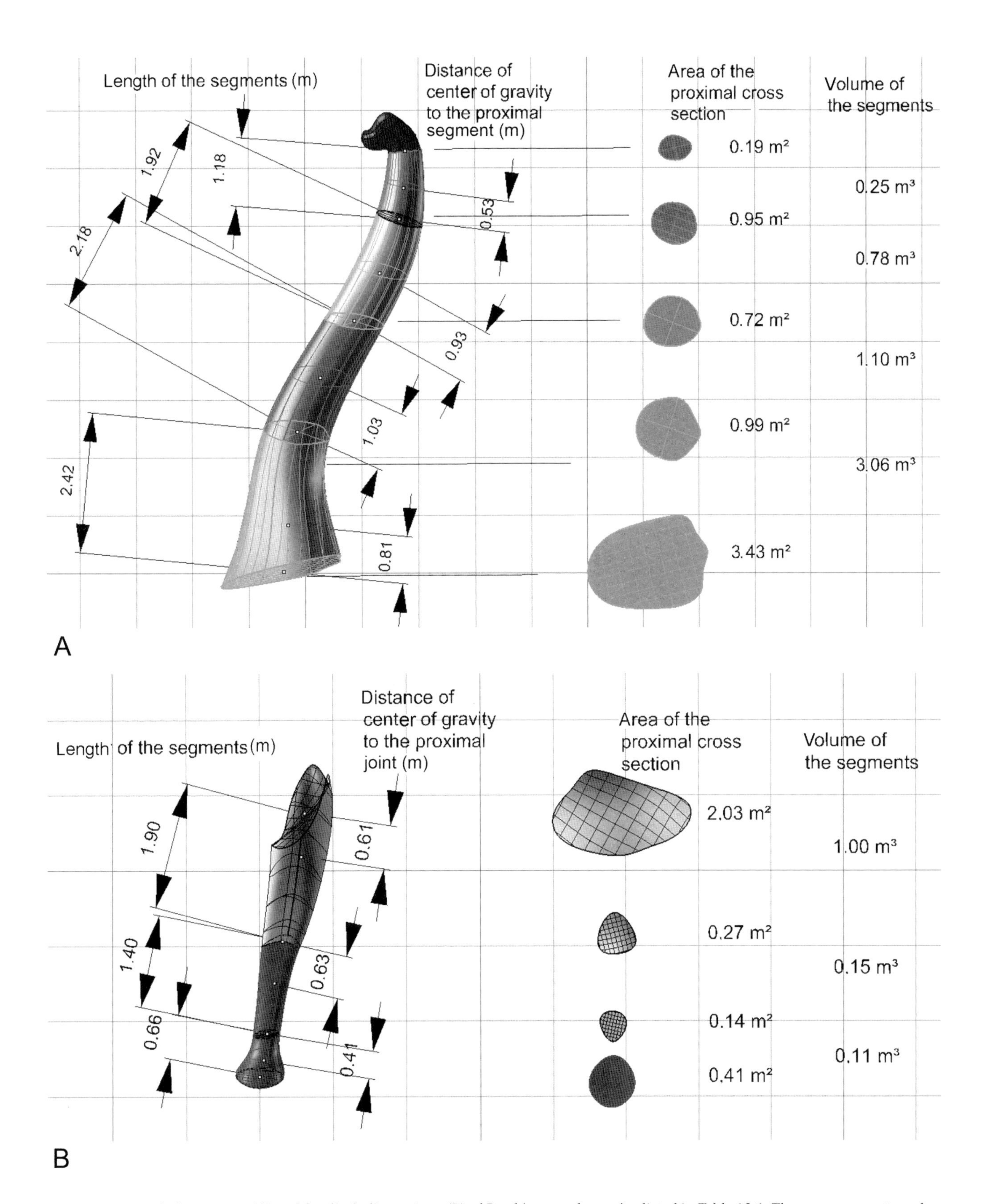

FIGURE 12.6. Neck dimensions (A) and forelimb dimensions (B) of *Brachiosaurus brancai* as listed in Table 12.1. The measurements and reconstructed volumes formed the basis for the calculations in this chapter.

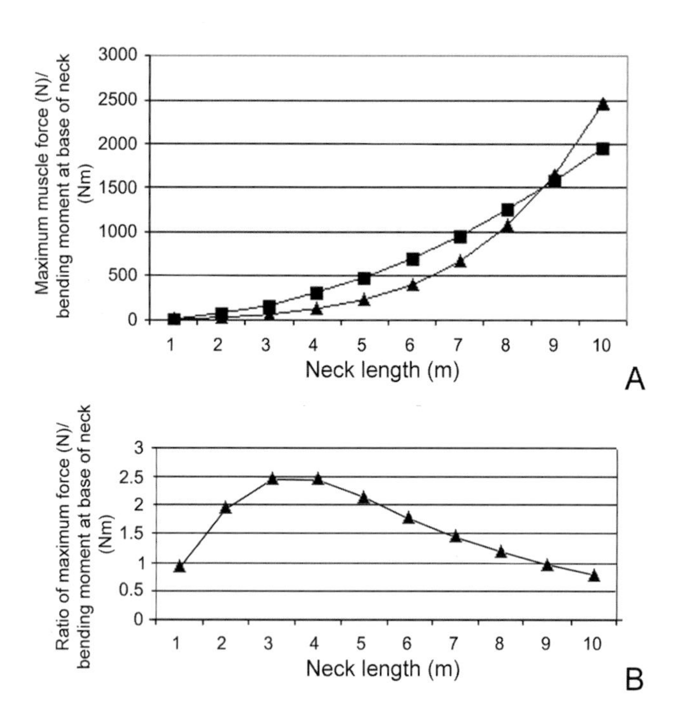

FIGURE 12.7. Biomechanical limitations to neck length. (A) The moments of load at the base of the neck (triangles) are proportional to neck weight and neck length. They increase by the fourth power, while muscle forces increase by the second power (squares) and therefore lag behind the increase of load moments. In neck lengths shorter than about 6 m the increase in muscle forces may be faster than the increase in load moments. This means first that neck lengths of less than 6 m are obviously advantageous, while longer necks are less so, and still longer necks become disadvantageous. It also means that generally these curves show the importance of the absolute values for neck length. (B) Based on the curves in (A), the ratio between muscle force and bending moments shows a maximum between 2 and 6 m neck length. Neck lengths of more than 7 m are not advantageous if neck diameters remain unchanged. This result, however, holds only true for the conditions assumed here (horizontal neck posture, given neck mass, and given muscle force). If the vertical neck diameter and neck mass are changed, this ratio changes as well, but in all cases, the fact remains that the advantage of a long neck becomes smaller with increasing length.

Beyond doubt, there is a considerable influence of neck mass on the shoulder–forelimb apparatus and on the postures of the forelimbs. Preuschoft & Gudo (2005) have shown that the long and heavy neck could not have been flexed to one side without moving one forelimb in the same direction. Likewise, a forceful movement of the heavy and long neck to one side required lateral movement of the opposite forelimb (see also Christian & Dzemski 2007) in order to support the body under the influence of mass moment inertia.

Body Size, Limb Length, and Locomotion

GAITS

Locomotion is a characteristic of animals. In vertebrates, the locomotor apparatus—that is, those parts of the body that take part in locomotion—determines the general shape of the body, although it is not the case in many invertebrates, for example, clams (Preuschoft et al. 1975; Preuschoft 1976), starfish, sea urchins, and corals. The majority of land-living vertebrates perform locomotion with the aid of extremities. Exceptions are snakes, some lizards, and a number of tailed amphibians. The number of limbs is principally four (reflected in the term *tetrapod*), and locomotion on four limbs is called *quadrupedalism*. However, in various groups (birds, kangaroos, some rodents, bats, and our own genus, *Homo*) the anterior pair of limbs is either reduced or has taken over special functions (for instance, flying in most birds and in bats, arm-swinging in gibbons, reaching and collecting food, or using tools in a variety of primates). In a remarkably high number of dinosaurs, the forelimbs are also reduced in comparison to the hindlimbs. In all of these cases, the hindlimbs take over the greater share or all of the locomotor tasks.

Quadrupedal locomotion can take place in various ways, which are called gaits. Most of the concepts concerning gaits are based on those quadrupeds we know best: dogs and horses. The latter in particular have long been bred to make use of their locomotor abilities, which is, in itself, a good reason to look at their gaits. This hippomorphic view of quadrupedal locomotion is still influenced by Muybridge's (1898) famous studies, but these have largely been superseded by a system based on absolute data in the work of Hildebrand (1960, 1965, 1966, 1985) and Hildebrand & Goslow (2001, 2004). Still, to this day, the horse gaits dominate terminology and discussion. The gaits of horses are the walk, the amble, the trot or pace, and the canter (which in small animals is often replaced by bound and half-bound). Although the walk is used by all tetrapods, the ways of running (trotting. pacing, and asymmetrical gaits like bounding, cantering, and galloping) vary (as mentioned above).

In the last decades, a major distinction between two general kinds of gaits has become increasingly apparent, namely between walking and running. In all gaits, the limbs perform cycles of movements, in which they alternatingly behave as a suspended pendulum (during the fore swing) and as an inverted pendulum (during the stance phase). The essence of walking (or striding) is that it is governed by the laws of the pendulum (Mochon & McMahon 1980a, 1980b; Witte et al. 1991; Preuschoft et al. 1992; Preuschoft & Witte 1993), whereas running is governed by the elastic stretching and recoiling of springs (Witte et al. 1995a, 1995b). These springs are usually the units of muscles and their tendons, which possess their own frequencies dictated by the material, and these frequencies fit to the repetition of locomotor cycles (Witte 1996; Witte et al. 1995b; Rao 1999). Walking per se is safe (because most of the time two or three limbs are on the ground), but slow. The faster gaits reduce the phases during which the body is safely supported on two or more limbs, increase the less stable phases of one-limb support, and prolong the phases of ballistic suspension in the air. Because the weight of the animal must be countered by ground reaction forces, shorter phases of support result in increasingly higher forces that must be exchanged between the ground and the animal. Slow walking can be performed by swinging forward one forelimb and the contralateral hindlimb nearly at the same time (as in many modern reptiles; Christian 1995; Preuschoft et al. 2007) or by swinging forward each limb in sequence (as in horses) (Fig. 12.8).

There are two basic methods for increasing speed: step elongation and step frequency increase. The former is produced by greater limb length and wider angles of excursion. However, step elongation is limited by the extreme torques about the hip and shoulder joints occuring in very long steps. If the excursion angle from touchdown to liftoff is much wider than the angle covered by the ground reaction forces, the latter deviate widely from the proximal joints (see Fig. 12.9) and therefore need excessive muscle force for balancing the joints. An increase in stride frequency also consumes energy because an additional input of muscle force is needed for accelerating the forward swing and shortening the stance phase. The shortening of the stance (=support) phase results in greater ground reaction forces (see above) and reduces the time during which more than one limb is on the ground to control and correct the movement of the body.

A way out is a phase of aerial suspension to increase stride length without wide excursion angles. This can be accomplished either while maintaining the symmetry of limb pairs, as in the trot and the pace, or by moving the limbs asymmetrically in the leaping gaits. The introduction of aerial suspension phases in the symmetrical gaits leads to a true trot in modern reptiles, and in mammals to unsupported, aerial suspension phases of one third or even one half of the stance phases in the trot or pace, both of which can be observed in horses and dogs. These phases of aerial suspension can also be disrupted for both hindlimbs and forelimbs. This happens at high speeds in the amble or running walk typical of some horse breeds (Preuschoft et al. 1994b) and of elephants (Gambaryan 1974; Christian et al. 1999b; Hutchinson et al. 2003). The running walk reduces the duration of the gait cycle (Fig. 12.8) and increases speed by increasing step lengths without changing the footfall sequence.

In locomotion, speed can be increased by increasing the frequency or by extending step lengths combined with reducing the duty factors (as in the ambling and trotting horse), which is the percentage of the step cycle during which the foot touches the ground, or by reducing the time lag between the touchdown of hind- and forelimb of the same side (as in the llama and some antelopes). Although most faster gaits require greater expenditure of muscle force, a pace-like walk, as in camels, seems to reduce the time needed for a full cycle without additional expenditure of energy. This option is also used in the trot and the pace (at the expense of safe supports on more than two limbs). In addition, in these gaits, the duty factor is below 50%, so that the animal briefly floats unsupported in the air.

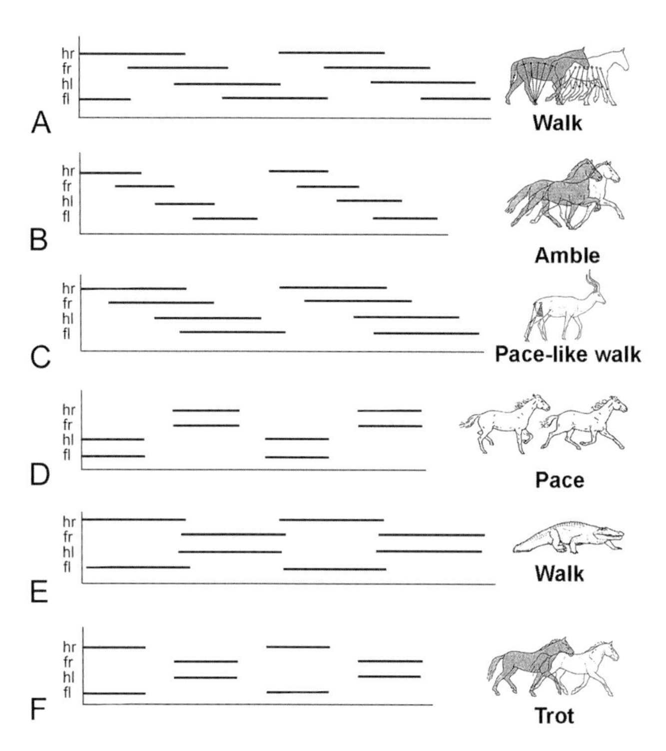

FIGURE 12.8. The most important symmetrical gaits in quadrupeds and the patterns of support, that is, the ground contact of the respective foot (horizontal bars). (A) Walking horse, lateral footfall sequence. The sketches on the right show the stance phase of hindlimb (gray) and the swing phase of forelimb (white). (B) Horse traveling in an amble or running walk. This is also the gait seen in fast-moving elephants. (C) Pace-like walking in some antelopes and especially in llamas and camels. (D) Pace in the strict sense (camels, some horses), which is similar to the true trot in horses. (E) Walk as seen in many recent reptiles with a diagonal footfall sequence using extended diagonal supports. This can easily be transformed into the true trot (F) with a fully suspended phase, as in horses. hr, right hindlimb; fr, right forelimb; hl, left hindlimb; fl, left forelimb.

In walking, the stance phase normally is longer than the

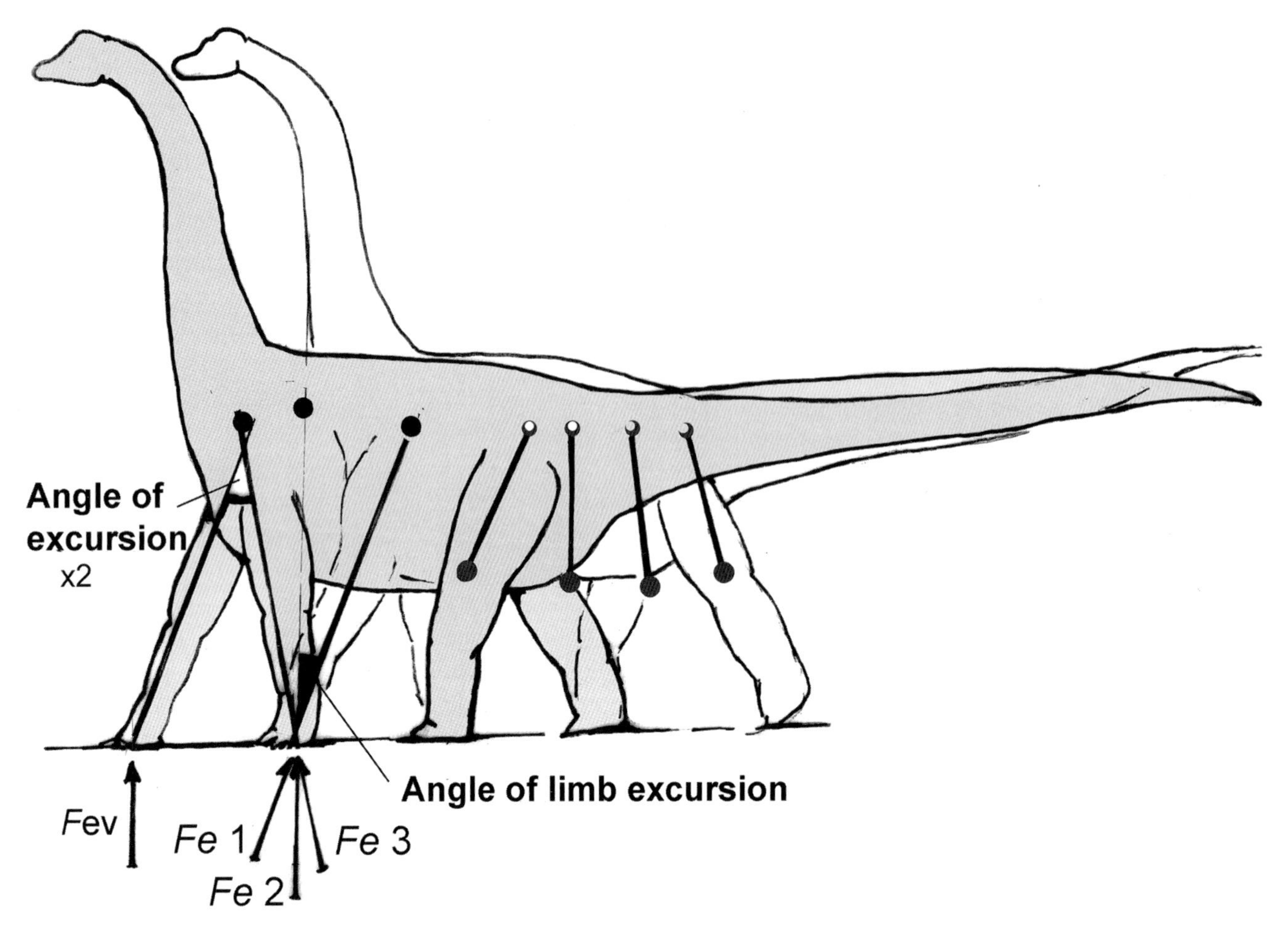

FIGURE 12.9. Walking mechanics, using a sauropod as an example. The positions of the limbs were only chosen to illustrate the mechanics of pendulous locomotion, not a particular gait. The left forelimb is shown as an inverted pendulum during the stance phase of locomotion. The pivot is in the autopodium, just above the point of ground contact. Ground reaction forces (*Fe*1–3) were estimated by analogy with a variety of living animals, with the length of the arrows indicating relative magnitude. Note that the angle of limb excursion is measured from the vertical upward. For the right forelimb, only the vertical component of the ground reaction force (*Fev*) is shown to illustrate its long lever arm at the shoulder. Balancing a limb in such a position requires enormous muscle forces and is therefore commonly avoided. The hindlimb is shown as a suspended pendulum during the forward swing. The dots in the shoulder region indicate the approximate positions of the most proximal part of the scapula (which is the pivot during the forward swing). The swing frequency of the hindlimb depends on the length of the pendulum chord, that is, the distance from the hip joint to the limb's center of mass.

swing phase of each limb; in other words, the duty factor is more than 50% of cycle duration. In running, the reverse holds true. The duty factor may be less than 50% in the amble and is less than 50% in the trot and the pace. Because the ground reaction forces must be equal to the continuously acting body weight, the ground reactions increase in all variants of running (2 *g* or more in large mammals), and lower in walking (slightly more than 1 *g*). As Christian et al. (1999a) have shown, the ground reaction forces in huge sauropods would have reached values beyond the strength of bone and joint cartilage if high speeds were assumed and no effective damping occurred by either movements of the joints or compliant sole pads (as in elephants; Hutchinson et al. 2006; Weissengruber et al. 2006; Miller et al. 2007). In addition, the time intervals needed for the fore swing of the extremely long limbs of sauropods do not allow cycle frequencies short enough to make use of the natural frequency (Preuschoft & Demes 1985; Preuschoft et al. 1994a; Preuschoft & Christian 1999; Figs. 12.9, 12.10). For short periods, most animals can run at frequencies higher than determined by the rules of the pendulum or by elastic properties. However, this requires a considerable input of muscular energy (Langmann et al. 1995), and therefore, it is not performed often or maintained over long distances. In the case of sauropods, both reasons—

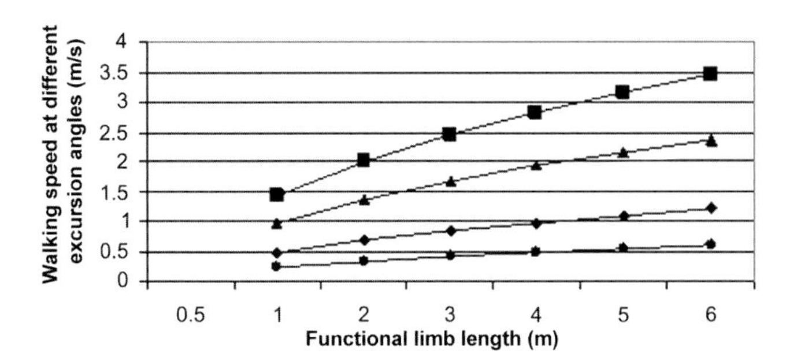

FIGURE 12.10. Increase of walking speed with increasing functional limb length at different excursion angles. Excursion angle determines step length and can be chosen by the animal. The curves are drawn for excursion angles of 5° (dots), 10° (diamonds), 20° (triangles), and 30° (squares). Note that the curves indicate an increase in speed following a square root function, and thus the increase in speed with increasing limb length is less than linear.

the tremendous, intolerable ground reaction forces and the long step cycles in comparison to the natural frequency of the legs—render all faster running, and especially all asymmetric gaits, unrealistic for these animals.

INFLUENCE OF SIZE ON SAUROPOD WALKING

Walking speed is dependent on the step length times the number of steps taken per unit time. Therefore, locomotion is positively influenced by the great step length of large animals. Step length is proportional to the length of the extremity if the excursion angles are kept constant.

Ground reaction forces are the result of the vertical component (gravitational force exerted by body mass) and horizontal braking or propulsive components, both of which usually increase with speed. In large animals, the vertical (weight) component grows faster than the horizontal components. Consequently, the larger the animal, the narrower its excursion angles, or, in other words, the shorter its step length in relation to limb length. The positive influence of limb length on walking speed is also limited by the reduced natural pendulous frequency of elongated limbs (see Fig. 12.10) during the forward swing of the leg, which is thus the major limitation of gait frequency. The time needed for one swing is defined by

$$T = 2\pi\sqrt{\frac{d}{g}}\ ,$$

where d is the distance from the pivot to the center of mass of the limb, which is proportional to limb length l. The comparison with a physical pendulum would be more realistic, but as detailed in Preuschoft & Demes (1984), the deviation is negligible at small excursion angles, and the more complicated method does not yield better results than the less complicated approximation we use here. The maximal speed of walking v therefore is:

$$v = \frac{4\,l \sin \alpha}{T}\ .$$

Speed increases slowly with body size because it increases by the square root of limb length (Preuschoft & Demes 1984; illustrated in Fig. 12.10). A sauropod with limbs 4 m in length has a swing period of 2.8 seconds, so it needs 1.4 seconds alone for each fore swing. Gait frequency is as low as 0.36/s. In comparison, horses have walking frequencies of little more than 1/s, and trotting frequencies of nearly 2/s (pers. obs.); elephants have frequencies of about 0.3–1.2/s (Christian et al. 1999b), or even 2/s (Hutchinson et al. 2006). Both animals obviously use the specific elastic properties of their limbs for locomotion. The points made above suggest rather convincingly that huge sauropods strictly preferred walking over running (i.e., all gaits involving aerial suspension of the entire body). The same points render all asymmetrical gaits unrealistic for sauropods. The aerial supension phases in these gaits inevitably produce higher ground reaction forces than gaits in which the body is continuously supported. These ground reaction forces exceed the strength of the limbs in huge animals.

Like modern cursorial mammals (Preuschoft & Witte 1991; Preuschoft et al. 1994a) and elephants, sauropods seem to have reduced pendulum length by reducing the mass of the distal limb segments, but in a different way: the phalanges are poorly developed, so that the animal walks more or less on its metacarpals. In fact, the anatomical construction of the sauropod forefoot may well have been similar to that of modern elephants (Weissengruber et al. 2006; Miller et al. 2007). The musculature of the forearm needed for operating the digits is reduced and allows the proximal shift of the center of mass of the entire limb (Fig. 12.9), with the effect of a shortened pendulum period and increased step frequency. The speed of walking under these conditions is somewhat higher than illustrated, but the curve illustrating the forces remains essentially the same. The great length of their limbs alone permitted large sauropods to achieve traveling speeds of 1.5–2.4 m/s (equal to 5.4–8.6 km/h) at the lowest possible level of energy expenditure. This is more than the 3–4 km/h estimated by Christian et al. (1999b). These authors took the usual locomotion of elephants into consideration, which often extends their stance phases, and tracks of sauropods, which record shorter step lengths than those assumed here for greater excursion angles (see also Alexander 1976, 1985, 1989). Both factors lead to lower traveling speeds.

In addition, the transport of a larger body mass simply requires more energy. According to Taylor (1977) and Taylor et al. (1982), this increase in energy requirement seems to follow a roughly linear relationship.

Body Mass and Muscle Force in Agonistic Encounters

BIOMECHANICS OF PREDATOR ATTACKS ON SAUROPODS

If the bodies of two animals are firmly connected (by embracing, clawing, biting), the smaller mass inevitably has to follow the larger mass. This means that if two animals are interlocked, the larger one will drag the smaller one down or push it aside. In addition to the influence of mass alone, the larger animal has more musculature. According to Gunga et al. (2007, 2008), the musculature of the prosauropod *Plateosaurus* amounts to 36% of body mass, and that of *Brachiosaurus* to 45%. These estimates are in general agreement with Grand's (1977, 1991) findings among mammals and with the experience of butchers: meat is usually some 45–50% of total body mass. Muscle force (F_m) is dependent on the physiological cross-sectional area (A_m) of the muscle and thus scales with a square function. The relationship between body mass (M_b) and cross-sectional area of a muscle is roughly

$$F_m \propto A_m \propto M_b^{2/3}.$$

Although muscle force indeed increases with body mass, it lags behind because mass scales to the third power.

The energy of a sudden impact during an antagonistic encounter (such as one animal ramming another or by jumping on its back) is dependent not only on the body mass of the impactor but also on the square of its speed ($E = 1/2mv^2$). Therefore, a smaller but more agile impactor may have the same effect as a larger, more sluggish one (Fig. 12.12A). At the given masses, the intervals between the curves in Fig. 12.12A are dependent on speed to the second power.

On impact, the force ($F = m\ a$) developed by the moving body acts only for a short time. The length of this time for the exchange of forces is dependent on the compliance of the two bodies in contact, which is difficult to estimate. An important function of muscles is to absorb the energy of an impact, but its effect can hardly be estimated in generalized form. The impulses ($m\ a/t$) contained in the bodies of animals increase linearly with mass and with speed ($m\ v$) if impact time is constant (Fig. 12.12B).

On the other hand, larger body mass and a greater body length result in a greater resistance against being moved—more specifically, greater mass moments of inertia, which must be overcome by muscle force. Imagine, for example, a defensive action of the Berlin *Brachiosaurus,* using its forelimb (Table 12.1). The density of the forelimb is assumed to be 1 because it consists mainly of bone and musculature. In con-

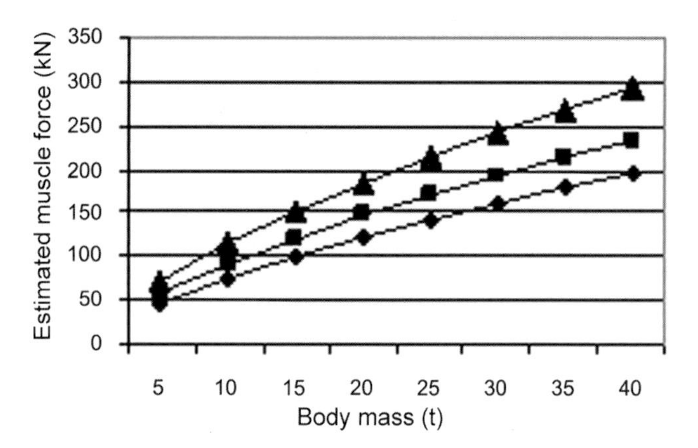

FIGURE 12.11. Rough estimates of the scaling of total muscle force with body mass The curves assume that different percentages (triangles 50%; squares 40%; diamonds 30%) of the cross-sectional area of the body consist of muscles.

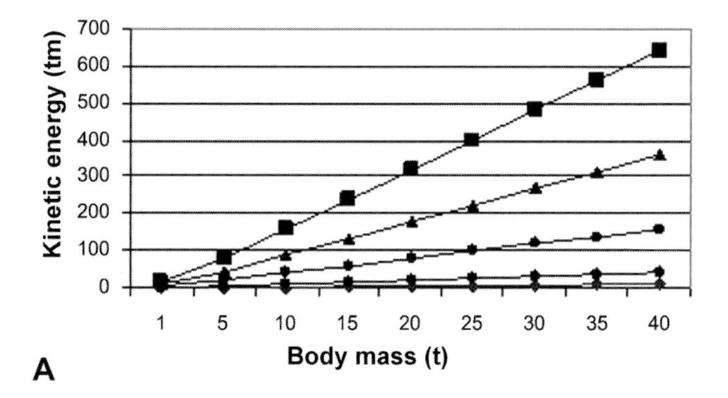

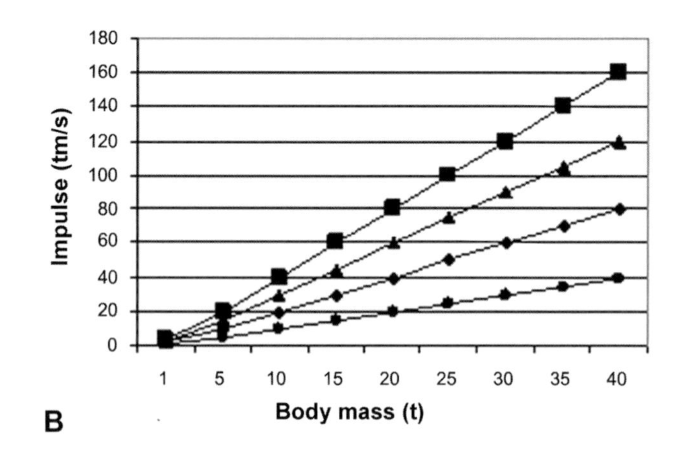

FIGURE 12.12. Scaling of kinetic energy and impulse with mass. (A) Increase of kinetic energy with body mass for different speeds of a dinosaur. Squares 4 m/s; triangles 3 m/s; diamonds 2 m/s; dots, 1 m/s. (B) Impulses contained in dinosaurs of varying mass; moving at the same speeds as in (A). At a given mass, for example, 15 metric tons, the kinetic energy contained in a moving body grows with the square of speed. On the other hand, the graph shows that the impulse contained in a giant of 35 metric tons moving at a speed of 1 m/s may easily be surpassed by an attacker of less than one third its mass but moving twice as fast.

trast, the density of the neck is assumed to be 0.67 (Gunga et al. 2008), while the head is assumed to have a density of 1, like the forelimb.

The segment masses and lever arm lengths for the forelimb and for the neck are listed in Table 12.1. In analogy to the results of Christian & Dzemski (2007) for neck movements in mammals, and for the sake of simplicity, we assume that the pivot of the movement is between the most proximal neck segment 1 and neck segment 2.

Mass moments of inertia J increase by length to the second power,

$$J = \sum (m_i l_i^2).$$

Their increase with increasing limb length is illustrated in Fig. 12.13. The impact that can be exerted by a defensive movement of a forelimb on the attacker depends on speed. After multiplication with the angular velocity ω, we obtain the impulse, or angular momentum D,

$$D = \sum (m_i l_i^2)\cdot\omega,$$

which is contained in a moving extremity or the moving neck (Fig. 12.13B). The angular velocity ω is measured here as $2\pi n$, where n is the number of rotations per second. Its dimension is $1/s$. If an object, or an enemy, is hit by a foot or distal neck segment, the impulse transferred D is $J\omega$, or $m\, v\, l$. Because we do not know the speed and the acceleration of movements in sauropods, we use values that have been observed in living mammals as a proxy (Preuschoft & Witzel 2005): 100°/s, 300°/s, 500°/s, and 700°/s. According to Ren et al. (2008), elephants use similar angular velocities. Because larger individuals move more slowly than smaller ones, we may assume that sauropods probably used angular velocities toward the lower end of this range. The large r of large animals, however, results in high peripheral velocities. Calculations of impulses generated by the forelimb and the neck upon impact are shown in Fig. 12.14. These values must be considered in comparison to the kinetic energy and impulses of the entire animals in Fig. 12.12. The impulses transferred by the moving neck and forelimb of a sauropod have the same order of magnitude as the impulse transferred by the entire body of a theropod (with a mass of less than 10 metric tons) as potential predator. It is easy to understand that an attacker struck by a limb or a neck of sauropod dimensions runs a considerable risk!

In view of the slow accelerations that must be assumed for dinosaurs (see below), the lower values shown in Fig. 12.14 seem to be more realistic. On the basis of the data in Table 12.1, the greatest lever arm of the shoulder retractors may be about 0.8 m. The muscle forces and lever arm lengths set limits to the acceleration of the body segments, either for defense action or for evasive movements. The acceleration that could have been exerted by the neck of a sauropod can be calculated from estimates of muscle force and lever arm length. The

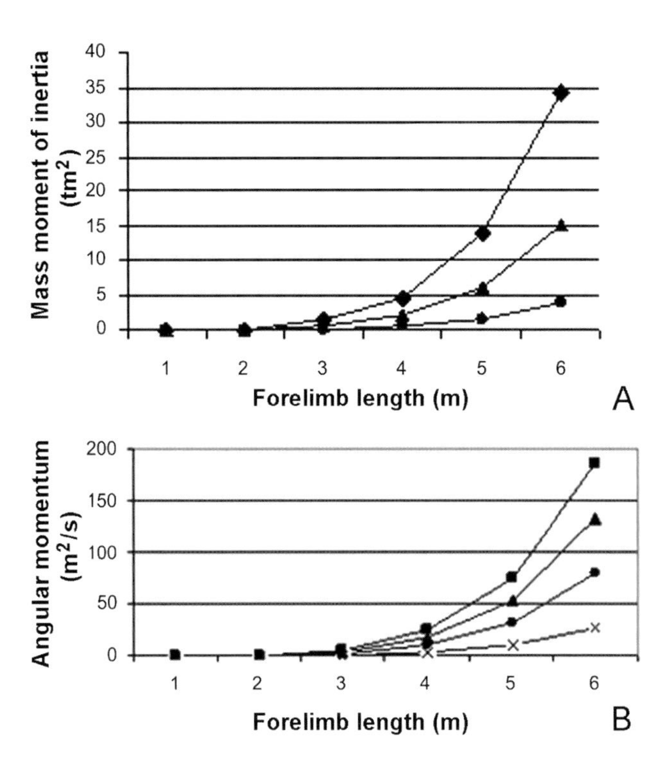

FIGURE 12.13. (A) Relationship between mass moments of inertia J and forelimb length in sauropods. Because the forelimb of *Brachiosaurus* was exceptionally massive (because of its long and heavy neck), we reduced its length–radius ratios of 13.3 (diamonds) to 20 (triangles) and to 40 (circles) to show the influence of limb diameter on limb mass. (B) Angular momentum D of the average forelimb length–radius ratio of 0.5 moving at angular velocities of 100°/s (crosses), 300°/s (circles), 500°/s (triangles), and 700°/s (squares). Note that both, mass moments of inertia and angular momentum, increase steeply in very large sauropods.

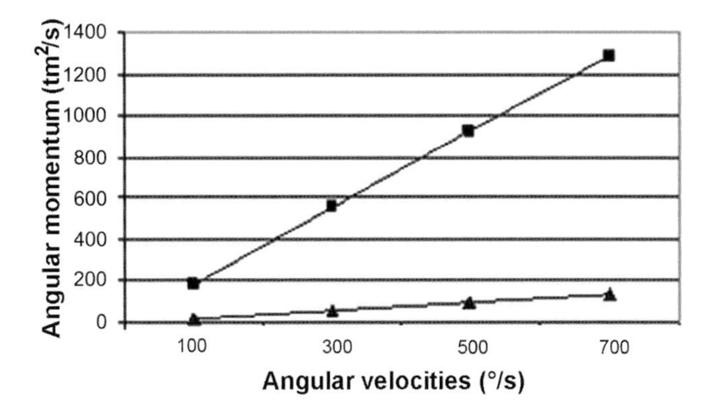

FIGURE 12.14. Rotational mpulses of the neck and head (squares) and forelimb (triangles) of a *Brachiosaurus* at different angular velocities.

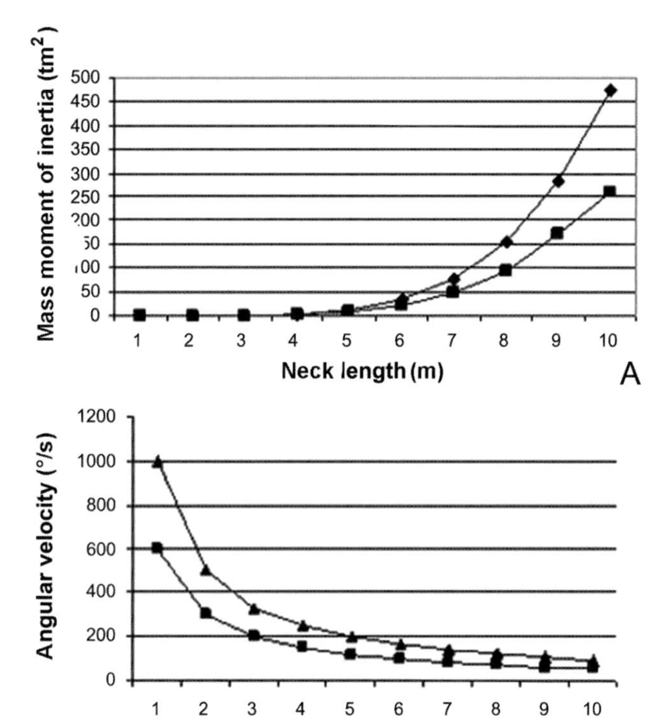

FIGURE 12.15. (A) Mass moments of inertia of a cylindrical neck increase with neck length. Two different length/radius ratios of 13.5 (diamonds) and of 10.5 (squares) were considered. The neck of *Brachiosaurus* varied between these limits. (B) Angular velociy of the neck in an evasive movement decreases with increasing neck length if muscle force remains constant. Calculations are based on the assumption that lever arms of muscles are constant, and muscle forces are proportional to 50% of the cross-sectional area (triangles) or 30% of the cross-sectional area (squares) of the neck, respectively. The fraction of neck cross sectional area determines the cross sectional area of the muscles and is thus a measure of the available muscle force. The greater the available force, the greater the angular velocity that can be reached.

nearly circular cross section of the neck between segments 1 and 2 is 0.99 m^2, and its radius is 0.56 m. Assuming that either one third or one half of the cross section was musculature, these cross sectional areas may have produced forces of 165 kN and 247.5 kN, respectively, if the muscles were parallel fibered and produced a force of 50 N/cm^2 (Bouteillier & Ulmer 2007, p. 930). The average lever arm length of the neck muscles is arbitrarily assumed to have been 0.45 m. The torques produced by the muscles therefore amount to 74.25 kNm and 111.38 kNm, respectively. The mass moments of inertia were estimated for a cylindrical neck with a length:diameter ratio of 5.025 and 6.725, respectively, because according to Gunga et al. (2008), the diameter of the *Brachiosaurus* neck along its length varied between these limits. The marked increase of mass moments of inertia with increasing neck length are shown in Fig. 12.15A.

The decrease of angular velocity of neck movements with increasing neck length is illustrated for *Brachiosaurus* in Fig. 12.15B. The peripheral velocity ω is defined by

$$\omega = \frac{mvr}{J},$$

where *r* is the lever arm length of the external force involved in hitting an enemy (as above) and *J* is the total mass moment of inertia of the neck. Assuming that the shortest neck (2 m) moved with a speed of 500°/s, the graph shows that a 10 m long neck would have needed as much as one second for sweeping through an arc of about 90°. If the initial movement was as fast as jaw closure in a snapping dog (about 700°/s; Preuschoft & Witzel 2005), the angular velocity of a long neck would not have exceeded 140°/s. We admit that the values for speed are rough estimates, but the asymptotic curves reliably show the inverse relationship between angular velocity and neck length. The same mathematical rules, of course, apply for movements of the limbs (as discussed above).

In an agonistic encounter, slowness of defense actions may well be fatal, especially if the enemy is smaller and therefore moves more quickly.

The close contact between two animals can involve claws or teeth inserted into the body of the victim. The same mechanical principle underlies both claws and teeth: the claw or tooth must be driven by muscle force and (perhaps) impact into the skin and deeper tissues.

An immediate effect of large size is that vulnerable organs, primarily the brain, spinal cord, nerves, tendons, large blood vessels, trachea, and intestines, are hidden deep below the skin (which itself becomes thicker with size). The distance from the surface to the respective organ increases linearly with body length. Obviously the depth of, for example, an artery below the surface of a leg is proportional to a fraction of limb diameter (Fig. 12.16). Limb diameters, of course, are on average thicker in larger animals than in smaller ones. Likewise, the intestines in a large animal are covered by a thicker layer of musculature than in a smaller one (Fig. 12.16; Grand 1977).

This can be illustrated by a predator with a given gape (measured between the bite points, which are defined by opening angle and length of the jaws) that attacks increasingly large sauropods (Fig. 12.17). The large diameter *r* of a limb, the neck, or of the trunk of the sauopod is equivalent to the long radius of curvature of the external outline (Fig. 12.17). The attacker's gape covers the distance between the two bite points on the external outline, that is, the bite line D to D′. The protective effect of size (=body diameter) is illustrated by the ratio *PA*/*r*, where *PA* is the depth of bite. Depth of bite is defined as the distance from the external outline, marked by point *A*, to the middle of the bite line, marked by point *P* (Fig. 12.17). The protective effect of size can also be illustrated by the ratio *r*/*ZP*, which describes the part of the victim's

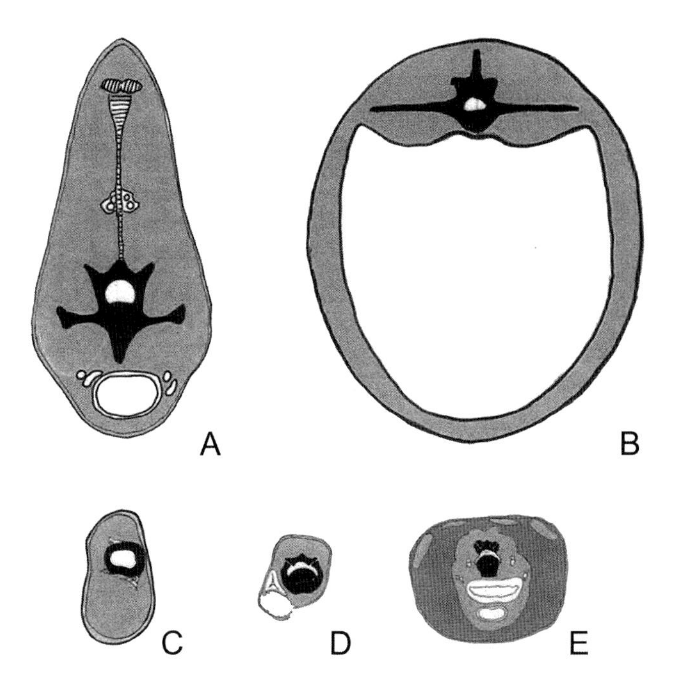

FIGURE 12.16. Schematic anatomical cross sections through the neck (A), abdomen (B), and forearm (C) of a horse, and the neck of an ostrich (D) and of a leatherback turtle (E) to illustrate the distance of the most vulnerable parts (white) from the body surface. Skeletal elements are depicted in black. All cross sections are drawn roughly to the same scale. *Modified after Nickel et al. (1968; A–C), Dzemski (2006; D), and Schumacher (1972; E).*

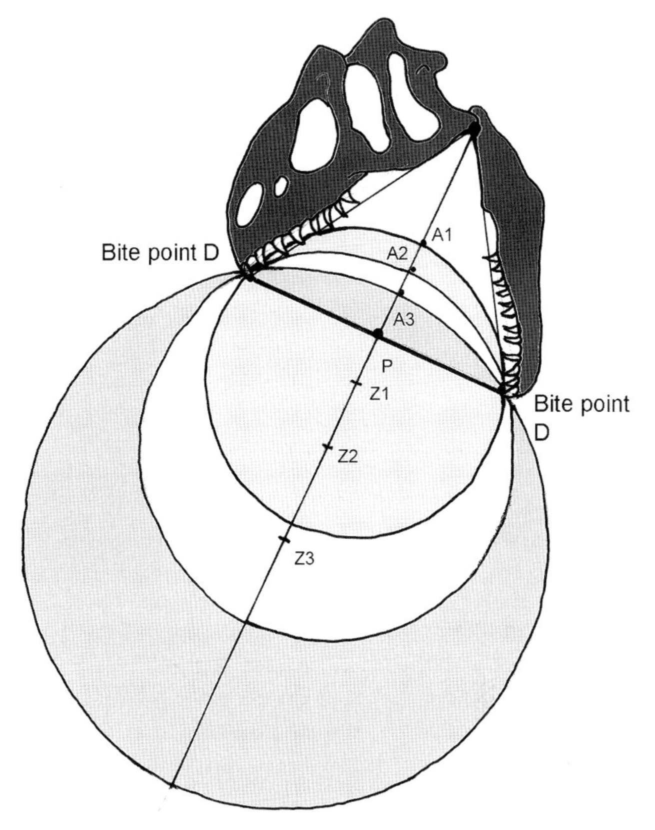

FIGURE 12.17. The protective effect of increasing body size modeled. An attacker (such as *Allosaurus*) delivers a bite to prey objects with three different radii (Z1–A1; Z2–A2; Z3–A3). Gape is defined as the distance between the bite points D and D′ that are connected by the bite line. If gape is constant, a greater part of the circumference of a smaller than of a larger prey object is covered. This can be expressed by the ratios *PA*/*r* and *r*/*ZP*, respectively, where *r* is prey object diameter, *PA* is the distance between the surface *A* and the middle of bite line *P*, and *ZP* is the distance between the center of the prey object *Z* and the middle of bite line *P*.

diameter not directly exposed to the teeth of the attacker (Fig. 12.17). *ZP* is the distance between the center of the body *Z* and *P*.

In the case of a small animal, gape may be large enough to gain a firm and effective tooth hold. The greater the diameters of limbs, neck, and trunk, the smaller the central angle (sensu Cartmill 1974) formed by the points of force application (bite points, D–D′ in Fig. 12.17) and the jaw joint, and the smaller the force components that can be applied at right angles to the jaws. Thus, the larger the prey, the harder it is for the predator to get more than just skin between its teeth. However, as can be seen in Fig. 12.18, the protective effect of size follows an asymptotic curve. This means that the advantage of large size on the part of the victim becomes smaller and smaller if the attacker does not increase its size as well! From a certain value onward, the advantage of size increase does not exceed the disadvantages incurred by increasing mass.

COEVOLUTION OF THEROPODS AND SAUROPODS

As noted above (Fig. 12.11), in a crash between two animals, the speed of movement can compensate for mass, because

$$F = m\,a,$$

and acceleration *a,* which in this context is negative, depends on speed. The kinetic energy contained in a moving body is

$$E_{kin} = \frac{1}{2} mv^2.$$

Therefore, even a smaller predator (like *Allosaurus*, which weighed less than 10 metric tons) may be able to bring down a much larger victim of more than 30 metric tons. However, a predator of much smaller size is threatened by the poorly aimed and relatively slow defensive movements of a huge victim, simply because of its sheer size (Figs. 12.11, 12.14). This seems to be a good argument for the predators to become larger as well to compensate for this disadvantage and to reduce their risk.

Also relevant in terms of coevolution is the problem of limited gape. In the predator, this can be overcome either by

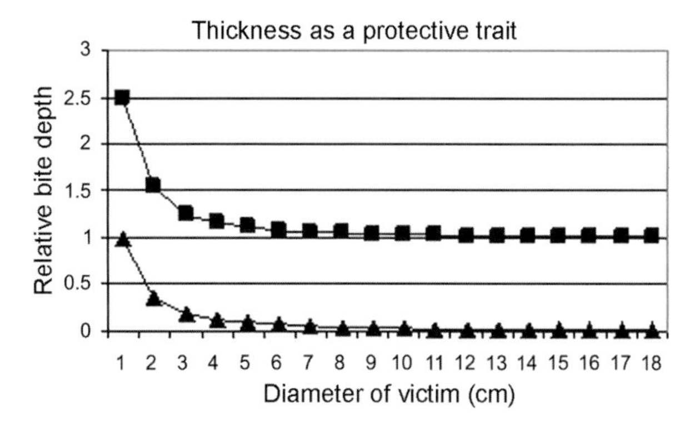

FIGURE 12.18. The protective effect of increasing body size quantified in two different ratios, *PA*/*r* (triangles) and *r*/*ZP* (squares). The part of the victim between the teeth of a predator becomes smaller with increasing size, that is, whole body diameter and body part diameters *r*. With increasing diameter, the bite depth *PA* (see Fig. 12.17 and text for explanation) drops rapidly from 1 to 0 (triangles). The distance from the center of the body *Z* to the middle of the bite line *P* (see Fig. 12.17) increases linearly with body size, resulting in a drop of the ratio *r*/*ZP* from 2.5 to 1 (squares). Both ratios thus approach an asymptotic limit, at which point the protective effect no longer increases with body size. Graph is assumed as constant.

an elongation of the jaws or by a wider opening angle. Both means have their limitations, as will be discussed below.

A wider opening angle (Fig. 12.19B) results in a decrease in the lever arm of the muscles that adduct the mandible, and this decrease follows a cosine function of the opening angle unless the muscles change their direction. One option would be a bony structure to redirect muscle force, but this would be heavily stressed and require much heavy bone, making the skull heavier. In addition, the bite forces exerted by upper and lower jaws combine and result in squeezing the victim out of the attacker's mouth. This squeezing increases with the sine of half the opening angle and is therefore much smaller in the other examples (Fig. 12.19A, C). Nevertheless, this consideration provides a reason why the teeth of predators are commonly recurved: the bitten object is prevented from being squeezed out of the mouth. More details on jaw muscles and their combined action are available in Demes et al. (1986), Preuschoft et al. (1986a, 1986b), and Witzel et al. (this volume).

Longer jaws have the consequence that the distance between the bite points and the jaw joint (that is, the load arms) becomes greater. The original bite force can only be maintained if the lever arms of the jaw-closing muscles also increase, or if the muscles become stronger (Fig. 12.19). Increase in lever arm length must result from a shift of muscle insertion toward the tips of the jaws by the same elongation factor as the jaws themselves. The forward shift of the muscles closes off the mouth posteriorly, but the proportions of the larger

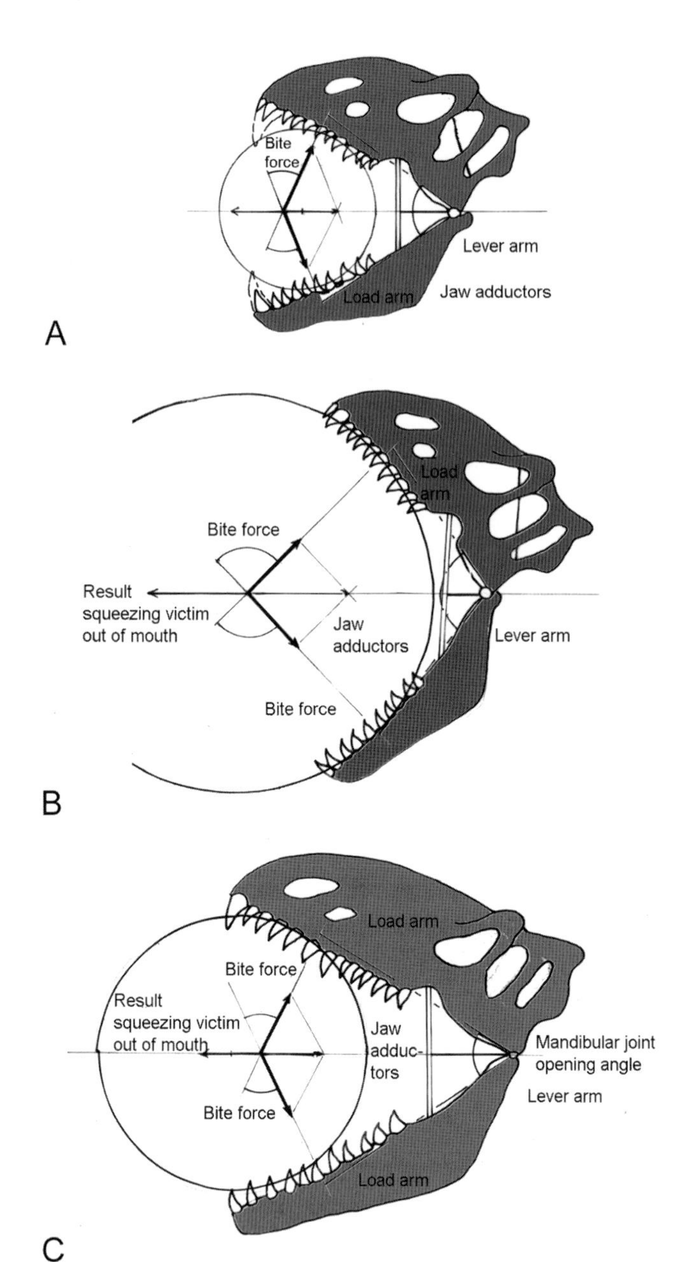

FIGURE 12.19. The relationship of prey size and the opening angle and length of predator jaws. (A) The jaws of an attacker, with *Allosaurus* taken as an example, can be opened to firmly grasp a victim not too voluminous. (B) To cope with a more voluminous victim, the jaws can be opened wider. This, however, reduces the lever arm of the jaw muscles. Note that this drawing does not show the jaw adductors, but rather their common resultant, which will squeeze the prey out of the predator's jaws. (C) Another means of getting hold of a larger prey is elongation of the jaws. Jaw elongation, however, requires greater strength of the jaws and stronger muscles—both resulting in increased mass.

skull will remain the same as in the smaller one. The other option, an increase in jaw muscle strength, is only possible by increasing the muscles' cross-sectional area. As a consequence, muscle volume increases, as does muscle mass. These increases are limited, however, because the linear increase in jaw length does not require more than a linear increase in cross-sectional area, and the resulting muscle mass increase follows a square root function. However, longer jaws incur greater bending moments induced by the bite forces, and thus require greater strength. Bending strength of the skull to resist a given moment of force (*Fl*) is defined by the material properties of bone (which are beyond the scope of this chapter and therefore taken as a constant factor) and the arrangement of the material in the cross section, which can be calculated as the area moment of inertia

$$J = \sum y^2 \Delta A,$$

where ΔdA is a small part of the cross-sectional area and y is its distance from the neutral line of bending (i.e., a fraction of skull height).

For the sake of simplicity, we use skull length as a proxy for jaw length and model the skull as a solid beam. Bending moments *Mb* grow linearly with skull length. In a body of equal strength, such as a skull, the stresses (σ, measured in Ncm) have the same value everywhere. Stresses are defined as Mb/W, where $W = 2I/h$ is a function of the moment of inertia (I) of the skull cross section, which depends on its diameter. Because I must correspond to the bending moments, the height and width of the beam that can sustain them can be calculated. If we keep skull width constant at 0.2 m, skull height (h, black diamonds in Fig. 12.20) increases with skull length following a root function ($h = \sqrt{6W/b}$; see Hohn et al. 2007). Alternatively, if we keep skull height constant at 0.2 m, skull width (b, solid squares in Fig. 12.20) grows linearly with skull length ($b = 6W/h^2$). Height and width, together with length, determine the volume of the skull and thus its mass. If height is taken as the dependent variable, skull mass increases linearly with skull length (open triangles in Fig. 12.20). However, if width is taken as the dependent variable, skull mass increases exponentially (open circles in Fig. 12.20). As can be seen from the intersection points of the curves in Fig. 12.20, at small skull sizes, mass increase may be nearly linear with size increase, depending on factors like specific material strength, specific density, and the length/diameter ratio of the beam. The intersection of the curves therefore may shift in both directions. In longer jaws and thus larger skulls, an increase in height is obviously more advantageous than an increase in width. In fact, most elongate skulls show an outline following the root function. As a result, all possible means to increase gape lead to an increase in head mass.

As can be observed in Fig. 12.17, the bite forces that occur between predator and a voluminous victim concentrate on the

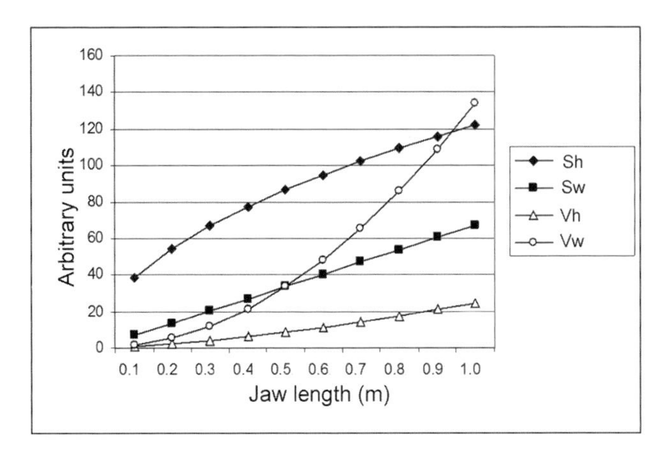

FIGURE 12.20. Dependence of skull shape and volume on increasing skull length, which is proportional to jaw length. Model caluclations are based on bending moments in a beam; see text for details. Jaw length is of interest because it determines gape (D–D' in Figs. 12.17–19). Note that if skull height (black diamonds) increases together with jaw length but width remains constant, skull volume (open triangles) and thus mass increases linearly. However, if skull height is kept constant and skull width (black squares) increases with jaw length, skull volume (open circles) and thus mass increases exponentially. This is why the snouts of large theropod dinosaurs are tall and narrow. Sk, skull height; Sw, skull width; Vh, volume at variable height; Vw, volume at variable width.

anterior teeth, providing a good reason to make them especially large and strong. If, however, a victim with a smaller diameter is bitten, the most anterior teeth do not come into contact with the victim unless the teeth are larger (Fig. 12.19A). This leads us to a discussion of the problems of increasing tooth size. As is apparent in Figs. 12.16, 12.17, and 12.19, larger prey requires teeth on the part of the predator that can be inserted deeper. When specializing on larger prey, predators therefore need longer teeth or longer claws to reach the more deeply hidden organs of the victim.

The teeth of a predator are exposed to compressive components of the bite force—which increases along the part of the tooth inserted into the victim, then remains constant—and to force components at right angles to compression (Fig. 12.21). These force components assume their greatest values along the jaws because these are the directions in which the attacker will pull its prey and in which the prey will try to escape. The force components exert bending moments on each tooth that increase from the tooth tip to the root. The increase is exponential in the part of the tooth inserted into tissue, but linear between the surface of victim and tooth root. Upon biting, the tip of the tooth slides between the randomly orientated bundles of collagenous fibers in the tough lower layer of the skin (the so-called corium, which is transformed into leather by the process of tanning; Fig. 12.21, left column). As the tooth is

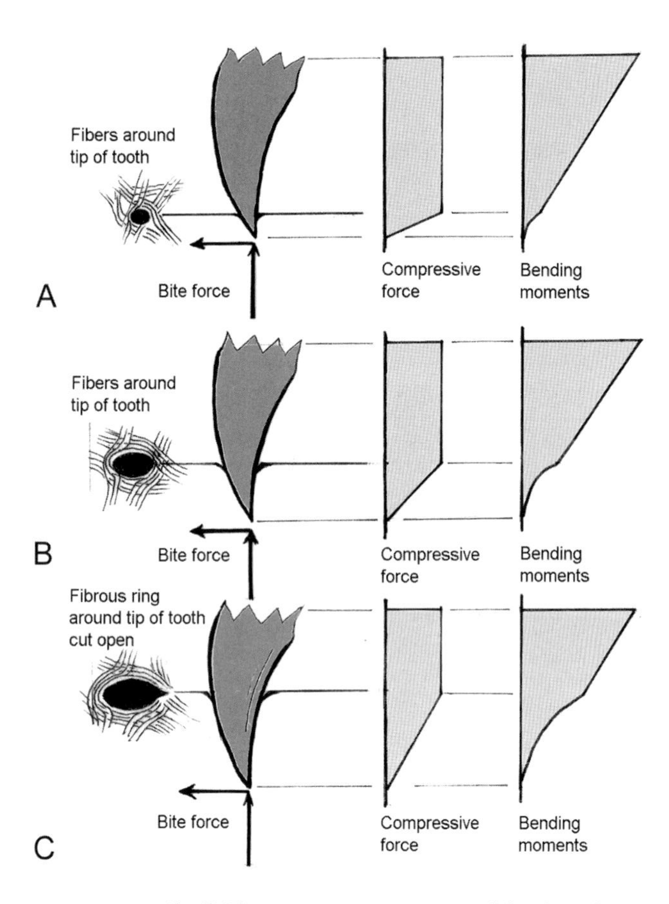

FIGURE 12.21. (A–C) Three steps in penetration of the skin of a victim by the tooth of a predatory dinosaur. The first column shows the anatomical interaction of tooth and skin. Initially, the tooth tip pushes the collagenous fibers of the tough deep layer of the skin (corium) out of the way (A) until these are arranged like a ring around the tooth (B). A cutting edge then permits deeper insertion (C). Bite forces have components along the tooth (second column) and at right angles, leading to bending moments. Consequently, the teeth are exposed to compression (third column) and to bending (fourth column).

inserted deeper, the fibers become more and more stretched, until they form a highly resistant ring around the tooth, which prevents deeper penetration. The posterior and sometimes anterior cutting edges, commonly bearing serrations, serve in cutting open the ring of fibers, thus permitting deeper tooth insertion. This tooth shape is called *ziphodont* and has been observed in many meat-eating archosaurs, lizards, basal synapsids, carnivorous mammals, and even primates. The details of their function were clarified for sharks by Preuschoft et al. (1974).

The primary mechanical danger for an individual tooth is bending moments leading to breakage, especially if hard parts of the victim are hit before a part of the tooth is embedded into the victim's soft tissue. Accordingly, the tooth must be able to resist these moments (M_{max} = longitudinal bite force times tooth height). Although tooth height and therefore the bending moments also increase linearly with body size, tooth thickness must increase as well, simply to resist bending and prevent mechanical failure. Technically speaking, the local maximum stress close to the roots of the teeth (σ_{max}), where bending moments are highest, must be kept constant. The conditions are the same as those explained for the skull.

Because mechanical stress is constant, the bending strength of a roughly cone-shaped tooth depends on the square root of its diameter. Because resistance to bending is offered by a nearly homogenous hard tissue (dentine), the resisting force will often be concentrated at one point, the tip, and the profile of a tooth therefore must increase parabolically to give its tip sufficient strength (Preuschoft et al. 1974). This means that if the tooth strength is to be maintained in a carnivorous dinosaur of increasing size, thickness must increase with length (or, more exactly, with height) of the tooth, but the increase in thickness is lower than the increase in height.

Inserting a tooth into the body of the prey animal requires force produced by the jaw adductor muscles of the predator. This force grows with the square of muscular diameter, as shown above for the neck and forelimb musculature, and leads to an increase in head mass. The resistance offered by the skin of the victim depends on its properties, that is, the existence of dermal armor and thickness and quality of the corium, which varies considerably among body regions. Although these factors are not directly related to our main topic, which is the advantages and consequences of size increase, the increasing cross-sectional areas of the tooth A, where $A = r^2\pi$, with increasing height results in an increased force required for penetration of prey skin. In Fig. 12.22B, the necessary diameters and the cross-sectional areas are shown.

The force required to insert the tooth, or the ratio of required strength of the tooth to available muscle force, therefore increases, but it does not change dramatically with increasing tooth height. Larger teeth also mean a heavier head, not only because of greater tooth volume, but also because larger teeth need stronger and heavier implantation in the jaws.

Conclusions

The results of this study can be summed up as follows: an increase in body size can be defined by either body mass or by linear dimensions of specific body parts, depending on the biological context under consideration. Increased size does offer biomechanical advantages. These advantages either follow a linear function (e.g., kinetic energy and the impulse of a defense movement), a root function (walking speed), or an asymptotic function (increased neck length for feeding and the protective effect of a large volume). This means that the advantages become smaller with increasing size or that they reach a point of diminishing return. These results were unexpected.

While the advantage of a long neck for harvesting a large

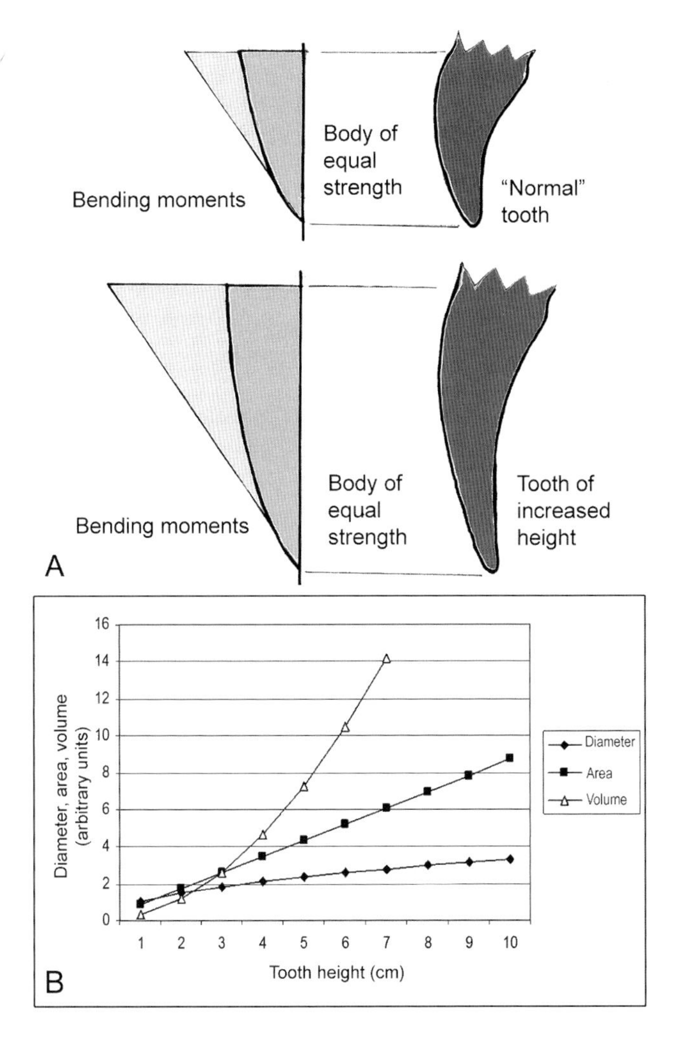

FIGURE 12.22. Shape change and underlying biomechanics in teeth of theropod dinosaurs of increasing size. (A) To resist bending moments (left column, light gray), the diameter ($2r$) of the tooth must increase with a square root function (left column, medium gray) to maintain strength with increasing height (Preuschoft et al. 1974). Indeed, all teeth, from fish to mammals, show this pattern. Generalized examples of conical teeth of different height are shown on the right in dark gray. (B) Diagram of the dependence of increase in tooth thickness on increase in tooth height under the condition that constant strength is maintained. Horizontal axis shows tooth height in centimeters. The vertical axis shows tooth diameter (diamonds), cross-sectional area at base of the crown (squares), and volume of the tooth (triangles) in arbitrary units. If bending strength is to remain constant with increasing tooth size, the diameter of the tooth must increase with increasing height, following a root function. This leads to a linearly increasing cross-sectional area of the tooth, which is a proxy for the compressive strength necessary for the increasing force required for prey skin penetration with increasing tooth height. The cross-sectional area and the height of the tooth determine its volume and weight. The weight of the adductor muscles must follow a curve proportional to cross-sectional area of the teeth to overcome the increasing resistance due to increasing tooth height.

volume increases by the third power, the potential gain is offset by the disadvantage of increased neck mass and—worse!—neck length. If the latter reaches a value of more than 8 m, the load grows faster than the force.

Longer, or, in anatomical terms, higher teeth in the jaws of a predator need greater diameters to maintain the same bending strength as smaller ones. However, this includes the disadvantage of a greater cross-sectional area, which increases the force required for penetrating the skin exponentially. This requires additional strength of the jaw-closing muscles, which thus increases their mass and that of the whole head.

Although the quadrupedal bauplan of the sauropods does not seem to limit size increases, there seems to be a limitation set by the slowness of movements in extremely large forms. The decrease in gait frequency with increasing size is obviously compensated for by greater step lengths, so that walking speed can become quite high with increasing size. However, the evasive movements of the limbs and neck are so slow that huge creatures become vulnerable—even in agonistic encounters with much smaller, but quicker, enemies. The slowness of movements may even lead to a point at which not enough food can be acquired within the available time.

Emphasis must be placed on quantifying the advantages and disadvantages in mathematical equations. Because limitations in the energy budget seem to exert a major influence on evolution, we conclude that the quantitatively defined advantages directly represent the selective pressures that have led to the acquisition of the morphology of sauropod dinosaurs.

Acknowledgments

We thank the members of the DFG Research Unit 533 for their stimulating discussions and two anonymous reviewers for their sensitive criticism and precise comments. This investigation was supported by the Deutsche Forschungsgemeinschaft and is contribution number 71 of the DFG Research Unit 533 "Biology of the Sauropod Dinosaurs: The Evolution of Gigantism."

References

Alexander, R. M. 1976. Estimates of speeds of dinosaurs.—*Nature* 261: 129–130.

Alexander, R. M. 1985. Mechanics of posture and gait in some large dinosaurs.—*Zoological Journal of the Linnean Society of London* 83: 1–25.

Alexander, R. M. 1989. *Dynamics of Dinosaurs and Other Extinct Giants.* Columbia University Press, New York.

Blankenhorn, W. U. 2000. The evolution of body size: what keeps organisms small?—*Quarterly Review of Biology* 75: 385–407.

Bouteillier, U. & Ulmer, H.-V. 2007. Sport- und Arbeitsphysiologie. *In* Schmidt, R. & Lang, F. (eds.). *Physiologie des Menschen.* Springer, Heidelberg: pp. 928–952.

Burness, G. P., Diamond, J. & Flannery, T. 2001. Dinosaurs,

dragons, and dwarfs: the evolution of maximal body size.—*Proceedings of the National Academy of Sciences of the United States of America* 98: 14518–14523.
Cartmill, M. 1974. Pads and claws in arboreal locomotion. *In* Jenkins, F. A., Jr. (ed.). *Primate Locomotion*. Academic Press, New York: pp. 45–83.
Christian, A. 1995. Zur Biomechanik der Lokomotion vierfüßiger Reptilien (besonders der Squamata).—*Courier Forschungs-Institut Senckenberg* 180: 1–58.
Christian, A. 2002. Neck posture and overall body design in sauropods.—*Mitteilungen aus dem Museum für Naturkunde in Berlin, Geowissenschaftliche Reihe* 5: 269–279.
Christian, A. & Dzemski, G. 2007. Reconstruction of the cervical skeleton posture of *Brachiosaurus brancai* Janensch, 1914, by an analysis of the intervertebral stress along the neck and a comparison with the results of different approaches.—*Fossil Record* 10: 38–49.
Christian, A. & Dzemski, G. This volume. Neck posture in sauropods. *In* Klein, N., Remes, K., Gee, C. T. & Sander, P. M. (eds.). *Biology of the Sauropod Dinosaurs: Understanding the Life of Giants*. Indiana University Press, Bloomington: pp. 251–260.
Christian, A. & Heinrich, W.-D. 1998. The neck posture of *Brachiosaurus*.—*Mitteilungen aus dem Museum für Naturkunde in Berlin, Geowissenschaftliche Reihe* 1: 73–80.
Christian, A., Heinrich, W.-D. & Golder, W. 1999a. Posture and mechanics of the forelimbs of *Brachiosaurus brancai* (Dinosauria: Sauropoda).—*Mitteilungen aus dem Museum für Naturkunde in Berlin, Geowissenschaftliche Reihe* 2: 63–73.
Christian, A., Müller, R. H. G., Christian, G. & Preuschoft, H. 1999b. Limb swinging in elephants and giraffes and implications for the reconstruction of limb movements and speed estimates in large dinosaurs.—*Mitteilungen aus dem Museum für Naturkunde in Berlin, Geowissenschaftliche Reihe* 2: 81–90.
Demes, B. 1991. Biomechanische Allometrie: Wie die Körpergröße Fortbewegung und Körperform der Primaten bestimmt.—*Courier Forschungs-Institut Senckenberg* 141: 1–84.
Demes, B., Creel, N. & Preuschoft, H. 1986. Functional significance of allometric trends in the hominoid masticatory apparatus. *In* Else, J. G. & Lee. P. C. (eds.). *Primate Evolution*. Cambridge University Press, Cambridge: pp. 229–237.
Dzemski, G. 2005. Funktionsmorphologische Betrachtung der Halsstellung bei Zoogiraffen.—*Der Zoologische Garten* 3: 189–201.
Dzemski, G. 2006. *Funktionsmorphologische Analysen langer Hälse bei rezenten terrestrischen Wirbeltieren zur Rekonstruktion der Stellung und Beweglichkeit langer Hälse prähistorischer Tiere*. Ph.D. Dissertation. University of Flensburg, Flensburg.
Dubbel, H. 1981. *Taschenbuch für den Maschinenbau*. Springer, Berlin.
Engesser, H. 1996. *Der Kleine Duden, Mathematik*. Dudenverlag, Mannheim.
Gambaryan, P. P. 1974. *How Animals Run*. Wiley & Sons, New York.
Gee, C. T. This volume. Dietary options for the sauropod dinosaurs from an integrated botanical and paleobotanical perspective. *In* Klein, N., Remes, K., Gee, C. T. & Sander, P. M. (eds.). *Biology of the Sauropod Dinosaurs: Understanding the Life of Giants*. Indiana University Press, Bloomington: pp. 34–56.
Grand, T. 1977. Body weight: its relation to tissue composition, segment distribution, and motor function. Part I: interspecific comparison.—*American Journal of Physical Anthropology* 47: 211–240.
Grand, T. 1991. Patterns of muscular growth in the African Bovidae.—*Applied Animal Behavior Science* 29: 471–482.
Gunga, H.-C., Kirsch, K. A., Baartz, F., Röcker, L., Heinrich, W. D., Lisowski, W., Wiedemann, A. & Albertz, J. 1995. New data on the dimensions of *Brachiosaurus brancai* and their physiological implications.—*Naturwissenschaften* 82: 189–192.
Gunga, H.-C., Kirsch, K. A, Rittweger, J., Clarke, A., Albertz, J., Wiedemann, A., Mokry, S., Suthau, T., Wehr, A., Heinrich, W.-D. & Schultze, H. P. 1999. Body size and body volume distribution in two sauropods from the Upper Jurassic of Tendaguru (Tanzania).—*Mitteilungen aus dem Museum für Naturkunde in Berlin, Geowissenschaftliche Reihe* 2: 91–102.
Gunga, H.-C., Suthau, T., Bellmann, A., Friedrich, A., Schwanebeck, T., Stoinski, S., Trippel, T., Kirsch, K. & Hellwich, O. 2007. Body mass estimates for *Plateosaurus engelhardti* using laser scanning and 3D reconstruction methods.—*Naturwissenschaften* 94: 623–630.
Gunga, H.-C., Suthau, T., Bellmann, A., Stoinski, S., Friedrich, A., Trippel, T., Kirsch, K. & Hellwich, O. 2008. A new body mass estimation of *Brachiosaurus brancai,* Janensch 1914, mounted and exhibited at the Museum of Natural History (Berlin, Germany).—*Fossil Record* 11: 28–33.
Hildebrand, M. 1960. How animals run.—*American Scientist* 202: 148–157.
Hildebrand, M. 1965. Symmetrical gaits of horses.—*Science* 150: 701–708.
Hildebrand, M. 1966. Analysis of the symmetrical gaits of tetrapods.—*Folia Biotheoretica* 13: 9–22.
Hildebrand, M. 1985. Walking and running. *In* Hildebrand, M., Bramble, D. M., Liem, K. F. & Wake, D. B. (eds.). *Functional Vertebrate Morphology*. Belknap Press of Harvard University, Cambridge: pp. 38–57.
Hildebrand, M. & Goslow, G. 2001. *Analysis of Vertebrate Structure*. 5th edition. John Wiley & Sons, New York.
Hildebrand, M. & Goslow, G. 2004. *Vergleichende und Funktionelle Anatomie der Wirbeltiere*. Springer, Berlin.
Hohn, B., Preuschoft, H. & Witzel, U. 2007. A new fossil, *Tiktaalik roseae,* and the biomechanical conditions for the evolution of the tetrapod bauplan, based on the basal tetrapod. *In 8th International Congress of Vertebrate Morphology*. Université Pierre et Marie Curie, Paris: p. 56.
Hummel, J. & Clauss, M. This volume. Sauropod feeding and digestive physiology. *In* Klein, N., Remes, K., Gee, C. T. & Sander, P. M. (eds.). *Biology of the Sauropod Dinosaurs: Understanding the Life of Giants*. Indiana University Press, Bloomington: pp. 11–33.
Hummel, J., Südekum, K.-H., Streich, W. & Clauss, M. 2006. Comparative in vitro fermentative behaviour of temperate forage plant classes—potential consequences for herbivore ingesta retention times.—*Functional Ecology* 20: 989–1002.

Hummel, J., Gee, C. T., Südekum, K.-H., Sander, P. M., Nogge, G. & Clauss, M. 2008. In vitro digestibility of fern and angiosperm foliage: implications for sauropod feeding ecology and diet selection.—*Proceedings of the Royal Society B: Biological Sciences* 275: 1015–1021.

Hutchinson, J. R., Famini, D., Lair, R. & Kram, R. 2003. Are fast-moving elephants really running?—*Nature* 422: 493–494.

Hutchinson, J. R., Schwerda, D., Famini, D., Dale, R. H. I., Fischer, M. S. & Kram, R. 2006. The locomotor kinematics of Asian and African elephants: changes with speed and size.—*Journal of Experimental Biology* 209: 3812–3827.

Langmann, V. A., Roberts, T. J., Black, J., Maloy, G. M. O., Heglund, N. C., Weber, J.-M., Kram, R. & Taylor, C. R. 1995. Moving cheaply: energetics of walking in the African elephant. —*Journal of Experimental Biology* 108: 629–232.

Lehmann, T. 1974–77. *Elemente der Mechanik, Bände 1–3*. Vieweg, Braunschweig.

Miller, C. E., Ren, L. & Hutchinson, J. R. 2007. An integrative analysis of elephant foot biomechanics. *In 8th International Congress of Vertebrate Morphology*. Université Pierre et Marie Curie, Paris: p. 83.

Mochon, S. & MacMahon, T. A. 1980a. Ballistic walking.—*Journal of Biomechanics* 3: 49–57.

Mochon, S. & MacMahon, T. A. 1980b. Ballistic walking: an improved model.—*Mathematical Biosciences* 52: 241–260.

Muybridge, E. 1898. *Animals in Motion*. Dover, New York.

Nickel, R., Schummer, A. & Seiferle, E. 1968. *Lehrbuch der Anatomie der Haustiere, Band I*. Parey-Verlag, Berlin.

Preuschoft, H. 1976. Funktionelle Anpassung evoluierender Systeme.—*Aufsätze und Reden der Senckenbergischen Naturforschenden Gesellschaft* 28: 98–117.

Preuschoft, H. 2010. Selective value of big size and sexual dimorphism in primates.—*Abstracts of the Congress of the International Primatological Society in Kyoto, Japan, September 2010*.

Preuschoft, H. & Christian, A. 1999. Statik und Dynamik bei Tetrapoden. *In* Gansloßer, U. (ed.). *Spitzenleistungen—was Tiere alles können*. Filander-Verlag, Fürth: pp. 89–130.

Preuschoft, H. & Demes, B. 1984. Biomechanics of brachiation. *In* Preuschoft, H., Brockelman, W. Y., Chivers, D. J. & Creel N. (eds.). *The Lesser Apes: Evolutionary and Behavioral Biology*. Edinburgh University Press, Edinburgh: pp. 96–118.

Preuschoft, H. & Demes, B. 1985. Influence of size and proportions on biomechanics of brachiation. *In* Jungers, W. L. (ed.). *Size and Scaling in Primate Biology*. Plenum Press, New York: pp. 383–398.

Preuschoft, H. & Gudo, M. 2005. Die Schultergürtel von Wirbeltieren: Biomechanische Überlegungen zu den Bauprinzipien des Wirbeltierkörpers und zur Fortbewegung von Tetrapoden.—*Zentralblatt für Geologie und Paläontologie, Teil II* 2005: 339–361.

Preuschoft, H. & Witte, H. 1991. Biomechanical reasons for the evolution of hominid body shape. *In* Coppens, Y. & Senut, B. (eds.). *Origine(s) de l'Homme*. CNRS, Paris: pp. 59–77.

Preuschoft, H. & Witte, H. 1993. Die Körpergestalt des Menschen als Ergebnis biomechanischer Erfordernisse. *In* Voland, E. (ed.). *Evolution und Anpassung—Warum die Vergangenheit die Gegenwart erklärt*. Hirzel Verlag, Stuttgart: pp. 43–74.

Preuschoft, H. & Witzel, U. 2005. The functional shape of the skull in vertebrates: which forces determine morphology? With a comparison of lower primates and ancestral synapsids.—*Anatomical Record Part A* 283A: 402–413.

Preuschoft, H., Reif, W. E. & Müller, W. H. 1974. Funktionsanpassungen in Form und Struktur an Haifischzähnen.—*Zeitschrift für Anatomie und Entwicklungsgeschichte* 143: 315–344.

Preuschoft, H., Fritz, M. & Krämer, T. 1975. Beziehungen zwischen mechanischer Beanspruchung und Gestalt bei Wirbellosen: Spannungsanalyse an Muschelklappen.—*Neues Jahrbuch der Geologie und Paläontologie, Abhandlungen* 150: 161–181.

Preuschoft, H., Demes, B., Meyer, M. & Baer, H. F. 1986a. The biomechanical principles realised in the upper jaw of long-snouted primates. *In* Else, J. G. & Lee, P. C. (eds.). *Primate Evolution*. Cambridge University Press, Cambridge: pp. 249–264.

Preuschoft, H., Demes, B., Meyer, M. & Baer, H. F. 1986b. The biomechanical principles realised in the upper jaw of long-snouted vertebrates. *In* Sakka, M. (ed.). *Définition et Origines de l'Homme*. CNRS, Paris: pp. 177–202.

Preuschoft, H., Hayama, S. & Günther, M. M. 1988. Curvature of the lumbar spine as a consequence of mechanical necessities in Japanese macaques trained for bipedalism.—*Folia Primatologica* 50: 42–58.

Preuschoft, H., Witte, H. & Demes, B. 1992. Biomechanical factors that influence overall body shape of large apes and humans. *In* Matano, S., Tuttle, R. H., Ishida, H. & Goodman, M. (eds.). *Topics in Primatology Vol. 3: Evolutionary Biology, Reproductive Endocrinology and Virology*. University of Tokyo Press, Tokyo: pp. 259–289.

Preuschoft, H., Lesch, C., Witte, H. & Loitsch, C. 1994a. Die biomechanischen Grundlagen der Gangarten, insbesondere des Galopps. *In* Knezevic, P. F. (ed.). *2. Internationaler Kongress für Orthopädie bei Huf-u. Klauentieren*. Schattauer, Wien: pp. 355–370.

Preuschoft, H., Witte, H., Christian, A. & Recknagel, S. 1994b. Körpergestalt und Lokomotion bei großen Säugetieren. *Verhandlungen der Deutschen Gesellschaft für Zoologie* 87: 147–163.

Preuschoft, H., Schulte, D., Distler, C., Witzel, U. & Hohn, B. 2007. Body shape and locomotion in monitor lizards.—*Mertensiella* 16: 58–79.

Putz, R. 1976. Zur Morphologie und Rotationsmechanik der kleinen Gelenke der Lendenwirbel.—*Zeitschrift für Orthopädie* 114: 902–912.

Putz, R. 1983. Zur Morphologie und Dynamik der Wirbelsäule.—*Radiologie* 23: 145–150.

Rao, G. J. 1999. *Über die Rolle der Elastizität in der Lokomotion. Das Kollagen-Myosin-Titin-Modell (CMT-Modell)*. Diploma Thesis. Ruhr-University of Bochum, Bochum.

Rauhut, O. W. M., Fechner, R., Remes, K. & Reis, K. This volume. How to get big in the Mesozoic: the evolution of the sauropodomorph body plan. *In* Klein, N., Remes, K., Gee, C. T. & Sander, P. M. (eds.). *Biology of the Sauropod Dinosaurs: Under-*

standing the Life of Giants. Indiana University Press, Bloomington: pp. 119–149.

Ren, L., Butler, M., Miller, D., Paxton, H., Schwerda, D., Fischer, M. S. & Hutchinson, J. R. 2008. The movements of limb segments and joints during locomotion in African and Asian elephants.—*Journal of Experimental Biology* 211: 2735–2751.

Sander, P. M., Christian, A., Clauss, M., Fechner, R., Gee, C. T., Griebeler, E. M., Gunga, H.-C., Hummel, J., Mallison, H., Perry, S., Preuschoft, H., Rauhut, O., Remes, K., Tütken, T., Wings, O. & Witzel, U. 2010. Biology of the sauropod dinosaurs: the evolution of gigantism.—*Biological Reviews of the Cambridge Philosophical Society*. doi: 10.1111/j.1469=185X.2010.00137.x.

Schumacher, G.-H. 1972. *Die Kopf- und Halsregion der Lederschildkröte, Dermochelys coriacea (Linnaeus 1766)*. Akademie-Verlag, Berlin.

Shipley, L. A., Spalinger, D. E., Gross, J. E., Thompson Hobbs, N. & Wunder, B. A. 1996. The dynamics and scaling of foraging velocity and encounter rate in mammalian herbivores.—*Functional Ecology* 10: 234–244.

Stevens, K. A. & Parrish, M. J. 1999. Neck posture and feeding habits of two Jurassic sauropod dinosaurs. *Science* 284: 798–800.

Stevens, K. A. & Parrish, M. J. 2005a. Reconstructions of sauropod dinosaurs and implications for feeding. *In* Curry Rogers, K. & Wilson, J. A. (eds.). *The Sauropods: Evolution and Paleobiology*. University of California Press, Berkeley: pp. 178–200.

Stevens, K. A. & Parrish, M. J. 2005b. Neck posture, dentition and feeding strategies in Jurassic sauropod dinosaurs. *In* Tidwell, V. & Carpenter, K. (eds.). *Thunder Lizards: The Sauropodomorph Dinosaurs*. Indiana University Press, Bloomington: pp. 212–232.

Stoinski, S., Suthau, T. & Gunga, H.-C. This volume. Reconstructing body volume and surface area of dinosaurs using laser scanning and photogrammetry. *In* Klein, N., Remes, K., Gee, C. T. & Sander, P. M. (eds.). *Biology of the Sauropod Dinosaurs: Understanding the Life of Giants*. Indiana University Press, Bloomington: pp. 94–104.

Taylor, C. R. 1977. The energetics of terrestrial locomotion and body size in vertebrates. *In* Pedley, T. J. (ed.). *Scale Effects in Animal Locomotion*. Academic Press, London: pp. 127–141.

Taylor, C. R., Heglund, N. C. & Maloy, G. M. O. 1982. Energetics and mechanics of terrestrial locomotion. I. Metabolic energy consumption as a function of speed and body size in birds and mammals.—*Journal of Experimental Biology* 97: 1–21.

Weissengruber, G. E., Egger, G. F., Hutchinson, J. R., Groenewald, H. B., Elsässer, L., Famini, D. & Forstenpointner, G. 2006. The structure of the cushions in the feet of African elephants (*Loxodonta africana*).—*Journal of Anatomy* 209: 781–792.

Witte, H. 1996. *Beiträge zur Anatomie und Biomechanik elastischer Elemente in Bewegungsapparaten*. Habilitation Thesis. Ruhr-University of Bochum, Bochum.

Witte, H., Preuschoft, H. & Recknagel, S. 1991. Human body proportions explained on the basis of biomechanical principles.—*Zeitschrift für Morphologie und Anthropologie* 78: 407–423.

Witte, H., Lesch, C., Preuschoft, H. & Loitsch, C. 1995a. Die Gangarten der Pferde: Sind Schwingungsmechanismen entscheidend? Pendelschwingungen der Beine bestimmen den Schritt.—*Pferdeheilkunde* 11: 199–206.

Witte, H., Lesch, C., Preuschoft, H. & Loitsch, C. 1995b. Die Gangarten der Pferde: Sind Schwingungsmechanismen entscheidend? Federschwingungen bestimmen den Trab und den Galopp.—*Pferdeheilkunde* 11: 265–272.

Witzel, U., Mannhardt, J., Goessling, R,. de Micheli, P. & Preuschoft, H. This volume. Finite element analyses and virtual syntheses of biological structures and their application to sauropod skulls. *In* Klein, N., Remes, K., Gee, C. T. & Sander, P. M. (eds.). *Biology of the Sauropod Dinosaurs: Understanding the Life of Giants*. Indiana University Press, Bloomington: pp. 171–181.

13

Plateosaurus in 3D: How CAD Models and Kinetic–Dynamic Modeling Bring an Extinct Animal to Life

HEINRICH MALLISON

CAD (COMPUTER-AIDED DESIGN) software combined with biomechanical considerations can be used to create extremely accurate skeletal reconstructions of dinosaurs and other extinct vertebrates. CAE (computer-aided engineering) methods that are based on such accurate models give insight into the way dinosaurs moved and behaved, and they greatly ease the task of calculating physical properties (such as position of the center of mass) compared to traditional methods. On the basis of a high-resolution 3D model of *Plateosaurus*, I show that this animal was an agile obligate biped with strong grasping hands. The assessment of possible postures and ranges of motions of the 3D model was done with a CAD program, while the total mass, mass distribution, and the position of the center of mass of the model were assessed with CAE software.

Introduction

Biomechanics deals with the function and structure of biological systems. This chapter will address certain aspects within this broad field of study, focusing on the mechanics of posture and motion of animals. The prosauropod dinosaur *Plateosaurus* will be used as a detailed example of how two different modern computer technologies can aid research on extinct animals. CAD (computer-aided design) programs can be applied to the study of large assemblies of objects, for example, bones in a skeletal mount of a dinosaur, without the bother of actually having to lift and support the many, and often heavy, elements. Digital bones, in contrast, have no weight and cannot break, and are easily combined into a virtual skeleton in a CAD program. A virtual skeleton of *Plateosaurus* is used to assess the posture and range of motion of this animal. Additional information on posture and on locomotion capabilities is derived from CAE (computer-aided engineering) modeling, using a CAD model of the living animal based on the virtual skeleton. The CAE modeling can be used to determine the position of the center of mass (COM) and its shift when the animal moves, as well as joint torques and many other important physical parameters. This approach to biomechanical modeling was termed *kinetic–dynamic modeling* by Mallison (2007) because it derives information on the kinetics—the movements of the modeled animal—from the dynamics—the forces that cause this movement—and vice versa.

COMPUTER SOFTWARE

There is a wide variety of CAD software available at low cost, and even some freeware programs. Models described in this chapter were all created in McNeel Associates Rhinoceros 4.0 NURBS Modeling for Windows. NURBS refers to nonuniform rational B splines, mathematical curves for creating and visualizing 2D curves and rectangles, but also for creating

3D freehand surfaces and 3D volume solids (see Stoinski et al., this volume).

Many CAE programs use NASTRAN (NASA Structural Analysis System), which is a general-purpose finite-element analysis program, originally developed by NASA and mainly written in the programming language FORTRAN. NASTRAN can model the motion of rigid-body systems, taking gravity, inertia, and various other physical forces and principles into account, and is thus capable of full kinetic–dynamic modeling. Various commercially available software packages based on NASTRAN offer advanced kinetic–dynamic modeling. The models described in this chapter were created and run in MSC.visualNASTRAN 4D by MSC software, which is described in detail below.

Virtual Skeletons

ADVANTAGES AND DISADVANTAGES

In order to understand the biomechanical adaptations of an extinct vertebrate animal, it is often helpful to create an accurate reconstruction of its skeleton. Drawings can be a great aid, but they do not show the 3D arrangement of the component bones. Modern computers have sufficient power to display large and complex 3D structures in great detail, and they can rotate such a display freely. If an extinct vertebrate animal is known from sufficient skeletal material, and if that material can be digitized at high resolution, a virtual skeletal mount can be created in CAD software. Such a virtual skeleton offers many benefits over drawings and even over real mounts. The process of creating a virtual mount is in some regards much easier than mounting a skeleton with real bones. Virtual bones will not suffer damage from handling, and they do not require an expensive and technically difficult structural support. In CAD software, large partial assemblies can be moved with a mouse click, and the entire animal can be reposed just as easily. Somewhat work-intensive but occasionally quite helpful is the option of deforming the virtual bones, for example, for removing taphonomic distortion and damage. Also, bones missing from the real skeleton can be mirrored from contralateral ones, or they can be replaced with those from other individuals. These may even be of a different size because scaling up or down is also easily accomplished. Similarly, symmetrical bones (e.g., vertebrae) can be restored if only half the bone is preserved.

A virtual skeletal mount can be created from any kind of 3D digital file, for example, a computed tomographic (CT) scan, a laser scan, or a file produced with a mechanical digitizer (Wilhite 2003; Mallison 2007; Mallison et al. 2009). The mount can be used in a variety of applications, ranging from biomechanical studies—for example, on the range of motion of the animal's limbs—to the planning of museum exhibits. It is also possible to digitally recreate the musculature and other soft tissues on the virtual skeleton. This process allows more accurate musculature models than reconstructions in 2D on paper, because 3D models visualize the actual paths of the muscles, highlighting possible conflicts in the third dimension. Also, volumetric estimates can be made, both from complete models and from sections.

The drawbacks of digitally mounting a skeleton are less obvious: although it is possible to rotate the virtual bones freely and display them from any desired angle, the lack of a stereo image makes it hard to judge the correct contact between curved surfaces. Also, because CAD objects can intersect freely, a correct fit cannot be created by making the bones touch and then pulling them apart a short distance to create the gap for the articular cartilage. Rather, the correct arrangement and distance must be judged from a perpendicular view—or, better, several views. These can be displayed simultaneously, but more views on the screen mean that each individual view window becomes smaller. Also, the depth and distance of objects is hard to estimate, so that overlapping bones often appear to be closer than they are (or nearly touch when they seem far apart), especially if there is another bone surface visible in the gap between them. Although assembling a large number of real bones manually is much slower than with digital files, bringing only two or three into correct articulation is much faster using the real physical objects.

BUILDING A VIRTUAL SKELETON

Creating a dinosaur mount, whether digital or real, requires knowledge of how the individual bones of the skeleton articulate, because some important areas of the dinosaur skeleton typically provide little reliable information about bone articulation. Vertebrae, finger bones, and skull elements are usually easy to fit together. Similar to the bones of mammals, they are normally well ossified, and their articular surfaces fit quite well, with little room for cartilage. Long limb bones, however, as well as ribs, some girdle bones (scapulae, coracoids), and axial elements (sternal plates, gastral ribs), remained cartilaginous at their margins and tips, giving the long bones the typical pitting on their articular surfaces. The preserved bone surface is not necessarily similar in shape to the surface that articulated with adjacent bones at a joint surface (Holliday et al. 2001, 2010; Bonnan et al. 2010). Thus, even large processes may have been lost during fossilization because they were cartilaginous, especially in juvenile and subadult individuals (Bonnan et al. 2010).

The thickness of the articular cartilage is another unknown variable. Although some researchers posit that it was only a thin cover (Stevens & Parrish 1999, 2005a, 2005b), others argue for a more massive layer (Holliday et al. 2010; Bonnan & Senter 2007; Mallison 2007; Schwarz et al. 2007a), at least in some sauropodomorphs. Such variations can influence overall limb length and the interpretation of possible limb motion. In the sole published case of articular cartilage preservation in a sauropod dinosaur, in a humerus of *Cetiosauriscus greppini* from Switzerland, the preserved cartilage does not include the uncalcified articular cartilage cap, but only the underlying calcified cartilage that defines the growth zone in living animals (Schwarz et al. 2007a). Often it is claimed that a thick cartilage cap is impossible because articular cartilage is avascular (lacks blood vessels) and thus cannot be adequately supplied with nutrients. Only a thin layer can absorb enough nutrients from the joint fluid. In fact, extant dinosaurs (birds) have avascular, hyaline cartilage in their joints, but this thin layer is underlain by a massive, vascularized mass of fibrous cartilage, so that nutrient transport is not problematic (Graf et al. 1993).

Additional help in arranging the bones comes from their function and its influence on their shape. For example, curvatures in the limb bone shaft indicate bending stresses (e.g., Biewener 1983), suggesting an inclined position either permanently or during activities that exert large forces on the bone. Cases in point are the curved femur shaft of most mammals and the longitudinal bending of human metatarsals. The usual zigzag posture of the hindlimb bends the femur, and the bone accordingly curves longitudinally to distribute the load better. Also, at midshaft, the femur cross section is almost circular because the magnitude of stresses is similar in all directions.

POSING THE SKELETON

The pose of a digitally mounted skeleton depends on its intended use. For the initial mounting and for locomotion simulation, an osteologically neutral pose (Stevens & Parrish 1999, 2005a, 2005b) is best. The alternative is a pose reflecting the habitual body, neck, and tail posture during locomotion, should the neutral and assumed habitual poses differ. This is usually the case, as demonstrated for the necks of extant mammals by Taylor et al. (2009). The limbs should be placed symmetrically for 3D model creation. Later, the skeleton can be posed differently by moving body parts in accordance with the motions permitted by the limb joints.

CREATING A CAD MODEL OF THE LIVING ANIMAL

For CAE modeling as described below, a 3D model of the shape of the living animal is required. This can be created by CAD with little effort on the basis of a virtual skeletal mount (Gunga et al. 2007; Mallison 2010a) or a laser scan of a mounted skeleton (Gunga et al. 2008), but accurate 2D drawings of the type standardized by Paul (1987, 1997) can be used as well. In this case, the resulting model will be more accurate if a set of corresponding views, for example, the top view and the cross section at the hips, are used alongside the standard lateral view. The CAD model can be built either by creating NURBS curves around it, from which a body can be derived (Bates et al. 2009), or by deforming ellipsoid bodies. A CAD model probably results in far more accurate shapes and volume estimates than digital or physical models based on drawings, because the actual skeleton and the model can be viewed from all direction and as cross sections at all times. The 3D model of *Plateosaurus* used here is intentionally overweight so that errors in NASTRAN modeling will lead to underestimating the animal's athletic capabilities, not overestimating them.

A Virtual Skeleton of *Plateosaurus engelhardti*

MATERIAL

The first complete dinosaur mounted digitally for DFG Research Unit 533 was *Plateosaurus engelhardti* Meyer 1837 from the Upper Triassic Löwenstein Formation of southern Germany (Mallison 2010a, b). *Plateosaurus* is one of the oldest dinosaurs known from good material, and numerous individuals have been found at several localities across Central Europe. The taxonomic history of *Plateosaurus* is complex and confusing, with many species having been erected in the past, but the recent consensus is that there were only two species: *P. engelhardti* and *P. gracilis,* previously *Sellosaurus gracilis* (Galton 1985; Sander 1992; Moser 2003; Yates 2003). In addition, all results described here are based on a single well preserved individual and thus possibly pertain only to that animal.

This skeleton, one of the best-preserved finds of *Plateosaurus,* is housed at the museum of the Institute of Geosciences of the University of Tübingen. It is a nearly complete individual of medium size excavated in the Trossingen quarry by von Huene between 1921 and 1923 and cataloged as GPIT 1/RE/7288 (short GPIT 1). Figure 13.1A shows the Tübingen mount by von Huene, and Figs. 13.1B–D show different views and poses of the digital mount of the same individual (Mallison 2010a, 2010b). All bones of this skeleton were separately scanned in a CT scanner, but some badly preserved bones, including the complete lower left arm and hand, were taken from another skeleton (GPIT Skelett 2, also from Trossingen). In GPIT 1, the skull is partly disarticulated and deformed, so the skull of SMNS 13200 was used instead. This is permissible because Galton (1985) showed that all skull material assigned to *Plateosaurus* is indeed monospecific. For the assembly of the virtual skeleton, it was also necessary to mirror some ribs and elements of the hands and feet. The gastral ribs are missing in the museum mount and in the virtual skeleton, as are the very small clavicles, although these elements were present in the field and collected. They remain embedded in rock matrix and could not be digitized.

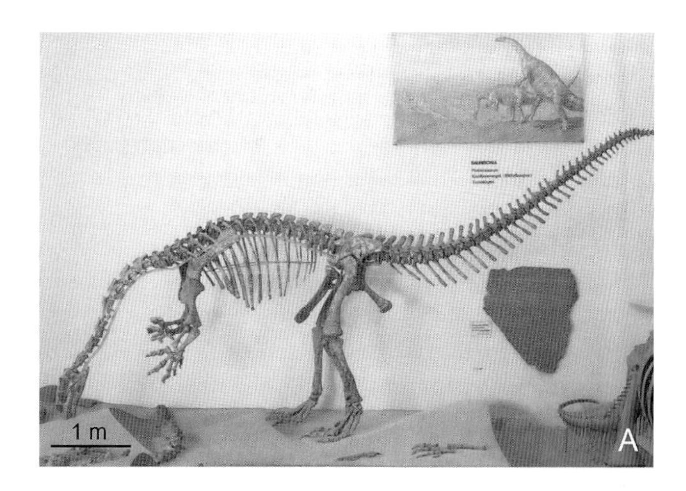

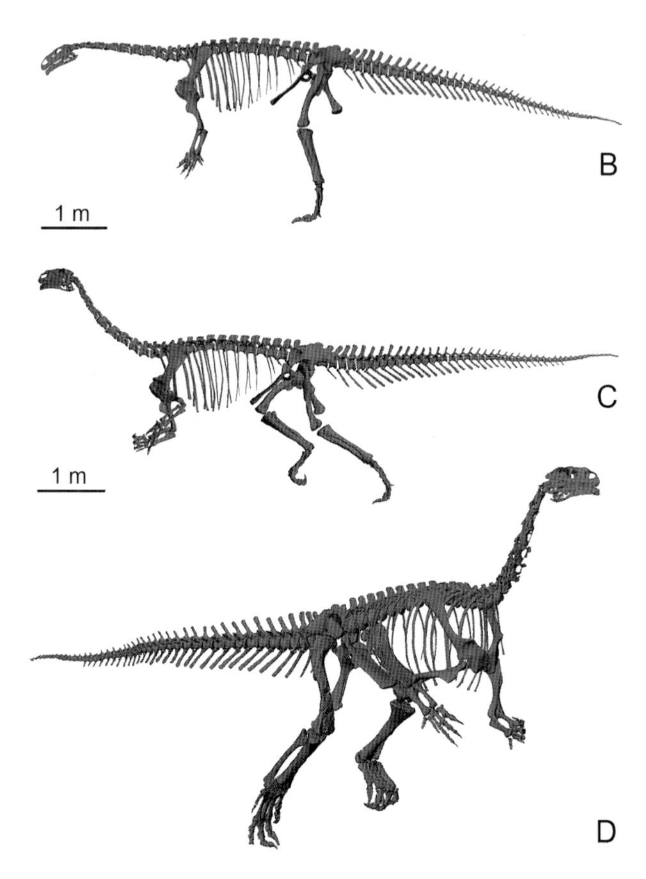

FIGURE 13.1. *Plateosaurus engelhardti* (GPIT 1). (A) Photograph of bipedal skeletal mount by von Huene on display at the Institute for Geosciences, University of Tübingen. (B–D) Virtual skeletal mount. (B) Lateral view of neutral articulation pose. (C, D) Lateral and anterolateral view of rapid running pose. Length of left femur = 597 mm.

ARTICULATING THE BONES

The femur of *Plateosaurus* has a curved shaft (Fig. 13.1) and a subcircular cross section; moreover, what is preserved of the distal condyles faces slightly backward. The shape of the distal condyles of the femur indicates that the knee could not be fully extended when bearing a full load, in contrast to the situation in humans. Altogether, these observations suggest that *Plateosaurus* held its hindlimbs flexed—how strongly flexed cannot exactly be ascertained unless the position of the COM is calculated. The flexion of the limb must result in a position that places the COM above the contact of the feet with the ground; otherwise any bipedal posture would not be stable.

The articulation of the ribs and dorsal vertebrae shows that the anterior body was narrow from side to side, but the exact position of the scapulae on the rib cage is difficult to determine. Dinosaur ribs have two heads and can rotate around the axis connecting the two articulations. Whether the rib motion enlarges the rib cage volume, for example, for breathing in as in humans, or whether rib motion has no influence on the body volume depends on the orientation of this axis. In Fig. 13.2A, the cervical and dorsal vertebrae are shown in lateral view with the axis of rib rotation marked on each vertebra. It is immediately apparent that the anterior body region, the area where the pectoral girdle envelops the rib cage, differs from both the neck and the posterior body. In the anterior body, the axes are nearly vertical, so that rib motion does not have any influence on body width (Fig. 13.2B). This is consistent with a pectoral girdle architecture that solidly connects the left and right girdle elements medially and ventrally to the rib cage, while a widening and narrowing rib cage would require the pectoral girdle to be flexible and separate along the midline. Moving backward, in the middle of the trunk, motion of the ribs articulating with dorsal 6 widens the body cavity from around 50–68 cm, and in the posterior part, the motion of the ribs of dorsal 10 almost doubles body cavity width, from 33–64 cm. This allowed for lung ventilation independent of limb motion, indicating that no liver piston system, as in crocodiles, was required for breathing in *Plateosaurus* (see also Perry et al., this volume).

If the virtual skeleton of *Plateosaurus* is placed in a resting posture on the ground, the shape of this rib cage means that the ends of the ribs would contact the ground and would have to bear some weight, unless the scapulae, coracoids, and sternal plates were arranged to form a brace on which the animal could have rested. This, however, is only feasible with steeply inclined scapular blades and an anteroventral position of the coracoids relative to the most anterior trunk ribs (Fig. 13.1; Remes 2008; see also Hohn, this volume).

In *Massospondylus,* a close relative of *Plateosaurus,* Yates & Vasconcelos (2005) found articulated clavicles that were ar-

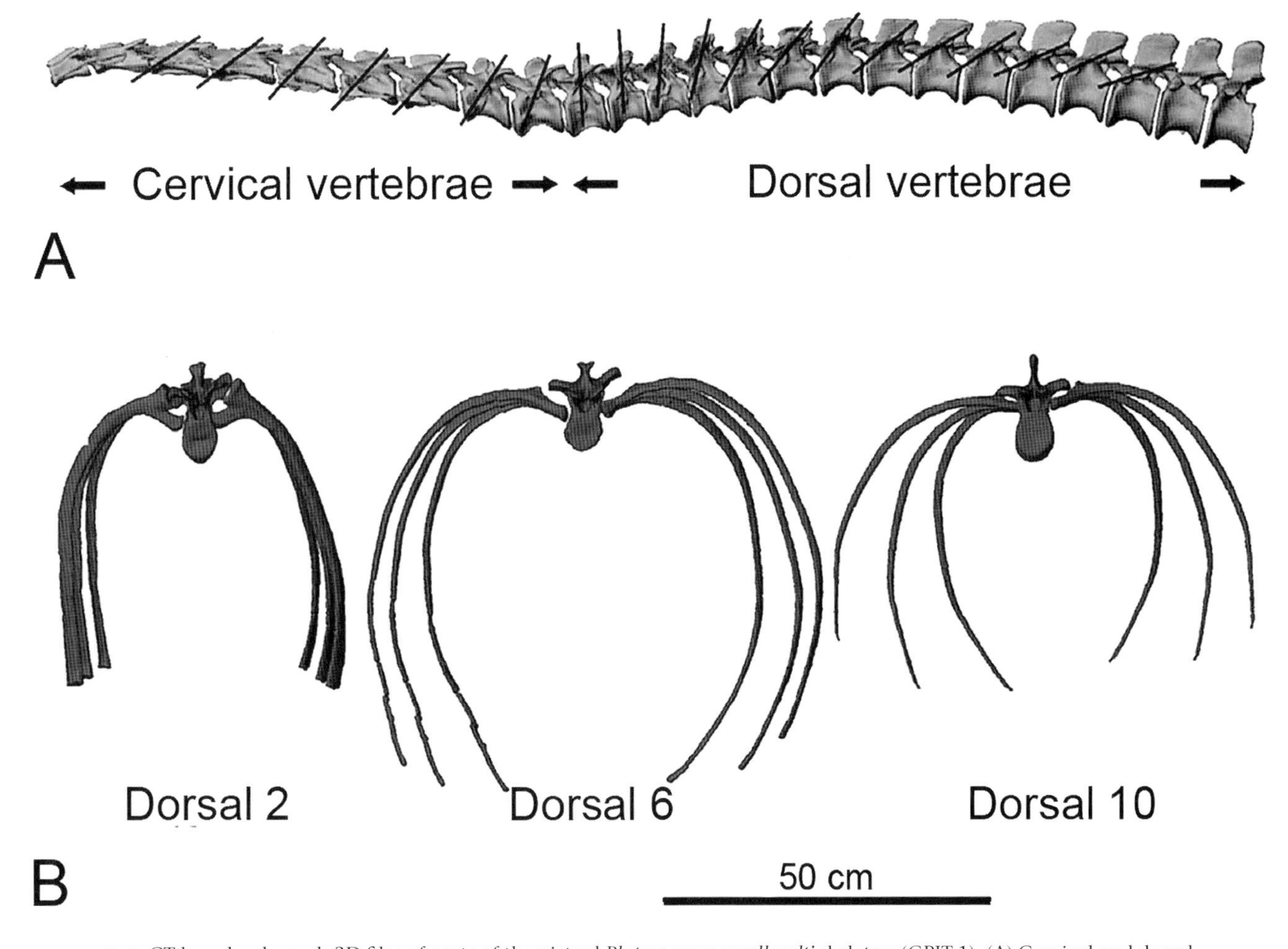

FIGURE 13.2. CT-based polymesh 3D files of parts of the virtual *Plateosaurus engelhardti* skeleton (GPIT 1). (A) Cervical and dorsal vertebral column in lateral view. Black lines indicate the inclination of the axis of rotation of the ribs. (B) Anterior view of dorsal vertebrae and ribs, the latter in three different positions to show lateral expansion of the rib cage during breathing. Note the minimal width change in anterior dorsal ribs, where the shoulder girdle is located, and the strong expansion toward the back.

ranged in a furcula-like manner, touching at the midline. This further confirms the narrow arrangement of the shoulder girdle. For *Plateosaurus,* touching but not co-ossified coracoids and clavicles were already observed by von Huene (1926) in several skeletons. In the virtual skeleton, the scapulae thus cannot be separated and shifted up and backward on the side of the rib cage. It is interesting to note that this tight arrangement of the clavicles may have been retained in derived sauropodomorphs, and that the clavicles even may have been fused into a furcula in some sauropods. A small potential furcula (or unusually curved clavicle) is preserved in the diplodocid baby sauropod "Toni" (specimen SMA 009) in the Sauriermuseum Aathal, Switzerland (Schwarz et al. 2007b).

CHOOSING A POSE

In Fig. 13.1B, the virtual *Plateosaurus* mount has been placed in a neutral pose, while in Figs. 13.1C–D, it is shown in a dynamic pose, with the animal running at high speed with its head held high. Note how much the pose influences the perception of the animal's agility: the neutral pose looks sluggish, while the running pose gives the animal a sleeker and more nimble appearance, even though this pose does not exhaust the possible range of motion of any body part.

RANGE OF MOTION OF INDIVIDUAL JOINTS

An important use of virtual skeletons, and a crucial step for CAE modeling of dinosaur motions, is determining the range of motion of all joints. The details of limb joint and vertebral column motion and their significance for locomotion will be described elsewhere. Here, only a short overview of those results important for determining the habitual posture of *Plateosaurus* is given. For any suggested locomotory mode and gait, the motion limits of all joints must allow an effective walking cycle. For example, a much restricted range of motion in the forelimb indicates that the limb was probably not used in rapid locomotion.

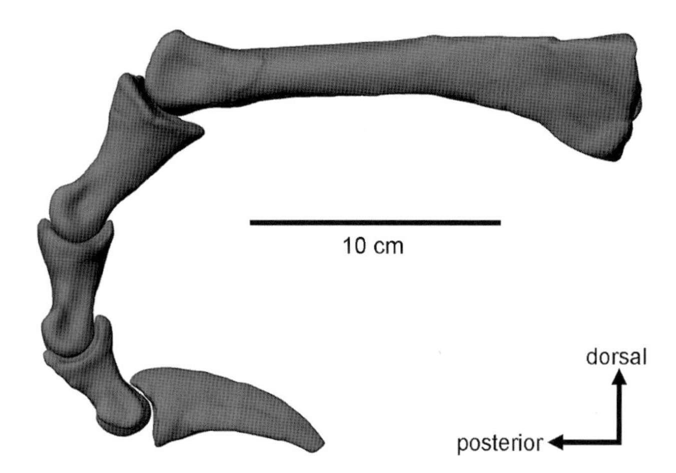

FIGURE 13.3. CT-based polymesh 3D files of toe III of the left foot of *Plateosaurus engelhardti* (GPIT 1). The toe is shown at maximum extension in medial view. Note that a far higher degree of hyperextension is possible than would be required for any gait. The angle between the metatarsal and the ungual is 180°, while that between the metatarsal and the second phalanx is 87°.

The toes of *Plateosaurus* have well developed claws that can be hyperextended to a surprising degree (Fig. 13.3). This is a fundamental requirement for a digitigrade runner, but less so for a plantigrade walker. The ankle also is sufficiently extendable for digitigrady. At full extension (straightening), the metatarsus can be angled approximately 110° from the horizontal, with the tibia and fibula at approximately 70°. This corresponds to an opening angle in the tarsus of approximately 140°. As noted, the knee cannot be fully straightened but must remain flexed at about 160°. The forward movement of the femur (protraction) is restricted by the pubis. With the femur shaft in a vertical position in anterior view, only a 35° forward protraction of the vertical position is possible. Greater protraction angles require a sideways movement (abduction) of the femur to approximately 20°. This abduction angle is so large because of the wide, tray-like shape of the pubis. Backward movement of the femur (retraction) is easily possible beyond the vertical (contra Paul 1997). Osteologically, retraction to a nearly horizontal position seems possible, but the soft tissues must have imposed lower limit. Any angle less than 50° between the long axis of the sacrum and the femur shaft seems unlikely. In such a position, the femur shaft would have already passed the ischia, so that the ischiofemoral musculature could not have aided in retraction, which thus solely would have been performed by the caudofemoralis muscle.

Forelimb mobility is more restricted (Mallison 2010b). Although the elbow has limits to its mobility similar to those of the knee (minimum flexion approximately 245°), the shoulder joint (glenoid) is a simple trough. This shape limits flexion and extension to angles between approximately 135° and approximately 55° relative to the long axis of the scapular blade. If the scapula is inclined posteriorly at an angle of 55° to the horizontal (a reasonable position: Schwarz et al. 2007c; Remes 2008), then the humerus cannot be protracted beyond vertical, while maximum retraction places the elbow at a level above the shoulder joint. The glenoid does not have a dorsal bony rim to constrain dorsal motions of the humerus head at large abduction angles. Therefore, abduction of the forelimb to angles greater than 40° was probably possible, but only when there was no significant compressive force acting on the shoulder joint.

The shapes of radius and ulna prevent rotation of the radius for manus pronation (inward rotation). 3D analysis (Mallison & Bachmann 2006; Mallison 2010b) confirms the results of Bonnan & Senter (2007) that *Plateosaurus* could not have placed the palms of its hands on the ground without strong abduction of the humerus.

The carpus is difficult to analyze as a result of the low degree of ossification of the carpals, with evidence coming from several specimens. Fortunately, a *Plateosaurus* skeleton from Halberstadt housed at the Museum für Naturkunde, Berlin (skeleton XXV) has five carpals preserved in its right hand. This previously mounted, nearly complete skeleton was damaged during World War II, and only parts of the skeleton remain. In the well preserved left wrist of GPIT 1 and the hand of skeleton XXV, there are a radiale and an ulnare shaped like flattened pyramids in dorsal view and a series of three distal carpals. One of these is plate-like and fits to the proximal end of metacarpal II; the other two are small and rounded, and they fit between the ulna and metacarpals III and IV. GPIT Skelett 2 also has a small, rounded distal carpal that was found proximal to metacarpal II. The wrist may have had as much mobility as in humans: strong extension and especially flexion were possible, as well as significant ulnar and radial deviation. In contrast to the toes, the fingers were capable of only limited hyperextension (Mallison 2010b).

An Introduction to Kinetic–Dynamic Modeling in NASTRAN

THE MODELING SOFTWARE

Kinetic–dynamic modeling is a new approach to modeling animals. As previously noted, the models presented here were created in MSC.visualNastran 4D. Figure 13.4 shows a screenshot of the work environment in which the important features are indicated. As with all NASTRAN implementations, the program exclusively works with rigid bodies. In the simulation software, rigid bodies do not deform or break apart during simulation runs, even though the physical object they represent may actually do so. In order to simulate a flexible body, it is necessary to connect a series of rigid bodies with appropriate constraints (joints) to approximate this flexible body.

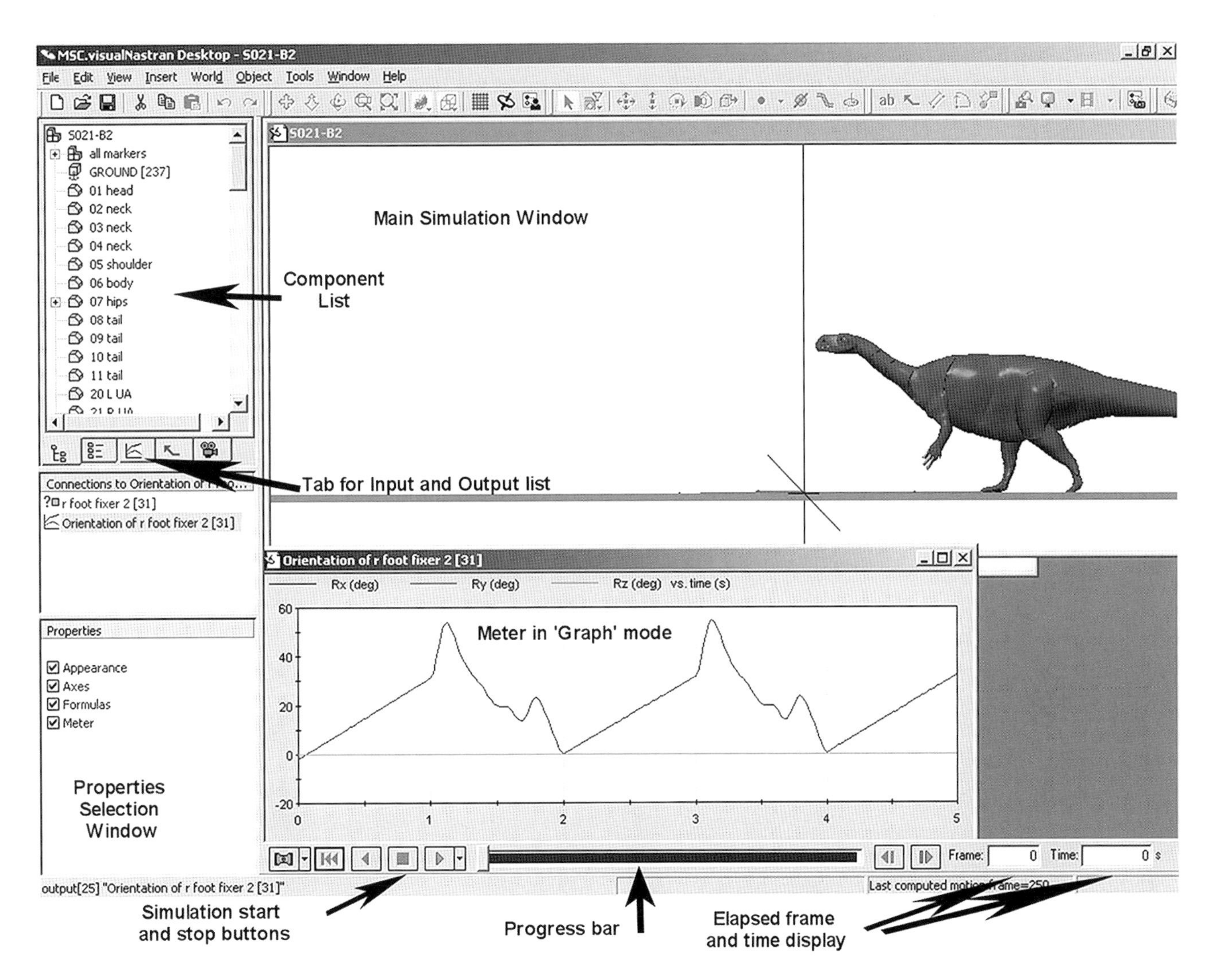

FIGURE 13.4. Annotated screenshot of the work environment of CAE modeling software MSC.visualNASTRAN 4D running a simulation of a walking *Plateosaurus*.

Problems are time discretized—that is, split into short steps of time, so that the program computes motions and accompanying forces while ensuring that all constraint conditions are satisfied. Both the fast and simple Euler integration with fixed integration intervals and the more accurate Kutta-Merson integration can be used. Kutta-Merson integration, which allows variable time steps, is described in detail in Fox (1962).

MODELING DINOSAURS IN NASTRAN

Mallison (2007) developed a method for modeling COM position and animal motions of extinct and extant vertebrates that allows rapid estimates of many biomechanical parameters and interspecific comparisons (cf. Mallison & Pfretzschner 2005; Mallison & Bachmann 2006).

A simulation model is first created by dividing the 3D CAD model into functional units. Usually this involves separating limb segments from each other and dividing neck, trunk, and tail into a small number of segments. Also, areas with strong average density differences should be divided into separate units; for example, in sauropodomorphs the dense skull should be separated from the pneumatized neck, even if no motion between them is planned for the simulation. Tissues of very low density (lungs, air sacs) can be cut out of the 3D solid in the CAD software. Theoretically, it is possible to create each tissue type or organ as a separate 3D body and assign the correct density in the simulation software, but the gain in accuracy is minimal while the added workload, computation time, and uncertainties are prohibitive. Adjusting densities in NASTRAN is, in comparison, much easier and faster, thus making it possible to test how different interpretations influence the modeling results.

In the next step of CAE simulation model building, the functional units must be connected with articulations to simulate the joints of the real animal. Many types of predefined articulations exist in NASTRAN, but it is also possible to define complex joints that adhere to up to six rules by defining appropriate formulae. For example, a joint between two parts can be set to rotate around a certain axis by applying torque values defined in a table, while simultaneously sliding freely along a different axis (i.e., based on the forces created within the simulation). Such an articulation can also be specified to dislocate when one of the parts achieves an angular velocity greater than, for example, 10°/s, simulating forces potentially resulting in injuries.

Next, the model is moved into its intended starting position and the simulation is prepared for the problem that is to be investigated. This can involve the creation of a prescribed motion sequence, or a sequence of forces and torques to be exerted at certain times in certain joints. Creating a complicated sequence often requires extensive corrections by a trial-and-error procedure, which can be a time-consuming effort. Finally, the motion sequence can be calculated and the results exported in a variety of formats.

For each functional unit or body, NASTRAN offers a wide range of parameters that can be measured. The program can directly calculate the mass, the COM, the position, orientation, linear and angular acceleration and velocity, the linear and angular momentum, and the kinetic and potential energy of this body. Thus, a wide range of research questions can be investigated using one basic model, without the need to tailor a new specific, complex model for each new question.

The first step in modeling must always be a test of the consistency of the input data, that is, whether the articulations of the functional units of the model work correctly and the animal is positioned in a balanced posture. This basic stability test automatically means that impossible postures (e.g., a cow standing on its hind legs) are excluded as well. If the animal is placed in an impossible posture to start with, or if the density of a body part has been set to a wrong value, all simulations run on the basis of such a model will be meaningless. Testing the motion of joints is simply a matter of entering different values for joint position and seeing whether the body parts actually move as desired (the motion test). Testing the balance —that is, the stability test—is best done by starting the simulation without any motion command. If the animal is unbalanced, it will tip over. Although this may seem trivial, accidentally mispositioning a model is an easy mistake to make. The differences in tissue density, and therefore in weight of the different model segments, tend to confuse the observer, and the COM is hardly ever located where one suspects it to be. Additionally, external factors in the simulation may be wrong, such as the formula for ground reaction conditions; or the value for the gravitational constant or some other value may have been entered incorrectly.

So far, different species pertaining to two sauropodomorph groups were modeled in NASTRAN using the approach described above. *Plateosaurus* was used to establish and validate modeling techniques for the assessment of postural stability and the development of locomotion cycles. Several sauropods, including *Brachiosaurus* and *Diplodocus,* were modeled to assess their ability of adopting a bipedal or tripodal pose and to evaluate some aspects of their locomotion (Mallison 2007, Chapter 14 in this volume).

VARIATIONS IN MASS DISTRIBUTION

Obviously, the 3D model of the animal's external shape that serves as starting data for a simulation must be as accurate as possible, and different densities of different body parts must be taken into account. Although published mass estimates for dinosaurs vary widely (e.g., from 15 metric tons [Russell et al. 1980] to 78 metric tons [Colbert 1962] for *Brachiosaurus;* see also Appendix), modern reconstructions that are based on reliable skeletal data show a relatively low range of variation (Henderson 1999, 2006; Seebacher 2001; Mazzetta et al. 2004). However, large variations in assumed muscle mass can influence the models profoundly. Still, even large variations in overall bulkiness of the animal lead to relatively small differences in the qualitative interpretation of the final modeling results as long as similar reconstruction methods were applied to all models. Tests with models of extant animals (Mallison, unpublished data) confirm that models that are based on sufficiently reliable fossil data (such as nearly complete skeletons) give useful and sufficiently accurate results. In any case, it is advisable to use not one model, but several versions with differing mass distribution. This can take the form of a sensitivity analysis, testing whether the assumed soft tissue distribution has a major influence on the modeling results.

ADVANTAGES OF NASTRAN MODELING

NASTRAN modeling is vastly superior to biomechanical pen-and-paper calculations because of its ability to handle large numbers of calculations quickly. Therefore, the models can be considerably more complex than simple calculations, resulting in greater accuracy. For example, the constant shift of the COM within the body during locomotion is taken into account by a NASTRAN model, while this is difficult to track on paper (Alexander 2003a). Equally, the complex shift in length of a limb during its swing forward continuously alters the position of the COM of the limb and the deviation of its path relative to the hip joint as compared to a simple pendulum swing. This has the effect of continuously altering the swing frequency of the limb. Rotational inertia is transferred from one segment to its neighboring segments whenever muscles act on them, a factor that cannot be considered in simple calculations, while in NASTRAN this is automatically included. Lateral motions of the thorax and abdomen can also be modeled because NASTRAN can be used for 3D simulation.

Another great advantage of NASTRAN modeling is the ease with which various animals or motion sequences (or versions of the same animal/motion sequence) can be compared, and the simplicity with which additional variables can be measured after completion of the model. Slimmer and fatter as well as longer and shorter versions of each 3D section can be exchanged quickly, and once a motion sequence has been developed, it can be quickly adapted for a new model of a geometrically similar animal. Additionally, many problems do not even require complicated motions but rather only (near-) static analysis, and results of model variants can be obtained within seconds.

In contrast to many other CAE tools, it is also possible in NASTRAN to view the results of each time step immediately and thus to build and improve a model step by step. In addition, it is even possible to include finite element stress analyses during the motion. Video and graph outputs make interpreting the results of modeling easy. Because of the easy modification of a basic model, models can be used to investigate scaling differences in ontogenetic and phylogenetic series and to compare species that are only distant relatives but that share overall morphological similarity as a result of convergent evolution (e.g., sauropods and elephants). Also, instead of a minimum and a maximum case for a force (or a weight, or weight distribution, etc.; e.g., Carpenter et al. 2005), many different combinations can be tested quickly and easily (e.g., Mallison 2010c). Although the NASTRAN method is not as detailed as, for example, interactive musculoskeletal modeling in the software package SIMM (Delp & Loan 1995, 2000), it is much faster. In this respect, NASTRAN is intermediate between simple biomechanical calculations and highly detailed modeling, and well suited to a range of tasks for which both much simpler and more detailed methods are not suited. Some of these are the rapid determination of the COM, rotational and linear inertia, and the kinetic and potential energy of complex and moving systems (e.g., in a swinging limb) or the assessment of collision conditions (e.g., a stegosaur tail spike impacting a theropod body; Mallison 2010c).

An important aspect of NASTRAN modeling is that the protocol for model creation as described in Mallison (2007) can be tested on extant animals. Motions of live animals can be captured on video, and the NASTRAN models can be set to perform the exact same motion over the same time in order to obtain force and torque values, which can then be compared to measurements from living animals. Also, the relative differences between the models of extinct and living animals can yield interesting results, as long as all data used in a study are created using the same protocol and therefore are affected by the same errors. Thus, their relative values can be interpreted (e.g., "twice the force is required, but the moment arm is twice as large, and thus the torques are equivalent"), even if the absolute values are meaningless.

DISADVANTAGES OF NASTRAN MODELING

NASTRAN modeling has various limitations that limit model accuracy. The software cannot model deformable bodies, and thus soft tissues, accurately. Many important biomechanical processes in animal locomotion involve deformable tissues, such as the contact between the foot and the ground, and the transfer and storage of energy in stretched muscles and tendons. Also, muscles constantly alter their shapes and paths during motion, wrapping around bones and around each other. Highly detailed modeling of muscle systems as in SIMM (Delp & Loan 1995, 2000; Hutchinson et al. 2005) are therefore not possible. For the same reason, ground reaction forces (the force exerted on the foot by the ground during contact) cannot be taken directly from the simulation but must be calculated by hand or as a separate program entity (a model component, e.g., a formula for a graph) on the basis of the weight and acceleration of the limb parts. Also, realistic collision conditions are difficult to set up in NASTRAN. Most commonly, this problem affects the simulation of foot-to-ground contact, which is treated as a collision in the software and leads to problems with ground friction. The easy solution—attaching the foot to the ground with a constraint—leads to massive artifacts for each new foot-to-ground contact and precludes interpretation of the ground reaction force. Other, more complex solutions exist, but these are usually difficult and increase calculation times considerably.

MODEL COMPLEXITY AND ACCURACY

An ideal model in any modeling method produces sufficiently accurate results from a minimal model building and calculation effort. The simpler the model, the easier it is to determine which of the characteristics is responsible for the observed effect. However, a more complex model usually allows more accurate results, and care must be taken to avoid oversimplifications that lead to fundamentally erroneous results. Alexander (2003b) discusses an example in which a simple model (Alexander 1989) fails to deliver sufficiently accurate results compared to a study that used a more complex model (Pandy et al. 1990). On the other hand, complexity always increases the model building time as well as the calculation time, whether the time is spent running computer software or doing the math by hand, and also the likelihood of typographical and other model building errors.

The main sources of errors in NASTRAN, however, are the

above-mentioned difficulties in accurately modeling collisions, which are inherent in the mathematical methods used, and the inability to use deformable materials. Here, it is especially necessary to validate modeling methods by applying them to extant taxa, so that gross errors will become apparent as significant differences between the modeling results and the known motion capabilities of the animals. Sensitivity analyses are an important part of validating results. In these, important parameters are varied to determine which of them influence the results markedly.

Kinetic–Dynamic Modeling of *Plateosaurus engelhardti*

PLATEOSAURUS POSTURE AND LOCOMOTION: STATE OF THE ART

A wide variety of gaits has been suggested for *Plateosaurus* in the literature: von Huene (1908, 1926) argued that *Plateosaurus* was an obligate digitigrade biped, much as the similar *Anchisaurus,* which had been described as bipedal and digitigrade by Marsh (1893a, 1893b). Von Huene based his reasoning on the striking discrepancy between forelimb and hindlimb length, and on adaptations of the hand, which he interpreted as a specialized grasping organ not suited for supporting the body during locomotion. Von Huene visualized his ideas in the famous and well executed mounts of two *Plateosaurus* skeletons from the Trossingen locality (GPIT 1 and GPIT Skelett 2), which were on display at the University of Tübingen (Weishampel & Westphal 1986). Other researchers have suggested practically all stances imaginable: Jaekel (1910, p. 276) proposed obligate quadrupedality and plantigrady, "like lizards." Later, he revised his opinion to conclude that a clumsy, kangaroo-like hop was the only possible mode of locomotion (Jaekel 1911). Fraas (1912, 1913), on the basis of the position of the skeletal finds in the field, argued for a sprawling gait. Later, researchers began to agree with von Huene on the notion of digitigrady, although plantigrady resurfaced in Sullivan et al. (2003). Weishampel & Westphal (1986) depict *Plateosaurus* running bipedally on digitigrade feet, but they argue for facultative quadrupedality. However, the wide spread of the metacarpals in their drawing is inconsistent with the hands bearing much weight during locomotion. Paul (1997) also argues for bipedality, but his outline skeletal drawing seems to imply permanent rather than facultative quadrupedality. He also proposed that *Plateosaurus* was capable of a bounding gallop (Paul 2000).

Among others, Galton (1971a, 1976, 1990) advocated quadrupedality in prosauropods on the basis of the hindlimb/trunk and hindlimb/forelimb ratios—the same character that had led von Huene to mount his plateosaurs bipedally. Wellnhofer (1994) also reconstructed *Plateosaurus* in a quadrupedal stance on the basis of characters seen in the tail of the material from Ellingen, Bavaria. He concluded that the strong downward curve of the tail would make a bipedal stance impossible, as it would run the tail into the ground. However, Moser (2003) showed that Wellnhofer's (1994) reconstruction was based on a few selected caudals that show stronger than normal keystoning, and that the tail of *Plateosaurus* was probably straight. In the first functional morphology study to address *Plateosaurus* locomotion, Christian & Preuschoft (1996) studied the resistance of the vertebral column to bending in various vertebrates in order to determine locomotory modes. *Plateosaurus* shows an intermediate pattern between obligate bipeds and obligate quadrupeds, exhibiting a medium peak of resistance to bending over its shoulders instead of either the small peak of bipeds or the large peak of quadrupeds, leading Christian & Preuschoft (1996) to suggest that the animal was facultatively bipedal at high speeds only. In their study on the shape of the pelvis and on hindlimb posture, Christian et al. (1996) argue for quadrupedal locomotion at walking speeds and conclude that the femur was held in a parasagittal, erect position.

This wide variety of suggested locomotion modes, combined with the availability of excellently preserved and complete material, appeared to make the *Plateosaurus* the ideal subject for kinetic–dynamic modeling for gaining insight into the development of quadrupedality in sauropodomorphs. However, the detailed analysis of the 3D digital skeleton (see above; Mallison 2007, 2010a, 2010b) later showed that this animal could not have walked on all fours, a finding that confirms the conclusion reached by Bonnan & Senter (2007) based on 2D data. Remes (2008) agreed with this conclusion because *Plateosaurus* has a narrower distal humerus than all other prosauropods, leaving insufficient space between radius and ulna for them to cross each other, thus preventing pronation. Therefore, additional evidence for obligate bipedality was sought through modeling (Mallison 2007, 2010a).

MODELING: INSTANTANEOUS CALCULATION OF THE CENTER OF MASS

As mentioned, the 3D shape reconstructed of an animal can have an important influence on simulation results. It is therefore advisable to test several variations of the model with different weights assigned to the model parts, or even differently shaped sections in the initial stability test. The COM of assemblages of bodies is not directly displayed by the software, while that of a single body is automatically calculated. In order to display the COM of the entire model, it is therefore necessary to find its position via the following formula:

$$v_{COM} = \frac{\Sigma \check{v}_{body} \cdot mass_{body}}{mass}, \qquad (1)$$

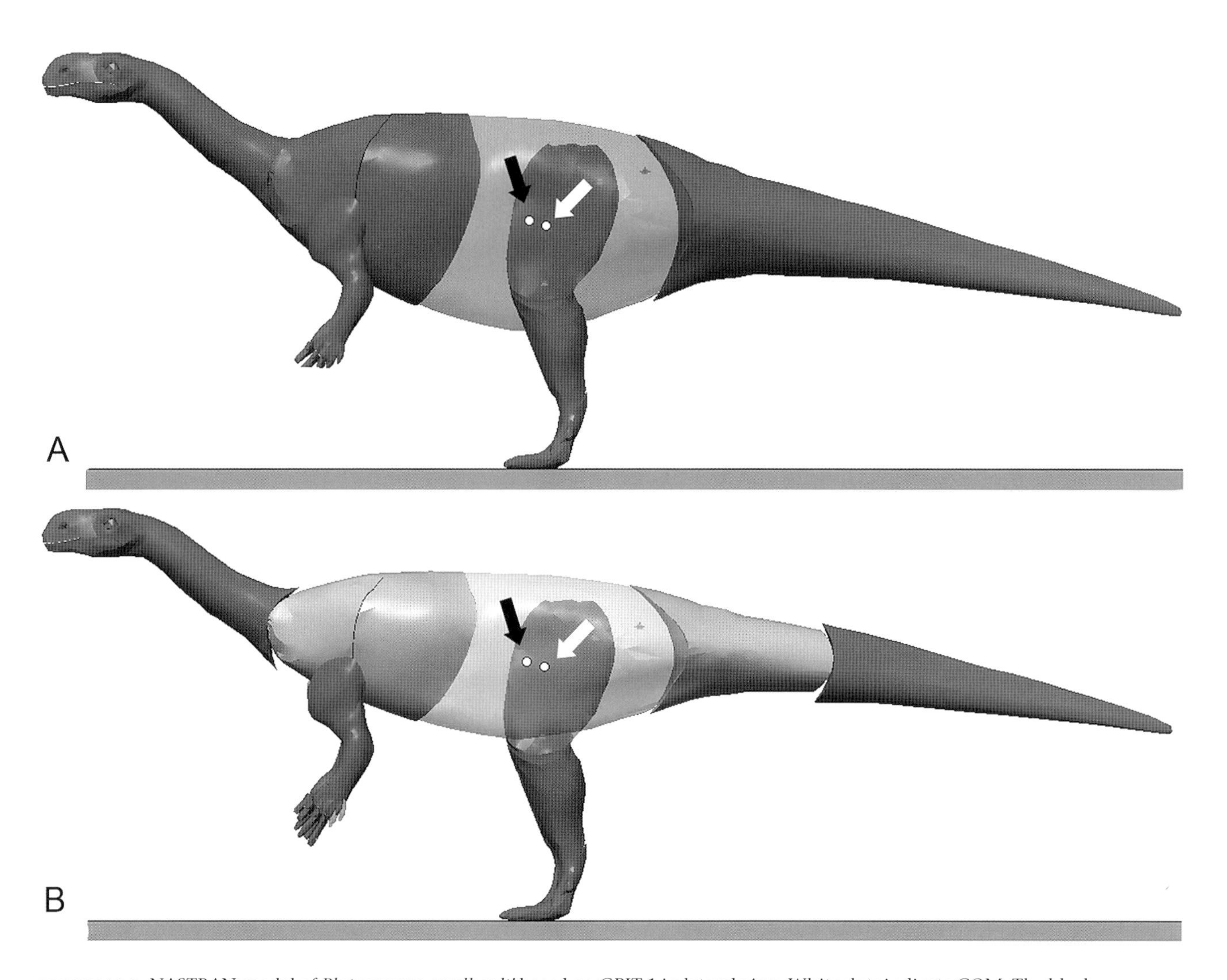

FIGURE 13.5. NASTRAN model of *Plateosaurus engelhardti* based on GPIT 1 in lateral view. White dots indicate COM. The black arrow points to the COM at uniform density, while the white arrow indicates the COM at adjusted density (i.e., different tissue types are taken into account). (A) Initial model. (B) Slimmed-down model. Neither the density adjustment nor the overall bulkiness of the model results in a shift of the COM that would be important when compared to the support area (i.e., the part of the foot that is in contact with the ground).

where $\check{v}_{COM}$ is the position vector of the COM, *Mass* is the mass of the entire assemblage, $\check{v}_{body}$ is the position vector, and $mass_{body}$ the mass of the individual part. For NASTRAN, this must be transcribed separately for each coordinate axis. The formulae for the x, y, and z position of the COM of the assemblage are used to assign motion to a small NASTRAN body, which moves to the position of the COM, and moves accordingly if the COM moves as a result of a motion of the whole model. Therefore, motions of the COM become almost instantaneously visible.

Figure 13.5A shows the *Plateosaurus* model and the position of the COM calculated by equation (1) for a uniform density of 1.0 kg/L. Additionally, the position of the COM is shown for a model version in which the densities of all body parts have been adjusted to take the respiratory organ into account, as well as the high percentage of bone in the tail and limbs, which lead to lower densities in the neck and anterior body, and higher densities in the limbs and tail. Figure 13.5B shows a version of the model in which the body and the anterior tail have been altered to produce a slimmer animal. This version was evaluated both for a uniform density of 1 kg/L and for the same density adjustments as the initial model. The influence of reducing the body outline on the COM is negligible; the COM is located slightly higher in the slimmer model. Adjusting the densities of the various body parts results in a caudal shift of the COM by about 9 cm on average and a downward shift of approximately 1.5 cm. Compared to the length of the part of the middle toe that is in contact with the ground (about 25 cm), this is also insignificant. Slightly stronger flexion of the ankle would move the COM back over the middle of

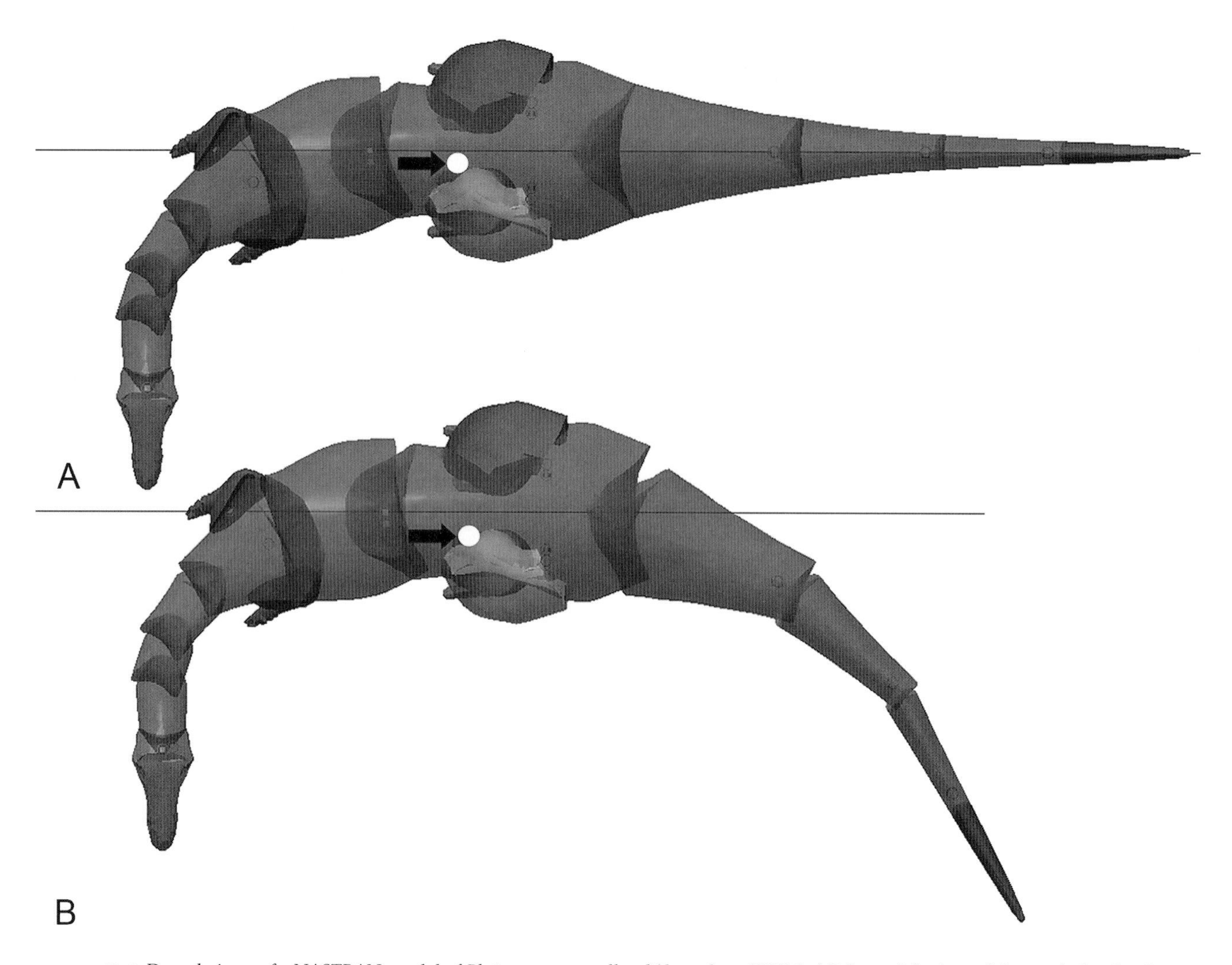

FIGURE 13.6. Dorsal views of a NASTRAN model of *Plateosaurus engelhardti* based on GPIT 1. (A) Lateral flexion of the neck that leads to maximal lateral displacement of the COM. (B) As (A), but with both neck and tail flexed. White dots mark the position of the COM. Line indicates plane of symmetry of model.

the toes. Mallison (2007) showed that even a rather unlikely redistributions of body weight, such as a ridiculously thin tail on an otherwise fat model, still would allow a bipedal posture for *Plateosaurus*.

MODELING: EXPANDED STABILITY TEST

The basic stability test can be expanded into an assessment of the stability of extreme poses. For example, a dinosaur with a long and heavy tail and a long and heavy neck may lose its balance when bending both the neck and tail to the same side without broadening its base support by abducting at least one limb. *Plateosaurus* remains stable even during strong lateral flexion of both tail and neck (Fig. 13.6), but it becomes unstable if both are moved into such a position quickly, the result of rotational inertia induced by the rapid motion. Sauropod models, in contrast, can swing their tails and necks rapidly without a marked influence on stability because of the wider stance in their hindlimbs, and especially the additional stability of the quadrupedal posture (Mallison, unpublished data).

MODELING OF LOCOMOTION

The *Plateosaurus* models were used to assess both bipedal and quadrupedal posture, although the latter appears unlikely on osteological grounds, as discussed earlier. A cyclical motion sequence for a moderately fast (approximately 1 m/s) bipedal walk of *Plateosaurus* was also developed (Mallison 2007, unpublished data), and the resulting forces were measured. A sequence of screenshots with the accompanying torque meter graphs is shown in Fig. 13.7. The sequence was then optimized to reduce overall energy consumption and compared to a sequence for quadrupedal walking. This research showed that the muscle forces required for *Plateosaurus* to move quickly in a bipedal gait are well within the limits set by the amount of muscles that can be placed on the skeleton, and not much

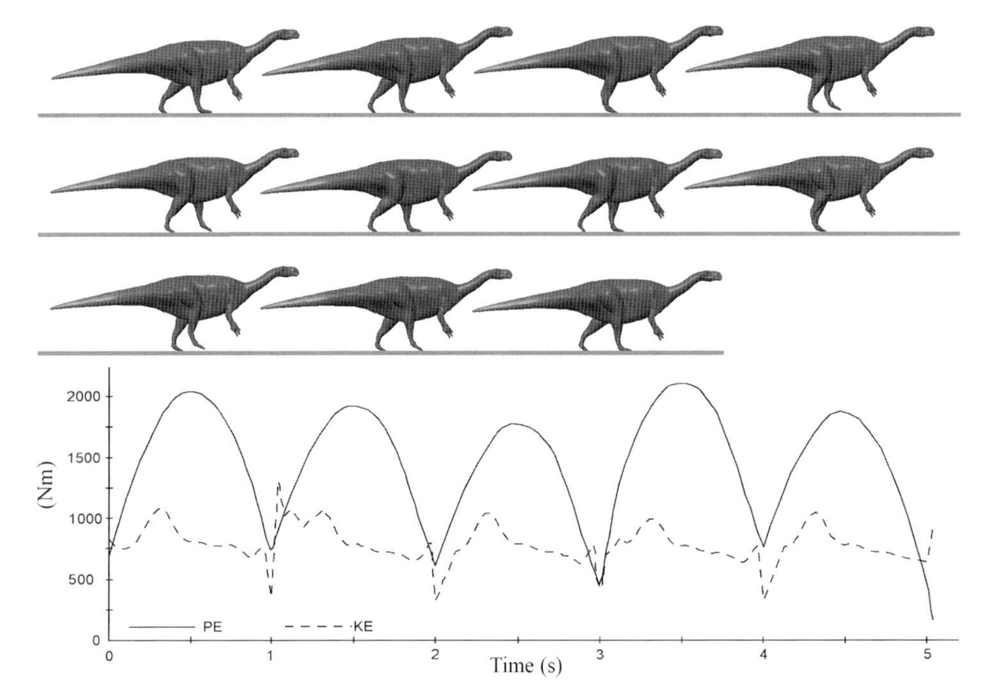

FIGURE 13.7. Walking cycle for a NASTRAN model of *Plateosaurus engelhardti* based on GPIT 1 at moderate speed (4 km/h). Time between frames is 0.2 seconds. The sequence runs left to right, top to bottom. Bottom right: graph of potential energy (continuous; PE) and kinetic energy (dashed line; KE) of the animal.

higher than for a quadrupedal walk—despite a far higher speed possible for the bipedal motion.

The model was also placed into a quadrupedal posture with almost vertically placed forelimbs (Fig. 13.8), forcibly pronating the hands. In this posture, the ratio between the effective limb lengths (ELL), measured as the height of the glenoid above the ground divided by the height of the acetabulum above the ground, is approximately 0.45, that is, the hindlimb has more than double the effective length (Mallison 2010a; Fig. 13.8A). Also, the back and tail point absurdly upward in the quadrupedal posture. The resulting vectors of gravity and forward motion compresses the forelimb, forcing the elbow to flex. This requires unrealistically large torques to counter.

If the hindlimb is flexed more to improve the ELL ratio to approximately 0.58 (Fig. 13.8B), hindlimb protraction becomes difficult, despite this ratio being still far from suitable for quadrupedal locomotion. Some sauropods have an ELL ratio of approximately 0.6, but their humeri can be brought much further forward, resulting in much greater relative forelimb stride lengths. To achieve an ELL ratio of approximately 0.7 (Fig. 13.8C), the femora of the *Plateosaurus* model must both be permanently brought up against the pubes, and limb protraction becomes nearly impossible, even if the feet are placed in a plantigrade position. This posture would also mean that limb retraction could not have been performed by the caudofemoralis muscle, similar to the situation in birds (Gatesy 1990). However, birds lack a fourth trochanter, the attachment site of the caudofemoralis, which is well developed in *Plateosaurus* (Fig. 13.1), and thus this assumption is proven wrong.

In any of these quadrupedal postures, locomotion is slow because the orientation of the glenoid, which faces posteroventrally, makes humerus protraction beyond the vertical almost impossible. Additionally, the bipedal posture was found to be as laterally stable as any reasonable quadrupedal one (Mallison & Bachmann 2006), suggesting that *Plateosaurus* had little to gain from walking on all fours. Simply stated, a quadrupedal gait offered no advantages to the animal and only slowed it down, placing large stresses on the forelimbs. Further, *Plateosaurus* shows a number of adaptations to cursoriality, such as elongated metatarsals, a digitigrade foot, and long hindlimbs in relation to body length. Assuming that the animal moved in a way that negated the advantages conveyed by these cursorial adaptations is counterintuitive.

Discussion

PALEOBIOLOGICAL IMPLICATIONS

The CAD and CAE modeling results have implications for the paleobiological interpretation of *Plateosaurus*. The specialized grasping hand indicates that powerful grasping played an important role in the biology of the animal. The claws, however, are not as strongly curved as those of predators, nor do they possess the characteristic large process for the flexor tendon. Mallison (2007) showed that, contrary to the position in the drawings in Galton (1971b) and Paul (1987, 1997), the first claw was not abducted to nearly 90° during motion, and that the angle between the basal phalanges of digits I and II was reduced from approximately 25° at maximum hyperextension to only approximately 13° at maximum flexion (Mallison 2010b). Thus, the thumb claw did not fully oppose the other digits for precision grasping, as the thumb does in humans. In addition, as pointed out by Galton (1976) for *Anchisaurus*, which has a similarly shaped hand, during flexion, the ungual

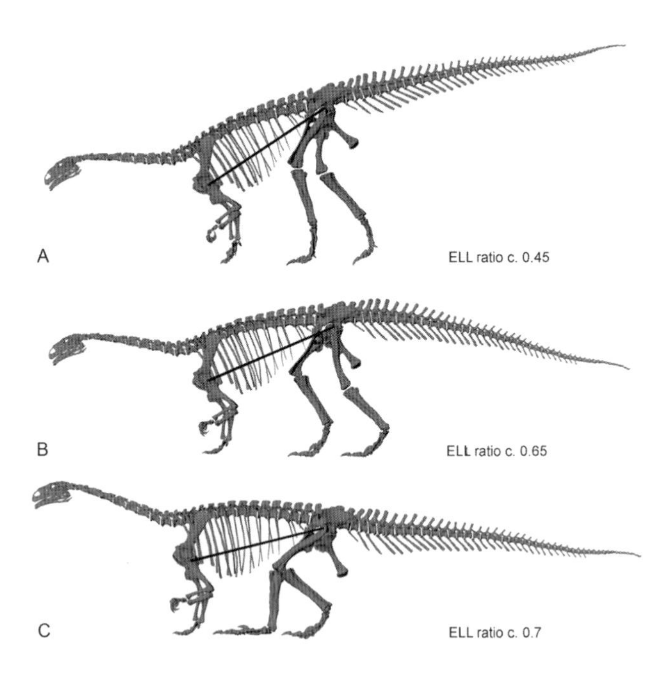

FIGURE 13.8. Lateral view of the virtual skeleton of *Plateosaurus engelhardti* (GPIT 1) in different quadrupedal poses. (A) Hindlimbs upright with digitigrade feet. (B) Hindlimbs positioned as in (A) but more flexed. (C) Hindlimb strongly flexed with plantigrade feet. Lines connect glenoid and acetabulum to illustrate effective limb length ratios (ELL). Typical large (>25 kg body weight) quadrupeds have an effective limb length ratio of approximately 1.0. Note that hindlimb protraction in (C) is barely possible.

phalanx rotated laterally as well, and thus nearly lined up with the second and third digits.

Although von Huene (1926) overemphasized the inward turn of the first finger, his assessment of the hand function appears essentially correct: the fourth and the widely splayed fifth finger with their reduced distal phalanges probably functioned as an abutment for the first to third fingers, allowing *Plateosaurus* to grasp and manipulate objects effectively.

On the basis of the CAD and NASTRAN models, the range of motion of the forelimb is surprisingly limited (Fig. 13.9; Mallison 2007, 2010b), which is also the case in some other bipedal dinosaurs (e.g., Senter & Robins 2005; Senter 2006; Longrich & Currie 2008), so that the hand cannot have reached the snout (unless *Plateosaurus* had a much more mobile neck than often assumed; see, e.g., Taylor et al. 2009 for data suggesting high neck mobility). Apparently, *Plateosaurus* was limited to grasping objects immediately beneath its trunk and in front of its feet, and to the left and right as far as the arms could be abducted. However, this lateral range depended on the height of the body above the ground: for a large range, the animal had to squat down into a semisitting position. What good is there in a strong grasping hand attached to a strong arm that had hardly any reach? Although intraspecific combat is possible (as suggested by Bakker, pers. comm., 2006), the acquisition of food appears more likely. It is possible that the animals dug out subterranean roots and shoots, ripping large pieces out of the ground, or pulled branches off trees. The strong and (during extension) everted first claw may have served to break open rotten logs. The range of motion appears to have been far too limited, however, to pull down foliage to feed on higher branches in a tree.

IMPLICATIONS FOR THE EVOLUTION OF SAUROPODOMORPH LOCOMOTION

Although *Plateosaurus* turned out to be an unsuitable model for the hypothesized facultatively bipedal stage in the evolution of sauropodomorph locomotion, some conclusions can nevertheless be drawn. In light of the research by Fechner (2006) and Remes (2006, 2008), it is obvious that the ratio of forelimb to hindlimb length alone is not a good indicator for the ability of an animal to walk in a quadrupedal gait. Rather, the ELL should be determined, but this is often difficult (e.g., see Gatesy et al. 2009 for an attempt to constrain limb flexure in *Tyrannosaurus*). The NASTRAN models of *Plateosaurus* show that the combination of short forelimbs and a limited range of motion in the forelimb joints makes a quadrupedal posture unlikely. Restrictions on manus pronation increases this problem because it would necessitate stronger abduction of the humeri in order for the palm to face the ground, further decreasing ELL. Considering the varying adaptations of the forearm and manus to pronation (and therefore locomotion) in other basal sauropodomorphs (Bonnan & Senter 2007; Bonnan & Yates 2007; Remes 2008; Rauhut et al., this volume), as well as evidence for only facultative bipedality in basal saurischians and early sauropodomorphs (Langer 2004, 2007; Fechner 2006; Remes 2006, 2008), it becomes evident that obligatory bipedality is derived in *Plateosaurus*.

CAE modeling also indicates that lateral stability is an important issue in early sauropodomorph locomotion (Mallison, unpublished data). Herbivores, especially large herbivores, have massive guts. This means that the thighs must be abducted appreciably to clear the body at large protraction angles. Such a posture leads either to a wide-gauge trackway, which is at odds with the ichnological record for plateosaurs, or somehow the foot must be brought close to the center line of the track even if the knee is placed far laterally from the body midline. This can be achieved in two ways. The limb can be placed in a zigzag posture in anterior view, or the body must move laterally or by rotation around the vertical and/or longitudinal axes through the hips during each stride. For the first solution, the outward–inward canting of the limb, there is no supporting evidence in the skeleton of *Plateosaurus* or any other prosauropod. The distal end of the hu-

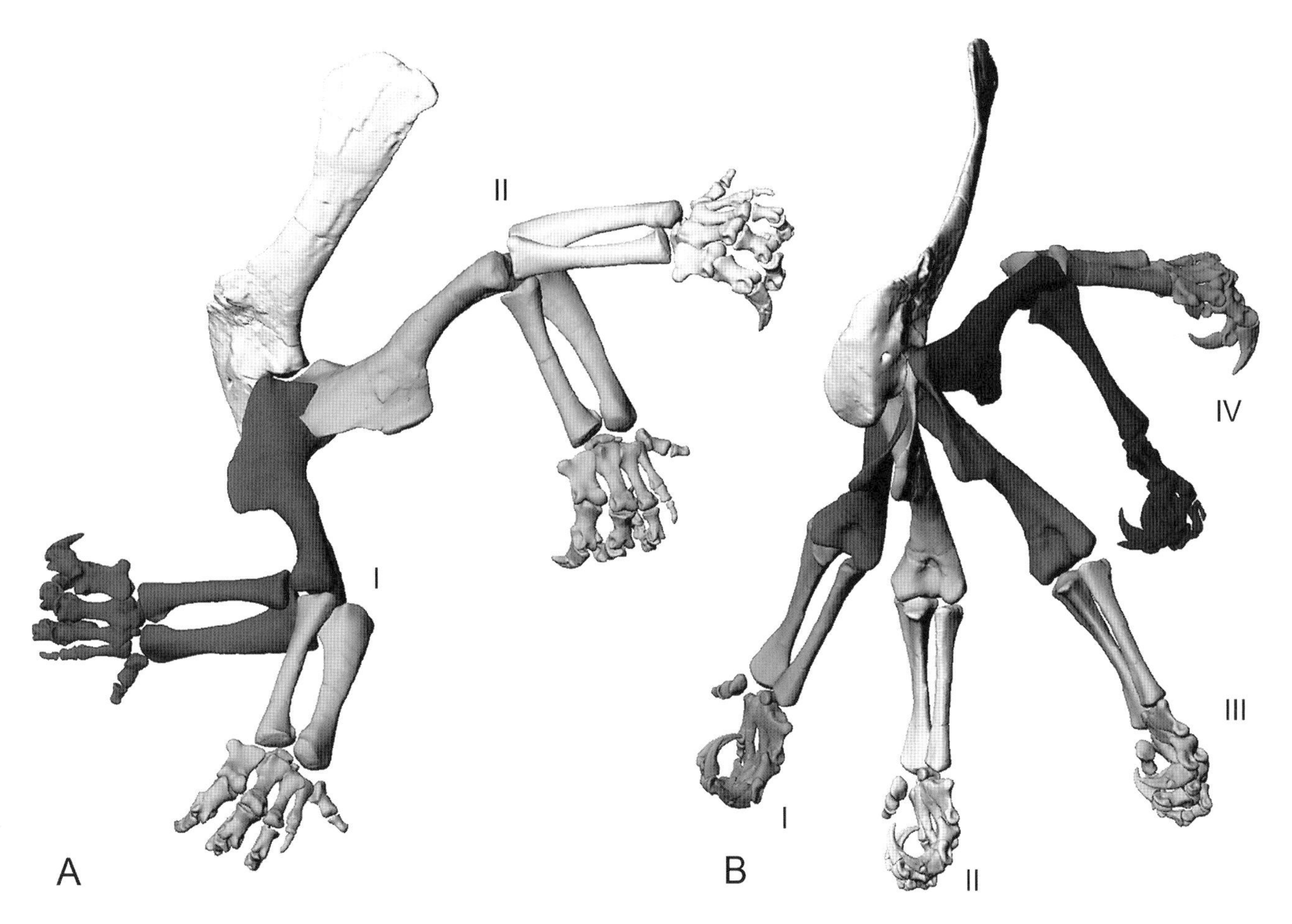

FIGURE 13.9. Virtual forelimb skeleton of *Plateosaurus engelhardti* (GPIT 1). (A) Lateral and (B) anterior view of the left scapulocoracoid and forelimb. Multiple poses of the limb show the maximum range of motion in 3D. (A) (I) Maximum protraction and (II) maximum retraction of the humerus, each with maximally extended and strongly flexed elbow. (B) (I) maximally adducted, (II) neutral, and (III) maximally abducted humerus at full protraction, corresponding to (A) (I), with maximally extended elbow. (IV) Maximally abducted humerus at maximum retraction, corresponding to (A) (II), with extended and strongly flexed elbow. Note the slightly flexed position of the hand.

merus is not markedly slanted inward, nor is the proximal end of the tibia beveled outward. On the contrary, the line across the distal condyles of the femur is tilted outward against the long axis of the femur. This means that it is horizontal, resulting in an even distribution of compressive forces across the knee joint, when the femur is strongly adducted, which places the foot under the body midline. However, the largest lateral forces on the knee would result when the limb is markedly abducted to clear the belly. The slight inward tilt makes the knee joint especially vulnerable to lateral displacement in such a position, and therefore makes a zigzag limb posture in anterior view improbable. Similarly, the ankle joint forms an almost perfect right angle with the long axis of the tibia, indicating that the lower leg was perfectly vertical when large compressive forces acted on it (i.e., during midstance). Therefore, it appears likely that motions of the body ensured that the COM passed over the supporting foot, or close enough to it that the inertia of the body was sufficient for energy-efficient forward travel. Strong lateral swaying is not an effective forward motion, as demonstrated by a cyclist riding too slowly, expending much energy moving from side to side instead of forward. Although some swaying certainly occurred, the broad pubes of *Plateosaurus* require abduction angles of >25°, widening the gauge to approximately 68 cm in the model. Strong adduction at high retraction angles can ameliorate this somewhat, but there is a practical limit imposed by the ischia and the caudofemoralis musculature. Therefore, rotation of the hips around both the vertical and longitudinal axes must have been involved in fast gaits of *Plateosaurus*. However, such rotation leads to large side-to-side swings of the anterior body and tail unless it is countered by undulating motions of the body and tail base. The vertebral column of *Plateosaurus* certainly allowed such motions (Mallison 2007, unpublished data).

Summary and Future Research

Virtual skeletons can be of tremendous help in paleontological research. The example of *Plateosaurus engelhardti* GPIT 1 shows how a virtual skeleton can be used to assess the posture and range of motion of an extinct animal, and how it can be used as a basis for detailed assessments of the body mass and mass distribution. Kinetic–dynamic modeling in a NASTRAN-based program has proven its usefulness for calculating the position of the COM of a moving model and for assessing limb joint forces during complex motions, with inertia and gravity taken into account.

CAD modeling suggests that *Plateosaurus* was an obligatory biped because it was not able to pronate its hands. This conclusion is supported by the results of the kinetic–dynamic modeling: the animal would have been stable on its hind feet while moving the long neck and tail, and bipedal progression required the least energy while minimizing joint torques. CAD modeling also shows that the forelimb of *Plateosaurus* possessed a grasping hand, but that the range of forelimb motion was rather limited.

Future models of sauropodomorphs should be created in greater detail to assess more complex combinations of movements. For example, models will have to allow rotation around three axes in all limb and axial joints. Such models should be used to clarify the exact limb segment motion sequences required for locomotion at various speeds, and to test how scaling influences them. Additionally, more models of both extinct and extant animals should be created to allow interspecific comparisons that can be cross-checked by measurements on living animals.

Acknowledgments

This project was supported by the Deutsche Forschungsgemeinschaft as part of Project PF-219/20 in the DFG Research Unit 533. Hans-Ulrich Pfretzschner (University of Tübingen) supervised my research on *Plateosaurus* biomechanics. The late Wolf-Ernst Reif (University of Tübingen) aided me during the final phases of my work. Helpful hints, ideas, and pointed questions that guided me in the right direction came from many collegues, for which I am very grateful. In particular, I would like to acknowledge the fruitful exchange of ideas with the following members of Research Unit 533: Holger Preuschoft, Ulrich Witzel, Bianca Hohn (all Ruhr-University of Bochum), Marcus Clauss (University of Zurich), Jürgen Hummel, Steven Perry, Martin Sander, Kristian Remes (all University of Bonn), and Oliver Rauhut (Bavarian State Collection for Paleontology and Geology, Munich). This chapter has greatly profited from extensive reviews by Ray Wilhite (Auburn University) and Mathew Bonnan (Western Illinois University, Macomb), as well as Martin Sander (University of Bonn). This is contribution number 72 of the DFG Research Unit 533 "Biology of the Sauropod Dinosaurs: The Evolution of Gigantism."

References

Alexander, R. M. 1989. Sequential joint extension in jumping.—*Human Movement Science* 8: 339–345.

Alexander, R. M. 2003a. *Principles of Animal Locomotion*. Princeton University Press, Princeton.

Alexander, R. M. 2003b. Modelling approaches in biomechanics.—*Philosophical Transactions of the Royal Society B: Biological Sciences* 358: 1429–1435.

Bates, K. T., Manning, P. L., Hodgetts, D. & Seller, W. I. 2009. Estimating mass properties of dinosaurs using laser imaging and 3D computer modeling.—*PloS ONE* 4: e4532.

Biewener, A. A. 1983. Allometry of quadrupedal locomotion: the scaling of duty factor, bone curvature and limb orientation to body size.—*Journal of Experimental Biology* 105: 147–171.

Bonnan, M. F. & Senter, P. 2007. Were the basal sauropodomorph dinosaurs *Plateosaurus* and *Massospondylus* habitual quadrupeds?—*Special Papers in Palaeontology* 77: 139–155.

Bonnan, M. F. & Yates, A. M. 2007. A new description of the forelimb of the basal sauropodomorph *Melanorosaurus:* implications for the evolution of pronation, manus shape and quadrupedalism in sauropod dinosaurs.—*Special Papers in Palaeontology* 77: 157–168.

Bonnan, M. F., Sandrik, J., Nishiwaki, T., Wilhite, D. R., Elsey, R. 2009. Calcified cartilage shape in extant archosaur long bones reflects overlying joint shape in load-bearing elements: implications for inferring dinosaur joint shape.—*Journal of Vertebrate Paleontology* 29 (3 Suppl.): 67A–68A.

Bonnan, M. F., Sandrik, J. L., Nishiwaki, T., Whilhite, D. R., Elsey, R. M. & Vittore, C. 2010. Calcified cartilage shape in archosaur long bones reflects overlying joint shape in stress-bearing elements: implications for nonavian dinosaur locomotion.—*The Anatomical Record*. doi: 10.1002/ar.21266.

Carpenter, K., Sanders, F., McWhinney, L. A. & Wood, L. 2005. Evidence for predator-prey relationships. Examples for *Allosaurus* and *Stegosaurus*. *In* Carpenter, K. (ed.). *The Carnivorous Dinosaurs*. Indiana University Press, Bloomington: pp. 325–350.

Christian, A. & Preuschoft, H. 1996. Deducing the body posture of extinct large vertebrates from the shape of the vertebral column.—*Palaeontology* 39: 801–812.

Christian, A., Koberg, D. & Preuschoft, H. 1996. Shape of the pelvis and posture of the hind limbs in *Plateosaurus*.—*Paläontologische Zeitschrift* 70: 591–601.

Colbert, E. H. 1962. The weights of dinosaurs.—*American Museum Novitates* 2076: 1–16.

Delp, S. L. & Loan, J. P. 1995. A graphics-based software system to develop and analyze models of musculoskeletal structures.—*Computers in Biology and Medicine* 25: 21–34

Delp, S. L. & Loan, J. P. 2000. A computational framework for simulating and analyzing human and animal movement.—*Computing in Science and Engineering* 2: 46–55

Fechner, R. 2006. Evolution of bipedality in dinosaurs.—*Journal of Vertebrate Paleontology* 26: 60A.

Fraas, E. 1912. Die schwäbischen Dinosaurier.—*Jahreshefte des Vereins für vaterländische Naturkunde in Württemberg* 68: 66–67.

Fraas, E. 1913. Die neuesten Dinosaurierfunde in der schwäbischen Trias.—*Naturwissenschaften* 45: 1097–1100.

Fox, L. 1962. *Numerical Solution of Ordinary and Partial Differential Equations*. Addison-Wesley, Munich.

Galton, P. M. 1971a. The prosauropod *Ammosaurus,* the crocodile *Protosuchus,* and their bearing on the age of the Navajo Sandstone of northeastern Arizona.—*Journal of Paleontology* 45: 781–795.

Galton, P. M. 1971b. Manus movements of the coelurosaurian dinosaur *Syntarsus rhodesiensis* and the opposability of the theropod hallux.—*Arnoldia* 5: 1–8.

Galton, P. M. 1976. Prosauropod dinosaurs of North America.—*Postilla* 169: 1–98.

Galton, P. M. 1985. Cranial anatomy of the prosauropod dinosaur *Plateosaurus* from the Knollenmergel (Middle Keuper, Upper Triassic) of Germany. II. All the cranial material and details of soft-part anatomy.—*Geologica et Palaeontologica* 19: 119–159.

Galton, P. M. 1990. Basal Sauropodomorpha-Prosauropoda. *In* Weishampel, D. B., Dodson, P. & Osmólska, H. (eds.). *The Dinosauria*. University of California Press, Berkeley: pp. 320–344.

Gatesy, S. M. 1990. Caudofemoral musculature and the evolution of theropod locomotion.—*Paleobiology* 16: 170–186.

Gatesy, S. M., Bäker, M. & Hutchinson, J. R. 2009. Constraint-based exclusion of limb poses for reconstructing theropod dinosaur locomotion.—*Journal of Vertebrate Paleontology* 29: 535–544.

Graf, J., Stofft, E., Freese, U. & Niethard, F. U. 1993. The ultrastructure of articular cartilage of the chicken's knee joint.—*International Orthopedics* 17: 113–119.

Gunga, H.-C., Suthau, T., Bellmann, A., Friedrich, A., Schwanebeck, T., Stoinski, S., Trippel, T., Kirsch, K. & Hellwich, O. 2007. Body mass estimations for *Plateosaurus engelhardti* using laser scanning and 3D reconstruction methods.—*Naturwissenschaften* 94: 623–630.

Gunga, H.-C., Suthau, T., Bellmann, A, Stoinski, S., Friedrich, A., Trippel, T., Kirsch, K. & Hellwich, O. 2008. A new body mass estimation of *Brachiosaurus brancai* Janensch 1914 mounted and exhibited at the Museum of Natural History (Berlin, Germany).—*Fossil Record* 11: 28–33.

Henderson, D. M. 1999. Estimating the masses and centers of mass of extinct animals by 3-D mathematical slicing.—*Paleobiology* 25: 88–106.

Henderson, D. M. 2006. Burly gaits: centers of mass, stability and the trackways of sauropod dinosaurs.—*Journal of Vertebrate Paleontology* 26: 907–921.

Hohn, B. This volume. Walking with the shoulder of giants: biomechanical conditions in the tetrapod shoulder girdle as a basis for sauropod shoulder reconstruction. *In* Klein, N., Remes, K., Gee, C. T. & Sander, P. M. (eds.). *Biology of the Sauropod Dinosaurs: Understanding the Life of Giants*. Indiana University Press, Bloomington: pp. 182–196.

Holliday, C. M., Ridgley, R. C., Sedlmayr, J. C. & Witmer, L. M. 2001. The articular cartilage of extant archosaur long bones: implications for dinosaur functional morphology and allometry.—*Journal of Vertebrate Paleontology* 21: 62A.

Holliday, C. M., Ridgley, R. C., Sedlmayr, J. C. & Witmer, L. M. 2010. Cartilaginous epiphyses in extant archosaurs and their implications for reconstructing limb function in dinosaurs.—*PLoS ONE* 5(9): e13120.

Hutchinson, J. R., Anderson, F. C., Blemker, S. & Delp, S. L. 2005. Analysis of hindlimb muscle moment arms in *Tyrannosaurus rex* using a three-dimensional musculoskeletal computer model: implications for stance, gait, and speed.—*Paleobiology* 31: 676–701.

Jaekel, O. 1910. Die Fussstellung und Lebensweise der grossen Dinosaurier.—*Zeitschrift der Deutschen Geologischen Gesellschaft, Monatsberichte* 62: 270–277.

Jaekel, O. 1911. *Die Wirbeltiere. Eine Übersicht über die fossilen und lebenden Formen*. Borntraeger, Berlin.

Langer, M. C. 2004. Basal Saurischia. *In* Weishampel, D. B., Dodson, P. & Osmólska, H. (eds.). *The Dinosauria. 2nd edition*. University of California Press, Berkeley: pp. 25–46.

Langer, M. C. 2007. The pectoral girdle and forelimb anatomy of the stem-sauropodomorph *Saturnalia tupiniquim* (Upper Triassic, Brazil).—*Special Papers in Palaeontology* 77: 113–137.

Longrich, N. R. & Currie, P. J. 2008. *Albertonykus borealis,* a new alvarezsaur (Dinosauria: Theropoda) from the early Maastrichtian of Alberta, Canada: implications for the systematics and ecology of the Alvarezsauridae.—*Cretaceous Research* 30: 239–252.

Mallison, H. 2007. *Virtual Dinosaurs—Developing Computer Aided Design and Computer Aided Engineering Modeling Methods for Vertebrate Paleontology*. Ph.D. Dissertation. Eberhardt-Karls-Universität, Tübingen.

Mallison, H. 2010a. The digital *Plateosaurus* I: body mass, mass distribution and posture assessed using CAD and CAE on a digitally mounted complete skeleton.—*Palaeonotologia Electronica* 13: 1–26.

Mallison, H. 2010b. The digital *Plateosaurus* II: an assessment of the range of motion of the limbs and vertebral column and of previous reconstructions using a digital skeletal mount.—*Acta Palaeontologica Polonica* 55: 433–458.

Mallison, H. 2010c. CAD assessment of the posture and range of motion of *Kentrosaurus aethiopicus* Hennig 1915.—*Swiss Journal of Geosciences* 103: 211–233.

Mallison, H. & Bachmann, E. 2006. Kinematical modeling in MSC.visualNASTRAN proves bipedality of *Plateosaurus*.—*Journal of Vertebrate Paleontology* 26: 94A.

Mallison, H. & Pfretzschner, H.-U. 2005. Walking with sauropods: modeling dinosaur locomotion in MSC.visualNASTRAN 4D.—*Journal of Vertebrate Paleontology* 25: 88A.

Mallison, H., Hohloch, A. & Pfretzschner, H.-U. 2009. Mechanical digitizing for paleontology—new and improved techniques.—*Palaeontologica Electronica* 12: 1–41.

Mallison, H. This volume. Rearing giants: kinetic-dynamic modeling of sauropod bipedal and tripodal poses. *In* Klein, N., Remes, K., Gee, C. T. & Sander, P. M. (eds.). *Biology of the Sauropod Dinosaurs: Understanding the Life of Giants*. Indiana University Press, Bloomington: pp. 237–250.

Marsh, O. C. 1893a. Restoration of *Anchisaurus*.—*American Journal of Science* 45: 169–170.

Marsh, O. C. 1893b. Restoration of *Anchisaurus, Ceratosaurus,* and *Claosaurus.—Geological Magazine* 10: 150–157.

Mazzetta, G. V., Christiansen, P. & Fariña, R. A. 2004. Giants and bizarres: body size of some southern South American Cretaceous dinosaurs.—*Historical Biology* 16: 71–83.

Meyer, H. v. 1837. Mitteilungen, an Professor Bronn gerichtet (*Plateosaurus engelhardti*).—*Neues Jahrbuch für Mineralogie, Geologie und Paläontologie* 1837: 817.

Moser, M. 2003. *Plateosaurus engelhardti* Meyer, 1837 (Dinosauria, Sauropodomorpha) aus dem Feuerletten (Mittelkeuper; Obertrias) von Bayern.—*Zitteliana B* 24: 1–186.

Pandy, M. G., Zajac, F. E., Sim, E. & Levine, W. S. 1990. An optimal control model for maximum-height human jumping.—*Journal of Biomechanics* 23: 1185–1198

Paul, G. S. 1987. The science and art of restoring the life appearance of dinosaurs and their relatives: a rigorous how-to guide. *In* Czerkas, S. J. & Olson, E. C. (eds.): *Dinosaurs Past and Present, Vol. II.* University of Washington Press, Seattle: pp. 5–49.

Paul, G. S. 1997. Dinosaur models: the good, the bad, and using them to estimate the mass of dinosaurs. *In* Wolberg, D. L., Stump, E. & Rosenberg, G. D. (eds.). *Dinofest International.* Academy of Natural Sciences, Philadelphia: pp. 129–154.

Paul, G. S. 2000. Restoring the life appearance of dinosaurs. *In* Paul, G. S. (ed.). *The Scientific American Book of Dinosaurs.* Byron Press and Scientific American, New York: pp. 78–106.

Perry, S. F., Breuer, T. & Pajor, N. This volume. Structure and function of the sauropod respiratory system. *In* Klein, N., Remes, K., Gee, C. T. & Sander, P. M. (eds.). *Biology of the Sauropod Dinosaurs: Understanding the Life of Giants.* Indiana University Press, Bloomington: pp. 57–79.

Rauhut, O. W. M., Fechner, R., Remes, K. & Reis, K. This volume. How to get big in the Mesozoic: the evolution of the sauropodomorph body plan. *In* Klein, N., Remes, K., Gee, C. T. & Sander, P. M. (eds.). *Biology of the Sauropod Dinosaurs: Understanding the Life of Giants.* Indiana University Press, Bloomington: pp. 119–149.

Remes, K. 2006. Evolution of forelimb functional morphology in sauropodomorph dinosaurs.—*Journal of Vertebrate Paleontology* 26: 115A.

Remes, K. 2008. *Evolution of the Pectoral Girdle and Forelimb in Sauropodomorpha (Dinosauria, Saurischia): Osteology, Myology, and Function.* Ph.D. Dissertation. Ludwig-Maximilians-Universität, München.

Russell, D. A., Beland, P. & McIntosh, J. S. 1980. Paleoecology of the dinosaurs of Tendaguru.—*Memoirs de la Societe Geologique de France* 139: 169–175.

Sander, P. M. 1992. The Norian *Plateosaurus* bonebeds of central Europe and their taphonomy.—*Palaeogeography, Palaeoclimatology, Palaeoecology* 93: 255–299.

Schwarz, D., Wings, O. & Meyer, C. A. 2007a. Super sizing the giants: first cartilage preservation at a sauropod limb joint.—*Journal of the Geological Society* 164: 61–65.

Schwarz, D., Ikejiri, T., Breithaupt, B. H., Sander, P. M. & Klein, N. 2007b. A nearly complete skeleton of an early juvenile diplodocid (Dinosauria: Sauropoda) from the Lower Morrison Formation (Late Jurassic) of north central Wyoming and its implications for early ontogeny and pneumaticity in sauropods.—*Historical Biology* 19: 225–253.

Schwarz, D., Frey, E. & Meyer, C. A. 2007c. Novel reconstruction of the orientation of the pectoral girdle in sauropods.—*Anatomical Record* 290: 32–47.

Seebacher, F. 2001. A new method to calculate allometric length-mass relationships of dinosaurs.—*Journal of Vertebrate Paleontology* 21: 51–60.

Senter, P. 2006. Comparison of forelimb function between *Deinonychus* and *Bambiraptor* (Theropoda: Dromaeosauridae).—*Journal of Vertebrate Paleontology* 26: 897–906.

Senter, P. & Robins, J. H. 2005. Range of motion in the forelimb of the theropod dinosaur *Acrocanthosaurus atokensis,* and implications for predatory behaviour.—*Journal of Zoology* 266: 307–318.

Stevens, K. A. & Parrish, J. M. 1999. Neck posture and feeding habits of two Jurassic sauropod dinosaurs.—*Science* 284: 798–800.

Stevens, K. A. & Parrish, J. M. 2005a. Digital reconstructions of sauropod dinosaurs and implications for feeding. *In* Curry Rogers, K. A. & Wilson, J. A. (eds.). *The Sauropods: Evolution and Paleobiology.* University of California Press, Berkeley: pp. 178–200.

Stevens, K. A. & Parrish, J. M. 2005b. Neck posture, dentition, and feeding strategies in Jurassic sauropod dinosaurs. *In* Carpenter, K. & Tidwell, V. (eds.). *Thunder-Lizards: The Sauropodomorph Dinosaurs.* Indiana University Press, Bloomington: pp. 212–232.

Stoinski, S., Suthau, T. & Gunga, H.-C. This volume. Reconstructing body volume and surface area of dinosaurs using laser scanning and photogrammetry. *In* Klein, N., Remes, K., Gee, C. T. & Sander, P. M. (eds.). *Biology of the Sauropod Dinosaurs: Understanding the Life of Giants.* Indiana University Press, Bloomington: pp. 94–104.

Sullivan, C., Jenkins, F. A., Gatesy, S. M. & Shubin, N. H. 2003. A functional assessment of the foot structure in the prosauropod dinosaur *Plateosaurus.—Journal of Vertebrate Paleontology* 23: 102A.

Taylor, M. P., Wedel, M. J. & Naish, D. 2009. Head and neck posture in sauropod dinosaurs inferred from extant animals.—*Acta Palaeontologica Polonica* 54: 213–220.

von Huene, F. 1908. Die Dinosaurier der europäischen Triasformation mit Berücksichtigung der aussereuropäischen Vorkommnisse.—*Geologische und Paläontologische Abhandlungen* 1: 1–419.

von Huene, F. 1926. Vollständige Osteologie eines Plateosauriden aus dem schwäbischen Keuper.—*Geologische und Palaeontologische Abhandlungen, Neue Folge* 15: 139–179.

Weishampel, D. B. & Westphal, F., 1986. *Die Plateosaurier von Trossingen.* Attempto, Tübingen.

Wellnhofer, P. 1994. Prosauropod dinosaurs from the Feuerletten (middle Norian) of Ellingen near Weissenburg in Bavaria.—*Revue de Paleobiologie, Volume Spéciale* 7: 263–271.

Wilhite, R. 2003. Digitizing large fossil skeletal elements for three-dimensional applications.—*Palaeontologia Electronica* 5: 1–10.

Yates, A. M. 2003. The species taxonomy of the sauropodomorph dinosaurs from the Löwenstein Formation (Norian, Late Triassic) of Germany.—*Palaeontology* 46: 317–337.

Yates, A. M. & Vasconcelos, C. C. 2005. Furcula-like clavicles in the prosauropod dinosaur *Massospondylus.—Journal of Vertebrate Paleontology* 25: 466–468.

14

Rearing Giants: Kinetic–Dynamic Modeling of Sauropod Bipedal and Tripodal Poses

HEINRICH MALLISON

BECAUSE OF THEIR LARGE BODY MASSES, sauropod dinosaurs must have required enormous amounts of plant matter to support their metabolism, even if one assumes a much lower metabolic rate in adults than in extant mammals and birds. Therefore, their methods of food acquisition are of interest, specifically how they procured a sufficient volume of food without expending unlikely large amounts of energy during feeding. Some, if not all, sauropods supposedly could rear up onto their hindlimbs to access food at heights beyond the reach of other herbivores, increasing their feeding envelopes without requiring energetically more costly locomotion. Kinetic–dynamic modeling in comparison with elephants indicates that at least diplodocids could rear easily and for prolonged times without significant exertion, while brachiosaurids were probably not capable of extended upright feeding. Modeling results also suggest that optimizing body shape for rearing by a posterior shift of the center of mass may be detrimental to locomotory abilities.

Introduction

Sauropod dinosaurs were the largest terrestrial animals of all time and were the dominant terrestrial herbivores for most of the Mesozoic (McIntosh 1990; McIntosh et al. 1997; Upchurch et al. 2004; Sander et al. 2010). Aside from the question, "How did they get so big?," it is also important to ask, "How did they maintain their bulk? How did they acquire sufficient energy?" Thus, we need to investigate how much energy sauropods required—how their metabolism functioned, what and how much fodder they required, how they fed and digested. Other chapters of this volume address some of these issues, such as metabolic rate and adaptations of the digestive tract. A special aspect of food acquisition, the question of whether sauropods could feed in a bipedal pose, shall be addressed here.

A simple way to increase feeding height is to rear onto the hindlimbs. Hatcher (1901) was the first to suggest this, pointing out that rearing would have granted sauropods an enormous height range for feeding. This notion has also been reiterated for defense and copulation, not only in the literature (e.g., Borsuk-Bialynicka 1977; Bakker 1978, 1986; Alexander 1985; Jensen 1988; Dodson 1990), but also in films and TV programs. In the entrance hall of the American Museum of Natural History (New York City, USA), a cast of *Barosaurus* was mounted in a rearing pose, defending its young against a predator attack. In contrast to copulation, and possibly defense, feeding in a bipedal or tripodal posture requires maintaining a near-upright position for a prolonged time without great exertion. The term *tripodal* in this regard means that the animal stands on its hindlimbs and uses its tail as an additional support, which has been suggested by Bakker (1986) for sauropods and stegosaurs. Upchurch & Barrett (2000) noted that the high-browsing (Bakker 1986) and low-browsing (Barrett & Upchurch 1994) models of *Diplodocus* rearing to increase feeding height, are not mutually exclusive. Rather, such variation in vertical feeding range is consistent with higher food selectivity and the specialized branch-stripping mode proposed for diplodocids (Barrett & Upchurch 1994; Upchurch & Barrett 2000).

Brachiosaurus, in contrast, with its adaptations to cutting through foliage instead of branch stripping, was assumed to have fed at heights not reached by other sauropods by using its near-vertical neck (Upchurch & Barrett 2000). Stevens & Parrish (2005a, 2005b) disagreed, and they reconstructed *Brachiosaurus* with the neck held at less than 45° to the horizontal. This means that browsing at heights of 10 to 13 m would only have been possible through rearing. However, Christian (2010), Christian & Dzemski (2007, this volume), and Sander et al. (2009, 2010) convincingly argued for higher mobility in the neck of sauropods, so *Brachiosaurus* was probably not only limited to low browsing.

Alexander (1985) calculated the position of the center of mass (COM) for *Diplodocus,* arguing that its extremely posterior position made rearing possible, but did not attempt to assess the forces and stresses involved. Henderson (2006) calculated the position of the COM for a variety of sauropods. Of these, *Diplodocus* has the most posteriorly placed COM (Henderson 2006, fig. 11). Although all sauropods could certainly rear to some degree and for a short time, as is necessary for most imaginable modes of copulation, it is questionable whether all could hold such a position without support. Some, however, could have possibly attained an upright position so easily that one must ask whether this position was frequently adopted, and for extended times. Although rearing for a short time may have aided in defense, the only realistic use of an upright position that is held for a long time is high browsing, as suggested by Bakker (1986). Moving the neck was comparatively cheap, despite the long lever arm, as it was lightly built (Wedel 2003, 2005; Sander et al. 2010) and supported by ligaments (Schwarz et al. 2007). Moving the entire animal probably took more energy (Preuschoft et al., this volume; Christian & Dzemski, this volume). This, however, is also true in low browsers with a long neck, which can cover a larger area while walking slowly or standing still than animals with a short neck.

Sauropods inherited a relatively long neck from early dinosaurs (Rauhut et al., this volume). In early dinosaurs, such a neck would have allowed the animals to lower their heads to the ground for drinking or feeding while maintaining an upright body posture ready to flee. However, most sauropod groups developed markedly longer necks (Parrish 2006; Rauhut et al., this volume). Did the long neck simply serve to extend the low browsing range and thus conserve energy, as suggested by, for example, Martin (1987), Barrett & Upchurch (1994), Upchurch & Barrett (2000), Stevens & Parrish (2005a), Sander et al. (2010) and Preuschoft et al. (this volume)? Did it serve to allow mid- to high-level browsing using dorsiflexion (e.g., Riggs 1904; Bakker 1971)? Or did it combine with rearing to open up a new food source—the high-up branches of large trees (e.g., Bakker 1986)? Some studies suggest vertical partitioning of feeding among neosauropods (Barrett & Upchurch 1994; Christiansen 2000), while others find an extensive overlap between the feeding envelopes (Stevens & Parrish 1999, 2005b).

In one respect, upright browsing on high-up vegetation is fundamentally different from low browsing: high browsing is limited to trees and tree-like plants (tree ferns etc.), while low browsing gives access to a wide variety of food plants, including herbaceous plants, shrubs, and small or young trees. A high browser that adopts a position making locomotion impossible (e.g., a rearing sauropod) must have extreme mobility in the neck, as it must be able to move the head in a maze of branches in order to browse a considerable volume. This requires the ability to bend consecutive segments of the neck into opposite directions, the way swans do today. In fact, it probably requires greater mobility than the necks of giraffes, which rarely push their heads deep between strong branches. Ground or midlevel browsing does not require such a flexible neck. It is sufficient to sweep the neck laterally to cover a large area and occasionally lift it over an obstacle, without having to expend energy for moving the body. It appears logical that a mobile neck weighs more than a stiff neck of the same length because it cannot be supported by lightweight ligaments and possibly long cervical ribs to the same degree a less mobile neck can. Rather, it must be held up by the action of muscles. In return, the high browser gains access to a greater food volume in three dimensions, while a low browser can harvest only from an (almost) 2D surface. Therefore, we should see greater neck mobility in sauropods with rearing abilities than in those that remained on all fours.

Indeed, there is a strong indicator that diplodocids had a more mobile neck, especially in the lateral direction: their cervical ribs do not extend across intervertebral joints. Those of brachiosaurids offer a marked contrast. Although they were reconstructed to be quite short by Janensch (1950), and even slightly shorter by Paul (1988), those reconstructions were based on damaged Tendaguru material. In fact, they were probably much longer, overlapping the succeeding two cervicals (Upchurch et al. 2004). This reduced mobility while adding a means of supporting the neck without adding much weight (Schwarz et al. 2007). Short cervical ribs are also seen in dicraeosaurids and in *Shunosaurus,* but not in the other major sauropod lineages (Upchurch et al. 2004).

It is important to remember that high browsing by dorsiflexion of the neck may have been limited in most sauropods, even in the laterally mobile neck of *Diplodocus.* The exception may be *Apatosaurus,* which was able to reach feeding heights similar to the mean head height of *Brachiosaurus* (Stevens & Parrish 1999, 2005a). Therefore, rearing may have been an important option for reaching new food sources, especially for those groups that had a subhorizontal neck. Even if the values for dorsiflexion in Stevens & Parrish (1999, 2005a, 2005b) are too conservative, rearing can add nearly the entire body length to feeding height, in addition to the elevation gained from a straight instead of a curved neck position.

Aside from the COM position, other factors could theoretically influence rearing ability. In the saltasaurid *Opisthocoelicaudia,* Borsuk-Bialynicka (1977) described several features that appear related to a bipedal posture. The strongly opisthocoelous centra of the anterior half of the tail are interpreted as an indication that the tail was used as support in rearing, but no biomechanical explanation was given. Additionally, the large range of motion in the acetabulum (nearly 180°), and the ability to extend the anterior tail are cited as adaptations to the regular use of a bipedal pose. Borsuk-Bialynicka (1977) additionally cited the extremely oblique position of the first five chevrons as indicative of rearing, but she admitted that this position may also have been required for laying eggs. Also, she noted that maximum extension seems possible at just the right distance from the hips so that the anterior part of the tail could have served as a prop and the posterior part could have rested on the ground (Borsuk-Bialynicka 1977).

Opisthocoelicaudia may have been able to produce more of a kink between the anterior and posterior tail sections, and thus keep the anterior tail less curved in a tripodal pose than would be required in other sauropods. However, the interpretation of the range of motion of sauropod tails, and especially of the femur in the acetabulum, is difficult, and even tails that seem much stiffer have sufficient mobility to allow a tripodal pose. Borsuk-Bialynicka (1977) further cited the fused pubic symphysis and the flaring ilia as potentially connected to rearing because they may have provided support to the viscera. However, a fused symphysis is not necessary for supporting the viscera. Wilson & Carrano (1999) interpreted a set of skeletal features in titanosauriforms as progressive adaptations to a wide-gauge posture, including the medial deflection of the proximal femur, increased femoral midshaft eccentricity, and beveled distal femoral condyles. Flared ilia, which are present in *Brachiosaurus* as well, may be linked to locomotion, not rearing.

One biomechanical study (Rothschild & Molnar 2005) claimed that sauropods could not have regularly used a tripodal pose. Among other evidence, Rothschild & Molnar (2005) compared the descent from a tripodal pose to the en pointe landing of ballet dancers. Ballet dancers are prone to repeated stress fractures in their feet from this action. Because such stress fractures were not observed in any sauropod manus bones, Rothschild & Molnar (2005) concluded that sauropods did not perform the associated motion, thus arguing against rearing.

To test the hypothesis that at least some sauropods could rear into a bipedal or tripodal pose, NASTRAN (NASA Structural Analysis System)-based CAE (computer-aided engineering) modeling was applied to the problem.

MATERIAL AND METHODS

The 3D models required for NASTRAN were all created in McNeel Associates Rhinoceros 4.0 NURBS Modeling for Windows and analyzed in MSC.visualNASTRAN 4D by MSC software. For a description of NASTRAN modeling, see Mallison & Pfretzschner (2005) and Mallison (2007, 2010, Chapter 13 in this volume).

The two extremes in body plan, *Diplodocus* and *Brachiosaurus,* are best suited for a comparative analysis of rearing ability. The required 3D models were created on the basis of 2D skeletal drawings by Scott Hartmann (http://www.skeletaldrawing.com/) for *Diplodocus carnegii,* and 2D skeletal drawings in lateral view and cross section at the pelvic girdle and pectoral girdle by Paul (2000) for *Brachiosaurus*. To reduce the risk of error inherent in drawings, many photographs of the new mounts of *Diplodocus carnegii* and *Brachiosaurus brancai* at the Berlin Museum für Naturkunde were taken and used to improve the models. The amount of soft tissue in the models is mostly congruent with the body outlines in the drawings, with *Brachiosaurus* slim and *Diplodocus* sturdier. The tail of *Diplodocus,* however, was further thickened because dinosaurs probably possessed much more musculature in the upper hindlimbs and tail than most previous reconstructions assumed (National Geographic Society 2007). Because the drawings of *Brachiosaurus* in Paul (2000) show a very slim animal, model variants differing in mass and mass distribution were created to account for a sturdier reconstruction (see below).

Modeling results should be validated by comparison to extant animals. For this purpose, a 3D model of an African elephant (*Loxodonta africana*) was created on the basis of the 2D drawing and photograph in Goldfinger (2004), aided by numerous photographs of living elephants in front and rear view taken by the author.

NASTRAN Modeling of Rearing

BIPEDAL/TRIPODAL MODEL POSITIONS

NASTRAN-based modeling of extinct vertebrates is described in detail in Mallison (2007), and a short introduction can be found in Chapter 13 of this volume. Because the ability to assume a bipedal posture is not the focus of this investigation on prolonged bipedal feeding, most modeling runs used a near–static approach, in which the animal was already placed in an upright pose, and the limb joint torques (hip, knee, ankle) required to maintain it were measured.

In addition, the stability of the upright pose was assessed, both while standing still and during rapid neck movements. Also, the forces required for returning the models to a quadrupedal pose were assessed, as well as the impact forces acting on the forelimbs upon touching the ground when returning from the bipedal pose.

Diplodocus

A motion sequence developed for *Diplodocus* illustrates the transition from a quadrupedal to a bipedal pose. This sequence is generally similar to the motion performed by a rearing elephant, but it differs in one important aspect: elephants push themselves up by flexing their forelimbs, then extending them rapidly. For *Diplodocus,* this push action was omitted, so that all power is created by the hindlimbs. This simplification of the motion process is justified because it avoids potentially overestimating the strength of the forelimbs, and it increases the ability to compare results with those from other models. Figure 14.1 shows a series of screenshots of this sequence, in which the tail provides only limited support during the motion and carries only sufficient weight in the final upright stance to stabilize the pose against toppling when the neck is moved. This position was chosen to allow a large, but not maximal, height gain at a minimal risk of loss of balance: the pose places the COM above the toes of both feet, and bends the knees a little. Also, the hindlimbs are not fully retracted, but only to 30°. As viewed in the *Diplodocus* mount in the Museum für Naturkunde, Berlin, the ischia form a much greater angle with the vertical than in, for example, *Brachiosaurus* and *Dicraeosaurus,* allowing for greater powered retraction angles, possibly as far as 55°.

The limb position of this model allows rapid and easy correction should the animal become unbalanced, avoiding backward toppling simply by resting the tail on the ground, or by slightly bending ankles and knees. Forward and lateral motions can be stopped by flexing and extending one or both knees and ankles. In this position, termed the *tripodal minimum* in Table 14.1, the back is inclined by approximately 48°. A more upright position would reduce bending moments in the femora and torques in the limb joints slightly, but would also reduce the ability to correct the posture if the model loses its balance. The sequence shown in Fig. 14.1 takes nearly 13 seconds overall to complete in the simulation software, but this long time is caused by modeling-related delays.

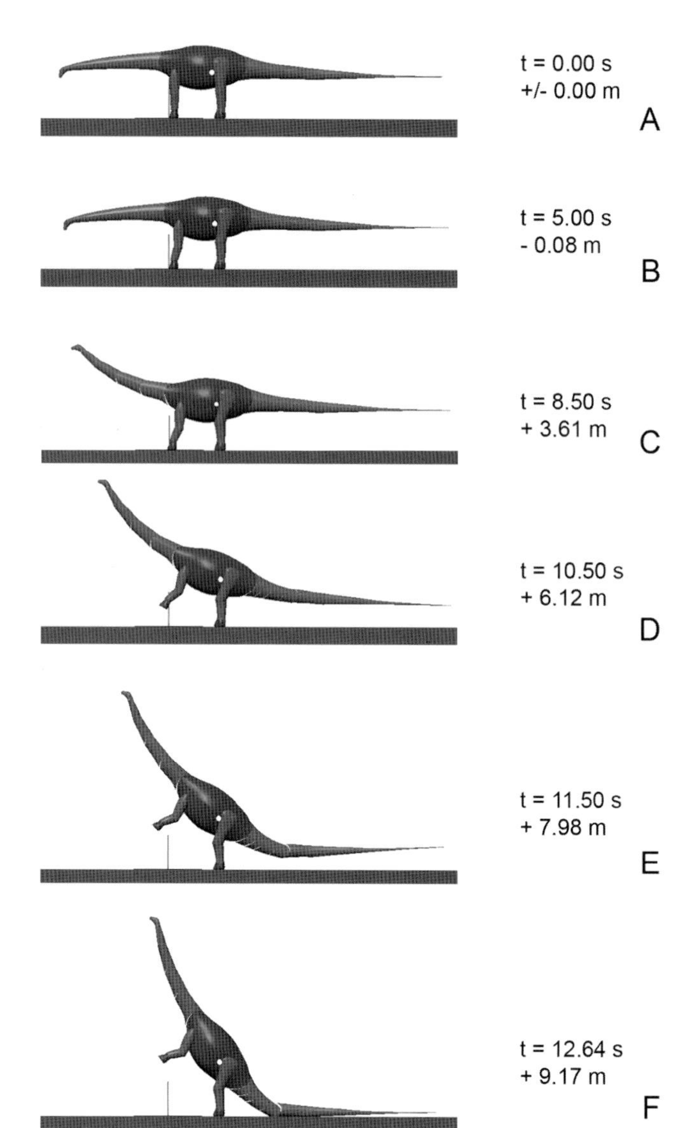

FIGURE 14.1. Rearing sequence for *Diplodocus* in the MSC.visualNASTRAN 4D model. The COM is marked by a white dot. Time and height gain of the snout are given for all frames. (A to B) The COM moves back over the hind feet. (B to C) Extension of neck reduces acceleration of the head during later phases. (C to D) Initial phase of rearing, where the trunk rotates upward while the neck returns to neutral position. (D to E) Further upward rotation of the trunk. Extension of the tail stops caudal motion of the COM. (E to F) The tail contacts the ground while the neck reaches the neutral pose. Note that the kink in the tail is an artifact of modeling. In reality, the tail would have bent smoothly and would have been well capable of producing the required extension.

In the beginning, the model stands still for nearly 2 seconds in order to avoid unwanted influences from the initial calculation of collisions between the feet and ground objects. A second delay of over 2 seconds takes place after the COM has been shifted over the hind feet (t = 2.8 seconds to t = 5 seconds), so that no inertia transfer from the COM shift to the rearing model can happen. In reality, *Diplodocus* probably used exactly such a transfer to ease the torque loads on the limb joints. Removing it from the calculation ensures that these torques are not underestimated. Between t = 5 seconds and t = 8.5 seconds, only the neck moves. At the end of the sequence, for almost 1.5 seconds, the body has already achieved the upright pose, and only the neck is moves slightly to maximize browsing height. The actual rearing motion takes

Table 14.1. Simulation Input Data for Rearing Modeling Files of *Loxodonta, Diplodocus,* and *Brachiosaurus*

Genus	Torque (Nm) Hip sum	Knee sum	Ankle sum	Total	Mass (kg)	$d = (\text{mass}^2)^{1/3}$	Torque/d $R(a)$	$R_n(a)$	PE difference (Nm)	PE/mass·g (m)
Loxodonta	46,804	6,984	1,288	55,076	7,400	379.74	145.04	1.00	+4.01	1.00
Diplodocus										
Bipedal	68,214	9,885	9,178	87,277	14,381	591.37	147.58	1.02	81,913.00	0.62
Tail touching	66,380	8,647	8,672	83,699			141.53	0.98		
Tripodal										
Max. support	170,209	8,227	8,100	186,536			194.07	1.34		
Med. support	117,681	8,713	9,314	135,708			141.19	0.97		
Min. support	98,091	7,843	6,852	112,786			117.34	0.81		
Brachiosaurus										
Upright	367,882	72,133	95,317	535,332	36,500	1,100.34	486.51	3.35	445,766.03	1.24
Reasonable	451,889	276,284	107,835	836,008			759.77	5.24	487,055.00	1.36
JP	392,779	31,645	17,337	441,761			401.48	2.77	429,153.45	1.20

Torques for the limb joints are given along with model mass and the correction factor *d* for mass vs. muscle cross-sectional growth. $R_n(a)$ indicates the rearing ability and is the standardized ratio of torque per factor *d*. The value for the elephant is 1. Higher values indicate that a relatively greater exertion is required to sustain an upright pose (*Brachiosaurus*), while lower values mean that rearing was easier (*Diplodocus*). PE difference indicates the gain in potential energy from rearing, showing that a rearing elephant must perform work equal to lifting its entire body by 1 m, while the work required by a rearing *Diplodocus* is much less. Although *Brachiosaurus* fares barely worse than an elephant, it has a far higher $R_n(a)$ value, indicating that its bauplan is much less suited for rearing.

less than 5 seconds, even in this conservative model. The use of a forelimb push-up and inertia transfers from the caudal COM shift would have allowed far more rapid rearing.

Various modifications of this model were created, testing different degrees of tripodal support. Lifting the tail off the ground minimally increases the torques in the limbs ("bipedal" pose in Table 14.1) and reduces stability when the neck is moved rapidly. Resting a larger part of the weight of the tail on the ground, to provide better support in a tripodal pose, actually increases the torques ("tripodal maximum support," "tripodal medium support" in Table 14.1) because it requires stronger limb flexion.

The *Diplodocus* model is potentially tail heavy because the 3D shape chosen is based on a muscular model, especially in the tail, and has ample air sacs and pneumatic spaces in the vertebral column in the anterior body cavity. In contrast, the posterior part, where no osteological evidence for air sacs exists except for some small pneumatic cavities in the sacral region, makes no provisions for weight-reducing structures. To study the effect of tail-heaviness, the densities of the model components in the tail were reduced to 0.8 kg/l, and an additional 500 kg weight was added to the anterior body. This resulted in an anterior shift of the COM of less than 0.1 m, which has practically no influence on the possible upright poses.

Brachiosaurus

Several different rearing poses were tested in *Brachiosaurus*. In the first ("upright," Fig. 14.2A), the body was simply rotated 55° upward, along with a slight extension at the base of the neck to achieve maximum browsing height. The vertebral column in this pose is inclined approximately 77° from the vertical, with the COM at a height of nearly 6 m, while the hip joint is 3.4 m above the ground. This position is quite risky: any backward shift of the weight puts tremendous stress on the tail, and recovery from such a motion is difficult, as flexion of the knees and extension of the hips would only increase the problem. Additionally, lateral stability is a problem. The angle between the vertical and the line connecting the COM and the outside of the foot is small at about 10°. This is roughly two thirds the angle in *Diplodocus*.

An attempt was then made of maximizing height gain while placing the animal in a pose that still allows some recovery capability ("reasonable," Fig. 14.2B). This still leads to an upright body pose, with the vertebral column inclined approximately 69°. This angle is still much greater than in the "bipedal" *Diplodocus* model, but it includes roughly the same amount of rotation from the quadrupedal pose because the back of *Brachiosaurus* is already angled up by about 22° from the horizontal in the quadrupedal pose. To increase lateral stability and retain a limited ability of preventing backward toppling, the limbs could not be placed in parallel. One foot was placed in front of the COM, and the other is placed behind it. If the COM moves backward, bending the knee of the forward leg keeps the foot of the posteriorly placed limb behind it, allowing recovery. A third model (Fig. 14.2C) was placed in a pose that copies the rearing pose shown in the

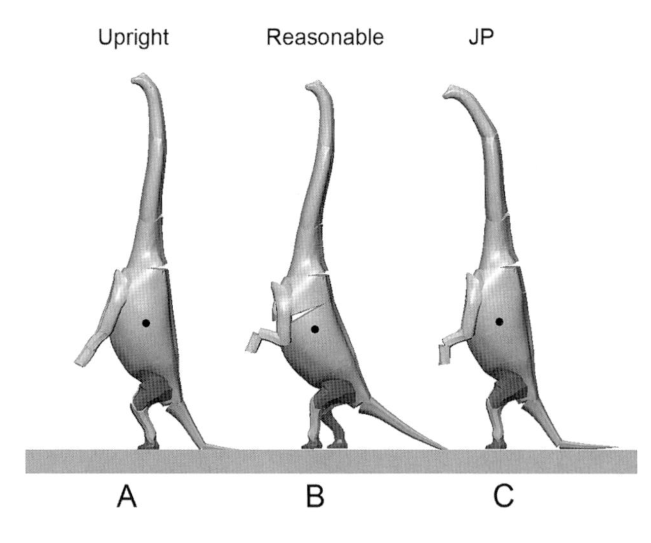

FIGURE 14.2. Rearing poses of *Brachiosaurus:* (A) "upright," (B) "reasonable," and (C) "JP" (Jurassic Park) position. Black dots mark position of COM. Note that the COM is located high above the hip joint, making any of the postures prone to instability.

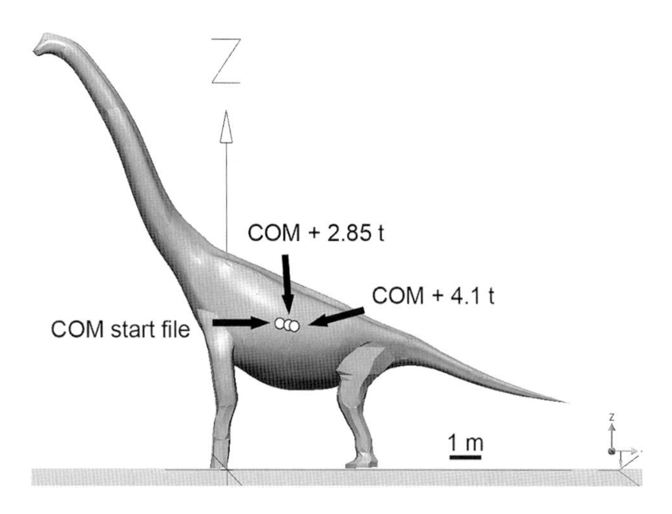

FIGURE 14.3. Mass variants of the *Brachiosaurus* model show that even markedly more tail-heavy models than used here do not have a COM position (white dots) more suitable to rearing.

movie *Jurassic Park* (termed "JP"). This pose is at first glance quite similar to "upright," but the hindlimbs are less flexed.

The *Brachiosaurus* model based on a reconstruction drawing in Paul (1988), is probably inaccurately proportioned and front heavy. Therefore, several variants were created, in which the weight of the "hips" segment and all tail segments was markedly increased. However, even a doubling or tripling of their masses did not result in a shift of the COM that appreciably altered the rearing poses. A weight increase by 2,862 kg (plus approximately 7.9% of the initial weight) from 36.323 metric tons to 39.183 metric tons shifted the COM a mere 0.29 m caudally and 0.07 m ventrally. A further weight increase to 40.435 metric tons (plus 4,112 kg, or 11% of the initial weight) resulted in a backward shift of the COM of 0.45 cm, displacing it ventrally by 0.11 cm (Fig. 14.3). Because the ventral shift of the COM cancels out part of the effect of the caudal shift in an upright pose, the balanced upright pose of the heavier models varies only minimally.

Results

For all reasonable variants of weight distribution, all positions modeled for *Brachiosaurus* are inherently far more unstable than those modeled for *Diplodocus*. This is not a result of the modeling per se, because an attempt was made to lower the torques in the limb joints of *Brachiosaurus* to the minimum possible. However, the body shape of *Brachiosaurus* invests the animal with a far greater risk of toppling sideways or backward, even in the "reasonable" model. This is caused by the great height that the COM is shifted during rearing as compared to hip height. More tail-heavy versions have the COM in a lower position, but there is still a marked contrast to *Diplodocus*. This animal was examined from a perspective of maximum safety. Therefore, the differences in required torque between the two animals reflects the absolute minimum theoretically possible, as any erroneous assumption in either model will tend to reduce the difference in resulting torques and not increase it. This approach ensures that *Diplodocus* is not erroneously found to be capable of prolonged rearing, or *Brachiosaurus* falsely found to be incapable of it. The opposite approach would be reducing the torques in *Brachiosaurus* to a level comparable to that in *Diplodocus* by altering the pose. However, the end result would be that only nonrearing poses would be created.

TORQUES

The dinosaur models were compared to that of an African elephant (*Loxodonta africana*), which was placed in a bipedal pose based on photographs of wild animals rearing. In each simulation of the three different animals, the torques in all limb joints were measured and added up. To put the efforts required by the models for rearing into relation to each other, the mass of each model was used to determine a factor that roughly correlates to the available muscle force. Volume increases by the cube and area by the square; therefore, an animal that has *n* times the mass of a geometrically similar one has roughly $n^* \sqrt[3]{mass^2}$ the muscle cross-sectional area available (which determines muscle forces). The importance of such scaling issues in biology is discussed in detail in numerous textbooks (e.g., Schmidt-Nielsen 1984; Brown & West 2000; Bonner 2006) and will not be elaborated on here. The torque estimate used here is a rough one, but given the inaccuracies inherent in the 3D models, the torques in the NASTRAN models must be seen primarily as qualitative result, and the only quantitative assessment possible is their relative, not their absolute, values. The sum of the limb torques for each model was divided by the factor $\sqrt[3]{m(a)^2}$ to derive an index for rearing ability:

$$\text{Relative rearing capability of animal } a\text{: } R(a) = \frac{\Sigma M(a)}{\sqrt[3]{m(a)^2}}, \quad (1)$$

where $M(a)$ is the sum of all torque values in the hindlimb joints and $m(a)$ is the total mass of the animal.

The resulting values were normalized by dividing them by the value $R_{(Loxodonta)}$ derived for the elephant model so that the value for the elephant is $R = 1$.

$$\text{Normalized rearing capability of animal } a\text{: } R_n(a) = \frac{R_a}{R_{Loxodonta}}. \quad (2)$$

Values of R_n smaller than $R_n = 1$ indicate that the dinosaur could have reared more easily than an elephant can, and values larger than $R_n = 1$ suggest that the dinosaur had to exert itself more than an elephant in order to sustain an upright posture. Note, however, that R_n indicates rearing ability relative to total body mass. Sauropods had proportionally stronger hindlimbs than mammals, so that a larger percentage of total muscle mass was available to produce the torques required for rearing. Thus, the above calculation method is rather conservative because it underestimates rearing ability in sauropods.

The models indicate that *Diplodocus* could rear as easily as an elephant [$R_{n(Diplodocus)} \approx 1.0$], despite its significantly greater weight, and that sustaining a bipedal pose was easier than a tripodal one. Taking the proportionally stronger hindlimbs of sauropods into account, *Diplodocus* could rear more easily than an elephant. In contrast, rearing required a much greater effort for *Brachiosaurus,* at least nearly three times as great as for an elephant (JP model, $R_{n[Brachiosaurus\ JP]} \approx 2.77$), and with a high risk of serious injury should the animal become unbalanced. Somewhat less risky, but more exhausting, is the "reasonable" pose ($R_{n[Brachiosaurus\ rears.]} \approx 5.24$). It is thus questionable whether *Brachiosaurus* could rear at all for a prolonged time without additional support.

While the overall proportions and the position of the COM are similar to elephants, the rearing ability of *Brachiosaurus* was severely curtailed by its great weight in addition to the high risk of toppling. In a pose similar to elephants, the forces are much higher, and altering the pose to ease the loads is difficult as a result of the limit on femoral retraction imposed by the anatomy of the pelvis. The femur could have been retracted quite far, but beyond an angle of 35°, there was no muscle that had an effective lever arm, so that forceful or prolonged retraction must have been difficult. Although we do not know the exact architecture of the muscles involved, data on limb retraction angles in archosaurs from Gatesy (1990) suggest that the caudofemoralis longus muscle had to be kept close to maximum contraction because of the low angle between the long axis of the vertebral column and the vertical, and therefore produced little force. The same is true for any muscle running from the ischium to the femur. Solely iliofemoral muscles could have further retracted the femur. This keeps the COM in front of the hips, and thus requires the knees to remain flexed, increasing torques in both hips and knees.

STABILITY

It is no surprise that all *Brachiosaurus* models become unstable from even relatively small motions of the neck. The COM is elevated too far above the hip joint during rearing, creating a large moment arm. Even relatively small neck motions require continuous high-torque correcting motions in the limbs to sustain the upright pose. The tail, as a result of its relative shortness, provides only minimal aid because the support triangle (both hind feet and the tail) is fairly small. In contrast, any pose with tail support in *Diplodocus* can be maintained with relative ease, even during speedy and large neck excursions. Because of the low position of the COM compared to the hip joints and the larger support triangle, most imbalances can be corrected for by bending the knees, at the cost of a moderate torque increase.

The differences in stability between the two dinosaurs modeled here have a twofold influence on the interpretation of their rearing ability. *Brachiosaurus* probably must have expended more energy for correcting motions. In addition, and more importantly, the inherent stability of a tripodal pose in *Diplodocus* ("bend the knees a bit and you'll be fine") contrasts markedly with the constant corrections that would have been required by *Brachiosaurus*. This is best explained by comparing a human sitting on a chair to one standing on a ladder. Sitting down means that one can make any rapid motion with one or both arms without concern for stability. In contrast, somebody standing on a ladder constantly has to ensure that sudden motions do not cause a fall. One cannot stand on a ladder and forget about balance while shifting attention to something else. Therefore, on the basis of stability, it also appears unlikely that *Brachiosaurus* engaged in prolonged rearing.

RETURN TO A QUADRUPEDAL POSE

Generally, returning to a quadrupedal pose in the models required less energy than rearing up. Still, there is a big difference between the *Diplodocus* and the *Brachiosaurus* models. These suggest that *Diplodocus* could have used a continuous fluid motion with an initially relatively high rotation speed and a marked slowdown to lessen the impact just before the forefeet touched the ground. In this motion sequence, the torques in the hindlimbs stay below or at the level required for sustaining the upright pose, and only marginally in them for the final slowdown. *Brachiosaurus,* however, must have either moved in continuous slow motion or risked extreme forces acting on the manus, which would have lead to injury (Rothschild & Molnar 2005). A rapid motion that was decelerated to avoid a hard impact would have required extraordinarily large torques in the hindlimb joints.

ACCURACY TEST

The results of the simulations were first tested for accuracy by calculating the difference in potential energy between the starting pose and the rearing pose in three model variants: *Loxodonta, Diplodocus* "bipedal," and *Brachiosaurus* "reasonable" (Table 14.1; Fig. 14.2). For *Brachiosaurus,* both the energy difference for the Stevens & Parrish (2005a) start pose and the Paul (1988) start pose were measured. Unsurprisingly, the values (given in Table 14.1) show that in relation to total body weight, *Diplodocus* has by far the easiest task: the increase in potential energy is equivalent to lifting slightly more than half of its body weight by 1 m, which corresponds to only 1.1 times the potential energy difference in the elephant model, while the latter must have produced the equivalent of lifting its total mass by 1 m. In *Brachiosaurus,* the average of the two tested variants is roughly six times that of the elephant, requiring forces that would suffice to lift 1.4 times the complete animal by 1 m. These values, which are proportional to the torque value differences, show that the simulation is internally consistent.

A different approach to testing simulation accuracy involves considerably more work. To judge whether the animal could sustain an upright pose, the required physiological diameter of the muscles can be calculated if the lever arm is known. Luckily, muscle forces are the sole forces that need to be considered; elastic processes, such as storage of energy in stretched tendons and ligaments, do not influence the question of sustaining an upright pose (as opposed to attaining it, in which they may play a role). If the calculated required muscle diameters do not fit on the skeleton, the animal could not have reared. This approach requires exact knowledge of the muscle's moment arm, and an accurate value for exactly how much force a given muscle cross section can produce. The former is hard to determine for extinct animals, especially for sauropods, as a result of their large cartilage caps on the long bones (e.g., Holliday et al. 2001, 2010; Bonnan et al. 2010). The moment arm also changes during motion. Here, a minimum and a maximum value are used that cover the range of probable moment arms. Literature values for the force a muscle can produce per cross-sectional area vary widely, from as low as 16 N/cm^2 to as high as 100 N/cm^2 (e.g., Fick 1911; Franke & Bethe 1919; Barmé 1964; Langenberg 1970). Carpenter et al. (2005) use two values to bracket the probable range, 39 N/cm^2 and 78 N/cm^2, based on data in Ikai & Fukunaka (1968). For the sake of simplicity, it is here assumed that the muscles in question all were parallel fibered, or if they were pinnate, that the angle between the direction of pull and the fibers was relatively small. A pinnate muscle would produce more force per cross section, so that the muscle diameters calculated would be somewhat too large.

Brachiosaurus

Starting with the same values as given in Carpenter et al. (2005) for the "reasonable" *Brachiosaurus* model (Fig. 14.2B), it is possible to calculate muscle diameter–moment arm data pairs for the hip joint:

Average torque $M_{\text{hip}} = 225{,}945$ Nm

Moment arm length for the caudofemoralis longus muscle, derived graphically from the 3D model, is

Minimum estimate: $l_{\text{min}} = 0.43$ m
Maximum estimate: $l_{\text{max}} = 0.83$ m

$$\text{Required muscle force } F_r\text{: } F_r = \frac{M}{l} \tag{3}$$

For $l = 0.35$ m$\rightarrow F_r = 525{,}453.49$ N
For $l = 0.64$ m$\rightarrow F_r = 272{,}222.89$ N

$$\text{Physiological muscle cross-section area: } A = \frac{F_r}{F_m}, \tag{4}$$

where F_m is the maximal muscle force per square centimeter.

For $F_m = 78$ N/cm$_2$ and $F_r = 272{,}222.89$ N$\rightarrow A \approx 3{,}490$ cm^2
For $F_m = 78$ N/cm^2 and $F_r = 525{,}453.49$ N$\rightarrow A \approx 6{,}737$ cm^2
For $F_m = 39$ N/cm^2 and $F_r = 272{,}222.89$ N$\rightarrow A \approx 6{,}980$ cm^2
For $F_m = 39$ N/cm^2 and $F_r = 645{,}557.14$ N$\rightarrow A \approx 13{,}473$ cm^2

Obviously, these values do not represent the exact cross-sectional area of the caudofemoralis longus muscle because other muscles participated in limb retraction. However, all of these other muscles have smaller moment arms, most of them significantly so, especially when the limb is retracted.

The most exact test of simulation accuracy would be to create a detailed 3D model of the hindlimb and tail musculature of *Brachiosaurus* to compare cross-sectional areas to the calculated values. This could be done in the musculoskeletal modeling software SIMM (see Hutchinson et al. 2005 for a description). Although the effort required would be large, it may not result in a significant increase in accuracy because of the many uncertainties inherent in muscle reconstruction in extinct animals.

However, from the 3D model used for the kinetic–dymanic model, it is also possible to estimate the maximal muscle cross-sectional area that fits onto the skeleton. Given the relatively low accuracy of this estimate, simple elliptical cross sections are sufficient, for which the enclosed area can be easily calculated. A generous estimate of the diameter of the caudofemoralis longus muscle, larger than the rather skinny model based on Paul (1988), is 0.79 m^2 or 7,900 cm^2. This means that rearing was impossible for *Brachiosaurus* at the smallest moment arm and lowest force value. Assuming an intermediate value, the cross section required for rearing is actually available. However, there is hardly any safety factor.

Normally, animals do not perform motions that take all their strength. Rather, muscles (as well as bones, ligaments, and tendons) are overengineered, much like engineers construct buildings and machines to standards much higher than the expected loads. This ensures that accidental overloading during normal use does not lead to damage, for example, a failure of a bridge span or a torn tendon. Such safety factors in animals often range from 50% to 100% (Biewener 1983, 1989). Therefore, *Brachiosaurus* should have possessed a hip muscle cross section somewhat larger than 9,500 cm^2 for an intermediate value. When the "reasonable" rearing model is used, the muscle cross section required for rearing becomes much larger, making it unlikely that *Brachiosaurus* reared for any length of time, if at all.

Diplodocus

The same calculations for *Diplodocus* ("tail touching" model; Fig. 14.1F) yields the following values:

Average torque M_{hip} = 33,190 Nm

Moment arm length for the caudofemoralis longus muscle, derived graphically from the 3D model, is:

Minimum estimate: l_{min} = 0.35 m
Maximum estimate: l_{max} = 0.75 m

Equation (3) gives:

For l = 0.35 m→ F_r = 94,828.57 N
For l = 0.75 m→ F_r = 44,253.33 N

Equation (4) gives:

For F_m = 78 N/cm^2 and F_r = 44,253.33 N→ A ≈ 567 cm^2
For F_m = 78 N/cm^2 and F_r = 94,828.57 N→ A ≈ 1,216 cm^2
For F_m = 39 N/cm^2 and F_r = 44,253.33 N→ A ≈ 1,134 cm^2
For F_m = 39 N/cm^2 and F_r = 94,828.57 N→ A ≈ 2,432 cm^2

Even the largest of these muscle cross sections easily fits into the body outline of the *Diplodocus* model, providing a large safety margin. It is therefore likely that *Diplodocus* could rear easily and for a prolonged time.

For comparison, the same calculations were repeated for *Loxodonta:*

Average torque M_{hip} = 23,402 Nm

Moment arm length for the gluteus muscle, derived graphically from the 3D model, is:

Minimum estimate: l_{min} = 0.20 m
Maximum estimate: l_{max} = 0.40 m

Equation (3) gives:

For l = 0.20 m→ F_r = 117,010 N
For l = 0.40 m→ F_r = 58,505 N

Equation (4) gives:

For F_m = 78 N/cm^2 and F_r = 58,505 N→ A ≈ 750 cm^2
For F_m = 78 N/cm^2 and F_r = 117,010 N→ A ≈ 1,500 cm^2
For F_m = 39 N/cm^2 and F_r = 58,505 N→ A ≈ 15,00 cm^2
For F_m = 39 N/cm^2 and F_r = 94,828.57 N→ A ≈ 3,000 cm^2

Such a large retractor musculature can be fitted onto an elephant's skeleton, although with less room to spare than in the *Diplodocus* model. Elephants thus have a lower safety factor for rearing, which fits well with their actual use of this movement: all elephants can rear up, and most rear up regularly, but rearing is not employed in daily food acquisition.

Discussion

FEASIBILITY OF REARING

The results of the kinetic–dynamic modeling clearly show that diplodocids could rear quickly, easily, and for a long time. Although their rearing ability factor is identical to that of elephants, their stronger hindlimbs mean that rearing was less of an exertion for *Diplodocus.* A tripodal pose would have allowed high browsing without compromising the ability to locomote, as regaining a quadrupedal pose was unproblematic. In contrast, *Brachiosaurus* would have expended considerably more energy, could not have attained a stable upright pose, and would have risked serious injury to its forefeet when descending too rapidly.

HEIGHT INCREASE THROUGH REARING

The height gained by rearing is impressive: the snout of *Diplodocus* moves nearly up 9 m, and that of *Brachiosaurus* gains between 3.86 m and 4.38 m for the "JP" and "upright" variants, respectively. Note, however, that these distances are based on the reconstruction by Paul (1988) and would increase to 5.19 and 7.71 m if the habitual posture reconstructed by Stevens & Parrish (1999) with a subhorizontal neck was used as a reference instead. Stevens & Parrish (2005a, 2005b) concluded that *Brachiosaurus* habitually held its head 6 m above the ground, but this is based on a slightly smaller model of *Brachiosaurus* than that of Paul (1988). At best, the upright pose for *Brachiosaurus* less than doubled the maximum feeding height (based on the model of Stevens & Parrish 2005a), while *Diplodocus* at least tripled, if not nearly quadrupled, feeding height (11.11 m versus approximately 3 m), depending on whether the neck was held level instead of slightly drooping in the quadrupedal position. This difference between the two dinosaurs is caused by the differing angles at which the neck articulates with the body, the orientation of the dorsal vertebral column, and the neck–body–forelimb length ratios. Simply stated, the short front limb and extremely long neck of *Diplodocus* mean that they would have gained more from rearing than *Brachiosaurus,* while at the same time these differences also would have made rearing easier and less risky by providing a more suitably placed COM.

POTENTIAL INFLUENCE OF REARING ABILITY ON LOCOMOTION

Acceleration

An additional result of the modeling effort presented here is that *Diplodocus* payed a potentially steep price for their excellent rearing abilities. The COM of *Diplodocus* is located so far posteriorly that rapid acceleration, from a standstill to a walk, or a slow walk to a rapid elephantine run, would have been difficult. Instead of propelling the body forward, hindlimb retraction in the model tends to rotate the body up. Additionally, the resulting vector from the acetabulum through the COM is directed at a point posterior to the contralateral shoulder, indicating that acceleration would have had a large lateral component (see below). This means that at least juvenile *Diplodocus* must have been vulnerable to predator attacks because this bauplan made them sitting ducks (see also Senter 2007 on vulnerable necks). Adult diplodocids may have been protected from predator attacks as a result of their size (Sander & Clauss 2008; Sander et al. 2010). Therefore, juvenile *Diplodocus* must have had some kind of defense reaction that did not require rapid flight. It is interesting to note that *Diplodocus* has lower strength indicators than, for example, *Apatosaurus,* which indicates that it was less athletic than more sturdily built dinosaurs (Alexander 1985).

Lateral Stability

An additional problem with the extreme posterior position of the COM in diplodocids is the issue of lateral stability during locomotion. In general, the further posteriorly the COM is situated, the more the vector from the acetabulum of the supporting limb to the COM is directed laterally across the midline of the animal. In diplodocids, the vector is directed so far laterally that it exits the body behind the front limbs, which means that lifting one hindlimb and retracting the other forcefully will result in acceleration with a strong lateral component, threatening to tip the animal over or turn it around the vertical axis. This could be countered by moving the tail and neck laterally, transferring large rotational inertia to them, and resulting in large swinging amplitudes. Such swinging amplitudes may have resulted in collisions between herd members if these sauropods lived in herds. It increases the risk of serious injury as well, when the neck collides with a solid object such as a tree trunk. An alternative solution to the problem of strong lateral acceleration would be to adduct the supporting hindlimb to the midline, but that would only have been possible in midstance. When the limb was protracted or retracted to any large extent, the pubes and ischia, respectively, would have made adduction to the midline difficult. Theoretically, sauropods could have moved with a high degree of lateral swaying and/or strong undulating motions of the back, but this appears energetically inefficient because much energy would have been wasted for motions that did not propel the animal forward. So far, however, the effect of an extremely posterior COM on locomotion has not been investigated to test this hypothesis.

PALEOBIOLOGICAL IMPLICATIONS

In their study of Late Jurassic vegetation, Rees et al. (2004) describe the paleoenvironment of the Morrison Formation as a savanna-like habitat, in which tall trees were rare, while herbaceous plants and short trees such as ginkgo and cycads dominated. In the Morrison, diplodocid fossils are vastly more abundant than brachiosaurids. In contrast, the Tendaguru Beds in Africa preserve an ecosystem rich in tall conifers, and the tall brachiosaurids were more common (Rees et al. 2004). Stevens & Parrish (2005a) argue that this difference in sauropod abundance was caused by diplodocids having been limited to low browsing, while *Brachiosaurus* and possibly *Camarasaurus* were the only true high browsers able to feed on tall conifers. The modeling results do not support this interpretation. By rearing, *Diplodocus* could have reached as high as *Brachiosaurus,* and therefore there is no difference in potential feeding height. The different abundance in sauropod taxa in geographically separate areas cannot be explained by different vegetation heights. Moreover, it is possible that it would have been easier for brachiosaurids with their stiffer necks to feed in savanna-like habitats by browsing on the tops of short, solitary trees, while diplodocids were able to feed effectively in denser tree stands.

Araucaria trees today routinely grow much taller than the assumed feeding heights of nonrearing sauropods, which would have also been out of reach of a rearing *Diplodocus,* or even a rearing *Brachiosaurus.* Their presence in the Morrison Formation and abundance in other sauropod-rich habitats, as well as their high energy content (e.g., Hummel et al. 2008; Gee, this volume), begs the question whether they were fed on by rearing, extra-long-necked diplodocids such as *Supersaurus* in the Morrison Formation, or supersized titanosaurs like *Argentinosaurus* in the Cretaceous.

OTHER USES OF A BIPEDAL OR TRIPODAL POSE

An alternative use that has been suggested for a bipedal or tripodal pose is self-defense (e.g., Bakker 1986, Jensen 1988), as visualized by the *Barosaurus* mount at the American Museum of Natural History. However, it is difficult to imagine how young or subadult herd members could be defended more easily in a bipedal stance because it limits the ability to locomote. For the rearing animal itself, it is also counterintuitive that rearing for a prolonged time would help in defense. Although it does protect the belly, because an approaching theropod risks being kicked by the front limbs, and removes the vulnerable neck (Senter 2007) from the predator's reach, it exposes the base of the tail and the axial musculature on the hips as well as the thigh muscles to attack. Severing these muscles would be a good way to cripple a sauropod. There is fossil evidence for one probable attack of a predator on a sauropod, a trackway found on the Paluxy River by R. T. Bird. Thomas & Farlow (1997) describe the possible attack sequence: apparently, the predator synchronized its stride with the intended prey, then attempted to strike at the thigh or tail base—exactly the areas that would have become more vulnerable by rearing.

Rearing has also been suggested to function in intraspecific fights (Bakker 1986). Obviously, a rearing animal is taller than one that remains in a quadrupedal pose, and today many animals use poses and motion that make them appear larger to frighten or impress competitors or predators. It is very possible that sauropods reared to impress; possibly even brachiosaurids pushed themselves up onto their hindlimbs for a moment to show off. If so, it was probably a good idea for competitors and enemies alike to stay clear of the area in front of them, because when coming down, they would certainly have done serious harm to any animal that got caught under them.

Rearing only works well for defense if the sauropod can rear up and get back down quickly, which in the case of *Brachiosaurus* probably resulted in prohibitive compressive forces on the front limbs. An alternative defensive strategy could have been coordinated herd behavior with several rearing animals forming a circle to cover each others' backs. However, the risk of one animal stepping on its neighbor's tail may have made such behavior impossible.

STRESS FRACTURES: WHY THEIR ABSENCE DOES NOT EXCLUDE REARING

Rothschild & Molnar (2005) claim that sauropods did not rear on the basis of the absence of stress fractures in their handbones. Although it is true that repeated excessive stress causes fractures, leaving typical marks on the bones, and while sauropods indeed lack such fracture marks, their claim that "the presence of such injuries indicates that the stance was adopted, and the absence of these injuries indicates that the stance was not adopted" (Rothschild & Molnar 2005, p. 382) is questionable. Rothschild & Molnar assume that "the resumption of a normal pose would potentially exert extreme forces"; however, this is most probably not universally true, given the extremely posterior position of the COM in *Diplodocus*. Returning to a quadrupedal pose slowly is simple, and it does not put excessive stress on the hands. The human example chosen by Rothschild & Molnar (2005) as a comparison (ballet dancers landing in an en pointe position) is far outside the norm of behavior for *Homo sapiens*. Adaptations that reduce the risk of injury from this motion are imaginable, but there is no indication that those were ever selected for. In contrast, sauropod rearing is viewed in the literature as an everyday activity, performed by all individuals of the species. It would have greatly improved the fitness of the individual by allowing easier mating, better defense, and/or access to more or more nutritious food. Selection for rearing abilities should therefore be strong. If a certain type of injury is commonly caused by an everyday motion, evolutionary processes should lead to adaptations that make the injury less likely. The literature is tellingly silent on common injuries in extant mammals caused by daily feeding motions. Therefore, the absence of injuries can mean that the motion was not performed, or that it was performed so regularly that the animal had adapted to it to a degree that guaranteed safety from injury during routine motions.

An additional case in point is the frequency of stress fractures in elephants: in contrast to wild elephants, circus elephants rear almost daily, sometimes even several times a day. If Rothschild & Molnar (2005) were correct, there should be a higher incidence of fractures in the metacarpals of circus elephants than in wild elephants or zoo elephants that do not perform shows for visitors. However, a cursory search of the veterinary literature did not return reports on stress fractures in the hands of circus elephants.

Summary and Future Research

Kinetic–dynamic modeling provides insights into the energy balance and forces required for rearing in sauropods and permits fast calculation of COM position during motion. Modeling suggests that *Diplodocus* could have easily reared up into a bipedal pose and maintained it for an extended period of time at moderate energy expenses and without the risk of toppling. Thus, diplodocids may have spent considerable time on their hindlimbs just for high browsing. *Brachiosaurus,* in contrast, was probably unable to use a bipedal or tripodal posture regularly and for an extended period of time. Although this dinosaur could certainly have reared up, for example during mating, this was probably a rare and short-lived event. Rearing for mating may not even have required a true bipedal posture, in which the COM rests over the feet, but only a short push-up leading to a final pose in which part of the weight was supported by the female. NASTRAN modeling also indicates that a bipedal or tripodal posture was stable for *Diplodocus,* but unstable for *Brachiosaurus.*

The speed of rearing into a bipedal pose and returning to a quadrupedal one has not yet been modeled. Additionally, more dinosaurs must be included in future studies to obtain a comprehensive picture of the evolution of rearing abilities in sauropods through time. The influence of biomechanical adaptations optimizing rearing performance on locomotion and on feeding envelope may provide insights into niche partitioning in sauropods. This will also require detailed models of the energy requirements for locomotion at a wide range of speeds.

Results from kinetic–dynamic modeling should ideally be cross-checked with high-detail modeling, for example in SIMM, of selected species representative of certain bauplan types.

Acknowledgments

This project was supported by the Deutsche Forschungsgemeinschaft as part of project MA 4249/1-1 in Research Unit 533. Although too many people to list have contributed to my research through extensive and helpful discussions, it is Ulrich Witzel, Holger Preuschoft (both Ruhr-University of Bochum), and Hans-Ulrich Pfretzschner (University of Tübingen) who have earned my special thanks for advice on biomechanical issues. The entire sauropod research group has furthered my work on sauropod rearing through helpful suggestions and critical questions. This chapter has greatly profited from extensive reviews by Ray Wilhite (Auburn University) and Matthew Bonnan (Western Illinois University, Macomb). This is contribution number 73 of the DFG Research Unit 533 "Biology of the Sauropod Dinosaurs: The Evolution of Gigantism."

References

Alexander, R. M. 1985. Mechanics of posture and gait of some large dinosaurs.—*Zoological Journal of the Linnean Society of London* 83: 1–25.

Bakker, R. T. 1971. The ecology of the brontosaurs.—*Nature* 229: 172–174.

Bakker, R. T. 1978. Dinosaur feeding behaviour and the evolution of flowering plants.—*Nature* 276: 661–663.

Bakker, R. T. 1986. *The Dinosaur Heresies.* William Morrow and Company, New York.

Barmé, M. 1964. *Kraft, Ermüdung und elektrische Aktivität der Kaumuskulatur und der Flexoren am Oberarm.* Diploma Thesis. University of Cologne, Cologne.

Barrett, P. M. & Upchurch, P. 1994. Feeding mechanisms of *Diplodocus.*—*GAIA* 10: 195–204.

Biewener, A. A. 1983. Allometry of quadrupedal locomotion: the scaling of duty factor, bone curvature and limb orientation to body size.—*Journal of Experimental Biology* 105: 147–171.

Biewener, A. A. 1989. Scaling body support in mammals: limb posture and muscle mechanics.—*Science* 245: 45–48.

Bonnan, M. F., Sandrik, J. L., Nishiwaki, T., Wilhite, D. R., Elsey, R. M. & Vittore, C. 2010. Calcified cartilage shape in archosaur long bones reflects overlying joint shape in stress-bearing elements: implications for nonavian dinosaur locomotion.—*The Anatomical Record.* doi: 10.1002/ar.21266.

Bonner, J. T. 2006. *Why Size Matters: From Bacteria to Blue Whales.* Princeton University Press, New Jersey.

Borsuk-Bialynicka, M. 1977. A new camarasaurid sauropod *Opisthocoelicaudia skarzynskii,* gen. n., sp. n. from the Upper Cretaceous of Mongolia.—*Palaeontologia Polonica* 37: 5–64.

Brown, J. H. & West, G. B. 2000. *Scaling in Biology.* Oxford University Press, Oxford.

Carpenter, K., Sanders, F., McWhinney, L. A. & Wood, L. 2005. Evidence for predator–prey relationships. Examples for *Allosaurus* and *Stegosaurus. In* Carpenter, K. (ed.). *The Carnivorous Dinosaurs.* Indiana University Press, Indiana: pp. 325–350.

Christian, A. 2010. Some sauropods raised their necks—evidence for high browsing in *Euhelopus zdanskyi.*—*Biology Letters* 6: 823–825.

Christian, A. & Dzemski, G. 2007. Reconstruction of the cervical skeleton posture of *Brachiosaurus brancai* Janensch 1914 by an analysis of the intervertebral stress along the neck and a comparison with the results of different approaches.—*Fossil Record* 10: 38–49.

Christian, A. & Dzemski, G. This volume. Neck posture in sauropods. *In* Klein, N., Remes, K., Gee, C. T. & Sander, P. M. (eds.). *Biology of the Sauropod Dinosaurs: Understanding the Life of Giants.* Indiana University Press, Bloomington, pp. 251–260.

Christiansen, P. 2000. Feeding mechanisms of the sauropod dinosaurs *Brachiosaurus, Camarasaurus, Diplodocus,* and *Dicraeosaurus.*—*Historical Biology* 14: 137–152.

Dodson, P. 1990. Sauropod paleoecology. *In* Weishampel, D. B., Dodson, P. & Osmólska, H. (eds.). *The Dinosauria.* University of California Press, Berkeley: pp. 402–407.

Fick, R. 1911. Anatomie und Mechanik der Gelenke unter Berücksichtigung der bewegenden Muskeln. *In* Bardeleben, K. v. (ed.). *Handbuch der Anatomie des Menschen, 2. Bd., Abt. 1, Teil 1–3.* Gustav Fischer Verlag, Jena.

Franke, F. & Bethe, A. 1919. Beiträge zum Problem der willkürlich beweglichen Armprothesen, IV. Kraftkurven.—*Münchner medizinische Wochenschrift* 66: 201–205.

Gatesy, S. M. 1990. Caudofemoral musculature and the evolution of theropod locomotion.—*Paleobiology* 16: 170–186.

Gee, C. T. This volume. Dietary options for the sauropod dinosaurs from an integrated botanical and paleobotanical perspective. *In* Klein, N., Remes, K., Gee, C. T. & Sander, P. M. (eds.). *Biology of the Sauropod Dinosaurs: Understanding the Life of Giants.* Indiana University Press, Bloomington: pp. 34–56.

Goldfinger, E. 2004. *Animal Anatomy for Artists: The Elements of Form.* Oxford University Press, Oxford.

Hatcher, J. B. 1901. *Diplodocus* (Marsh): its osteology, taxonomy, and probable habits, with a restoration of the skeleton.—*Memoirs of the Carnegie Museum* 1: 1–61.

Henderson, D. M. 2006. Burly gaits: centers of mass, stability and the trackways of sauropod dinosaurs.—*Journal of Vertebrate Paleontology* 26: 907–921.

Holliday, C. M., Ridgley, R. C., Sedlmayr, J. C. & Witmer, L. M. 2001. The articular cartilage of extant archosaur long bones: implications for dinosaur functional morphology and allometry.—*Journal of Vertebrate Paleontology* 21: 62A.

Holliday, C. M., Ridgley, R. C., Sedlmayr, J. C. & Witmer, L. M. 2010. Cartilaginous epiphyses in extant archosaurs and their implications for reconstructing limb function in dinosaurs.—*PLoS ONE* 5(9): e13120.

Hummel, J., Gee, C. T., Südekum, K.-H., Sander, P. M., Nogge, G. & Clauss, M. 2008. In vitro digestibility of fern and gymnosperm foliage: implications for sauropod feeding ecology and diet selection.—*Proceedings of the Royal Society B: Biological Sciences* 275: 1015–1021.

Hutchinson, J. R., Anderson, F. C., Blemker, S., Delp, S. L. 2005. Analysis of hindlimb muscle moment arms in *Tyrannosaurus rex* using a three-dimensional musculoskeletal computer model: implications for stance, gait, and speed.—*Paleobiology* 31: 676–701.

Ikai, M. & Fukunaka, T. 1968. Calculation of muscle strength per unit cross-sectional area of human muscle by means of ultrasonic measurement.—*Internationale Zeitschrift für Angewandte Physiologie* 26: 26–32.

Janensch, W. 1950. Die Wirbelsäule von *Brachiosaurus brancai*.—*Palaeontographica Supplement* 7 (3): 27–93.

Jensen, J. A. 1988. A fourth new sauropod from the Upper Jurassic of the Colorado Plateau and sauropod bipedalism.—*The Great Basin Naturalist* 48: 121–145.

Langenberg, W. 1970. Morphologie, physiologischer Querschnitt und Kraft des M. erector spinae in Lumbalbereich des Menschen.—*Zeitschrift für Anatomie und Entwicklungsgeschichte* 132: 158–190.

Mallison, H. 2007. *Virtual Dinosaurs—Developing Computer Aided Design and Computer Aided Engineering Modeling Methods for Vertebrate Paleontology*. Ph.D. Dissertation. Eberhardt-Karls-University, Tübingen.

Mallison, H. 2010. The digital *Plateosaurus* I: body mass, mass distribution and posture assessed using CAD and CAE on a digitally mounted complete skeleton.—*Palaeontologia Electronica* 13: 1–26.

Mallison, H. This volume. *Plateosaurus* in 3D: how CAD models and kinetic–dynamic modeling bring an extinct animal to life. *In* Klein, N., Remes, K., Gee, C. T. & Sander, P. M. (eds.). *Biology of the Sauropod Dinosaurs: Understanding the Life of Giants*. Indiana University Press, Bloomington, pp. 219–236.

Mallison, H. & Pfretzschner, H.-U. 2005. Walking with sauropods: modeling dinosaur locomotion in MSC.visualNASTRAN 4D.—*Journal of Vertebrate Paleontology* 25: 88A.

Martin, J. 1987. Mobility and feeding of *Cetiosaurus:* why the long neck?—*Occasional Papers of the Tyrell Museum of Palaeontology* 3: 150–155.

McIntosh, J. 1990. Sauropoda. *In* Weishampel, D. B., Dodson, P. & Osmólska, H. (eds.). *The Dinosauria*. University of California Press, Berkeley: pp. 345–401.

McIntosh, J., Brett-Surman, M. K. & Farlow, J. O. 1997. Sauropods. *In* Farlow, J. O. & Brett-Surman, M. K. (eds.). *The Complete Dinosaur.* Indiana University Press, Bloomington: pp. 264–290.

National Geographic Society. 2007. Dinosaur mummy found with fossilized skin and soft tissues. December 3.—*ScienceDaily.* Available at: http://www.sciencedaily.com/releases/2007/12/071203103349.html. Accessed June 29, 2008.

Parrish, J. M. 2006. The origin of high browsing and the effect of phylogeny and scaling on neck length in sauropodomorphs. *In* Carrano, M., Blob, R., Gaudin, T. & Wible, J. (eds.). *Amniote Paleobiology: Phylogenetic and Functional Perspectives on the Evolution of Mammals, Birds, and Reptiles*. University of Chicago Press, Chicago: pp. 201–224.

Paul, G. S. 1988. The brachiosaur giants of the Morrison and Tendaguru with the description of a new subgenus, *Giraffatitan,* and a comparison of the world's largest dinosaurs.—*Hunteria* 2: 1–14.

Paul, G. S. 2000. Restoring the life appearance of dinosaurs. *In* Paul, G. S. (ed.). *The Scientific American Book of Dinosaurs*. Byron Press, New York: pp. 78–106.

Preuschoft, H., Hohn, B., Stoinski, S. & Witzel, U. This volume. Why so huge? Biomechanical reasons for the acquisition of large size in sauropod and theropod dinosaurs. *In* Klein, N., Remes, K., Gee, C. T. & Sander, P. M. (eds.). *Biology of the Sauropod Dinosaurs: Understanding the Life of Giants*. Indiana University Press, Bloomington: pp. 197–218.

Rauhut, O. W. M., Fechner, R., Remes, K. & Reis, K. This volume. How to get big in the Mesozoic: the evolution of the sauropodomorph body plan. *In* Klein, K., Remes, K., Gee, C. T. & Sander, P. M. (eds.). *Biology of the Sauropod Dinosaurs: Understanding the Life of Giants*. Indiana University Press, Bloomington: pp. 119–149.

Rees, P. M., Noto, C. R., Parrish, J. M. & Parrish, J. T. 2004. Late Jurassic climates, vegetation, and dinosaur distributions.—*Journal of Geology* 112: 643–653.

Riggs, E. S. 1904. Structure and relationship of the opisthocoelian dinosaurs. Part II: Brachiosauridae.—*Publications of the Field Columbian Museum, Geological Series* 2: 229–248.

Rothschild, B. & Molnar, R. 2005. Sauropod stress fractures as clues to activity. *In* Tidwell, V. & Carpenter, K. (eds.). *Thunder Lizards: The Sauropodomorph Dinosaurs*. Indiana University Press, Bloomington: pp. 381–392.

Sander, P. M. & Clauss, M. 2008. Sauropod gigantism.—*Science* 322: 200–201.

Sander, P. M., Christian, A. & Gee, C. T. 2009. Sauropods kept their heads down. Response.—*Science* 323: 1671–1672.

Sander, P. M., Christian, A., Clauss, M. Fechner, R., Gee, C. Griebeler, E. M., Gunga, H.-C., Hummel, J., Mallison, H., Perry, S., Preuschoft, H., Rauhut, O., Remes, K., Tütken, T., Wings, O. & Witzel, U. 2010. Biology of the sauropod dinosaurs: the evolution of gigantism.—*Biological Reviews of the Cambridge Philosophical Society.* doi: 10.1111/j.1469=185X.2010.00137.x.

Schmidt-Nielsen, K. 1984. *Scaling: Why Is Animal Size So Important?* Cambridge University Press, Cambridge.

Schwarz, D., Frey, E. & Meyer, C. A. 2007. Pneumaticity and soft-tissue reconstructions in the neck of diplodocid and dicraeosaurids sauropods.—*Acta Palaeontologica Polonica* 52: 167–188.

Senter, P. 2007. Necks for sex: sexual selection as an explanation for sauropod dinosaur neck elongation.—*Journal of Zoology* 271: 45–53.

Stevens, K. A. & Parrish, J. M. 1999. Neck posture and feeding habits of two Jurassic sauropod dinosaurs.—*Science* 284: 798–800.

Stevens, K. A. & Parrish, J. M. 2005a. Digital reconstructions of sauropod dinosaurs and implications for feeding. *In* Curry Rogers, K. A. & Wilson, J. A. (eds.). *The Sauropods: Evolution and Paleobiology.* University of California Press, Berkeley: pp. 178–200.

Stevens, K. A. & Parrish, J. M. 2005b. Neck posture, dentition, and feeding strategies in Jurassic sauropod dinosaurs. *In* Tidwell, V. & Carpenter, K. (eds.). *Thunder Lizards: The Sauropodomorph Dinosaurs.* Indiana University Press, Bloomington: pp. 212–232.

Thomas, D. A. & Farlow, J. A. 1997. Tracking a dinosaur attack.—*Scientific American* 277: 48–53.

Upchurch, P. & Barrett, P. M. 2000. The evolution of sauropod feeding mechanisms. *In* Sues, H.-D. (ed.). *Evolution of Herbivory in Terrestrial Vertebrates: Perspectives from the Fossil Record.* Cambridge University Press, Cambridge: pp. 79–122.

Upchurch, P., Barrett, P. M. & Dodson, P. 2004. Sauropoda. *In* Weishampel, D. B., Dodson, P. & Osmólska, H (eds.). *The Dinosauria. 2nd edition.* University of California Press, Berkeley: pp. 259–322.

Wedel, M. J. 2003. Vertebral pneumaticity, air sacs, and the physiology of sauropod dinosaurs.—*Paleobiology* 29: 243–255.

Wedel, M. J. 2005. Postcranial pneumaticity in sauropods and its implication for mass estimates. *In* Curry Rogers, K. A. & Wilson, J. A. (eds.). *The Sauropods: Evolution and Paleobiology.* University of California Press, Berkeley: pp. 201–228.

Wilson, J. A. & Carrano, M. T. 1999. Titanosaurs and the origin of "wide-gauge" trackways; a biomechanical and systematic perspective on sauropod locomotion.—*Paleobiology* 25: 252–267.

15

Neck Posture in Sauropods

ANDREAS CHRISTIAN AND GORDON DZEMSKI

THE NECK POSTURE IN sauropod dinosaurs is a crucial feature that affects their biomechanics, physiology, ecology, and evolution. Yet neck posture and utilization in sauropods are still controversial topics. In this chapter, we use a biomechanical approach to reconstruct the habitual neck posture of sauropods. The analysis is based on a comparison of stresses on the intervertebral cartilage along the vertebral column of the neck. In previous studies on extant animals with long necks, this method has shown to yield reliable results. The habitual neck posture is shown to differ considerably among sauropods. At least in some sauropod species, the long sauropod neck was biomechanically capable of both feeding at great heights and sweeping over a large feeding area without moving much of the body. Differences in neck posture indicate that the feeding strategy varied among sauropods.

Introduction

A long neck is a characteristic feature of almost all sauropod dinosaurs (McIntosh 1990; but see Rauhut et al. 2005). The necks of some sauropods, such as *Brachiosaurus, Barosaurus, Diplodocus,* and *Mamenchisaurus*, reach twice or even more the length of the trunk (e.g., Janensch 1950a, 1950b; Bonaparte 1986; McIntosh 1990). Neck posture is a crucial feature for understanding the ecology, physiology, biomechanics, and evolution of sauropods. Yet the neck posture continues to be a highly controversial subject (Figs. 15.1, 15.2). The long neck has been interpreted as either a means for high vertical browsing (e.g., Bakker 1987; Paul 1987, 1988) or for increasing the horizontal feeding range (e.g., Martin 1987). Taking a single species, *Brachiosaurus brancai* for example, the range of neck postures suggested extends from horizontal (Frey & Martin 1997; Berman & Rothschild 2005; Stevens & Parrish 2005a, 2005b), to forwardly inclined (Janensch 1950b; Christian & Dzemski 2007), to nearly vertical (Bakker 1987; Paul 1987, 1988; Christian & Heinrich 1998; Christian 2002) (Figs. 15.1, 15.2).

Some physiologists doubt that the very long necks of some sauropods, such as *Brachiosaurus,* could have been held in a more or less vertical position for long periods of time, because if the head was elevated approximately 10 m above the heart, an unlikely high blood pressure would be required to perfuse the brain (Hohnke 1973; Seymour 1976, 2009a; Hargens et al. 1987; Pedley 1987; Dodson 1990; Badeer & Hicks 1996; Seymour & Lillywhite 2000). According to Seymor (2009b), energy expenditures due to higher blood pressure increase greatly with feeding height. To avoid imposing dangerous stresses on the cardiovascular system, sauropods with extreme neck length may have habitually fed at moderate heights, and browsed at relatively greater heights for shorter periods (Dodson 1990). However, the possibility of mechanisms that would have enabled sauropods to cope with the physiological problems associated with a greatly elevated head cannot be excluded. Such mechanisms have been described for giraffes

(Dagg & Foster 1976; Hargens et al. 1987; Pedley 1987). Christian (2010) demonstrated that despite an increased metabolic rate, high browsing was worthwhile for a sauropod if resources were far apart.

There is more agreement on the neck posture in other sauropods such as *Diplodocus* and its close relatives, which are usually reconstructed with low- or medium-height neck postures (McIntosh 1990; Fastovsky & Weishampel 1996; Stevens & Parrish 1999; Berman & Rothschild 2005). Neck posture might have varied among species (Dodson 1990). Differences in neck posture and feeding strategy among sauropods appear reasonable from an ecological point of view and correspond well to the diversity of jaw and tooth morphologies observed among sauropods (e.g., Upchurch & Barrett 2000; Sereno & Wilson 2005). Adult individuals of sympatric species of sauropods may have reduced competition for food by exploiting sources at different heights.

According to some studies, sauropod necks tended to be rather stiff (Stevens & Parrish 1999). Extensive comparative studies, including those on living animals with long necks, however, indicate that even the neck of the presumably low-browsing *Diplodocus* was flexible enough for raising the head high above the shoulders (Dzemski & Christian 2007).

One difficulty in reconstructing the habitual neck posture of a sauropod is the possibility that a single individual may have employed different neck postures during different activities, as has been observed in some extant terrestrial vertebrates with long necks such as ostriches (Dzemski 2006; Christian & Dzemski 2007; Dzemski & Christian 2007). This is one likely explanation for why different methods used for reconstructing sauropod neck posture sometimes yield different results, for example, in *Brachiosaurus brancai*. The models proposed by Stevens & Parrish (1999, 2005a, 2005b) for this sauropod are based on proper articulation between the neck vertebrae, especially the zygapophyses. Their "zygapophyseal alignment" or "best-fit" postures of the neck tend to indicate a rather straight and low orientation. Berman & Rothschild (2005) also proposed a horizontal neck posture for *Brachiosaurus brancai* and other sauropods with very long necks based on an analysis of the internal structure of the vertebral centra that, according to these authors, indicates a predominance of bending moments rather than compressive forces along the neck. Gunga & Kirsch (2001) concluded from studies of the inner ear that the neck posture in *Brachiosaurus* varied from a vertical to a more horizontal posture during feeding. The overall body design of *Brachiosaurus,* with long forelimbs and a short tail, suggests at least a short-term elevated position of the neck during feeding (e.g., Paul 1987, 1988; Christian 2002, 2010; Remes et al., this volume).

FIGURE 15.1. The flexibility of sauropod necks is still controversial.

FIGURE 15.2. Neck postures for *Brachiosaurus brancai* suggested in the scientific literature. Depicted are the two extremes of all possible postures, a horizontal neck (left) and a vertical neck (right).

In our study, the neck posture of different sauropods is reconstructed by comparing the stress in the intervertebral joints of the neck. This method is first demonstrated in detail for *Brachiosaurus brancai,* then applied to the sauropods *Apatosaurus louisae, Diplodocus carnegii,* and *Euhelopus zdanskyi.* On the basis of the results of these analyses, possible feeding strategies of sauropods are discussed.

Methodology

The method presented here is based on the analysis of the stress pattern in the intervertebral cartilage along the neck. This method was first proposed by Preuschoft (1976) and is therefore referred to as the Preuschoft Method. A similar approach was used by Alexander (1985). The Preuschoft Method has been shown to be a robust and reliable means for reconstructing the posture of long necks of extinct and extant vertebrates (Christian 2002).

The necks of sauropods experience forces and torques that are a function of the posture and the distribution of neck mass. Bending moments act primarily in a sagittal plane except during rapid lateral accelerations. They are usually highest at the base of the neck and decrease toward the head. In a vertical position, bending moments are low, and weight forces predominate.

As long as the neck is not swept backward over the trunk, bending moments are counteracted by tension in muscles,

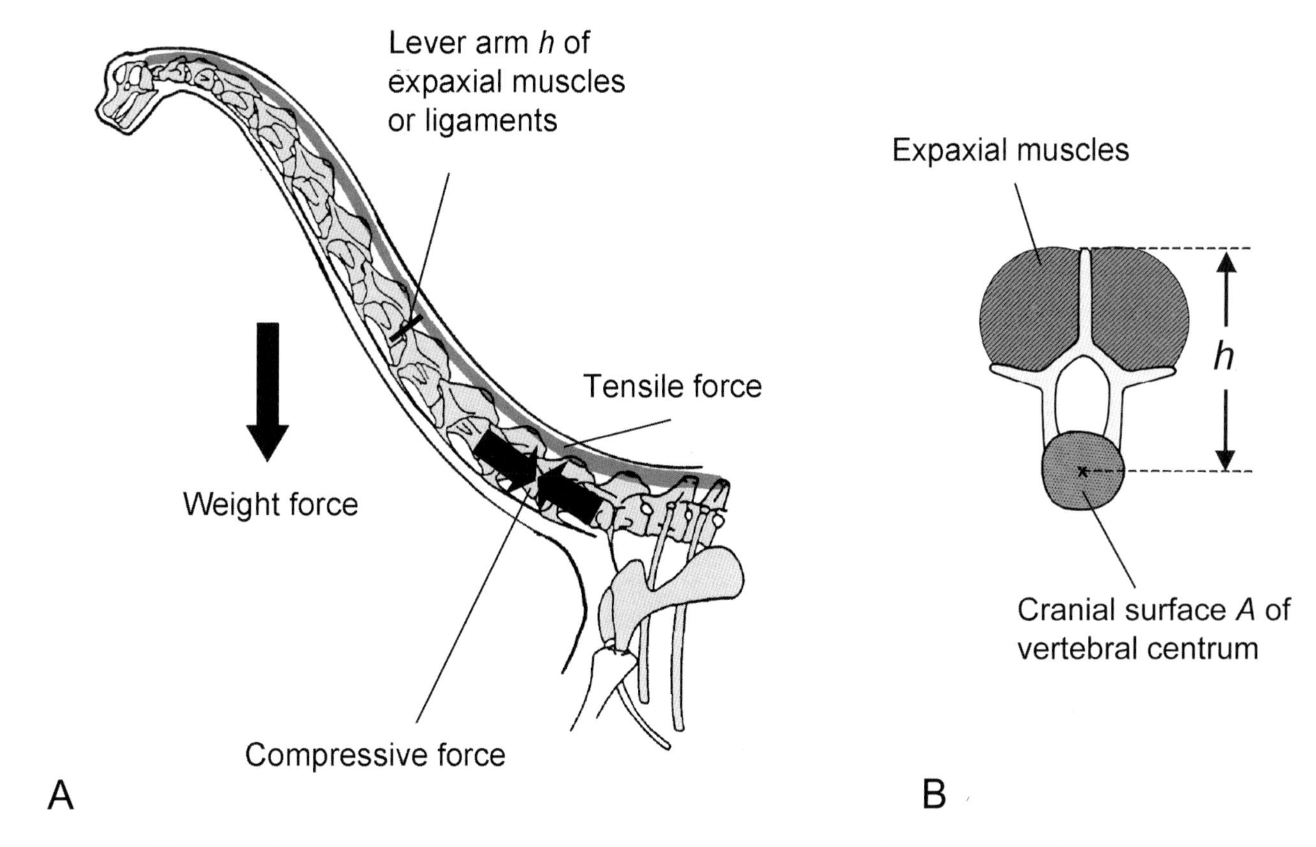

FIGURE 15.3. (A) Forces acting in and on the neck. (B) Schematic cross section of the neck with the lever arm of the epaxial muscles estimated as the distance h (see text for further explanation).

tendons, or ligaments located dorsal to the vertebral centra (Fig. 15.3; Preuschoft 1976; Alexander 1985, 1989; Christian & Preuschoft 1996). A muscle, tendon, or ligament force F_m acting in a distance h above the transverse axis of an intervertebral joint produces a torque $F_m \times h$ about that joint, where h is the lever arm of the force F_m about the intervertebral joint. The transverse axis of an intervertebral joint can be assumed to pass through the center of the intervertebral joint (Preuschoft 1976; Alexander 1985). The lever arms of the epaxial muscles can be estimated as the vertical distances between the centers of the intervertebral joints and a line that connects the tips of the neural spines (Alexander 1985; Christian 2002). Difficulties in estimating lever arms arise if epaxial muscles, tendons, or ligaments are located far above the spinal processes in parts of the neck. This is common in the necks of mammals (Preuschoft & Fritz 1977; Preuschoft & Günther 1994) but unlikely in most regions of the sauropod neck, with the exception of the basal neck region (e.g., Alexander 1985; Paul 1988; Christian & Heinrich 1998; Christian 2002; Christian & Dzemski 2007; for the reconstruction of muscles and ligaments in sauropod necks, see also Wedel & Sanders 2002; Tsuihiji 2004).

The tensile force, F_m, of epaxial muscles or ligaments produces a compressive force of the same magnitude in the intervertebral cartilage between the vertebral centra (Preuschoft 1976; Alexander 1985; Christian & Preuschoft 1996). At a given position in the vertebral column, the force F_m can be calculated as the bending moment in the sagittal plane, M, at this position divided by the lever arm h of the tensile force; that is, $F_m = M/h$ (Preuschoft 1976; Alexander 1985; Christian & Preuschoft 1996). Additionally, the intervertebral cartilage is compressed by the weight force, F_g, of the neck cranial to the position investigated multiplied by the cosine of the angle φ between the plane of the intervertebral joint and the horizontal plane (Fig. 15.3; Preuschoft 1976; Christian & Preuschoft 1996; Christian & Heinrich 1998; Christian 2002). Thus the total force F acting on the intervertebral cartilage is $F = F_m + F_g \times \cos \varphi$.

Effects from static or quasistatic forces are neglected, assuming that forces due to accelerations or other activities are not predominant. This assumption appears reasonable for sauropod necks except at the distal part of the neck close to the head, where additional forces are generated during feeding, as discussed below (Christian 2002; Christian & Dzemski 2007; Dzemski & Christian 2007). The assumption of predominantly static or quasistatic forces acting along long necks has been corroborated in studies on long-necked mammals (giraffes and camels; Christian 2002), despite the occasional use of the head for combat in these animals. Under the assumption of equal safety factors, the highest regularly occurring compressive forces, F, acting on the intervertebral cartilage along the neck should be proportional to the cross sections, A, of the intervertebral joints, so that the stress on the joint cartilage is approximately constant along the neck (Preuschoft 1976; Christian & Preuschoft 1996; Christian 2002). The stress on the intervertebral cartilage is given by the force F divided by the cross-sectional area A of the intervertebral joint. The assumption of constant stress on the intervertebral cartilage in the habitual neck posture has also been shown to be a reasonable assumption (Christian 2002). The cross-sectional area of the intervertebral cartilage is estimated by assuming an elliptical shape of the joint, with the transverse and dorsoventral diameters of the cranial surface of the adjacent vertebral centrum used as the major axes. At any given section in a sauropod neck, the caudal and cranial surfaces of adjacent centra are approximately proportional, so that two sources of measurements are available in case the cranial surface of a vertebral centrum is not well preserved.

Several hypothetical neck postures were tested in each sauropod studied. To calculate the compressive force F on the intervertebral cartilage, the distribution of neck mass needed to be determined. Reconstructions of the mass distribution in sauropod bodies, however, may differ considerably from one another (e.g., Gunga et al. 1995, 1999, 2008; Seebacher 2001; Henderson 1999, 2004; Wedel 2005; Appendix). Therefore, different models for the distribution of the head and neck mass were used for the sauropods under investigation. The mass of the head was estimated by the dimensions of the skull and an assumed density of the head of 1.0 g/cm^3. Head mass estimates were successfully tested on some recent vertebrates (ostriches, horses, and giraffes). The density of the neck in sauropods was probably much lower than that of the head as a result of air sacs that were presumably present in the necks of sauropods (Henderson 1999, 2004; Seebacher 2001; Wedel 2003a, 2003b, 2005; Schwarz & Fritsch 2006; Schwarz et al. 2007; Sander et al. 2010; Perry et al., this volume). The mass of different neck segments was estimated from scaled plaster models of sauropods, as well as from the dimensions of the vertebrae.

With the exception of *Apatosaurus,* a neck density of 0.67 g/cm^3 without the tracheal air volume was used for the calculations. For *Apatosaurus,* as explained further below, a lower density of the neck had to be used. For most species, additional calculations were performed with neck densities between 0.33 and 1.0 g/cm^3 in order to test the effect of variation in neck mass on the calculated stress patterns. An even wider range of neck mass was tested in *Brachiosaurus*. The actual density is considerably lower than these values above because the cavity of the trachea was not taken into account.

For all hypothetical neck postures, the compressive force, F, was calculated along the neck and divided by the cross-sectional area, A, of the intervertebral joints. The stress, F/A, is expected to be more or less constant for habitual postures. Hypothetical neck postures were rejected in which the stress was not more or less constant along the neck.

Even when the distribution of mass along the head and neck and the lever arms of the neck muscles and ligaments are only roughly estimated, the Preuschoft Method can be applied because it is not affected by systematic errors in estimates of segment masses, lever arms, muscle forces, or cross-sectional areas of intervertebral discs (Dzemski & Taylor, forthcoming). Therefore, the Preuschoft Method tolerates inaccuracies in the estimates of the head and neck weights based on reconstructions and the lever arms or cross-sectional areas, as long as the errors are consistent for different segments along the neck. However, the results are affected by the relationship between head and neck mass. Therefore, relatively high differences in the estimates for the neck mass result in visible differences in the reconstructed neck posture. This is the reason why a wide range of apparently reasonable neck densities was tested.

MATERIAL

Four sauropods with well preserved neck skeletons were studied: *Apatosaurus louisae* (CM 3018), *Brachiosaurus brancai* (specimens MFN S I, S II for the reconstruction of the neck, and specimen MFN MB.R.2223 [field number t 1] for the reconstruction of the head), *Diplodocus carnegii* (CM 84), and *Euhelopus zdanskyi* (PMU R233). Additional data were also collected from a specimen of *Diplodocus* (SMA H.Q. 1) on exhibit in the Sauriermuseum Aathal, Switzerland. Measurements of neck vertebra dimensions were taken from the original skeletons, casts, scaled illustrations, or photographs. Damaged vertebrae were either reconstructed or excluded from the calculations, depending on the degree of damage. For *Brachiosaurus brancai,* only the cranial 4.8 m of the neck was studied because the neural spines are not preserved in the caudal part of the neck.

Results

THE NECK POSTURE OF *BRACHIOSAURUS BRANCAI*

Hypothetical neck postures of *Brachiosaurus* were tested that differed in inclination and in the shape of the neck. These postures are illustrated in Figs. 15.4 and 15.5, along with the calculated stress values along the well preserved first 4.8 m of the neck. The density was assumed to be 0.67 g/cm^3 without taking into account the air volume of the trachea. The assumed mass distribution of the neck is given in Table 15.1.

In all postures tested with the neck held straight but not in a vertical position, the stress on the intervertebral cartilage increased toward the base of the neck. In a vertical neck posture, the stress is more or less constant along the neck section examined, except at the joint between the axis and third ver-

Table 15.1. Basic Estimates for Segment Mass of the Head and the First 4.8 m of Neck of *Brachiosaurus brancai*

Segment	Length (m)	Mass (kg)
Head	–	100
Neck 1	0.42	60
Neck 2	0.40	50
Neck 3	0.60	75
Neck 4	0.72	91
Neck 5	0.83	132
Neck 6	0.86	219
Neck 7	0.96	328

From Christian & Dzemski (2007).

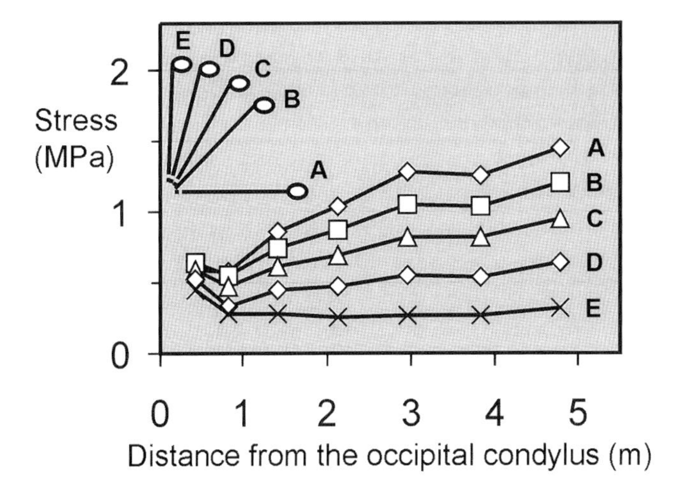

FIGURE 15.4. The calculated stress on the intervertebral joints along the foremost 4.8 m of the neck in *Brachiosaurus brancai* for five hypothetical straight neck postures A–E. The horizontal neck posture A results in the highest stress values; a vertical neck E leads to minimal stress values. A steeply inclined posture D leads to minimal variation in the intervertebral stress along the neck.

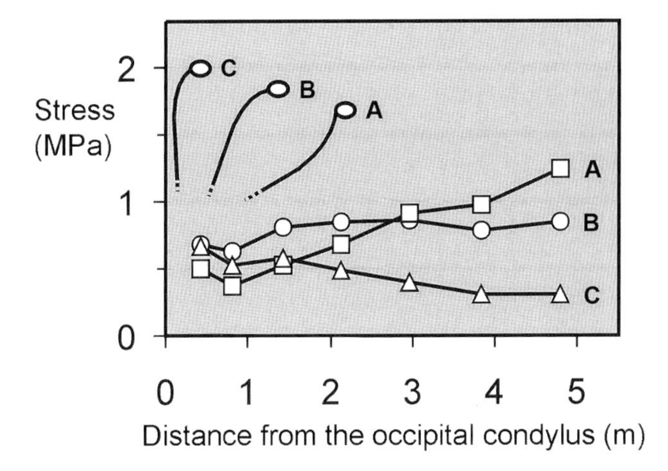

FIGURE 15.5. The calculated stress on the intervertebral joints along the foremost 4.8 m of the neck in *Brachiosaurus brancai* for three hypothetical bent neck postures A–C. A cranially bent distal part of a vertical neck leads to minimal stress values C, while an inclined, dorsally bent neck produces much higher stress values in the distal intervertebral joints A. The least amount of stress is observed in posture B.

Table 15.2. Standard Deviations of Calculated Stress Values

Neck posture	SD/MS[a]
Horizontal (Fig. 15.4A)	0.34
Horizontal double mass	0.46
Horizontal half mass	0.21
Inclined (Fig. 15.4B)	0.28
Inclined (Fig. 15.4C)	0.23
Inclined (Fig. 15.4D)	0.19
Vertical (Fig. 15.4E)	0.22
Flexed (Fig. 15.5A)	0.42
Flexed (Fig. 15.5B)	0.12
Flexed (Fig. 15.5C)	0.29

[a]SD/MS is the standard deviation divided by mean stress for all joints between the second and eighth vertebral centra. *From Christian & Dzemski (2007).*

tebra where the stress is comparatively high (Fig. 15.4). More or less constant stress values were also obtained with an inclined neck that is flexed ventrally at its distal end (Fig. 15.5). In this posture, the stress is low in the joints close to the head. Variation in stress is considerably higher in the other neck postures tested. Variation in the stress values for different neck postures can be compared by dividing the standard deviations by the mean values (Table 15.2).

On the basis of these results, two neck postures yielding more or less constant stress values are plausible: a vertical and an inclined neck posture with a ventral flexion close to the head. In the inclined posture, however, the calculated stress values are comparatively low close to the head, whereas they are comparatively high in this region in a vertical posture. Because of additional forces incurred during repositioning and moving the head while feeding, the calculated static stress values should be comparatively lower close to the head. Therefore, the vertical posture must be rejected, despite the more or less constant stress values along the neck, whereas the inclined posture with a ventrally flexed anterior part of the neck fits this prediction well.

In the neck skeleton of *Brachiosaurus brancai,* a marked increase in the heights of the neural spines can be observed between the sixth and seventh vertebrae. This increase in the lever arms of epaxial muscles also indicates a ventral flexion of the portion of the neck cranial to the seventh vertebra (Christian & Dzemski 2007; see also Wedel et al. 2000 for a similar observation in *Sauroposeidon*). The posture of the basal section

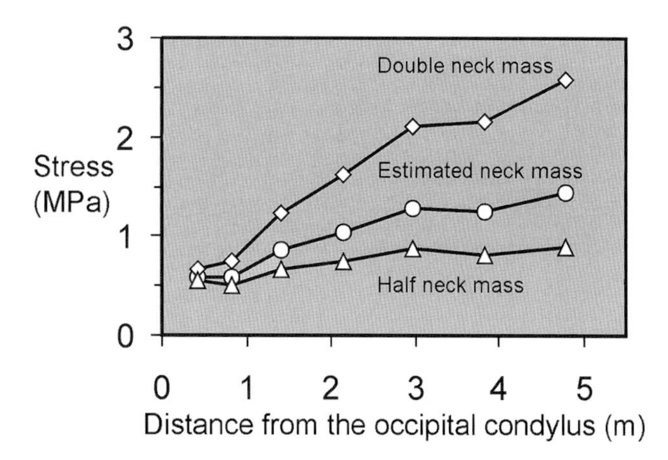

FIGURE 15.6. The influence of neck mass on the calculated stress values demonstrated for a horizontal neck posture in *Brachiosaurus brancai* with three different mass estimates. The mass of the head is not altered. Heavier necks lead to increasingly high stress values in the distal intervertebral joints.

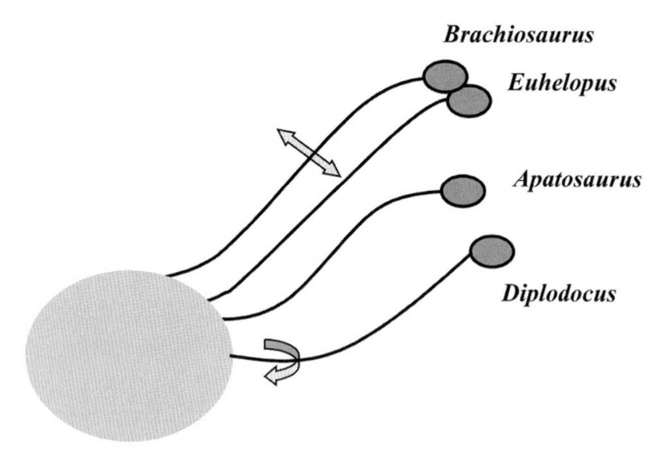

FIGURE 15.7. The habitual neck posture reconstructed for some sauropods at rest based on minimal stress values in the intervertebral joints. Arrows indicate neck movements during feeding. All sauropods were able to move the neck sideways during feeding. *Brachiosaurus* and *Euhelopus* also employed vertical movements of the neck.

of the neck cannot be reconstructed reliably because the neural spines are not preserved. Given the inclined posture of the middle section of the neck, however, there must have been an upward bend in the basal part of the neck when *Brachiosaurus* was standing at rest (Christian & Dzemski 2007).

Variation in neck mass relative to that of the head does change the stress pattern along the neck. With decreasing neck mass, the variation of stress along a horizontal neck is reduced. In general, models with lighter necks result in lower postures of the neck, but even in extremely light necks, stress values are not constant in a horizontal posture (Fig. 15.6). Therefore, in contrast to the results of Stevens & Parrish (2005a, 2005b), it appears unlikely that *Brachiosaurus brancai* regularly employed a horizontal neck posture while standing at rest.

On the basis of the mass distribution given in Table 15.1, the Preuschoft Method yields an inclination of the neck of roughly 60° from the horizontal. This calculation is already based on the assumption of a lightly built neck. The neck mass was possible even lower, yielding a less inclined habitual posture of the neck during standing. With a neck density between 0.4 and 0.5 g/cm^3, an inclination angle of 45° would be in accordance with the results of the Preuschoft Method.

Extant animals with long necks typically reduce the height of the head during locomotion (Dzemski & Christian 2007). In *Brachiosaurus,* a reduction in the neck inclination of 20° or greater can be expected during locomotion, so that the inclination of the neck did not differ much from the inclination of the trunk. This means that the entire presacral vertebral column was almost straight during locomotion (Christian & Dzemski 2007).

VARIATION OF NECK POSTURE AMONG SAUROPODS

The Preuschoft Method yields a similar but slightly more inclined neck posture in *Euhelopus zdanskyi* than in *Brachiosaurus brancai* (Fig. 15.7). For *Diplodocus carnegii,* as well as for the *Diplodocus* (H.Q. 1) in the Sauriermuseum Aathal collections, a considerably lower neck posture is obtained. In contrast to the reconstruction by Stevens & Parrish (1999), the habitual posture of the neck in *Diplodocus* at rest appears to have been not unlike the neck posture of a camel (Fig. 15.7).

When the most probable habitual neck posture for several sauropods and large extant animals with long necks are compared, similar absolute stress values in the intervertebral joints show up, which usually range between 0.5 and 0.9 MPa. However, for *Apatosaurus louisae,* and its more voluminous neck, such stress values could not be obtained unless models with a very low density—clearly below 0.5 g/cm^3—were chosen. This strongly suggests that the neck in *Apatosaurus* contained especially large air sacs. Under the assumption of a very lightly built neck with a density between 0.3 and 0.5 g/cm^3, the reconstructed neck posture in *Apatosaurus* was probably similar to that of a camel with an additional downward bend in the foremost part of the neck (Fig. 15.7).

Discussion

NECK POSTURE AT REST

The habitual posture of the neck of a *Brachiosaurus brancai* and *Euhelopus zdanskyi* at rest appears to have been not only less upright than has been proposed by several workers (e.g., Bakker 1987; Paul 1988; Christian & Heinrich 1998), but also not as horizontal as proposed by Frey & Martin (1997), Berman & Rothschild (2005), and Stevens & Parrish (2005a, 2005b). The "zygapophyseal alignment" or "best-fit" postures proposed by Stevens and Parrish (1999, 2005a, 2005b) yield comparatively straight necks that are held low. However, comparative studies on living vertebrates with long necks indicate that such best-fit models tend to yield neck postures lower than the habitual neck postures during rest, but approximate the posture during locomotion (Dzemski 2006; Christian & Dzemski 2007; Dzemski & Christian 2007). In contrast to most other reconstructions of *Brachiosaurus* (e.g., Christian & Dzemski 2007), in the model proposed by Stevens and Parrish (1999, 2005a, 2005b), the vertebral column of the trunk is not inclined upward immediately behind the neck. Berman & Rothschild (2005) concluded that the neck was held horizontally in *Brachiosaurus* and other sauropods with long necks because the internal structure of the vertebral centra indicates a predominance of bending moments rather than compressive forces along the neck. However, more detailed mechanical studies including finite element analyses are necessary for a better understanding of the reinforcement structure of the centra. In any case, bending moments along the neck are predominant, even when the neck posture deviates by less than 20° from the vertical. The argument of Berman & Rothschild (2005) allows a wide range of possible postures and does not exclude the possibility that even a fully vertical posture was temporarily adopted (e.g., during feeding).

In the light of recent mass estimates (e.g., Henderson 1999, 2004; Seebacher 2001; Wedel 2005), the Preuschoft Method yields a habitual neck posture in *Brachiosaurus brancai* and *Euhelopus zdanskyi* that is similar to the early reconstruction of *Brachiosaurus brancai* by Janensch (1950b) which proposed an upward bend at the base of the neck, a rather straight and inclined middle section of the neck, followed by a more or less pronounced downward bend in the section close to the head. Depending on the estimated neck mass, the Preuschoft Method yields an inclination of the neck between 45° and 60° from the horizontal in these sauropods.

For *Diplodocus* and *Apatosaurus,* the results obtained by the Preuschoft Method indicate a medium high habitual neck posture during standing that was not unlike that of a camel except for a downward flexion in the neck of *Apatosaurus* closest to the head. In *Apatosaurus* and *Diplodocus,* the vertebral column of the neck was less inclined than in *Brachiosaurus,* especially at the base, according to the reconstructions by Stevens and Parrish (1999, 2005a, 2005b). In contrast to Stevens and Parrish (1999, 2005a, 2005b), however, the Preuschoft Method yields an upward bend between the middle and the base of the neck, so that the head was held well above the shoulders while standing at rest. For ostriches, camels, and giraffes, Dzemski & Christian (2007) found a correlation between the dorsoventral flexibility and the posture of the neck. Applied to *Diplodocus carnegii,* these results corroborate the assumption of an upward bend in the neck of *Diplodocus* (Dzemski & Christian 2007). Even if sauropods such as *Apatosaurus* and *Diplodocus* fed at low levels, an elevated position of the head at rest appears reasonable from an ecological point of view with regard to a better overview of the surrounding area.

NECK POSTURE DURING OTHER ACTIVITIES

According to the findings by Dzemski (2006) and by Dzemski & Christian (2007) for all sauropods studied, it can be assumed that during locomotion, the entire neck was held at a lower position than at rest. During feeding, the neck posture could have differed considerably from that at rest. In all sauropods studied, the section of the neck close behind the head was comparatively flexible (see also Upchurch & Barrett 2000), so that moderate vertical movements of the head could have been performed with little energy expenditure without greatly altering the height of the center of gravity of the neck. Lateral movements of the head without altering the height of the neck were performed by lateral flexion at the base of the neck (Christian & Dzemski 2007; Dzemski & Christian 2007; Christian 2010).

In sauropods with comparatively straight and upward-inclined, giraffe-like necks, such as *Brachiosaurus* and *Euhelopus,* pronounced changes in feeding heights were possible by slow upward and downward movements of the entire neck. In order to maximize accessibility to food sources, the neck could have been raised into a vertical position or swung in wide lateral arcs at medium heights. Therefore, the very long neck of these sauropods was both a means of browsing at great heights and a means of increasing feeding volume without moving the body (Christian & Dzemski 2007).

The more camel-like necks of sauropods such as *Apatosaurus* and *Diplodocus* could have been easily flexed downward so that the head reached the ground with little displacement of the center of gravity of the neck (Dzemski & Christian 2007). This supports the widespread view that sauropods such as *Diplodocus* and *Apatosaurus* kept the neck at low or medium heights in order to sweep over a large feeding area (e.g., Dodson 1990; McIntosh 1990; Stevens & Parrish 1999; Christian 2002; Dzemski & Christian 2007). Both species possessed a very long tail that may have been useful as passive and active counterweight when the neck was moved sideways. Compared to *Brachiosaurus,* the forelimbs in these species were short and forceful (Christiansen 1997) and therefore suitable for counteracting torques and sideways forces that were transmitted by the neck and tail to the trunk. The results presented here do not exclude the possibility that the diplodocids occasionally adopted upright neck postures or that *Diplodocus*

FIGURE 15.8. Variation of neck posture of *Brachiosaurus brancai* during different activities. The posture at rest (dark colored, elevated neck) probably was considerably higher than the posture during locomotion (dark colored, low neck). Different neck postures can be assumed during feeding (light colored postures).

employed a tripodal stance on the hindlimbs, with the tail serving as a third limb to reach high food sources (Bakker 1987; Paul 1988; but see also Rothschild & Berman 1991).

Conclusions

Despite the similarities in the overall body construction, different sauropods employed different feeding strategies (see also Hummel & Clauss, this volume). High browsers such as *Brachiosaurus* are characterized by a rather straight, upward-inclined neck, similar to that of giraffes, which appears to have been primarily used for feeding at medium and great heights. In forms like *Diplodocus,* the neck appears to have been more suitable for feeding at low levels. However, in all species investigated, the neck was flexible enough for large vertical movements of the head, so that food sources at different heights could have been exploited. This might be especially true if a tripodal stance was possible in those forms with an otherwise comparatively low posture of the neck, such as that in *Diplodocus.* For the same species, different reconstructions of the neck posture may be correct because different neck postures were employed during different activities, as is observed in extant vertebrates with long necks such as giraffes, ostriches, and camels. While standing at rest, low browsers probably raised the head well above the shoulders in a camel-like fashion, allowing a better overview of the surrounding area for potential predators. During locomotion, a lower position of the head would be expected. It appears that the neck was generally kept under more tension during locomotion when compared with standing at rest. It can be concluded that the neck posture of sauropods varied considerably interspecifically, reflecting adaptation to different ecological niches, as well as intraspecifically, reflecting different behavioral activities of individual sauropods.

Acknowledgments

We are obliged to the Bayerische Staatssammlung für Paläontologie und Geologie in Munich, Germany, the Sauriermuseum Aathal in Switzerland, the Museum für Naturkunde in Berlin, Germany, and Uppsala University in Sweden, for their support in collecting data. Hanns-Christian Gunga (Free University, Berlin) and Albert Wiedemann supplied original data for segment masses along the necks of sauropods analyzed in this study. Gundula Christian and Jan-Thomas Möller (University of Flensburg) assisted in collecting data. We thank David S. Berman (Carnegie Museum, Pittsburgh) and Paul Upchurch (University College, London) for critical and constructive reviews of the manuscript. This is contribution number 74 of the DFG Research Unit 533 "Biology of the Sauropod Dinosaurs: The Evolution of Gigantism."

References

Alexander, R. M. 1985. Mechanics of posture and gait of some large dinosaurs.—*Zoological Journal of the Linnean Society of London* 83: 1–25.

Alexander, R. M. 1989. *Dynamics of Dinosaurs and Other Extinct Giants.* Columbia University Press, New York.

Badeer, H. S. & Hicks, J. W. 1996. Circulation to the head of *Barosaurus* revised: theoretical considerations.—*Comparative Biochemistry and Physiology A* 114: 197–203.

Bakker, R. 1987. *The Dinosaur Heresies.* Longman Scientific & Technical, Harlow, UK.

Berman, D. S. & Rothschild, B. M. 2005. Neck posture of sauropods determined using radiological imaging to reveal three-dimensional structure of cervical vertebrae. *In* Tidwell, V. & Carpenter, K. (eds.). *Thunder Lizards: The Sauropodomorph Dinosaurs.* Indiana University Press, Bloomington: pp. 233–247.

Bonaparte, J. F. 1986. The early radiation and phylogenetic relationship of the Jurassic sauropod dinosaurs, based on vertebral anatomy. *In* Padian, K. (ed.). *The Beginning of the Age of Dinosaurs.* Cambridge University Press, Cambridge: pp. 247–258.

Christian, A. 2002. Neck posture and overall body design in sauropods.—*Mitteilungen aus dem Museum für Naturkunde in Berlin, Geowissenschaftliche Reihe* 5: 269–279.

Christian, A. 2010. Some sauropods raised their necks—evidence for high browsing in *Euhelopus zdanskyi*.—*Biology Letters* 6: 823–825.

Christian, A. & Dzemski, G. 2007. Reconstruction of the cervical skeleton posture of *Brachiosaurus brancai* Janensch, 1914 by an analysis of the intervertebral stress along the neck and a comparison with the results of different approaches.—*Fossil Record* 10: 37–48.

Christian, A. & Heinrich, W.-D. 1998. The neck posture of *Brachiosaurus brancai*.—*Mitteilungen aus dem Museum für Naturkunde in Berlin, Geowissenschaftliche Reihe* 1: 73–80.

Christian, A. & Preuschoft, H. 1996. Deducing the body posture of extinct large vertebrates from the shape of the vertebral column.—*Palaeontology* 39: 801–812.

Christiansen, P. 1997. Locomotion in sauropod dinosaurs.—*Gaia* 14: 45–75.

Dagg, A. I. & Foster, J. B. 1976. *The Giraffe: Its Biology, Behavior, and Ecology.* Van Nostrand Reinhold, New York.

Dodson, P. 1990. Sauropod paleoecology. *In* Weishampel, D. B., Dodson, P. & Osmólska, H. (eds.). *The Dinosauria.* University of California Press, Berkeley: pp. 402–407.

Dzemski, G. 2006. *Funktionsmorphologische Analysen langer Hälse bei rezenten terrestrischen Wirbeltieren zur Rekonstruktion der Stellung und Beweglichkeit langer Hälse prähistorischer Tiere.* Ph.D. Dissertation. University of Flensburg, Flensburg.

Dzemski, G. & Christian, A. 2007. Flexibility along the neck of the ostrich (*Struthio camelus*) and consequences for the reconstruction of dinosaurs with extreme neck length.—*Journal of Morphology* 268: 701–714.

Dzemski, G. & Taylor, M. Forthcoming. Intervertebral disc space in mammals and sauropods necks.—*Journal of Morphology.*

Fastovsky, D. E. & Weishampel, D. B. 1996. *The Evolution and Extinction of the Dinosaurs.* Cambridge University Press, Cambridge.

Frey, E. & Martin, J. 1997. Long necks of sauropods. *In* Currie, P. J. & Padian, K. (eds.). *Encyclopedia of Dinosaurs.* Academic Press, San Diego: pp. 406–409.

Gunga, H.-C. & Kirsch, K. 2001. Von Hochleistungsherzen und wackeligen Hälsen.—*Forschung* 2: 4–9.

Gunga, H.-C., Kirsch, K. A., Baartz, F., Röcker, L., Heinrich, W.-D., Lisowski, W., Wiedemann, A. & Albertz, J. 1995. New data on the dimensions of *Brachiosaurus brancai* and their physiological implications.—*Naturwissenschaften* 82: 189–192.

Gunga, H.-C., Kirsch, K., Rittweger, J., Röcker, L., Clarke, A., Albertz, J., Wiedemann, A. Mokry, S., Suthau, T., Wehr, A., Heinrich, W.-D. & Schultze, H.-P. 1999. Body size and body volume distribution in two sauropods from the Upper Jurassic of Tendaguru (Tanzania).—*Mitteilungen aus dem Museum für Naturkunde in Berlin, Geowissenschaftliche Reihe* 2: 91–102.

Gunga, H.-C., Suthau, T., Bellmann, A., Stoinski, S., Friedrich, A., Trippel, T., Kirsch, K. & Hellwich, O. 2008. A new body mass estimation of *Brachiosaurus brancai* Janensch, 1914 mounted and exhibited at the Museum of Natural History (Berlin, Germany).—*Fossil Record* 11: 33–38.

Hargens, A. R., Millard, R. W., Pettersson, K. & Johansen, K. 1987. Gravitational haemodynamics and oedema prevention in the giraffe.—*Nature* 329: 59–60.

Henderson, D. M. 1999. Estimating the masses and centers of mass of extinct animals by 3-D mathematical slicing.—*Paleobiology* 25: 88–106.

Henderson, D. M. 2004. Tipsy punters: sauropod dinosaur pneumaticity, buoyancy and aquatic habits.—*Proceedings of the Royal Society B: Biological Sciences* 271: 180–183.

Hohnke, L. A. 1973. Haemodynamics in the Sauropoda.—*Nature* 244: 309–310.

Hummel, J. & Clauss, M. This volume. Sauropod feeding and digestive physiology. *In* Klein, N., Remes, K., Gee, C. T. & Sander, P. M. (eds.). *Biology of the Sauropod Dinosaurs: Understanding the Life of Giants.* Indiana University Press, Bloomington: pp. 11–33.

Janensch, W. 1950a. Die Wirbelsäule von *Brachiosaurus brancai.*—*Palaeontographica Supplement* 7 (3): 27–92.

Janensch, W. 1950b. Die Skelettrekonstruktion von *Brachiosaurus brancai.*—*Palaeontographica Supplement* 7 (3): 95–103.

Martin, J. 1987. Mobility and feeding of *Cetiosaurus* (Saurischia: Sauropoda)—why the long neck? *In* Curry, P. J. & Koster, E. H. (eds.). *4th Symposium on Mesozoic Terrestrial Ecosystems, Short Papers.* Royal Tyrell Museum of Palaeontology, Drumheller: pp. 154–159.

McIntosh, P. 1990. Sauropod Paleoecology. *In* Weishampel, D. B., Dodson, P. & Osmólska, H. (eds.). *The Dinosauria.* University of California Press, Berkeley: pp. 345–401.

Paul, G. S. 1987. The science and art of restoring the life appearance of dinosaurs and their relatives. *In* Czerkas, S. J. & Olsen, E. C. (eds.). *Dinosaurs Past and Present, Vol. 2.* Natural History Museum of Los Angeles County, Los Angeles: pp. 5–49.

Paul, G. S. 1988. The brachiosaur giants of the Morrison and Tendaguru with a description of a new subgenus, *Giraffatitan,* and a comparison of the world's largest dinosaurs.—*Hunteria* 2: 1–14.

Pedley, T. J. 1987. How giraffes prevent oedema.—*Nature* 329: 13–14.

Perry, S. F., Breuer, T. & Pajor, N. This volume. Structure and function of the sauropod respiratory system. *In* Klein, N., Remes, K., Gee, C. T. & Sander, P. M. (eds.). *Biology of the Sauropod Dinosaurs: Understanding the Life of Giants.* Indiana University Press, Bloomington: pp. 57–79.

Preuschoft, H. 1976. Funktionelle Anpassung evoluierender Systeme.—*Aufsätze und Reden der Senckenbergischen Naturforschenden Gesellschaft* 28: 98–117.

Preuschoft, H. & Fritz, M. 1977. Mechanische Beanspruchung im Bewegungsapparat von Springpferden.—*Fortschritte der Zoologie* 24: 75–98.

Preuschoft, H. & Günther, M. M. 1994. Biomechanics and body shape in primates compared with horses.—*Zeitschrift für Morphologie und Anthropologie* 80: 149–165.

Rauhut, O. W. M., Remes, K., Fechner, R., Cladera, G. & Puerta, P. 2005. Discovery of a short-necked sauropod dinosaur from the Late Jurassic period of Patagonia.—*Nature* 435: 670–672.

Remes, K. R., Unwin, D. M., Klein, N., Heinrich, W.-D. & Hampe, O. This volume. Skeletal reconstruction of *Brachiosaurus brancai* in the Museum für Naturkunde, Berlin: summarizing 70 years of sauropod research. *In* Klein, N., Remes, K., Gee, C. T. & Sander, P. M. (eds.). *Biology of the Sauropod Dinosaurs: Understanding the Life of Giants.* Indiana University Press, Bloomington: pp. 305–316.

Rothschild, B. M. & Berman, D. S. 1991. Fusion of caudal vertebrae in Late Jurassic sauropods.—*Journal of Vertebrate Paleontology* 11: 29–36.

Sander, P. M., Christian, A. & Gee, C. T. 2009. Sauropods kept their heads down. Response.—*Science* 323: 1671–1672.

Sander, P. M., Christian, A., Clauss, M., Fechner, R., Gee, C. T., Griebeler, E. M., Gunga, H.-C., Hummel, J., Mallison, H.,

Perry, S., Preuschoft, H., Rauhut, O., Remes, K., Tütken, T., Wings, O. & Witzel, U. 2010. Biology of the sauropod dinosaurs: the evolution of gigantism—*Biological Reviews of the Cambridge Philosophical Society.* doi: 10.1111/j.1469=185X .2010.00137.x.

Schwarz, D. & Fritsch, G. 2006. Pneumatic structures in the cervical vertebrae of the Late Jurassic (Kimmeridgian-Tithonian) Tendaguru sauropods *Brachiosaurus brancai* and *Dicraeosaurus.*—*Eclogae Geologicae Helvetiae* 99: 65–78.

Schwarz, D., Frey, E. & Meyer, C. A. 2007. Pneumaticity and soft-tissue reconstructions in the neck of diplodocid and dicraeosaurid sauropods.—*Acta Palaeontologica Polonica* 52: 167–188.

Seebacher, F. 2001. A new method to calculate allometric length-mass relationships of dinosaurs.—*Journal of Vertebrate Paleontology* 21: 51–60.

Sereno, P. C. & Wilson, J. A. 2005. Structure and evolution of a sauropod tooth battery. *In* Wilson, J. A. & Curry Rogers, K. (eds.). *The Sauropods: Evolution and Paleobiology.* University of California Press, Berkeley: pp. 157–177.

Seymour, R. S. 1976. Dinosaurs, endothermy and blood pressure.—*Nature* 262: 207–208.

Seymour, R. D. 2009a. Sauropods kept their heads down.—*Science* 323: 1671.

Seymour, R. S. 2009b. Raising the sauropod neck: it costs more to get less.—*Biology Letters* 5: 317–319.

Seymour, R. S. & Lillywhite, H. B. 2000. Hearts, neck posture, and metabolic intensity of sauropod dinosaurs.—*Proceedings of the Royal Society B: Biological Sciences* 267: 1883–1887.

Stevens, K. A. & Parrish, M. J. 1999. Neck posture and feeding habits of two Jurassic sauropod dinosaurs.—*Science* 284: 798–800.

Stevens, K. A. & Parrish, M. J. 2005a. Digital reconstructions of sauropod dinosaurs and implications for feeding. *In* Wilson, J. A. & Curry Rogers, K. (eds.). *The Sauropods: Evolution and Paleobiology.* University of California Press, Berkeley: pp. 178–200.

Stevens, K. A. & Parrish, M. J. 2005b. Neck posture, dentition and feeding strategies in Jurassic sauropod dinosaurs. *In* Tidwell, V. & Carpenter, K. (eds.). *Thunder Lizards: The Sauropodomorph Dinosaurs.* Indiana University Press, Bloomington: pp. 212–232.

Tsuihiji, T. 2004. The ligament system in the neck of *Rhea americana* and its implication for the bifurcated neural spines of sauropod dinosaurs.—*Journal of Vertebrate Paleontology* 24: 165–172.

Upchurch, P. & Barrett, P. M. 2000. The evolution of sauropod feeding mechanisms. *In* Sues, H.-D. (ed.). *Evolution of Herbivory in Terrestrial Vertebrates: Perspectives from the Fossil Record.* Cambridge University Press, Cambridge: pp. 79–122.

Wedel, M. J. 2003a. Vertebral pneumacity, air sacs, and the physiology of sauropod dinosaurs.—*Paleobiology* 29: 243–255.

Wedel, M. J. 2003b. The evolution of vertebral pneumacity in sauropod dinosaurs.—*Journal of Vertebrate Paleontology* 23: 344–357.

Wedel, M. J. 2005. Postcranial skeletal pneumaticity in sauropods and its implications for mass estimates. *In* Wilson, J. A. & Curry Rogers, K. (eds.). *The Sauropods: Evolution and Paleobiology.* University of California Press, Berkeley: pp. 201–228.

Wedel, M. J. & Sanders, R. K. 2002. Osteological correlates of cervical musculature in Aves and Sauropoda (Dinosauria: Saurischia), with comments on the cervical ribs of *Apatosaurus.*—*PaleoBios* 22: 1–6.

Wedel, M. J., Cifelli, R. I. & Sanders, R. K. 2000. *Sauroposeidon proteles,* a new sauropod from the Early Cretaceous of Oklahoma. —*Journal of Vertebrate Paleontology* 20: 109–114.

PART FOUR

GROWTH

16

The Life Cycle of Sauropod Dinosaurs

EVA MARIA GRIEBELER AND JAN WERNER

BECAUSE SAUROPOD DINOSAURS ARE EXTINCT, it might seem impossible to fully reconstruct their life cycles. Nevertheless, information on reproduction, reproductive behavior, growth in body size, and sexual maturity can be indirectly derived from the fossil record. In addition, we can also use living, phylogenetically related taxa as models for these extinct animals in order to support and expand our knowledge on sauropod life cycles. Predictions from life history theory on the relationship between reproductive traits and body size as well as the analyses of life cycle characteristics of extant reptiles, birds, and mammals are also appropriate. In the present chapter, we utilize this complex approach for the reconstruction of sauropod life cycles. We summarize the information on eggs, clutches, nests, hatching, adolescence, and growth in body size that has been derived from the fossil record. In addition, we try to fill the gaps in our knowledge concerning the reproductive behavior, the total reproductive output of animals, and the mortality during the life cycle using information from extant phylogenetic brackets or predictions of life history theory. Finally, we discuss hypotheses explaining gigantism of sauropods based on their life cycles.

Geographic Distribution of Eggshells, Intact Eggs, and Eggs with Embryonic Bone or Integument

As with all extant archosaurs such as birds and crocodilians, the life of a sauropod began in an egg. Dinosaur eggshell fragments were first found by Jean-Jacques Pouech in the rugged, egg-rich foothills of the French Pyrenees in 1859 (Buffetaut & Le Loeuff 1994). Matheron (1869) attributed similar eggshells to the sauropod *Hypselosaurus* on the basis of the thickness of the eggshell and the presence of this species in the same formation.

These large, round eggs are distributed nearly worldwide, occur exclusively in the Upper Cretaceous, and belong to the oofamily Megaloolithidae (Plate 16.1). Eggs identical to those first found in France are also known from the Tremp Basin in northeastern Spain (Sander et al. 1998, 2008; López-Martinéz 2000). During the Late Cretaceous and before the uplift of the Pyrenees, a continuous facies belt existed between the Spanish and French egg sites. Eggs in the Tremp Formation in northern Spain are preserved in paleosols that were developed on alluvial to marginal marine deposits (Sander et al. 2008). The Megaloolithidae include over 20 oospecies, and the most thoroughly documented localities occur in Argentina, northern Spain, southern France, and India; however, other occurrences are known from Romania, Korea, and Africa (Sander et al. 2008; Grellet-Tinner & Fiorelli 2010).

Substantial diversity within the Megaloolithidae is also documented in the Lameta Formation in north-central India (Mohabey 2001, 2005). The egg-bearing strata occur in the Deccan Traps, a rugged landscape containing lava flows that record immense volcanism during Late Cretaceous (Maastrichtian) time. The sauropods inhabited this region between periods of volcanic eruptions. Thousands of eggshells and unhatched eggs have been recovered, but none preserves embryonic remains. Although most of the egg material from the Lameta Formation is assigned to the Megaloolithidae (Sahni et al. 1994), detailed observations on egg occurrence have not been published (Sander et al. 2008).

In contrast to these Indian and European sites, the Auca Mahuevo locality in Argentina produced the first eggs containing sauropod embryonic bones and integument in 1997. Cranial characters of the in ovo embryos made it possible to identify the eggs as those of titanosaur sauropods (Chiappe et al. 1998, 2001; Salgado et al. 2005). This spectacular nesting site is located about 120 km north of the city Neuquén in the Neuquén Province. Egg-bearing layers occur in uniform mudstones representing overbank deposits on a fluvial plain. Several egg-bearing beds have also been reported from Barreales Norte and Barreales Escondido, two adjacent nesting sites 15 and 22 km south of Auca Mahuevo, respectively. Auca Mahuevo nesting sites include thousands of in situ titanosaur fossil eggs referable to *Megaloolithus patagonicus* (Calvo et al. 1997). Jackson et al. (2004) also documented clutches containing abnormal eggs from this locality.

Reisz et al. (2005) reported the oldest eggs referable to a sauropodomorph that contain embryos. This cluster of six subspherical eggs assigned to the prosauropod *Massospondylus carinatus* came from the Lower Jurassic Elliot Formation of South Africa. Because the eggshell structure of these eggs is unidentifiable, most likely as a result of recrystallization (Jackson, pers. comm.), the eggs are not assignable to any oofamily.

Grigorescu et al. (1994) described megaloolithid eggs from a site near the Romanian village of Tustea. They assigned them to the hadrosaur *Telmatosaurus* on the basis of embryonic and hatchling bones found in the mudstone that contained the eggs. These taxonomic assignments, however, are controversial (Carpenter 1999), because assignments of an egg to a taxon on any evidence other than embryonic remains within the egg have been proven fallacious (Horner & Weishampel 1988; Norell et al. 1994).

Morphology and Parataxonomy of Eggs

Assigning fossil eggs and eggshells to a specific dinosaur species is difficult and often speculative. Positive identifications rely on the presence of an embryo inside an egg, but even this does not necessarily guarantee a correct assignment. There can be considerable morphological difference between an embryo and the adult animal. Individuals can change characteristics during ontogeny, and various embryos may look similar to one another, even across broad categories of taxa. Bones of embryos or hatchlings are rare. Small bones are less likely to be preserved because they are more easily destroyed by environmental and geological forces or even by trampling by other animals. Nevertheless, the classification of dinosaur eggs and eggshells based on macrostructural (e.g., size, shape, shell thickness, and ornamentation) and microstructural characteristics (e.g., shell units and pores) can provide important insights into the reproductive biology of dinosaurs and their taxonomy.

Because all living archosaurs (birds, crocodilians) have hard-shelled eggs, it is assumed that the eggs laid by all sauropodomorph dinosaurs were hard-shelled too. This was corroborated by the basal sauropodomorph *Massospondylus* from the Lower Jurassic of South Africa, of which hard-shelled eggs with embryos were discovered (Reisz et al. 2005).

Eggs and eggs with embryos from the Auca Mahuevo locality vary little in size, shape, microstructure (Fig. 16.1), and surface ornamentation. They are spherical to subspherical, with a diameter between 13 and 15 cm and an average volume of 1.5 l (Chiappe et al. 1998; Jackson et al. 2008). Shell thickness of eggs ranges from 1.00 to 1.78 mm with a mean of 1.40 mm (Chiappe et al. 1998). Surface ornamentation consists of domed tubercles with an average diameter of 0.45 mm (Chiappe et al. 1998). Pore canals enabling gas exchange are unevenly distributed between the tubercles. Shell units exhibit a radial–tubular ultrastructure.

Eggs from southern Europe are round to slightly oval, ranging in size from about 14 × 27 cm to 18 × 22 cm (Plate 16.1; Peitz 2000; Sander et al. 2008). Eggshell thickness is rather variable between localities and between clutches; however, the average thickness among the eggs of a single clutch is rather constant (Peitz 2000; Sander et al. 2008). Average thickness of eggshells studied by Peitz (2000) at two Catalonian localities ranged from 1.0 mm to about 4.5 mm. In the egg with the thinnest shell, it ranges from 1.6–2.4 mm, while in the egg with the thickest shell, it ranges from 3.0–4.5 mm. The European megaloolith eggs have a much higher number of pores than the Auca Mahuevo eggs (Fig. 16.1; Sander et al. 2008) and hence higher water vapor conductance rates (Jackson et al. 2008). The diameter of shell units ranges from 0.25 to 1.31 mm, with high variability within a single egg (0.48–0.81 mm). Despite the high variability in eggshell attributes, Peitz (1999, 2000) and Sander et al. (2008) assigned these eggs to a single oospecies, *Megaloolithus mammilare,* and assumed that they were produced by *Ampelosaurus atacis* (Plate 16.2; Sander et al. 2008). The conclusion of Peitz (1999, 2000) and

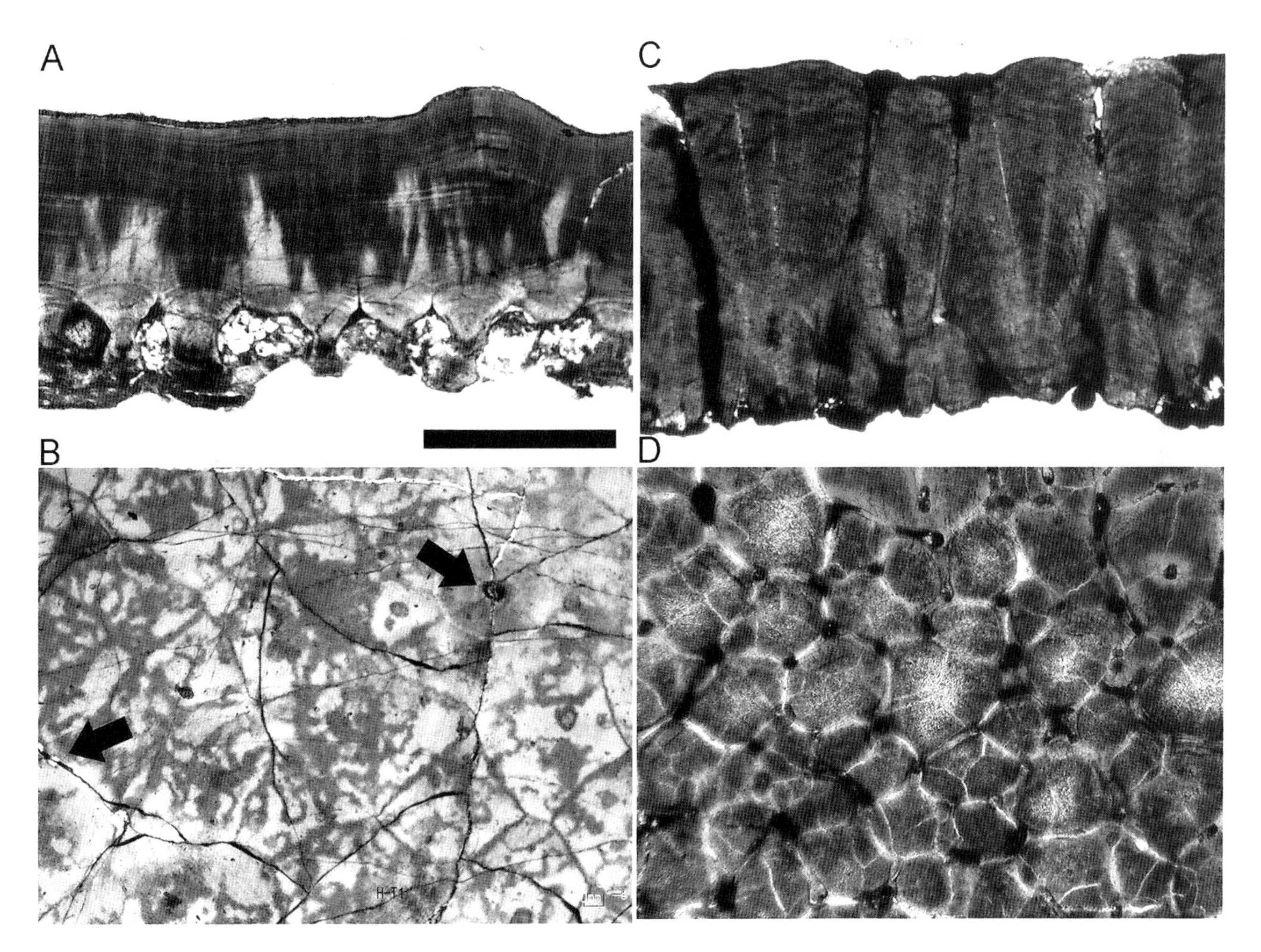

FIGURE 16.1. Comparison of shell microstructure and porosity. (A) *Megaloolithus patagonicus* from Auca Mahuevo, Argentina: radial cross section, specimen PVPH 113, an egg containing an embryo. (B) Tangential section of the egg figured in (A). Note that there are only two pores visible in this section (arrows), with a third one possibly located between the arrows. (C) *Megaloolithus mammillare* from Coll de Nargó, Spain: radial cross section, specimen XXII/4. (D) Tangential section of egg in (C). Note the numerous pores located at the junction of the shell units. The shell samples figured in (A) and (B) are deposited in the collections at the Museo Carmen Funes, Plaza Huincul, Argentina, and those figured in (C) and (D) at the Steinmann Institute, Division of Paleontology, University of Bonn, Germany. *Reproduced from Sander et al. (2008).*

Sander et al. (2008) about the parataxonomy of eggs from southern Europe differs from those of French and Spanish paleontologists who have assigned them to numerous oospecies (Vianey-Liaud & López-Martinéz 1997; Bravo et al. 2000; López-Martinéz 2000; López-Martinéz et al. 2000; Panadés I Blas 2002, 2005).

Indian eggs of the Megaloolithidae vary in diameter between 12 and 20 cm (Vianey-Liaud et al. 2003). They are spherical, and shell thickness ranges from 1.17 to 3.60 mm. Shell units show high morphological variation, ranging from cylindrical to fan-shaped. As in the European eggs, the Indian eggs have a much higher number of pores than the Argentinian eggs. Because of their varying shell microstructure, eggs from India have been assigned to numerous oospecies (Vianey-Liaud et al. 1987; Khosla & Sahni 1995; Mohabey 1998, 2001).

Egg Clutches, Nests, and Prehatching Parental Care

EGG CLUTCHES AND NESTS

Clutch size at Auca Mahuevo ranges from about 15 to nearly 40 eggs (Chiappe et al. 2004; Jackson et al. 2004). Eggs are in direct contact with one another or loosely packed with a random distribution in single or multiple layers; up to three layers have been observed so far (Jackson et al. 2004). Although most of these clutches are preserved in red mudstone, a few occur in large, subcircular to subelliptical to kidney-shaped depressions in sandstone. The interstitial spaces between eggs in these clutches are filled in with mudstone. These clutches preserve evidence of nest architecture because the eggs truncate cross bedding, and an elevated rim of structureless sandstone surrounds the clutch. The fact that the eggs in these nests are entombed by mudstone resulting from flooding events in this alluvial system suggests that the titanosaurs from Auca Mahuevo did not bury their eggs (Chiappe et al. 2004). The low porosity of eggshells limited gas conductance, thus providing further support for the absence of substrate burial in the Auca Mahuevo eggs (Jackson et al. 2008; Sander et al. 2008).

Clutches from Europe and India are much smaller than those at the Argentinian site (but see Vila et al. 2010). For example, clutches discovered at the Spanish and French localities did not exceed eight eggs, and some contained as few as three eggs (Sander et al. 2008). Clutches found in the Indian Lameta Formation show greater variability in size (Mohabey 1990, 1996, 2000, 2001; Sahni et al. 1994), but the maximum number of eggs per clutch, which is eight eggs, appears to be similar to that of the European clutches (Sander et al. 2008). European and Indian eggs occur in a single layer (but see Vila et al. 2010) and are more closely spaced than those at the Argentinian site; however, eggs in all of the clutches are irregularly and loosely arranged. On the basis of comparisons with extant amniote eggs, the high eggshell porosity documented in European and Indian eggs indicates that incubation occurred in a vegetation mound or that the eggs were buried in a substrate (Seymour 1979; Deeming 2006). The eggs were most likely deposited in a shallow depression and covered with soil and/or vegetation (Sander et al. 2008).

NESTING BEHAVIOR AND SITE FIDELITY

Maps showing the spatial distribution of eggs and clutches, the stratigraphic distribution of egg layers, and the sedimentological context in which they are preserved yield inferences on the nesting behavior and site fidelity of sauropods. The close spacing of clutches (approximately 3 m), their high density, and the continuity of clutches observed at Auca Mahuevo suggest gregarious nesting behavior among these titanosaurs (Chiappe et al. 2000). However, it remains unclear whether the clutches were laid at precisely the same time. Among the European and Indian sites, clutch spacing can only be estimated with confidence at the Spanish locality of Coll de Nargó, Tremp Formation (Sander et al. 2008). Here, but also at the other European and Indian sites, spacing is wide and irregular. This differs from the Argentinian site, and thus gregarious nesting of these titanosaurs seems unlikely. In contrast, this arrangement suggests that a small number of females may have deposited clutches in this area at any one time. Nevertheless, a long occupation time of over 10,000 years estimated for the Bastur locality documents a repeated use of this nesting site by titanosaurs (Sander et al. 1998).

PREHATCHING PARENTAL CARE?

Many extant birds and reptiles show nest-guarding behavior that reduces the risk of nest predation (Plate 16.3). The titanosaurs of Auca Mahuevo, however, might have offered little or no parental care of their clutches for several reasons. The eggs from Argentina were most likely not buried. Although this argues for nest guarding or attendance, the large size of adults and the close proximity of clutches contradict this conclusion. Inadvertent trampling of eggs by adults would have resulted in eggshell fragments around the clutches; the absence of eggshells between clutches, therefore, suggests that movement within the area was limited once the eggs were laid. Nest guarding may have been restricted to the periphery of the nesting ground, however (Chiappe & Dingus 2001). That clutches from Europe and India were most probably deposited in a shallow depression and covered with soil and/or vegetation does not completely contradict nest guarding or attendance because exactly this kind of parental care is observed in modern crocodilians (Plate 16.3) and in a few turtles (Somma 2003), which both bury their eggs. However, potential differences in the parental care for eggs between the Auca Mahuevo titanosaurs and those from Europe and India can also be the result of intraspecies variation, as is observed in living animal species.

EGG DEVELOPMENT

The amount of time necessary for a dinosaur embryo to mature to the hatching stage may never be known with certainty, but it can be at least roughly estimated by a model developed by Rahn & Ar (1974) for birds. On the basis of comparisons with extant birds that have, in contrast to modern reptiles, a rather constant incubation temperature of about 40°C, a dinosaur egg of 1.5 kg—the size of an ostrich egg—would require an incubation time of about 60 days to hatch. Temperature, however, can greatly lengthen or shorten incubation time in extant species. The incubation temperature depends on the body temperature of the breeding species. Further, the temperature in a nest that is not incubated by parents, for example, in a depression filled with sand or vegetation, or a mound (modern turtles or crocodilians, Plate 16.3) varies within the environment at the local scale. The female optimizes incubation temperature by nest site selection and the method of nest construction. For example, the mean temperature in the nests of the American crocodile (*Crocodylus acutus*) is 30.9°C in June and 34.3°C in August when eggs hatch after an incubation period of about 11 weeks (Lutz & Dunbar-Cooper 1984), and the daily temperature fluctuations in the nest range between 2 and 4°C (Packard & Packard 1988).

Hatching, Adolescence, Growth in Body Size, and Sexual Maturity

HATCHING

Fossil eggs are considered to be hatched when the upper proportion of the egg is missing and the inside is filled with the sediment that surrounds the egg (Carpenter 1999; Mueller-Töwe et al. 2002). Carpenter (1999) hypothesized that a dinosaur hatchling cracks the eggshell by pecking with an egg tooth (a small, sharp, cranial protuberance present in most birds and reptiles) and pushing its legs and shoulders against the shell. Although hatched eggs have not been documented at Auca Mahuevo, García (2007) has identified an egg tooth-like structure for in ovo embryos of these titanosaurs, the occurrence of which had been predicted by Mueller-Töwe et al. (2002). Hatched eggs of the Megaloolithidae are frequent in clutches in Europe (Sander et al. 1998) and India (Mohabey 2001), however.

PRECOCIAL VERSUS ALTRICIAL

Extant reptiles and many ground-nesting birds are precocial. They hatch fully developed and are able to care for themselves immediately after hatching. In modern altricial birds such as passerines, parents provide their nestlings with food and guard their nest to reduce the risk of predation of their young. Synchronization of hatching is a strategy that is applied by many precocial species to significantly reduce the predation risk of hatchlings, because predators are unable to eat the oversupply of prey (predator satiation). Whether sauropods were precocial or altricial, or whether they displayed synchronization in hatching is unknown. The possibly low or absent parental care in the nest phase, however, favors the notion that they might have been precocial, as is observed in modern turtles, reptiles, and some birds. The low fracturing of the hatched eggs in the Indian oospecies suggests that these hatchlings left the nests immediately after hatching, providing further support that at least these oospecies might have been precocial (Mohabey 2005). The lack of hatched material at the Auca Mahuevo sites does not lead to any conclusion on whether these sauropods were also precocial.

Studies on hatchling and adult skeletons (Horner & Weishampel 1988) and on the structure of the eggshell (Bond et al. 1988) are currently applied by paleontologists to test the hypothesis that sauropods were precocial (Horner 2000). These refer to differences in skeletons and eggshells that exist in extant precocial and altricial species. Contrary to altricial species, the hatchlings of precocial species are fully developed, and their skeleton shows a high level of ossification (Lack 1968). In extant birds, eggs yielding altricial young are usually smaller and are more quickly formed than eggs of precocial hatchlings (Lack 1968). Precocial birds have eggs that do not lose water rapidly as result of low porosity, and they are insulated against temperature change as a result of thick eggshells, resulting in a long incubation time (Rahn & Ar 1974).

Evidence for altriciality in *Maiasaura* was presented by Horner & Weishampel (1988), which was based on characteristics of the skeleton of hatchlings found in a nest. More recent work supports the contrary hypothesis that *Maiasaura* had been precocial and includes evidence such as a well ossified skeleton, no underdeveloped knee joints, and a well developed pelvis (Horner & Weishampel 1988). For sauropods, however, there is no direct evidence so far that they were altricial, but there are some hints that they were probably precocial, such as modern turtles, reptiles, and some birds.

ADOLESCENCE AND GROWTH IN BODY SIZE

Sauropod hatchlings may have been much shorter than a meter from head to tail (Chiappe et al. 1998, 2001) and weighed much less than 10 kg. The maximum volume of an egg is 4.5 l in those from southern Europe and about 1.5 l in those from Argentinia (Breton et al. 1986; Weishampel & Horner 1994; Jackson et al. 2008). As adults, the estimated lengths of these sauropods reached more than 30 m, and their weight ranged from 30–80 metric tons (Appendix and references cited therein).

Growth patterns in dinosaurs have been assessed by bone histology (summarized in Chinsamy-Turan 2005 and Erickson 2005). Growth lines comparable to the annual rings in trees can be counted in vertebrate long bones and are used to estimate the longevity of individuals of the same species (i.e., Castanet 1994; summarized in Erickson 2005; Klein & Sander 2007). Growth rates have been difficult to quantify in sauropods because histological growth marks are rare and at best appear late in ontogeny (Sander 2000; Klein & Sander 2008). The few growth mark records that are available suggest that longevity was more than four decades (Curry 1999; Sander 1999, 2000; Sander & Tückmantel 2003; Wings et al. 2007; Sander et al., this volume). Juvenile growth was rapid in sauropods because the bones of juveniles consist of highly vascularized bone (Sander 2000; Klein & Sander 2008). The lower limit in the exponential phase of growth that is documented in long bones during late ontogeny was about 2 metric tons per year (Wings et al. 2007; Sander et al., this volume; but see Lehman & Woodward 2008 for lower estimates). The qualitative growth record also suggests that sauropods reached sexual maturity well before maximum body size (Sander 2000; Klein & Sander 2008) and that this pattern is consistent with that of other nonavian dinosaurs (Erickson et al. 2007; Klein & Sander 2007; Lee & Werning 2008; de Ricqlès et al. 2008). Sauropods seem to have reached the age at sexual maturity in the second decade of life (Sander 2000; Sander & Tückmantel 2003). This observation coincides with the upper theoretical limit for the age at the onset of sexual maturity derived by Dunham et al. (1989). After a series of simulation studies of demographically imposed constraints to the life cycle, they showed that life histories in which the onset of reproduction occurs after the 20th year of life are unlikely. Because such an old age at sexual maturity would theoretically lead to juvenile and adult survival rates that have not been reported in populations of extant birds and reptiles, the authors concluded that such life cycles probably did not exist, not even in the dinosaurs.

We can only speculate about dinosaur behavior during the adolescent and adult stages. Sauropod track sites indicate that sauropods sometimes traveled in groups (Wright 2005). It has been suggested that hatchlings and juveniles did not stay with the adults because of the exceptional differences in body size and mass (Paul 1994). Juveniles might have been overlooked by adults and trampled; in addition, juveniles required food plants of lower height than adults. Trackway and bonebed evidence indicates that only juveniles of at least one third of adult size joined adults (Paul 1994). Myers & Storrs (2007) found clear age partitioning of a diplodocid herd into only juvenile and subadult individuals at the Mother's Day Quarry in Montana. Martin (1994), however, described *Phuwiangosaurus* material from northern Thailand that included both juveniles less than 2 m long and adults. This last observation suggests that at least in this species, parents might have attended or guarded their young, as is done by many extant herd-living herbivores. Living in herds has also been discussed for other sauropods (Martin 1994; Wright 2005) and would have been a simple form of defense against predators.

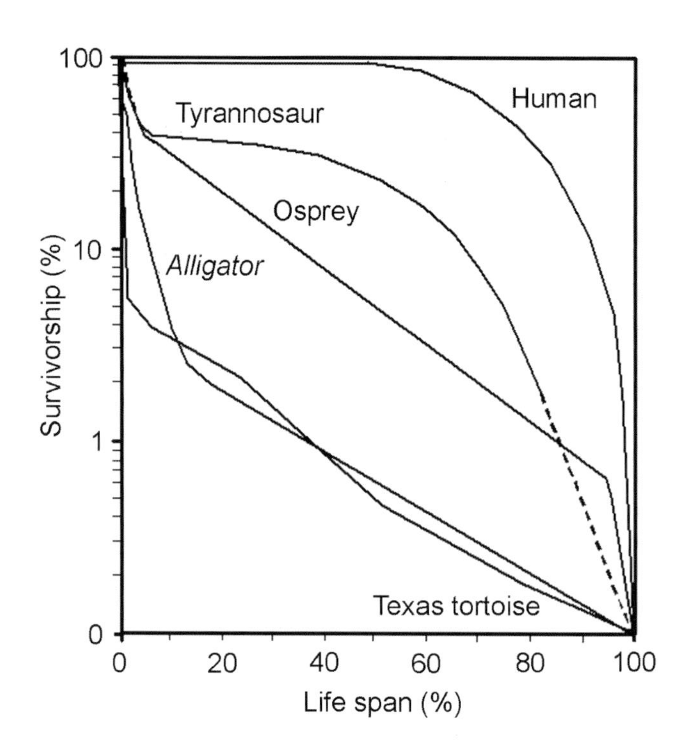

FIGURE 16.2. Survivorship curves. Curves for humans (industrialized country, representing a typical type 1 species with high juvenile survival), tyrannosaur (*Albertosaurus sarcophagus*), osprey (a type 2 species with a constant mortality), and alligator (a typical type 3 species with a high mortality in the early juvenile stage) are adapted from Erickson et al. (2006); the curve for the Texas tortoise (a type 3 species with a high mortality after hatching) is based on survival estimates obtained in a field study by Hellgren et al. (2000).

Mortality during Adolescence and Adulthood

Ecologists distinguish between three basic idealized curves to describe the age dependency of survival in animal species (examples in Fig. 16.2; Pearl & Minor 1935). Many survivorship curves show components of these three generalized types at different times in a species' life history. The curve of type 1 is convex and assumes a high survival before individuals reach sexual maturity, followed by a period of increasing rates of mortality caused by senescence-driven die-offs. This survivorship curve is characteristic for species that are extremely safe from predation and that experience low neonate mortality, such as humans living in highly industrialized countries (Fig. 16.2) or populations of captive animals. In contrast, the survival of species showing a linear survivorship curve (type 2) is characterized by a constant mortality risk throughout their life, which is dominated by predation (including cannibalism). Small, short-lived birds, mammals, and lizards have a linear survivorship curve. Species with a concave curve (type 3) experience high mortalities at the beginning of their life, which is mainly caused by predation, followed by low mortalities before reaching their maximal life span.

Large, long-lived reptiles show a concave pattern in survival (Fig. 16.2, alligator and Texas tortoise). Erickson et al. (2006) reported a sigmoidal survivorship curve based on type 1 survival for tyrannosaurs (Fig. 16.2), which suggests that these nonavian dinosaurs were minimally susceptible to predation throughout their life, with the exception of a higher mortality during the first two years. These authors derived their curves from an analysis of the age structure of four tyrannosaur populations (*Albertosaurus sarcophagus*, n = 22; *Daspletosaurus* sp., n = 14; *Gorgosaurus libratus*, n = 39; *Tyrannosaurus rex*, n = 30) and used growth lines in fibulae and metatarsals to estimate the age of individuals. Ages of individuals included in their analyses ranged from two years to a maximal age that differed between species. The survival pattern observed in tyrannosaurs greatly differs from curves found for birds (Fig. 16.2, osprey), crocodiles (Fig. 16.2, alligator), and turtles (Fig. 16.2, Texas tortoise), but conforms much better to large, long-lived mammals with a long period of parental care that increases juvenile survival (Erickson et al. 2006).

Predation and competition for resources are the main agents of mortality in populations. Sinclair et al. (2003) analyzed the patterns of predation and competition for resources of a highly diverse mammal community in the Serengeti ecosystem. They found that smaller-bodied ungulates experienced predation from many predators, and their predation rate was higher than in large ungulates that were preyed on by fewer species. Very large herbivores such as elephant, rhinoceros, and hippopotamus almost never experienced predation as adults and only rarely as juveniles. The relationship between ungulate body size and predation rate was not linear. Instead, there was a threshold body size above which the proportion of annual adult mortality caused by predators declined rapidly as body size increased. After having passed this size, resource limitation strongly controlled herbivore populations. When extrapolating their results to sauropod communities, we can expect that small-bodied sauropod species and small juveniles of sauropods might have experienced a strong predation pressure, while the predation pressure on large-bodied adult sauropods was probably low, similiar to that on modern megaherbivores (Sinclair et al. 2003; Owen-Smith & Mills 2008).

Growth rates in body size were exceptionally fast in sauropods (Sander 2000; Klein & Sander 2008). Rapid growth is advantageous when survival rate increases with body size (Arendt 1997), as is observed in ungulates in the Serengeti ecosystem. Rapid growth minimizes the amount of time spent in stages in which individuals are vulnerable to predation. This benefit of rapid growth was predicted by a model of Case (1978). In his model, growth rate was fastest in species with a high ratio of juvenile to adult mortality; in the case of sauropods, this means that large adult sauropods are virtually safe from predation. Cooper et al. (2008) found evidence for a relation between predation and growth rate in the hadrosaur *Hypacrosaurus stebingeri*. After analyzing sequences of lines of arrested growth in tibiae and femora of this prey species and its predators, they showed that the hadrosaur was able to avoid predators in part by growing. For sauropods, in general, the high juvenile mortality resulting from predation might be reflected in the low numbers of small and of even larger juveniles in the fossil record. Alternatively, the rarity of these individuals could be explained by the fact that small bones are less likely to be preserved.

However, the tyrannosaur species studied by Erickson et al. (2006) show stronger relative increases in body size than do sauropods (defined as a ratio of growth rate in the exponential phase and adult body mass). Their growth rate estimates in the exponential phase are as follows: *Tyrannosaurus rex*, approximately 0.6 metric tons per year (Erickson et al. 2006), and sauropods, approximately 2 metric tons per year (Wings et al. 2007; Sander et al., this volume). For estimated adult body mass, values are as follows: *Tyrannosaurus rex*, approximately 6 metric tons (Erickson et al. 2006), and sauropods, approximately 30–80 metric tons (Appendix and references cited therein). This corroborates the conclusion of Erickson et al. (2006) that tyrannosaurs had a long period of parental care that increased juvenile survival, while sauropods were at least most probably precocial and experienced a higher predation risk. Altricial and semialtrical young of extant species that are attended or guarded and fed by parents are able to allocate the saved energy that might otherwise be spent acquiring food or protecting themselves to grow more rapidly than precocial species that have to care for themselves after hatching (Case 1978).

In short, we conclude that the survivorship curve of sauropods was most likely characterized by a phase in which predation dominated survival, as in modern birds (Fig. 16.2, osprey), because sauropods were most probably precocial and the high growth rates in body size suggest a significant predation risk for juveniles. This was followed by a phase in which sauropods were more or less safe from predation due to their large body size, as is the case in turtles (Fig. 16.2, Texas tortoise), crocodiles (Fig. 16.2, alligator), and large mammals (Fig. 16.2, humans). Our interpretation of growth curves with respect to survivorship curves (Sander et al. 2010) contradicts Lockley (1994), who derived convex survivorship curves from the fossil footprints of sauropod herds. This may be explained by the fact that herds did not cover all age classes of animals, ranging from hatchlings to adults (Paul 1994).

Reproductive Output of Sauropod Females

The deaths of the individuals in a population always have to be compensated for by the births of new individuals to enable long-term population survival. Thus, high reproductive outputs of females are needed to balance the high mortality rate before reaching sexual maturity. The reproductive output of females is manifested in two variables: the frequency of reproduction, and the clutch and litter sizes. In extant reptiles, both variables increase with increasing body size (Thorbjarnarson 1996). However, body size also affects egg and offspring size. Mammals and birds increase neonate size to a much larger extent than reptiles in regard to increasing body size. Modern species with high juvenile mortalities therefore have high clutch sizes (i.e., most crocodiles, Thorbjarnarson 1996) and/or multiple clutches per breeding season (most turtles, Ernst & Barbour 1989; passerine birds, Ricklefs 2000). Because the clutch size of sauropods, which is 8 for European and Indian sauropods (but see Vila et al. 2010) and less than 40 for Argen-

Table 16.1. Life Cycles of Extant Turtles, Crocodiles, and Birds versus Sauropods

Characteristic	Tortoises	Crocodiles	Ratites	Sauropods
Clutch/nest	Buried	Buried	Open nest	Buried/open nest
Clutch size	Medium	Large	Small	Small/medium
Multiple broods per year	Yes	Yes	Yes	Probable
Nest guarding before hatching	Rare	All species	All species	Unlikely to limited
Parental care after hatching	Precocial	Precocial	Precocial	Probably precocial
Juvenile mortality	High	High	High	High
Adult mortality	Low	Low	Low	Low
Sexual maturity	After full growth	After full growth	After full growth	Considerably before full growth
Longevity	High	High	Low to medium	Medium

We include here only tortoises (Family Testudinidae) among the turtles because of their exclusively terrestrial lifestyle. Ratites were chosen as representatives for birds because of their inability to fly. Both taxa are phylogenetically closely related to sauropods. Data for tortoises from Ernst & Barbour (1989), for crocodiles from del Hoyo et al. (1992), for ratites from Thorbjarnarson (1996), and for sauropods from Somma (2003).

tinian sauropods, is low in comparison to that of most extant reptiles (Table 16.1), their eggs are small in relation to the size of fully grown adults. As a result of the physiological limitation of clutch size imposed by respiration of embryos if the eggs are buried (Seymour 1979), multiple clutches per breeding season are a likely option (Sander et al. 2008). Multiple clutches allow a female to spread the predation risk over multiple nests, which thus increases the hatching rate of eggs. Furthermore, if sauropods had multiple clutches, hatchlings must have been abundant (Sander et al. 2008).

HYPOTHESES EXPLAINING GIGANTISM IN SAUROPODS BASED ON THEIR LIFE CYCLE

One of the most obvious features of the living world is that organisms greatly differ in their life cycles. Species traits such as adult body size, longevity, age at sexual maturity, mortality, and fecundity show a wide variation and are the result of evolutionary adaptation; these processes are not mutually exclusive.

GIGANTISM AND INCREASED FECUNDITY

Classical life history theory approaches body size from the age at sexual maturity: age at sexual maturity increases with adult body size, because "one must grow for a longer time to get larger" (Stearns 1992, p. 127). Most models making predictions of age at sexual maturity rest on assumptions about trade-offs between reproductive benefits and costs of long growth periods, for example, through mortality. They predict that either fecundity increases or offspring mortality decreases with increasing age at maturity (Stearns 1992). Thus, since the offspring of sauropod dinosaurs must have experienced significant mortality (Sander et al. 2010; Wilson et al. 2010) and these animals do not show particularly early sexual maturity because reproduction presumably started in the second decade of life (Sander 2000; Sander & Tückmantel 2003), an increase in fecundity seems to be the most probable benefit of their increased body size.

This conclusion coincides with predictions from allometric scaling. The amount of space inside the body cavity available for eggs increases as the cube of the length of the species and thus the body mass of the species, presuming that egg size is constant and there is no change in the shape of animals with changing body sizes (Roff 2002). This positive relationship between clutch size and body size is well documented for many ectotherms, in particular for Crustacea, Mollusca, Annelida, Arachnida, Insecta, Tunicata, fish, amphibians, and reptiles (reviewed in Roff 2002). Endothermic birds and mammals, however, do not comply with the predictions from allometric scaling (Janis & Carrano 1992). Janis & Carrano (1992) found no correlation between body size and clutch size (or annual fecundity) for birds (see below). Positive allometric relationships between litter size and body size are rare in mammalian taxa (Millar & Hickling 1991; Roff 2002). In contrast, large mammals tend to have one to two young per year, whereas small mammals such as cricetine rodents are frequently quite fecund (Roff 2002). This bimodal distribution of litter size leads to an overall decrease in litter size and number of offspring per year with increasing body size for mammalian vertebrates (as noted by Janis & Carrano 1992; see below).

Whether the hypothesis of increased fecundity resulting from an increased age at sexual maturity (body size) is valid for sauropods is unclear and difficult to prove. Reptiles and birds show no consistent pattern in both fecundity and clutch size in regard to body mass. Reptiles point to an increase in sauropod fecundity and clutch size with increasing body mass, whereas birds show no relationship between fecundity or clutch size and body mass. Although clutch size can be determined from the fossil record, the frequency of reproduction determining fecundity cannot.

GIGANTISM AND THE RISK OF POPULATION EXTINCTION

The body size and the density of a population reflect whether the energy available in the area inhabited is allocated to either many small-bodied or fewer large-bodied individuals. Body size and density also strongly influence the chance of long-term survival of populations (Blackburn & Gaston 2001). Crucial factors limiting body size and the number of supported individuals depend on the bauplan of the organism, which reflects the evolutionary history of the species, its physiology (including mechanisms enabling energy efficiency, as well as its herbivorous or carnivorous mode of nutrition), the availability of resources, and the ecological interactions of the species with other species (Blackburn & Gaston 2001).

Janis & Carrano (1992) put forth a hypothesis that links the life cycle of dinosaurs to ecological processes. After analyzing large data sets for the number of offspring versus body mass for mammals heavier than 1 kg and terrestrial nonpasserine birds, they posited that the oviparous mode of reproduction was the most important factor contributing to gigantism in dinosaurs. The authors found no significant change with respect to adult body mass in the number of offspring per clutch, number of clutches per year, and total number of offspring per year for oviparous birds. In contrast, viviparous mammals showed a significant decrease of all of these life cycle characteristics with increasing body mass. They proposed that the oviparous reproduction strategy of dinosaurs—producing a greater number of offspring, but experiencing higher juvenile mortality than mammals—has a great advantage over evolutionary time. This is because dinosaurs have a greater potential for reproductive turnover than similarly sized mammals, making them less vulnerable to population extinction under environmental perturbations. High intrinsic growth rates allow species to rebuild populations more rapidly after severe declines in individual numbers than species with lower rates (e.g., mammals). This improves their chance of long-term survival and enables populations that have, on average, fewer large-bodied individuals. This hypothesis of Janis & Carrano (1992) was supported by Paul (1994, 1997) and was based on a statistical analysis of annual fecundity and body size of reptiles, birds, monotremes, marsupials, and placentals.

Macroecological analyses of oviparous taxa reveal further support for the success of the reproduction mode of dinosaurs. Increasing clutch sizes with increasing body sizes are documented for turtle species (Frazer 1986; Hailey & Loumbourdis 1988; Iverson 1992), for crocodiles (Thorbjarnarson 1996), for snakes (Ford & Seigel 1989), and for reptiles in general (Blueweiss et al. 1978; King 2000). No correlation between body size and clutch size was found in galliform birds (Kolm et al. 2007), which accords with the results obtained by Janis & Carrano (1992). The difference between extant reptiles and birds may be explained by their internal thermal conditions (Shine 2005) and the different mode of incubation of eggs in the taxa.

The hypothesis of Janis & Carrano (1992) assumes a selective advantage of an oviparous reproduction mode in comparison to the viviparous mode of mammals, which enables larger body sizes, including gigantism. As pointed out by Farlow et al. (1995), this hypothesis predicts that multiton birds should have evolved in the Tertiary, but this did not actually happen. Other factors may have prevented the birds and other oviparous taxa from becoming multiton animals because the evolution of any life history trait of an organism is always constrained. These may be any ecological, morphological, or physiological factor in general or, in the case of the Tertiary birds, competition from mammals or their evolutionary history, including their obligatory bipedalism. Large, bipedal, flightless birds evolved from birds that were able to fly, which definitely influenced their bauplan. In contrast, large mammals evolved from nonvolant animals adapted to a terrestrial life. Nevertheless, relative gigantism seems to be well represented by island birds from New Zealand, the moa, and from Madagascar, the elephant birds (although elephant birds and moa had small clutch sizes, most likely because they inhabited islands), and by predatory birds in mainland ecosystems that lacked efficient mammalian predators, for example, in the Early Tertiary of the Northern Hemisphere and throughout the Tertiary of South America (Mazzetta et al. 2004). This is also known from extant tortoises inhabiting several islands such as *Dipsochelys* from the Seychelles and *Chelonoidis nigra* from the Galapagos. Fossils of the giant crocodyliform *Sarcosuchus imperator* from the Cretaceous of Africa (Sereno et al. 2001) also corroborate the hypothesis of Janis & Carrano (1992).

Summary

The life of a sauropod started in an egg with a hard shell. After hatching, it grew quickly in body size and reached sexual maturity in the second decade of life. Parental care in the juvenile stage was most likely absent or low, and juvenile mortality must have been significant, as suggested by the high growth rates of juveniles and as observed in small extant herbivores. Clutches in Europe and India were buried in the substrate, whereas those in Argentina were deposited in an open-nest structure and remained uncovered by sediment or plant material. Gregarious nesting behavior is likely for the Argentinian titanosaurs, whereas repeated use of the nesting sites over a long period is probable for those from Europe and India.

The clutch size was small—8 eggs in Europe and India and less than 40 eggs in Argentina—in comparison to those of large extant reptiles. Eggs were small in comparison to adult size, which, along with absent or little parental care during incubation, suggest that sauropods had multiple clutches each season. This oviparous reproductive strategy might have contributed to the gigantism of dinosaurs in general because it enabled populations to recover rapidly after crashes in population size and to prevent population extinction. The life cycle of sauropods is similar in several aspects to the life cycle that has evolved in birds, but is also comparable to that of extant turtles and crocodiles

Acknowledgments

We are grateful to Frankie Jackson (Montana State University, Bozeman) and Chris Carbone (Zoological Society of London) for their critical reading of this work. Their valuable comments significantly improved the manuscript. This chapter is part of the Ph.D. dissertation of J.W. We thank Georg Oleschinski (University of Bonn) for providing Plate 16.1. Our research was funded by the Deutsche Forschungsgemeinschaft (DFG). This chapter is contribution number 75 of the DFG Research Unit 533 "Biology of the Sauropod Dinosaurs: The Evolution of Gigantism."

References

Arendt, J. D. 1997. Adaptive intrinsic growth rates: an integration across taxa.—*Quarterly Review of Biology* 72: 149–177.

Blackburn, T. & Gaston, K. J. 2001. Linking patterns in macroecology.—*Journal of Animal Ecology* 70: 338–352.

Blueweiss, L., Fox, H., Kudzma, V., Nakashima, D., Peters, R. & Sams, S. 1978. Relationships between body size and some life history parameters.—*Oecologia* 37: 257–272.

Bond, G. M., Board, R. G. & Scott, V. D. 1988. A comparative study of changes in the fine structure of avian eggshells during incubation.—*Zoological Journal of the Linnean Society of London* 92: 105–113.

Bravo, A. M., Moratalla, J. J., Santafé, J. V. & Santisteban, C. De. 2000. Faidella, a new Upper Cretaceous Nesting Site from the Tremp Basin (Lérida Province, Spain). *In* Bravo, A. M. & Reyes, T. (eds.). *First International Symposium on Dinosaur Eggs and Babies, Extended Abstracts*. Isona i Conda Dellà, Spain: pp. 15–22.

Breton, G., Fourniet, R. & Watté, J.-P. 1986. Le lieu de ponte de dinosaurs de Rennes-le-Château (Aude): Premiers résultates de la camoagne de fouilles 1984.—*Annales du Muséum du Havre* 32: 1–13.

Buffetaut, E. & Le Loeuff, J. 1994. The discovery of dinosaur eggshells in nineteenth-century France. *In* Carpenter, K., Hirsch, K. & Horner, J. R. (eds.). *Dinosaur Eggs and Babies*. Cambridge University Press, Cambridge: pp. 31–34.

Calvo, J. O., Engelland, S., Heredia, S. E. & Salgado, L. 1997. First record of dinosaur eggshells (?Sauropoda-Megaoolithidae) from Neuquen, Patagonia, Argentina.—*Gaia* 14: 23–32.

Carpenter, K. 1999. *Eggs, Nests, and Baby Dinosaurs: A Look at Dinosaur Reproduction*. Indiana University Press, Bloomington.

Case, T. J. 1978. On the evolution and adaptive significance of postnatal growth rates in terrestrial vertebrates.—*Quarterly Review of Biology* 53: 243–282.

Castanet, J. 1994. Age estimation and longevity in reptiles.—*Gerontology* 40: 174–192.

Chiappe, L. M. & Dingus, L. 2001. *Walking on Eggs*. Scribner, New York.

Chiappe, L. M., Coria, R. A., Dingus, L., Jackson, F., Chinsamy, A. & Fox, M. 1998. Sauropod dinosaur embryos from the Late Cretaceous of Patagonia.—*Nature* 396: 258–261.

Chiappe, L. M., Dingus, L., Jackson, F., Grellet-Tinner, G., Aspinall, R., Clarke, J., Coria, R. A., Garrido, A. & Loope, D. 2000. Sauropod eggs and embryos from the Late Cretaceous of Patagonia. *In* Bravo, A. M. & Reyes, T. (eds.). *First International Symposium on Dinosaur Eggs and Babies, Extended Abstracts*. Isona i Conda Dellà, Spain: pp. 23–29.

Chiappe, L. M., Salgado, L. & Coria, R. A. 2001. Embryonic skulls of titanosaur sauropod dinosaurs.—*Science* 293: 2444–2446.

Chiappe, L. M., Schmitt, J. G., Jackson, F. D., Garrido, A., Dingus, L. & Grellet-Tinner, G. 2004. Nest structure of sauropods: sedimentary criteria for recognition of dinosaur nesting traces.—*Palaios* 19: 89–95.

Chinsamy-Turan, A. 2005. *The Microstructure of Dinosaur Bone*. Johns Hopkins University Press, Baltimore.

Cooper, L. N., Lee, A. H., Taper, M. L. & Horner, J. R. 2008. Relative growth rates of predators and prey dinosaurs reflect effects of predation.—*Proceedings of the Royal Society B: Biological Sciences* 275: 2609–2615.

Curry, K. A. 1999. Ontogenetic histology of *Apatosaurus* (Dinosauria: Sauropoda): new insights on growth rates and longevity.—*Journal of Vertebrate Paleontology* 19: 654–665.

Deeming, D. C. 2006. Ultrastructural and functional morphology of eggshells supports the idea that dinosaur eggs were incubated buried in a substrate.—*Paleontology* 49: 171–185.

Del Hoyo, J., Elliot, A. & Sargata, L. J. 1992. *Handbook of the Birds of the World, Vol. 1: Ostrich to Ducks*. Lynx Editions, Barcelona.

de Ricqlès, A., Padian, K., Knoll, F. & Horner, J. R. 2008. On the origin of high growth rates in archosaurs and their ancient relatives: complementary histological studies on Triassic archosauriforms and the problem of a "phylogenetic signal" in bone histology.—*Annalés de Paléontologie* 94: 57–76.

Dunham, A. E., Overall, K. L., Porter, W. P. & Forster, C. A. 1989. Implications of ecological energetics and biophysical and developmental constraints for life-history variation in dinosaurs. *In* Farlow, J. O. (ed.). *Paleobiology of the Dinosaurs. GSA Special Paper 238*. Geological Society of America, Boulder: pp. 1–21.

Erickson, G. M. 2005. Assessing dinosaur growth patterns: a microscopic revolution.—*Trends in Ecology and Evolution* 20: 677–684.

Erickson, G. M., Currie, P. J., Inouye, B. D. & Winn, A. A. 2006. Tyrannosaur life tables: an example of nonavian dinosaur population biology.—*Science* 313: 213–217.

Erickson, G. M., Curry Rogers, K., Varricchio, D. J., Norell, M. A. & Xu, X. 2007. Growth pattern in brooding dinosaurs reveals the timing of sexual maturity in non-avian dinosaurs and genesis of the avian condition.—*Biology Letters* 3: 558–561.

Ernst, C. H. & Barbour, R. W. 1989. *Turtles of the World*. Smithsonian Institution Press, Washington, D.C.

Farlow, J. O., Dodson, P. & Chinsamy, A. 1995. Dinosaur biology.—*Annual Review of Ecology and Systematics* 26: 445–471.

Ford, N. B. & Seigel, R. A. 1989. Relationships among body size, clutch size, and egg size in three species of oviparous snakes.—*Herpetologia* 45: 75–83.

Frazer, N. B. 1986. The relationship of clutch size and frequency of body size in loggerhead turtles, *Caretta caretta*.—*Journal of Herpetology* 20: 81–84.

García, R. A. 2007. An "egg-tooth"-like structure in titanosaurian sauropod embryos.—*Journal of Vertebrate Paleontology* 27: 247–252.

Grellet-Tinner, G. & Fiorelli, L. E. 2010. A new Argentinean nesting site showing neosauropod dinosaur reproduction in a Cretaceous hydrothermal environment.—*Nature Communications* 1. doi: 10.1038/ncomms1031.

Grigorescu, D., Weishampel, D., Norman, D., Seclamen, M., Rusu, M., Baltres, A. & Teodorescu, V. 1994. Late Maastrichtian dinosaur eggs from the Hateg Basin (Romania). *In* Carpenter, K., Hirsch, K. & Horner, J. R. (eds.). *Dinosaur Eggs and Babies*. Cambridge University Press, Cambridge: pp. 75–87.

Hailey, A. & Loumbourdis, N. S. 1988. Egg size and shape, clutch dynamics, and reproductive effort in European tortoises.—*Canadian Journal of Zoology* 66: 1527–1536.

Hellgren, E. C., Kazmaier, R. T., Ruthven III, D. C. & Synatzske, D. R. 2000. Variation in tortoise life history: demography of *Gopherus berlandieri*.—*Ecology* 81: 1297–1310.

Horner, J. R. 2000. Dinosaur reproduction and parenting.—*Annual Review of Earth and Planetary Sciences* 28: 19–45.

Horner, J. R. & Weishampel, D. B. 1988. A comparative embryological study of two ornithischian dinosaurs.—*Nature* 332: 256–257.

Iverson, J. B. 1992. Correlates of reproductive output in turtles (Order Testudines).—*Herpetological Monographs* 6: 25–42.

Jackson, F. D., Garrido, A., Schmitt, J. G., Chiappe, L. M., Dingus, L. & Loope, D. 2004. Abnormal, multilayered titanosaur eggs from in situ clutches (Dinosauria: Sauropoda) at the Auca Mahuevo locality, Neuquen Province, Argentinia.—*Journal of Vertebrate Paleontology* 24: 913–922.

Jackson, F. D., Varricchio, D. J., Jackson, R. A., Vila, B. & Chiappe, L. M. 2008. Comparison of water vapor conductance in a titanosaur egg from the Upper Cretaceous of Argentina and a *Megaloolithus siruguei* egg from Spain.—*Paleobiology* 34: 229–246.

Janis, C. M. & Carrano, M. 1992. Scaling of reproductive turnover in archosaurs and mammals: why are large terrestrial mammals so rare?—*Acta Zoologica Fennica* 28: 201–206.

King, R. B. 2000. Analyzing the relationship between clutch size and female body size in reptiles.—*Journal of Herpetology* 34: 148–150.

Khosla, A. & Sahni, A. 1995. Parataxonomic classifications of Late Cretaceous dinosaur eggshells from India.—*Journal of the Palaeontological Society of India* 40: 87–102.

Klein, N. & Sander, M. 2007. Bone histology and growth of the prosauropod dinosaur *Plateosaurus engelhardti* (von Meyer 1837) from the Norian bonebeds of Trossingen (Germany) and Frick (Switzerland).—*Special Papers in Palaeontology* 77: 169–206.

Klein, N. & Sander, M. 2008. Ontogenetic stages in the long bone histology of sauropod dinosaurs.—*Paleobiology* 34: 247–263.

Kolm, N., Stein, R. W., Mooers, A., Verspoor, J. J. & Cunningham, E. J. A. 2007. Can sexual selection drive female life histories? A comparative study on galliform birds.—*Journal of Evolutionary Biology* 20: 627–638.

Lack, D. 1968. *Ecological Adaptations for Breeding in Birds*. Methuen, London.

Lee, A. H. & Werning, S. 2008. Sexual maturity in growing dinosaurs does not fit reptilian growth models.—*Proceedings of the National Academy of Sciences of the United States of America* 105: 582–587.

Lehman, T. M. & Woodward, H. N. 2008. Modeling growth rates for sauropod dinosaurs.—*Paleobiology* 34: 264–281.

Lockley, M. G. 1994. Dinosaur ontogeny and population structure: interpretations and speculations based on fossil footprints. *In* Carpenter, K., Hirsch, K. & Horner, J. R. (eds.). *Dinosaur Eggs and Babies*. Cambridge University Press, Cambridge: pp. 347–365.

López-Martinéz, N. 2000. Eggshell sites from the Cretaceous-Tertiary transition in the South-Central Pyrenees (Spain). *In* Bravo, A. M. & Reyes, T. (eds.). *First International Symposium on Dinosaur Eggs and Babies, Extended Abstracts*. Isona i Conda Dellà, Spain: pp. 95–115.

López-Martinéz, N., Moratalla, J. J. & Sanz, J. L. 2000. Dinosaur nesting on tidal flats.—*Paleogeography, Paleoclimatology, Paleoecology* 160: 253–163.

Lutz, P. L. & Dunbar-Cooper, A. 1984. The nest environment of the American crocodile (*Crocodylus acutus*).—*Copeia* 1984: 155–161.

Martin, V. 1994. Baby sauropods from the Sao Khua Formation (Lower Cretaceous) in Northeastern Thailand.—*Gaia* 10: 147–153.

Matheron, P. 1869. Notice sur les reptiles fossiles des dépôts fulvio-lacustres crétaces du bassin à lignite de Fuveau: note on the fossil reptiles from the fluvio-lacustrine deposits of the Fuveau lignite basin.—*Bulletin de la Société Géologique de France, Série 2* 26: 781–795.

Mazzetta, G. V., Christiansen, P. & Fariña, R. A. 2004. Giants and bizzarres: body size of some southern South American Cretaceous dinosaurs.—*Historical Biology* 16: 1–13.

Millar, J. S. & Hickling, G. J. 1991. Body size and the evolution of mammalian life histories.—*Functional Ecology* 5: 588–593.

Mohabey, D. M. 1990. Dinosaur eggs from Lameta Formation of western and central India: their occurrence, and nesting behaviour. *In* Sahni, A. & Jolly, A. (eds.). *Cretaceous Event Stratigraphy and the Correlation of the Nonmarine Strata. Contributions from the Seminar cum Workshop Contribution. IGCP 216 and 245*. Panjab University, Chandigarh, India: pp. 86–89.

Mohabey, D. M. 1996. A new oospecies, *Megaloolithus matleyi*, from the Lameta Formation (Upper Cretaceous) of Chandrapur district, Maharashtra, India, and general remarks on the palaeoenvironment and nesting behaviour of dinosaurs.—*Cretaceous Research* 17: 183–196.

Mohabey, D. M. 1998. Systematics of Indian Upper Cretaceous

dinosaur and chelonian eggshells.—*Journal of Vertebrate Paleontology* 18: 348–362.

Mohabey, D. M. 2000. Indian Upper Cretaceous (Maastrichtian) dinosaur eggs: their parataxonomy and implications in understanding the nesting behaviour. *In* Bravo, A. M. & Reyes, T. (eds.). *First International Symposium on Dinosaur Eggs and Babies, Extended Abstracts*. Isona i Conda Dellà, Spain: pp.139–153.

Mohabey, D. M. 2001. Indian dinosaur eggs: a review.—*Journal of the Geological Society of India* 58: 479–508.

Mohabey, D. M. 2005. Late Cretaceous (Maastrichtian) nests, eggs, and dung mass (coprolites) of sauropods (titanosaurs) from India. *In* Tidwell, V. & Carpenter, K. (eds.). *Thunder Lizards: The Sauropodomorph Dinosaurs*. Indiana University Press, Bloomington: pp. 466–489.

Mueller-Töwe, I., Sander, P. M., Schüller, H. & Thies, D. 2002. Hatching and infilling of dinosaur eggs as revealed by computed tomography.—*Palaeontographica Abt. A* 267: 119–168.

Myers, T. S. & Storrs G. W. 2007. Taphonomy of the Mother's Day Quarry, Upper Jurassic Morrison Formation, south-central Montana, USA.—*Palaios* 22: 651–666.

Norell, M. A., Clark, J. M., Demberelyin, D., Barsbold, R., Chiappe, L. M., Davidson, A. R., McKenna, M. C., Altangerel, P. & Novacek J. 1994. A theropod dinosaur embryo and the affinities of the Flaming Cliffs dinosaur eggs.—*Science* 266: 779–782.

Owen-Smith, N. & Mills, M. G. L. 2008. Predator-prey size relationships in an African large-mammal food web.—*Journal of Animal Ecology* 77: 173–183.

Packard, G. C. & Packard, M. J. 1988. The physiological ecology of reptilian eggs and embryos. *In* Gans, C. & Huey, R. B. (eds.). *Biology of the Reptilia, Vol. 16. Ecology B: Defense and Life History*. Academic Press, New York: pp. 523–605.

Panadés I Blas, X. 2002. Does diversity of eggshells mean diversity of dinosaurs?—*Journal of Vertebrate Paleontology* 22: 94A.

Panadés I Blas, X. 2005. Diversity versus variability in megaloolithid dinosaur eggshells.—*PalArch's Journal of Vertebrate Palaeontology* 2: 1–13.

Paul, G. S. 1994. Dinosaur reproduction in the fast lane: implications for size, success, and extinction. *In* Carpenter, K., Hirsch, K. & Horner, J. R. (eds.). *Dinosaur Eggs and Babies*. Cambridge University Press, Cambridge: pp. 244–255.

Paul, G. S. 1997. Reproductive behavior and rates. *In* Currie, P. J. & Padian, K. (eds.). *Encyclopedia of Dinosaurs*. Academic Press, San Diego: pp. 630–637.

Pearl, R. & Minor, J. R. 1935. Experimental studies on the duration of life. XIV. The comparative mortality of certain lower organisms.—*Quarterly Reviews in Biology* 10: 60–79.

Peitz, C. 1999. Parataxonomic implications of some megaloolithid dinosaur eggs from Catalunya, Spain. *In* Bravo, A. M. & Reyes, T. (eds.). *First International Symposium on Dinosaur Eggs and Babies, Extended Abstracts*. Isona i Conda Dellà, Spain: pp. 49–50.

Peitz, C. 2000. Megaloolithid dinosaur eggs from Maastrichtian of Catalunya (NE-Spain): parataxonomic implications and stratigraphic utility. *In* Bravo, A. M. & Reyes, T. (eds.). *First International Symposium on Dinosaur Eggs and Babies, Extended Abstracts, Extended Abstracts*. Isona i Conda Dellà, Spain: pp. 155–159.

Rahn, H. & Ar, A. 1974. The avian egg: incubation time and water loss.—*Condor* 76: 147–152.

Reisz, R. R., Scott, D., Sues, H.-D., Evans, D. C. & Raath, M. A. 2005. Embryos of an Early Jurassic prosauropod dinosaur and their evolutionary significance.—*Science* 309: 761–764.

Ricklefs, R. E. 2000. Density dependence, evolutionary optimization, and the diversification of avian life histories.—*Condor* 102: 9–22.

Roff, D. A. 2002. *Life History Evolution*. Sinauer Associates, Sunderland, Mass.

Sahni, A., Tandon, S. K., Jolly, A., Bajpai, S., Sood, A. & Srinivasan, S. 1994. Upper Cretaceous dinosaur eggs and nesting sites from the Deccan volcano-sedimentary province of peninsular India. *In* Carpenter, K., Hirsch, K. & Horner, J. R. (eds.). *Dinosaur Eggs and Babies*. Cambridge University Press, Cambridge: pp. 204–226.

Salgado, L., Coria, R. A. & Chiappe, L. M. 2005. Osteology of the sauropod embryos from the Upper Cretaceous of Patagonia.—*Acta Palaeontologica Polonica* 50: 79–92.

Sander, P. M. 1999. Life history of Tendaguru sauropods as inferred from long bone histology.—*Mitteilungen aus dem Museum für Naturkunde in Berlin, Geowissenschaftliche Reihe* 2: 103–112.

Sander, P. M. 2000. Long bone histology of the Tendaguru sauropods: implications for growth and biology.—*Paleobiology* 26: 466–488.

Sander, P. M. & Tückmantel, C. 2003. Bone lamina thickness, bone apposition rate, and age estimates in sauropod humeri and femora.—*Paläontologische Zeitschrift* 76: 161–172.

Sander, P. M., Klein, N., Stein, K. & Wings, O. This volume. Sauropod bone histology and its implications for sauropod biology. *In* Klein, N., Remes, K., Gee, C. T. & Sander, P. M. (eds.). *Biology of the Sauropod Dinosaurs: Understanding the Life of Giants*. Indiana University Press, Bloomington: pp. 276–302.

Sander, P. M., Peitz, C., Gallemi, J. & Cousin, R. 1998. Dinosaurs nesting on a red beach?—*Comptes Rendus de l'Académie des Sciences de Paris: Sciences de la Terre et des Planètes* 327: 67–74.

Sander, P. M., Peitz, C., Jackson, F. D. & Chiappe, L. M. 2008. Upper Cretaceous titanosaur nesting sites and their implications for sauropod dinosaur reproductive biology.—*Palaeontographica Abt. A* 284: 69–107.

Sander, P. M., Christian, A., Clauss, M. Fechner, R., Gee, C. T., Griebeler, E. M., Gunga, H.-C., Hummel, J., Mallison, H., Perry, S., Preuschoft, H., Rauhut, O., Remes, K., Tütken, T., Wings, O. & Witzel, U. 2010. Biology of the sauropod dinosaurs: the evolution of gigantism.—*Biological Reviews of the Cambridge Philosophical Society*. doi: 10.1111/j.1469=185X.2010.00137.x.

Sereno, P. C., Larsson, H. C. E., Sidor, C. A. & Gado, B. 2001. The giant crocodyliform *Sarcosuchus* from the Cretaceous of Africa.—*Science* 294: 1516–1519.

Seymour, R. S. 1979. Dinosaur eggs: gas conductance through the shell, water loss during incubation and clutch size.—*Paleobiology* 5: 1–11.

Shine, R. 2005. Life-history evolution in reptiles.—*Annual Review of Ecology, Evolution and Systematics* 36: 23–46.

Sinclair, A. R. E., Mduma, S. & Brashares, J. S. 2003. Patterns of predation in a diverse predator-prey system.—*Nature* 425: 288–290.

Somma, L. A. 2003. Reptilian parental behaviour. *Linnean* 19: 42–46.

Stearns, S. C. 1992. *The Evolution of Life Histories*. Oxford University Press, New York.

Thorbjarnarson, J. B. 1996. Reproductive characteristics of the order Crocodylia.—*Herpetologica* 52: 8–24.

Vianey-Liaud, M. & López-Martinéz, N. 1997. Late Cretaceous dinosaur eggshells from the Tremp Basin, southern Pyrenees, Lleida, Spain.—*Journal of Paleontology* 71: 1157–1171.

Vianey-Liaud, M., Jain, S. L. & Sahni, A. 1987. Dinosaur eggshells (Saurischia) from the Late Cretaceous Intertrappean and Lameta Formations (Deccan, India).—*Journal of Vertebrate Paleontology* 7: 408–242.

Vianey-Liaud, M., Khosla, A. & García, G. 2003. Relationships between European and Indian dinosaur eggs and eggshells of the oofamily Megaloolithidae.—*Journal of Vertebrate Paleontology* 23: 575–585.

Vila, B., Jackson, F. D., Fortuny, J., Sellés, A. G. & Galobart, Á. 2010. 3-D modelling of megaloolithid clutches: insights about nest construction and dinosaur behaviour.—*PLoS ONE* 5(5): e10362.

Weishampel, D. B. & Horner, J. R. 1994. Life history syndromes, heterochrony, and the evolution of Dinosauria. *In* Carpenter, K., Hirsch, K. & Horner, J. R. (eds.). *Dinosaur Eggs and Babies*. Cambridge University Press, Cambridge: pp. 229–243.

Wilson, J. A., Mohabey, D. M., Peters, S. E. & Head, J. J. 2010. Predation upon hatchling dinosaurs by a new snake from the Late Cretaceous of India.—*PLos Biology* 8(3): e1000322.

Wings, O., Sander, P. M., Tütken, T., Fowler, D. W. & Sun, G. 2007. Growth and life history of Asia's largest dinosaur.—*Journal of Vertebrate Paleontology* 27: 167A.

Wright, J. L. 2005. Steps in understanding sauropod biology. *In* Curry Rogers, K. A. & Wilson, J. (eds.). *The Sauropods: Evolution and Paleobiology*. Academic Press, San Diego: pp. 252–280.

17

Sauropod Bone Histology and Its Implications for Sauropod Biology

P. MARTIN SANDER, NICOLE KLEIN, KOEN STEIN, AND OLIVER WINGS

BONE HISTOLOGY HAS EMERGED as the major source of information on life history of dinosaurs, and sauropodomorphs are one of the best-sampled clades. The large long bones (humerus and femur) preserve the most complete growth record, which allows inference on life history, thermometabolism, and other aspects of sauropod biology.

Basal sauropodomorphs have fibrolamellar bone interrupted by regularly spaced growth marks, and termination of growth is recorded in an external fundamental system (EFS). However, in the best-studied basal sauropodomorph, *Plateosaurus engelhardti,* growth rate deduced from growth mark counts and termination of growth are highly variable (developmental plasticity). Growth series of many taxa of sauropods also show fibrolamellar bone exclusively but differ in that growth marks appear only late in life, or in most taxa only in the EFS. Growth rate and final size are taxon specific, not variable, and genetically predetermined.

As indicated by their highly vascularized fibrolamellar bone (but also by estimates from rare growth marks and estimates from local bone tissue apposition rates), growth rates of sauropods were comparable only to those of large herbivorous mammals and birds, suggesting that sauropods were tachymetabolic endotherms.

Evolution of large body size in sauropods was brought about by an increase in growth rate and evolution of tachymetabolic endothermy, which was only incompletely developed in basal sauropodomorphs. The small- to medium-sized titanosaurs *Phuwiangosaurus* and *Ampelosaurus* show a reduced growth rate compared to other neosauropods. The dwarfed island forms *Europasaurus* and *Magyarosaurus* evolved their diminutive body size by a decrease in growth rate and, in the case of *Europasaurus,* a shortening of the active phase of growth.

Introduction

Sauropod dinosaurs were the largest animals ever to evolve in the terrestrial realm. Late Jurassic neosauropods, such as the diplodocoids *Diplodocus, Barosaurus,* and *Apatosaurus,* grew easily to 20–30 m body lengths and reached masses of 15–35 metric tons (e.g., Seebacher 2001; Mazzetta et al. 2004; Foster 2007; Appendix). The Macronaria, the other successful group of sauropods, consisting primarily of the familiar basal forms *Brachiosaurus* and *Camarasaurus* and the Titanosauria, were more massively built, and similar in body size but heavier than the Diplodocoidea (Seebacher 2001; Mazzetta et al. 2004; Appendix). For example, the most recent body mass estimate for *Brachiosaurus* is 38 metric tons for the mounted Berlin skeleton (Gunga et al. 2008; Ganse et al., this volume). This mount is approximately 23 m long and lifts its head over 13 m above the museum floor (Remes et al., this volume). However, some sauropods (e.g., *Supersaurus, Sauroposeidon, Argentinosaurus*) are believed to have reached nearly 35 m in body length and body mass approaching 70 metric tons (Appendix).

Thus, sauropod body size is without precedent in the terrestrial realm today or in the past; other dinosaurs remained a magnitude smaller as well (Sander & Clauss 2008; Sander et al. 2010). The closest extant relatives of sauropods—crocodiles and birds—are minute in direct comparison, and both are highly adapted to their particular lifestyles (amphibious, volant). The African elephant, the largest living land mammal, which resembles sauropods in its graviportal stance with columnar legs, is still only half as heavy as a small sauropod (with the exception of island dwarf sauropods). This exceptional size of sauropods raises the question of their life history and biology.

Paleohistology, the study of the microstructure of fossil bone, offers insights into these questions. It is the most direct way of studying the growth and life history of sauropods and other extinct animals. The study of dinosaur bone histology, which was pioneered by Armand de Ricqlès (1968a, 1968b, 1980; de Ricqlès et al. 1991; Francillon-Vieillot et al. 1990), has yielded elementary new data that have fundamentally changed our understanding of dinosaurs as living animals (summarized in Erickson 2005 and Chinsamy-Turan 2005).

The long bone histology of basal sauropodomorphs so far has only been studied in detail in two taxa, *Massospondylus carinatus* (Chinsamy 1993; Chinsamy-Turan 2005) and *Plateosaurus engelhardti* (Sander & Klein 2005; Klein & Sander 2007). Other taxa studied are *Thecodontosaurus* (Cherry 2002), one of the basalmost sauropodomorphs (Upchurch et al. 2007; Yates 2007), and *Euskelosaurus browni* (de Ricqlès 1968a). However, *Euskelosaurus* is a nomen dubium (Galton & Upchurch 2004), and the material must be assigned to basal Sauropodomorpha indet. Sauropod dinosaurs, on the other hand, are now a histologically well sampled group, at least with respect to the shaft of their long bones (Rimblot-Baly et al. 1995; Curry 1999; Sander 1999, 2000; Sander & Tückmantel 2003; Sander et al. 2004, 2006; Curry Rogers & Erickson 2005; Woodward 2005; Klein & Sander 2008; Lehman & Woodward 2008; Klein et al. 2009a; Woodward & Lehman 2009). Gaps in sampling primarily exist in basal taxa from the Early and Middle Jurassic and among the great diversity of Titanosauria.

It is the purpose of this chapter to review and synthesize the paleohistological information available for sauropodomorph dinosaurs. Most of these data come from long bones and are based on our own research, but published observations will be used for comparison and discussion.

Bone Histology in Fossil Vertebrates

In extant animals, growth and other life history data can be easily obtained by field studies (mark–release–recapture studies, observation) or laboratory work. For extinct animals, these options are obviously not available. The only direct way to obtain information about past events in the life of an individual of an extinct species is from the growth record preserved in its hard tissues. For vertebrates, these are the teeth and bones.

Skeletochronology, the count of growth marks in skeletal tissues, is a standard method for determining the age of extant fish and ectothermic vertebrates (Castanet & Smirina 1990; Castanet et al. 1993). It can also address important questions regarding the life history of extinct vertebrates. How long did it take an approximately 1 m long sauropod hatchling weighing around 5 kg to become an adult of 25 m body length and a mass of over 20 metric tons? How many years did sauropods need to reach their enormous sizes? At what age did they reach sexual maturity? How old was a specific individual at death? What was the life expectancy of sauropods (i.e., their longevity)? And what can bone histology tell us about the conditions during the life of an individual (Klein et al. 2009b)?

Other questions are important from a more evolutionary perspective: How did sauropods evolve their enormous body size, and what limited it? What kind of physiology was needed to support their rapid growth? Was the physiology of sauropods more similar to that of modern ectothermic reptiles or more similar to that of modern endothermic mammals?

PRINCIPLES OF GROWTH AND LIFE HISTORY

Growth in vertebrates means an increase in size and mass over time in an individual's ontogeny. Growth rate decreases gradually during ontogeny. In most modern mammals and birds, the attainment of sexual maturity coincides roughly with the attainment of maximum body size and a final growth stop, or growth continues only for a short period after sexual maturity. Living large reptiles have a completely different life history and growth patterns because they continue to grow moderately fast long after attaining sexual maturity. Contrary to popular belief about large reptiles such as crocodiles and boid snakes, there is no such thing as true indeterminate growth in which an individual continues to grow at noticeable rates throughout life. Instead, growth in length and mass is asymptotic. Once maximum size is approached, growth decreases rapidly and finally stops. Small reptiles, such as some lizards and snakes, may show a pattern more akin to that of mammals, in which growth stops rather early in life.

During their phase of growth, ectothermic and endothermic vertebrates are both affected by endogenous and exogenous factors, resulting in individual growth rate variations and in differences in growth rate between individuals. These variations in growth rate are much more pronounced in ectothermic vertebrates and are interpreted as adaptive under the concept of developmental plasticity (West-Eberhard 2003).

Endogenous factors influencing growth rate are genetic differences and incubation temperature, while exogenous factors are climate, food supply, and inter- and intraspecific competition (Farmer 2000, 2003). Endotherms show comparatively little developmental plasticity (Stark & Chinsamy 2002).

PRINCIPLES OF SKELETAL GROWTH AND RATIONALE FOR SAMPLING LONG BONES

An increase in size and mass during growth is accompanied by shape changes and allometric growth. This is also why different bones of a single skeleton grow at different rates. In addition, because of the different sizes of the bones in a skeleton, local rates of bone deposition vary greatly, leading to variation in histology within a single skeleton. For example, a large weight-bearing long bone (e.g., a humerus or a femur) will grow much more rapidly in circumference compared to a smaller bone, such as a skull bone or a rib (Francillon-Vieillot et al. 1990; Chinsamy-Turan 2005). Thus, in comparative bone histological studies, it is most important to sample bones with similar growth patterns (e.g., long bones vs. flat bones). Limb bones (long bones like humerus and femur) are best suited for studying the growth record because of their appositional growth (Fig. 17.1). Because of the principles of bone growth, that is, that long bones generally grow in length at both ends at roughly the same rate (Francillon-Vieillot et al. 1990), the most complete growth record is preserved in their midshaft region (Sander 1999, 2000), also called the neutral region. The neutral region might not lie exactly in the middle of the shaft, however, because of slight differences in the rate of growth at the ends of the bone or due to a differential cessation of growth in older individuals in which one end continues to grow after the other has stopped.

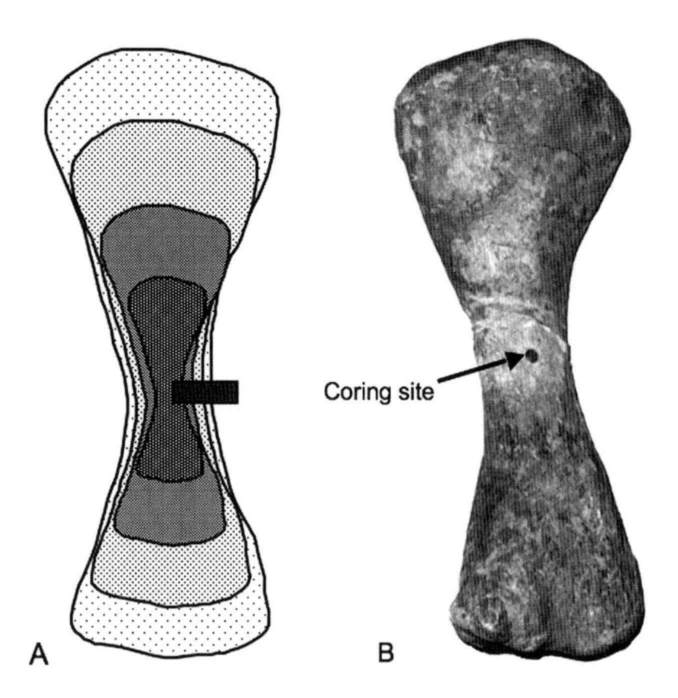

FIGURE 17.1. Optimal sampling location in long bones. (A) Sketch of the appositional growth of a sauropod humerus, showing four growth stages and the optimal sampling location (black bar) in the midshaft region that captures the most complete growth record. (B) Right humerus of the holotype of *Tornieria africana* (MFN MB.R.2672, field no. A1) in posterior view. The arrow marks the borehole left over from histological sampling by core drilling. Length of bone is 99.0 cm.

PRINCIPLES OF BONE HISTOLOGY

Principles of Bone Tissue Formation

The following is an introduction to the principles of bone tissue formation. More detailed reviews can be found in Francillon-Vieillot et al. (1990) and Reid (1997). Bone is best described at different hierarchical levels, from the morphological level to the crystallite level. Levels in between these two are, from highest to lowest, the microanatomical level, the histological or bone tissue level, the bone type level, and the cellular (cytological) level. We focus our description on the histology of the bone wall or cortex, and we note that the bone tissue produced at the growing ends of the long bone is formed differently. This kind of bone tissue is much less studied paleohistologically than cortical bone (see also Dumont et al., this volume).

The formation of bone tissue (osteogenesis) is initiated by specialized cells called osteoblasts, which lay down the collagenous matrix and bone mineral at the surface of the bone, or inside canals and cavities in the already formed bone (Fig. 17.2A). Some of the osteoblasts become incorporated into the bone tissue and are then called osteocytes. Unlike in tooth enamel, the shells of invertebrates, and arthropod exoskeletons, the bone cells stay active, making bone a living tissue. In addition—again, unlike other animal hard tissues—bone undergoes internal destruction (resorption) and redeposition, a process called remodeling. Bone is resorbed by another kind of specialized cells, the osteoclasts. Most bone is vascularized and continuously nourished by blood vessels that occupy vascular canals. Because bone is a living tissue, it can change throughout the entire life of an individual, which is crucial for the growth of complex shapes and structures.

The most important distinction in bone tissues is that between primary bone and secondary bone. Primary bone tissue is deposited during the growth of the bone. However, early in ontogeny, resorption and remodeling processes are initiated in the medullary cavity and progress outward. The bone resulting from these processes is called secondary bone, and secondary bone replaces primary bone by remodeling. This process is generally slower than the deposition of primary bone and thus lags behind the deposition of primary bone. How-

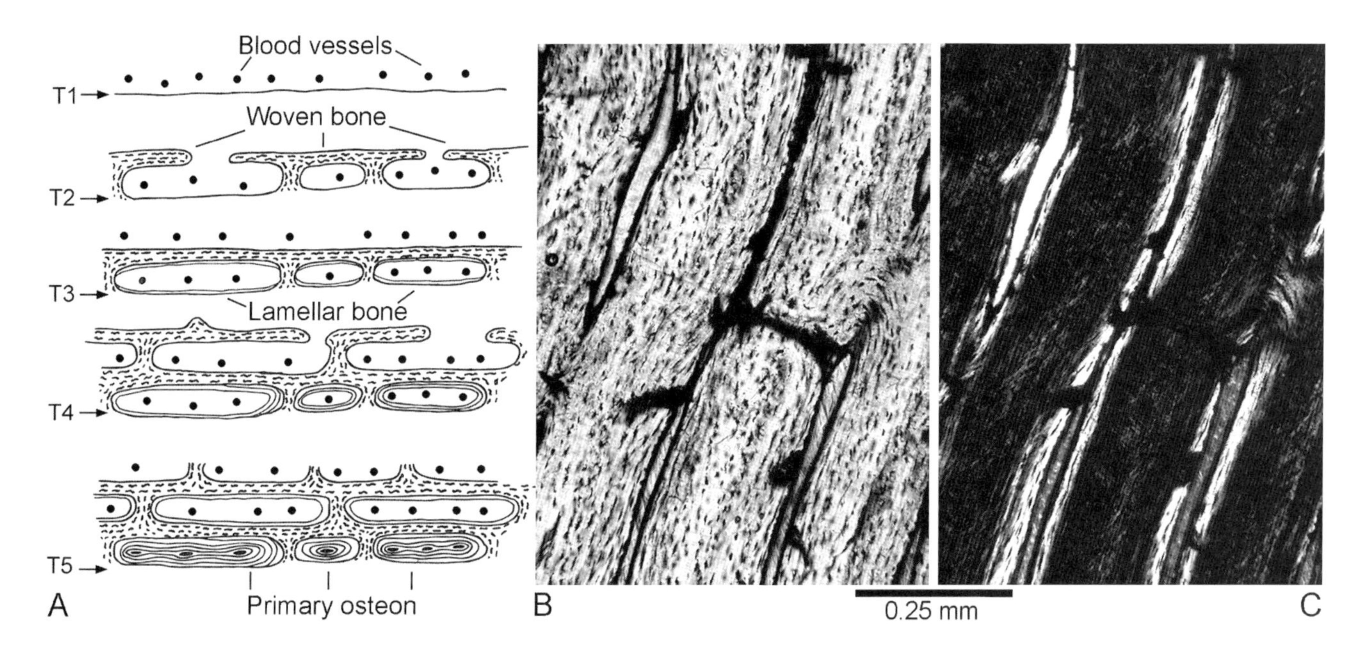

FIGURE 17.2. Laminar fibrolamellar bone tissue and its formation. (A) Formation of bone tissue traced over time (T1–T5 equals five time intervals). The horizontal arrow marks the same location in the bone through time. At time T1, a network of blood vessels covers the bone surface, around which at T2, a framework or matrix of woven bone is laid down, forming a lamina. At T3, the infilling of the vascular spaces with lamellar bone is initiated. At the same time, the next network of blood vessels is formed on the bone surface, initiating the formation of the next lamina. T4 repeats the situation at T2, but primary osteons continue to form. At T5, a third network of blood vessels is formed on the bone surface; the network from T3 is now fully surrounded by a framework of woven bone. In the lamina from T1, primary osteon formation is completed at T5. The apposition rate of laminar fibrolamellar bone can vary greatly, from a few to many laminae being laid down per year. Illustration from Currey (2002). (B) Fully formed laminar fibrolamellar bone tissue of a neosauropod dinosaur in a thin section under the microscope in normal light. Note the network of vascular canals. The small dark spots are the spaces for the osteocytes. (C) Same image as in (B) but in polarized light. The highlighted areas are the primary osteons consisting of lamellar bone, which has gradually filled in the vascular canals. The dark areas with diffuse extinction are the woven bone matrix. Note that compared to (A), the images in (B) and (C) are rotated counterclockwise by 80°.

ever, the trigger and function of this replacement is not entirely understood. Some authors suggest a biomechanical (loading) background, while others suggest that the cause is an aging process of bone (review in Currey 2002); a function in mineral storage and remobilization is also likely. A link with basal metabolic rate is suggested by the observation that mainly large endothermic animals show complete remodeling of their primary bone (de Ricqlès 1980). Any given histological section will thus reflect all these activities and changes in the bone microstructure, at least until it is completely remodeled by secondary bone.

Bone Tissue Types, Bone Types, and Vascularization

The primary bone tissue of all tetrapods is made up of one or more of the three main bone types: lamellar bone, parallel-fibered bone, and fibrous or woven bone (Fig. 17.3). In lamellar bone, the collagen fibers and apatite crystallites are highly organized, the result of relatively slow growth. Fibrous or woven bone is generally deposited quickly and, accordingly, is not well organized. Parallel-fibered bone is in its degree of organization and rate of deposition intermediate between the two others. In addition to its primary bone tissues, a bone type is characterized by the presence, density, and arrangement of blood vessels, preserved as vascular canals in fossil bone. For example, most living reptiles show lamellar-zonal bone tissue (Castanet et al. 1993), in which the primary bone tissue consists of lamellar bone or parallel-fibered bone. In this tissue type, the bone tissue is regularly interrupted by growth marks (Fig. 17.4), which in most cases are annual. Vascularization consists mainly of longitudinally arranged vascular canals. During ontogeny, when growth rate decreases, vascularization also decreases and finally disappears when growth stops completely. The bone tissue is then avascular.

The bone microstructure of birds, large herbivorous mammals, and most dinosaurs, including sauropods, primarily consists of fibrolamellar bone tissue (Fig. 17.2). Fibrolamellar bone is also called the fibrolamellar complex as a result of its 3D structure and the sequential deposition of different bone types to form a complex tissue. In a first step, a matrix or framework of woven (or fibrous) bone is deposited quickly around a large vascular canal, and only later is the vascular canal filled in centripetally by lamellar bone (Fig. 17.2A). This

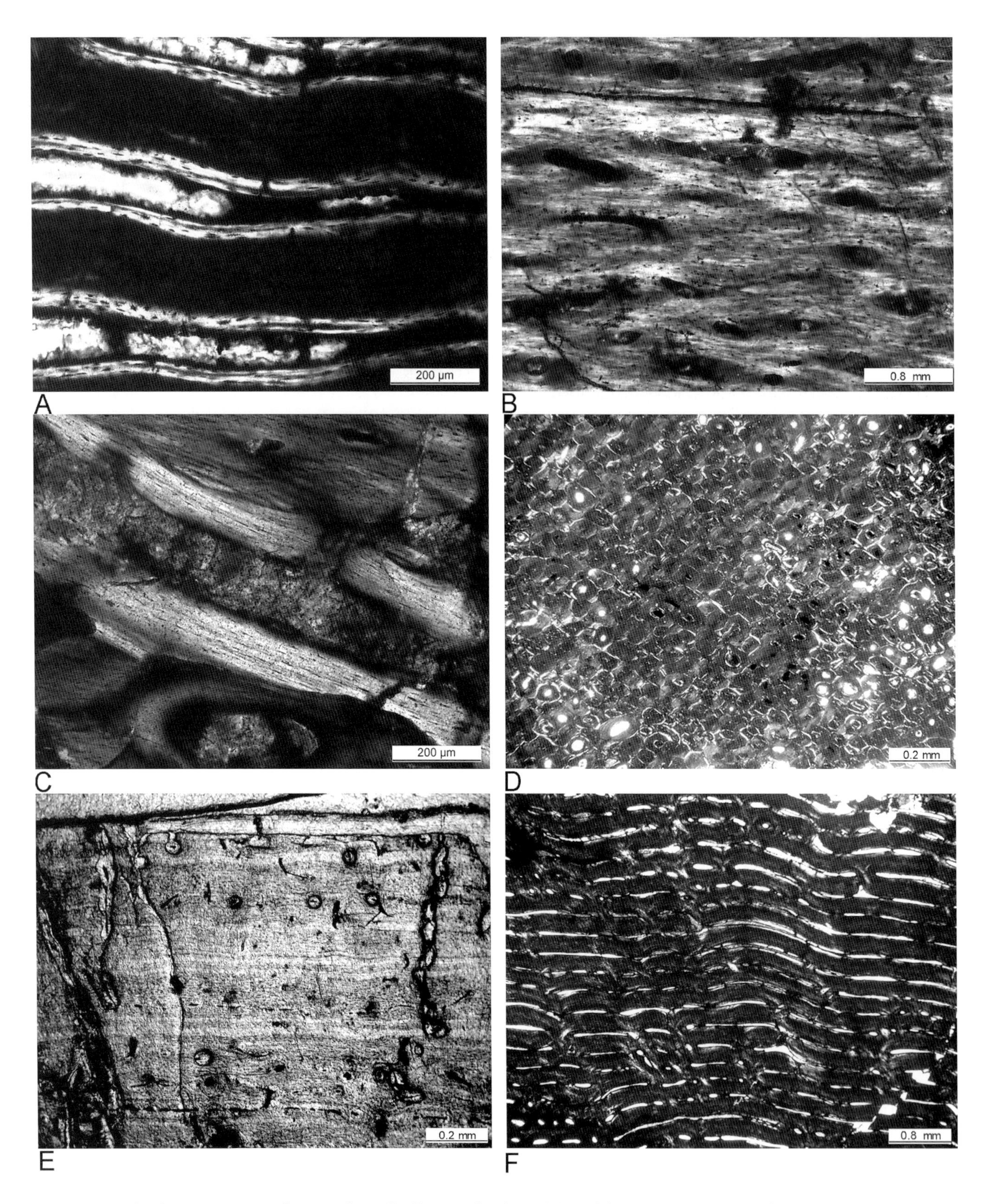

FIGURE 17.3. The three main types of primary bone (A–C), secondary bone (D), and the two most important bone tissue types (E, F) in the cortex of amniotes. (A) Woven bone (also called fibrous bone) in polarized light. The woven bone is in the dark region between the primary osteons (vascular canals lined with lamellar bone). This bone type lacks a clear pattern of extinction because the crystallites are arranged randomly in 3D. Detail of a humerus thin section of *Apatosaurus* (CM 21715). (B) Parallel-fibered bone in polarized light. In this bone type, the crystallites are oriented in 2D. Detail of a humerus thin section of *Ampelosaurus* (MDE C31139). (C) Lamellar bone

infill is called a primary osteon. Fibrolamellar bone is highly vascularized, and its vascular system is dominated by circumferential vascular canals, and less often by radial longitudinal or irregularly arranged vascular canals. Fibrolamellar bone tissue with circumferential vascular canals is also called laminar fibrolamellar bone, or laminar bone for short.

Amprino (1947) recognized that there is a connection between bone tissue type and rate of deposition. This relationship is now known as Amprino's rule (de Margerie et al. 2002; Montes et al. 2007, 2010; Cubo et al. 2008), although it has been difficult to quantify (Castanet et al. 2000; de Margerie et al. 2002; Montes et al. 2007).

Growth Cycles and Growth Marks

The lamellar-zonal bone tissue of modern and fossil ectothermic reptiles is characterized by regularly spaced annual growth cycles (Francillon-Vieillot et al. 1990; de Ricqlès et al. 1991; Castanet et al. 1993; Castanet 1994). Typically, well developed annual growth cycles are divided into a fast-growing zone and a slower-growing annulus, and then are completed by a type of growth mark called a line of arrested growth (LAG) (Fig. 17.4). The LAG indicates a temporary cessation of growth that has been shown to be deposited annually in extant vertebrates (Castanet & Smirina 1990; Castanet et al. 1993, 2004).

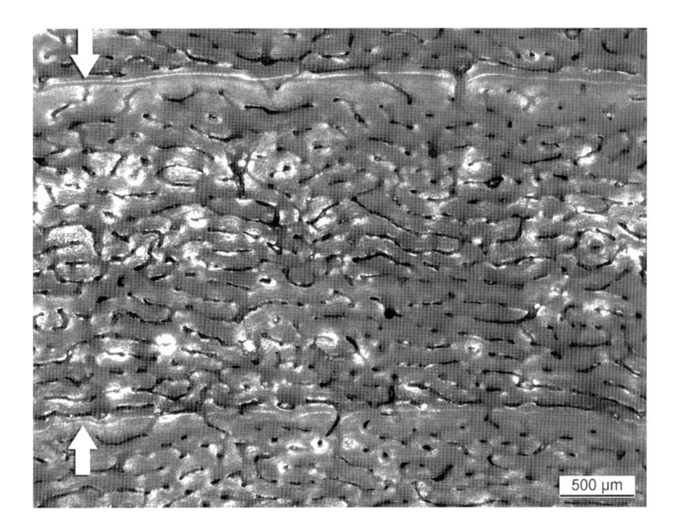

FIGURE 17.4. A well developed annual growth cycle in fibrolamellar bone, starting with a LAG (lower arrow), then divided into a wide, well vascularized zone, a narrow, poorly vascularized annulus, and finally a LAG (upper arrow). Polarized light image of femur IFG/GPIT 192.1 (74.0 cm) of *Plateosaurus engelhardti*.

LAGs are not restricted to the lamellar zonal bone tissue but can also occur in all other bone tissue types. Thus, they also appear in the fibrolamellar tissue of the long bones of mammals (Klevezal 1996; Horner et al. 1999; Sander & Andrássy 2006), birds (Turvey et al. 2005), and nonavian and nonsauropod dinosaurs (Erickson 2005), but only rarely in sauropods (Curry 1999; Sander 1999, 2000; Klein & Sander 2008).

Other types of growth marks are polish lines (Sander 2000; Sander & Tückmantel 2003; Klein & Sander 2007, 2008) and modulations (de Ricqlès 1983; Curry 1999). Polish lines are subtle cessations in growth that are not apparent in thin sections as LAGs, but are visible only in polished sections under reflected light. Modulations are changes in vascularity and have primarily been observed in sauropod flat bones (scapulae of *Apatosaurus;* Curry 1999) and long bones of sauropods (Sander & Tückmantel 2003).

Although the annual nature of growth marks in dinosaurs is still controversial (Chinsamy & Hillenius 2004; Padian & Horner 2004), regularly spaced growth marks were shown to be annual in modern tetrapods (e.g., Castanet et al. 1993). As already noted, there is no convincing explanation why their annularity should not also apply to extinct tetrapods such as dinosaurs. Furthermore, growth marks are found in endothermic animals as well (e.g., Sander & Andrássy 2006), suggesting that they are caused by endogenous cycles, rather than representing only the result of seasonal variations.

In large adult sauropods and other dinosaurs, the outermost cortex may show several closely spaced LAGs in avascular bone. This histological feature is known as an external fundamental system (EFS) and is an unambiguous indicator that the animal in question was fully grown. However, the annual nature and the number of the LAGs in the EFS is often difficult to ascertain because of their narrow spacing and variable distinctiveness.

Estimates of Individual Age and Life Span

Growth marks, if present, can be used to estimate the age of an individual at death, and the greatest number are generally found in the largest individuals, providing an indication of

Caption continued from opposite page in polarized light. The crystallites are all oriented in the same direction, leading to a strong extinction pattern. In this image, the lamellar bone is that of a primary osteon with a vascular canal in the center. Detail of a femur thin section of Diplodocinae indet. (MFN Ki 2). (D) Secondary bone in polarized light consisting of densely packed secondary osteons. This bone tissue type is also called Haversian bone. Detail of a humerus thin section of *Ampelosaurus* (MDE C3602). (E) Lamellar-zonal bone tissue in incident light. This bone tissue type consists of cyclically deposited lamellar or parallel-fibered bone with the cycles ending in LAGs. Detail of a humerus thin section of *Plateosaurus* (NAA F 88/B640). (F) Laminar fibrolamellar bone tissue in polarized light. This type of fibrolamellar bone tissue is called laminar bone because the vascular canals are oriented predominantly circumferentially, giving the tissue a layered appearance in cross section. Note, however, that the laminae do not represent growth cycles. Detail of a femur thin section of Diplodocinae indet. (MFN NW4).

the maximum life span of the species. However, expansion of the medullary cavity and remodeling destroy the inner growth marks, and the number of these missing growth marks must be estimated to arrive at a reliable age estimate (for different methods, see Erickson et al. 2004; Bybee et al. 2006; Klein & Sander 2007).

In sauropods, with their poor record of annual growth marks, another approach has been applied (Curry 1999; Sander & Tückmantel 2003): estimates of individual ages that are based on the thickness of the cortical tissue. In this approach, known or estimated tissue apposition rates (usually expressed in micrometers per day) are divided by the measured thickness of the cortex. Tissue apposition rates can be obtained by applying Amprino's rule by using rates for specific bone tissue types observed in living animals and applying them to the same bone tissue type in fossils (Curry 1998). Alternatively, apposition rates can be estimated by dividing the thickness of a growth cycle in the fossil by the numbers of days in a year (Curry 1998). These rates are then applied to parts of the cortex that lack growth marks, or to other individuals of the same taxon that lack growth marks. The rates obtained by measuring cycle thickness are minimal rates because the growth marks represent an interruption of growth of unknown duration (Sander & Tückmantel 2003). However, all these methods fail to take into account that growth in juveniles is much faster than that of adults.

Growth Curves and Growth Rates

In dinosaur taxa for which good growth mark records are available, for example, in theropods (Erickson et al. 2004, 2007; Horner & Padian 2004; Bybee et al. 2006) and psittacosaurs (Erickson & Tumanova 2000; Erickson et al. 2009), growth can be quantified and expressed in a growth curve. These growth curves generally plot body mass increase versus time, either for a growth series of several individuals or for a single individual. Although body length increase versus time is suitable for a growth curve, length-based growth curves are rarely used in the study of dinosaurs for two reasons: it is usually much easier to estimate the mass of a dinosaur than its length, and mass-based growth curves are more common in biology, allowing the comparison of dinosaur growth curves with those of extant amniotes. As an alternative to body mass, the length of a bone that represents body mass and body length, such as the femur, can be used. Growth curves based on the length of a single bone are useful for the comparison with extant reptiles because herpetologists commonly use length-based growth curves. The first step in constructing a growth curve is counting growth cycles and estimating the number of missing cycles (see above). Next, the number of growth cycles is plotted versus body mass or a length measurement.

Body mass can be obtained in a number of different ways, the discussion of which is beyond the scope of this chapter (for a recent review, see Foster 2007; but see also Stoinski et al., this volume; Ganse et al., this volume). If the growth curve is based on a growth series of several individuals, the number of cycles for each individual is estimated, and the body mass of the individual is determined. Different methods can be employed on the basis of the size of the sampled bone or skeleton. Mass estimates may be obtained for each individual from some measurement such as femur length (Mazzetta et al. 2004) or long bone circumference (Anderson et al. 1985), or by up- and downscaling from one individual of the growth series for which body mass can be reliably estimated (Erickson & Tumanova 2000; Erickson et al. 2001; Bybee et al. 2006; Lehman & Woodward 2008). Most commonly, the mass of the largest individual is estimated and set as 100%. The mass of smaller individuals in the growth series is then calculated from a percentage of a linear dimension (e.g., humerus length or femur length) compared to that of the largest individual (e.g., Erickson & Tumanova 2000; Lehman & Woodward 2008).

However, growth curves can be constructed from the growth mark record of single individuals as well (e.g., Wings et al. 2007; Lehman & Woodward 2008), if the increments of the growth marks can be transformed into increments of mass gain. The simplest way is to equate the percentage increase in cortex thickness per growth cycle with a percentage increase in body mass. Again, the size of the individual at death is expressed as percentage of the largest known individual of the species (whether this preserves growth marks or not), for example, 60%, so that the individual preserving growth marks may record a size increase of, for instance, 30–60% of maximum size. This method also assumes that there is a linear relationship between an increase in long bone cortex thickness and an increase in body mass. At least for sauropods, this appears reasonable because sauropod long bones have been shown to grow isometrically not only to the rest of the body, but also in proportions within the bone, for example, length versus circumference (Bonnan 2004; Kilbourne & Makovicky 2010).

In the third step, the growth curve based either on a growth series or on single individuals is obtained by fitting the plot of cycles versus body mass with one of several growth models (Erickson & Tumanova 2000; Erickson et al. 2001, 2004; Bybee et al. 2006; Cooper et al. 2008; Lehman & Woodward 2008). The most commonly used growth model is the von Bertalanffy equation (Reiss 1989; Erickson et al. 2001; Lehman & Woodward 2008). Such growth curves are informative in several ways: they indicate the age when the fastest growth occured and the age when full size was attained. Finally, they can be used to make inferences on evolutionary ecology (Erickson et al. 2006; Cooper et al. 2008).

Histological Ontogenetic Stages

In addition to a quantitative growth record, bone histology provides information on the ontogenetic status of an individual. This is particularly useful in taxa that lack a good growth mark record, such as sauropods (Sander 2000; Klein & Sander 2008). The method is rather simple and is based on the observation that in the growth series of a tetrapod taxon, a specific histology correlates with a specific body size. In addition, it is an application of Amprino's rule, namely, that a certain bone tissue type is laid down at a specific rate. The correlation between body size and histology works particularly well in taxa that have a well constrained growth trajectory, that is, little developmental plasticity. Most dinosaurs, birds, and mammals fit this criterion. In addition, if histological change in ontogeny is sufficiently discontinuous, histological ontogenetic stages can be established. Sauropods are the first dinosaur group for which this has been done (Klein & Sander 2008), but the concept has also been applied to stegosaurs (Hayashi et al. 2009). As in embryology, the histological ontogenetic stages make the recognition of evolutionary change in ontogeny possible (Scheyer et al. 2010).

MATERIAL BASIS

Our research has focused on long bones because they are the most abundant remains of basal sauropodomorphs and sauropods, they are well preserved, they are the least affected by crushing as a result of their massive bone walls, and the size of the animal is comparatively easy to estimate from long bones. One drawback to long bones is their often generalized morphology, making assignment of isolated long bones to a specific taxon difficult.

Our histological database consists of 238 samples (Table 17.1) derived from 2 "prosauropods" (*Plateosaurus engelhardti* and *Thecodontosaurus* sp.) and 17 sauropod taxa (Table 17.1, Fig. 17.5). The material comes from different localities in North America, East Africa, Europe (Switzerland, Germany, England, France, Romania), and Asia (China, Thailand). It was obtained during several trips to different collections over the last decade. The identifications of the material are based on the collection labels and in some cases on more current published information. All samples were obtained and studied firsthand. For more detailed information, see Klein & Sander (2007, 2008), Table 17.1, and the Appendix.

METHODS

The majority of samples were obtained by core drilling at a prescribed location in the midshaft region of the bones (Fig. 17.6), but several cross sections of entire bones were cut as well. Sampling by core drilling is similar to performing a medical biopsy in scope and philosophy because it minimizes tissue damage but provides only a limited view in return. This is why sampling homologous locations across individuals is crucial. Cores were obtained with an adjustable electrical drill and a hollow diamond drill bit (Sander 1999, 2000; Klein & Sander 2007, 2008; Stein & Sander 2009; Fig. 17.6). Cores of a diameter of about 16 mm (13 and 4 mm for smaller bones) were taken out of the midshaft region of humeri and femora. The cores were then embedded in epoxy resin and cut perpendicular to the long axis of the bone, which is also perpendicular to the growth direction (Fig. 17.6). Half of the core sample was processed into a standard petrographic thin section and the other half into a polished section. The complete and partial cross sections were also cut from the midshaft region and processed as described above.

The thin sections were photographed with a digital camera (Nikon D1) using bellows, resulting in a field of view of about 3 cm. Images of sections that were larger than this were created by merging two or three photographs by Adobe Photoshop software. To observe the histology in greater detail, all thin sections were examined by standard light microscopic techniques (normal transmitted light, polarized light) with a Leica DMLP compound microscope (16× to 400× magnification). Polished sections were studied in incident light in dark field and bright field illumination with the Leica DMLP compound microscope and also with a binocular microscope (Sander 2000). Under higher magnification, bone types, vascularization, primary and secondary bone tissue, and even bone cell lacunae become visible.

If at all possible, the cores were drilled on the anterior side of the middle of the femur shafts and on the posterior side of the middle of the humerus shafts. It is crucial that core samples are taken from these standardized locations because otherwise comparability is compromised. However, in a few specimens, the bone surface was damaged at the prescribed location or the bone was reconstructed in plaster, necessitating sampling at a different location in the midshaft region. Some girdle and a few limb bones were drilled completely through, thus sampling the medial and lateral, or anterior and posterior cortex, respectively.

Evolution of Sauropodomorph Bone Histology

BASAL SAUROPODOMORPHA: *PLATEOSAURUS ENGELHARDTI*

The prosauropod *Plateosaurus engelhardti* is the most abundant dinosaur from the Upper Triassic of Central Europe. At up to 10 m in body length, basal sauropodomorphs such as *Plateosaurus* were the dominant herbivores in faunas of this age worldwide, and the first to evolve the large body size generally attributed to dinosaurs. *Plateosaurus* is also one of the best-studied dinosaurs, including its bone histology (Sander & Klein 2005; Klein & Sander 2007).

Plateosaurus long bones are characterized by a large medul-

Table 17.1. Histological Database

Taxon	n	Stratigraphy	Locality and repository	No. and size range of specimens
Thecodontosaurus sp. (Benton et al. 2000)	1	Rhaetic Fissure Fills, Upper Triassic	Tytherington, IPB	1 indet. long bone
Plateosaurus engelhardti (summarized in Klein & Sander 2007)	44	Trossingen Fm. and equivalents, Upper Triassic	Trossingen, SMNS & GPIT Frick, SMF	13 femora (approximately 50.0–99.0 cm) 8 tibiae (approximately 51.0–66.0 cm 6 fibulae (46.5–59.0 cm) 6 humeri (41.0–53.0 cm) 6 scapulae (36.5–49.5 cm) 5 pubes (48.0–53.5 cm)
cf. *Isanosaurus* sp. (Buffetaut et al. 2002)	1	Nam Phong Formation, Upper Triassic	Phu Nok Khian, PC.DMR	1 humerus (105.0 cm)
Mamenchisaurus sp. (Young 1954)	1	Shishugou Formation, Junggar Basin, Middle Jurassic	Shishugou Formation, GPIT	1 ulna (> 96.0 cm)
Dicraeosaurus sattleri, *D. hansemanni* (Janensch 1961)	6	Tendaguru Beds, Upper Jurassic	Tendaguru Beds, MFN	3 humeri (58.0–62.0 cm) 3 femora (98.0–114.0 cm)
Apatosaurus spp. (Carpenter & McIntosh 1994 for the juveniles)	21	Morrison Formation, Upper Jurassic	Cactus Park, BYU Dinosaur National Monument, CM Dry Mesa Quarry, BYU Howe Ranch, SMA Jensen/Jensen Quarry, BYU Kenton Quarry, OMNH	4 humeri (25.8–98.0 cm) 10 femora (34.0–180.0 cm) 2 tibiae (20.05 cm; > 64.0 cm) 5 scapulae (35.5–145.5 cm)
Diplodocus spp. (Foster 2007)	11	Morrison Formation, Upper Jurassic	Cactus Park, BYU Dinosaur National Monument, CM Dry Mesa Quarry, BYU Kenton Quarry, OMNH	2 humeri (90.0–106.0 cm) 9 femora (61.0–142.0 cm)
Barosaurus sp.	1	Morrison Formation, Upper Jurassic	Dinosaur National Monument, CM	1 humerus (101.0 cm)
Tornieria africana (Remes 2006)	1	Tendaguru Beds, Upper Jurassic	Tendaguru Beds, MFN	1 humerus (99.0 cm)
Diplodocinae indet. (*Barosaurus africanus* of Janensch 1961)	13	Tendaguru Beds, Upper Jurassic	Tendaguru Beds, MFN	6 femora (79.0–135.0 cm) 5 humeri (43.5–80.5 cm) 1 tibia (84.0 cm) 1 fibula (96.0 cm)
Diplodocinae indet. (Ayer 2000; Schwarz et al. 2007)	17	Morrison Formation, Upper Jurassic	Howe Ranch, SMA	6 femora (24.8–149.0 cm) 4 humeri (71.5–92.5 cm) 3 tibiae (86.5–101.0 cm) 2 ulnae (57.5–74.5 cm) 1 scapula (138.0 cm) 1 ischium (82.0 cm)
Camarasaurus spp. (Foster 2007)	21	Morrison Formation, Upper Jurassic	Dinosaur National Monument, Freezeout Hills, both CM Kenton Quarry, OMNH Howe Ranch, SMA Dry Mesa Quarry, BYU	9 humeri (22.7–120.4 cm) 7 femora (55.0–156.6 cm) 1 tibia (61.5 cm) 1 scapula (49.0 cm) 3 ischia (84.0–108.0 cm)
Europasaurus holgeri (Sander et al. 2006)	11	Lower Saxony Basin, Upper Jurassic	Langenberg, Oker, DFMMh/FV	4 femora (16.5–51.0 cm) 3 tibiae (11.9–30.0 cm)

Table 17.1. *continued*

Taxon	n	Stratigraphy	Locality and repository	No. and size range of specimens
Brachiosaurus brancai (MFN specimens), *B. altithorax* (BYU specimen)	16	Tendaguru Beds, Upper Jurassic Morrison Formation, Upper Jurassic	Dry Mesa Quarry, BYU Tendaguru Beds, MFN	7 humeri (69.0–176.0 cm) 7 femora (69.0–219.0 cm) 1 tibia (85.0 cm) 1 ulna (90.0 cm)
Janenschia robusta (Janensch 1961; Bonaparte et al. 2000)	2	Tendaguru Beds, Upper Jurassic	Tendaguru Beds, MFN	1 femur (127.0 cm) 1 humerus (89.0 cm)
Phuwiangosaurus sirindhornae (Martin et al. 1999)	23	Sao Khua Formation, Lower Cretaceous	Phu Singha, Phu Pratu Teema, Wat Sakawan, Ban Na Khrai all PC.DMR	5 humeri (71.0–110.0 cm) 13 femora (38.5–112.0 cm) 2 tibiae (> 34.0 cm) 3 pubes (> 64.0 cm)
Ampelosaurus atacis (Martin et al. 1999; Le Loeuff 2005)	27	Marnes Rouges Infèrieures Formation, Marnes de la Maurine member, Aude department, Upper Cretaceous	Bellevue & others closely by MDE & LMC	12 humeri (approximately 12.0–approximately 70.0 cm) 15 femora (48.0–100.0 cm)
Alamosaurus sanjuanensis (Gilmore 1922)	3	Javelina Formation, Upper Cretaceous	Big Bend National Park	3 humeri (46.0–130.0 cm)
Magyaorosaurus dacus (Nopsca 1914)	22	Hateg Basin, Upper Cretaceous	Sînpetru Formation, FGGUB, MAFI Densuş-Ciula Formation, FGGUB Vălioara-Budurone, MAFI	4 femora (34.6–54.5 cm) 8 humeri (22.25–48.8 cm) 5 tibiae (35.45–45.0 cm) 2 ulnae (21.9 cm; 33.7 cm) 2 fibulae (38.45 cm; 38.8 cm)
Magyarosaurus hungaricus (von Huene 1932)	1	Hateg Basin, Upper Cretaceous	Vălioara, MAFI	1 humerus (91.4 cm)

The taxa are listed in systematic order according to the phylogeny of Wilson (2002). The size range of *Ampelosaurus, Europasaurus, Phuwiangosaurus,* and *Magyarosaurus* bones are partially based on estimates because of the fragmentary preservation of the bones (for more details about the database, see the appendix in Klein & Sander 2008).

lary cavity and relatively thin bone walls. The cortex is sharply set off from the medullary cavity with a few scattered or no secondary osteons. The primary bone of the cortex is dominated by growth cycles of fibrolamellar bone, ending in a LAG (Fig. 17.4). Vascular canals are primarily circumferential, and vascularity decreases toward the LAG. Although the predominance of fibrolamellar bone in the long bone cortex of *Plateosaurus* indicates that it grew at the fast rates typical for dinosaurs, qualitative (growth stop, EFS) and quantitative (growth-mark count) features of its histology are poorly correlated with body size (Sander & Klein 2005). This indicates a variable life history and dependence on environmental factors, as is typical for modern ectothermic reptiles, but not for mammals, birds, and other dinosaurs. This strong developmental plasticity of *Plateosaurus* may be a reversal to the plesiomorphic condition of basal archosaurs (de Ricqlès et al. 2008) and contrasts with all of the more derived dinosaurs.

Both *Massospondylus* (Chinsamy 1993; Chinsamy-Turan 2005) and *Thecodontosaurus* (Cherry 2002; M.S., N.K., pers. obs.) have a bone tissue similar to that of *Plateosaurus*. However, none of the specimens of these taxa shows a change from fibrolamellar bone to lamellar-zonal bone in the outer cortex. This may reflect a rather different growth strategy or, more likely, it may indicate that no fully grown individuals were sampled (Chinsamy-Turan 2005). A growth series has only been sampled in *Massospondylus*, but unlike *P. engelhardti*, this taxon shows a close correlation between age and body size (Chinsamy 1993; Chinsamy-Turan 2005). The bone histology of *"Euskelosaurus browni"* as described by de Ricqlès (1968a) is also consistent with our observations on *P. engelhardti*.

BASAL SAUROPODA: CF. *ISANOSAURUS*

The primary bone tissue in a humerus of the stratigraphically oldest sauropod that exhibits gigantism, cf. *Isanosaurus* (Buffetaut et al. 2002) from the Late Triassic of Thailand, entirely consists of laminar fibrolamellar bone tissue (Sander et al. 2004). As opposed to prosauropods, growth marks are not developed. The medullary cavity is small and consists of cancellous bone. The inner cortex contains scattered secondary

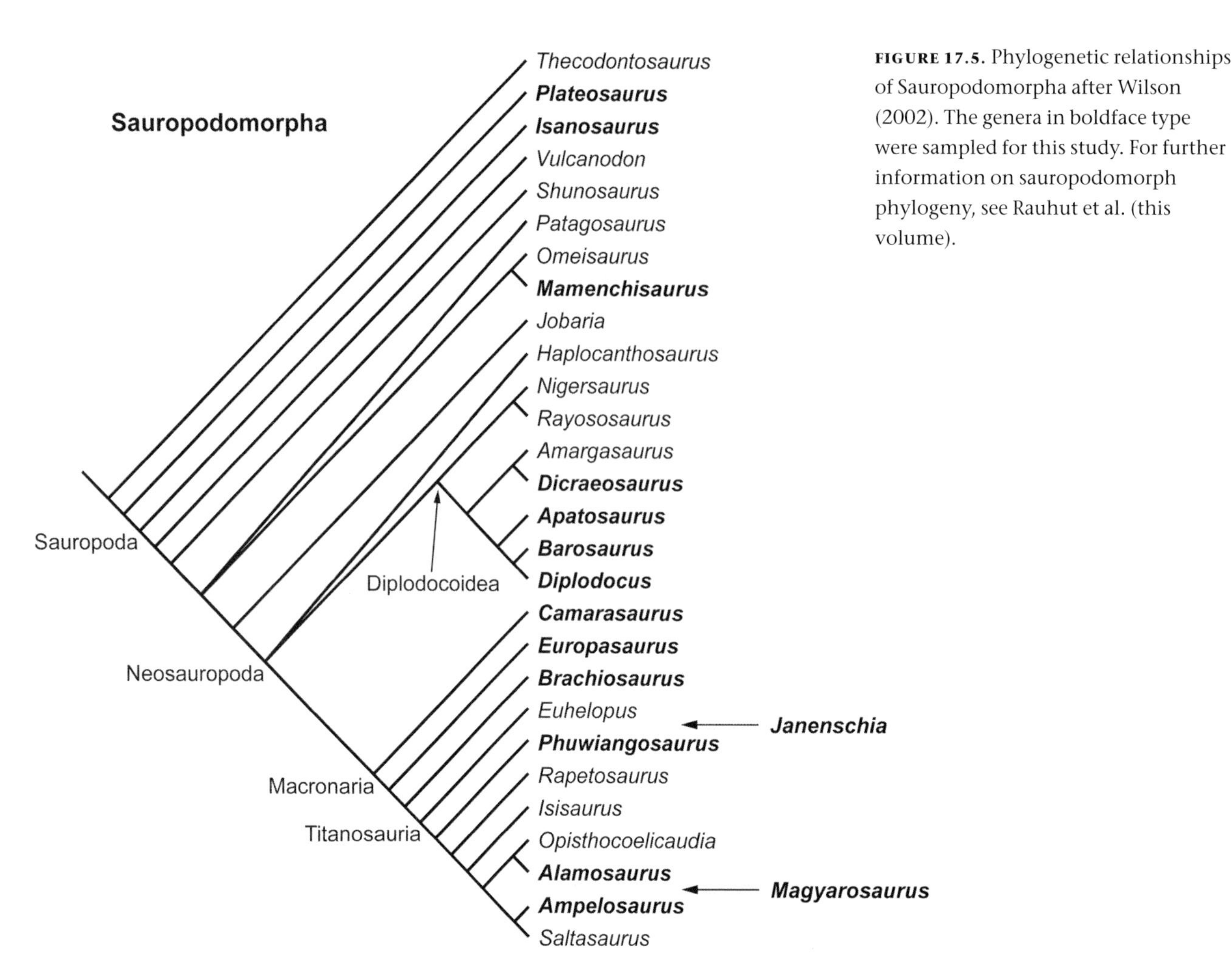

FIGURE 17.5. Phylogenetic relationships of Sauropodomorpha after Wilson (2002). The genera in boldface type were sampled for this study. For further information on sauropodomorph phylogeny, see Rauhut et al. (this volume).

osteons. The histology of the cortex thus documents the rapid and uninterrupted growth typical for later sauropods (Sander et al. 2004). We used the histological ontogenetic stages erected for sauropods by Klein & Sander (2008) to identify the specimen as a subadult or young adult individual.

NEOSAUROPODA

All neosauropod long bones studied (Table 17.1, Fig. 17.5) grew fairly uniformly, laying down large amounts of laminar fibrolamellar bone tissue in their cortex (Fig. 17.7). Growth marks of any kind are rare and not consistently present in any taxon sampled from growth series except for *Europasaurus*. Differences in primary bone tissue types mainly concern the organization of the vascular system, the degree of vascularization, and the presence and degree of development of primary osteons (i.e., the amount of lamellar bone deposited centripetally in the vascular canals). Vascularization, and therefore growth rate, decreases gradually from young to fully grown individuals. We used these differences in primary bone and the degree of remodeling by secondary osteons were used to establish histological ontogenetic stages for sauropods (Klein & Sander 2008). These work best with diplodocoid (*Apatosaurus, Dicraeosarus, Diplodocus,* indet. Diplodocinae) and basal macronarian taxa (*Camarasaurus, Brachiosaurus*), which are most uniform histologically. Only the dwarf basal macronarian *Europasaurus* has a different histology, which will be discussed below in the section on island dwarf sauropods.

Titanosauria show more variation in their histology. Although the two titanosaurs *Ampelosaurus* and *Phuwiangosaurus* went through these histological ontogenetic stages as well (Klein & Sander 2008; Klein et al. 2009a), details in their bone histology differ from other neosauropods. They also show clear differences to the most basal titanosaur *Janenschia* (Sander 2000; Sander & Tückmantel 2003) and the large derived titanosaur *Alamosaurus* (Woodward 2005; Woodward & Lehman 2009). Even in early histological ontogenetic stages, *Ampelosaurus* and *Phuwiangosaurus* exhibit an unusually high amount of parallel-fibered bone in the matrix of the fibrolamellar complex, and also strong remodeling by secondary osteons (Klein et al. 2009a). The high amount of the parallel-fibered bone tissue, even in early ontogenetic stages, indicates

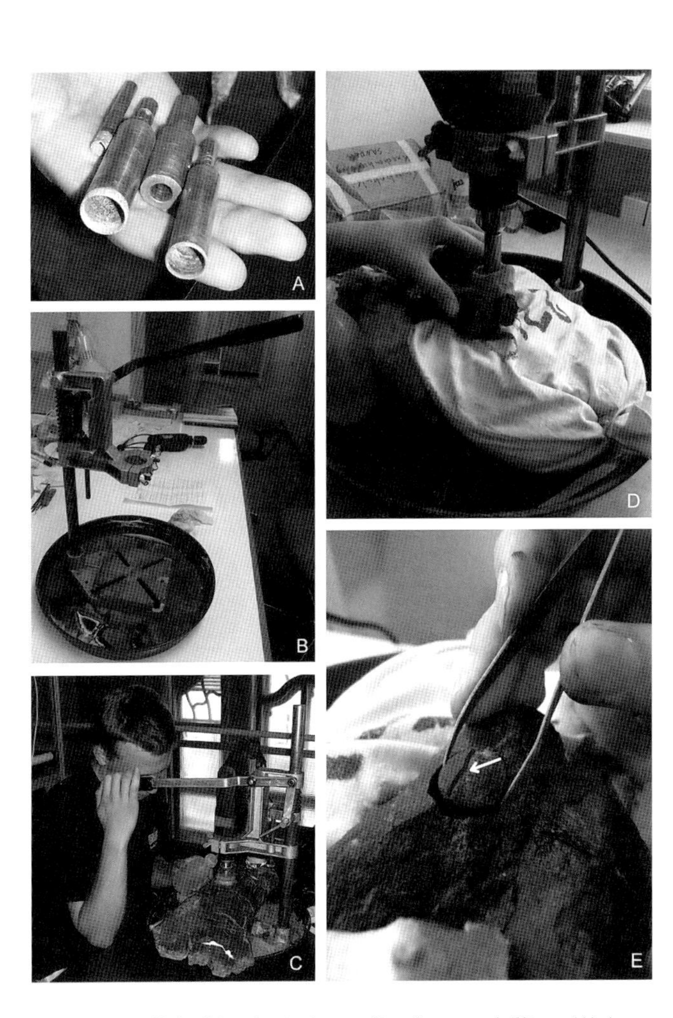

FIGURE 17.6. Paleohistological sampling by core drilling. (A) A selection of core drill bits ranging in diameter from 4–16 mm, and an extension piece for the 16 mm bit (second from right). (B) Drill press set in a flowerpot dish. (C) Drilling in progress with an electric drill mounted in the drill press. (D) The drill bit needs to be lubricated with water or oil, which is contained by a small plasticine dam set up around the drill hole. Note the sandbag on which the bone specimen is resting. (E) Before the core sample is carefully removed with tweezers, a mark indicating the bone long axis (arrow) is applied.

a generally slower growth rate in *Ampelosaurus* and *Phuwiangosaurus* than in diplodocoid and basal macronarian sauropods. This high amount of parallel-fibered bone in the matrix should not be confused with particularly well developed primary osteons that would have developed only after the deposition of the matrix (see Fig. 17.2A for the formation of fibrolamellar bone.). It is the rate of deposition of the matrix that determines the rate of size increase of the bone; the infilling of the matrix by primary osteons occurs later. The ontogenetically earlier occurrence of complete remodeling by secondary osteons in these two titanosaur taxa suggests a different life history (Klein & Sander 2008), which possibly includes a shortened growth period. The meaning of this is not clear yet, but it could be connected to the relative small size of both *Phuwiangosaurus* and *Ampelosaurus* compared to the large *Alamosaurus* and the Diplodocoidea and basal Macronaria.

ISLAND DWARF SAUROPODS

Although virtually all sauropods are very large animals and small individuals represent juveniles, there are two species with a small adult stature (body mass <1,000 kg): *Europasaurus holgeri* and *Magyarosaurus dacus*. They evolved their diminutive body size on paleoislands (Nopcsa 1914; Sander et al. 2006; Stein et al. 2010). *Europasaurus holgeri* is a basal macronarian from Kimmeridgian marine sediments of the Lower Saxony Basin in northern Germany (Sander et al. 2006), while *Magyarosaurus dacus* is a titanosaur from Maastrichtian terrestrial sediments of Romania (Stein et al. 2010).

The long bone histology of *Europasaurus* resembles that of large Jurassic sauropods except for the regularly spaced LAGs that occur in even small individuals. That the largest known individuals of *Europasaurus* were fully grown is indicated by an EFS in their outermost cortex (Sander et al. 2006). The regular occurrence of growth marks suggests that bone apposition rate was lower than in the large Jurassic sauropods (Sander et al. 2006), while the growth mark counts suggests that the phase of active growth was shortened in *Europasaurus*. *Europasaurus* dwarfing thus evolved both through a reduction in growth rate and a shortening of ontogeny.

The histology of *Magyarosaurus* is unique among sauropods so far because even the smallest individuals of 45% maximum size show nearly complete replacement of the primary cortex by secondary Haversian bone (Stein et al. 2010). Expressed in histological ontogenetic stage (HOS), the smallest specimen is nearly completely remodeled and shows a late histological ontogenetic stage (HOS 12.5), while all others represent the last stages, HOS 13 and HOS 14 (Table 17.2), which otherwise are only seen in the oldest and largest sauropod bones, such as a femur of *Apatosaurus* of 1.8 m length. The intense remodeling of the *Magyarosaurus* cortex suggests an extremely reduced growth rate compared to ancestral large titanosaurs (Stein et al. 2010). Whether this was accompanied by a reduction in metabolic rate is unclear at present.

IMPLICATIONS FOR SAUROPOD LIFE HISTORY AND PALEOBIOLOGY

Histological Ontogenetic Stages and Biological Ontogenetic Stages

The establishment of histological ontogenetic stages raises the question whether we can correlate these stages with biological ontogenetic stages and specific life history events. Although we are aware that biological ontogenetic stages are

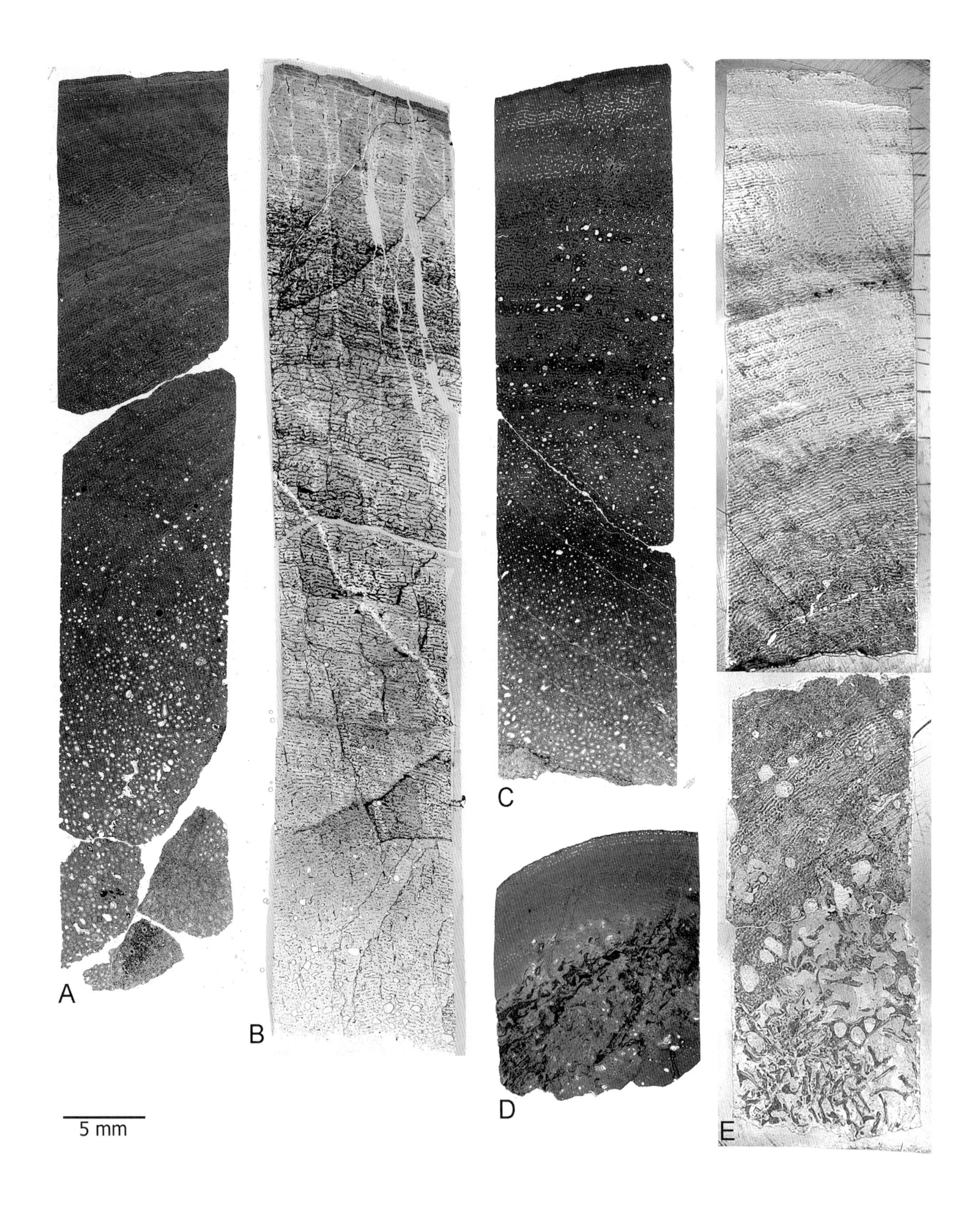
A
B
C
D
E
5 mm

FIGURE 17.7. Thin sections from drill cores of a variety of neosauropod long bones. The outer bone surface is at the top. Most samples extend from the cortex into the cancellous bone of the medullary region. All show variations of the fibrolamellar bone type typical for neosauropods (see Klein & Sander 2008). (A) *Diplodocus* sp. SMA Max (femur, 149.0 cm). (B) *Janenschia robusta* MFN Nr. 22 (femur, 127.0 cm). (C) *Apatosaurus* sp. SMA 0014 (tibia, 64.0 cm). (D) *Europasaurus holgeri* DFMMh/FV 495.5 (tibia, 28.6 cm). (E) Diplodocidae indet. MFN NW 4 (femur, 135.0 cm). (F) *Brachiosaurus brancai* MFN XV (femur, 219.0 cm). (G) *Camarasaurus* sp. CM 36664 (humerus, 117.2 cm). (H) *Dicraeosaurus* MFN M 1b (femur, 112.0 cm). (I) *Phuwiangosaurus sirindhornae* PC.DMR K4–366 (femur, 93.0 cm).

difficult to determine in extinct taxa, we tentatively suggest here interpretations of histological ontogenetic stages in terms of biological ontogenetic stages (Table 17.2). Terms such as *embryo, hatchling, juvenile, subadult,* and *adult,* which originate from the study of life history in recent amniotes, are in common use for extinct taxa as well. Because there is no uniform definition of biological ontogenetic stages in zoology, especially for the earliest stages, and additionally because biological ontogenetic stages are variably applied from group to group in modern tetrapods, we first provide a detailed definition of biological ontogenetic stages as we apply them to sauropod dinosaur ontogeny. The definitions are mainly based on the life history of recent oviparous reptiles and on the consensus in the herpetological literature, although no single reference defines them. Early ontogenetic stages are also similar to what is observed in recent birds. We recognize a total of seven biological ontogenetic stages (plus substages, Table 17.2).

Table 17.2. Ontogenetic Stages in Sauropods as Reflected in Long Bone Histology

Bone tissue type	Histological ontogenetic stage	Biological ontogenetic stage
Cortex consists of type A bone tissue.	HOS 1	Embryo
Cortex consists primarily of type A bone tissue with type B bone tissue laid down in the outer cortex.	HOS 2	Embryo to hatchling
Cortex consists primarily of type B bone tissue while in the inner cortex, remains of type A bone tissue can be preserved.	HOS 3	Hatchling
Cortex consists primarily of type B bone tissue with type C bone tissue laid down in the outer cortex.	HOS 4	Hatchling to juvenile
Cortex consists primarily of type C bone tissue while in the inner cortex, remains of type B bone tissue can be preserved.	HOS 5	Juvenile
Cortex consists primarily of type C bone tissue with type D bone tissue laid down in the outer cortex.	HOS 6	Juvenile to subadult
Cortex consists primarily of type D bone tissue while in the inner cortex, remains of type C bone tissue can be preserved.	HOS 7	Subadult
Cortex consists primarily of type D bone tissue with type E bone tissue laid down in the outer cortex.	HOS 8	Sexual maturity
Cortex consists primarily of type E bone tissue while in the inner cortex, remains of type D bone tissue can be preserved.	HOS 9	Adult I
Cortex consists primarily of type E bone tissue with type F bone tissue laid down in the outer cortex.	HOS 10	Adult I to adult II
Cortex consists primarily of type F bone tissue while in the inner cortex, remains of type E bone tissue can be preserved.	HOS 11	Adult II
Cortex consists primarily of type F bone tissue while in the outer cortex, an EFS is deposited.	HOS 12	Adult II to adult III
Cortex consists of type G bone tissue, which means that it is nearly completely remodeled by secondary osteons.	HOS 13 HOS 14	Adult III

Ontogenetic variation in long bone histology is subdivided into 6 ontogenetic bone tissue types and 14 histological ontogenetic stages. For the definition of the bone tissue types A to G, see Klein & Sander (2008). EFS, external fundamental system.

An *embryo* is an individual inside the egg, before hatching. This stage can be determined unequivocally in the fossil record of dinosaurs. However, among sauropodomorphs, there is currently only evidence for titanosaur and *Massospondylus* embryos (Chiappe et al. 2005; Reisz et al. 2005; Wilson et al. 2010). We do not have any evidence for embryonic material in our sample. In recent oviparous reptiles and birds, *hatchling* refers to an individual from the time when the embryo starts to hatch to the first few days after hatching. After having forced the egg membranes and the outer eggshell open, some hatchlings stay inside the egg from several hours to up to a few days to recover from the effort. The term *hatchling* is still applied after the individual leaves the egg, before the yolk is fully resorbed, and the hatchling starts to feed by itself, which can also last some days.

The hatchling stage is followed by the *juvenile* stage. Often *juvenile* is used for an "individual not sexually mature," but we prefer further splitting up of the early ontogenetic stages. We define the end of the juvenile stage as the time when all elements in the skeleton have ossified and adult skeletal proportion and shapes are attained, with only linear growth after this point. The juvenile phase can last from several months up to a few years in recent sauropsids. When shape changes are completed, the *subadult* stage begins. In recent reptiles (e.g., monitor lizards, turtles, agamids), often a change in diet (e.g., from insectivorous to omnivorous; from omnivorous to herbivorous) accompanies the transition from juvenile to subadult. In diet, appearance, and behavior, subadults are comparable to adults. They only grow in size and mass, but are not sexually mature. In many extant reptiles, this subadult stage does not last long—only a few months—before they start mating in the next season.

The *adult* stage starts with the first reproduction and describes sexually mature individuals. Most mammals, birds, and some reptiles reach their full size at nearly the same time as sexual maturity, but some others do not and continue to grow for several years afterward. Among extant reptiles, this is seen in turtles, crocodiles, boid snakes, and some lizards. Significant continued growth is also typical for the males of some large herbivorous mammals (e.g., African elephants and giraffes) and for primates. In life histories with significant growth after sexual maturity, we divide the adult stage into three substages. The first includes sexually mature but still significantly growing animals (adult I), the second substage includes adults in which growth slows down and eventually ceases (adult II), and finally the senescent stage in which growth has stopped but the individual continues to live for several years. Because of the continuing increase in size, adult I stage can clearly be observed. In contrast, adult II and the senescent stage are not obvious from a size increase and morphology. These last two ontogenetic stages are best distinguished by bone histology. However, both stages, and especially the senescent stage, are generally rare because of the greatly decreased likelihood of survival in the wild. Note that our definition of *adult* differs from the usage in a recent paper on sauropod growth curves by Lehman & Woodward (2008). These authors equated *adult* with having attained maximum size (our adult II), thereby implying that sexual maturity coincided with the termination of growth. We consider this unlikely for reasons discussed below.

In our correlation (Table 17.2), HOS 1 corresponds to the embryonic stage, HOS 3 to the hatchling stage, HOS 5 to the juvenile stage, HOS 7 to the subadult stage, HOS 9 to adult stage I, HOS 11 to adult stage II, and finally HOS 14 to the senescent stage. Sexual maturity would correlate with HOS 8.

Sexual Maturity

Because sexual maturity is one of the major events in life history, our correlation of HOS 8 with sexual maturity in sauropods warrants further discussion. Determining sexual maturity in extinct animals is rather difficult (Curry 1999; Horner et al. 1999; Sander 2000; Chinsamy-Turan 2005; Erickson 2005; Erickson et al. 2007; Klein & Sander 2007, 2008) because it remains hypothetical until clear evidence, such as medullary bone, is discovered for a sauropod taxon. Medullary bone is known from recent ovulating birds (Chinsamy-Turan 2005) but not from crocodiles. Medullary bone has recently been discovered in both theropod (*Tyrannosaurus,* Schweitzer et al. 2005; *Allosaurus,* Lee & Werning 2008) and ornithopod dinosaurs (*Tenontosaurus,* Lee & Werning 2008; *Dysalotosaurus,* T. Hübner, pers. comm.), which suggests that medullary bone is a character of all dinosaurs. In our large sauropodomorph sample (Klein & Sander 2007, 2008; Table 17.1), no medullary bone was observed. This may be because medullary bone as a storage tissue was not needed in sauropods because of their small clutch mass compared to the adult (Sander et al. 2008; Sander et al. 2010; Griebeler & Werner, this volume).

Sander (2000) hypothesized that sexual maturity in sauropods is recorded in their histological growth record, but this has not met with universal agreement (e.g., Erickson 2005; Chinsamy-Turan 2005). Although the onset of sexual maturity has not been identified with certainty in the bone histology of extinct vertebrates, one can argue the pros and cons for several possible moments in an individual's histological growth record as indicators of sexual maturity. Sander's (2000) reasoning was that the onset of sexual maturity would have resulted in a decrease in growth rate as a result of the shift in resource allocation from growth to reproduction as a sauropod became sexually mature. This is the pattern seen in virtually all living tetrapods, and it presumably applied to sauropods as well. Bone histology in sauropods documents several decreases in linear growth rate, and the question arises which of these might reflect the onset of sexual maturity.

One possibility, favored by Curry (1999), is that sexual maturity coincided with attainment of nearly final size, that is, the asymptotic phase of growth, as reflected by the deposition of an EFS or a similar bone tissue type. This possibility can be evaluated by using the limited quantitative life history data available for sauropods (see also below). Sander (2000) used growth cycles to estimate that an individual of the basal titanosaur *Janenschia* required at least 26 years to reach nearly final size. However, an individual of *Apatosaurus* of 91% maximum size (SMA 0014; Table 17.3) shows 20 growth cycles in the outer cortex, suggesting that it was at least 28 years old (including the resorbed cycles). Unfortunately, the outermost cortex is incomplete in this sample, and an exact cycle count within the EFS cannot be given. In this case, sexual maturity would not have been attained until the end of the third decade of life, which was considered highly unlikely by Dunham et al. (1989) for any dinosaur.

This is why we favor correlating HOS 8 with sexual maturity. As Erickson et al. (2007) and Lee & Werning (2008) have recently shown with different method, in nonavian theropods and ornithischians, sexual maturity also occurred well before full adult size was reached. Our *Apatosaurus* sample (Klein & Sander 2008) suggests a body size of around 50% of maximum adult size at HOS 8. This would correspond to a femur length of 90.0 cm, or 11.25 m body length. Attainment of sexual maturity at about half of maximum body size is consistent with the life histories of modern large reptiles, such as crocodiles and turtles, but not with those of large herbivorous mammals.

Estimates of Life Span and Longevity

Even in extant reptiles, life span is most easily estimated skeletochronologically (Castanet et al. 1993; Castanet 1994), and this method can be applied to fossil tetrapods as well. However, as noted above, growth marks are rare in Neosauropoda, and their record is largely restricted to the final phase of growth (Klein & Sander 2008). Apparently juvenile sauropods grew too fast, and thus local bone apposition rates

Table 17.3. Longevity, Maximal Growth Rate (kg/yr), and Mass Calculations of Some Sauropods[a]

Taxon	A	B	C	D	E	F	G	H
Plateosaurus engelhardti	GPIT/IFG 192.1	Femur (74.0 cm) 1,587 kg	71% 52.0 cm 575 kg	12	3	16	NC	253 kg
Mamenchisaurus sp.	GPIT SGP 2006/10	Ulna (>96.0 cm) 31,000 kg	NC	32	10	43	NC	2,006 kg
Apatosaurus sp.	SMA 0014	Femur (164.0 cm) 21,023 kg	54% 89.0 cm 3405 kg	~20	7	28	NC	2,531 kg
Apatosaurus sp.	BYU-601-17328	Femur (158.0 cm) 21,024 kg	61% 97.0 cm 4927 kg	21	19	41	NC	962 kg
Diplodocinae indet. ("*Barosaurus africanus*")	MFN NW 4	Femur (135.0 cm) 11,273 kg	52% 70.0 cm 1614 kg	23	8	32	26.8	934 kg
Diplodocinae indet. ("*Barosaurus africanus*")	MFN MB.R. 2685 (H 4)	Humerus (61.0 cm) 2,247 kg	74% 63.0 cm 928 kg	11	19	41	NC	202 kg
Dicraeosaurus sp.	MFN 02	Femur (98.0 cm) 3,802 kg	50.7% 49.7 cm 495 kg	~4	3	8	14.9	
Camarasaurus	CM 36664	(Femur 142.5 cm) 9,882 kg	80% 114.0 cm 5151 kg	14	25	40	NC	886 kg
Janenschia robusta	MFN Nr. 22	Femur (127.0 cm) 7,943 kg	57% 72.0 cm 1,473 kg	38	16	55	24.3	447 kg
Europasaurus holgeri	DFMMh/FV 415	Femur (c. 60.0 cm) 690 kg	65% 39.0 cm 189.5 kg	6	4	11	NC	353 kg

Mass calculations are based on Anderson et al. (1985) and Seebacher (2001). All percentages are based on the largest known individual of each taxon (Klein & Sander 2008). The number of resorbed cycles is based on an extrapolation method explained in Klein & Sander (2007). The total age is based on the sum of the visible cycles and the number of the calculated resorbed cycles. Age estimates on the basis of cortical thickness measured in the middle of the shaft with a presumed local bone apposition rate of 7.7 μm/day (following Sander & Tückmantel 2003). The maximum growth rate per cycle is obtained by the method described in the main text. Note that the maximum growth rate per cycle for *Janenschia* is a clear underestimate as a result of the poor mass correlation, and therefore the age estimate is an overestimate (see text).

[a]A, specimen number; B, bone, bone length and body mass estimate; C, percentage maximum size at onset of growth record and corresponding mass; D, number of visible growth marks; E, estimated number of resorbed growth marks; F, estimated total age in years; G, age estimates based on cortical thickness; H, maximum growth rate per year; NC, not calculated.

were too high, for annual interruptions to be recorded. In old adult sauropods, the growth marks have already been obliterated by secondary bone. Because of the lack of long records of regularly deposited growth marks in most sauropod specimens and taxa (Klein & Sander 2008), skeletochronology cannot be applied with confidence to sauropod dinosaurs.

Nevertheless, there are three approaches to estimating longevity in sauropods. The simplest is to count growth marks in the large specimens that show them and estimate the number of resorbed marks. A specimen of *Apatosaurus* sp. (SMA 0014) of 91% maximum size and of HOS 10 preserves 20 cycles in its outer cortex with an estimated 7 cycles missing, resulting in an age estimate of 28 years at death. A *Janenschia robusta* (MFN Nr. 22) that was fully grown (HOS 12) preserves 26 cycles in its outer cortex (Sander 2000), with an estimated 16 cycles missing. The EFS maximally preserves another 12 cycles (Sander 2000), which, if they are annual, would indicate a minimum age of 38 and a maximum age of 55 years. The largest known ulna of *Mamenchisaurus* sp. preserves 32 cycles, with an estimated 10 cycles missing, resulting in an age estimate of 43 years at death (Wings et al. 2007). These are the three most complete growth records available for sauropods, with the exception of the dwarf sauropod *Europasaurus,* in which all individuals show regularly spaced growth marks. For example, the largest individual preserves 6 cycles in its cortex with an estimated 4 cycles missing, resulting in an age estimate of 11 years at death (Table 17.3).

The second method is to estimate longevity on the basis of

cortical thickness (Sander & Tückmantel 2003). This can either be done by calculating apposition rates from growth cycles of sauropods, or by using apposition rates from modern animals. Age estimates range from about 15 years to over 30 years, depending on taxon (Table 17.3).

The third method, fitting of growth curves to growth mark records as done by Lehman & Woodward (2008), also results in longevity estimates, which represent the asymptotic age of the growth curve. However, asymptotic age estimates also involve a number of assumptions and depend on the growth model used. Lehman & Woodward (2008) calculated an age of 70 years at 90% maximal size for *Apatosaurus,* 45 years at 90% maximal size for *Alamosaurus,* and between 18 and 29 years at 90% maximal size for the *Janenschia robusta* specimen described by Sander (2000). They also calculated an age of between 50 and 75 years at 90% maximal size for an indeterminate sauropod from England described by Reid (1981), on the basis of growth marks in the pubis. Except for *Alamosaurus,* Lehman & Woodward (2008) did not observe the specimens firsthand, and their calculations involve many uncertainties. Interestingly, the specimen with the best growth mark record in their study, the *Janenschia robusta* femur mentioned above, also shows the best agreement between the longevity estimate based on growth mark counts (first method) and that based on growth curve fitting (third method). The other ages calculated by Lehman & Woodward (2008) using curve fitting appear to be substantial overestimates that resulted from incomplete growth mark records and uncertainties in estimating maximum size of the taxon (see section on growth rate estimates).

All three methods thus at least partially depend on the interpretation of the growth mark record and share two kinds of errors. One stems from the fact that part of the growth record is obliterated by resorption and remodeling, necessitating retrocalculation of the number of missing growth marks or the amount of tissue deposited. The other source of error stems from the fact that bone histology can only record life history while the animal is growing and depositing primary bone. The time from completion of growth to death is not recorded in primary bone but only in the progress of remodeling, that is, in the formation of Haversian bone. The rate of formation of Haversian bone has not been quantified in sauropods, however, and it is not known whether it is constant. Because the sign of the two errors is opposite (retrocalculation provides an upper limit to the number of resorbed cycles, while the entire preserved growth record provides a lower limit to the true age of the animal), longevity estimates for sauropods appear reasonably well supported. Generalizing from these estimates, it thus appears that sauropods lived for several decades but did not reach the centennial mark, despite their gigantic size.

This is also consistent with anecdotal evidence based on a few specimens of uncertain affinities published by de Ricqlès (1980), Reid (1981, 1990), and Rimblot-Baly et al. (1995). The first study resulted in an estimate of 43 years at full size, on the basis of a half-grown humerus for *Bothriospondylus madagascariensis* (de Ricqlès 1980). The second was based on similar material of the same taxon (called *Lapparentosaurus madagascariensis* in that study), with a humerus of 40% maximal length showing a minimum of 13 growth cycles (Rimblot-Baly et al. 1995). The taxonomic affinity of *Bothriospondylus madagascariensis* is uncertain at present (Upchurch et al. 2007), although earlier workers assumed brachiosaurid affinities. The last study documented 23 cycles in an indeterminate sauropod pubis from England (Reid 1981), with the total being estimated to have been only 28 or 29 (Reid 1990), contrasting with the much higher estimate of Lehman & Woodward (2008) for the same specimen.

Growth Rate Estimates for Sauropods

Because of the rare occurrence of growth mark records in sauropod long bones, there is as yet no reliable growth curves for any sauropod taxon, and those that have been published are controversial and poorly constrained. Erickson et al. (2001) obtained a maximum growth rate of >5,000 kg/yr for *Apatosaurus.* This extremely high growth rate is based on an age estimate of 15 years for a fully grown *Apatosaurus,* for which no further documentation was provided. First doubts about these high growth rates were raised by the description of an *Apatosaurus* specimen by Sander & Tückmantel (2003) that preserves 26 growth cycles in its outer cortex alone (see also Table 17.3). A reanalysis by Lehman & Woodward (2008) of the *Apatosaurus* growth series published by Curry (1999) led to a greatly reduced maximum growth rate for *Apatosaurus* of 520 kg/yr. The high rate in the analysis of Erickson et al. (2001; 5,000 kg) appears to suffer from an error in the analysis (Erickson, pers. comm.) and must be incorrect (Lehman & Woodward 2008). On the other hand, the low rate of Lehman & Woodward (2008) is poorly constrained because it is the only published one to be based on the growth record found in scapulae and not in large limb bones such as humeri and femora.

Although the assumption of isometric growth of limb bones relative to the rest of the skeleton appears well supported (see discussion in Lehman & Woodward 2008 and Bonnan 2004; Kilbourne & Makovicky 2010), as is the isometry of limb bone length versus circumference, that of the isometry of shape of the scapula remains untested for sauropods. More precisely, we do not know how the increase in cortical thickness of the medial surface of the shaft of the scapula (where the growth record was found by Curry 1999) translates into body mass increase. We consider this translation done by Lehman & Woodward (2008) particularly problematical because sauropod scapulae have a distinctive medial concavity, and their growth record will be strongly influenced by shape changes of the bone during growth.

Lehman & Woodward (2008) computed growth rates for

FIGURE 17.8. Polished cross section from the midshaft region of a *Mamenchisaurus* ulna (IFG/GPIT SGP 2006/10, length >96.0 cm) from the Upper Jurassic Shishugou Formation. Because of little remodeling and a small medullary cavity, the sample preserves an unusually long growth record in its primary cortex. It consists of 26 clearly visible cycles developed in fibrolamellar bone and six closely spaced cycles seen in an EFS in the outermost cortex.

three other taxa, *Janenschia robusta, Alamosaurus sanjuanensis,* and an indeterminate sauropod from England. The growth rates for the latter were based on the published information (Reid 1981) on a pubis of a single individual. They are the lowest (180–260 kg/yr) in the study of Lehman & Woodward (2008), but we think that they are much too poorly constrained to warrant further consideration because the body mass of the individual cannot be reliably estimated from a pubis, nor can the incremental increase in cortex thickness of the pubis be transformed into size increase of the individual.

The maximum growth rates for *Janenschia* of between 623 and 993 kg/yr were calculated by Lehman & Woodward (2008) from the information and illustrations published by Sander (2000) for the femur MFN Nr. 22. They are thus based on the growth record from a single individual. This estimate again is rather poorly constrained (although not as poorly as that for the indeterminate sauropod from England) because there are no associated humeri and femora known for *Janenschia* that would permit estimation of body mass of the individual on the basis of long bone circumference (Anderson et al. 1985). Instead, Lehman & Woodward (2008) estimated the circumference of the humerus of MFN Nr. 22 based on "proportions similar to those of other titanosaurs" (Lehman & Woodward 2008, caption to fig. 3) and arrived at a body mass estimate of 14 metric tons. This approach is problematic because it is not known how constant such proportions are among titanosaurs, even if *Janenschia* is indeed the most basal titanosaur, which has not been firmly established yet. An alternative approach to estimating body mass in the *Janenschia* individual MFN Nr. 22 would be to use the regression equation of Mazzetta et al. (2004), which predicts body mass from femur length. However, because of the unusual robustness of *Janenschia* bones (with the species name *robusta!*), this regression yields an unrealistically low body mass estimate of only

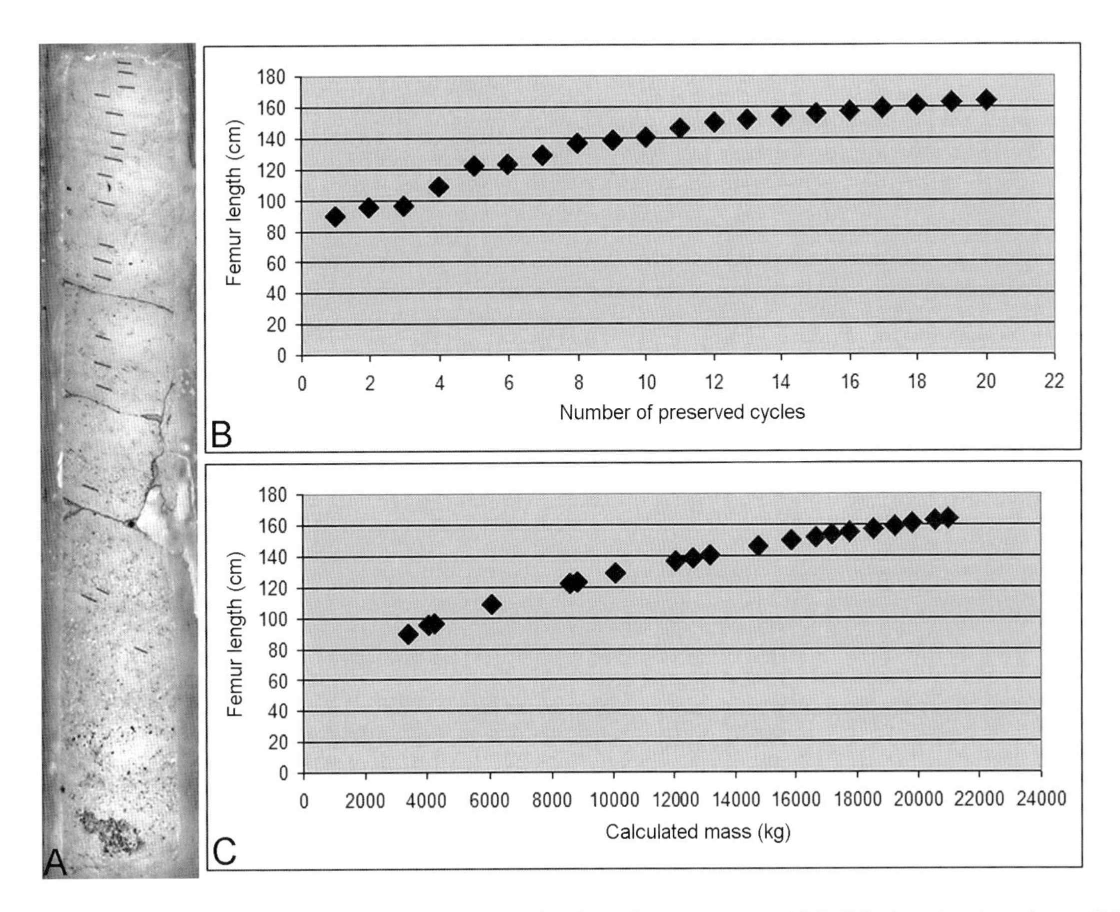

FIGURE 17.9. Quantification of the growth record of a large individual of *Apatosaurus* sp. (A) Polished section from femur SMA 0014 (length 164.0 cm) with well preserved growth cycles. The growth marks are marked here by additional black lines. (B) Graph showing the corresponding femur length for each visible cycle of this individual. (C) Graph showing the calculated body mass for a given femur length. See text for details of the calculations. The greatest mass gain was between cycles 4 and 5 and corresponds to approximately 2,500 kg.

8.5 metric tons for MFN Nr. 22, and thus unrealistically low growth rates (Table 17.3) were it to be used in constructing a growth curve.

In the study of Lehman & Woodward (2008), only the growth rate estimate for *Alamosaurus sanjuanensis* appears well constrained, which ranges from 940–1,160 kg/yr. We consider this estimate reliable because this taxon is known from entire skeletons, and at least some specimens preserve good histological growth records, which were studied first-hand by Lehman & Woodward (2008).

The best growth record of any sauropod discovered so far is that in an ulna (preserved length 96.0 cm) of a large *Mamenchisaurus* sp. (Wings et al. 2007). Cross-scaling with other long bones results in a body mass estimate of 31,000 kg (Wings et al. 2007). The growth record in the bone, consisting of well developed and regularly spaced growth cycles each terminated by a LAG, begins at 40% full size and continues all the way to the outer cortex (Fig. 17.8). Assuming that increase in cortex thickness is isometric with body mass gain, the animal attained its highest growth rate of 2,006 kg/yr between preserved cycle 13 and 14. Because only the inner 40% of the growth record is missing, the animal cannot have been much older than 20 years.

Some individuals provide a good growth record for the latter part of their life history (see above), rendering growth rate estimates possible (Table 17.3). The highest growth rate of 2.5 metric tons is here observed between the fourth and fifth preserved cycle (corresponding to an age of 12 years) of *Apatosaurus* sp. (SMA 0014; Fig. 17.9; Table 17.3).

As for the longevity estimates, Amprino's rule can be applied to the problem, and bone tissue apposition rates observed in modern animals can be used to estimate overall growth rates in dinosaurs. There are considerable difficulties with such estimates of appositional growth rates in any extinct tetrapods (Castanet et al. 1996, 2000; de Margerie et al. 2002; Sander & Tückmantel 2003). One difficulty is that the minimal and maximal known apposition rates of a particular bone tissue type may vary by a factor of 5 or more. In addition, the experimentally obtained growth rates for the fibrolamellar bone tissue originate from living birds and small mammals (Castanet et al. 1996, 2000; de Margerie et al. 2002; Montes et al. 2007; Cubo et al. 2008), and it is questionable whether they should be applied to sauropods because of the enormous differences in body size, physiology, and life history between these groups. On the other hand, the values of Curry (1999) obtained for *Apatosaurus* (10.5 μm/day) and Sander & Tückmantel (2003) for other sauropods (by dividing growth cycles thickness by the number of days per year) are consistent with the few known tissue apposition rates of large extant mammals (Curry 1999; Sander & Tückmantel 2003).

In conclusion, the few reliable estimates based on growth marks for maximum growth rates for sauropods we have (Table 17.3) range from about 1,000–2,000 kg per year. However, they are inferred from the growth record of subadults or adults, and nothing is known of juvenile growth rates. However, these rates compare well with those observed in modern mammalian megaherbivores (Case 1978a). In addition, the approach of using tissue apposition rates also suggests that sauropod dinosaurs grew at rates similar to extant large mammals.

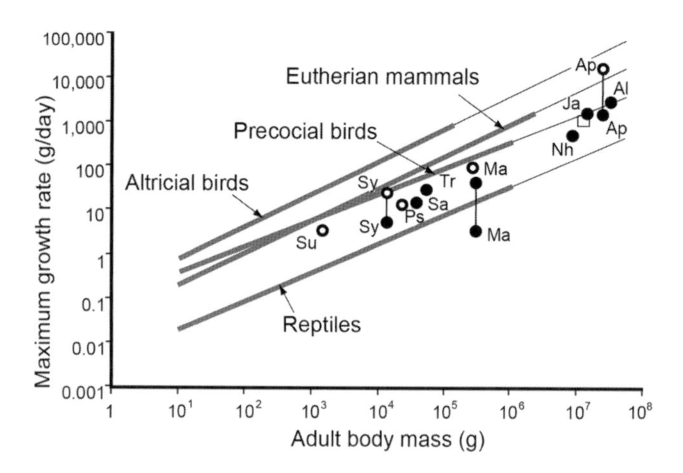

FIGURE 17.10. Relationship between adult body mass and maximum growth rate in extant amniotes as compiled by Case (1978a). Dinosaurs were added to this plot by several authors (e.g., Erickson et al. 2001; Lehman & Woodward 2008). On the basis of our data and our calculation methods, we found *Apatosaurus* SMA 0014 to have increased in body mass at a maximum growth rate of approximately 700 g/day (white rectangle in the graph). See text for details of calculation. *Modified after Lehman & Woodward (2008).*

Thermophysiology

Bone histological data have great implications for constraining possible thermophysiological strategies of sauropod dinosaurs, as was already recognized by de Ricqlès (1968a, 1980). Bone histological evidence for ectothermy in sauropods was seen in lamellar-zonal bone showing LAGs in the pelvis of a sauropod dinosaur (Reid 1981). However, this work has been superseded by the in-depth studies of sauropod long bone histology reviewed above. These document the overwhelming abundance of fibrolamellar bone tissue in long bones (Rimblot-Baly et al. 1995; Curry 1999; Sander 1999, 2000; Erickson et al. 2001; Sander & Tückmantel 2003; Sander et al. 2004, 2006; Erickson 2005; Klein & Sander 2007, 2008; Lehman & Woodward 2008; Woodward & Lehman 2009; Klein et al. 2009a). Laminar fibrolamellar bone as the main bone tissue type throughout 90% of the sauropod ontogeny, the strong remodeling by secondary osteons that finally lead to Haversian bone, and the lack of regularly deposited growth marks in early ontogenetic stages all point to tachymetabolic endothermy in sauropods. This is in agreement with the maximum growth rate of 1–2 metric tons per year. Such growth rates are not seen in any living ectotherm and cannot be reconciled with the basal metabolic rate of modern bradymetabolic terrestrial vertebrates but compare to those of large mammals (Case 1978a; Fig. 17.10). As Case (1978a, p. 243) put it, "the evolution of endothermy was a key factor in lifting physiological constraints upon growth rates."

A possible secondary reduction in metabolic rate may be observed in the small titanosaurs *Phuwiangosaurus* and *Ampelosaurus* because of their more slowly growing cortical bone. On the other hand, the slower local growth rate may result from their relatively small adult body size, because absolute growth rates are lower in smaller animals (Case 1978a). The strong remodeling seen in these two taxa and in the island dwarf sauropod *Magyarosaurus* argues against a reduction in metabolic rate as a reason for the decreased growth rate. Such remodeling is largely restricted today to endothermic animals such as large mammals and birds (de Ricqlès 1980). If metabolic rate had decreased evolutionarily in these small titanosaurs, this would have reintroduced the "physiological constraint upon growth rates" of Case (1978a), and we would expect to observe lamellar-zonal tissue instead. The evolution of bone tissues in pseudosuchians (crurotarsan archosaurs), including living crocodiles, suggests that a reduction in metabolic rate is indeed a possible evolutionary pathway (Padian & Horner 2004; de Ricqlès et al. 2003, 2008).

Bone histological evidence for endothermy also consists in the loss of developmental plasticity in the sauropodomorph lineage, namely that there is a tight correlation between body size and ontogenetic age in sauropods, and that terminal body size is not variable (Sander & Klein 2005; Klein & Sander 2007). In the basal sauropodomorph *Plateosaurus*, on the other hand, developmental plasticity was still present but in combination with fibrolamellar bone, and this may represent an early stage in the evolution of endothermy in sauropodomorphs (Sander & Klein 2005; Klein & Sander 2007). Bone histology also shows that evolutionary body size increase from basal sauropodomorphs to large sauropods was brought about by a strong increase in growth rate, for which the evolution of tachymetabolic endothermy may have been the prerequisite (Sander et al. 2004; Sander & Klein 2005; Sander &

Clauss 2008). The bone histology of *Plateosaurus* appears close to that of basal archosaurs (de Ricqlès et al. 2008).

The argument by Reid (1981) and others (Chinsamy & Hillenius 2004; Chinsamy-Turan 2005) that the presence of the rare LAGs in sauropods reflects cyclical growth, and hence a low metabolic rate, has also been weakened by the recognition that LAGs are common in the bones of mammals (Klevezal 1996; Horner et al. 1999; Sander & Andrássy 2006). Although fibrolamellar bone has repeatedly been ascribed to recent wild alligators which have moderate growth rates (Chinsamy & Hillenius 2004; Chinsamy-Turan 2005; Tumarkin-Deratzian 2007), this occurrence has not been documented in sufficient detail, such as high-magnification photomicrographs and polarized light images, to substantiate the claim.

Discussion

Bone histology is potentially the most powerful source of information about sauropod growth and physiology. However, recent seemingly contradictory results on sauropod dinosaur growth rates as deduced from bone histology are in danger of jeopardizing the recognition of the importance of bone histological evidence. This is why we will discuss the various sources of error in the interpretation of such evidence. It must be kept in mind, however, that in the interpretation of fossil bone histology and the quantification of any aspect of dinosaur biology, higher margins of error apply than in the study of recent animals. On the other hand, given that many aspects of sauropod dinosaur biology and life history are only accessible through bone histological studies, an error of 20% or more is acceptable if the only thing that matters is the order of magnitude. For example, it does not matter whether estimated the age of 20 years for a particular sauropod individual is off by plus or minus 4 years if the initial hypothesis was that sauropods took over 100 years to mature (e.g., Case 1978b). It also has to be kept in mind that bone histology is often the only line of evidence we have.

Although affected by diagenesis to a varying extent, fossil bone is generally well preserved at the tissue level, and there are no diagenetic artifacts that would alterate the in vivo signal. For example, fibrolamellar bone can always be recognized as such, and LAGs will not be obliterated or newly formed during diagenesis. The chemical composition of fossil bone differs greatly from that of fresh bone, and geochemical study of bone is always faced with issues of diagenetic alteration.

Errors introduced during sampling may be a problem as well. As noted earlier, ideally, cross sections of entire bones are taken. Because this will inevitably restrict the number of samples, parts of cross sections may be substituted (e.g., by coring) with the assumption that they are representative for the entire cross section. Although this can be tested in a growth series for a single species by calibrating the sampling location using entire bone cross sections, uncertainty remains, especially when the sampling location is applied to a different species without calibration. As the results described in this chapter show, the benefits of only sampling part of the cross section, namely, the large sample size, outweigh the disadvantages.

The problems of interpreting sauropod bone start with quantification, such as the reconstruction of the number of growth marks and estimates of tissue apposition rates. There are two major uncertainties in interpreting the growth mark record in sauropods: identifying subtle growth marks and retrocalculating resorbed growth marks. The former problem is best approached by two or more observers counting the growth marks in the same sample and coming up with a consensus number, as is done in skeletochronology of recent reptiles (e.g., Zug 1990). Although there will be variation between observers, this tends to be limited. Considering the other sources of error, counting errors are not very problematic.

Any of the several methods of retrocalculation (e.g., Chinsamy 1993; Erickson & Tumanova 2000; Sander & Klein 2005; Klein & Sander 2007) of missing growth marks will introduce errors, the nature of which depends on the assumptions underlying the method of retrocalculation (see above). A specific discussion is beyond the scope of this chapter. A general principle is that bone apposition rate is highest early in life, and as a result, the inner growth marks are more widely spaced than the outer ones, which means that the number of growth marks lost to resorption is relatively low. Growth mark records thus provide a reliable minimum estimate for a given animal. Growth mark records that start late in life, as is the case in most sauropods, should not be used in longevity estimates.

Similarly, the several methods for reconstructing growth rates carry their own suite of errors, as discussed earlier. One not mentioned so far derives from the error inherent in body mass estimates, which may vary widely, again depending on underlying assumptions (e.g., Henderson 1999; Seebacher 2001; Ganse et al., this volume; Appendix, this volume). The best way to deal with these errors is to carry out a sensitivity analysis, as was done by Lehman & Woodward (2008), who showed that even seemingly major differences in body mass estimates had a limited effect on their growth rate estimates.

We close this discussion with a particularly obvious example of how a poorly constrained growth rate may lead to erroneous conclusions. This is the study by Gillooly et al. (2006), in which the authors conclude that sauropod dinosaurs could not have had the metabolic rates of an endotherm because of the danger of overheating. This conclusion appears seriously flawed because they calculated a metabolic rate for *Apatosaurus* on the basis of the exaggerated growth rate of Erickson et al. (2001) and thus arrived at a much higher heat production (resulting in a body temperature of 41°C in a 13,000 kg

Apatosaurus) than would be the case with the growth rate of about 1,000 kg/yr suggested by our data. We predict that if realistic growth rates were entered in the formula of Gillooly et al. (2006), the resulting body temperatures would be well within the realm of modern endothermic animals.

Future Perspectives

The bone histology of sauropod dinosaurs offers a wealth of information about their growth, life history, and biology. Despite the advances reviewed in this chapter, there are obvious limitations and unexplored avenues. For one, the paucity of the growth mark record in the cortex of the large long bones means that sauropod growth rates are poorly constrained compared to other dinosaurs, especially considering the large sample base available. Work on the quantification of the few growth mark records from long bone is in progress. Other elements of the sauropod skeleton also need to be surveyed for growth mark records. So far, this has only been done for scapulae, but this record has been useful only for a few specimens (Curry 1999), while several others do not preserve a good growth record (P.M.S., N.K., pers. obs.). As indicated by a preliminary study (Waskow & Sander, 2010), ribs also offer potential, but only if they are from individuals for which one of the large long bones is also preserved because body mass is otherwise difficult to estimate. For reconstructing growth curves from ribs, a growth series of several individuals will be necessary because, unlike in long bones, growth cycle thickness in ribs cannot be translated into body mass increase. Vertebrae offer less potential because their complex morphology and pneumaticity makes preservation of a long growth mark record unlikely.

Another area of insufficient taxon sampling is basal sauropodomorphs as well as basal sauropods outside Neosauropoda. Although the already sampled taxa (*Thecodontosaurus, Plateosaurus,* cf. *Isanosaurus*) tell a coherent story of the evolution of giant body size by an increase in growth rate and presumably metabolic rate, one would like to see this evolutionary trajectory to be confirmed by denser sampling.

One aspect that has not been considered at all in this review is the potential use of histology in phylogenetic analyses: can discrete histological characters be extracted from obvious differences in long bone histology between taxa, as first noted by Sander (2000) for the Tendaguru sauropods? Differences of varying subtlety between other taxa are present as well, as noted in the systematic section. The field of extracting systematic information from dinosaur bone histology remains largely unexplored.

Similary unexplored are features at lower hierarchical levels, such as the quantification of tissue parameters such as lamina thickness in laminar fibrolamellar bone (for a first attempt see Sander & Tückmantel 2003), osteocyte lacunae morphology and morphometry (but see Rensberger & Watabe 2000; Organ et al. 2007, 2009), and the meaning of secondary osteons. Bone histology thus will continue to offer exciting and unique insights into the biology and evolution of extinct tetrapod groups for the foreseeable future.

Acknowledgments

Our foremost thanks go to the numerous curators who made sampling of sauropodomorph long bones possible. These are David Berman (Carnegie Museum, Pittsburgh), Eric Buffetaut (CNRS, Paris), Richard Cifelli and Kyle Davies (Oklahoma Museum of Natural History, Norman), Zoltan Csiki (University of Bucharest), Rainer Foelix (Naturama, Aarau), Wolf-Dieter. Heinrich (Museum für Naturkunde, Berlin), Laszlo Kordos (Geological Survey of Hungary, Budapest), Wann Langston (University of Texas, Austin), Jean LeLoeuff (Musée des Dinosaures, Espéraza), Nils Knötschke (Dinosaur Museum Münchehagen), Hans-Ulrich Pfretzschner (University of Tübingen), Heidi Birrer (Sauriermuseum Frick), Hans-Peter Schultze (Museum für Naturkunde, Berlin), Ken Stadtman (Brigham Young University, Provo), Hans-Jakob Siber (Sauriermuseum Aathal), Varavudh Suteethorn (Department of Mineral Ressources, Khon Kaen), and Rupert Wild (State Museum of Natural History, Stuttgart). Olaf Dülfer (University of Bonn) cut the hundreds of thin sections that form the basis of this study, while Georg Oleschinski (University of Bonn) aided us greatly in imaging the sections. Sebastian Marpmann (University of Bonn) compiled the data on sauropod masses used in this study and included in the Appendix of this book. We thank Torsten Scheyer (University of Zurich) for discussion throughout this research, and Hartmut Haubold (Martin-Luther-University Halle-Wittenberg) and an anonymous reviewer for the speedy but thorough review. Our research was generously funded by the Deutsche Forschungsgemeinschaft (DFG) through a variety of grants. This is contribution number 76 of the DFG Research Unit 533 "Biology of the Sauropod Dinosaurs: The Evolution of Gigantism."

References

Amprino, R. 1947. La structure du tissu osseux envisagée comme expression de différences dans la vitesse de l'accroisement.—*Archives de Biologie* 58: 315–330.

Anderson, J. F., Hall-Martin, A. & Russell, D. A. 1985. Long bone circumference and weight in mammals, birds, and dinosaurs.—*Journal of Zoology Series A* 207: 53–61.

Ayer, J. 2000. *The Howe Ranch Dinosaurs*. Sauriermuseum Aathal, Zürich.

Benton, M. J., Juul, L., Storrs, G. W. & P. M. Galton. 2000. Anatomy and systematics of the prosauropod dinosaur *Thecodontosaurus antiquus* from the Upper Triassic of southwest England.—*Journal of Vertebrate Paleontology* 20: 77–108.

Bonaparte, J. F., Heinrich, W.-D. & Wild, R. 2000. Review of

Janenschia Wild, with the description of a new sauropod from the Tendaguru Beds of Tanzania and a discussion on the systematic value of procoelus caudal vertebrae in the Sauropoda.—*Palaeontographica Abt. A* 256: 25–76.

Bonnan, M. F. 2004. Morphometric analysis of humerus and femur shape in Morrison sauropods: implications for functional morphology and paleobiology.—*Paleobiology* 30: 444–470.

Buffetaut, E., Suteethorn, V., Le Loeuff, J., Cuny, C., Tong, H. & Khansubha, S. 2002. The first giant dinosaurs: a large sauropod from the Late Triassic of Thailand.—*Comptes Rendus Palevol* 1: 103–109.

Bybee, P. J., Lee, A. H. & Lamm, E.-T. 2006. Sizing the Jurassic theropod dinosaur *Allosaurus:* assessing growth strategy and evolution of ontogenetic scaling of limbs.—*Journal of Morphology* 267: 347–359.

Carpenter, K. & McIntosh, J. 1994. Upper Jurassic sauropod babies from the Morrison Formation. *In* Carpenter, K., Hirsch, K. & Horner, J. (eds.). *Dinosaur Eggs and Babies*. Cambridge University Press, Cambridge: pp. 265–278.

Case, T. J. 1978a. On the evolution and adaptive significance of postnatal growth rates in the terrestrial vertebrates.—*Quarterly Review of Biology* 53: 243–282.

Case, T. J. 1978b. Speculations on the growth rate and reproduction of some dinosaurs.—*Paleobiology* 4: 320–328.

Castanet, J. 1994. Age estimation and longevity in reptiles.—*Gerontology* 40: 174–192.

Castanet, J. & Smirina, E. 1990. Introduction to the skeletochronological method in amphibians and reptiles.—*Annales des Sciences Naturelles Zoologie* 13: 191–196.

Castanet, J., Francillon-Vieillot, H., Meunier, F. J. & de Ricqlès, A. 1993. Bone and individual aging. *In* Hall, B. K. (ed.). *Bone. Vol. 7: Bone Growth*. CRC Press, Boca Raton: pp. 245–283.

Castanet, J., Grandin, A., Abourachid, A. & de Ricqlès, A. 1996. Expression de la dynamique de croissance dans la structure de l'os périostique chez *Anas platyrhynchos*.—*Comptes Rendus de l'Académie des Sciences de Paris: Science de la Vie* 319: 301–308.

Castanet, J., Curry Rogers, K., Cubo, J. & Boisard, J.-J. 2000. Periosteal bone growth rates in extant ratites (ostrich and emu). Implications for assessing growth in dinosaurs.—*Comptes Rendus de l'Académie des Sciences de Paris: Science de la Terre et des Planètes* 323: 543–550.

Castanet, J., Croci, S., Aujard, F., Perret, M., Cubo, J. & de Margerie, E. 2004. Lines of arrested growth in bone and age estimation in a small primate: *Microcebus murinus*.—*Journal of Zoology* 263: 31–39.

Cherry, C. 2002. *Bone Histology of the Primitve Dinosaur* Thecodontosaurus antiquus. M. Sc. Thesis. University of Bristol, Bristol.

Chiappe, L. M., Jackson, F. D., Coria, R. A. & Dingus, L. 2005. Nesting titanosaurs from Auca Mahuevo and adjacent sites: understanding sauropod reproductive behavior and embryonic development. *In* Curry Rogers, K. A. & Wilson, J. A. (eds.). *The Sauropods: Evolution and Paleobiology.* University of California Press, Berkeley: pp. 285–302.

Chinsamy, A. 1993. Bone histology and growth trajectory of the prosauropod dinosaur *Massospondylus carinatus* (Owen).—*Modern Geology* 18: 319–329.

Chinsamy, A. & Hillenius, W. J. 2004. Physiology of non-avian dinosaurs. *In* Weishampel, D. B., Dodson, P. & Osmólska, H. (eds.). *The Dinosauria. 2nd edition*. University of California Press, Berkeley: pp. 643–659.

Chinsamy-Turan, A. 2005. *The Microstructure of Dinosaur Bone*. Johns Hopkins University Press, Baltimore.

Cooper, L. N., Lee, A. H., Taper, M. L. & Horner, J. R. 2008. Relative growth rates of predator and prey dinosaurs reflect effects of predation.—*Proceedings of the Royal Society B: Biological Sciences* 275: 2609–2615.

Cubo, J., Legendre, P., de Ricqlès, A., Montes, L., de Margerie, E., Castanet, J. & Desdevises, Y. 2008. Phylogenetic, functional, and structural components of variation in bone growth rate of amniotes.—*Evolution and Development* 10: 217–277.

Currey, J. D. 2002. *Bones: Structure and Mechanics*. Princeton University Press, Princeton.

Curry, K. 1998. Histological quantification of growth rates in *Apatosaurus*.—*Journal of Vertebrate Paleontology* 18: 36A.

Curry, K. A. 1999. Ontogenetic histology of *Apatosaurus* (Dinosauria: Sauropoda): new insights on growth rates and longevity.—*Journal of Vertebrate Paleontology* 19: 654–665.

Curry Rogers, K. & Erickson, G. M. 2005. Sauropod histology: microscopic views on the lives of giants. *In* Curry Rogers, K. & Wilson, J. A. (eds.). *The Sauropods: Evolution and Palaeobiology.* University of California Press, Berkeley: pp. 303–326.

de Margerie, E., Cubo, J. & Castanet. J. 2002. Bone typology and growth rate: testing and quantifying "Amprino's rule" in the mallard (*Anas platyrhynchos*).—*Comptes Rendus de l'Académie des Sciences de Paris: Biologies* 325: 221–230.

de Ricqlès, A. 1968a. Recherches paléohistologiques sur les os longs des tétrapodes I.—Origine du tissu osseux plexiforme des dinosauriens sauropodes.—*Annales de Paléontologie* 54: 133–145.

de Ricqlès, A. 1968b. Quelques observations paléohistologiques sur le dinosaurien sauropode *Bothriospondylus*.—*Annales de l'Université de Madagascar* 6: 157–209.

de Ricqlès, A. 1980. Tissue structures of dinosaur bone. Functional significance and possible relation to dinosaur physiology. *In* Thomas, R. D. K. & Olson, E. C. (eds.). *A Cold Look at the Warm-Blooded Dinosaurs*. Westview Press, Boulder, Colo.: pp. 103–139.

de Ricqlès, A. 1983. Cyclical growth in the long limb bones of a sauropod dinosaur.—*Acta Palaeontologica Polonica* 28: 225–232.

de Ricqlès, A., Meunier, F. J., Castanet, J. & Francillon-Vieillot, H. 1991. Comparative microstructure of bone. *In* Hall, B. K. (ed.). *Bone. Vol. 3: Bone Matrix and Bone Specific Products*. CRC Press, Boca Raton: pp. 1–78.

de Ricqlès, A., Padian, K. & Horner, J. R. 2003. On the bone histology of some Triassic pseudosuchian archosaurs and related taxa.—*Annales de Paléontologie* 89: 67–101.

de Ricqlès, A., Padian, K., Knoll, F. & Horner, J. R. 2008. On the origin of high growth rates in archosaurs: complementary histological studies on Triassic archosauriforms and the problem of a "phylogenetic signal" in bone histology.—*Annales de Paléontologie* 94: 57–76.

Dumont, M., Borbély, A., Kostka, A., Sander, P. M. & Kaysser-Pyzalla, A. This volume. Characterization of sauropod bone structure. *In* Klein, N., Remes, K., Gee, C. T. & Sander, P. M. (eds.). *Biology of the Sauropod Dinosaurs: Understanding the Life of Giants.* Indiana University Press, Bloomington: pp. 150–170.

Dunham, A. E., Overall, K. L., Porter, W. P. & Forster, C. A. 1989. Implications of ecological energetics and biophysical and developmental constraints for life-history variation in dinosaurs. *In* Farlow, J. O. (ed.). *Paleobiology of the Dinosaurs.* GSA Special Paper 238. Geological Society of America, Boulder: pp. 1–21.

Erickson, G. M. 2005. Assessing dinosaur growth patterns: a microscopic revolution.—*Trends in Ecology and Evolution* 20: 677–684.

Erickson, G. M. & Tumanova, T. A. 2000. Growth curve of *Psittacosaurus mongoliensis* Osborn (Ceratopsia: Psittacosauridae) inferred from long bone histology.—*Zoological Journal of the Linnean Society of London* 130: 551–566.

Erickson, G. M., Curry Rogers, K. & Yerby, S. A. 2001. Dinosaurian growth patterns and rapid avian growth rates.—*Nature* 412: 429–433.

Erickson, G. M., Currie, P. J., Inouye, B. D. & Winn, A. S. 2004. Gigantism and comparative life-history parameters of tyrannosaurid dinosaurs.—*Nature* 430: 772–775.

Erickson, G. M., Mackovicky, P. J., Currie, P. J., Norell, M. A., Yerby, S. A. & Brochu, C. A. 2006. Tyrannosaur life tables: an example of nonavian dinosaur population biology.—*Science* 313: 213–217.

Erickson, G. M., Curry Rogers, K., Varricchio, D. K., Norell, M. A. & Xu, X. 2007. Growth patterns in brooding dinosaurs reveals the timing of sexual maturity in non-avian dinosaurs and the genesis of the avian condition.—*Biology Letters* 3: 558–561.

Erickson, G. M., Makovicky, P. J., Inouye, B. D., Chang-Fu, Z. & Gao, K.-Q. 2009. A life table for *Psittacosaurus lujiatunensis*: initial insights into ornithischian population biology.—*Anatomical Record* 292: 1514–1521.

Farmer, C. G. 2000. Parental care: the key to understanding endothermy and other convergent features in birds and mammals.—*American Naturalist* 155: 326–334.

Farmer, C. G. 2003. Reproduction: the adaptive significance of endothermy.—*American Naturalist* 162: 826–840.

Francillon-Vieillot, H., de Buffrénil, V., Castanet, J., Géraudie, J., Meunier, F. J., Sire, J. Y., Zylberberg, L. & de Ricqlès, A. 1990. Microstructure and mineralization of vertebrate skeletal tissues. *In* Carter, J. G. (ed.). *Skeletal Biomineralization: Patterns, Processes and Evolutionary Trends. Vol. 1.* Van Nostrand Reinhold, New York: pp. 471–530.

Foster, J. R. 2007. *Jurassic West. The Dinosaurs of the Morrison Formation and their World.* Indiana University Press, Bloomington.

Galton, P. M. & Upchurch, P. 2004. Prosauropoda. *In* Weishampel, D. B., Dodson, P. & Osmólska, H. (eds.). *The Dinosauria. 2nd edition.* University of California Press, Berkeley: pp. 232–258.

Ganse, B., Stahn, A., Stoinski, S., Suthau, T. & Gunga, H.-C. This volume. Body mass estimation, thermoregulation, and cardiovascular physiology of large sauropods. *In* Klein, N., Remes, K., Gee, C. T. & Sander, P. M. (eds.). *Biology of the Sauropod Dinosaurs: Understanding the Life of Giants.* Indiana University Press, Bloomington: pp. 105–115.

Gillooly, J. F., Allen, A. P. & Charnov, E. L. 2006. Dinosaur fossils predict body temperatures.—*PLoS Biology* 4: 1467–1469.

Gilmore, C. W. 1925. A nearly complete articulated skeleton of *Camarasaurus*, a saurischian from the Dinosaur National Monument, Utah.—*Memoirs of the Carnegie Museum* 10: 347–384.

Griebeler, E. M. & Werner, J. This volume. The life cycle of sauropod dinosaurs. *In* Klein, N., Remes, K., Gee, C. T. & Sander, P. M. (eds.). *Biology of the Sauropod Dinosaurs: Understanding the Life of Giants.* Indiana University Press, Bloomington: pp. 263–275.

Gunga, H.-C., Suthau, T., Bellmann, A., Stoinski, S., Friedrich, A., Trippel, T., Kirsch, K. & Hellwich, O. 2008. A new body mass estimation of *Brachiosaurus brancai* Janensch, 1914 mounted and exhibited at the Museum of Natural History (Berlin, Germany).—*Fossil Record* 11: 33–38.

Hayashi, S., Carpenter, K. & Suzuku, D. 2009. Different growth patterns between the skeleton and osteoderms of *Stegosaurus* (Ornithischia: Thyreophora).—*Journal of Vertebrate Paleontology* 29: 123–131.

Henderson, D. M. 1999. Estimating the masses and centers of mass of extinct animals by 3-D mathematical slicing.—*Paleobiology* 25: 88–106.

Horner, J. R. & Padian, K. 2004. Age and growth dynamics of *Tyrannosaurus rex*.—*Proceedings of the Royal Society B: Biological Sciences* 271: 1875–1880.

Horner, J. R., Padian, K. & de Ricqlès, A. 1999. Variation in dinosaur skeletochronology indicators: implications for age assessment.—*Paleobiology* 25: 49–78.

Janensch, W. 1961. Die Gliedmaßen und Gliedmaßengürtel der Sauropoden der Tendaguru-Schichten.—*Palaeontographica Abt. A* 3: 177–235.

Kilbourne, B. M. & Makovicky, P. J. 2010. Limb bone allometry during postnatal ontogeny in non-avian dinosaurs.—*Anatomical Record* 217: 135–152.

Klein, N. & Sander, P. M. 2007. Bone histology and growth of the prosauropod *Plateosaurus engelhardti* Meyer, 1837 from the Norian bonebeds of Trossingen (Germany) and Frick (Switzerland).—*Special Papers in Palaeontology* 77: 169–206.

Klein, N. & Sander, P. M. 2008. Ontogenetic stages in the long bone histology of sauropod dinosaurs.—*Paleobiology* 34: 247–263.

Klein, N., Sander, P. M. & Suteethorn, V. 2009a. Bone histology and its implications for the life history and growth of the Early Cretaceous titanosaur *Phuwiangosaurus sirindhornae*.—*Geological Society Special Publications* 315: 217–228.

Klein, N., Scheyer, T. & Tütken, T. 2009b. Skeletochronology and isotopic analysis of an individual specimen of *Alligator mississippiensis* Daudin, 1802.—*Fossil Record* 12: 121–131.

Klevezal, G. A. 1996. *Recording Structures of Mammals. Determination of Age and Reconstruction of Life History.* A. A. Balkema, Rotterdam.

Lee, A. H. & Werning, S. 2008. Sexual maturity in growing dinosaurs does not fit reptilian growth models.—*Proceedings of the*

National Academy of Sciences of the United States of America 105: 582–587.

Le Loeuff, J. 2005. Osteology of *Ampelosaurus atacis* (Titanosauria) from Southern France. *In* Tidwell, V. & Carpenter, K. (eds.). *Thunder-lizards*. Indiana University Press, Bloomington: pp. 115–137.

Lehman, T. M. & Woodward, H. N. 2008. Modeling growth rates for sauropod dinosaurs.—*Paleobiology* 34: 264–281.

Mazzetta, G. V., Christiansen, P. & Farina, R. A. 2004. Giants and bizarres: body size of some southern South American Cretaceous dinosaurs.—*Historical Biology* 16: 1–13.

Martin, V., Suteethorn, V. & Buffetaut, E. 1999. Description of the type and referred material of *Phuwiangosaurus sirindhornae* Martin, Buffetaut and Suteethorn, 1994, a sauropod from the Lower Cretaceous of Thailand.—*Oryctos* 2: 39–91.

Montes, L., Le Roy, N., Perret, M., de Buffrénil, V., Castanet, J. & Cubo, J. 2007. Relationships between bone growth rate, body mass and resting metabolic rate in growing amniotes: a phylogenetic approach.—*Biological Journal of the Linnean Society* 92: 63–76.

Montes, L., Castanet, J. & Cubo, J. 2010. Relationship between bone growth rate and bone tissue organization in amniotes: first test of Amprino's rule in a phylogenetic context.—*Animal Biology* 60: 25–41.

Nopcsa, F. 1914. Über das Vorkommen der Dinosaurier in Siebenbürgen.—*Verhandlungen der Zoologisch-Botanischen Gesellschaft* 54: 12–14.

Organ, C. L., Shedlock, A. M., Meade, A., Pagel, M. & Edwards, S. V. 2007. Origin of avian genome size and structure in non-avian dinosaurs.—*Nature* 446: 180–184.

Organ, C. L., Brusatte, S. L., and & Stein, K. 2009. Sauropod dinosaurs evolved moderately sized genomes unrelated to body size.—*Proceedings of the Royal Society B: Biological Sciences* 276: 4303–4308.

Padian, K. & Horner, J. R. 2004. Dinosaur physiology. *In* Weishampel, D. B., Dodson, P. & Osmólska, H. (eds.). *The Dinosauria. 2nd edition*. University of California Press, Berkeley: pp. 660–671.

Rauhut, O. W. M., Fechner, R., Remes, K. & Reis, K. This volume. How to get big in the Mesozoic: the evolution of the sauropodomorph body plan. *In* Klein, N., Remes, K., Gee, C. T. & Sander, P. M. (eds.). *Biology of the Sauropod Dinosaurs: Understanding the Life of Giants*. Indiana University Press, Bloomington: pp. 119–149.

Reid, R. E. H. 1981. Lamellar-zonal bone with zones and annuli in the pelvis of a sauropod dinosaur.—*Nature* 292: 49–51.

Reid, R. E. H. 1990. Zonal 'gowth rings' in dinosaurs.—*Modern Geology* 15: 19–48.

Reid, R. E. H. 1997. How dinosaurs grew. *In* Farlow, J. O. & Brett-Surman, M. K. (eds.). *The Complete Dinosaur*. Indiana University Press, Bloomington: pp. 403–413.

Reiss, M. J. 1989. *The Allometry of Growth and Reproduction*. Cambridge University Press, Cambridge.

Reisz, R. R., Scott, D., Sues, H.-D., Evans, D. C. & Raath, M. A. 2005. Embryos of an Early Jurassic prosauropod dinosaur and their evolutionary significance.—*Science* 309: 761–764.

Remes, K. 2006. Revision of the Tendaguru sauropod dinosaur *Tornieria africana* Fraas and its relevance for sauropod paleobiogeography.—*Journal of Vertebrate Paleontology* 26: 651–669.

Rensberger, J. & Watabe, M. 2000. Fine structure of bone in dinosaurs, birds and mammals.—*Nature* 406: 619–622.

Remes, K., Unwin, D. M., Klein, N., Heinrich, W.-D. & Hampe, O. This volume. Skeletal reconstruction of *Brachiosaurus brancai* in the Museum für Naturkunde, Berlin: summarizing 70 years of sauropod research. *In* Klein, N., Remes, K., Gee, C. T. & Sander, P. M. (eds.). *Biology of the Sauropod Dinosaurs: Understanding the Life of Giants*. Indiana University Press, Bloomington: pp. 305–316.

Rimblot-Baly, F., de Ricqlès, A. & Zylberberg, L. 1995. Analyse paléohistologique d'une série de croissance partielle chez *Lapparentosaurus madagascariensis* (Jurassique moyen): Essai sur la dynamique de croissance d'un dinosaure sauropode.—*Annales de Paléontologie (Invertébrés)* 81: 49–86.

Sander, P. M. 1999. Life history of the Tendaguru sauropods as inferred from long bone histology.—*Mitteilungen aus dem Museum für Naturkunde in Berlin, Geowissenschaftliche Reihe* 2: 103–112.

Sander, P. M. 2000. Long bone histology of the Tendaguru sauropods: implications for growth and biology.—*Paleobiology* 26: 466–488.

Sander, P. M. & Tückmantel, C. 2003. Bone lamina thickness, bone apposition rates, and age estimates in sauropod humeri and femora.—*Paläontologische Zeitschrift* 76: 161–172.

Sander, P. M., Klein, N., Buffetaut, E., Cuny, G., Suteethorn, V. & Le Loeuff, J. 2004. Adaptive radiation in sauropod dinosaurs: bone histology indicates rapid evolution of giant body size through acceleration.—*Organisms, Diversity and Evolution* 4: 165–173.

Sander, P. M. & Klein, N. 2005. Developmental plasticity in the life history of a prosauropod dinosaur.—*Science* 310: 1800–1802.

Sander, P. M. & Andrássy, P. 2006. Lines of arrested growth and long bone histology in Pleistocene large mammals from Germany: what do they tell us about dinosaur physiology?—*Palaeontographica Abt. A* 277: 143–159.

Sander, P. M., Mateus, O., Laven, T. & Knötschke, N. 2006. Bone histology indicates insular dwarfism in a new Late Jurassic sauropod dinosaur.—*Nature* 441: 739–741.

Sander, P. M. & Clauss, M. 2008. Sauropod gigantism.—*Science* 322: 200–201.

Sander, P. M., Peitz, C., Jackson, F. D. & Chiappe, L. M. 2008. Upper Cretaceous titanosaur nesting sites and their implications for sauropod dinosaur reproductive biology.—*Palaeontographica Abt. A* 284 (4–6): 69–107.

Sander, P. M., Christian, A., Clauss, M., Fechner, R., Gee, C. T., Griebeler, E. M., Gunga, H.-C., Hummel, J., Mallison, H., Perry, S., Preuschoft, H., Rauhut, O., Remes, K., Tütken, T., Wings, O. & Witzel, U. 2010. Biology of the sauropod dinosaurs: the evolution of gigantism.—*Biological Reviews of the Cambridge Philosophical Society*. doi: 10.1111/j.1469=185X.2010.00137.x.

Scheyer, T. M., Klein, N. & Sander, P. M. 2010. Developmental palaeontology of Reptilia as revealed by histological studies.—*Seminars in Cell and Developmental Biology* 21: 462–470.

Schwarz, D., Ikejiri, T., Breithaupt, B. H., Sander, P. M. & Klein, N. 2007. A nearly complete skeleton of an early juvenile diplodocid (Dinosauria: Sauropoda) from the Lower Morrison Formation (Late Jurassic) of North Central Wyoming and its implications for early ontogeny and pneumaticity in sauropods.—*Historical Biology* 19: 225–253.

Schweitzer, M. H., Wittmeyer, J. L. & Horner, J. R. 2005. Gender-specific reproductive tissue in ratites and *Tyrannosaurus rex*.—*Science* 308: 1456–1460.

Seebacher, F. 2001. A new method to calculate allometric length-mass relationships of dinosaurs.—*Journal of Vertebrate Paleontology* 21: 51–60.

Stark, J. M. & Chinsamy, A. 2002. Bone microstructure and developmental plasticity in birds and other dinosaurs.—*Journal of Morphology* 254: 232–246.

Stein, K. & Sander, P. M. 2009. Histological core drilling: a less destructive method for studying bone histology. *In* Brown, M. A., Kane, J. F. & Parker, W. G. (eds.). *Methods in Fossil Preparation: Proceedings of the First Annual Fossil Preparation and Collections Symposium*. Available at: http://fossilprep.org/Stein%20and%20Sander%202009.pdf.

Stein, K., Csikib, Z., Curry Rogers, K., Weishampel, D. B., Redelstorff, R., Carballidoa, J. L. & Sander, P. M. 2010. Small body size and extreme cortical bone remodelling indicate phyletic dwarfism in *Magyarosaurus dacus* (Sauropoda: Titanosauria).—*Proceedings of the National Academy of Sciences of the United States of America* 107: 9258–9263.

Stoinski, S., Suthau, T. & Gunga, H.-C. This volume. Reconstructing body volume and surface area of dinosaurs using laser scanning and photogrammetry. *In* Klein, N., Remes, K., Gee, C. T. & Sander, P. M. (eds.). *Biology of the Sauropod Dinosaurs: Understanding the Life of Giants*. Indiana University Press, Bloomington: pp. 94–104.

Tumarkin-Deratzian, A. R. 2007. Fibrolamellar bone in wild adult *Alligator mississippiensis*.—*Journal of Herpetology* 41: 341–345.

Turvey, S. T., Green, O. R. & Holdaway, R. N. 2005. Cortical growth marks reveal extended juvenile development in New Zealand moa.—*Nature* 435: 940–943.

Upchurch, P., Barrett, P. M., & Dodson, P. 2004. Sauropoda. *In* Weishampel, D. B., Dodson, P., & Osmólska, H. (eds.). *The Dinosauria. 2nd edition*. University of California Press, Berkeley, pp. 259-322.

Upchurch, P., Barrett, P. M. & Galton, P. M. 2007. A phylogenetic analysis of basal sauropodmorph relationships: implications for the origin of sauropod dinosaurs.—*Special Papers in Palaeontology* 77: 57–90.

von Huene, F. 1932. Die fossile Reptil-Ordnung Saurischia, ihre Entwicklung und Geschichte.—*Monographie für Geologie und Palaeontologie*, Serie 7, 4: 1–361.

Waskow, K. & Sander, P. M. 2009. Growth marks in sauropod ribs from the Upper Jurassic Morrison Formation, Tendaguru and Lower Cretaceous of Niger.—*Journal of Vertebrate Paleontology* 29: 198A.

West-Eberhard, M. J. 2003. *Developmental Plasticity and Evolution*. Oxford Unversity Press, Oxford.

Wilson, J. A. 2002. Sauropod dinosaur phylogeny: critique and cladistic analysis.—*Zoological Journal of the Linnean Society of London* 136: 217–276.

Wilson, J. A., Mohabey, D. M., Peters, S. E. & Head, J. J. 2010. Predation upon hatchling dinosaurs by a new snake from the Late Cretaceous of India.—*PloS Biology* 8(3): e100032.

Wings, O., Sander, P. M., Tütken, T., Fowler, D. W. & Sun, G. 2007. Growth and life history of Asia's largest dinosaur.—*Journal of Vertebrate Paleontology* 27: 167A.

Woodward, H. 2005. Bone histology of the titanosaurid sauropod *Alamosaurus sanjuanensis* from the Javelina Formation, Texas.—*Journal of Vertebrate Paleontology* 25: 132A.

Woodward, H. & Lehman, T. 2009. Bone histology and microanatomy of *Alamosaurus sanjuanensis* (Sauropoda: Titanosauria) from the Maastrichtian of Big Bend National Park, Texas.—*Journal of Vertebrate Paleontology* 29: 807–821.

Yates, A. M. 2007. The first complete skull of the Triassic dinosaur *Melanorosaurus* Haughton (Sauropodomorpha: Anchisauria).—*Special Papers in Palaeontology* 77: 9–55.

Young, C. C. 1954. On a new sauropod from Yiping, Szechuan.—*China. Acta Palaeontologica Sinica* 2: 355–369.

Zug, G. R. 1990. Age determination of long-lived reptiles: some techniques for seaturtles.—*Annales des Sciences Naturelles Zoologie* 11: 219–222.

PART FIVE

EPILOGUE

18

Skeletal Reconstruction of *Brachiosaurus brancai* in the Museum für Naturkunde, Berlin: Summarizing 70 Years of Sauropod Research

KRISTIAN REMES, DAVID M. UNWIN, NICOLE KLEIN, WOLF-DIETER HEINRICH, AND OLIVER HAMPE

THE SKELETAL RECONSTRUCTION OF *Brachiosaurus brancai* displayed in the Museum für Naturkunde, Berlin, is the largest mounted dinosaur skeleton in the world that incorporates original fossil material. Found during the course of the German Tendaguru expedition from 1909 to 1913, a composite skeleton of *B. brancai* was first mounted in 1938, and although it was demounted and remounted several times, it remained unchanged until the renovation of the Berlin dinosaur exhibition hall in 2005–2007. Here we describe the scientific progress, technical solutions, and specific decisions that led to the new mount, which has been on display since 2007. The new mount differs in a number of points from the old mount, including improved models of the presacral vertebrae and head, the posture of the neck, the shape of the torso, the orientation of the pectoral girdle and forelimbs, and the posture of the tail. Overall, the *Brachiosaurus* skeleton now looks livelier, evoking the impression of an active, relatively agile animal and symbolizing developments in our understanding of sauropods since the first mounting of the skeleton.

Introduction

In July 2007, the famous Dinosaur Hall of the Museum für Naturkunde in Berlin reopened after two years of reconstruction and renovation, returning one of the world's most famous dinosaur mounts to public view. The original reconstruction of *Brachiosaurus brancai* by Werner Janensch (1937; Fig. 18.1) has been emblematic of sauropod gigantism since the late 1930s, and pictures of the Berlin *Brachiosaurus* can be found in countless textbooks, popular articles, children's books, and posters around the world. However, research on sauropod dinosaurs has made substantial progress since Janensch's time, and the complete renovation of parts of the Museum für Naturkunde's exhibitions, which began in 2004, provided a unique opportunity to update the Berlin reconstruction according to our current understanding of sauropod paleontology. Discoveries made by the DFG Research Unit 533, as described elsewhere in this volume, had a substantial influence on the new *Brachiosaurus* mount, but would not have been possible without the tremendous research efforts of Richard McNeill Alexander, Robert Bakker, Paul Barrett, José Bonaparte, Eric Buffetaut, Jorge Calvo, Matt Carrano, Per Christiansen, Peter Dodson, John Foster, John Hutchinson, Martin Lockley, John McIntosh, Leonardo Salgado, Paul Sereno, Paul Upchurch, Mark Wedel, Jeff Wilson, C. C. Young, Dong Zhiming, and many others. Debates on sauropod anatomy, posture, and paleobiology continue today, and not every sauropod researcher will agree with all aspects of the new reconstruction now on display. Therefore, as an epilogue to the issues discussed earlier in this book, we describe the history and science that led to the new *Brachiosaurus* mount on show in Berlin, and how the results of studies by our research group influenced individual decisions.

FIGURE 18.1. Original mount of *Brachiosaurus brancai* in the Museum für Naturkunde, Berlin in 1938. Note the elbow-out position of the forelimbs, the bent knees, and the tail dragging on the ground.

History and Components of the Mount

All the original dinosaur specimens in the dinosaur hall of the Museum für Naturkunde, Berlin, were discovered in the Late Jurassic (Kimmeridgian–Tithonian) Tendaguru Beds of southern Tanzania, East Africa. From 1909 to 1913, Werner Janensch led one of the most productive paleontological excavation campaigns in history to the area around Tendaguru Hill, about 60 km northwest of the port town of Lindi (Janensch 1914; Maier 2003). More than 250 metric tons of fossils were transported back to Berlin, including two partial skeletons from Tendaguru site "S" that form the main part of the *Brachiosaurus* mount today (Janensch 1937). After World War I, preparation proceeded slowly, and it was only in November 1937 that the reconstruction of *Brachiosaurus* was opened to the public (Janensch 1937, 1950; Fig. 18.1).

The mount of *Brachiosaurus brancai* is a composite because no complete skeleton was found by the German Tendaguru expedition of 1909–1913 (Janensch 1950). Janensch (1950) described in detail the source of the individual bones and the rationale behind his mount. We repeat here the complete information on the provenance of the bones to make it accessible to a wider audience. The majority of the skeletal elements included in the mount are from skeleton S II, recovered from the Middle Saurian Beds (Tendaguru site "S"). The tail skeleton is derived from another individual of similar size found in the Upper Saurian Beds at Tendaguru site "no." In addition, skeletal elements of comparable dimensions to S II, obtained from different sites in the Tendaguru area, were also included in the original mounting, both as originals and as reconstructions.

The skull of the original mount was modeled in plaster on the basis of skull fragments, including lower and upper jaws from skeleton S II. Missing parts were reproduced from the remarkably well preserved complete skull t 1. Skeleton S II provided the presacral vertebral column, including 11 cervical and 11 dorsal vertebrae, but because of their extreme fragility, only plaster copies were incorporated into the mount. The sacrum was completely modeled in plaster on the basis of two specimens found at Tendaguru sites "Aa" and "T." Skeleton "no" supplied a series of 50 articulated caudal vertebrae for the tail, at the tip of which four small pieces were added as freehand reconstructions in plaster. The missing first caudal vertebra and most of the chevrons were also modeled in plaster. Cervical ribs were reconstructed in plaster on the basis of incomplete examples from skeleton S II. With the exception of four plaster reconstructions, the dorsal ribs are originals from skeleton S II.

The right scapula is a plaster reconstruction that used the original left scapula as a guide. The right coracoid and forelimb, with the exception of a carpal bone in plaster, consists of originals from S II; also originals are the left scapula, the coracoids, sternal plates, and the left humerus, radius, and ulna. The left manus was constructed entirely from plaster using the original manus from the opposite site as a guide.

The right ilium is an original from Tendaguru site "Ma," while the left ilium is a reconstruction in plaster mirrored from the right ilium. Both pubes are S II originals. The right ischium came from Tendaguru locality "L," and the left ischium was modeled in plaster as a mirror image of the latter. The hindlimbs are composites of bones of S II and other skeletons, partly original and partly modeled in plaster. The left femur is a fragmentary S II original completed with plaster, and the right femur an original from Tendaguru site "Ni." The right tibia and fibula are S II originals, whereas those from the left side are derived from Tendaguru site "Bo." The ankle bones are plaster imitations modeled from skeleton "Bo" originals, while the remaining elements of the hind feet are composites, mainly modeled in plaster, of badly preserved foot bones from skeleton S II and other finds.

Preliminary work on the mount commenced under Janensch in 1934 (Maier 2003) and took advantage of experience gained from mounting skeletal reconstructions of *Kentrosaurus aethiopicus* (1924), *Elaphrosaurus bambergi* (1926), and *Dicraeosaurus hansemanni* (1930/1931), all of which were also from Tendaguru. Unfortunately, no sketches of the mounting of *Brachiosaurus brancai* are available in the archives of the Museum für Naturkunde. There is, however, an extended photographic record produced by the *New York Times* GmbH Berlin that documents the mounting in detail (Fig. 18.2).

First, a scale model of the mount, about 1 m high and 1 m long, was produced in plaster. This was followed by a full-scale mock-up (Maier 2003), which allowed precise measurements and adjustments of the skeletal elements. Next, a metal armature to hold the bones in the desired position was constructed. Holes were drilled in the heavy bones, and steel tubes of about 5 cm in diameter were inserted that attached the skeletal elements to the metal armature. Because the armature was largely hidden, the bones were rendered highly visible. Finally, the skeleton was supported by two vertical T bars that were anchored in a basal platform.

FIGURE 18.2. The mount of *Brachiosaurus brancai* shortly before completion in late 1938. The ribs had not yet been mounted at the time the picture was taken. Note the extensive wooden scaffolding and the still-uncolored plaster models of the presacral vertebrae.

The skeleton was reconstructed with a neck that sloped steeply upward, giving *Brachiosaurus brancai* a giraffe-like appearance. This clearly distinguished it from most other sauropod dinosaur mounts of the early 20th century (e.g., Hatcher 1901; Osborn & Mook 1919; Gilmore 1936). On the basis of detailed comparative anatomical examinations, Janensch reconstructed the huge forelimbs in an elbow-out position.

The mounted skeleton was of superlative size. It stood about 12 m high, was approximately 23 m long, and became the tallest dinosaur skeleton on display anywhere in the world. Overall, the resulting skeletal reconstruction was an outstanding masterpiece and a milestone in mounting huge dinosaur skeletons (Maier 2003).

During World War II, the skeleton of *Brachiosaurus brancai* was taken down for safekeeping and stored in the museum basement from 1943 onward. In the spring of 1953, the remounted skeleton was put on public view once again. It remained on display until 2005, apart from a short interval during 1984 when it was disassembled for an exhibition in Japan and then remounted in Berlin that same year.

Technical Solutions for Remounting

The dismantling and remounting of all Tendaguru skeletons in the 2005–2007 period was performed by Research Casting International, a Canadian company that specializes in conserving, casting, and mounting fossils, including large dinosaur skeletons. The dismantling of the *Brachiosaurus* skeleton was carried out during April and May 2005.

Before taking the skeleton apart, a thorough inspection of each fossil bone was conducted. This inspection determined which bones needed additional stabilization or other special attention before dismantling and ensured that each specimen was stable enough to be disarticulated and moved. All of the disarticulated skeletal sections were packed and placed in drawers or, in the case of larger elements, firmly secured to custom pallets or metal frames. Some of the smaller *Brachiosaurus* bones were removed from their armatures by hand; larger and heavier bones and sections were lowered using rigging methods. Equipment used included two articulated man lifts, scaffolding with a movable crane, and other metal working equipment including welding machines. The specimens were crated and transported in a specialized air-ride moving truck to an 800 m^2 storage facility about 4 km from the museum, where the preparation took place.

The original sculpted plaster vertebrae were replaced by vacuum-formed epoxy. Carbon fiber casts were taken from new molds of original bones from the museum collection. All specimens were cleaned, breaks were repaired, and all bones were stabilized before remounting. In contrast to the old mount, the internal steel tubes were not used as supportive elements. Instead, individual external steel struts and clamps were welded and shaped to fit smoothly against the surface of the bone (Fig. 18.3). Because of the organic appearance of the ultralight metal armature, this mounting method resulted in an elegant appearance that avoided damage to the specimens (e.g., by drilling through the fossils). Furthermore, all clamps can be individually removed, allowing any single bone to be separately dismounted (e.g., for scientific research) without the need for disassembling the entire skeleton.

In April 2007, the remounting of *Brachiosaurus* was begun that used the same set of scaffolding and equipment as had been employed for the dismantling.

As mentioned above, in the old *Brachiosaurus* mount, the head and presacral vertebral column were modeled in plaster, but the original elements were not suitable for reuse in the mount for a number of reasons: the vertebral laminae and individual elements of the skull are delicate and easily broken; the fossil specimens are heavy and would have required complex (and presumably unaesthetic) steelwork to support them securely; most of the original presacral vertebrae are slightly deformed and would not articulate comfortably with one another; and if mounted between 5 and 13 m above floor level, the specimens would have been out of reach for scientific examination. Therefore, in the new mount, models were used again in place of the original presacral vertebral column and head, but took advantage of modern techniques and materials. The 1937 models were rather clumsy sculptures that bore only a superficial resemblance to the original fossil material. Thus, on this occasion, we sought to produce replicas that were as similar as possible to the originals. Although the old models were made of plaster, the new models were constructed from carbon fiber. This material has many advantages: it is more lightweight than plaster, it is tougher, and complex surface structures can be more easily modeled. For

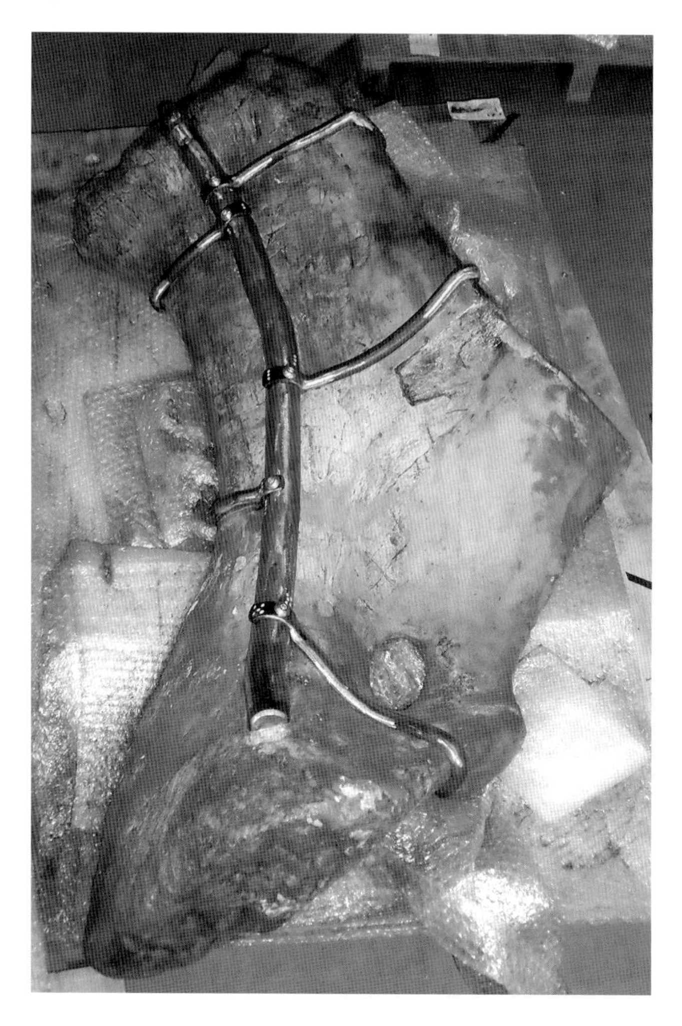

FIGURE 18.3. A steel strut with individually welded and shaped clamps on the left pubis of *Brachiosaurus brancai*. The clamps were screwed to the strut, enabling easy demounting of individual elements for scientific study.

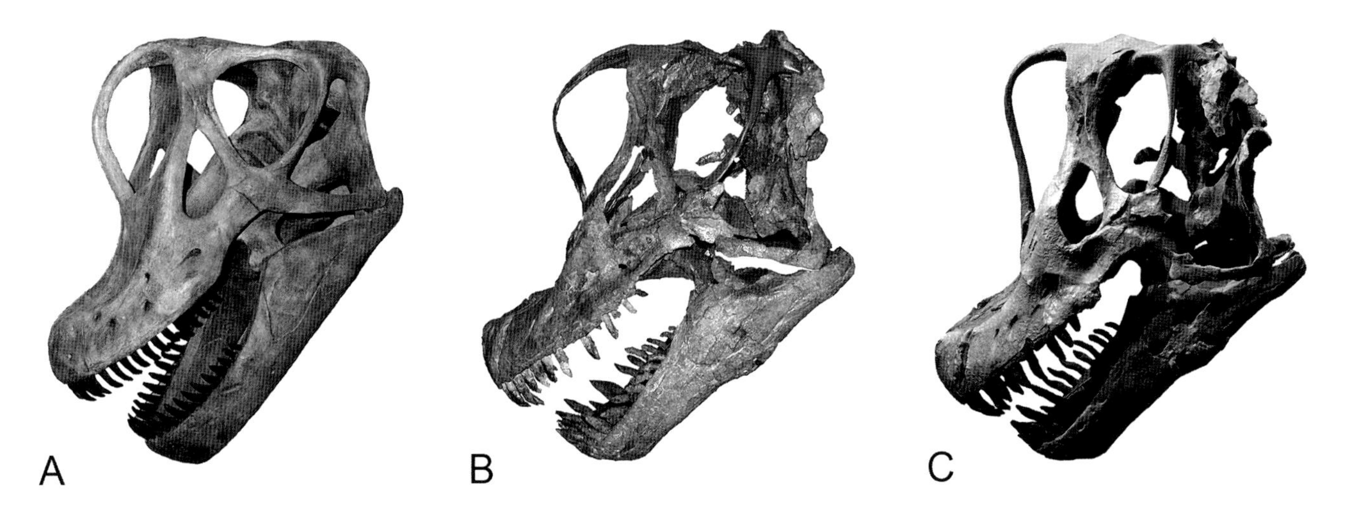

FIGURE 18.4. Comparison of the old skull model of *Brachiosaurus brancai* (A) to the original specimen t 1 (B) and the new model (C) created with the help of 3D laser scanning and rapid prototyping. The new model is much closer to the original.

the vertebral column, the original vertebrae were first molded using the techniques described above. Subsequently, carbon fiber models of three groups of articulated vertebrae (anterior cervicals, posterior cervicals, dorsals), which could be mounted comparatively easily and quickly, were manufactured together with their supporting armature.

Compared to the vertebral column, it was much more difficult to produce an accurate model of the skull. The old skull model (Fig. 18.4A) was unsatisfactory by modern standards, but fabricating a new, improved version proved to be a quite complex undertaking. It appeared that the skull elements of skeleton t 1 were assembled into an elaborate skull reconstruction in 1934. Because of the complexity of this construction and the fragility of the individual bones, it was deemed an unacceptable risk to deconstruct the skull and produce molds from its various components. Therefore, Research Casting International used a Minolta/Konica Vivid 9i 3D color laser scanner mounted on a moveable crane to capture more than 1,000 images of the original skull reconstruction from various perspectives. These images were used to construct a complete virtual 3D model of the skull to be printed out by a rapid prototyping printer. This allowed geometric transformations to be applied before printing, such as size scaling and replacement of missing bones by mirror imaging where contralateral bones were available. For example, the positions of the jugals were corrected at this stage. Because this data set was too large to be handled by most computers, it was divided into smaller parts, each of which was then replicated using a 3D printer. The printed models were reunited to form a complete skull, which was then molded. The mold was used to produce two skull casts, one mounted in the actual skeleton 13 m above ground level and the other on display in front of the mount of the entire skeleton (Figs. 18.4C, 18.5).

Scientific Rationale for Remounting *Brachiosaurus brancai*

As for all the dinosaurs in the refurbished hall, the principal idea behind remounting *Brachiosaurus* was to present the skeleton in a dynamic, lively pose that was set strictly within the limits of likely postures predicted by scientific studies such as those reported on in this volume. At the same time, it was deemed important to retain the majestic appearance of Janensch's classical mount in the center of the hall, which is why the museum chose the concept of a "dinosaur trek," with the skeletons mounted in poses showing them moving slowly toward the visitor as they enter the hall (Fig. 18.5).

Not surprisingly, the remounting of *Brachiosaurus* was by far the most challenging. The Berlin mount is the world's largest skeletal mount with original material, and it has now obtained its rightful place in the *Guinness Book of Records* (Anonymous 2008). A single femur weighs more than 300 kg, and the entire construction, including steel, fossilized bones, and carbon fiber models, adds up to about 5 metric tons. In addition, the schedule for the remounting of the skeleton was tight because of delays in the architectural restructuring of the exhibition halls, leaving no possibility for a test mount and only limited opportunities to correct errors.

The basic design of the *Brachiosaurus* mount is that of a giraffe-like animal with a fully erect neck, resulting in a skull located more than 13 m above the level of the feet (Fig. 18.5). The limbs are mounted in a walking pose, while the tail is held clear of the ground and is slightly curved distolaterally. In the following, potentially controversial issues of the new mount are addressed consecutively.

ARCHAEOPTERYX
DIE WELT IM OBEREN JU

NECK POSTURE

The neck posture of *Brachiosaurus* has been one of the most hotly debated issues in dinosaur paleontology in the last decade (e.g., Sander et al. 2009, 2010; Seymour 2009a, 2009b; Taylor et al. 2009; Christian & Dzemski, this volume). In Janensch's original reconstruction, the head was positioned high above the level of the shoulders, but instead of a vertical orientation, the neck was inclined cranially at about 30° from the vertical. On the basis of computer simulations of intervertebral articulations, Stevens & Parrish (1999, 2005a, 2005b) argued that the neutral pose of the neck of *Brachiosaurus* was closer to the horizontal plane; they rejected the giraffe-like position of earlier reconstructions (Bakker 1986; Paul 1988; Christian & Heinrich 1998). Other workers also agreed with a more horizontal orientation (Frey & Martin 1997; Berman & Rothschild 2005). However, in a series of Research Unit 533 papers on biomechanical calculations of the distribution of forces within the neck and in comparisons with extant long-necked animals, Christian & Dzemski (2007; this volume) showed that the neck could indeed reach a near-vertical orientation (80–85° above the horizontal plane), as is reconstructed in the new mount. During locomotion, forces would have been minimized when the neck was inclined about 30° cranially, corresponding closely to Janensch's original reconstruction. The individual in the exhibition has a walking pose and an almost fully vertical neck. This is less ergonomic than a more inclined neck, but it is still anatomically and behaviorally feasible (Christian 2010; Christian & Dzemski, this volume, for further discussion of this issue). The major argument against the neck raised high is the strain that this would place on the cardiovascular system (Seymour 2009a). In any case, the decision to adopt a fully erect neck posture maximizes the visual impact of the mount and also supports the educational aims of the new exhibition in terms of displaying dinosaurs as active and versatile animals.

RIB CAGE AND STERNUM

The new, anatomically accurate models of the dorsal vertebrae (see above, Technical Solutions for Remounting) had a profound impact on the shape of the rib cage. The ribs formed a rather barrel-shaped trunk in the old mount, but improvements in the accuracy of the position of the rib heads with the diapophyses and parapophyses on the newly modeled vertebrae resulted in a markedly slimmer profile of the trunk (compare Figs. 18.5, 18.6A). In dorsal view, the trunk is teardrop-shaped, reaching its widest point at the level of the fourth dorsal rib and then gradually tapering toward the hips. As a result, the entire animal appears much more slender and elegant, with a considerably reduced volume enclosed by the rib cage. In the exhibition, the mass of *Brachiosaurus* individual S II is cited as up to 50 metric tons, which is based on the mean of previously published estimates (Anderson et al. 1985; Gunga et al. 1995, 1999), preliminary laser scanning measurements, and physiological calculations (Gunga, pers. comm. 2006), as well as on 3D kinematic modeling (Mallison, pers. comm. 2006). However, shortly after the exhibition opened, Gunga et al. (2008) published a refined model that was based on research completed by Research Unit 533, which recalculated the mass of *Brachiosaurus* (still employing the old reconstruction) as approximately 38 metric tons. Consequently, the likely true body mass of a living animal with the same dimensions as that mounted in the new exhibition would have been considerably less and is currently being reassessed (Ganse et al., this volume; Stoinski et al., this volume).

After the rib cage was mounted, we encountered an unexpected problem: the sternal plates were too large to fit between the ends of the anterior ribs when an attempt was made to mount them in a horizontal plane. Such an arrangement is usually preferred for dinosaur sternals and is consistent with the usual orientation in most extant tetrapods. However, because the cartilaginous sternal ribs do not normally fossilize and because complete, undistorted sauropod rib cages with sternal plates in their original position have not been found to date, there is no direct evidence to support a strictly horizontal orientation of the sternal plates in sauropods. In birds, the contralateral halves of the sternum often stand at an angle to the horizontal plane, giving the dorsal side of the sternum a distinct concavity. Hence, we decided to mount the sternals of *Brachiosaurus* in a similar way, with a slight V-like orientation when viewed from the front (Fig. 18.6A). This is also corroborated by the observation that the estimated line of action of m. pectoralis (which runs from the ventral surface of the sternum to the medial side of the deltopectoral crest on the humerus; Remes 2008; Rauhut et al., this volume) would have been more effective with the sternals placed in this position. Alternatively, the problem of accommodating the sternals and rib cage may reflect their origin from two different individuals. However, there is no evidence to support this in the taphonomic data in the records of Janensch's expeditions (Heinrich 1999).

TAIL

In the original exhibition, Janensch opted for restorations in which all sauropods, and even the bipedal dinosaurs *Dysaloto-*

FIGURE 18.5. *Opposite page*: Right anterolateral view of the new mount of *Brachiosaurus brancai,* on display since July 2007. Note the walking pose, the vertical forelimbs with backwardly directed elbows, and the tail held clear of the ground.

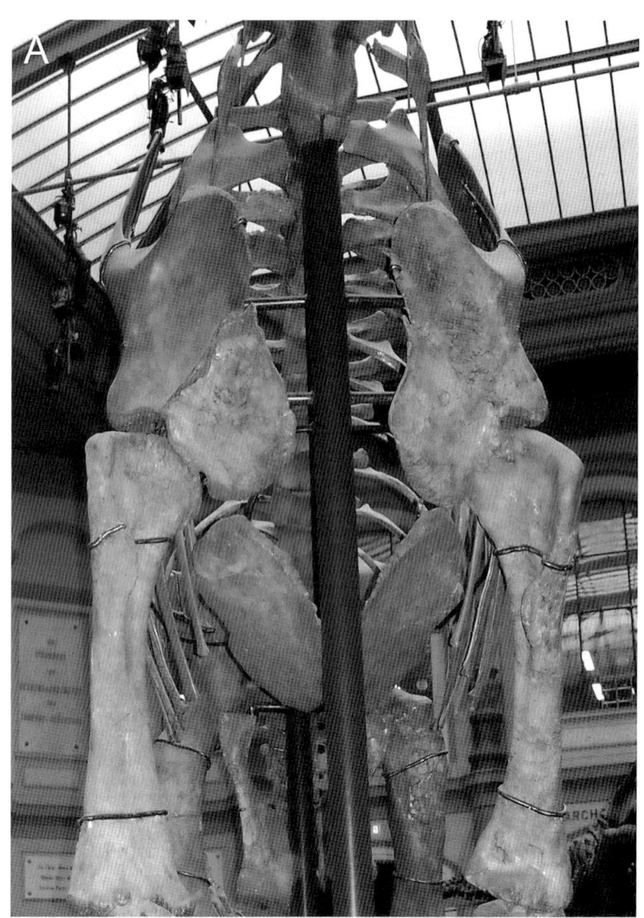

FIGURE 18.6. Close-up views of the shoulder girdle of the new mount of *Brachiosaurus brancai.* (A) Cranial view. (B) Left lateral view. The scapular blades are inclined about 60° from the horizontal. Note the space between the rib cage and the scapula, illustrating the volume of the serratus and subscapularis muscles. Because the improved models of the dorsal vertebrae led to a narrower trunk, the sternal plates had to be arranged in a V-like manner to fit between the distal ends of the ribs. See text for further explanations.

saurus, Plateosaurus, and *Elaphrosaurus,* were reconstructed with their tails resting on the ground. This implied that the tail was dragged behind the animal during locomotion in a fashion similar to that of modern lepidosaurs. Such a reconstruction was widely used during the first half of the 20th century and reflected contemporary interpretations of dinosaurs, especially sauropods, as sluggish, lizard-like creatures (e.g., Hay 1908, 1910, 1911; Tornier 1909; Abel 1910; Holland 1910; Matthew 1910). However, examination of modern archosaurs with a long tail (i.e., crocodilians) reveals that even these forms carry their tail above the ground, with only its distal tip occasionally contacting the substrate. Since the time of the Tendaguru expedition, thousands of examples of dinosaur tracks and traces have been discovered and recorded (e.g., Lockley 1991). Many of these have been attributed to sauropods, and almost without exception, they lack any evidence of tail drags. However, it was only fairly recently that the significance of these tracks was fully appreciated (e.g., Bakker 1971, 1986) and incorporated into modern reconstructions of dinosaurs, which now are almost always shown with the tail held in the air in a near-horizontal position. This arrangement has a number of biomechnical advantages: the tail can serve as a counterbalance for the anterior half of the animal and may facilitate rearing (Mallison, Chapter 14, this volume). Moreover, the principal muscle serving to propel the animal during terrestrial locomotion, the m. caudofemoralis longus (Gatesy 1990, 1995; Hutchinson & Gatesy 2000), which originated from the anterior one third of the tail, would have been most effective with the tail held horizontally. Otherwise, the distance between origin and insertion of this muscle would have been too short for efficient locomotion (Carrano 1998, 2000, 2005; Rauhut et al., this volume; Mallison, Chapter 14, this volume). Hence, there is both ichnological and biomechanical evidence to support the idea that sauropods held their tails clear of the ground, and this is reflected in the newly mounted *Brachiosaurus* and other dinosaurs in the exhibition hall.

FIGURE 18.7. Comparison of the originally planned walking pose (A) and the actual mount (B). Because the anterior steel column proved to be too short during the mounting process, the left forelimb had to be retracted further in order to fit between glenoid and floor level. This makes the animal appear to be walking faster. See text for further explanations.

SHOULDER GIRDLE

The position of the shoulder girdle in sauropods has become a matter of debate in recent years (Schwarz et al. 2007). Observations of partially articulated skeletons (Gilmore 1925) and impressions of the rib cage have led to suggestions that in life, the sauropod scapula was subhorizontally oriented (Parrish & Stevens 2002), a position already incorporated by Janensch in the original mount. However, the position of the scapula, even in fully articulated skeletons, may have been altered by postmortem desiccation of muscle tissue (Schwarz et al. 2007). Impressions of the scapula on dorsal ribs are most likely due to diagenetic effects because the scapula does not articulate with the rib cage in extant tetrapods but instead is widely separated from the ribs by muscle tissue (m. subscapularis, mm. serrati). On the basis of functional considerations, Schwarz et al. (2007) argued that the scapula was rather steeply inclined caudodorsally at an angle of about 50–60° to the horizontal. Remes (2006, 2008) came to the same conclusion and added a phylogenetic perspective, demonstrating the absence of any rotation toward a subhorizontal orientation throughout sauropodomorph evolution. Therefore, the scapula of *Brachiosaurus* was mounted at an inclination of about 60° to the horizontal, with the scapular part of the glenoid facing ventrally. This brought the glenoid into a position distinctly ventral to that in the old mount, which resulted in a steeper inclination of the back and a somewhat higher position for the base of the neck.

LIMBS

The position of the limbs is crucial for conveying the impression of a skeletal reconstruction in a static stance or in a dynamic pose. A walking position was therefore the natural choice for the new *Brachiosaurus* mount and consistent with the concept of the exhibition. However, mounting the limbs also caused the majority of problems.

Initially, it was planned to mount *Brachiosaurus* as if it were walking at a slow speed (Fig. 18.7A). The body would have been supported by the right forelimb and left hindlimb in contact with the ground, while the left forelimb and the right hindlimb would have been elevated in the phase of protraction. This arrangement was inspired by the lateral footfall pattern used by elephants, which is identical to the gait known as *toelt* in, for example, Icelandic ponies (Hutchinson et al. 2006). In such a configuration, there may have been enough space between the fore- and hindlimbs to allow visitors to walk beneath the rib cage, a novel exhibition feature that was discussed during the development phase of the exhibition. However, in the course of the mounting process, it was found that as a result of a miscalculation, the steel tube forming the front supporting pillar was too short. To correctly increase the length of the support, the already mounted elements (at this point in time the dorsal vertebrae, the sacrum, and the hindlimbs) would have had to be dismounted again followed by welding an extension to the front pillar. In addition, the sacrum and ilia would have had to be redesigned because extending the front pillar would have steepened the angle of the dorsal vertebral column. Because the schedule for mounting the skeletons was tight, this was not an option. It was decided instead to further retract the left forelimb so that it could be accommodated between the base of the mount and the glenoid joint. As a result, the entire mount became even more dynamic in appearance, primarily because the increase in step length of the forelimbs suggests a rapid walk rather than a slow toelt gait. This mode of locomotion resembles a pace (e.g., as in camels), a gait probably not typical for *Brachiosaurus;* however, this had no negative effect on the realistic appearance of the mount in the eyes of the museum visitors. Another consequence of this repositioning was a reduction in the gap between the forelimbs and the hindlimbs, excluding any possibility of a passageway beneath the mount for visitors.

A second problem was encountered during the mounting of the limbs. This also stemmed from the improved reconstruction of the rib cage and shoulder girdle, which imposed a bilateral symmetry on the cranial half of the torso fixing the positions of the glenoid joints. However, the limb elements themselves were not entirely symmetrical; the damaged left humerus had been supplemented with plaster for the original mount. This resulted in humeri of different lengths, with the complete right humerus being about 5 cm shorter than the partially reconstructed left humerus. As a consequence, there was sufficient room to mount the right limb with a substantial gap between the distal humeral condyles and the lower limb, representing space for the articular cartilage, while on the left side the humerus, radius, and ulna were mounted in contact with one another.

Ultimately, the problems encountered, and their solutions, had only a minor impact on the overall appearance of the animal, and are only likely to be detected after prolonged study of the mount. Despite its size, from all perspectives, the new *Brachiosaurus* mount is suggestive of an elegant, giraffe-like animal, captured in a dynamic stance and gait that contrasts quite sharply with the more static reconstruction from 1938.

Acknowledgments

We thank all participants in the ERDF (European Regional Development Fund) exhibition project for the excellent manner in which the collaboration was conducted, and without which the remounting of *Brachiosaurus* and the success of the new exhibition would not have been possible. We also thank all those institutions that provided funding, including the European Regional Development Fund, the Lotto Foundation, and the state of Berlin. K.R. thanks the Deutsche Forschungsgemeinschaft for funding his research through project RA 1012/2. This is contribution number 77 of the DFG Research Unit 533 "Biology of the Sauropod Dinosaurs: The Evolution of Gigantism."

References

Abel, O. 1910. Die Rekonstruktion des *Diplodocus.—Abhandlungen der kaiserlichen und königlichen Zoologisch-Botanischen Gesellschaft in Wien* 5: 1–60.

Anderson, J. F., Hall-Martin, A. & Russell, D. A. 1985. Long bone circumference and weight in mammals, birds and dinosaurs.—*Journal of Zoology A* 207: 53–61.

Anonymous. 2008. *Guinness World Records 2009.* Bibliographisches Institut, Mannheim.

Bakker, R. L. 1971. Ecology of the brontosaurs.—*Nature* 229: 172–174.

Bakker, R. L. 1986. *The Dinosaur Heresies.* William Morrow, New York.

Berman, D. S. & Rothschild, B. M. 2005. Neck posture of sauropods determined using radiological imaging to reveal three-dimensional structure of cervical vertebrae. *In* Tidwell, V. & Carpenter, K. (eds.). *Thunder Lizards: The Sauropodomorph Dinosaurs.* Indiana University Press, Bloomington: pp. 233–247.

Carrano, M. T. 1998. Locomotion in non-avian dinosaurs: integrating data from hindlimb kinematics, in vivo strains, and bone morphology.—*Paleobiology* 24: 450–469.

Carrano, M. T. 2000. Homoplasy and the evolution of dinosaur locomotion.—*Paleobiology* 26: 489–512.

Carrano, M. T. 2005. The evolution of sauropod locomotion: morphological diversity of a secondarily quadrupedal radiation. *In* Curry Rogers, K. & Wilson, J. A. (eds.). *The Sauropods: Evolution and Paleobiology.* University of California Press, Berkeley: pp. 229–251.

Christian, A. 2010. Some sauropods raised their necks—evidence for high browsing in *Euhelopus zdanskyi.—Biology Letters* 6: 823–825.

Christian, A. & Dzemski, G. 2007. Reconstruction of the cervical skeleton posture of *Brachiosaurus brancai* Janensch, 1914 by an analysis of the intervertebral stress along the neck and a comparison with the results of different approaches.—*Fossil Record* 10: 38–49.

Christian, A. & Dzemski, G. This volume. Neck posture in sauropods. *In* Klein, N., Remes, K., Gee, C. T. & Sander, P. M. (eds.). *Biology of the Sauropod Dinosaurs: Understanding the Life of Giants*. Indiana University Press, Bloomington: pp. 251–260.

Christian, A. & Heinrich, W.-D. 1998. The neck posture of *Brachiosaurus brancai*.—*Mitteilungen aus dem Museum für Naturkunde in Berlin, Geowissenschaftliche Reihe* 1: 73–80.

Frey, E. & Martin, J. 1997. Long necks of sauropods. *In* Currie, P. J. & Padian, K. (eds.). *Encyclopedia of Dinosaurs*. Academic Press, San Diego: pp. 406–409.

Ganse, B., Stahn, A., Stoinski, S., Suthau, T. & Gunga, H.-C. This volume. Body mass estimation, thermoregulation, and cardiovascular physiology of large sauropods. *In* Klein, N., Remes, K., Gee, C. T. & Sander, P. M. (eds.). *Biology of the Sauropod Dinosaurs: Understanding the Life of Giants*. Indiana University Press, Bloomington: pp. 105–115.

Gatesy, S. M. 1990. Caudofemoral musculature and the evolution of theropod locomotion.—*Paleobiology* 16: 170–186.

Gatesy, S. M. 1995. Functional evolution of the hindlimb and tail from basal theropods to birds. *In* Thomason, J. J. (ed.). *Functional Morphology in Vertebrate Paleontology*. Cambridge University Press, Cambridge: pp. 219–234.

Gilmore, C. W. 1925. A nearly complete articulated skeleton of *Camarasaurus*, a saurischian dinosaur from the Dinosaur National Monument, Utah.—*Memoirs of the Carnegie Museum* 10: 347–384.

Gilmore, C. W. 1936. Osteology of *Apatosaurus*, with special reference to specimens in the Carnegie Museum.—*Memoirs of the Carnegie Museum* 11: 175–271.

Gunga, H.-C., Kirsch, K. A., Baartz, A., Röcker, L., Heinrich, W.-D., Lisowski, W., Wiedemann, A. & Albertz, J. 1995. New data on the dimensions of *Brachiosaurus brancai* and their physiological implications.—*Naturwissenschaften* 82: 190–192.

Gunga, H.-C., Kirsch, K. A., Rittweger, J., Röcker, L., Clarke, A., Albertz, J., Wiedemann, A., Makry, S., Suthau, T., Wehr, A., Heinrich, W.-D. & Schultze, H.-P. 1999. Body size and body volume distribution in two sauropods from the Upper Jurassic of Tendaguru.—*Mitteilungen aus dem Museum für Naturkunde in Berlin, Geowissenschaftliche Reihe* 2: 91–102.

Gunga, H.-C., Suthau, T., Bellmann, A., Stoinski, S., Friedrich, A., Trippel, T., Kirsch, K. & Hellwich, O. 2008. A new body mass estimation of *Brachiosaurus brancai* Janensch, 1914 mounted and exhibited at the Museum of Natural History (Berlin, Germany).—*Fossil Record* 11: 33–38.

Hatcher, J. B. 1901. *Diplodocus* (Marsh): its osteology, taxonomy, and probable habits, with a restoration of the skeleton.—*Memoirs of the Carnegie Museum* 1: 1–63.

Hay, O. P. 1908. On the habits and the pose of the sauropodous dinosaurs, especially of *Diplodocus*.—*American Naturalist* 42: 672–681.

Hay, O. P. 1910. On the manner of locomotion of the dinosaurs especially *Diplodocus*, with remarks on the origin of birds.—*Proceedings of the Washington Academy of Science* 12: 1–25.

Hay, O. P. 1911. Further observations on the pose of the sauropodous dinosaurs.—*American Naturalist* 45: 398–412.

Heinrich, W.-D. 1999. The taphonomy of dinosaurs from the Upper Jurassic of Tendaguru (Tanzania) based on field sketches of the German Tendaguru expedition (1909–1913).—*Mitteilungen aus dem Museum für Naturkunde in Berlin, Geowissenschaftliche Reihe*: 2:25–61.

Holland, W. J. 1910. A review of some recent criticisms of the restorations of sauropod dinosaurs existing in the museums of the United States, with special reference to that of *Diplodocus carnegiei* in the Carnegie Museum.—*American Naturalist* 44: 259–283.

Hutchinson, J. R. & Gatesy, S. M. 2000. Adductors, abductors, and the evolution of archosaur locomotion.—*Paleobiology* 26: 734–751.

Hutchinson, J. R., Schwerda, D., Famini, D., Dale, R. H. I., Fischer, M. S. & Kram, R. 2006. The locomotor kinematics of Asian and African elephants: changes with speed and size.—*Journal of Experimental Biology* 209: 3812–3827.

Janensch, W. 1914. Bericht über den Verlauf der Tendaguru-Expedition.—*Archiv für Biontologie* 3: 17–58.

Janensch, W. 1937. Skelettrekonstruktion von *Brachiosaurus brancai* aus den Tendaguru-Schichten Deutsch-Ostafrikas.—*Zeitschrift der Deutschen Geologischen Gesellschaft* 89: 550–552.

Janensch, W. 1950. Die Skelettrekonstruktion von *Brachiosaurus brancai*.—*Palaeontographica Supplement* 7: 97–103.

Lockley, M. G. 1991. *Tracking Dinosaurs: A New Look at an Ancient World*. Cambridge University Press, Cambridge.

Maier, G. 2003. *African Dinosaurs Unearthed: The Tendaguru Expeditions*. Indiana University Press, Bloomington.

Mallison, H. This volume. Rearing giants: kinetic–dynamic modeling of sauropod bipedal and tripodal poses. *In* Klein, N., Remes, K., Gee, C. T. & Sander, P. M. (eds.). *Biology of the Sauropod Dinosaurs: Understanding the Life of Giants*. Indiana University Press, Bloomington: pp. 237–250.

Matthew, W. D. 1910. The pose of sauropodous dinosaurs.—*American Naturalist* 44: 547–560.

Osborn, H. F. & Mook, C. C. 1919. Characters and restoration of the sauropod genus *Camarasaurus* Cope.—*Proceedings of the American Philosophical Society* 58: 386–396.

Parrish, M. J. & Stevens, K. A. 2002. Rib angulation, scapular position, and body profiles in sauropod dinosaurs.—*Journal of Vertebrate Paleontology* 22: 95A.

Paul, G. S. 1988. The brachiosaur giants of the Morrison and Tendaguru with a description of a new subgenus, *Giraffatitan*, and a comparison of the world's largest dinosaurs.—*Hunteria* 2: 14.

Rauhut, O. W. M., Fechner, R., Remes, K. & Reis, K. This volume. How to get big in the Mesozoic: the evolution of the sauropodomorph body plan. *In* Klein, N., Remes, K., Gee, C. T. & Sander, P. M. (eds.). *Biology of the Sauropod Dinosaurs: Understanding the Life of Giants*. Indiana University Press, Bloomington: pp. 119–149.

Remes, K. 2006. Evolution of forelimb functional morphology in sauropodomorph dinosaurs.—*Journal of Vertebrate Paleontology* 26: 115A.

Remes, K. 2008. *Evolution of the Pectoral Girdle and Forelimb in Sau-*

ropodomorpha (Dinosauria, Saurischia): Osteology, Myology, and Function. Ph.D. Dissertation. Ludwig-Maximilians-Universität, Munich.

Sander, P. M., Christian, A. & Gee, C. T. 2009. Sauropods kept their heads down. Response.—*Science* 323: 1671.

Sander, P. M., Christian, A., Clauss, M., Fechner, R., Gee, C. T., Griebeler, E. M., Gunga, H.-C., Hummel, J., Mallison, H., Perry, S., Preuschoft, H., Rauhut, O., Remes, K., Tütken, T., Wings, O. & Witzel, U. 2010. Biology of the sauropod dinosaurs: the evolution of gigantism.—*Biological Reviews of the Cambridge Philosophical Society*. doi: 10.1111/j.1469=185X.2010.00137.x.

Schwarz, D., Frey, E. & Meyer, C. A. 2007. Novel reconstruction of the orientation of the pectoral girdle in sauropods.—*Anatomical Record* 290: 32–47.

Seymour, R. S. 2009a. Sauropods kept their heads down.—*Science* 323: 1671.

Seymour, R. S. 2009b. Raising the sauropod neck: it costs more to get less.—*Biology Letters* 5: 317–319.

Stevens, K. A. & Parrish, J. M. 1999. Neck posture and feeding habits of two Jurassic sauropod dinosaurs.—*Science* 284: 798–800.

Stevens, K. A. & Parrish, J. M. 2005a. Digital reconstructions of sauropod dinosaurs and implications for feeding. *In* Curry Rogers, K. & Wilson, J. A. (eds.). *The Sauropods: Evolution and Paleobiology*. University of California Press, Berkeley: pp. 178–200.

Stevens, K. A. & Parrish, J. M. 2005b. Neck posture, dentition, and feeding strategies in Jurassic sauropod dinosaurs. *In* Tidwell, V. & Carpenter, K. (eds.). *Thunder-Lizards: The Sauropodomorph Dinosaurs*. Indiana University Press, Bloomington: pp. 212–232.

Stoinski, S., Suthau, T. & Gunga, H.-C. This volume. Reconstructing body volume and surface area of dinosaurs using laser scanning and photogrammetry. *In* Klein, N., Remes, K., Gee, C. T. & Sander, P. M. (eds.). *Biology of the Sauropod Dinosaurs: Understanding the Life of Giants*. Indiana University Press, Bloomington: pp. 94–104.

Taylor, M. P., Wedel, M. J. & Naish, D. 2009. Head and neck posture in sauropod dinosaurs inferred from extant animals.—*Acta Palaeontologica Polonica* 54: 213–220.

Tornier, G. 1909. Wie war der *Diplodocus carnegii* wirklich gebaut?—*Sitzungsberichte der Gesellschaft naturforschender Freunde zu Berlin* 1909: 193–209.

Compilation of Published Body Mass Data for a Variety of Basal Sauropodomorphs and Sauropods

In the appendix, published body mass data for a variety of basal sauropodomorphs and sauropods have been compiled. Note the differing results for some taxa, mainly depending on the method used for body mass reconstruction. The different methods used (as described by the authors cited) are coded in the table as follows: 0, method not given; 1, von Bertalanffy equation; 2, 3D mathematical slicing; 3, polynomial technique and volume figures; 4, log transformed (base 10) database consisting of model-based body estimates and measurements of bone dimensions; 5, bone measuring, midshaft femora, and/or humeri circumference; 6, ontogenetic growth curves of dinosaur species, estimated from data on the scaling of maximum growth rates for reptiles and mammals; 7, scale model, water displacement, and volume of living animals scaled up from the model; 8, weighing scale models in air and water, recalculation using a slightly lower overall density (950 kg/m^3); 9, 3D stereophotogrammetry, laser scanning of mounted skeletons; 10, estimating cubic meters; 11, considering pneumaticity, reduced neck and tail volume; 12, laser stereophotogrammetry, laser scanning, and 3D reconstruction methods; 13, plasticine scale models, following a skeletal restoration; 14, estimated by personal opinion; 15, "gathered data"; 16, modern skeletal reconstructions, numerical estimates of centers of mass.

Published Body Mass Data for a Variety of Basal Sauropodomorphs and Sauropods

Taxon	Reference	Mass (kg)	Body size (m) as height (h) or length (l) (error ±1 dm)	Method of mass reconstruction
Alamosaurus sp.	Lehman & Woodward (2008)	32,663		1 and 15
Amargasaurus cazaui	Mazzetta et al. (2004)	2,600	9.1 (l)	2
Amargasaurus cazaui	Seebacher (2001)	6,852	10.3 (l)	3
Amphicoelias altus	Paul (1998)	11,000–12,000	23.0–28.0 (l)	0
Amphicoelias fragillimus	Paul (1998)	90,000–150,000	40.0–60.0 (l)	0
Anchisaurus sinensis	Seebacher (2001)	84	2.8 (l)	3
Antarctosaurus giganteus	Mazzetta et al. (2004)	69,000		4
Antarctosaurus giganteus	Van Valen (1969)	80,000		0
Antarctosaurus sp.	Alexander (1998)	75,000		0
Antarctosaurus sp.	Benton (1989)	80,000	30.0 (l)	0
Antarctosaurus wichmannianus	Mazzetta et al. (2004)	25,797–47,611		4
Antarctosaurus wichmannianus	Mazzetta et al. (2004)	34,000		4
Antarctosaurus wichmannianus	Mazzetta et al. (2004)	33,410		4
Antarctosaurus wichmannianus	Mazzetta et al. (2004)	24,617		4
Apatosaurus alenquerensis	Anderson et al. (1985)	26,424		5
Apatosaurus excelsus	Foster (2007)	24,247		0
Apatosaurus excelsus	Gillooly et al. (2006)	25,952		6
Apatosaurus excelsus	Gillooly et al. (2006)	25,000		6
Apatosaurus excelsus	Gillooly et al. (2006)	12,979		6
Apatosaurus excelsus (*Brontosaurus*)	Foster (2007)	<10,000		0
Apatosaurus louisae	Christiansen (1997)	19,500	22.8 (l)	7
Apatosaurus louisae	Foster (2007)	34,035	21.0–25.0 (l)	0
Apatosaurus louisae	Foster (2007)	17,017		0
Apatosaurus louisae	Foster (2007)	10,000	17.0 (l)	0
Apatosaurus louisae	Mazzetta et al. (2004)	20,600	22.8 (l)	8
Apatosaurus louisae	Seebacher (2001)	22,407.2	21.0 (l)	3
Apatosaurus louisae	Paul (1998)	17,500	28.8 (l)	0
Apatosaurus louisae	Paul (1998)	11,000		0
Apatosaurus louisae	Alexander (1989)	34,000–35,000		7
Apatosaurus louisae	Anderson et al. (1985)	37,500		5
Apatosaurus louisae	Anderson et al. (1985)	35,000		5
Apatosaurus louisae	Anderson et al. (1985)	34,842		5
Apatosaurus louisae	Anderson et al. (1985)	30,000		5
Apatosaurus louisae	Colbert (1962)	33,500		7
Apatosaurus louisae	Colbert (1962)	32,420		7
Apatosaurus louisae	Colbert (1962)	27,870		7
Apatosaurus louisae (juvenile)	Foster (2007)	4,254	11.0 (l)	0
Apatosaurus sp.	Lehman & Woodward (2008)	25,952		1
Apatosaurus sp.	Paul (1998)	23,000–28,000		0
Apatosaurus sp.	Paul (1998)	19,100	23.0 (l)	0
Apatosaurus sp.	Paul (1998)	18,000		0
Apatosaurus sp.	Paul (1998)	13,000		0
Apatosaurus sp.	Colbert (1962)	27,000–32,000		0
Argentinosaurus huinclensis	Mazzetta et al. (2004)	72,936		4
Argentinosaurus huinclensis	Burness et al. (2001)	73,000		15
Argentinosaurus huinclensis	Hokkanen (1986)	100,000		0
Barosaurus lentus	Foster (2007)	11,957	24.0 (l), 4.6 (h)	0
Barosaurus lentus	Paul (1998)	11,600	27.9 (l)	0
Barosaurus sp.	Béland & Russell (1980)	8,300		5
Barosaurus sp.	Seebacher (2001)	20,039	26.0 (l)	3
Barosaurus sp.	Peczkis (1994)	10,000–40,000		5

Appendix Table *continued*

Taxon	Reference	Mass (kg)	Body size (m) as height (h) or length (l) (error ±1 dm)	Method of mass reconstruction
Brachiosaurus altithorax	Foster (2007)	43,896	18.0 (l), 9.4 (h)	0
Brachiosaurus altithorax	Seebacher (2001)	28,264	21.0 (l)	3
Brachiosaurus altithorax	Paul (1998)	35,000	20.0–21.0 (l), 12.0–13.0 (h)	0
Brachiosaurus altithorax	Paul (1998)	32,000		7
Brachiosaurus altithorax (*Giraffatitan*)	Paul (1988)	50,000		7
Brachiosaurus brancai	Christiansen (1997)	37,400	21.8 (l)	7
Brachiosaurus brancai	Alexander (1985)	46,600		7
Brachiosaurus brancai	Janensch (1938)	40,000		14
Brachiosaurus brancai	Gunga et al. (2008)	38,000		9
Brachiosaurus brancai	Mazzetta et al. (2004)	78,300		0
Brachiosaurus brancai	Mazzetta et al. (2004)	39,500	21.8 (l)	8
Brachiosaurus brancai	Gunga et al. (1999)	74,420		9
Brachiosaurus brancai	Alexander (1989)	32,000–87,000		7
Brachiosaurus brancai	Alexander (1989)	47,000		7
Brachiosaurus brancai	Paul (1988)	45,000–50,000	25.0 (l)	7
Brachiosaurus brancai	Anderson et al. (1985)	31,600		5
Brachiosaurus brancai	Anderson et al. (1985)	29,335		5
Brachiosaurus brancai	Colbert (1962)	87,000		0
Brachiosaurus brancai	Colbert (1962)	80,000		7
Brachiosaurus brancai	Colbert (1962)	78,260		7
Brachiosaurus macintoshi (*Ultrasauros*)	Paul (1998)	45,000–50,000	24.0–25.0 (l), 15.0–16.0 (h)	0
Brachiosaurus sp.	Reid (1997)	30,000		0
Brachiosaurus sp.	Béland & Russell (1980)	14,900		5
Brachiosaurus sp.	Janensch (1935–36)	25,000–32,000		10
Brachiosaurus sp.	Janensch (1935–36)	40,000		10
Brachiosaurus sp.	Christian et al. (1999)	63,000		11
Brachiosaurus sp.	Paul (1998)	10,000		0
Brachiosaurus sp.	Peczkis (1994)	10,000–40,000		5
Brachiosaurus sp.	Benton (1989)	40,000–78,000	23.0–27.0 (l), 12.0 (h)	0
Brachiosaurus sp.	Paul (1988)	29,000		7
Camarasaurus grandis	Foster (2007)	18,413	15.0 (l)	0
Camarasaurus grandis	Foster (2007)	14,000	15.0 (l)	0
Camarasaurus grandis	Foster (2007)	9,321	15.0 (l)	0
Camarasaurus lentus	Paul (1998)	14,200	15.3 (l), 7.0 (h)	0
Camarasaurus lentus	Paul (1998)	640	5.0 (l), 2.2 (h)	0
Camarasaurus lewisi	Seebacher (2001)	11,652	15.4 (l)	3
Camarasaurus sp.	Paul (1998)	20,000–25,000	18.0 (l)	0
Camarasaurus sp.	Paul (1998)	3,800	10.0 (l)	0
Camarasaurus supremus	Christiansen (1997)	8,800	13.8 (l)	7
Camarasaurus supremus	Foster (2007)	47,000	15.0–23.0 (l), 2.0–5.0 (h)	0
Camarasaurus supremus	Mazzetta et al. (2004)	9,300	13.8 (l)	8
Cetiosaurus oxoniensis	Mazzetta et al. (2004)	15,900	16.5 (l)	8
Dicraeosaurus hansemanni	Christiansen (1997)	5,400	14.2 (l)	7
Dicraeosaurus hansemanni	Mazzetta et al. (2004)	5,700	14.2 (l)	8
Dicraeosaurus hansemanni	Gunga et al. (1999)	12,800		9
Dicraeosaurus hansemanni	Peczkis (1994)	10,000–40,000		5
Dicraeosaurus sp.	Béland & Russell (1980)	3,300		5
Dicraeosaurus sp.	Seebacher (2001)	4,421	12.0 (l)	3
Dicraeosaurus sp.	Gunga et al. (1995)	12,800		12
Diplodocus carnegiei	Paul (1998)	6,000–19,000		0
Diplodocus carnegiei	Alexander (1985)	18,500	20.3 (l)	7

Appendix Table *continued*

Taxon	Reference	Mass (kg)	Body size (m) as height (h) or length (l) (error ±1 dm)	Method of mass reconstruction
Diplodocus carnegiei	Mazzetta et al. (2004)	16,000	25.6 (l)	8
Diplodocus carnegiei	Seebacher (2001)	19,654	25.7 (l)	3
Diplodocus carnegiei	Anderson et al. (1985)	5,800		5
Diplodocus carnegiei	Colbert (1962)	11,700		7
Diplodocus carnegii	Christiansen (1997)	15,200	25.6 (l)	7
Diplodocus carnegii	Foster (2007)	12,657		0
Diplodocus carnegii	Foster (2007)	12,000	28.0 (l), 0.8–4.3 (h)	0
Diplodocus carnegii	Paul (1998)	11,400	24.8 (l)	0
Diplodocus carnegii	Paul (1998)	6,000		0
Diplodocus longus	Foster (2007)	12,000	24.0 (l)	0
Diplodocus sp.	Henderson (1999)	13,421	22.4 (l)	16
Diplodocus sp.	Paul (1998)	10,000		0
Diplodocus sp.	Benton (1989)	18,500	27.0 (l)	0
Diplodocus sp.	Anderson et al. (1985)	15,000		5
Diplodocus sp.	Anderson et al. (1985)	5,000		5
Diplodocus sp.	Colbert (1962)	10,560	18.0 (l)	7
Diplodocus sp. (*Supersaurus*)	Paul (1998)	80,000		7
Euhelopus zdanskyi	Mazzetta et al. (2004)	3,800	10.5 (l)	13
Europasaurus holgeri	Stein et al. 2010	800	6.2 (l)	5
Haplocanthosaurus delfsi	Foster (2007)	21,000	21.0 (l)	0
Haplocanthosaurus delfsi	Paul (1998)	12,700	14.0–15.0 (l)	0
Haplocanthosaurus priscus	Foster (2007)	10,500		0
Haplocanthosaurus priscus	Foster (2007)	7,000		0
Haplocanthosaurus priscus	Mazzetta et al. (2004)	12,800	14.8 (l)	13
Haplocanthosaurus sp.	Foster (2007)	10,000		0
Haplocanthosaurus sp.	Seebacher (2001)	14,528	14.0 (l)	3
Haplocanthosaurus sp.	Paul (1998)	5,000	10.0–11.0 (l)	0
Janenschia sp.	Lehman & Woodward (2008)	14,029		1
Janenschia sp.	Peczkis (1994)	10,000–40,000		5
Lufengosaurus huenei	Seebacher (2001)	1,193	6.2 (l)	3
Magyarosaurus dacus	Stein et al. 2010	900		5
Mamenchisaurus hochuanensis	Christiansen (1997)	14,300	20.4 (l)	7
Mamenchisaurus hochuanensis	Mazzetta et al. (2004)	15,100	20.4 (l)	8
Mamenchisaurus hochuanensis	Seebacher (2001)	18,169	21.0 (l)	3
Massospondylus sp.	Seebacher (2001)	136	4.0 (l)	3
Omeisaurus tianfunensis	Christiansen (1997)	9,800	20.2 (l)	7
Omeisaurus tianfunensis	Mazzetta et al. (2004)	9,800	20.2 (l)	8
Omeisaurus tianfunensis	Seebacher (2001)	11,796	20.0 (l)	3
Opisthocoelicaudia skarzynskii	Mazzetta et al. (2004)	8,400	11.3 (l)	13
Opisthocoelicaudia skarzynskii	Seebacher (2001)	10,522	11.4 (l)	3
Opisthocoelicaudia skarzynskii	Anderson et al. (1985)	22,000		5
Paralititan stromeri	Burness et al. (2001)	59,000		15
Patagosaurus sp.	Seebacher (2001)	9,435	15.0 (l)	3
Plateosaurus engelhardti	Gunga et al. (2007)	912		12
Plateosaurus engelhardti	Gunga et al. (2007)	630		12
Plateosaurus engelhardti	Seebacher (2001)	1,072	6.5 (l)	3
Riojasaurus sp.	Seebacher (2001)	3,038	10.0 (l)	3
Sauroposeidon proteles	Burness et al. (2001)	55,000		0
Sauroposeidon proteles	Wedel (2000)	50,000–60,000		0
Seismosaurus halli	Paul (1997)	30,000	32.0–35.0 (l)	0
Seismosaurus halli	Seebacher (2001)	49,275	40.0 (l)	3

Appendix Table *continued*

Taxon	Reference	Mass (kg)	Body size (m) as height (h) or length (l) (error ±1 dm)	Method of mass reconstruction
Seismosaurus halli	Gillette (1994)	100,000	50.0 (l)	0
Seismosaurus hallorum	Foster (2007)	42,500	36.0 (l)	0
Seismosaurus hallorum	Paul (1998)	25,000–30,000	28.0–34.0 (l)	0
Seismosaurus sp.	Benton (1989)	80,000+	30.0–36.0 (l)	0
Shunosaurus lii	Christiansen (1997)	3,400	9.9 (l)	7
Shunosaurus lii	Mazzetta et al. (2004)	3,600	9.9 (l)	8
Shunosaurus lii	Seebacher (2001)	4,793	8.7 (l)	3
Supersaurus sp.	Jensen (1985)	50,000		0
Supersaurus sp.	Paul (1998)	30,000–40,000	30.0–40.0 (l)	0
Supersaurus sp.	Benton (1989)	75,000–100,000	24.0–30.0 (l), 15.0 (h)	0
Supersaurus vivianae	Foster (2007)	40,200	38.0 (l)	0
Supersaurus viviane	Paul (1998)	45,000–55,000	35.0–45.0 (l)	0
Thecodontosaurus antiquus	Seebacher (2001)	24	2.6 (l)	3
Ultrasaurus sp.	Benton (1989)	100,000–140,000	30.0–35.0 (l), 16.0–17.0 (h)	0

References

Alexander, R. M. 1985. Mechanics of posture and gait of some large dinosaurs.—*Zoological Journal of the Linnean Society of London* 83: 1–25.

Alexander, R. M. 1989. *Dynamics of Dinosaurs and Other Extinct Giants*. Columbia University Press, New York.

Alexander, R. M. 1998. All-time giants: the largest animals and their problems.—*Palaeontology* 41: 1231–1245.

Anderson, J. F., Hall-Martin, A. & Russell, D. A. 1985. Long bone circumference and weight in mammals, birds, and dinosaurs.—*Journal of Zoology A* 207: 53–61.

Béland, P. & Russell, D. A. 1980. Dinosaur metabolism and predator/prey ratios in the fossil record. *In* Thomas, R. D. K. & Olson, E. C. (eds.). *A Cold Look at the Warm-Blooded Dinosaurs*. Westview Press, Boulder.

Benton, M. J. 1989. Evolution of large size. *In* Briggs, D. E. G. & Crowther, P. R. (eds.). *Palaeobiology: A Synthesis*. Blackwell Scientific, Oxford: pp. 147–152.

Burness, G. P., Diamond, J. & Flannery, T. 2001. Dinosaurs, dragons, and dwarfs: the evolution of maximal body size.—*Proceedings of the National Academy of Sciences of the United States of America* 98: 14518–14523.

Christian, A., Heinrich, W.-D. & Golder, W. 1999. Posture and mechanics of the forelimbs of *Brachiosaurus brancai* (Dinosauria: Sauropoda).—*Mitteilungen aus dem Museum für Naturkunde in Berlin, Geowissenschaftliche Reihe* 2: 63–73.

Christiansen, P. 1997. Locomotion in sauropod dinosaurs.—*Gaia* 14: 45–75.

Colbert, E. H. 1962. The weights of dinosaurs.—*American Museum Novitates* 2076: 1–16.

Foster, J. R. 2007. *Jurassic West: The Dinosaurs of the Morrison Formation and Their World*. Indiana University Press, Bloomington.

Gillette, D. D. 1994. *Seismosaurus: The Earth Shaker*. Columbia University Press, New York.

Gillooly, J. F., Allen, A. P. & Charnov, E. L. 2006. Dinosaur fossils predict body temperatures.—*PLoS Biology* 4: 1467–1469.

Gunga, H.-C., Kirsch, K., Rittweger, J., Röcker, L., Clarke, A., Albertz, J., Wiedemann, A., Mokry, S., Suthau, T., Wehr, A., Heinrich, W.-D. & Schultze, H.-P. 1999. Body size and body volume distribution in two sauropods from the Upper Jurassic of Tendaguru (Tanzania).—*Mitteilungen aus dem Museum für Naturkunde in Berlin, Geowissenschaftliche Reihe* 2: 91–102.

Gunga, H.-C., Suthau, T., Bellmann, A., Friedrich, A., Schwanebeck, T., Stoinski, S., Trippel, T., Kirsch, K. & Hellwich, O. 2007. Body mass estimations for *Plateosaurus engelhardti* using laser scanning and 3D reconstruction methods.—*Naturwissenschaften* 94: 623–630.

Gunga, H.-C., Suthau, T., Bellmann, A., Stoinski, S., Friedrich, A., Trippel, T., Kirsch, K. & Hellwich, O. 2008. A new body mass estimation of *Brachiosaurus brancai* Janensch, 1914 mounted and exhibited at the Museum of Natural History (Berlin, Germany).—*Fossil Record* 11: 33–38.

Gunga, H.-C., Kirsch, K., Baartz, F., Röcker, L., Heinrich, W.-D., Lisowski, W., Wiedemann, A. & Albertz, J. 1995. New data on the dimensions of *Brachiosaurus brancai* and their physiological implications.—*Naturwissenschaften* 82: 190–192.

Henderson, D. M. 1999. Estimating the masses and centers of mass of extinct animals by 3-D mathematical slicing.—*Paleobiology* 25: 88–106.

Hokkanen, J. E. I. 1986. The size of the biggest land animal.—*Journal of Theoretical Biology* 118: 491–499.

Janensch, W. 1935–36. Die Schädel der Sauropoden *Brachiosaurus, Barosaurus* und *Dicraeosaurus* aus den Tendaguru-Schichten Deutsch-Ostafrikas.—*Palaeontographica Supplement* 7 (2): 147–298.

Janensch, W. 1938. Gestalt und Größe von *Brachiosaurus* und anderen riesenwüchsigen Sauropoden.—*Der Biologe* 7: 130–134.

Jensen, J. A. 1985. Three new sauropod dinosaurs from the Upper Jurassic of Colorado.—*Great Basin Naturalist* 45: 697–709.

Lehman, T. M. & Woodward, H. N. 2008. Modeling growth rates for sauropod dinosaurs.—*Paleobiology* 34: 264–281.

Mazzetta, G. V., Christiansen, P. & Farina, R. A. 2004. Giants and bizarres: body size of some southern South American Cretaceous dinosaurs.—*Historical Biology* 16: 1–13.

Paul, G. S. 1997. Dinosaur models: the good, the bad, and using them to estimate the mass of dinosaurs. *In* Wolberg, D. L., Stump, E. & Rosenberg, G. D. (eds.). *DinoFest International.* Academy of Natural Sciences, Philadelphia: pp. 129–154.

Paul, G. S. 1998. Terramegathermy and Cope's rule in the land of titans.—*Modern Geology* 23: 179–217.

Peczkis, J. 1994. Implications of body-mass estimates for dinosaurs.—*Journal of Vertebrate Paleontology* 14: 520–533.

Reid, R. E. H. 1997. Dinosaurian physiology: the case for "intermediate" dinosaurs. *In* Farlow, J. O. & Brett-Surman, M. K. (eds.). *The Complete Dinosaur.* Indiana University Press, Bloomington: pp. 449–473.

Seebacher, F. 2001. A new method to calculate allometric length–mass relationships of dinosaurs.—*Journal of Vertebrate Paleontology* 21: 51–60.

Stein, K., Csiki, Z., Curry Rogers, K., Weishampel, D. B., Redelstorff, R., Carballido, J. L. & Sander, P. M. 2010. Small body size and extreme cortical bone remodeling indicate phyletic dwarfism in *Magyarosaurus dacus* (Sauropoda: Titanosauria).—*Proceedings of the National Academy of Sciences of the United States of America* 107: 9258–9263.

Van Valen, L. 1969. What was the largest dinosaur?—*Copeia* 3: 624–626.

Wedel, M. J., Cifelli, R. L. & Sanders, R. K. 2000. *Sauroposeidon proteles,* a new sauropod from the Early Cretaceous of Oklahoma. —*Journal of Vertebrate Paleontology* 20: 109–114.

INDEX

Page numbers in italics refer to figures.

ABOUT THE EDITORS

Nicole Klein
is a vertebrate paleontologist at the University of Bonn who specializes in sauropodomorph dinosaur bone histology and marine reptiles from the Middle Triassic Muschelkalk deposits of Central Europe. She has done extensive fieldwork in many parts around the world, including Alaska and Nevada in the United States, and Ethiopia.

Kristian Remes
has studied sauropodomorph anatomy, functional morphology, and phylogeny. He played a major role in the remounting of the famous *Brachiosaurus* skeleton in the newly renovated Dinosaur Hall at the Museum für Naturkunde in Berlin. He is now a program director at the German Research Foundation (DFG).

Carole T. Gee,
a senior research scientist at the University of Bonn, has worked on the Mesozoic flora for the last 25 years. She is the Research Unit's paleobotanist and answers questions on sauropod herbivory and the Mesozoic vegetation. Her research applies the knowledge of living plants and their ecological preferences to the interpretation of fossil plants and their habitats, and also includes studies on Eocene mangroves, Tertiary fruits and seeds, and plant taphonomy.

P. Martin Sander
is a professor of vertebrate paleontology at the University of Bonn and head of the DFG Research Unit 533 "Biology of the Sauropod Dinosaurs: The Evolution of Gigantism." His research interests are the major events in the evolution of tetrapod vertebrates and how the fossil record helps us to understand them. His core expertise is the microstructure of dinosaur bone and the diversity and evolution of marine reptiles.